# Functional Equations in Mathematical Analysis

# Springer Optimization and Its Applications

VOLUME 52



*Aims and Scope*

Optimization has been expanding in all directions at an astonishing rate during the last few decades. New algorithmic and theoretical techniques have been developed, the diffusion into other disciplines has proceeded at a rapid pace, and our knowledge of all aspects of the field has grown even more profound. At the same time, one of the most striking trends in optimization is the constantly increasing emphasis on the interdisciplinary nature of the field. Optimization has been a basic tool in all areas of applied mathematics, engineering, medicine, economics and other sciences.

The series *Springer Optimization and Its Applications* publishes undergraduate and graduate textbooks, monographs and state-of-the-art expository works that focus on algorithms for solving optimization problems and also study applications involving such problems. Some of the topics covered include nonlinear optimization (convex and nonconvex), network flow problems, stochastic optimization, optimal control, discrete optimization, multi-objective programming, description of software packages, approximation techniques and heuristic approaches.

For further volumes:
http://www.springer.com/series/7393

Themistocles M. Rassias • Janusz Brzdęk
Editors

# Functional Equations in Mathematical Analysis

Springer

*Editors*
Themistocles M. Rassias
Department of Mathematics
National Technical University of Athens
Zografou Campus
15780 Athens
Greece
trassias@math.ntua.gr

Janusz Brzdęk
Department of Mathematics
Pedagogical University
30084 Krakow
Poland
jbrzdek@ap.krakow.pl

ISSN 1931-6828
ISBN 978-1-4614-0054-7 e-ISBN 978-1-4614-0055-4
DOI 10.1007/978-1-4614-0055-4
Springer New York Dordrecht Heidelberg London

Library of Congress Control Number: 2011935375

Mathematics Subject Classification (2010): 39-XX, 46-XX, 33-XX



Printed on acid-free paper

Springer is part of Springer Science+Business Media (www.springer.com)

*Dedicated to the memory of*
*Stanisław Marcin Ulam (1909–1984)*
*on the occasion of the 100th anniversary*
*of his birth*

# Preface

The volume consists of articles written by eminent scientists from the international mathematical community, who present important research works in the field of Mathematical Analysis and related subjects, in particular, in Functional Equations and Inequalities. These works provide an insight in a progress in the study of various problems of nonlinear character.

Several of these results have been influenced by the work of the well-known mathematician and physicist Stanisław Marcin Ulam (April 3, 1909 to May 13, 1984). An emphasis is given to one of his questions concerning approximate homomorphisms.

The book is dedicated to the memory of Ulam on the occasion of the 100th anniversary of his birth.

It is divided into two parts. Part I focuses on several aspects of the Ulam stability theory. Part II contains papers on various subjects of Mathematical Analysis.

It is a pleasure to express our deepest thanks to all of the mathematicians who, through their works, participated in this volume.

We would also wish to acknowledge the superb assistance that the staff of Springer has provided in the preparation of this publication.

Athens and Cracow

Themistocles M. Rassias
Janusz Brzdęk

# Contents

# Contributors

**Eisa Al-Said** Mathematics Department, College of Science, King Saud University, Riyadh, Saudi Arabia, eisasaid@ksu.edu.sa

**Roman Badora** Institute of Mathematics, Silesian University, Bankowa 14, 40-007 Katowice, Poland, robadora@math.us.edu.pl

**H. Baghani** Department of Mathematics, Semnan University, P. O. Box 35195-363, Semnan, Iran, h.baghani@gmail.com

**Bogdan Batko** Department of Mathematics, WSB – NLU, Zielona 27, 33-300 Nowy Sącz, Poland, bbatko@wsb-nlu.edu.pl

Department of Mathematics, Pedagogical University, Podchorążych 2, 30-084 Kraków, Poland, bbatko@ap.krakow.pl

**Liviu Cădariu** Department of Mathematics, "Politehnica" University of Timişoara, Piaţa Victoriei no. 2, 300006 Timişoara, Romania, liviu.cadariu@mat.upt.ro;lcadariu@yahoo.com

**A. Charifi** Department of Mathematics, Faculty of Sciences, Ibn Tofail University, 14000 Kenitra, Morocco, charifi2000@yahoo.fr

**Wing-Sum Cheung** Department of Mathematics, The University of Hong Kong, Pokfulam Road, Hong Kong, wscheung@hku.hk

**Jacek Chmieliński** Instytut Matematyki, Uniwersytet Pedagogiczny w Krakowie, Podchorążych 2, 30-084 Kraków, Poland, jacek@up.krakow.pl

**Antoni Chronowski** Department of Mathematics, Pedagogical University, Podchorążych 2, 30-084 Kraków, Poland, chron@up.krakow.pl

**Małgorzata Chudziak** Department of Mathematics, University of Rzeszów, Rejtana 16 C, 35-959 Rzeszów, Poland, mchudziak@univ.rzeszow.pl

**Jae-Young Chung** Department of Mathematics, Kunsan National University, Kunsan 573-701, Korea, jychung@kunsan.ac.kr

**Krzysztof Ciepliński** Department of Mathematics, Pedagogical University, Podchorążych 2, 30-084 Kraków, Poland, kc@up.krakow.pl

**Iulia Costin** Department of Computer Sciences, Technical University of Cluj-Napoca, Cluj-Napoca, Romania, Iulia.Costin@cs.utcluj.ro

**Stefan Czerwik** Department of Mathematics, Silesian University of Technology, Kaszubska 23, 44-100 Gliwice, Poland, stefan.czerwik@polsl.pl

**Sever S. Dragomir** Department of Mathematics, School of Enginering & Science, Victoria University, PO Box 14428, Melbourne, VIC 8001, Australia, sever.dragomir@vu.edu.au

**Konstantinos Drakakis** Complex and Adaptive Systems Laboratory (UCD CASL), University College Dublin, Belfield, Dublin 4, Ireland

The School of Electronic, Electrical & Mechanical Engineering, University College Dublin, Dublin, Ireland, Konstantinos.Drakakis@ucd.ie

**Elhoucien Elqorachi** Department of Mathematics, Faculty of Sciences, University Ibn Zohr, Agadir, Morocco, elqorachi@hotmail.com

**M. Eshaghi-Gordji** Department of Mathematics, Semnan University, P. O. Box 35195-363, Semnan, Iran, madjid.eshaghi@gmail.com

**Jinmei Gao** Department of Mathematics, Qingdao University, Qingdao 266071, China

Department of Mathematics, Nankai University, Tianjin 300071, China, gjinmei1979@163.com

**Laura Găvruţa** Department of Mathematics, "Politehnica" University of Timişoara, Piaţa Victoriei 2, 300006 Timişoara, Romania, Laura.Gavruta@mat.upt.ro

**Paşc Găvruţa** Department of Mathematics, "Politehnica" University of Timişoara, Piaţa Victoriei 2, 300006 Timişoara, Romania, pgavruta@yahoo.com

**Takahiro Hayata** Department of Informatics, Graduate School of Science and Engineering, Yamagata University, Yonezawa 992-8510, Japan, hayata@yz.yamagata-u.ac.jp

**Go Hirasawa** Faculty of Engineering, Ibaraki University, Hitachi 316-8511, Japan, gou@mx.ibaraki.ac.jp

**Eliza Jabłońska** Department of Mathematics, Rzeszów University of Technology, W. Pola 2, 35–959 Rzeszów, Poland, elizapie@prz.edu.pl

**Wojciech Jabłoński** Department of Mathematics, University of Rzeszów, ul. Rejtana 16 A, 35–310 Rzeszów, Poland, wojciech@univ.rzeszow.pl

**Kil-Woung Jun** Department of Mathematics, Chungnam National University, Taejon 305-764, Republic of Korea, kwjun@cnu.ac.kr

**S. Kabbaj** Department of Mathematics, Faculty of Sciences, Ibn Tofail University, 14000 Kenitra, Morocco, samkabbaj@yahoo.fr

**Hans-Heinrich Kairies** Institut für Mathematik, Technische Universität Clausthal, Erzstraße 1, 38678 Clausthal-Zellerfeld, Germany, kairies@math.tu-clausthal.de

**H. Khodaei** Department of Mathematics, Semnan University, P. O. Box 35195-363, Semnan, Iran, khodaei.hamid.math@gmail.com

**Hark-Mahn Kim** Department of Mathematics, Chungnam National University, 79 Daehangno, Yuseong-gu, Daejeon 305-764, Korea, hmkim@cnu.ac.kr

**George Kokolakis** Department of Mathematics, National Technical University of Athens, Zografou Campus, 15780 Athens, Greece, kokolakis@math.ntua.gr

**George Kouvaras** Department of Mathematics, National Technical University of Athens, Zografou Campus, 15780 Athens, Greece, gkouv@math.ntua.gr

**Yang-Hi Lee** Department of Mathematics Education, Kongju National University of Education, Kongju 314-711, Republic of Korea, yanghi2@hanmail.net

**Youssef Manar** Department of Mathematics, Faculty of Sciences, University Ibn Zohr, Agadir, Morocco, manaryoussefr32@hotmail.fr

**Ricardo Pérez Marco** CNRS, LAGA UMR 7539, Université Paris XIII, 99, Avenue J.-B. Clément, 93430-Villetaneuse, France, ricardo@math.univ-paris13.fr

**Janusz Matkowski** Faculty of Mathematics Computer Science and Econometrics, University of Zielona Góra, Podgórna 50, 65-246 Zielona Góra, Poland

Institute of Mathematics, Silesian University, Bankowa 14, 42-007 Katowice, Poland, J.Matkowski@wmie.uz.zgora.pl

**Krzysztof Misztal** Institute of Computer Science, Jagiellonian University, Łojasiewicza 6, 30-348 Kraków, Poland, krzysztof.misztal@ii.uj.edu.pl

**Vladimir V. Mityushev** Department of Computer Science and Computer Methods, Pedagogical University, Podchorążych 2, 30-084 Kraków, Poland, vmityu@yahoo.com

**Takeshi Miura** Department of Applied Mathematics and Physics, Yamagata University, Yonezawa 992-8510, Japan, miura@yz.yamagata-u.ac.jp

**Vicente Muñoz** Facultad de Matemáticas, Universidad Complutense de Madrid, Plaza Ciencias 3, 28040 Madrid, Spain, vicente.munoz@mat.ucm.es

**Anna Mureńko** Department of Mathematics, University of Rzeszów, Rejtana 16 A, 35–959 Rzeszów, Poland, aniam@univ.rzeszow.pl

**Abbas Najati** Department of Mathematics, Faculty of Sciences, University of Mohaghegh Ardabili, Ardabil, Iran, a.nejati@yahoo.com

**Prem Nath** Department of Mathematics, University of Delhi, Delhi 110007, India, pnathmaths@gmail.com

**Muhammad Aslam Noor** Mathematics Department, COMSATS Institute of Information Technology, Islamabad, Pakistan

Mathematics Department, College of Science, King Saud University, Riyadh, Saudi Arabia, mnoor.c@ksu.edu.sa

**Khlaida Inayat Noor** COMSATS Institute of Information Technology, Islamabad, Pakistan, khalidanoor@hotmail.com

**Choonkil Park** Department of Mathematics, Hanyang University, Seoul 133-791, Republic of Korea, baak@hanyang.ac.kr

**Dorian Popa** Department of Mathematics, Technical University of Cluj-Napoca, Str. Memorandumului 28, 400114 Cluj-Napoca, Romania, Popa.Dorian@math.utcluj.ro

**Vladimir Yu. Protasov** Department of Mechanics and Mathematics, Moscow State University, Vorobyovy Gory, Moscow 119992, Russia, v-protassov@yandex.ru

**Viorel Radu** Faculty of Mathematics and Computer Science, West University of Timişoara, Bv. Vasile Pârvan 4, 300223 Timişoara, Romania, radu@math.uvt.ro; viorelvradu@yahoo.com

**M. Ramezani** Department of Mathematics, Semnan University, P. O. Box 35195-363, Semnan, Iran, ramezanimaryam873@gmail.com

**Themistocles M. Rassias** Department of Mathematics, National Technical University of Athens, Zografou Campus, 15780 Athens, Greece, trassias@math.ntua.gr

**A. Roukbi** Department of Mathematics, Faculty of Sciences, Ibn Tofail University, 14000 Kenitra, Morocco, rroukbi.a2000@gmail.com

**Ioan A. Rus** Department of Applied Mathematics, Babeş–Bolyai University, Kogălniceanu Street, No. 1, 400084 Cluj-Napoca, Romania, iarus@math.ubbcluj.ro

**Prasanna K. Sahoo** Department of Mathematics, University of Louisville, Louisville, KY 40292 USA, sahoo@louisville.edu

**Dhiraj Kumar Singh** Department of Mathematics, University of Delhi, Delhi 110007, India, dksingh@maths.du.ac.in; dhiraj426@rediffmail.com

**Jiae Son** Department of Mathematics, Chungnam National University, 79 Daehangno, Yuseong-gu, Daejeon 305-764, Korea, hand0627@cnu.ac.kr

**Jacek Tabor** Institute of Computer Science, Jagiellonian University, Łojasiewicza 6, 30-348 Kraków, Poland, tabor@ii.uj.edu.pl

**Józef Tabor** Institute of Mathematics, University of Rzeszów, Rejtana 16A, 35-310 Rzeszów, Poland, tabor@univ.rzeszow.pl

**Sin-Ei Takahasi** Department of Applied Mathematics and Physics, Graduate School of Science and Engineering, Yamagata University, Yonezawa 992-8510, Japan, sin-ei@emperor.yz.yamagata-u.ac.jp

**Gheorghe Toader** Department of Mathematics, Technical University of Cluj-Napoca, Cluj-Napoca, Romania, Gheorghe.Toader@math.utcluj.ro

**Silvia Toader** Department of Mathematics, Technical University of Cluj-Napoca, Cluj-Napoca, Romania, Silvia.Toader@math.utcluj.ro

**Zhihua Wang** School of Science, Hubei University of Technology, Wuhan, Hubei 430068, PR China, matwzh2000@126.com

**Bicheng Yang** Department of Mathematics, Guangdong Education Institute, Guangzhou 510303, China, bcyang@pub.guangzhou.gd.cn

**G. Zamani Eskandani** Faculty of Mathematical Sciences, University of Tabriz, Tabriz, Iran, zamani@tabrizu.ac.ir

**D. Zeglami** Department of Mathematics, Faculty of Sciences, Ibn Tofail University, 14000 Kenitra, Morocco, zeglamidriss@yahoo.fr

**Chang-Jian Zhao** Department of Information and Mathematics Sciences, College of Science, China Jiliang University, Hangzhou 310018, PR China, chjzhao@yahoo.com.cn; chjzhao@163.com; chjzhao@cjlu.edu.cn

**Wanxiong Zhang** College of Mathematics and Statistics, Chongqing University, Chongqing 401331, PR China, cqumatzwx@163.com

# Part I
# Stability in Mathematical Analysis

# Chapter 1
# Stability Properties of Some Functional Equations

**Roman Badora**

*Dedicated to the memory of S.M. Ulam*
*on the 100th anniversary of his birth*

**Abstract** We prove stability results for a family of functional equations.

**Keywords** Functional equation • Stability

**Mathematics Subject Classification (2000):** Primary 39B82

## 1.1 Introduction

This paper deals with the stability properties of functional equations which are joint generalizations of some classical equations (e.g., of Cauchy, Jensen, d'Alembert, Wilson, exponential and quadratic). For information concerning the stability we refer to Forti [10], Ger [11] and Hyers et al. [13].

Throughout this note $\mathbb{N}$, $\mathbb{R}$ and $\mathbb{C}$ stand, as usual, for the set of positive integers, reals and complex numbers, respectively. Moreover, $\mathbb{K}$ is either the field $\mathbb{R}$ or $\mathbb{C}$, $X$ is an abelian group and $\Lambda$ is a finite subgroup of the automorphism group of $X$ (the action of $\lambda \in \Lambda$ on $x \in X$ is denoted by $\lambda x$) and $N$ is the cardinality of $\Lambda$.

The stability problem for the functional equation

$$\frac{1}{N}\sum_{\lambda\in\Lambda} f(x+\lambda y) = f(x)g(y), \tag{1.1}$$

where $f,g : X \to \mathbb{K}$, was posed and solved by Badora in [3] (see also [2]). Equation (1.1) is a joint generalization of the exponential functional equation ($\Lambda = \{Id\}$, $g = f$), d'Alembert's equation ($\Lambda = \{Id, -Id\}$, $g = f$) and Wilson's

R. Badora (✉)
Institute of Mathematics, Silesian University, Bankowa 14, PL-40-007 Katowice, Poland
e-mail: robadora@math.us.edu.pl

Th.M. Rassias and J. Brzdęk (eds.), *Functional Equations in Mathematical Analysis*, Springer Optimization and Its Applications 52, DOI 10.1007/978-1-4614-0055-4_1,

equation ($\Lambda = \{Id, -Id\}$). In [4], Badora gave a generalization of his results to the case of the following equation

$$\int_{\Lambda} f(x+\lambda y)\mathrm{d}\mu(\lambda) = f(x)g(y), \tag{1.2}$$

where $X$ is a topological abelian group, $\Lambda$ is a compact subgroup of authomorphisms of $X$ with the normalized Haar measure $\mu$ and $f, g : X \to \mathbb{K}$. Since then, stability problems concerning (1.2) and its generalizations have been studied (see [7–9]). In 2005, Badora during the 10th ICFEI (see [5]) presented a simple solution of the stability problem for the following functional equation

$$\frac{1}{N} \sum_{\lambda \in \Lambda} f(x+\lambda y) = f(x) + h(y),$$

where $f, h : X \to \mathbb{K}$, which covers Jensen's equation ($\Lambda = \{Id, -Id\}$, $h = 0$), Cauchy's equation ($\Lambda = \{Id,\}$, $h = f$) and the functional equation of the square of the norm ($\Lambda = \{Id, -Id\}$, $h = f$). Later, this result was published (with a different proof and $h = f$) by Ait Sibaha et al. in [1] and generalized by Charifi et al. in [6].

In this paper, we study more interesting problem concerning the stability of the following functional equations

$$\frac{1}{N} \sum_{\lambda \in \Lambda} f(x+\lambda y) = f(x)g(y) + h(y); \tag{1.3}$$

$$\frac{1}{N} \sum_{\lambda \in \Lambda} f(x+\lambda y) = f(y)g(x) + h(x), \tag{1.4}$$

where $f, g, h : X \to \mathbb{K}$. These type of equations was studied by Stetkaer in [15]. In the case when $\Lambda = \{Id\}$ the problem of the stability of our equation was solved by Székelyhidi in [14].

## 1.2 Stability of (1.3)

First, we prove that if the function $f$ is bounded then (1.3) (and, with the same proof, equation (1.4)) is superstable with respect to the function $g$ and stable in the Hyers–Ulam sense with respect to functions $f$ and $h$.

**Theorem 1.1.** *If $f, g, h : X \to \mathbb{K}$, the function*

$$X \times X \ni (x,y) \mapsto \frac{1}{N} \sum_{\lambda \in \Lambda} f(x+\lambda y) - f(x)g(y) - h(y) \in \mathbb{K} \tag{1.5}$$

*is bounded and the function $f$ is bounded, then there exist $F,H : X \to \mathbb{K}$ such that*

$$\frac{1}{N}\sum_{\lambda\in\Lambda} F(x+\lambda y) = F(x)g(y)+H(y), \quad x,y\in X,$$

*and $f-F$, $h-H$ are bounded.*

*Proof.* If $f$ is bounded, then the function

$$X\times X \ni (x,y) \mapsto f(x)g(y)+h(y)\in\mathbb{K} \tag{1.6}$$

is also bounded.

First, we assume that $h$ is a bounded function. Then, we can take $F=0$ and $H=0$.

If $h$ is an unbounded function, then $f$ is a non-zero function and $g$ is also unbounded. For a given $x_1,x_2\in X$ let $a=f(x_1)$ and $b=f(x_2)$. Moreover, let $(y_n)$ be a sequence in $X$ with $\mid h(y_n)\mid\to\infty$. Then, $\mid g(y_n)\mid\to\infty$. By (1.6) the sequences

$$(g(y_n)a+h(y_n)) \quad \text{and} \quad (g(y_n)b+h(y_n))$$

are bounded. Hence, the sequence $(g(y_n)(a-b))$ is bounded which means that $a=b$ and $f$ is a constant function $f=c$, for some $c\in\mathbb{K}$. Then, $F=c$ and $H=-cg+c$ have all required properties. □

Since now we can assume that the function $f$ is unbounded.

First, we prove that (1.3) is superstable with respect to $g$.

**Theorem 1.2.** *If $f,g,h : X\to\mathbb{K}$, the function (1.5) is bounded on $X\times X$ and the function $f$ is unbounded, then the function $g$ satisfies the equation*

$$\frac{1}{N}\sum_{\lambda\in\Lambda} g(x+\lambda y) = g(x)g(y), \quad x,y\in X. \tag{1.7}$$

*Proof.* Assume that the absolute values of the function (1.5) is bounded by $M\geq 0$. Since $f$ is unbounded, there exists a sequence $(z_n)$ such that

$$0\neq\mid f(z_n)\mid\to\infty.$$

Then our assumption implies that

$$\left|\frac{1}{N}\sum_{\lambda\in\Lambda}\frac{f(z_n+\lambda y)}{f(z_n)} - g(y) - \frac{h(y)}{f(z_n)}\right| \leq \frac{M}{\mid f(z_n)\mid}, \quad y\in X,$$

which gives

$$g(y) = \lim_{n\to\infty}\frac{1}{N}\sum_{\lambda\in\Lambda}\frac{f(z_n+\lambda y)}{f(z_n)}, \quad y\in X. \tag{1.8}$$

Next, we get

$$\left|\frac{1}{N^2}\sum_{\kappa\in\Lambda}\sum_{\lambda\in\Lambda} f(z_n+\kappa x+\lambda y) - \frac{1}{N}\sum_{\kappa\in\Lambda} f(z_n+\kappa x)g(y)-h(y)\right| \le \frac{1}{N}\sum_{\kappa\in\Lambda} M = M$$

for all $x,y \in X$. Consequently, using the boundedness of the function (1.5), we obtain

$$\left|\frac{1}{N}\sum_{\lambda\in\Lambda}\left(\frac{1}{N}\sum_{\kappa\in\Lambda}\frac{f(z_n+\kappa(x+\lambda y))}{f(z_n)}\right) - \left(\frac{1}{N}\sum_{\kappa\in\Lambda}\frac{f(z_n+\kappa x)}{f(z_n)}\right)g(y)-\frac{h(y)}{f(z_n)}\right| \le \frac{M}{|f(z_n)|}.$$

Now, by (1.8) we get that the function $g$ satisfies (1.7). □

The next result shows that (1.3) is stable in the Hyers–Ulam sense with respect to $h$.

**Theorem 1.3.** *If $f,g,h : X \to \mathbb{K}$, the function (1.5) is bounded on $X \times X$ and the function $f$ is unbounded, then there exists a function $H : X \to \mathbb{K}$ fulfilling the equation*

$$\frac{1}{N}\sum_{\lambda\in\Lambda} H(x+\lambda y) = H(x)g(y)+H(y), \quad x,y \in X \tag{1.9}$$

*and such that $h-H$ is bounded.*

*Proof.* Let $x \in X$ be fixed. Then our assumption implies that the function

$$X \ni z \mapsto \frac{1}{N}\sum_{\lambda\in\Lambda} f(z+\lambda x) - f(z)g(x) \in \mathbb{K}$$

is bounded. Let $m$ be an invariant mean on the space $B(X,\mathbb{K})$ (see [12]). We define the map $H : X \to \mathbb{K}$ by the formula:

$$H(x) = m_z\left(\frac{1}{N}\sum_{\lambda\in\Lambda} f(z+\lambda x) - f(z)g(x)\right), \quad x \in X,$$

where the subscript $z$ next to $m$ indicates that the mean $m$ is applied to a function of the variable $z$. Then the difference $h-H$ is bounded. Moreover, for $x,y \in X$, from the linearity of $m$ and Theorem 1.2, we obtain

$$\begin{aligned}
\frac{1}{N}\sum_{\lambda\in\Lambda}H(x+\lambda y) &= \frac{1}{N}\sum_{\lambda\in\Lambda}m_z\left(\frac{1}{N}\sum_{\kappa\in\Lambda}f(z+\kappa x+\kappa\lambda y)-f(z)g(x+\lambda y)\right)\\
&= m_z\left(\frac{1}{N^2}\sum_{\lambda\in\Lambda}\sum_{\kappa\in\Lambda}f(z+\kappa x+\kappa\lambda y)-f(z)\sum_{\lambda\in\Lambda}g(x+\lambda y)\right)\\
&= m_z\left(\frac{1}{N^2}\sum_{\kappa\in\Lambda}\sum_{\lambda\in\Lambda}f(z+\kappa x+\kappa\lambda y)-f(z)g(x)g(y)\right)\\
&= m_z\left(\frac{1}{N^2}\sum_{\kappa\in\Lambda}\sum_{\lambda\in\Lambda}f(z+\kappa x+\lambda y)-f(z)g(x)g(y)\right)\\
&= m_z\left(\frac{1}{N}\sum_{\kappa\in\Lambda}\left(\frac{1}{N}\sum_{\lambda\in\Lambda}f(z+\kappa x+\lambda y)-f(z+\kappa x)g(y)\right)\right.\\
&\quad\left.+\left(\frac{1}{N}\sum_{\kappa\in\Lambda}f(z+\kappa x)-f(z)g(x)\right)g(y)\right)\\
&= m_z\left(\frac{1}{N}\sum_{\kappa\in\Lambda}\left(\frac{1}{N}\sum_{\lambda\in\Lambda}f(z+\kappa x+\lambda y)-f(z+\kappa x)g(y)\right)\right)\\
&\quad+m_z\left(\frac{1}{N}\sum_{\kappa\in\Lambda}f(z+\kappa x)-f(z)g(x)\right)g(y)\\
&= \frac{1}{N}\sum_{\kappa\in\Lambda}m_z\left(\frac{1}{N}\sum_{\lambda\in\Lambda}f(z+\kappa x+\lambda y)-f(z+\kappa x)g(y)\right)\\
&\quad+H(x)g(y).
\end{aligned}$$

The left invariance of $m$ yields

$$\begin{aligned}
\frac{1}{N}\sum_{\lambda\in\Lambda}H(x+\lambda y) &= \frac{1}{N}\sum_{\kappa\in\Lambda}m_z\left(\frac{1}{N}\sum_{\lambda\in\Lambda}f(z+\kappa x+\lambda y)-f(z+\kappa x)g(y)\right)\\
&\quad+H(x)g(y)\\
&= \frac{1}{N}\sum_{\kappa\in\Lambda}m_z\left(\frac{1}{N}\sum_{\lambda\in\Lambda}f(z+\lambda y)-f(z)g(y)\right)+H(x)g(y)\\
&= \frac{1}{N}\sum_{\kappa\in\Lambda}H(y)+H(x)g(y)=H(x)g(y)+H(y).
\end{aligned}$$

□

Now, we will look after the function $f$. First, we observe the following.

*Remark 1.1.* If $M \geq 0$, functions $f, g, h : X \to \mathbb{K}$ satisfy

$$\left| \frac{1}{N} \sum_{\lambda \in \Lambda} f(x+\lambda y) - f(x)g(y) - h(y) \right| \leq M, \quad x, y \in X$$

and

$$\phi = \frac{1}{N} \sum_{\lambda \in \Lambda} f \circ \lambda,$$

then $\phi, g, h$ satisfy

$$\left| \frac{1}{N} \sum_{\lambda \in \Lambda} \phi(x+\lambda y) - \phi(x)g(y) - h(y) \right| \leq M, \quad x, y \in X$$

and $\phi \circ \lambda = \phi$ for all $\lambda \in \Lambda$.

*Proof.* Our inequality implies that

$$\begin{aligned} &\left| \frac{1}{N^2} \sum_{\gamma \in \Lambda} \sum_{\lambda \in \Lambda} f(\gamma x + \lambda y) - \frac{1}{N} \sum_{\gamma \in \Lambda} f(\gamma x)g(y) - h(y) \right| \\ &\leq \frac{1}{N} \sum_{\gamma \in \Lambda} \left| \frac{1}{N} \sum_{\lambda \in \Lambda} f(\gamma x + \lambda y) - f(\gamma x)g(y) - h(y) \right| \leq M, \quad x, y \in X. \end{aligned}$$

On the other hand, for $x, y \in X$, we get

$$\begin{aligned} &\left| \frac{1}{N^2} \sum_{\gamma \in \Lambda} \sum_{\lambda \in \Lambda} f(\gamma x + \lambda y) - \frac{1}{N} \sum_{\gamma \in \Lambda} f(\gamma x)g(y) - h(y) \right| \\ &= \left| \frac{1}{N} \sum_{\lambda \in \Lambda} \frac{1}{N} \sum_{\gamma \in \Lambda} f(\gamma(x + \lambda y)) - \frac{1}{N} \sum_{\gamma \in \Lambda} f(\gamma x)g(y) - h(y) \right| \\ &= \left| \frac{1}{N} \sum_{\lambda \in \Lambda} \phi(x+\lambda y) - \phi(x)g(y) - h(y) \right|. \end{aligned}$$

□

The main result concerning the stability of (1.3), in the case when the function $f$ is unbounded, reads as follows.

**Theorem 1.4.** *If $f, g, h : X \to \mathbb{K}$, the function (1.5) is bounded on $X \times X$, the function $f$ is unbounded and $f \circ \lambda = f$, for all $\lambda \in \Lambda$, then there exist functions $F, H : X \to \mathbb{K}$ fulfilling the following equation*

$$\frac{1}{N} \sum_{\lambda \in \Lambda} F(x+\lambda y) = F(x)g(y) + H(y), \quad x, y \in X$$

*and such that $f - F$, $h - H$ are bounded.*

*Proof.* From Theorem 2 we obtain the existence of a function $H : X \to \mathbb{K}$ such that the function $h - H$ is bounded. By our assumption, putting $x = 0$, we have that the function

$$X \ni y \mapsto \frac{1}{N} \sum_{\lambda \in \Lambda} f(\lambda y) - f(0)g(y) - h(y) = f(y) - f(0)g(y) - h(y) \in \mathbb{K}$$

is bounded. Let $F = c \cdot g + H$, where $c = f(0)$. Then the difference $F - f = (F - cg - h) + (h - H)$ is bounded. Moreover, using the fact that the function $g$ satisfies (1.7) and the function $H$ satisfies (1.9), for all $x, y \in X$, we get

$$\begin{aligned} \frac{1}{N} \sum_{\lambda \in \Lambda} F(x + \lambda y) &= c \frac{1}{N} \sum_{\lambda \in \Lambda} g(x + \lambda y) \\ &\quad + \frac{1}{N} \sum_{\lambda \in \Lambda} H(x + \lambda y) \\ &= cg(x)g(y) + H(x)g(y) + H(y) \\ &= F(x)g(y) + H(y). \end{aligned}$$

□

*Remark 1.2.* Without the assumption $f \circ \lambda = f$, for all $\lambda \in \Lambda$, by Remark 1.1, we only get the existence of the function $F$ which is close to the function

$$\frac{1}{N} \sum_{\lambda \in \Lambda} f \circ \lambda.$$

## 1.3 Stability of (1.4)

Passing to the stability of (1.4), similarly to the proof of Theorem 2, we get the following superstability result on $g$.

**Theorem 1.5.** *If $f, g, h : X \to \mathbb{K}$, the function*

$$X \times X \ni (x, y) \mapsto \frac{1}{N} \sum_{\lambda \in \Lambda} f(x + \lambda y) - f(y)g(x) - h(x) \in \mathbb{K} \qquad (1.10)$$

*is bounded on $X \times X$ and the function $f$ is unbounded, then the function $g$ satisfies (1.7).*

In the proof of the next result, we use the following fact concerning invariant means.

*Remark 1.3.* Let $\Lambda$ be a left amenable semigroup of the automorphism group of an amenable group $X$. Then there exists an invariant mean $M$ on the space $B(X,\mathbb{K})$ such that

$$M(f\circ\lambda)=M(f),\quad f\in B(X,\mathbb{K}),\ \lambda\in\Lambda. \tag{1.11}$$

*Proof.* Let $m$ be a left invariant mean on the space $B(X,\mathbb{K})$ and let $\mu$ be a left invariant mean on $B(\Lambda,\mathbb{K})$. Then the function

$$\Lambda\ni\lambda\mapsto m_x((f\circ\lambda)(x))\in\mathbb{K}$$

is bounded for each $f\in B(X,\mathbb{K})$. So, we can define the functional $M$ on the space $B(X,\mathbb{K})$ by

$$M(f)=\mu(m(f\circ\lambda)),\quad f\in B(X,\mathbb{K}).$$

Then $M$ is a mean. Moreover, $M$ is left invariant: for $f\in B(X,\mathbb{K})$ and $a\in X$ we have

$$\begin{aligned}
M({}_af) &= \mu_\lambda(m_x((({}_af)\circ\lambda)(x)))\\
&= \mu_\lambda(m_x({}_af(\lambda x)))\\
&= \mu_\lambda(m_x(f(a+\lambda x)))\\
&= \mu_\lambda(m_x(f(\lambda(\lambda^{-1}a+x))))\\
&= \mu_\lambda(m_x(f(\lambda x)))=M(f).
\end{aligned}$$

Next, by the left left invariance of $\mu$, $M$ has property (1.11):

$$\begin{aligned}
M(f\circ\lambda) &= \mu_\kappa(m_x((f\circ\lambda\circ\kappa)(x)))\\
&= \mu_\kappa(m_x((f\circ\kappa)(x))=M(f).
\end{aligned}$$

□

Now, for a function $h$ now we have the following.

**Theorem 1.6.** *If $f,g,h:X\to\mathbb{K}$, the function (1.10) is bounded and the function $f$ is unbounded, then there exists a function $H:X\to\mathbb{K}$ fulfilling the equation*

$$\frac{1}{N}\sum_{\lambda\in\Lambda}H(x+\lambda y)=H(y)g(x)+H(x),\quad x,y\in X$$

*and such that $h-H$ is bounded.*

*Proof.* Let $x\in X$ be fixed. If the function (1.10) is bounded, then the function

$$X\ni z\mapsto\frac{1}{N}\sum_{\lambda\in\Lambda}f(z+\lambda x)-f(z)g(x)\in\mathbb{K}$$

is also bounded. We can define the map $H : X \to \mathbb{K}$ by the following formula:

$$H(x) = M_z\left(\frac{1}{N}\sum_{\lambda\in\Lambda} f(x+\lambda z) - f(z)g(x)\right), \quad x \in X,$$

where $M$ is an invariant mean on $B(X,\mathbb{K})$ satisfying condition (1.11) which existence is guaranteed by Remark 1.3. Then the difference $h - H$ is bounded. Moreover, for $x, y \in X$, from the linearity of $M$ and Theorem 1.5, we get

$$\begin{aligned}
&\frac{1}{N}\sum_{\lambda\in\Lambda} H(x+\lambda y)\\
&= \frac{1}{N}\sum_{\lambda\in\Lambda} M_z\left(\frac{1}{N}\sum_{\kappa\in\Lambda} f(x+\lambda y+\kappa z) - f(z)g(x+\lambda y)\right)\\
&= M_z\left(\frac{1}{N^2}\sum_{\lambda\in\Lambda}\sum_{\kappa\in\Lambda} f(x+\lambda y+\kappa z) - f(z)\sum_{\lambda\in\Lambda} g(x+\lambda y)\right)\\
&= M_z\left(\frac{1}{N^2}\sum_{\lambda\in\Lambda}\sum_{\kappa\in\Lambda} f(x+\lambda y+\lambda\kappa z) - f(z)g(x)g(y)\right)\\
&= M_z\left(\frac{1}{N^2}\sum_{\lambda\in\Lambda}\sum_{\kappa\in\Lambda} f(x+\lambda(y+\kappa z))\right.\\
&\qquad -\frac{1}{N}\sum_{\kappa\in\Lambda} f(y+\kappa z)g(x)\\
&\qquad \left. + \left(\frac{1}{N}\sum_{\kappa\in\Lambda} f(y+\kappa z) - f(z)g(y)\right)g(x)\right).
\end{aligned}$$

Applying property (1.11) and the left invariance of $M$ we get

$$\begin{aligned}
&\frac{1}{N}\sum_{\lambda\in\Lambda} H(x+\lambda y)\\
&= M_z\left(\frac{1}{N^2}\sum_{\lambda\in\Lambda}\sum_{\kappa\in\Lambda} f(x+\lambda(y+z)) - \frac{1}{N}\sum_{\kappa\in\Lambda} f(y+z)g(x)\right)\\
&\quad + M_z\left(\frac{1}{N}\sum_{\kappa\in\Lambda} f(y+\kappa z) - f(z)g(y)\right)g(x)
\end{aligned}$$

$$= \frac{1}{N} \sum_{\kappa \in \Lambda} M_z \left( \frac{1}{N} \sum_{\lambda \in \Lambda} f(x + \lambda z) - f(z) g(x) \right) + H(y) g(x)$$

$$= \frac{1}{N} \sum_{\kappa \in \Lambda} H(x) + h(y) g(x) = H(y) g(x) + H(x). \qquad \square$$

The main result of this part of the paper reads as follows.

**Theorem 1.7.** *If $f, g, h : X \to \mathbb{K}$, the function (1.10) is bounded on $X \times X$ and the function $f$ is unbounded, then there exist functions $F, H : X \to \mathbb{K}$ satisfying*

$$\frac{1}{N} \sum_{\lambda \in \Lambda} F(x + \lambda y) = F(y) g(x) + H(x), \quad x, y \in X \tag{1.12}$$

*and such that $f - F$, $h - H$ are bounded.*

*Proof.* By Theorem 1.6 there exists a function H which satisfies (1.11) and such that the difference $h - H$ is a bounded function. Let $F = c \cdot g + H$, where $c = f(0)$. Then our assumption (with $y = 0$) and the boundedness of $h - H$ imply that the function $f - F$ is bounded. Moreover, for $x, y \in X$, we have

$$\begin{aligned} \frac{1}{N} \sum_{\lambda \in \Lambda} F(x + \lambda y) &= c \frac{1}{N} \sum_{\lambda \in \Lambda} g(x + \lambda y) \\ &\quad + \frac{1}{N} \sum_{\lambda \in \Lambda} H(x + \lambda y) \\ &= c g(x) g(y) + H(y) g(x) + H(x) \\ &= F(y) g(x) + H(x). \end{aligned}$$

$\square$

From Theorem 1.1 and Theorem 1.7 we have the following stability result concerning (1.4).

**Corollary 1.1.** *If $f, g, h : X \to \mathbb{K}$ and the function (1.10) is bounded on $X \times X$, then there exist functions $F, H : X \to \mathbb{K}$ satisfying (1.12) and such that $f - F$, $h - H$ are bounded.*

## References

1. Sibaha, A., Bouikhalene, B., Elqorachi, E.: Hyers–Ulam–Rassias stability of the $K$-quadratic functional equation. J. Ineq. Pure Appl. Math. **8**, article 89 (2007)
2. Badora, R.: Stability of $K$-spherical functions. In: Report of Meeting, The 34th International Symposium on Functional Equations (June 10–19, 1996, Wisła-Jawornik, Poland), p. 164. Aequationes Math. **53** (1997)

3. Badora, R.: On the stability of a functional equation for generalized trigonometric functions. In: Th.M. Rassias (ed.) Functional Equations and Inequalities, pp. 1–5. Kluwer Academic Publishers (2000)
4. Badora, R.: On Hyers–Ulam stability of Wilson's functional equation. Aequationes Math. **60**, 211–218 (2000)
5. Badora, R.: On the stability of some functional equations. In: Report of Meeting, 10th International Conference on Functional Equations and Inequalities (September 11-17, 2005, Będlewo, Poland), p.130. Ann. Acad. Paed. Cracoviensis Studia Math. **5** (2006)
6. Charifi, A., Bouikhalene, B., Elqorachi, E.: Hyers–Ulam–Rassias stability of a generalized Pexider functional equation. Banach J. Math. Anal. **1**, 176–185 (2007)
7. Elqorach, E., Akkouchi, M.: The stability of the generalized d'Alembert and Wilson functional equations. Aequationes Math. **66**, 241–256 (2003)
8. Elqorach, E., Akkouchi, M.: On Hyers–Ulam stability of Cauchy and Wilson equations. Georgian Math. J. **11**, 69–82 (2004)
9. Elqorach, E., Akkouchi, M.: On Hyers–Ulam stability of the generalized Cauchy and Wilson equations. Publ. Math. Debrecen **66**, 283–301 (2005)
10. Forti, G.L.: Hyers–Ulam stability of functional equations in several variables. Aequationes Math. **50**, 143–190 (1995)
11. Ger, R.: A survey of recent results on stability of functional equations. In: Proc. of the 4th International Conference on Functional Equations and Inequalities (Cracow), pp. 5–36. Pedagogical University of Cracow, Poland (1994)
12. Greenleaf, F.P.: Invariant means on topological groups and their applications. Van Nostrand Mathematical Studies 16, New York–Toronto–London–Melbourne (1969)
13. Hyers, D.H., Isac, G., Rassias, Th.M.: Stability of functional equations in several variables. Birkhäuser, Boston–Basel–Berlin (1998)
14. Székelyhidi, L.: Stability properties of functional equations describing the scientific laws. J. Math. Anal. Appl. **150**, 151–158 (1990)
15. Stetkaer, H.: Functional equations and matrix-valued spherical functions. Aequationes Math. **69**, 271–292 (2005)

# Chapter 2
# Note on Superstability of Mikusiński's Functional Equation

**Bogdan Batko**

*Dedicated to the memory of S.M. Ulam*
*on the 100th anniversary of his birth*

**Abstract** We show superstability of Mikusiński's functional equation

$$f(x+y)\,(f(x+y)-f(x)-f(y))=0.$$

**Keywords** Stability • Superstability • Conditional Cauchy equation • Mikusiński's equation

**Mathematics Subject Classification (2000):** Primary 39B82; Secondary 39B55

## 2.1 Introduction

Certain geometrical considerations have led J. Mikusiński to the functional equation

$$f(x+y)((f(x+y)-f(x)-f(y))=0, \tag{2.1}$$

for the continuous real-valued function $f$ of the real variable. Equation (2.1) is usually written in the conditional form

$$f(x+y)\neq 0 \Longrightarrow f(x+y)=f(x)+f(y), \tag{2.2}$$

which enables us to deal with it more generally – in structures endowed only with one binary operation.

B. Batko (✉)
Department of Mathematics, WSB – NLU, Zielona 27, 33-300 Nowy Sącz, Poland

Department of Mathematics, Pedagogical University, Podchorążych 2, 30-084 Kraków, Poland
e-mail: bbatko@wsb-nlu.edu.pl; bbatko@ap.krakow.pl

Th.M. Rassias and J. Brzdęk (eds.), *Functional Equations in Mathematical Analysis*, Springer Optimization and Its Applications 52, DOI 10.1007/978-1-4614-0055-4_2,

The general sulution of Mikusiński's equation (2.2) is described in [4]. Stability of Mikusiński's equation in both forms (2.1) and (2.2) is proved in [1].

We show that the method for proving superstability of conditional Cauchy equations, proposed in [6] (see also [2, 3]), is applicable to Mikusiński's equation.

## 2.2 Superstability

We use [1, Theorem 2] in order to prove superstability of Mikusiński's equation (2.1).

**Theorem 2.1.** *Let* $(G,+)$ *be an Abelian group and a function* $f : G \to \mathbb{C}$ *satisfy*

$$|f(x+y)(f(x+y) - f(x) - f(y))| \leq \varepsilon \quad \text{for } x,y \in G, \tag{2.3}$$

*with some* $\varepsilon \geq 0$. *Then* $f$ *is additive, or bounded with* $|f(x)| \leq 2\sqrt{6\varepsilon}$ *for* $x \in G$.

*Proof.* By [1, Theorem 2] there exists an additive function $a : G \to \mathbb{C}$ with

$$| f(x) - a(x) | \leq 2\sqrt{6\varepsilon} \qquad \text{for } x \in G. \tag{2.4}$$

If $f$ is bounded, then $a = 0$ and $| f(x) | \leq 2\sqrt{6\varepsilon}$. Thus, let us consider $f$ unbounded. According to (2.4) $a$ is nontrivial and there is a bounded function $b$ such that $f = a + b$. Taking into account this representation and (2.3) one can easily see that the function

$$G \ni y \longmapsto a(y)(b(x+y) - b(x) - b(y))$$

is bounded for an arbitrary $x \in G$. This implies that

$$b(x) = \lim_{n \to +\infty} (b(x+y_n) - b(y_n)) \qquad \text{for } x \in G,$$

where $(y_n)_{n \in \mathbb{N}}$ is an arbitrary sequence in $G$ with $| a(y_n) | \to +\infty$ (such a sequence exists since $a$ is nontrivial). Now, using the same argumentation as in [3] one can check that $b$ is additive, and consequently has to be trivial, which yields $f = a$. □

*Remark 2.1.* It is to be noticed that Moszner [5] proposed another general method for proving superstability of some functional equations which is also applicable to Mikusiński's equation.

## References

1. Batko, B.: On the stability of Mikusiński's equation. Publ. Math. Debrecen **66**, 17–24 (2005)
2. Batko, B.: On the stability of an alternative functional equation. Math. Inequal. Appl. **8**, 685–691 (2005)

3. Batko, B.: Stability of Dhombres' equation. Bull. Austral. Math. Soc. **70**, 499–505 (2004)
4. Dubikajtis, L., Ferens, C., Ger, R., Kuczma, M.: On Mikusiński's functional equation. Ann. Polon. Math. **28**, 39–47 (1973)
5. Moszner, Z.: On stability of some functional equations and topology of their target spaces. Result. Math. (submitted)
6. Schwaiger, J.: Remark 13. In: Report of Meeting, The 41st International Symposium on Functional Equations, p. 309. Aequationes Math. **67** (2004)

# Chapter 3
# A General Fixed Point Method for the Stability of Cauchy Functional Equation

**Liviu Cădariu and Viorel Radu**

*Dedicated to the memory of S.M. Ulam*
*on the 100th anniversary of his birth*

**Abstract** In this note, we extend the ideas in [12] to obtain some general stability results for additive Cauchy functional equations in $\beta$-normed spaces. It is worth noting that two fixed point alternatives together with the error estimations for generalized contractions of type Bianchini–Grandolfi and Matkowski are pointed out, and then used as fundamental tools. Some examples which emphasize the very general hypotheses, are also given.

**Keywords** Additive Cauchy functional equation • Fixed points • Stability • Generalized contractions

**Mathematics Subject Classification (2000):** Primary 39B62,39B72,39B82,47H09

## 3.1 Introduction

The study of stability problems for functional equations is related to a question of S. M. Ulam concerning the stability of group homomorphisms (see [33]):

*Let* $(G_1, \circ)$ *be a group and* $(G_2, *)$ *a metric group with a metric* $d(\cdot,\cdot)$. *Given* $\varepsilon > 0$, *does there exist a* $\delta > 0$ *such that if* $f : G_1 \to G_2$ *satisfies*

$$d(f(x \circ y), f(x) * f(y)) \le \delta, \qquad \forall x, y \in G_1,$$

*then there exists a homomorphism* $h : G_1 \to G_2$ *with* $d(f(x), h(x)) \le \varepsilon$ for $x \in G_1$?

L. Cădariu (✉)
Department of Mathematics, "Politehnica" University of Timişoara,
Piaţa Victoriei no. 2, 300006 Timişoara, Romania
e-mail: liviu.cadariu@mat.upt.ro; lcadariu@yahoo.com

V. Radu
Faculty of Mathematics and Computer Science, West University of Timişoara,
Bv. Vasile Pârvan 4, 300223 Timişoara, Romania
e-mail: radu@math.uvt.ro; viorelvradu@yahoo.com

Th.M. Rassias and J. Brzdęk (eds.), *Functional Equations in Mathematical Analysis*,
Springer Optimization and Its Applications 52, DOI 10.1007/978-1-4614-0055-4_3,

Hyers provided in [21] a first partial answer to Ulam's problem. Subsequently, the Hyers result was extended and generalized in several ways [17, 22, 24, 29], for different types of functional equations including the stability problem with unbounded differences (see e.g. [2, 6, 16, 19, 20, 28]). For the sake of convenience we note here the following partial result.

**Theorem 3.1.** *Suppose that X is a real normed space, Y is a real Banach space and $f: X \to Y$ is a given function, such that the following condition holds*

$$(\mathbf{C_p}) \qquad \|f(x+y)-f(x)-f(y)\|_Y \leq \theta(\|x\|_X^p + \|y\|_X^p), \qquad \forall x, y \in X,$$

*for some $p \in [0,\infty)\backslash\{1\}$. Then there is a unique additive function $a: X \to Y$ with*

$$(\mathbf{Est_p}) \qquad \|f(x)-a(x)\|_Y \leq \frac{2\theta}{|2-2^p|}\|x\|_X^p, \qquad \forall x \in X.$$

Almost all proofs used the idea conceived by Hyers. Namely, the additive function $a: X \to Y$ is explicitly constructed and directly dependent on the approximate solution-function $f$ by the following formulae

$$(\mathbf{2_{p<1}}) \qquad a(x) = \lim_{n\to\infty} \frac{1}{2^n} f(2^n x), \quad \text{if } p < 1;$$

$$(\mathbf{2_{p>1}}) \qquad a(x) = \lim_{n\to\infty} 2^n f\left(\frac{x}{2^n}\right), \quad \text{if } p > 1.$$

We now recall some necessary notions and results, used in the sequel. Let $X$ be a vector space over the real or the complex field $\mathbb{K}$, $\alpha \in \mathbb{R}_+$, and $\beta \in (0,1]$.

**Definition 3.1.** A mapping $\|.\|_\alpha : X \to \mathbb{R}_+$ is called an *h-functional of order* $\alpha$ *(or a sub-homogenous functional of order* $\alpha$*)* if it has the property

$$(h_\alpha): \quad \|\lambda x\|_\alpha \leq |\lambda|^\alpha \|x\|_\alpha, \quad \text{for all } \lambda \in \mathbb{K}, x \in X.$$

As usual, $X$ is identified with $X \times \{0\}$ in $X \times X$, so that $\|x\|_\alpha = \|(x,0)\|_\alpha$ for all $x \in X$ and each h-functional of order $\alpha$ on $X \times X$.

**Definition 3.2.** A mapping $\|\cdot\|_\beta : E \to \mathbb{R}_+$ is called a *$\beta$-norm* if it has the following properties:

$n_\beta^I$: $\|x\|_\beta = 0 \Longleftrightarrow x = 0$;
$n_\beta^{II}$: $\|\lambda x\|_\beta = |\lambda|^\beta \|x\|_\beta$, for all $x \in X$;
$n_\beta^{III}$: $\|x+y\|_\beta \leq \|x\|_\beta + \|y\|_\beta$, for all $x, y \in X$.

For more details see, e.g., [31]. We will consider two vector spaces (over the same field) $X$ and $Y$, and suppose we are given an *h-functional of order* $\alpha$ on $X \times X$, and a *$\beta$-norm on* $Y$, with $\alpha \neq \beta$. It is to be noted that, for $d_\beta(x,y) := \|x-y\|_\beta$, one has a metric space $(Y, d_\beta)$, which we suppose to be complete, that is $Y$ is a *complete $\beta$-normed space.*

On can define $N^1_{\alpha\beta} : Y^X \to \overline{\mathbb{R}}_+$ and $N^2_{\alpha\beta} : Y^{X\times X} \to \overline{\mathbb{R}}_+$ by

$$N^1_{\alpha\beta}(f) = \inf\{C : \|f(x)\|_\beta \leq C \cdot \|\overline{x}\|_\alpha, x \in X\}$$

and

$$N^2_{\alpha\beta}(g) = \inf\{C : \|g(z)\|_\beta \leq C \cdot \|z\|_\alpha, \forall z \in X \times X\},$$

for given $f : X \to Y$, respectively, $g : X \times X \to Y$. Although not necessarily with finite values, $N^1_{\alpha\beta}$ and $N^2_{\alpha\beta}$ have the properties of a $\beta$-norm.

In what follows we will use the linear operators $L, M : Y^X \to Y^{X\times X}$, given by

$$Lf(x,y) = f(x+y), \forall (x,y) \in X \times X,$$
$$Mf(x,y) = f(x) + f(y), \forall (x,y) \in X \times X.$$

The following stability result, proved in [7] by the direct method, can be seen as an appropriate answer to the problem of Ulam, mentioned in the beginning of the paper.

**Theorem 3.2.** *Given* $\varepsilon > 0$, *there exists* $\delta(\varepsilon)$ *such that for any mapping* $f : X \to Y$ *satisfying*

$$N^2_{\alpha\beta}((L-M)f) \leq \delta(\varepsilon),$$

*there exists a unique mapping* $a : X \to Y$, *with the properties*

$$La = Ma$$

*and*

$$N^1_{\alpha\beta}(f - a) \leq \varepsilon.$$

*Remark 3.1.* If we consider in the above definitions of $N^1_{\alpha\beta}$ and $N^2_{\alpha\beta}$, the *h-functional of order p* on $X \times X$ given by $\|(x,y)\|_p := \delta(\varepsilon)\ (\|x\|^p + \|y\|^p)$ and the 1-*norm* on $Y$ defined by $\|z\|_1 := \|z\|$, then we obtain Theorem 3.1, for $p \in [0,\infty)\backslash\{1\}$, with $\delta(\varepsilon) = \frac{\varepsilon}{2}\ |2 - 2^p|$.

As it has been observed in [27], both *the existence of* the mapping $a$ and *the estimation* ($\text{Est}_p$) in Theorem 3.1 can really be obtained by applying the fixed point alternative for a strictly contractive operator defined on a suitable generalized metric (function) space:

$$Jh(x) := \frac{1}{r} h(rx).$$

As a matter of fact, apart from the *contractivity of* $J$, the condition $(C_p)$ does ensure, on the one hand, the first two successive approximations starting from $f$ to be at a *finite distance* and, on the other hand, the fixed point function of $J$ to be an *additive* one.

We used the fixed point method to obtain generalized Ulam–Hyers stability results for Jensen functional equation [8] and for additive Cauchy equation in

$\beta$-normed spaces [9], as well as for quadratic, cubic and quartic equations (cf. [10, 11, 13, 14]; see also [1, 3, 18]).

In this paper, we extend some generalized stability results for additive Cauchy functional equations in $\beta$-normed spaces. Two fixed point theorems as well as the error estimations for *generalized contractions* are used to prove these results. Namely, we used the fixed point alternatives of type Bianchini–Grandolfi and Matkowski, respectively.

Some examples which emphasize the very general hypotheses are given. Moreover, the main theorem and a corollary in [9] are obtained as applications of our results.

## 3.2 Generalized Contractions and Stability Properties of the Cauchy Equation

**Definition 3.3.** A nondecreasing function $c : \mathbb{R}_+ \to \mathbb{R}_+$ is said to be a *comparison function* [4, 30] or *Matkowski gauge function* [23] if $\{c^n(t)\}$ converges to 0 for all $t \geq 0$.

A nondecreasing function $c : \mathbb{R}_+ \to \mathbb{R}_+$ is said to be a *c-comparison function* [4] or *Bianchini–Grandolfi gauge function* [5] if

$$\sum_{k=0}^{\infty} c^k(t) \text{ is convergent,} \qquad \forall t \in (0,\infty),$$

where $c^k(t)$ denotes the $k$-th iteration of $c$.

It is easy to see that any *c-comparison function* is a *comparison function.* Notice that there are known various different conditions for comparison functions (see [30] or [4]). For example, the mapping $c_1 : \mathbb{R}_+ \to \mathbb{R}_+, c_1(t) = Lt$, with $L \in (0,1)$, is a *c-comparison function*, but the function $c_2 : \mathbb{R}_+ \to \mathbb{R}_+, c_2(t) = t/(t+1)$, is a *comparison function* without being *c-comparison function.*

**Definition 3.4.** A selfmapping $A$ of the metric space $(X,d)$, for which there exists a *comparison function* $c$ such that

$$(1) \qquad d(Ax,Ay) \leq c(d(x,y)), \qquad \forall x,y \in X,$$

is called *Matkowski contraction* [23]. Similarly, a selfmapping $A$ of the metric space $(X,d)$, for which there exists a *c-comparison function* $c$ such that relation (1) holds, is called *Bianchini–Grandolfi contraction.*

It is worth noting that the above contractions are only two examples of *gauge contractions* or *generalized contractions* [4, 25, 30].

The following result is a fixed point alternative of the famous theorem of Bianchini and Grandolfi [5]. We shall use it to prove our stability results for additive Cauchy functional equations.

**Lemma 3.1 (cf. [8, 15, 32] for some particular cases).** *Let $(X,d)$ be a complete **generalized** metric space, i.e. one for which $d$ may assume infinite values, and $A : X \to X$ a Bianchini–Grandolfi contraction. Then, for each given element $x \in X$, **either***

$(A_1)$ $d(A^n x, A^{n+1}x) = +\infty$ *for all* $n \geq 0$,

***or***

$(A_2)$ *there exists $k$ such that* $d(A^n x, A^{n+1}x) < +\infty$ *for all* $n \geq k$.

*Further, if $(A_2)$ holds, then*

$(A_{21})$ *the sequence $(A^n x)$ is convergent to a fixed point $y^*$ of $A$;*

$(A_{22})$ *$y^*$ is the unique fixed point of $A$ in* $\mathscr{Y} := \{y \in X,\ d(A^k x, y) < +\infty\}$;

$(A_{23})$ $d(y, y^*) \leq \sum_{k=0}^{\infty} c^k(d(y, Ay))$ *for all* $y \in Y$.

In the following definition, we introduce the notion of an *admissible pair*. Let us consider the family $\mathscr{M}$ of all mappings $m : X \times [0,\infty) \to [0,\infty)$ with the properties:

$(i)$ $m_x$ is continuous at 0, for each $x \in X$;

$(ii)$ $m_x$ is superadditive, that is $m_x(t+s) \geq m_x(t) + m_x(s)$ for $t, s \in [0,\infty)$, where $m_x := m(x,\cdot)$ for each $x \in X$.

Let us consider a *comparison function* $c$, an element $m \in \mathscr{M}$ and the numbers

$$r_j = 2^{1-2j} = \begin{cases} 2, & j = 0; \\ 1/2, & j = 1. \end{cases}$$

**Definition 3.5.** We say that $(m,c)$ is a *j-admissible pair of order* $\beta$ ($j \in \{0,1\}$, $\beta \in (0,1]$)) if

$$(\mathbf{A_j}) \qquad m(r_j\, x, t) \leq r_j^\beta\, m(x, c(t)), \qquad \forall t \in [0,\infty),\ x \in X.$$

### 3.2.1 A Stability Result for Cauchy Functional Equation with Bianchini–Grandolfi Fixed Point Alternative

We shall use the fixed point result in Lemma 3.1 and the estimation of the form $(A_{23})$ to prove the following very general stability theorem for the additive Cauchy functional equation:

**Theorem 3.3.** *Let us consider a real linear space $X$, a complete $\beta$-normed space $Y$, a c-comparison function $c$ and a j-admissible pair $(m,c)$, of order $\beta$, with $j \in \{0,1\}$. Suppose that a mapping $f : X \to Y$, with $f(0) = 0$, satisfies the inequality*

$$(\mathbf{C}_\varphi) \qquad \|f(x+y)-f(x)-f(y)\|_\beta \le \varphi(x,y), \qquad \forall x,y\in X,$$

*where* $\varphi : X\times X\to[0,\infty)$ *is a given function, and there exists* $\delta>0$ *such that*

$$(\mathbf{M}_\delta) \qquad \varphi\left(\frac{x}{2},\frac{x}{2}\right)\le m(x,\delta), \quad \forall x\in X.$$

*Then there exists a unique mapping* $a : X\to Y$ *such that*

$$\left(\mathbf{Est}_{\mathbf{j}}^{c}\right) \qquad \|f(x)-a(x)\|_\beta \le m\left(x,\sum_{k=0}^{\infty}c^{k+1-j}(\delta)\right), \quad \forall x\in X.$$

*Moreover, if* $\varphi$ *has the property*

$$\left(\mathbf{H}_{\mathbf{j}}^{*}\right) \qquad \lim_{n\to\infty}\frac{\varphi\left(r_j^n x, r_j^n y\right)}{r_j^{n\beta}}=0, \quad \forall x,y\in X,$$

*then the mapping a is additive.*

*Proof.* Let $\mathscr{E} := \{g : X\to Y, g(0)=0\}$ and introduce the mapping

$$d(g,h)=d_m(g,h)=\inf\left\{k\in\mathbb{R}_+,\ \|g(x)-h(x)\|_\beta\le m_x(k),\ \forall x\in X\right\}.$$

Then, $d_m$ is a *generalized complete metric* on $\mathscr{E}$ (as usual, $\inf\emptyset=\infty$). The triangle inequality results using the superadditivity of $m_x$. The fact that $(\mathscr{E},d)$ *is complete* follows from the continuity condition on $m_x$.

Now, consider the operator

$$(\mathbf{OP_j}) \qquad J:\mathscr{E}\to\mathscr{E}, \quad Jg(x):=\frac{g(r_j x)}{r_j}.$$

and notice that $r_j=2^{1-2j}$, $j\in\{0,1\}$.

Step I. Using the admissibility relation $(A_j)$, we show that $J$ satisfies the relation $(1)$. We can write, for any $g,h\in\mathscr{E}$ :

$$\begin{aligned} d(g,h)<k &\Longrightarrow \|g(x)-h(x)\|_\beta\le m_x(k), \quad \forall x\in X\\ &\Longrightarrow \left\|\frac{1}{r_j}g(r_jx)-\frac{1}{r_j}h(r_jx)\right\|_\beta \le \frac{1}{r_j^\beta}m_{r_j\cdot x}(k), \quad \forall x\in X\\ &\Longrightarrow \left\|\frac{1}{r_j}g(r_jx)-\frac{1}{r_j}h(r_ix)\right\|_\beta \le m_x(c(k)), \quad \forall x\in X\\ &\Longrightarrow d(Jg,Jh)\le c(k). \end{aligned}$$

Using the definition of the c-comparison function $c$ we see that

$$d\left(Jg,Jh\right)\le c(d\left(g,h\right)),\quad \forall g,h\in\mathscr{E},$$

hence, $J$ is a *Bianchini–Grandolfi contractive operator* on $\mathscr{E}$.

STEP II. We show that $d\left(f,Jf\right)<\infty$. For $j=0$, we replace $y$ by $x$ in the condition $\left(\mathbf{C}_\varphi\right)$, and we see that

$$\left\|f(x)-\frac{1}{2}f(2x)\right\|_\beta\le\frac{1}{2^\beta}\varphi(x,x),\quad \forall x\in X.$$

Using the hypothesis $(\mathbf{M}_\delta)$ and the admissibility condition $(\mathbf{A}_0)$ it results

$$\left\|f(x)-\frac{1}{2}f(2x)\right\|_\beta\le\frac{1}{2^\beta}\varphi(x,x)\le\frac{1}{2^\beta}m_{2x}(\delta)\le m_x(c(\delta)),\quad \forall x\in X,$$

that is $d\left(f,Jf\right)\le c(\delta)<\infty$.

For $j=1$, we replace $x$ and $y$ by $x/2$ in the condition $\left(\mathbf{C}_\varphi\right)$, and we see (by using again the relation $(\mathbf{M}_\delta)$) that

$$\left\|2f\left(\frac{x}{2}\right)-f\left(x\right)\right\|_\beta\le\varphi\left(\frac{x}{2},\frac{x}{2}\right)\le m_x(\delta),\quad \forall x\in X,$$

whence $d\left(f,Jf\right)\le\delta<\infty$.

STEP III. In both cases, we can apply Lemma 3.1 on the complete metric space $\mathscr{F}=\{g\in\mathscr{E},\ d\left(f,g\right)<\infty\}$, and we obtain the existence of a mapping $a:X\to Y$ with the following properties.

- $a$ is a fixed point of $J$, that is

$$a(2x)=2a(x),\quad \forall x\in X. \tag{3.1}$$

  The mapping $a$ is the unique fixed point of $J$ in the set $\mathscr{F}$. This says that $a$ is the unique mapping with *both* the properties (3.1) and (3.2), where

$$\exists k\in(0,\infty)\text{ such that }\|a\left(x\right)-f\left(x\right)\|_\beta\le m_x(k),\quad \forall x\in X. \tag{3.2}$$

- $d(J^nf,a)\xrightarrow[n\to\infty]{}0$, which implies the equality

$$\lim_{n\to\infty}\frac{f\left(r_j^nx\right)}{r_j^n}=a(x),\quad \forall x\in X. \tag{3.3}$$

By using $(A_{23})$, we obtain

$$d(a,f) \leq \sum_{k=0}^{\infty} c^k(d(f,Jf)).$$

Therefore,

$$\left(\mathbf{Est}_{\mathbf{j}}^{c}\right) \qquad \|f(x)-a(x)\|_{\beta} \leq m\left(x, \sum_{k=0}^{\infty} c^{k+1-j}(\delta)\right), \quad \forall x \in X.$$

STEP IV. The fact that $a$ is an additive mapping follows immediately from $\left(\mathbf{C}_{\varphi}\right)$, (3.3) and $\left(\mathbf{H}_{\mathbf{j}}^{*}\right)$. Namely, if in $\left(\mathbf{C}_{\varphi}\right)$ we replace $x$ by ${r_j}^n x$ and $y$ by ${r_j}^n y$, then we obtain

$$\left\|\frac{f({r_j}^n(x+y))}{{r_j}^n} - \frac{f({r_j}^n(x))}{{r_j}^n} - \frac{f({r_j}^n(y))}{{r_j}^n}\right\|_{\beta} \leq \frac{\varphi({r_j}^n x, {r_j}^n y)}{{r_j}^{n\beta}},$$

for all $x,y \in X$. Letting $n \to \infty$, we get

$$a(x+y)-a(x)-a(y)=0, \quad \forall x,y \in X.$$

□

As a direct consequence of Theorem 3.3, we have the following

**Corollary 3.1.** *Let $X$ be a real (or complex) linear space, $Y$ a complete $\beta$-normed space and let us consider a superadditive function $\mu : \mathbb{R}_+ \to \mathbb{R}_+$ which is continuous at 0 and not identically 0 (whence $\Delta := \{\delta > 0 : \mu(\delta) \geq 1\} \neq \emptyset$), and a monotone nondecreasing function $\gamma : \mathbb{R}_+ \to \mathbb{R}_+$. Suppose that the mapping $f : X \to Y$ satisfies the inequality*

$$(\mathbf{C}_{\varphi}) \qquad \|f(x+y)-f(x)-f(y)\|_{\beta} \leq \varphi(x,y), \quad \forall x,y \in X,$$

*where $\varphi : X \times X \to [0,\infty)$ is a given function. If there exists a c-comparison function c such that*

$$(\mathbf{A}_{\mathbf{j}}^{\mu,\gamma}) \qquad \mu(t)\cdot(\gamma(t)+\varphi(r_j x, r_j x)) \leq r_j^{\beta}\cdot\mu(c(t))\cdot(\gamma(c(t))+\varphi(x,x)),$$

*for $t \in [0,\infty)$, $x \in X$, $r_j = 2^{1-2j}, (j \in \{0,1\})$, then the following two statements hold.*

*(i) There exists a unique mapping $a : X \to Y$ which satisfies*

$$\left(\mathbf{Est}_{\mathbf{j}}^{\mu,\gamma}\right) \qquad \inf_{\delta\in\Delta} \mu\left(\sum_{k=0}^{\infty} c^{k+1-j}(\delta)\right)\left(\gamma\left(\sum_{k=0}^{\infty} c^{k+1-j}(\delta)\right)+\varphi\left(\frac{x}{2},\frac{x}{2}\right)\right)$$
$$\geq \|f(x)-a(x)\|_{\beta}\,, \qquad \forall x \in X.$$

*(ii) The mapping a is additive whenever $\varphi$ has the property $\left(\mathbf{H}_{\mathbf{j}}^{*}\right)$.*

*Proof.* If we take $m : X \times [0,\infty) \to [0,\infty)$, $m(x,t) = \mu(t)\ (\gamma(t)+\varphi(x/2,x/2))$, then it is easy to see that $m \in \mathcal{M}$. Since $(\mathbf{A}_{\mathbf{j}}^{\mu,\gamma})$ holds, we immediately obtain that $(m,c)$ is an admissible pair of order $\beta$. The condition $(\mathbf{M}_{\delta})$ is equivalent to the inequality $\mu(\delta) \geq 1$, that is $\delta \in \Delta$. So we can apply Theorem 3.3. □

*Example 3.1.* In the above corollary, consider the mappings $c : \mathbb{R}_{+} \to \mathbb{R}_{+}$, $c(t) = Lt$, with $L \in (0,1)$, and $m : X \times [0,\infty) \to [0,\infty)$, $m(x,t) = \mu(t)\cdot\varphi(x/2,x/2)$. Clearly, $c$ is a *c-comparison function* and $m \in \mathcal{M}$. Since $(\mathbf{A}_{\mathbf{j}}^{\mu,\mathbf{0}})$ holds, then we immediately obtain that $(m,c)$ is an admissible pair of order $\beta$ and we get the following result.

Let $X$ be a real linear space, $Y$ a complete $\beta$-normed space and $\mu : \mathbb{R}_{+} \to \mathbb{R}_{+}$ a superadditive function which is continuous at 0 and not identically 0. Suppose that the mapping $f : X \to Y$ satisfies the inequality

$$(\mathbf{C}_{\varphi}) \qquad \|f(x+y)-f(x)-f(y)\|_{\beta} \leq \varphi(x,y)\,, \qquad \forall x,y \in X,$$

where $\varphi : X \times X \to [0,\infty)$ is a given function. If there exists $L \in (0,1)$ such that

$$(\mathbf{A}_{\mathbf{j}}^{\mu,\mathbf{0}}) \qquad \mu(t)\ \varphi(r_j x, r_j x) \leq r_j^{\beta}\ \mu(Lt)\ \varphi(x,x)\,, \qquad \forall t \in [0,\infty)\,,\ x \in X,$$

for $r_j = 2^{1-2j}, (j \in \{0,1\})$, then the following two statements are valid.

(i) There exists a unique mapping $a : X \to Y$ which satisfies the inequality

$$\left(\mathbf{Est}_{\mathbf{j}}^{\mu}\right) \qquad \|f(x)-a(x)\|_{\beta} \leq \inf_{\delta\in\Delta} \mu\left(\frac{L^{1-j}\,\delta}{1-L}\right)\ \varphi\left(\frac{x}{2},\frac{x}{2}\right), \qquad \forall x \in X.$$

(ii) The mapping $a$ is *additive* whenever $\varphi$ has the property $\left(\mathbf{H}_{\mathbf{j}}^{*}\right)$.

*Remark 3.2.* As of matter of fact, if $\widetilde{L} := \inf_{t>0} \mu(Lt)/\mu(t)$, then the condition $(\mathbf{A}_{\mathbf{j}}^{\mu,\mathbf{0}})$ is equivalent with the following condition which appears in [9, Theorem 2.5]:

$$\varphi(r_j x, r_j x) \leq \widetilde{L}\, r_j^{\beta}\ \varphi(x,x)\,, \qquad \forall x \in X.$$

*Example 3.2.* Let us consider, in the above example, the mapping $\mu(t) = t^{s}$, with $s \geq 1$ fixed. We have the following result.

*Let $X$ be a (real or complex) linear space and $Y$ a complete $\beta$-normed space. Suppose that the mapping $f : X \to Y$ satisfies the inequality*

$$(\mathbf{C}_\varphi) \qquad \|f(x+y) - f(x) - f(y)\|_\beta \le \varphi(x,y), \qquad \forall x,y \in X,$$

*where $\varphi : X \times X \to [0,\infty)$ is a given function. If there exists $L \in (0,1)$ such that*

$$(\mathbf{A_j^L}) \qquad \varphi(r_j x, r_j x) \le r_j^\beta\, L^s\, \varphi(x,x), \qquad \forall t \in [0,\infty),\ x \in X,$$

*for $r_j = 2^{1-2j}$, with fixed $j \in \{0,1\}$, then the following two statements are valid.*

*(i) There exists a unique mapping $a : X \to Y$ which satisfies the inequality*

$$\left(\mathbf{Est_j^{\mathit{L}}}\right) \qquad \|f(x) - a(x)\|_\beta \le \left(\frac{L^{1-j}}{1-L}\right)^s \varphi\left(\frac{x}{2}, \frac{x}{2}\right), \qquad \forall x \in X.$$

*(ii) The mapping $a$ is additive whenever $\varphi$ has the property $(\mathbf{H_j^*})$.*

We note only that the estimation $\left(\mathbf{Est_j^{\mathit{L}}}\right)$ is obtained for $\delta = 1$.

*Remark 3.3.* For $s = 1$ in the Example 3.2, we obtain the generalized Ulam–Hyers stability theorem for additive Cauchy functional equation ([9], Theorem 2.5).

Another application is obtained by taking in the Example 3.2

$$\varphi(x,y) := \delta(\varepsilon)\ \|(x,y)\|_\alpha, \qquad x,y \in X.$$

*Example 3.3.* Let $X$ and $Y$ two linear spaces over the same (real or complex) field. Suppose that we are given an h-functional of order $\alpha$ on $X \times X$ and $Y$ is a complete $\beta$-normed space, with $\alpha \neq \beta$. In these conditions, we have the following stability property.

*For each $\varepsilon > 0$, there exists*

$$\delta(\varepsilon) := \varepsilon \left|2^{\beta/s} - 2^{\alpha/s}\right|^s > 0$$

*such that, for every mapping $f : E_1 \to E_2$ which satisfies*

$$(\mathbf{C}_{\alpha\beta}) \qquad \|f(x) + f(y) - f(x+y)\|_\beta \le \delta(\varepsilon)\ \|(x,y)\|_\alpha, \qquad x,y \in E_1,$$

*there exists a unique mapping $a : X \to Y$, with the properties*

$$a(x+y) = a(x) + a(y), \qquad \forall x,y \in X,$$

and

$$(\mathbf{Est}_{\alpha\beta}) \qquad \|f(x) - a(x)\|_\beta \le \varepsilon\ \|(x,x)\|_\alpha, \qquad \forall x \in X.$$

*Proof.* Indeed, $(H_j^*)$ is true, because

$$\frac{\varphi\left(r_j{}^n x, r_j{}^n y\right)}{r_j{}^{n\beta}} \leq \delta(\varepsilon)\, r_j{}^{n(\alpha-\beta)}\, \|(x,y)\|_\alpha \to 0.$$

Moreover, it is easy to see that either $(A_0^L)$ holds with $L = 2^{(\alpha-\beta)/s} < 1$ or $(A_1^L)$ holds with $L = 2^{(\beta-\alpha)/s} < 1$. Therefore, there is a *unique additive* mapping $a : X \to Y$ such that either

$$(\mathbf{Est}_0^L) \qquad \|f(x) - a(x)\|_\beta \leq \left(\frac{L}{1-L}\right)^s \theta(x), \qquad \forall x \in X.$$

holds with $L = 2^{(\alpha-\beta)/s}$, or

$$(\mathbf{Est}_1^L) \qquad \|f(x) - a(x)\|_\beta \leq \left(\frac{1}{1-L}\right)^s \theta(x), \qquad \forall x \in E_1,$$

holds with $L = 2^{(\beta-\alpha)/s}$. Consequently, the inequality $\left(\mathrm{Est}_{\alpha\beta}\right)$ holds true for $\delta(\varepsilon) = \varepsilon\ \left(2^{\beta/s} - 2^{\alpha/s}\right)^s$ and for $\delta(\varepsilon) = \varepsilon\ \left(2^{\alpha/s} - 2^{\beta/s}\right)^s$, respectively. □

*Remark 3.4.* For $s = 1$ in the above example, we obtain [9, Corollary 2.6].

*Remark 3.5.* As in Remark 3.1, taking the *h-functional of order p* on $X \times X$ given by $\|(x,y)\|_p := \delta(\varepsilon)\ (\|x\|^p + \|y\|^p)$ and the 1-*norm* on $Y$ defined by $\|z\|_1 := \|z\|$, we obtain Theorem 3.1, for $p \in [0,\infty)\backslash\{1\}$.

### 3.2.2 A Stability Result for Cauchy Functional Equation with Matkowski Fixed Point Alternative

In this section, we shall prove the stability result for additive Cauchy functional equations by using the fixed point alternative of type Matkowski [26].

**Lemma 3.2 (cf. [8,15,32]).** *Let $(X,d)$ be a complete generalized metric space, i.e., one for which d may assume infinite values, and $A : X \to X$ a Matkowski contraction. Then, for each given element $x \in X$,* ***either***

$(A_1)$ $d(A^n x, A^{n+1}x) = +\infty$ *for all* $n \geq 0$,

***or***

$(A_2)$ *There exists k such that* $d(A^n x, A^{n+1}x) < +\infty$ *for all* $n \geq k$.
*Further, if* $(A_2)$ *holds, then the following three statements are valid.*

$(A_{21})$ *The sequence* $(A^n x)$ *is convergent to a fixed point* $y^*$ *of* $A$.
$(A_{22})$ $y^*$ *is the unique fixed point of* $A$ *in* $y := \left\{y \in X,\ d\left(A^k x, y\right) < +\infty\right\}$.

$(A'_{23})$ *If the mapping* $t \to 1_{\mathbb{R}_+} - c(t)$ *is a bijection, then*

$$d(y,y^\star) \leq (1_{\mathbb{R}_+} - c)^{-1}(d(y,Ay)), \qquad \forall y \in Y.$$

We can use the above result and the estimation of the form $(A'_{23})$ to prove the following variant of general stability:

**Theorem 3.4.** *Let us consider a real linear space* $X$, *a complete* $\beta$*-normed space* $Y$, *a comparison function* $c$ *and a* $j$*-admissible pair* $(m,c)$, *of order* $\beta$, *with* $j \in \{0,1\}$. *Let us suppose that the mapping* $t \to 1_{\mathbb{R}_+} - c(t)$ *is an increasing bijection and that the mapping* $f : X \to Y$, *with* $f(0) = 0$, *satisfies the inequality*

$$(\mathbf{C}_\varphi) \qquad \|f(x+y) - f(x) - f(y)\|_\beta \leq \varphi(x,y), \qquad \forall x,y \in X,$$

*where* $\varphi : X \times X \to [0,\infty)$ *is a given function.*

*If there exists* $\delta > 0$ *such that*

$$(\mathbf{M}_\delta) \qquad \varphi\left(\frac{x}{2},\frac{x}{2}\right) \leq m(x,\delta), \qquad \forall x \in X,$$

*then the following two statements are valid.*

*(i) There exists a unique mapping* $a : X \to Y$ *such that*

$$\left(\mathbf{Est}_{\mathbf{j}}^{\prime c}\right) \qquad \|f(x) - a(x)\|_\beta \leq m\left(x, (1_{\mathbb{R}_+} - c)^{-1}(c^{1-j}(\delta))\right), \qquad \forall x \in X.$$

*(ii) If* $\varphi$ *has the property*

$$\left(\mathbf{H}_{\mathbf{j}}^*\right) \qquad \lim_{n\to\infty} \frac{\varphi\left(r_j^n x, r_j^n y\right)}{r_j^{n\beta}} = 0, \qquad \forall x,y \in X,$$

*then the mapping* $a$ *is additive.*

The proof is similar to that of Theorem 3.3. The estimation $\left(\mathrm{Est}_j^{\prime c}\right)$ follows from $(A'_{23})$.

*Remark 3.6.* As a direct consequence of the Theorem 3.4, for an admisible function $m$ as in the Corollary 3.1, we obtain a result that is similar as in the above mentioned corollary, with the the following estimation:

$$\|f(x) - a(x)\|_\beta \leq \inf_{\delta\in\Delta} \mu\left(1_{\mathbb{R}_+} - c\right)^{-1}(c^{1-j}(\delta)) \left(\gamma\left(\sum_{k=0}^{\infty} c^{k+1-j}(\delta)\right) + \varphi\left(\frac{x}{2},\frac{x}{2}\right)\right).$$

*Remark 3.7.* For $c(t) = Lt$, $L \in (0,1)$, $\mu(t) = t$ and $\gamma \equiv 0$ we obtain again our [9, Theorem 2.5].

## References


1. Agarwal, R.P., Xu, B., Zhang, W.: Stability of functional equations in single variable. J. Math. Anal. Appl. **288**, 852–869 (2003)
2. Aoki, T.: On the stability of the linear transformation in Banach spaces. J. Math. Soc. Japan **2**, 64–66 (1950)
3. Baker, J.A.: The stability of certain functional equations. Proc. Amer. Math. Soc. **112**, 729–732 (1991)
4. Berinde, V.: Iterative Approximation of Fixed Points. Lecture Notes in Math. **1912** (2007)
5. Bianchini, R.M., Grandolfi, M.: Transformazioni di tipo contracttivo generalizzato in uno spazio metrico. Atti Accad. Naz. Lincei Cl. Sci. Fis. Mat. Natur. Rend. Lincei (9) Mat. Appl. **45**, 212–216 (1968)
6. Bourgin D.G.: Classes of transformations and bordering transformations. Bull. Amer. Math. Soc. **57**, 223–237 (1951)
7. Cădariu, L., Radu, V.: A general theorem of stability for the Cauchy's equation. Bul. Ştiinţ. Univ. Politeh. Timiş. Ser. Mat. Fiz. **47(61)**, no. 2, 14–28 (2002)
8. Cădariu, L., Radu, V.: Fixed points and the stability of Jensen's functional equation. JIPAM. J. Inequal. Pure Appl. Math. **4(1)** (2003), Art. 4. http://jipam.vu.edu.au
9. Cădariu, L., Radu, V.: On the stability of Cauchy functional equation: a fixed points approach. In: Sousa Ramos, J., Gronau, D., Mira, C., Reich, L., Sharkovsky, A. (eds.) Iteration Theory (ECIT '02), Proceedings of European Conference of Iteration Theory, Evora, Portugal, September 1–7, 2002, pp. 43–52. Grazer Math. Ber. **346**, (2004)
10. Cădariu, L., Radu, V.: Fixed points and the stability of quadratic functional equations. An. Univ. Vest Timiş. Ser. Mat.-Inform. **41(1)**, 25–48 (2003)
11. Cădariu, L., Radu, V.: Fixed points in generalized metric spaces and the stability of a quartic functional equation. Bul. Ştiinţ. Univ. Politeh. Timiş. Ser. Mat. Fiz. **50(64)**, no. 2, 25–34 (2005)
12. Cădariu, L., Radu, V.: A general fixed point method for the stability of Jensen functional equation. Bul. Ştiinţ. Univ. Politeh. Timiş. Ser. Mat. Fiz. **51(65)**, no. 2, 63–72 (2006)
13. Cădariu, L., Radu, V.: Fixed points in generalized metric spaces and the stability of a cubic functional equation. In: Cho, Y.-J., Kim, J.-K., Kang, S.-M. (eds.) Fixed Point Theory and Applications **7**, pp. 53–68. Nova Science Publ. (2007)
14. Cădariu, L., Radu, V.: Stability results for a functional equation of quartic type. Automat. Comput. Appl. Math. **15**, 7–21 (2006)
15. Diaz, J.B., Margolis, B.: A fixed point theorem of the alternative for contractions on a generalized complete metric space. Bull. Amer. Math. Soc. **74**, 305–309 (1968)
16. Forti, G.L.: An existence and stability theorem for a class of functional equations. Stochastica **4**, 23–30 (1980)
17. Forti, G.L.: Hyers–Ulam stability of functional equations in several variables. Aequationes Math. **50**, 143–190 (1995)
18. Forti, G.L.: Comments on the core of the direct method for proving Hyers–Ulam stability of functional equations. J. Math. Anal. Appl. **295**, 127–133 (2004)
19. Gajda, Z.: On stability of additive mappings. Internat. J. Math. Math. Sci. **14**, 431–434 (1991)
20. Găvruţa, P.: A generalization of the Hyers Ulam Rassias stability of approximately additive mappings. J. Math. Anal. Appl. **184**, 431–436 (1994)
21. Hyers, D.H.: On the stability of the linear functional equation. Prod. Natl. Acad. Sci. USA **27**, 222–224 (1941)
22. Hyers, D.H., Isac, G., Rassias, Th.M.: Stability of Functional Equations in Several Variables. Birkhauser, Basel (1998)
23. Jachymski, J., Jóźwik, I.: Nonlinear contractive conditions: a comparison and related problems. In: Jachymski, J. et al. (eds.) Fixed Point Theory and Its Applications - Proc. of the Internat. Conf. Będlewo, Poland, August 1–5, 2005, pp. 123–146. Polish Academy of Sciences, Institute of Mathematics, Banach Center Publications **77**, (2007)

24. Jung, S.-M.: Hyers–Ulam-Rassias Stability of Functional Equations in Mathematical Analysis. Hadronic Press, Palm Harbor, Florida (2002)
25. Kirk, W.A., Sims, B. (eds.): Handbook of Metric Fixed Point Theory. Kluwer Academic Publishers, Dordrecht (2001)
26. Matkowski, J.: Integrable solutions of functional equations. Dissertationes Math. **127**, 63 pp. (1975)
27. Radu, V.: The fixed point alternative and the stability of functional equations. Fixed Point Theory Cluj-Napoca **4(1)**, 91–96 (2003)
28. Rassias, Th.M.: On the stability of the linear mapping in Banach spaces. Proc. Amer. Math. Soc. **72**, 297–300 (1978)
29. Rassias, Th.M.: On the stability of functional equations and a problem of Ulam. Acta Appl. Math. **62**, 23–130 (2000)
30. Rus, I.A.: Generalized Contractions and Applications. Cluj University Press, Cluj-Napoca (2001)
31. Simons, S.: Boundedness in linear topological spaces. Trans. Amer. Math. Soc. **113**, 169–180 (1964)
32. Zeidler, E.: Nonlinear Functional Analysis and Its Applications. I. Fixed-Point Theorems. Springer-Verlag, New York (1986)
33. Ulam, S.M.: A Collection of Mathematical Problems. Interscience Publ., New York, (1960); reprinted as: Problems in Modern Mathematics. Wiley, New York (1964)

# Chapter 4
# Orthogonality Preserving Property and its Ulam Stability

**Jacek Chmieliński**

*Dedicated to the memory of S.M. Ulam*
*on the 100th anniversary of his birth*

**Abstract** We survey the results concerning the preservation (exact and approximate) of various types of orthogonality relations. We focus on the stability of the orthogonality preserving property. Our considerations are carried out in spaces with inner product structure as well as in normed spaces. Some related topics are also discussed.

**Keywords** Orthogonality • Birkhoff orthogonality • Isosceles-orthogonality • Approximate orthogonality • Orthogonality preserving property • Right-angle preserving property • Linear preservers • Stability • Orthogonality equation • Wigner equation • Inner product spaces • Hilbert modules • Normed spaces • Semi-innner product • Norm derivatives • Isometric mappings • Approximate isometry

**Mathematics Subject Classification (2010):** Primary 46B20, 47B49, 15A86, 39B52, 39B82

## 4.1 Introduction

The study on *linear orthogonality preserving mappings* can be considered as a part of the theory of *linear preservers*. In the simplest case, for $X$ and $Y$ being real or complex inner product spaces with the standard orthogonality relation $\perp$, a mapping $T\colon X \to Y$ which satisfies

$$x \perp y \Longrightarrow f(x) \perp f(y), \qquad x, y \in X$$

---

J. Chmieliński (✉)
Instytut Matematyki, Uniwersytet Pedagogiczny w Krakowie,
Podchorążych 2, 30-084 Kraków, Poland
e-mail: jacek@up.krakow.pl

Th.M. Rassias and J. Brzdęk (eds.), *Functional Equations in Mathematical Analysis*, Springer Optimization and Its Applications 52, DOI 10.1007/978-1-4614-0055-4_4,

is called *orthogonality preserving* (o.p.). Now, if we replace the exact orthogonality $\perp$ by somehow defined *approximate orthogonality* $\perp^\varepsilon$ we obtain a larger class of *approximately orthogonality preserving* (a.o.p.) mappings defined by

$$x \perp y \Longrightarrow f(x) \perp^\varepsilon f(y), \qquad x, y \in X.$$

It can be proved that linear o.p. mappings are just similarities whereas linear a.o.p. ones are, in a sense, approximate similarities. Moreover, it can be shown that each linear a.o.p. mapping can be approximated by a linear o.p. one. Following the idea of the stability of functional equations, stemming from the celebrated Ulam's problem (cf. [48]), we may speak about *stability of the orthogonality preserving property*.

The problem can be easily generalized from the realm of inner product spaces to normed spaces, where the norm not necessarily comes from an inner product and the orthogonality relation may be defined in various ways. Another direction of generalization is to replace the scalar-valued inner product by an inner product taking values in some $C^*$-algebra, i.e., in the realm of inner product (Hilbert) modules.

In this paper we survey the results concerning the orthogonality preserving property and its stability in various settings. Because of an expository character of the paper, usually we merely present the results, referring for their proofs to original sources.

## 4.2 Orthogonalities and Approximate Orthogonalities

### 4.2.1 Inner Product and Semi-Inner Product Orthogonality

Let $X$ be a normed space over the scalar field $\mathbb{K} \in \{\mathbb{R}, \mathbb{C}\}$. If the norm $\|\cdot\|$ in $X$ comes from an inner product $\langle\cdot|\cdot\rangle$, then the natural orthogonality relation is the standard one: $x \perp y \Leftrightarrow \langle x|y\rangle = 0$. For $\varepsilon \in [0,1)$, we define *approximate orthogonality* ($\varepsilon$-orthogonality) of vectors $x$ and $y$:

$$x \perp^\varepsilon y \iff |\langle x|y\rangle| \le \varepsilon \|x\| \, \|y\|.$$

Due to Lumer [38] and Giles [25], each norm in $X$ admits a functional $[\cdot|\cdot] : X \times X \to \mathbb{K}$ satisfying:

(s1) $[\lambda x + \mu y|z] = \lambda [x|z] + \mu [y|z], \ x, y, z \in X, \ \lambda, \mu \in \mathbb{K}$;
(s2) $[x|\lambda y] = \bar{\lambda} [x|y], \ x, y \in X, \ \lambda \in \mathbb{K}$;
(s3) $[x|x] = \|x\|^2, \ x \in X$;
(s4) $|[x|y]| \le \|x\| \, \|y\|, \ x, y \in X$,

called the *semi-inner product* (s.i.p.) in a normed space $X$ (generating the given norm). There could be many different semi-inner products in $X$. The uniqueness is equivalent to the smoothness of the norm (understood as the existence, in each point of the unit sphere $S$, of a unique supporting hyperplane or, equivalently, the Gâteaux differentiability of the norm on $S$ – cf. [19]). If the norm in $X$ comes from an inner product, then the inner product itself is the unique semi-inner product.

Now, for a given semi-inner product and vectors $x,y \in X$ we may define:

- semi-orthogonality

$$x \perp_s y \iff [y|x] = 0$$

  and
- approximate semi-orthogonality ($\varepsilon$-semi-orthogonality)

$$x \perp_s^{\varepsilon} y \iff |[y|x]| \leq \varepsilon \|x\| \, \|y\|.$$

It is also possible to consider the orthogonality relation in more general inner product structures. Let $\mathscr{A}$ be a $C^*$-algebra. We may consider $\mathscr{A}$ as a subalgebra of an algebra $\mathscr{B}(\mathscr{H})$ of all linear bounded operators on a Hilbert space $\mathscr{H}$. By $\mathscr{K}(\mathscr{H})$ we denote the subalgebra of compact operators.

Let $\mathscr{X}$ be a complex linear space with additional multiplication $\mathscr{X} \times \mathscr{A} \to \mathscr{X}$ so that $\mathscr{X}$ is an algebraic right $\mathscr{A}$-module and compatible with scalar multiplication, i.e., $(\lambda x)a = x(\lambda a) = \lambda(xa)$ for all $x \in \mathscr{X}$, $a \in \mathscr{A}$, $\lambda \in \mathbb{C}$. Then $\mathscr{X}$ is called a *(right) inner product $\mathscr{A}$-module* if there exists an $\mathscr{A}$-valued inner product, i.e., a mapping $\langle \cdot|\cdot \rangle : \mathscr{X} \times \mathscr{X} \to \mathscr{A}$ satisfying

(i) $\langle x|x \rangle \geq 0$ (positive element of $\mathscr{A}$) and $\langle x|x \rangle = 0$ if and only if $x = 0$;
(ii) $\langle x|\lambda y + z \rangle = \lambda \langle x|y \rangle + \langle x|z \rangle$;
(iii) $\langle x|ya \rangle = \langle x|y \rangle a$;
(iv) $\langle y|x \rangle = \langle x|y \rangle^*$,

for all $x,y,z \in \mathscr{X}$, $a \in \mathscr{A}$, $\lambda \in \mathbb{C}$ (cf. [34]).

The orthogonality relation in $\mathscr{X}$ can be naturally defined by $x \perp y \Leftrightarrow \langle x|y \rangle = 0$ and the $\varepsilon$-orthogonality by $x \perp^{\varepsilon} y \Leftrightarrow \| \langle x|y \rangle \| \leq \varepsilon \|x\| \, \|y\|$.

### 4.2.2 Isosceles, Birkhoff and ρ-Orthogonalities

In a normed space $X$ one can consider, among other, the following orthogonality relations:

- isosceles orthogonality ($i$-orthogonality, cf. [28])

$$x \perp_i y \iff \|x+y\| = \|x-y\|;$$

- Birkhoff (Birkhoff–James) orthogonality (cf. [5,29])

$$x \perp_B y \iff \forall \alpha \in \mathbb{K}: \ \|x + \alpha y\| \geq \|x\|.$$

For properties of the above and other orthogonalities in normed spaces – see [1,2].

Some of the considered in this paper orthogonalities are related to *norm derivatives* (cf. [3,4])

$$\rho'_{\pm}(x,y) = \lim_{t\to 0^{\pm}} \frac{\|x+ty\|^2-\|x\|^2}{2t}, \qquad x,y\in X,\ t\in\mathbb{R}$$

and the so-called M-semi-inner product (cf. [22,39])

$$\langle y|x\rangle_g := \frac{1}{2}\left(\rho'_+(x,y)+\rho'_-(x,y)\right).$$

Remark that the mapping $\langle\cdot|\cdot\rangle_g$ need not be additive with respect to the first variable. If it is the case, i.e., if

$$\langle x+y|z\rangle_g = \langle x|z\rangle_g + \langle y|z\rangle_g, \qquad x,y,z\in X$$

we say that $X$ is *semi-smooth* (smoothness implies semi-smoothness but not conversely). In this case $\langle\cdot|\cdot\rangle_g$ is a semi-inner product in Lumer–Giles' sense (cf. [22]).

The related orthogonalities are defined as follows.

$$x\perp_{\rho_+}y \iff \rho'_+(x,y)=0;$$
$$x\perp_{\rho_-}y \iff \rho'_-(x,y)=0$$

and

$$x\perp_{\rho}y \iff \langle y|x\rangle_g=0 \iff \rho'_+(x,y)+\rho'_-(x,y)=0.$$

Relations $\perp_{\rho_+}$, $\perp_{\rho_-}$ and $\perp_{\rho}$ are generally (unless $X$ is smooth) incomparable (cf. [18]).

For each orthogonality relation listed above we may define its approximate extension. Let us start with the $i$-orthogonality. For $\varepsilon\in[0,1)$ and $x,y\in X$ we define:

$$x\perp_i^{\varepsilon}y \iff \left|\,\|x+y\|^2-\|x-y\|^2\right| \le 4\varepsilon\|x\|\,\|y\|$$

and

$$x\,{}^{\varepsilon}\!\!\perp_i y \iff \left|\,\|x+y\|-\|x-y\|\,\right| \le \varepsilon(\|x+y\|+\|x-y\|)$$
$$\iff \frac{1-\varepsilon}{1+\varepsilon}\|x-y\| \le \|x+y\| \le \frac{1+\varepsilon}{1-\varepsilon}\|x-y\|.$$

Obviously, for $\varepsilon=0$ both approximate orthogonalities coincide with $i$-orthogonality. It is a simple observation that the second definition of $\varepsilon$-$i$-orthogonality is weaker than the first one, i.e., that for an arbitrary $\varepsilon\in[0,1)$

$$x\perp_i^{\varepsilon}y \implies x\,{}^{\varepsilon}\!\!\perp_i y, \qquad x,y\in X,$$

but not conversely (see an example below). It is also easy to check that in the case where the norm comes from a real valued inner product, we have

$$x\perp_i^{\varepsilon}y \iff |\langle x|y\rangle| \le \varepsilon\|x\|\,\|y\| \iff x\perp^{\varepsilon}y$$

and

$$x\,{}^{\varepsilon}\!\perp_i y \iff |\langle x|y\rangle| \le \frac{\varepsilon}{1+\varepsilon^2}(\|x\|^2+\|y\|^2).$$

Thus, the first (stronger) approximate $i$-orthogonality coincides with the standard notion of approximate orthogonality for inner product spaces.

*Example 4.1.* Let $X$ be an inner product space, $x \in X\setminus\{0\}$ and $y=\lambda x$ for some $\lambda > 0$. Then

$$\frac{|\langle x|y\rangle|}{\|x\|^2+\|y\|^2} = \frac{\lambda}{1+\lambda^2} \to 0 \qquad (\text{as } \lambda\to\infty)$$

and, for all $\lambda$,

$$\frac{|\langle x|y\rangle|}{\|x\|\,\|y\|} = 1.$$

Thus, for an arbitrary $x \ne 0$ and $\varepsilon \in [0,1)$ there exists a number $\lambda$ large enough so that $x\,{}^{\varepsilon}\!\perp_i \lambda x$ whereas $x\perp_i^{\varepsilon}\lambda x$ does not hold for any $\varepsilon\in[0,1)$.

The following definition of the approximate Birkhoff orthogonality has been introduced by Dragomir [21]:

$$x\underset{\varepsilon}{\perp}_B y \iff \forall\lambda\in\mathbb{K}:\ \|x+\lambda y\| \ge (1-\varepsilon)\|x\|. \tag{4.1}$$

For inner product spaces $x\underset{\varepsilon}{\perp}_B y \Leftrightarrow x\perp^{\delta}y$ with $\delta := \sqrt{(2-\varepsilon)\varepsilon}$. In order to have this equivalence with $\delta=\varepsilon$, one can modify (4.1):

$$x\,{}^{\varepsilon}\!\perp_B y \iff \forall\lambda\in\mathbb{K}:\ \|x+\lambda y\| \ge \sqrt{1-\varepsilon^2}\|x\|. \tag{4.2}$$

Another definition of the approximate Birkhoff orthogonality (generally, not equivalent to (4.2)) has been given by the author [8]:

$$x\perp_B^{\varepsilon}y \iff \forall\lambda\in\mathbb{K}:\ \|x+\lambda y\|^2 \ge \|x\|^2 - 2\varepsilon\|x\|\|\lambda y\| \tag{4.3}$$

(cf. [8] and also [40] for comparison of ${}^{\varepsilon}\!\perp_B$ and $\perp_B^{\varepsilon}$).

Now, we introduce an approximate $\rho_\pm$ and $\rho$-orthogonality. For an $\varepsilon\in[0,1)$ we define:

$$x\perp_{\rho_+}^{\varepsilon}y \iff |\rho'_+(x,y)| \le \varepsilon\|x\|\,\|y\|;$$

$$x\perp_{\rho_-}^{\varepsilon}y \iff |\rho'_-(x,y)| \le \varepsilon\|x\|\,\|y\|$$

and

$$x\perp^{\varepsilon}_{\rho}y \iff |\langle y|x\rangle_g| \le \varepsilon\|x\|\,\|y\| \iff |\rho'_+(x,y)+\rho'_-(x,y)| \le 2\varepsilon\|x\|\,\|y\|.$$

Obviously, if the norm in $X$ comes from an inner product, then

$$\perp^{\varepsilon}_{\rho_+} = \perp^{\varepsilon}_{\rho_-} = \perp^{\varepsilon}_{\rho} = \perp^{\varepsilon}.$$

For inner product spaces $\perp_i = \perp_B = \perp_s = \perp$. It is known that in a smooth space, $\perp_B = \perp_s$. On the other hand, equality $\perp_i = \perp_B$ (and even each of the inclusions $\perp_i \subset \perp_B$ and $\perp_i \supset \perp_B$) holds true only if the norm comes from an inner product.

In inner product space we have also $\perp^{\varepsilon}_s = \perp^{\varepsilon}_B = {}^{\varepsilon}\!\perp_B = \perp^{\varepsilon}$. Moreover, the following can be shown.

**Theorem 4.1 ([8]).** *If $X$ is smooth, then $\perp^{\varepsilon}_B = \perp^{\varepsilon}_s$. In general case only the inclusion $\perp^{\varepsilon}_s \subset \perp^{\varepsilon}_B$ holds.*

**Theorem 4.2 ([18]).** *If $X$ is a real smooth space, then $\perp^{\varepsilon}_B = \perp^{\varepsilon}_{\rho}$. In general case only the inclusion $\perp^{\varepsilon}_{\rho} \subset \perp^{\varepsilon}_B$ holds.*

## 4.3 Orthogonality Preserving Mappings

In the theory of linear preservers the main goal is (see [41]) to describe transformations of a given space preserving a quantity attached to the space, some set of elements of the space or a given relation among those elements. We consider the latter case. However, in our considerations some approximate preservation is allowed.

### *4.3.1 Mappings Which Exactly Preserve Orthogonality*

We start with a mapping which exactly preserves orthogonality, i.e., which transforms any two orthogonal vectors into orthogonal ones.

#### 4.3.1.1 Inner Product Spaces

Let $X$ and $Y$ be (real or complex) inner product spaces. We say that $T\colon X \to Y$ *preserves orthogonality* iff

$$x\perp y \implies T(x)\perp T(y), \qquad x,y\in X. \tag{4.4}$$

Such mappings can be irregular, far from being continuous or linear. Thus, in considerations that follows we restrict ourselves to linear mappings only.

**Theorem 4.3.** *For a nonzero linear mapping $T\colon X \to Y$ the following conditions are equivalent with some $\gamma > 0$:*

(i) *$T$ preserves orthogonality;*
(ii) $\|T(x)\| = \gamma\|x\|,\ x \in X;$
(iii) $\langle T(x)|T(y)\rangle = \gamma^2\langle x|y\rangle,\ x,y \in X;$
(iv) $|\langle T(x)|T(y)\rangle| = \gamma^2|\langle x|y\rangle|,\ x,y \in X.$

*Proof.* Implication (i)⇒(ii) can be proved elementarily (see [9]). (ii)⇒(iii) follows from polarization formulas and implications (iii)⇒(iv)⇒(i) are obvious. □

Notice that it follows from the above result that a linear mapping $T$ preserves orthogonality if and only if it preserves orthogonality in both directions, i.e.,

$$x \perp y \iff T(x) \perp T(y), \qquad x,y \in X.$$

Let us make a note about the functional equations which appeared in (iii) and (iv). For $\gamma = 1$ and an arbitrary mapping $f\colon X \to Y$ the first one takes the form

$$\langle f(x)|f(y)\rangle = \langle x|y\rangle\,, \qquad x,y \in X \tag{4.5}$$

and is called an *orthogonality equation*. Notice, that each solution of (4.5) must be linear. Obviously, each solution of (4.5) satisfies also the celebrated *Wigner equation*

$$|\langle f(x)|f(y)\rangle| = |\langle x|y\rangle|, \qquad x,y \in X. \tag{4.6}$$

We say that mappings $f,g\colon X \to Y$ are *phase-equivalent* iff $f(x) = \sigma(x)g(x),\ x \in X$ with some $\sigma\colon X \to \mathbb{C},\ |\sigma(x)| = 1, x \in X$. Wigner's theorem (cf. [50]) states that each solution of (4.6) has to be phase equivalent to a linear or conjugate-linear isometry.

#### 4.3.1.2 Hilbert Modules

Now, let $\mathscr{A}$ be a $C^*$-algebra and let $V,W$ be inner product $\mathscr{A}$-modules. Ilišević and Turnšek [27] proved validity of Theorem 4.3 in this setting.

**Theorem 4.4 ([27], Theorem 3.1).** *For a $C^*$-algebra $\mathscr{A}$ with $\mathscr{K}(\mathscr{H}) \subset \mathscr{A} \subset \mathscr{B}(\mathscr{H})$ and a nonzero $\mathscr{A}$-linear mapping $T\colon V \to W$ the following conditions are equivalent (with some $\gamma > 0$):*

(i) *$T$ is orthogonality preserving;*
(ii) $\|Tx\| = \gamma\|x\|, \quad x \in V;$
(iii) $\langle Tx|Ty\rangle = \gamma^2\langle x|y\rangle\,, \quad x,y \in V.$

For an arbitrary $C^*$-algebra, (ii) and (iii) are equivalent and each of them implies (i). But in general, (i) is not equivalent neither to (ii) nor (iii) (see [27, Example 2.4]). Recently, Leung, Ng and Wong [35–37] have made a further progress in this topic. Further generalizations of Theorem 4.4 can be found in [24].

#### 4.3.1.3 Normed Spaces

For $X$ and $Y$ being normed spaces, the following result can be shown immediately for the isosceles-orthogonality.

**Theorem 4.5.** *Let $X$ and $Y$ be normed spaces. For a linear mapping $f : X \to Y$ the following conditions are equivalent:*

(a) $\exists \gamma > 0 \quad \|fx\| = \gamma\|x\|,\ x \in X;$
(b) $x \perp_i y \implies fx \perp_i fy, \quad x, y \in X;$
(c) $x \perp_i y \iff fx \perp_i fy, \quad x, y \in X.$

*Proof.* It is visible that (a)⇒(c) and (c)⇒(b) is obvious. For the proof of (b)⇒(a) notice that (b) is equivalent to the condition $\|x\| = \|y\| \ \Rightarrow\ \|fx\| = \|fy\|$. Let $\gamma := \|f(u)\|$ for an arbitrarily fixed unit vector $u$. Then, for $x \neq 0$ we have $\|f(x)\| = \|f((1/\|x\|)x)\|\,\|x\| = \gamma\|x\|$. □

Koehler i Rosenthal [31] showed that a linear operator from a normed space $X$ into itself is an isometry if and only if it preserves some semi-inner product. This can be reformulated to a slightly more general statement (cf. [14]).

**Theorem 4.6.** *Let $X$ and $Y$ be normed spaces and let $f : X \to Y$ be a linear operator. Then $f$ is a similarity (i.e., for some $\gamma > 0$, $\|fx\| = \gamma\|x\|$, $x \in X$), if and only if there exist semi-inner products $[\cdot|\cdot]_X$ and $[\cdot|\cdot]_Y$ in $X$ and $Y$, respectively, such that*

$$[fx|fy]_Y = \gamma^2\,[x|y]_X\,, \qquad x, y \in X. \tag{4.7}$$

*Moreover, if $X = Y$ (with the same norm), the assertion holds with one semi-inner product.*

*Proof.* The sufficiency is obvious. If $X$ and $Y$ are different spaces (at least the norms are different) we choose an arbitrary s.i.p. $[\cdot|\cdot]_Y$ in $Y$ and define

$$[x|y]_X := \frac{1}{\gamma^2}\,[fx|fy]_Y\,, \qquad x, y \in X$$

to obtain a s.i.p. in $X$ such that (4.7) is satisfied. However, this reasoning does not guarantee that if $X = Y$ and the norm is the same, $[\cdot|\cdot]_X = [\cdot|\cdot]_Y$ (unless $X$ is smooth which yields the uniqueness of s.i.p.). In this case, one can apply the proof of Koehler and Rosenthal (with a slight modification concerning the constant $\gamma$). □

Koldobsky [32] showed that a linear mapping from a real normed space into itself, preserving the Birkhoff orthogonality must be a similarity. Blanco and Turnšek [6] extended it to complex spaces.

**Theorem 4.7 ([6], Theorem 3.1).** *Let $X$ and $Y$ be (real or complex) normed spaces and let $f : X \to Y$ be a linear operator. Then $f$ preserves the Birkhoff orthogonality, i.e.,*

$$x \perp_B y \implies fx \perp_B fy, \qquad x, y \in X \tag{4.8}$$

*if and only if, for some $\gamma > 0$, $\|fx\| = \gamma\|x\|$, $x \in X$.*

Blanco and Turnšek remarked also that their proof of Theorem 4.7 can be easily adapted to the case where the Birkhoff orthogonality is replaced by a s.i.p.-orthogonality. Namely, we have the following result.

**Theorem 4.8 ([6], Remark 3.2).** *Let $X$ and $Y$ be (real or complex) normed spaces and let $f : X \to Y$ be a linear operator which preserves the semi-orthogonality related to some s.i.p. $[\cdot|\cdot]_X$ and $[\cdot|\cdot]_Y$ in $X$ and $Y$, respectively, i.e.,*

$$x \perp_s y \Longrightarrow fx \perp_s fy, \qquad x, y \in X. \tag{4.9}$$

*Then, for some $\gamma > 0$, $\|fx\| = \gamma\|x\|$, $x \in X$.*

The reverse implication follows from Theorem 4.6.

Now, we are able to collect the quoted results and show the equivalency of some conditions connected with the preservation of orthogonality relations (we generalize Theorem 2.5 in [14]).

**Theorem 4.9.** *Let $X$ and $Y$ be normed spaces. For a linear operator $f : X \to Y$ and some $\gamma > 0$ the following conditions are equivalent (we assume them for all $x, y \in X$):*

(a) $\|fx\| = \gamma\|x\|$;
(b) $[fx|fy]_Y = \gamma^2 [x|y]_X$;
(c) $x \perp_s y \Longleftrightarrow fx \perp_s fy$;
(d) $x \perp_s y \Longrightarrow fx \perp_s fy$;
(e) $x \perp_B y \Longleftrightarrow fx \perp_B fy$;
(f) $x \perp_B y \Longrightarrow fx \perp_B fy$;
(g) $x \perp_i y \Longleftrightarrow fx \perp_i fy$;
(h) $x \perp_i y \Longrightarrow fx \perp_i fy$.

*Conditions* (b)–(d) *should be understood that they hold with some semi-inner products $[\cdot|\cdot]_X$ and $[\cdot|\cdot]_Y$ in $X$ and $Y$, respectively.*

*Proof.* (a)$\Rightarrow$(b) follows from Theorem 4.6; implications (b)$\Rightarrow$(c)$\Rightarrow$(d) are trivial; implication (d)$\Rightarrow$(a) follows from Theorem 4.8. This proves that (a)–(d) are equivalent. It is easy to show that (a)$\Rightarrow$(e), (e)$\Rightarrow$(f) is obvious and (f)$\Rightarrow$(a) follows from Theorem 4.7, which proves the equivalency of (a), (e) and (f). Equivalency of (a), (g), (h) follows from Theorem 4.5. □

*Remark 4.1.* Notice that, in particular, the orthogonality preserving properties for linear mappings with respect to isosceles, Birkhoff and semi-orthogonalities are equivalent, although the relations $\perp_B$, $\perp_s$ and $\perp_i$ themselves are generally not equivalent.

*Remark 4.2.* For $X = Y$ the above results remain true with the same semi-inner products for arguments and values (cf. the remark in Theorem 4.6).

Taking $X = Y$ and the identity mapping we get from Theorem 4.9 the following corollary.

**Corollary 4.1.** *Let $\|\cdot\|_1$ and $\|\cdot\|_2$ be two norms in a vector space $X$ (with associated semi-inner products $[\cdot|\cdot]_1$ and $[\cdot|\cdot]_2$, semi-orthogonality relations $\perp_{s1}, \perp_{s2}$ and isosceles and Birkhoff orthogonalities $\perp_{B1}, \perp_{B2}$, $\perp_{i1}, \perp_{i2}$, respectively). The following conditions, with some $\gamma > 0$, are equivalent:*

(a) $\|x\|_2 = \gamma\|x\|_1, \quad x \in X;$

(b) $[x|y]_2 = \gamma^2 [x|y]_1, \quad x, y \in X;$

(c) $\perp_{s1} = \perp_{s2};$

(d) $\perp_{s1} \subset \perp_{s2};$

(e) $\perp_{B1} = \perp_{B2};$

(f) $\perp_{B1} \subset \perp_{B2};$

(g) $\perp_{i1} = \perp_{i2};$

(h) $\perp_{i1} \subset \perp_{i2}.$

As for the $\rho$ and $\rho_\pm$-orthogonality preserving mapping the following characterization has been given in [18].

**Theorem 4.10.** *Let $X, Y$ be real normed spaces, $f\colon X \to Y$ a nonzero, linear mapping. Then, the following conditions are equivalent (with some $\gamma > 0$):*

*(a) $f$ preserves $\rho_+$-orthogonality;*

*(b) $f$ preserves $\rho_-$-orthogonality;*

*(c) $\|fx\| = \gamma\,\|x\|, \quad x \in X;$*

*(d) $\rho'_+(fx, fy) = \gamma^2\, \rho'_+(x, y), \quad x, y \in X;$*

*(e) $\rho'_-(fx, fy) = \gamma^2\, \rho'_-(x, y), \quad x, y \in X;$*

*(f) $\langle fx|fy\rangle_g = \gamma^2\, \langle x|y\rangle_g, \quad x, y \in X.$*

*Moreover, each of the above conditions implies that*

*(g) $f$ preserves $\rho$-orthogonality.*

If $X$ is smooth (or at least semi-smooth), (g) is equivalent to (a)–(f). Actually, P. Wójcik has proved recently that (g) is equivalent to (a)–(f) in general.

If the mapping $f$ exactly preserves one of the above considered orthogonalities, $f$ must be then a linear similarity. Thus, the spaces $X$ and $Y$ have to share some geometrical properties. In particular, the modulus of convexity $\delta_X$ and modulus of smoothness $\rho_X$ must be preserved, i.e., $\delta_X = \delta_{f(X)}$ and $\rho_X = \rho_{f(X)}$. As a consequence, we have the following theorem.

**Theorem 4.11.** *Let $X$ be a normed space. Suppose that there exists a normed space $Y$ which is uniformly convex (uniformly smooth), strictly convex, or an inner product space and a nontrivial linear mapping $f$ from $X$ into $Y$ (or from $Y$ onto $X$) such that $f$ preserves the isosceles (Birkhoff, semi-, $\rho_+$, $\rho_-$) orthogonality. Then $X$ is, respectively, a uniformly convex (uniformly smooth), strictly convex, an inner product space as well.*

*Proof.* 1. Uniform convexity (smoothness) follows from $\delta_X = \delta_{f(X)}$ and $\rho_X = \rho_{f(X)}$.

2. We have $\|fx\| = \gamma\|x\|$, $x \in X$ for some $\gamma > 0$. Suppose that $Y$ is strictly convex. Let $x, y \in X$ and $\|x+y\| = \|x\| + \|y\|$. Then also $\|fx+fy\| = \|fx\| + \|fy\|$ whence $fx = cfy$ for some $c > 0$. Since $f$ is injective, $x = cy$, thus $X$ is strictly convex.
3. If $Y$ is an inner product space, since $f$ is a similarity, the parallelogram identity is preserved and the assertion follows. It can be also derived from the fact that $\delta_X = \delta_{f(X)}$ and that the equality $\delta_X = \delta_{\mathscr{H}}$ characterizes inner product spaces. □

## 4.3.2 Mappings Which Approximately Preserve Orthogonality

It may be of some interest for applications to consider mappings which transform orthogonal vectors to approximately orthogonal ones.

### 4.3.2.1 Inner Product Spaces and Hilbert Modules

Let $X$ be an inner product space. We say that $f\colon X \to Y$ is an *approximately orthogonality preserving* (a.o.p.) mapping iff

$$x \perp y \Longrightarrow f(x) \perp^{\varepsilon} f(y), \qquad x, y \in X \tag{4.10}$$

with some $\varepsilon \in [0,1)$. We know that linear o.p. mappings are similarities, thus, it should be expected that linear a.o.p. mappings are, in some sense, approximate similarities. Indeed, the following result was proved in [9] (cf. also [47]).

**Theorem 4.12.** *Let $f\colon X \to Y$ be a nonzero linear mapping satisfying* (4.10) *with $\varepsilon \in [0,1)$. Then $f$ is injective, continuous and, with some $\gamma > 0$, $f$ satisfies the functional inequality*

$$|\langle f(x)|f(y)\rangle - \gamma\langle x|y\rangle| \le \frac{4\varepsilon}{1+\varepsilon} \min\{\gamma\|x\|\,\|y\|, \|f(x)\|\,\|f(y)\|\}, \qquad x, y \in X.$$

*Conversely, if $f\colon X \to Y$ satisfies*

$$|\langle f(x)|f(y)\rangle - \gamma\langle x|y\rangle| \le \varepsilon \min\{\gamma\|x\|\,\|y\|, \|f(x)\|\,\|f(y)\|\}, \qquad x, y \in X$$

*with $\varepsilon \ge 0$ and with $\gamma > 0$, then $f$ is a quasi-linear mapping, approximately orthogonality preserving. More precisely, $f$ satisfies*

$$\|f(x+y) - f(x) - f(y)\| \le 2\sqrt{\varepsilon}\sqrt{\gamma}(\|x\| + \|y\|), \qquad x, y \in X;$$
$$\|f(\lambda x) - \lambda f(x)\| \le 2\sqrt{\varepsilon}\sqrt{\gamma}|\lambda|\,\|x\|, \qquad x \in X, \lambda \in \mathbb{K}$$

*and*

$$x \perp y \Longrightarrow f(x) \perp^{\varepsilon} f(y), \quad f(x) \perp f(y) \Longrightarrow x \perp^{\varepsilon} y, \qquad x, y \in X.$$

For Hilbert modules an analogous result has been obtained by Ilišević and Turnšek [27].

**Theorem 4.13.** *Let $\mathscr{A}$ be a $C^*$-algebra such that $\mathscr{K}(\mathscr{H}) \subset \mathscr{A} \subset \mathscr{B}(\mathscr{H})$. Let $T\colon V \to W$ be an $\mathscr{A}$-linear mapping $\varepsilon$-approximately orthogonality preserving. Then $T$ is bounded and*

$$\left\| \langle Tx|Ty\rangle - \|T\|^2 \langle x|y\rangle \right\| \leq \frac{4\varepsilon}{1+\varepsilon} \|T\|^2 \|x\| \|y\|, \qquad x,y \in V.$$

#### 4.3.2.2 Normed Spaces

Assume now that $X$ is a real normed space. We start with $i$-orthogonality and approximately $i$-orthogonality preserving mappings. What we mean is that for an arbitrary pair of $i$-orthogonal arguments $x,y$ their images $fx, fy$ are approximately $i$-orthogonal, with respect to ${}^{\varepsilon}\!\perp_i$ or ${}^{\varepsilon}\!\perp_i$ (cf. definitions on page 36). Also in this case, it can be shown that such mappings are approximate similarities.

Let $f\colon X \to Y$ be a linear and continuous operator. Define

$$[f] := \inf\{\|fx\| : \|x\| = 1\} = \sup\{M \geq 0 : \|fx\| \geq M\|x\|,\ x \in X\}.$$

The following result was proved in [17]:

**Theorem 4.14 ([17], Theorem 3.6).** *Let $\varepsilon \in [0,1)$ and let $f\colon X \to Y$ be a nonzero linear mapping. Then the following conditions are equivalent:*

$$x \perp_i y \quad \Longrightarrow \quad f(x)\,{}^{\varepsilon}\!\perp_i f(y), \qquad x,y \in X; \tag{4.11}$$

$$\frac{1-\varepsilon}{1+\varepsilon} \|f\| \|x\| \leq \|fx\| \leq \frac{1+\varepsilon}{1-\varepsilon} [f] \|x\|, \qquad x \in X; \tag{4.12}$$

$$\frac{1-\varepsilon}{1+\varepsilon} \gamma\|x\| \leq \|fx\| \leq \frac{1+\varepsilon}{1-\varepsilon} \gamma\|x\|, \qquad x \in X,\ \gamma \in [[f], \|f\|];$$

$$\|f\| \leq \frac{1+\varepsilon}{1-\varepsilon} [f];$$

$$\|fx\| \|y\| \leq \frac{1+\varepsilon}{1-\varepsilon} \|fy\| \|x\|, \qquad x,y \in X.$$

For $\varepsilon = 0$, from (4.12) one gets

$$\|f\| \|x\| \leq \|fx\| \leq [f]\|x\| \leq \|f\| \|x\|, \qquad x \in X,$$

thus $\|fx\| = \gamma\|x\|$, $x \in X$ with $\gamma = \|f\| = [f]$; $f$ is a similarity.

Instead of the condition (4.11) one can consider a stronger one:

$$x \perp_i y \quad \Longrightarrow \quad f(x) \perp_i^{\varepsilon} f(y), \qquad x,y \in X. \tag{4.13}$$

Let $\|\cdot\|_1$ and $\|\cdot\|_2$ be two norms in $X$ and let $\perp_{i1}$, $\perp_{i2}$ denote $i$-orthogonality relations with respect to the first or the second norm, respectively. Applying Theorem 4.14 for identity mapping from $(X,\|\cdot\|_1)$ to $(X,\|\cdot\|_2)$ one gets that if $\perp_{i1}\subset {}^{\varepsilon}\!\perp_{i2}$ (i.e., $x\perp_{i1}y \Longrightarrow x\,{}^{\varepsilon}\!\perp_{i2}y$, $x,y\in X$), then for all $\gamma$ such that

$$\inf_{\|x\|_1=1}\|x\|_2 \le \gamma \le \sup_{\|x\|_1=1}\|x\|_2,$$

there is

$$\frac{1-\varepsilon}{1+\varepsilon}\gamma\|x\|_1 \le \|x\|_2 \le \frac{1+\varepsilon}{1-\varepsilon}\gamma\|x\|_1, \qquad x\in X.$$

In a recent paper, Mojškerc and Turnšek [40] considered the class of linear mappings approximately preserving the Birkhoff orthogonality. They showed that each such a mapping must be an approximate similarity. Namely, they proved the following result, the proof of which is by no means elementary.

**Theorem 4.15 ([40], Theorem 3.5).** *Let* $X,Y$ *be normed spaces,* $\varepsilon\in[0,1/2)$ *and let* $T: X\to Y$ *satisfies*

$$x\perp_B y \Longrightarrow Tx\perp^{\varepsilon}_B Ty, \qquad x,y\in X. \tag{4.14}$$

*Then*

$$(1-16\varepsilon)\|T\|\,\|x\| \le \|Tx\| \le \|T\|\,\|x\|. \tag{4.15}$$

For $\mathbb{K}=\mathbb{R}$ the constant $1-16\varepsilon$ can be replaced by $1-8\varepsilon$.

Now, let us consider the class linear mappings $f: X\to X$ which preserve approximately the $\perp_\rho$ ($\perp_{\rho\pm}$) orthogonality; namely, which satisfy:

$$x\perp_\rho y \Longrightarrow fx\perp^{\varepsilon}_\rho fy, \qquad x,y\in X;$$
$$x\perp_{\rho+} y \Longrightarrow fx\perp^{\varepsilon}_{\rho+} fy, \qquad x,y\in X$$

or

$$x\perp_{\rho-} y \Longrightarrow fx\perp^{\varepsilon}_{\rho-} fy, \qquad x,y\in X.$$

It is easy to see that the latter two conditions are equivalent. The natural problem is to describe those classes of approximately orthogonality preserving mappings. In case of smooth space $X$, we have $\perp_\rho=\perp_B$ and $\perp^{\varepsilon}_\rho=\perp^{\varepsilon}_B$ for all $\varepsilon\in[0,1)$. Thus, if $X$ is smooth, then the class of (approximate) $\rho$-orthogonality preserving linear mappings coincide with that of (approximate) Birkhoff orthogonality preserving and hence it follows from Theorem 4.15 that linear approximately $\rho$-orthogonality preserving mappings are approximate similarities in sense of (4.15). The case of non-smooth spaces generally remains an open problem, however, a positive answer has been obtained for some particular spaces by Wójcik (unpublished yet).

## 4.4 Stability Problems

The theory of *stability of functional equations* originated in the question posed by S. Ulam: "when is it true that the solution of an equation differing slightly from a given one, must of necessity be close to the solution of the given equation?" (see [48, p. 63]) The expressions "differing slightly" and "be close" may have various meanings, which leads to a great number of different kinds of stability to dealt with. To get acquainted with the theory of the stability of functional equations we refer to monographs [26, 30] as well as to survey papers, e.g., [23, 42].

### *4.4.1 Stability of Isometric Operators*

For $\varepsilon \in [0,1)$ consider an $\varepsilon$-*isometric* operator, i.e., a linear operator $f\colon X \to Y$ satisfying

$$\big|\,\|fx\| - \|x\|\,\big| \le \varepsilon\|x\|, \qquad x \in X.$$

By a stability problem for isometric operators we mean here a question whether each linear $\varepsilon$-isometric operator admits a uniform approximation by a linear isometric one. More precisely, given a pair of normed spaces $(X,Y)$, does there exist a mapping $\delta\colon [0,1) \to \mathbb{R}_+$ satisfying $\lim_{\varepsilon\to 0}\delta(\varepsilon) = 0$ and such that for any $\varepsilon$-isometric mapping $f\colon X \to Y$ (with $0 \le \varepsilon < 1$) there exists a linear isometric operator $I\colon X \to Y$ such that

$$\|f - I\| \le \delta(\varepsilon)?$$

Let $\mathfrak{A}$ denote the class of all pairs of normed spaces for which the above problem has an affirmative solution.

One can also define a broader class $\mathfrak{B}$ of all pairs $(X,Y)$ of normed spaces for which the following condition holds: for each $\delta > 0$ there exists $\varepsilon > 0$ such that for each linear $\varepsilon$-isometry $f\colon X \to Y$ there exists a linear isometry $I\colon X \to Y$ such that

$$\|f - I\| \le \delta.$$

Obviously, $\mathfrak{A} \subset \mathfrak{B}$. It follows from the definition that if $(X,Y) \in \mathfrak{B}$, then there exist an $\varepsilon_0 \in (0,1]$ and a mapping $\delta\colon [0,\varepsilon_0) \to \mathbb{R}_+$ such as required in the definition of $\mathfrak{A}$. The constant $\varepsilon_0$ may be smaller than 1 and the reason is that some pairs of (different) spaces $(X,Y)$ do not admit isometries between them, although they do admit $\varepsilon$-isometries if only $\varepsilon$ is large enough (see Example 4.2 below). On the other hand, in the case $X = Y$, for an arbitrary normed space $X$, we have

$$(X,X) \in \mathfrak{A} \iff (X,X) \in \mathfrak{B}$$

(for details see [17]).

If the dimension of the domain $X$ is greater than the dimension of $Y$, then the pair $(X,Y)$ belongs to $\mathfrak{A}$ simply because there are no $\varepsilon$-isometries in this case. Ding [20] remarked that each pair $(X,Y)$ of finite-dimensional normed spaces belongs to $\mathfrak{B}$. The following example shows that it does not necessarily belong to $\mathfrak{A}$.

*Example 4.2.* Consider $\mathbb{R}^2$ with norms $\|\cdot\|_1$ and $\|\cdot\|_2$ where

$$\|x\|_1 := |x_1| + |x_2|, \quad \|x\|_2 := \sqrt{x_1^2 + x_2^2}, \qquad x = (x_1, x_2) \in \mathbb{R}^2.$$

Then for $X = (\mathbb{R}^2, \|\cdot\|_1)$, $Y = (\mathbb{R}^2, \|\cdot\|_2)$ and $f = \mathrm{id}$ we have $(\sqrt{2}/2)\|x\|_1 \leq \|x\|_2 \leq \|x\|_1$ for $x \in \mathbb{R}^2$ whence, with $\varepsilon := 1 - \sqrt{2}/2$,

$$(1-\varepsilon)\|x\|_1 \leq \|fx\|_2 \leq (1+\varepsilon)\|x\|_1, \qquad x \in X.$$

Thus, there exists a linear $\varepsilon$-isometry $f : X \to Y$ which cannot be approximated by a linear isometry $I : X \to Y$ since there is no one between considered spaces $X$ and $Y$.

It is proved in [20] that for a Hilbert space $\mathscr{H}$ the pair $(\mathscr{H}, \mathscr{H})$ belongs to $\mathfrak{A}$ with $\delta(\varepsilon) = \varepsilon$. It is also proved that $(m,m), (c,c), (c_0, c_0) \in \mathfrak{A}$. Generally, a pair $(X,Y)$ of normed spaces does not have to belong to $\mathfrak{B}$; neither if $X$ and $Y$ are different nor in the, more difficult, case $X = Y$.

Dealing with the problem of the stability of isometric mappings one can also restrict to surjective mappings, so that another classes of pairs of normed spaces can be introduced. For a much more comprehensive study on the stability of the isometric mappings, including some new results and examples, we refer the reader to the paper of Protasov [44], in the same volume.

A similar stability problem can be stated for the orthogonality equation (4.5) as well as for its generalization: the Wigner equation (4.6). Various approaches to the stability of these equations were considered both in Hilbert spaces (cf. a survey paper [13]) and in Hilbert modules (cf. [15, 16]). The following, selected, result remains in some connection with our present considerations.

**Theorem 4.16 ([10], Theorem 2 and [47], Theorem 2.6).** *Let $X$ and $Y$ be inner product spaces and let $X$ be finite-dimensional. Then, there exists a continuous mapping $\delta : [0,1) \to \mathbb{R}_+$ such that $\lim_{\varepsilon \to 0^+} \delta(\varepsilon) = 0$ and satisfying the following property. For each mapping $f : X \to Y$ satisfying*

$$|\langle f(x)|f(y)\rangle - \langle x|y\rangle| \leq \varepsilon\|x\|\,\|y\|, \qquad x, y \in X$$

*there exists a linear isometry $I : X \to Y$ such that*

$$\|f(x) - I(x)\| \leq \delta(\varepsilon)\|x\|, \qquad x \in X.$$

Turnšek [47] presented somewhat different method of the proof of the theorem than the one given in [10] and he was able to give an explicit formula for $\delta(\varepsilon)$. Namely

$$\delta(\varepsilon) := \min\left\{1 - \sqrt{1-\varepsilon} + \sqrt{8\varepsilon(n + 2\sqrt{n} - 1)}, 1 + \sqrt{1+\varepsilon}\right\},$$

where $n$ denotes the dimension of the domain $X$. Unfortunately, the answer for infinite-dimensional case is not known. In the paper of Kong and Cao [33], the stability is proved for arbitrary separable Hilbert, however, under some additional, restrictive, assumption on the considered mappings.

### 4.4.2 Stability of the Orthogonality Preserving Property

#### 4.4.2.1 Inner Product Spaces and Hilbert Modules

We know from Theorem 4.12 that each linear approximately orthogonality preserving mapping is an approximate solution of a modified orthogonality equation. Knowing that for finite-dimensional spaces the later equation is stable (Theorem 4.16) the author in [10] was able to obtain the stability of the orthogonality preserving property. However, it turned out that the finite-dimensionality assumption can be omitted. Turnšek in [47], applying different method of the proof, based on the operator theory for Hilbert spaces, was able to obtain the following result.

**Theorem 4.17 ([10], Theorem 4; [47], Theorem 2.3).** *Let $X,Y$ be Hilbert spaces. For a linear, $\varepsilon$-approximately orthogonality preserving mapping $f\colon X\to Y$ there exists a linear orthogonality preserving operator $T\colon X\to Y$ such that*

$$\|f-T\| \leq \left(1-\sqrt{\frac{1-\varepsilon}{1+\varepsilon}}\right)\min\{\|f\|,\|T\|\}. \tag{4.16}$$

*Proof.* Let us describe the sketch of the proof. First, it can be shown quite elementarily (see [9,47]) that for an arbitrary $x\in X$ we have

$$\sqrt{\frac{1-\varepsilon}{1+\varepsilon}}\|T\|\,\|x\| \leq \|Tx\| \leq \|T\|\,\|x\|, \qquad x\in X.$$

Next, we apply some operator theory. Let $|T|\in\mathscr{L}(X,Y)$ denote the positive square root of $T^*T$ and let $U\in\mathscr{L}(X,Y)$ be a partial isometry such that the *polar decomposition* $T=U|T|$ holds (see, e.g., [43]). The main idea from [47] was to prove, that assuming

$$m\|x\| \leq \|Tx\| \leq M\|x\|, \qquad x\in X$$

for some $m,M>0$, $U$ must be an isometry and

$$\|T-U\| \leq \max\{|M-1|,|m-1|\}.$$

Applying this for the operator $(1/\|T\|)\,T$ and the constants $m=\sqrt{(1-\varepsilon)/(1+\varepsilon)}$ and $M=1$ one gets (4.16). □

Ilišević and Turnšek [27] generalized Theorem 4.17 to Hilbert modules as follows.

**Theorem 4.18 ([27], Theorem 4.4).** *Let $\mathscr{A} = \mathscr{K}(\mathscr{H})$ and let $V, W$ be Hilbert $\mathscr{A}$-modules. If $T\colon V \to W$ is an $\mathscr{A}$-linear, $\varepsilon$-approximately orthogonality preserving mapping, then there exists an $\mathscr{A}$-linear isometry $U : V \to W$ such that*

$$\|T - \|T\|U\| \leq \varepsilon\|T\|.$$

It is an open problem if the above result is valid for an arbitrary $C^*$-algebra $\mathscr{A}$ such that $\mathscr{K}(\mathscr{H}) \subset \mathscr{A} \subset \mathscr{B}(\mathscr{H})$.

#### 4.4.2.2 Normed Spaces

Seemingly, the first most comprehensive answer to the stability problem in normed spaces has been given for the isosceles-orthogonality in real normed spaces. Assume that a linear mapping $f$ is, in a sense, close to a linear i-o.p. mapping $g$. Then one can show that $f$ is an a.o.p. mapping. Namely, if $g : X \to Y$ is a linear i-o.p. mapping, $f : X \to Y$ is linear and, with some $\varepsilon \in [0,1)$,

$$\|f - g\| \leq \varepsilon\|g\|,$$

then $f$ satisfies (4.11). Now, we would like to know if the reverse result is true, i.e., whether an approximately $i$-orthogonality preserving mapping has to be close to an $i$-orthogonality preserving one.

Applying Theorem 4.14, Chmieliński and Wójcik [17] were able to prove that the stability of the $i$-orthogonality preserving property holds true for pairs of spaces from the class $\mathfrak{A}$ (or $\mathfrak{B}$) and only for those classes.

**Theorem 4.19 ([17], Theorem 5.2).** *Let $X$ and $Y$ be real normed spaces such that $(X,Y) \in \mathfrak{A}$ with a suitable mapping $\delta\colon [0,1) \to \mathbb{R}_+$ satisfying $\lim_{\varepsilon\to 0}\delta(\varepsilon) = 0$ from the definition of $\mathfrak{A}$. Let $\varepsilon \in [0,1)$ and let $f\colon X \to Y$ be a linear mapping which satisfies* (4.11)*. Then there exists a linear mapping $g\colon X \to Y$ preserving i-orthogonality, i.e., a linear similarity, such that*

$$\|f - g\| \leq \delta(\varepsilon)\ \min\{\|f\|, \|g\|\}.$$

Similarly, assuming that $(X,Y) \in \mathfrak{B}$ we can get a weaker form of stability.

**Theorem 4.20 ([17], Theorem 5.3).** *Suppose that $X$ and $Y$ are real normed spaces and $(X,Y) \in \mathfrak{B}$. Then for an arbitrary $\delta > 0$, there exists an $\varepsilon > 0$ such that for any linear mapping $f\colon X \to Y$ satisfying* (4.11) *there exists a linear mapping $g\colon X \to Y$ preserving i-orthogonality such that*

$$\|f - g\| \leq \delta\ \min\{\|f\|, \|g\|\}.$$

Now assume that $X$ and $Y$ are real normed spaces for which the $i$-orthogonality preserving property is stable, in sense of Theorem 4.19. We will show that $(X,Y) \in \mathfrak{A}$. For let $f\colon X \to Y$ be a linear $\varepsilon$-isometry (with $\varepsilon \in [0,1)$). Then, in particular, $(1-\varepsilon) \leq [f] \leq \|f\| \leq (1+\varepsilon)$ and it is easy to observe that (4.12) holds, whence $f$ is a linear a.o.p. mapping (it satisfies (4.11)). Applying the stability assumption, there exists a linear o.p. mapping $g\colon X \to Y$, or equivalently, there exist a linear isometry $I\colon X \to Y$ and a constant $\gamma > 0$, such that

$$\|f-g\| = \|f-\gamma I\| \leq \delta(\varepsilon)\ \min\{\|f\|, \gamma\}.$$

We have $\|f\| \leq 1+\varepsilon$ and it can be verified that $|\gamma - 1| \leq \varepsilon + (1+\varepsilon)\delta(\varepsilon)$. Therefore

$$\|f-I\| \leq \|f-\gamma I\| + \|\gamma I - I\| \leq \varepsilon + 2(1+\varepsilon)\delta(\varepsilon)$$

and $\delta'(\varepsilon) := \varepsilon + 2(1+\varepsilon)\delta(\varepsilon) \to 0$ (as $\varepsilon \to 0$). This proves that $(X,Y) \in \mathfrak{A}$.

Similarly, one can prove that the stability of o.p. mappings in sense of Theorem 4.20 yields that $(X,Y) \in \mathfrak{B}$ and we can formulate the following result.

**Theorem 4.21 ([17], Theorem 5.5).** *If the stability of the i-orthogonality preserving property, in sense of Theorem 4.19 or 4.20, can be proved for a pair of real normed spaces $(X,Y)$, then $(X,Y) \in \mathfrak{A}$ or $(X,Y) \in \mathfrak{B}$, respectively.*

It follows from the above that the stability of $i$-orthogonality preserving property cannot be proved for all pairs of normed spaces. It cannot be proved even for some spaces which are close to Hilbert ones. The following counterexample is based on a construction communicated to the author by Protassov and included in his paper [44, Theorem 1].

*Example 4.3.* Let $\alpha = (\alpha_k)_{k\in\mathbb{N}}$ be an increasing sequence of positive numbers with

$$\sum_{k=1}^{\infty}(1-\alpha_k^2) < 1$$

(hence $\alpha_k \nearrow 1$). In the Hilbert space $l^2$ we introduce an equivalent norm

$$\|x\|_\alpha := \sup\left\{\|x\|_{l^2}, \left|\frac{x_1}{\alpha_1}\right|, \left|\frac{x_2}{\alpha_2}\right|, \dots\right\}$$

and we denote by $H_\alpha$ the space $l^2$ with the norm $\|\cdot\|_\alpha$.

We define a linear operator $A_k$ ($k \in \mathbb{N}$) by

$$A_k(e_k) = e_{k+1},\ A_k(e_{k+1}) = e_k \text{ and } A_k(e_j) = e_j, \text{ for } j \in \mathbb{N}\setminus\{k,k+1\}.$$

With $\delta_k := 1/\alpha_k - 1/\alpha_{k+1}$ ($\delta_k \searrow 0$ as $k \to \infty$) it can be proved that

$$\big|\,\|A_k x\|_\alpha - \|x\|_\alpha\big| \leq \delta_k\|x\|_\alpha, \qquad x \in H_\alpha.$$

We have from it

$$(1-\delta_k)\|x\|_\alpha \le \|A_k x\|_\alpha \le (1+\delta_k)\|x\|_\alpha$$

whence, for $\|x\|_\alpha = \|y\|_\alpha$, we get

$$|\,\|A_k x\|_\alpha - \|A_k y\|_\alpha\,| \le \delta_k(\|x\|_\alpha + \|y\|_\alpha) \le \varepsilon_k(\|A_k x\|_\alpha + \|A_k y\|_\alpha).$$

Thus, with $\varepsilon = \varepsilon_k := \delta_k/(1-\delta_k)$ we have

$$\|x\|_\alpha = \|y\|_\alpha \quad \Longrightarrow \quad |\,\|A_k x\|_\alpha - \|A_k y\|_\alpha\,| \le \varepsilon_k(\|A_k x\|_\alpha + \|A_k y\|_\alpha).$$

This is equivalent to (4.11) with the constant $\varepsilon_k$ instead of $\varepsilon$ (and $\varepsilon_k \searrow 0$ as $k \to \infty$).

Let $0 < \delta < 1$ be arbitrarily chosen. Suppose that the assertion of Theorem 4.20 holds true for $H_\alpha$. Then for $k$ large enough (so that $\varepsilon_k$ is sufficiently small) there should exist a linear isometry $I_k \colon H_\alpha \to H_\alpha$ and a constant $\gamma_k > 0$:

$$\|A_k - \gamma_k I_k\|_\alpha \le \delta \|\gamma_k I_k\|_\alpha = \delta \gamma_k. \tag{4.17}$$

It can be proved (see [44, Lemma 1]) that each linear isometry $I \colon H_\alpha \to H_\alpha$ has to be coordinate symmetry, i.e., $I(e_i) = \pm e_i$, $i \in \mathbb{N}$. Thus we have for an arbitrary $\gamma > 0$

$$\begin{aligned}
\|A_k - \gamma I\|_\alpha &\ge \alpha_{k+1}\|(A_k - \gamma I)e_{k+1}\|_\alpha = \alpha_{k+1}\|e_k \pm \gamma e_{k+1}\|_\alpha \\
&= \alpha_{k+1} \sup\left\{ \|e_k \pm \gamma e_{k+1}\|_{l^2}, 0, \dots, 0, \frac{1}{\alpha_k}, \frac{\gamma}{\alpha_{k+1}}, 0, \dots \right\} \\
&= \max\left\{ \alpha_{k+1}\sqrt{1+\gamma^2}, \frac{\alpha_{k+1}}{\alpha_k}, \gamma \right\} \ge \gamma.
\end{aligned}$$

Thus

$$\|A_k - \gamma_k I_k\|_\alpha \ge \gamma_k,$$

which contradicts (4.17) (since $\delta < 1$).

The answer for the analogous stability problem for mappings preserving Birkhoff orthogonality has been recently given by Mojškerc and Turnšek [40]. Using our notation their result reads as follows.

**Theorem 4.22 ([40], Theorem 4.1).** *Let $X, Y$ be normed spaces such that $(X,Y) \in \mathfrak{A}$ and let $T \colon X \to Y$ be a linear mapping satisfying* (4.14), *i.e., such that*

$$x \perp_B y \Longrightarrow Tx \perp_B^{\varepsilon} Ty, \qquad x, y \in X.$$

*Then, there exists an orthogonality preserving mapping $U \colon X \to Y$ (a scalar multiple of an isometry), close to $T$, i.e., such that*

$$\|T - U\| \le \delta(\varepsilon)\|T\| \quad \textit{and} \quad \lim_{\varepsilon \to 0^+} \delta(\varepsilon) = 0.$$

As for the other definition of approximate Birkhoff orthogonality $\underset{\varepsilon}{\perp}_B$ (see (4.1)), the same assertion was obtained for $L_p$ spaces and for finite-dimensional ones (cf. [40, Proposition 4.2, 4.3]).

Like in the Example 4.3 above, a similar construction was presented ([40, Example 4.4]) so that there exist approximate Birkhoff orthogonality preserving mappings (in both considered sences) which are far from isometries.

Recently, Wójcik has obtained some results concerning the stability of the orthogonality preserving property with respect to some wide class of orthogonalities. He showed that stability holds true in all finite-dimensional spaces and also in some classical Banach spaces. However, generally stability does not hold. Yet another construction of a suitable space was given.

## 4.5 Generalizations, Related Results and Open Problems

### 4.5.1 The Class $\mathscr{L}'$

In the above considerations concerning orthogonality preserving mappings, we restricted ourselves to linear mappings. This class can be, however, slightly enlarged. For given inner product spaces $X,Y$ let $\mathscr{L}' = \mathscr{L}'(X,Y)$ denote the class of all mappings $f\colon X \to Y$ which are phase-equivalent with linear or conjugate-linear mappings $T\colon X \to Y$.

It is possible to strengthen Theorem 4.12.

**Theorem 4.23 ([12], Theorem 2.3).** *Let $f\colon X \to Y$ be a nonzero mapping satisfying* (4.10) *with some $\varepsilon \in [0,1)$ and let $f \in \mathscr{L}'$. Then there exists $\gamma > 0$ such that*

$$\big|\,|\langle f(x)|f(y)\rangle| - \gamma|\langle x|y\rangle|\,\big| \le \frac{4\varepsilon}{1+\varepsilon}\min\{\gamma\|x\|\,\|y\|, \|f(x)\|\,\|f(y)\|\}, \qquad x,y \in X.$$

Taking $\varepsilon = 0$ one gets that for a nonzero orthogonality preserving mapping $f \in \mathscr{L}'(X,Y)$, there exists $\gamma > 0$ such that

$$|\langle f(x)|f(y)\rangle| = \gamma|\langle x|y\rangle|, \qquad x,y \in X.$$

Then it follows from Wigner's theorem that $f$ is phase-equivalent to a linear or conjugate-linear isometry multiplied by a positive constant.

As for the stability, the following holds true.

**Theorem 4.24.** *Let $X,Y$ be inner product spaces. Then, for an arbitrary $\delta > 0$ there exists an $\varepsilon > 0$ such that each $\varepsilon$-a.o.p. mapping $f\colon X \to Y$ from the class $\mathscr{L}'$ can be $\delta$-approximated by an o.p. mapping $T\colon X \to Y$ from $\mathscr{L}'$, i.e.,*

$$\|f(x) - T(x)\| \le \delta\min\{\|f(x)\|, \|T(x)\|\}, \qquad x \in X.$$

The above result has been explicitly stated and proved in [12, Theorem 3.3, Corollary 3,4] for finite-dimensional space $X$. But one can give the proof based on Turnšek's reasoning in the proof of Theorem 4.17 for an arbitrary space.

Similarly, in Theorem 4.7 linearity of $f$ can be replaced by its conjugate-linearity (cf. [6, Remark 3.3]) thus also (since the Birkhoff orthogonality is homogeneous) by the assumption $f \in \mathscr{L}'$.

### 4.5.2 *Right-Angle Preserving Property*

Kestelman and Tissier (see [45]) introduced a property similar to the orthogonality preserving one. We say that $f$ has the *right-angle preserving* (r.a.p.) property iff

$$x - z \perp y - z \Longrightarrow f(x) - f(z) \perp f(y) - f(z), \qquad x, y, z \in X. \tag{4.18}$$

It is known (see [45]) that in real normed spaces each r.a.p. mapping must be an affine, continuous similarity (with respect to some point). It is easily seen that if $f$ satisfies (4.18) then so does $f_0 := f - f(0)$ (and additionally $f_0(0) = 0$). It follows that for real spaces (cf. [11]) $f$ satisfies (4.18) and $f(0) = 0$ if and only if $f$ is orthogonality preserving and linear (actually, additivity suffices).

It seems natural to consider the approximate right-angle preserving (a.r.a.p.) property:

$$x - z \perp y - z \Longrightarrow f(x) - f(z) \perp^{\varepsilon} f(y) - f(z), \qquad x, y, z \in X. \tag{4.19}$$

It is easily seen that if $f$ is additive and approximately orthogonality preserving (satisfies (4.10)), then $f$ satisfies (4.19) and $f(0) = 0$. Conversely, if $f$ satisfies (4.19) with $\varepsilon < 1/8$ and $f(0) = 0$, then $f$ satisfies (4.10) and is in some sense almost additive (see [11, Theorem 2.2]). It can be also shown that a homogeneous mapping $f$ satisfying (4.19) is an approximate similarity (cf. [11, Theorem 3.2]).

Finally, we can prove the stability of the right-angle preserving property.

**Theorem 4.25.** *Let $X, Y$ be real inner product spaces. For each linear mapping $g \colon X \to Y$ satisfying* (4.19)*, there exists $f : X \to Y$ satisfying* (4.18) *and such that*

$$\|f - g\| \leq \left(1 - \sqrt{\frac{1-\varepsilon}{1+\varepsilon}}\right) \min\{\|f\|, \|g\|\}. \tag{4.20}$$

*Proof.* If $g$ is linear and satisfies (4.19), then $g$ is linear and satisfies (4.10). It follows then from Theorem 4.17 that there exists a linear and orthogonality preserving mapping $f$ (which must be also right-angle-preserving), such that (4.20) holds. □

### 4.5.3 $(\delta,\varepsilon)$-Orthogonality Preserving Mappings

In our considerations, a linear $\varepsilon$-approximate orthogonality preserving mapping $f$ can be replaced by a more general, linear $(\delta,\varepsilon)$-a.o.p. one, i.e., such that

$$x\perp^{\delta} y \quad\Longrightarrow\quad fx\perp^{\varepsilon} fy, \qquad x,y\in X. \tag{4.21}$$

Definitely, the class of mappings satisfying the condition (4.21) is wider than the class of a.o.p mappings (those with $\delta=0$). Such mappings, and stability problems connected with such a class of a.o.p. mappings were investigated in [33].

### 4.5.4 Preservation of the Given Value of the Inner Product

The following considerations can be found in [11]. In this section $X$ and $Y$ are inner product spaces over the field $\mathbb{K}$ of real or complex numbers. Suppose that, for a fixed number $c\in\mathbb{K}$, a function $f\colon X\to Y$ preserves this particular value of the inner product, i.e.,

$$\langle x|y\rangle = c \Longrightarrow \langle f(x)|f(y)\rangle = c, \qquad x,y\in X. \tag{4.22}$$

If $c=0$, the condition (4.22) simply means that $f$ preserves orthogonality. Let us discuss a stability problem. For fixed $0\neq c\in\mathbb{K}$ and $\varepsilon\geq 0$ we consider the condition

$$\langle x|y\rangle = c \Longrightarrow |\,\langle f(x)|f(y)\rangle - c\,|\leq\varepsilon, \qquad x,y\in X. \tag{4.23}$$

**Theorem 4.26.** *For a finite-dimensional inner product space $X$ and an arbitrary inner product space $Y$ there exists a continuous mapping $\delta:\mathbb{R}_+\to\mathbb{R}_+$ satisfying $\lim_{\varepsilon\to 0^+}\delta(\varepsilon)=0$ and such that for each linear mapping $f:X\to Y$ satisfying* (4.23) *there exists a linear isometry $I:X\to Y$ such that*

$$\|f-I\|\leq\delta(\varepsilon).$$

For $\varepsilon=0$ we obtain from the above result (we can omit the assumption concerning the dimension of $X$ in this case, considering a subspace spanned on given vectors $x,y\in X$):

**Corollary 4.2.** *Let $f:X\to Y$ be linear and satisfy* (4.22)*, with some $0\neq c\in\mathbb{K}$. Then $f$ satisfies the orthogonality equation* (4.5)*.*

Now, let us replace the condition (4.23) by

$$\langle x|y\rangle = c \Longrightarrow |\,\langle f(x)|f(y)\rangle - c\,|\leq\varepsilon\,\|f(x)\|\,\|f(y)\|, \qquad x,y\in X. \tag{4.24}$$

For $c=0$, (4.24) means that $f$ is $\varepsilon$-approximately orthogonality preserving.

**Theorem 4.27.** *If $X$ is a finite-dimensional inner product space and $Y$ an arbitrary inner product space, then there exists a continuous mapping $\delta : \mathbb{R}_+ \to \mathbb{R}_+$ with* $\lim_{\varepsilon \to 0^+} \delta(\varepsilon) = 0$ *such that for a linear mapping $f : X \to Y$ satisfying* (4.24)*, with $c \neq 0$, there exists a linear isometry $I : X \to Y$ such that*

$$\|f - I\| \leq \delta(\varepsilon).$$

A converse theorem is also true, even with no restrictions concerning the dimension of $X$. Let $I : X \to Y$ be a linear isometry and $f : X \to Y$ a mapping, not necessarily linear, such that

$$\|f(x) - I(x)\| \leq \delta \|x\|, \qquad x \in X$$

with $\delta = \sqrt{(1+2\varepsilon)/(1+\varepsilon)} - 1$ (for a given $\varepsilon \geq 0$). It can be shown that

$$|\langle f(x)|f(y)\rangle - \langle x|y\rangle| \leq \varepsilon \|f(x)\| \, \|f(y)\|.$$

Thus, $f$ satisfies (4.24) with an arbitrary $c$.

### 4.5.5 Other Results

Obviously, there are many other related results. Let us just mention some of them, without going into details.

As it was said, our considerations can be treated as a part of theory of linear preservers (see monograph [41]). Some part of this theory is devoted to *orthogonality preservers* and just a sample examples are papers of Turnšek [46], Burgos et al. [7].

*Bi-orthogonality preserving mappings* and their stability were considered by Vestfrid [49].

Wójcik in his recent results was considering a wider class of relations in $X$ (not necessarily with any geometrical background justifying regarding them as orthogonality relations). On introducing associate approximate relations he was able to obtain some stability results.

### 4.5.6 Open Problems

The above state-of-the-art description of the topic show that it is far from being exhausted. Many open questions remain and some of them could be derived from the text. Let us list some of them briefly.

- Stabilities of the orthogonality preserving property with respect to other mentioned orthogonalities and for possibly most general class of orthogonalities (or just relations).
- Description of the classes $\mathfrak{A}$ and $\mathfrak{B}$ of spaces which admit stability of isometric mappings.
- Stability of the orthogonality and Wigner equation for the infinite-dimensional spaces.

## References


1. Alonso, J., Benitez, C.: Orthogonality in normed linear spaces: A survey. Part I: Main properties. Extracta Math. **3**, 1–15 (1988). Part II: Relations between main orthogonalities. Extracta Math. **4**, 121–131 (1989)
2. Alonso, J.: Some properties of Birkhoff and isosceles orthogonality in normed linear space. In: Rassias Th.M. (ed.) Inner product spaces and applications, pp. 1–11. Pitman Res. Notes Math. Ser. 376, Addison Wesley Longman, Harlow (1997)
3. Alsina, C., Sikorska, J., Tomás, M.S.: Norm Derivatives and Characterizations of Inner Product Spaces. World Scientific, Hackensack, NJ (2009)
4. Amir, D.: Characterization of Inner Product Spaces. Birkhäuser, Basel–Boston–Stuttgart (1986)
5. Birkhoff, G.: Orthogonality in linear metric spaces. Duke Math. J. **1**, 169–172 (1935)
6. Blanco, A., Turnšek, A.: On maps that preserve orthogonality in normed spaces. Proc. Roy. Soc. Edinburgh Sect. A **136**, 709–716 (2006)
7. Burgos, M., Fernández-Polo, F., Garcés, J.J., Martínez Moreno J., Peralta A.M.: Orthogonality preservers in $C^*$-algebras, $JB^*$-algebras and $JB^*$-triples. J. Math. Anal. Appl. **348**, 220–233 (2008)
8. Chmieliński, J.: On an $\varepsilon$-Birkhoff orthogonality. J. Inequal. Pure Appl. Math. **6**(3), Art. 79 (2005)
9. Chmieliński, J.: Linear mappings approximately preserving orthogonality. J. Math. Anal. Appl. **304**, 158–169 (2005)
10. Chmieliński, J., Stability of the orthogonality preserving property in finite-dimensional inner product spaces. J. Math. Anal. Appl. **318**, 433–443 (2006)
11. Chmieliński, J.: On some approximate functional relations stemming from orthogonality preserving property. J. Inequal. Pure Appl. Math. **7**, no. 3, Article 85, 1–14 (2006)
12. Chmieliński, J.: Orthogonality preserving property, Wigner equation and stability. J. Inequal. Appl. **2006**, Article ID 76489, 1–9 (2006)
13. Chmieliński, J.: Stability of the Wigner equation and related topics. Nonlinear Funct. Anal. Appl. **11**, 859–879 (2006)
14. Chmieliński, J.: Remarks on orthogonality preserving mappings in normed spaces and some stability problems. Banach J. Math. Anal. **1**, no. 1, 117–124 (2007)
15. Chmieliński, J., Moslehian, M.S.: Approximately $C^*$-inner product preserving mappings. Bull. Korean Math. Soc. **45**, 157–167 (2008)
16. Chmieliński, J., Ilišević, D., Moslehian M.S., Sadeghi Gh.: Perturbation of the Wigner equation in inner product $C^*$-modules. J. Math. Phys. **49**, 033519 (2008)
17. Chmieliński, J., Wójcik, P.: Isosceles-orthogonality preserving property and its stability. Nonlinear Anal. **72**, 1445-1453 (2010)
18. Chmieliński, J., Wójcik, P.: On a $\rho$-orthogonality. Aequationes Math. **80**, 45–55 (2010)
19. Day, M.M.: Normed Linear Spaces. Springer-Verlag, Berlin – Heidelberg – New York (1973)

20. Ding, G.G.: The approximation problem of almost isometric operators by isometric operators. Acta Math. Sci. (English Ed.) **8**, 361–372 (1988)
21. Dragomir, S.S.: On approximation of continuous linear functionals in normed linear spaces. An. Univ. Timişoara Ser. Ştiinţ. Mat. **29**, 51–58 (1991)
22. Dragomir, S.S.: Semi-Inner Products and Applications. Nova Science Publishers, Inc., Hauppauge, NY (2004)
23. Forti, G.L.: Hyers–Ulam stability of functional equations in several variables. Aequationes Math. **50**, 143–190 (1995)
24. Frank, M., Mishchenko, A.S., Pavlov, A.A.: Orthogonality-preserving, $C^*$-conformal and conformal module mappings on Hilbert $C^*$-modules. J. Funct. Anal. **260**, 327–339 (2011)
25. Giles, J.R.: Classes of semi-inner-product spaces. Trans. Amer. Math. Soc. **129**, 436–446 (1967)
26. Hyers, D,H., Isac, G., Rassias, Th.M.: Stability of Functional Equations in Several Variables. Progress in Nonlinear Differential Equations and Their Applications vol. 34, Birkhäuser, Boston-Basel-Berlin (1998)
27. Ilišević, D., Turnšek, A.: Approximately orthogonality preserving mappings on $C^*$-modules. J. Math. Anal. Appl. **341**, 298–308 (2008)
28. James, R.C: Orthogonality in normed linear linear spaces. Duke Math. J. **12**, 291–301 (1945)
29. James, R.C.: Orthogonality and linear functionals in normed linear spaces. Trans. Amer. Math. Soc. **61**, 265–292 (1947)
30. Jung, S.-M.: Hyers–Ulam–Rassias Stability of Functional Equations in Mathematical Analysis. Hadronic Press, Inc., Palm Harbor (2001)
31. Koehler, D., Rosenthal, P.: On isometries of normed linear spaces. Studia Math. **36**, 213–216 (1970)
32. Koldobsky, A.: Operators preserving orthogonality are isometries. Proc. Roy. Soc. Edinburgh Sect. A **123**, 835–837 (1993)
33. Kong, L., Cao, H.: Stability of orthogonality preserving mappings and the orthogonality equation (Chinese). J. Shaanxi Norm. Univ. Nat. Sci. Ed. **36**, 10–14 (2008)
34. Lance, E.C.: Hilbert $C^*$-Modules. London Math. Soc. Lecture Note Series **210**, Cambridge University Press, Cambridge (1995)
35. Leung, C.W., Ng, C.K., Wong, N.C.: Linearorthogonality preservers of Hilbert $C^*$-modules over $C^*$-algebras with real rank zero. Preprint, arXiv: 0910.2335
36. Leung, C.W., Ng, C.K., Wong, N.C.: Linear orthogonality preservers of Hilbert boundles. Preprint, arXiv:1005.4502
37. Leung, C.W., Ng, C.K., Wong, N.C.: Linear orthogonality preservers of Hilbert $C^*$-modules over general $C^*$-algebras. J. Aust. Math. Soc. **89**, 245–254 (2010)
38. Lumer, G.: Semi–inner–product spaces. Trans. Amer. Math. Soc. **100**, 29-43 (1961)
39. Miličić, P.M.: Sur la G-orthogonalité dans les espaces normés. Mat. Vesnik **39**, 325–334 (1987)
40. Mojškerc, B., Turnšek, A.: Mappings approximately preserving orthogonality in normed spaces. Nonlinear Anal. **73**, 3821–3831 (2010)
41. Molnár, L.: Selected Preserver Problems on Algebraic Structures of Linear Operators and on Function Spaces. Lecture Notes in Mathematics 1895, Springer, Berlin (2007)
42. Moszner, Z.: On the stability of functional equations. Aequationes Math. **77**, 33–88 (2009)
43. Pedersen, G.K.: Analysis Now. Grad. Texts in Math. **118**, Springer-Verlag, New-York (1989)
44. Protasov, V.Yu.: On stability of isometries in Banach spaces. In: Rassias Th.M., Brzdęk J. (eds) Functional Equations in Mathematical Analysis, Springer (the paper is to be published in the same volume.)
45. Tissier, A.: A right-angle preserving mapping (a solution of a problem proposed in 1983 by H. Kestelman). Advanced Problem 6436, Amer. Math. Monthly **92**, 291–292 (1985)
46. Turnšek, A.: On operators preserving James' orthogonality. Lin. Algebra Appl. **407**, 189–195 (2005)
47. Turnšek, A.: On mappings approximately preserving orthogonality. J. Math. Anal. Appl. **336**, 625–631 (2007)

48. Ulam, S.M.: Problems in Modern Mathematics (Chapter VI, Some Questions in Analysis: §1, Stability). Science Editions, John Wiley & Sons, New York (1964)
49. Vestfrid, I.A.: Linear approximation of approximately biorthogonality preserving maps. J. Math. Anal. Appl. **336**, 418–424 (2007)
50. Wigner, E.P.: Gruppentheorie und ihre Anwendungen auf die Quantenmechanik der Atomspektren. Friedr. Vieweg und Sohn Akt.-Ges. (1931)

# Chapter 5
# On the Hyers–Ulam Stability of Functional Equations with Respect to Bounded Distributions

**Jae-Young Chung**

*Dedicated to the memory of S.M. Ulam*
*on the 100th anniversary of his birth*

**Abstract** We consider the Hyers–Ulam stability of the Cauchy, Jensen, Pexider, Pexider–Jensen equations with respect to bounded distributions. We also consider the Hyers–Ulam–Rassias stability problem for the quadratic functional equation in the space of Fourier hyperfunctions.

**Keywords** Bounded distribution • Fourier hyperfunction • Cauchy equation • Pexider equation • Jensen equation • Quadratic functional equation • Heat kernel • Hyers–Ulam stability

**Mathematics Subject Classification (2000):** Primary 39B82; Secondary 46F15

## 5.1 Introduction

The Hyers–Ulam stability problems of functional equations was originated by Ulam in 1940 when he proposed the following question [33]:

*Let $f$ be a mapping from a group $G_1$ to a metric group $G_2$ with metric $d(\cdot,\cdot)$ such that*

$$d(f(xy), f(x)f(y)) \le \varepsilon, \quad x, y \in G_1.$$

*Then does there exist a group homomorphism $h$ and $\delta_\varepsilon > 0$ such that*

$$d(f(x), h(x)) \le \delta_\varepsilon, \quad x \in G_1?$$

J.-Y. Chung (✉)
Department of Mathematics, Kunsan National University, Kunsan, 573-701 Korea
e-mail: jychung@kunsan.ac.kr

Th.M. Rassias and J. Brzdęk (eds.), *Functional Equations in Mathematical Analysis*, Springer Optimization and Its Applications 52, DOI 10.1007/978-1-4614-0055-4_5,

One of the first assertions to be obtained is the following result, essentially due to Hyers [20], that gives an answer for the question of Ulam.

**Theorem 5.1.** *Suppose that $S$ is an additive semigroup, $Y$ is a Banach space, $\varepsilon \geq 0$, and $f : S \to Y$ satisfies the inequality*

$$\|f(x+y) - f(x) - f(y)\| \leq \varepsilon \tag{5.1}$$

*for all $x, y \in S$. Then there exists a unique function $A : S \to Y$ satisfying*

$$A(x+y) = A(x) + A(y)$$

*for which*

$$\|f(x) - A(x)\| \leq \varepsilon$$

*for all $x \in S$.*

In 1949–1951, this result was generalized by Aoki [2] and Bourgin [5, 6]. In 1978 Th.M. Rassias generalized the Hyers' result to approximately linear mappings [31]. Since then the stability problems have been investigated in various directions for many other functional equations [3, 4, 7–9, 15, 22–25, 27–31].

Recently, the above stability problem (5.1) and the following three inequalities that are related to it:

$$\left\|2f\left(\frac{x+y}{2}\right) - f(x) - f(y)\right\| \leq \varepsilon, \tag{5.2}$$

$$\|f(x+y) - g(x) - h(y)\| \leq \varepsilon, \tag{5.3}$$

$$\left\|2f\left(\frac{x+y}{2}\right) - g(x) - h(y)\right\| \leq \varepsilon \tag{5.4}$$

have been considered in various spaces of generalized functions such as the space $\mathscr{S}'(\mathbb{R}^n)$ of tempered distributions of Schwartz, the space $\mathscr{F}'(\mathbb{R}^n)$ of Fourier hyperfunctions (see [7–9]). For example, a distribution version of the inequality (5.1) has been reformulated for generalized functions $u$ as

$$u \circ A - u \circ P_1 - u \circ P_2 \in L^\infty_\varepsilon(\mathbb{R}^{2n}),$$

where $\circ$ denotes the pullback, $A(x,y) = x+y$, $P_1(x,y) = x$, $P_2(x,y) = y$ for $x, y \in \mathbb{R}^n$, and $L^\infty_\varepsilon(\mathbb{R}^{2n})$ denotes the space of bounded measurable functions $\phi$ on $\mathbb{R}^{2n}$ such that $\|\phi\|_{L^\infty} \leq \varepsilon$. Due to Schwartz [32] the space $L^\infty$ of bounded measurable functions has been generalized to the space $\mathscr{D}'_{L^\infty}$ of bounded distributions which is a subspace of tempered distributions and later the space $\mathscr{D}'_{L^\infty}$ was further generalized to the space $\mathscr{A}'_{L^\infty}$ of bounded hyperfunctions which is a subspace of Sato hyperfunctions.

In this paper, we generalize the stability problems (5.1)–(5.4) to the space of Fourier hyperfunctions and consider the stability problem when the differences

$$C(u) := u \circ (x+y) - u \circ (x) - u \circ (y),$$

$$J(u) := 2u \circ \left(\frac{x+y}{2}\right) - u \circ (x) - u \circ (y),$$

$$P(u,v,w) := u \circ (x+y) - v \circ (x) - w \circ (y),$$

$$PJ(u,v,w) := 2u \circ \left(\frac{x+y}{2}\right) - v \circ (x) - w \circ (y)$$

belong to the space of bounded distributions, which is a very natural generalization of the classical Hyers–Ulam stability problem to the spaces of distributions. In the space of bounded distributions, however, the validity of the bound $\varepsilon > 0$ is deprived. Thus, we consider the stability problems

$$u \circ A - u \circ P_1 - u \circ P_2 \in \mathscr{D}'_{L^\infty}(\mathbb{R}^{2n}) \text{ [resp. } \mathscr{A}'_{L^\infty}(\mathbb{R}^{2n})], \tag{5.5}$$

$$2u \circ \frac{A}{2} - u \circ P_1 - u \circ P_2 \in \mathscr{D}'_{L^\infty}(\mathbb{R}^{2n}) \text{ [resp. } \mathscr{A}'_{L^\infty}(\mathbb{R}^{2n})], \tag{5.6}$$

$$u \circ A - v \circ P_1 - w \circ P_2 \in \mathscr{D}'_{L^\infty}(\mathbb{R}^{2n}) \text{ [resp. } \mathscr{A}'_{L^\infty}(\mathbb{R}^{2n})], \tag{5.7}$$

$$2u \circ \frac{A}{2} - v \circ P_1 - w \circ P_2 \in \mathscr{D}'_{L^\infty}(\mathbb{R}^{2n}) \text{ [resp. } \mathscr{A}'_{L^\infty}(\mathbb{R}^{2n})]. \tag{5.8}$$

We also consider the Hyers–Ulam–Rassias stability of the quadratic functional equation

$$\|u \circ A + u \circ B - 2u \circ P_1 - 2u \circ P_2\| \le \varepsilon(|x|^p + |y|^p) \tag{5.9}$$

in the space of Fourier hyperfunctions, where $B(x,y) = x - y$. As results we prove that all the solutions of the above stability problems (5.5)–(5.8) are additive functions up to bounded distributions. We also prove that let $u \in \mathscr{F}'$ satisfy the inequality (5.9), then there exists a unique quadratic function

$$q(x) = \sum_{1 \le j \le k \le n} a_{jk} x_j x_k$$

such that

$$\|u - q(x)\| \le \frac{2\varepsilon}{|4 - 2^p|} |x|^p$$

for $0 < p < 2, 4 < p$, and, for $2 < p \le 4$,

$$\|u - q(x)\| \le \frac{\varepsilon(2^p + 2)}{2^p - 4} |x|^p.$$

## 5.2 The Heat Kernel Method in Distributions and Hyperfunctions

We first introduce the space $\mathscr{F}'$ of hyperfunctions which is a natural generalization of the space $\mathscr{S}'$ of tempered distributions (see [18, 19] for these spaces). We use the notations: $|\alpha| = \alpha_1 + \cdots + \alpha_n$, $\alpha! = \alpha_1! \cdots \alpha_n!$, $x^\alpha = x_1^{\alpha_1} \cdots x_n^{\alpha_n}$ and $\partial^\alpha = \partial_1^{\alpha_1} \cdots \partial_n^{\alpha_n}$, for $x = (x_1, \cdots, x_n) \in \mathbb{R}^n$, $\alpha = (\alpha_1, \ldots, \alpha_n) \in \mathbb{N}_0^n$, where $\mathbb{N}_0$ is the set of non-negative integers and $\partial_j = \frac{\partial}{\partial x_j}$.

**Definition 5.1 ([10, 18]).** We denote by $\mathscr{F}$ or $\mathscr{F}(\mathbb{R}^n)$ the space of all infinitely differentiable functions $\varphi$ in $\mathbb{R}^n$ such that

$$\|\varphi\|_{h,k} = \sup_{x\in\mathbb{R}^n, \alpha, \beta\in\mathbb{N}_0^n} \frac{|x^\alpha \partial^\beta \varphi(x)|}{h^{|\alpha|} k^{|\beta|} \alpha! \beta!} < \infty$$

for some $h, k > 0$. We say that $\varphi_j \longrightarrow 0$ as $j \to \infty$ if $\|\varphi_j\|_{h,k} \longrightarrow 0$ as $j \to \infty$ for some $h, k$, and denote by $\mathscr{F}'$ the dual space of $\mathscr{F}$ and call its elements Fourier hyperfunctions.

Following Schwartz [32] we introduce the space $\mathscr{D}'_{L^\infty}$ of bounded distributions.

**Definition 5.2.** We denote by $\mathscr{D}_{L^1}(\mathbb{R}^n)$ the space of smooth functions on $\mathbb{R}^n$ such that $\partial^\alpha \varphi \in L^1(\mathbb{R}^n)$ for all $\alpha \in \mathbb{N}_0^n$ equipped with the topology defined by the countable family of seminorms

$$\|\varphi\|_m = \sum_{|\alpha| \le m} \|\partial^\alpha \varphi\|_{L^1}, \quad m \in \mathbb{N}_0.$$

We denote by $\mathscr{D}'_{L^\infty}$ the dual space of $\mathscr{D}_{L^1}$ and call its elements bounded distributions.

Generalizing bounded distributions the space $\mathscr{A}'_{L^\infty}$ of bounded hyperfunctions has been introduced [11] as a subspace of hyperfunctions.

**Definition 5.3.** We denote by $\mathscr{A}_{L^1}$ the space of smooth functions on $\mathbb{R}^n$ satisfying

$$\|\varphi\|_h = \sup_\alpha \frac{\|\partial^\alpha \varphi\|_{L^1}}{h^{|\alpha|} \alpha!} < \infty$$

for some constant $h > 0$. We say that $\varphi_j \to 0$ in $\mathscr{A}_{L^\infty}$ as $j \to \infty$ if there is a positive constant $h$ such that

$$\sup_\alpha \frac{\|\partial^\alpha \varphi_j\|_{L^1}}{h^{|\alpha|} \alpha!} \to 0 \quad \text{as } j \to \infty.$$

We denote by $\mathscr{A}'_{L^\infty}$ the dual space of $\mathscr{A}_{L^1}$.

It is well known that the following topological inclusions hold:

$$\mathscr{F} \hookrightarrow \mathscr{S} \hookrightarrow \mathscr{D}_{L^1}, \qquad D'_{L^\infty} \hookrightarrow \mathscr{S}' \hookrightarrow \mathscr{F}'$$
$$\mathscr{F} \hookrightarrow \mathscr{A}_{L^1} \hookrightarrow \mathscr{D}_{L^1}, \quad \mathscr{D}'_{L^\infty} \hookrightarrow \mathscr{A}'_{L^\infty} \hookrightarrow \mathscr{F}'.$$

It is easy to see that the $n$-dimensional heat kernel $E_t(x)$ given by

$$E_t(x) = (4\pi t)^{-n/2}\exp(-|x|^2/4t),\ t > 0$$

belongs to the space $\mathscr{F}(\mathbb{R}^n)$ for each $t > 0$.

## 5.3 Stability of Pexider Equations with Respect to Bounded Distributions

The main tool of the proofs of the results is *the heat kernel method* initiated by Matsuzawa [26] which represents the generalized functions in some class as the initial values of solutions of the heat equation with appropriate growth conditions [11, 26]. Making use of the heat kernel method we can convert (5.5)–(5.8) to the following classical Hyers–Ulam type stability problems; there exist $C > 0$, $N > 0$ [ resp. for every $\varepsilon > 0$ there exists $C_\varepsilon > 0$ ] such that

$$|f(x+y,t+s) - g(x,t) - h(y,s)| \le C\left(\frac{1}{t}+\frac{1}{s}\right)^N \left[\text{resp. } C_\varepsilon e^{\varepsilon(1/t+1/s)}\right]$$

for all $x, y \in \mathbb{R}^n$, $0 < t, s < 1$, where $f, g, h : \mathbb{R}^n \times (0,\infty) \to \mathbb{C}$ are the corresponding solutions of the heat equation. Thus, we first consider the above stability problem in a more general setting: Let $G$ be a group, $S$ an semigroup divisible by 2 and $\psi : S \times S \to [0,\infty)$.

**Theorem 5.2.** *Let $f, g, h : G \times S \to \mathbb{C}$ satisfy*

$$|f(x+y,t+s) - g(x,t) - h(y,s)| \le \psi(t,s) \tag{5.10}$$

*for all $x, y \in G$, $t, s \in S$. Then exists an additive function $A : G \to \mathbb{C}$ such that*

$$\left|f(x,t) - A(x) - g\left(0,\tfrac{t}{2}\right) - h\left(0,\tfrac{t}{2}\right)\right| \le 3\psi\left(\tfrac{t}{2},\tfrac{t}{2}\right), \tag{5.11}$$

$$|g(x,t) - A(x) - g(0,t)| \le 4\psi(t,t), \tag{5.12}$$

$$|h(x,t) - A(x) - h(0,t)| \le 4\psi(t,t), \tag{5.13}$$

*for all $(x,t) \in G \times S$.*

*Proof.* It follows from (5.10) that

$$|f(x,2t)-g(x,t)-h(0,t)| \le \psi(t,t), \tag{5.14}$$
$$|f(y,2s)-h(y,s)-g(0,s)| \le \psi(s,s) \tag{5.15}$$

for all $(x,t)\in G\times S$. Using triangle inequality with (5.10), (5.14) and (5.15) we have

$$\begin{aligned}|f(x+y,t+s) \;-\; & f(x,2t)-f(y,2s)-g(0,s)-h(0,t)| \\ & \le \psi(t,s)+\psi(t,t)+\psi(s,s),\end{aligned} \tag{5.16}$$

for all $x,\,y\in G,\,t,s\in S$. Putting $y=x,\,s=t$ in (5.16) we have

$$|f(2x,2t)-2f(x,2t)-g(0,t)-h(0,t)| \le 3\psi(t,t) \tag{5.17}$$

for all $(x,t)\in G\times S$. Fixing $t>0$ and using the well known induction argument of Hyers–Ulam [20] with respect to $x$ it is easy to see that the mapping $A(x,t) := \lim_{n\to\infty} 2^{-n} f(2^n x,t)$ satisfies

$$A(x+y,t+s)-A(x,2t)-A(y,2s)=0, \tag{5.18}$$

and

$$|f(x,2t)-A(x,2t)-g(0,t)-h(0,t)| \le 3\psi(t,t). \tag{5.19}$$

It follows from (5.18) that $A(0,2t)=0$, $A(x,2t)=A(x,t+s)=A(x,s+t)=A(x,2s)$ and $A(x,t+s)=A(x,s+t)$ for all $x\in G,\,t,s\in S$. Since $S$ is divisible by 2, $A(x,t)$ is independent of $t\in S$. If we denote $A(x,t)$ by $A(x)$, $A$ is an additive function on $G$. Thus the inequality (5.11) follows. Now (5.12) follows from (5.14) and (5.19), and (5.13) follows from (5.15), (5.19). □

The following structure theorem for bounded distributions and bounded hyperfunctions will be useful.

**Lemma 5.1 ([11,32]).** (i) *Every $u\in\mathscr{D}'_{L^\infty}(\mathbb{R}^n)$ can be expressed as*

$$u=\sum_{|\alpha|\le m}\partial^\alpha f_\alpha \tag{5.20}$$

*for some $m\in\mathbb{N}_0$ where $f_\alpha$ are bounded continuous functions on $\mathbb{R}^n$.*

(ii) *Every $u\in\mathscr{A}'_{L^\infty}(\mathbb{R}^n)$ can be expressed by*

$$u=\left(\sum_{k=0}^{\infty} a_k\Delta^k\right)g+h, \tag{5.21}$$

*where $\Delta$ denotes the Laplacian, $g,h$ are bounded continuous functions on $\mathbb{R}^n$ and $a_k$, $k=0,1,2,\ldots$, satisfy the estimates: for every $L>0$ there exists $C>0$ such that*

$$|a_k|\le CL^k/k!^2$$

*for all $k=0,1,2,\ldots$.*

**Lemma 5.2.** (i) *Let $u := u(\xi,\eta) \in \mathscr{D}'_{L^\infty}(\mathbb{R}^{2n})$. Then we have the estimate: there exist positive constants $C, N$ and $d$ such that*

$$\left|[u * (E_t(\xi)E_s(\eta))](x,y)\right| \le C\,(1/t+1/s)^N, \quad x \in \mathbb{R}^n,\ 0<t,s<1. \tag{5.22}$$

(ii) *Let $u = u(\xi,\eta) \in \mathscr{A}'_{L^\infty}(\mathbb{R}^{2n})$. Then we have the estimate: for every $\varepsilon > 0$ there exists $C_\varepsilon > 0$ such that*

$$|[u * (E_t(\xi)E_s(\eta))](x,y)| \le C_\varepsilon\, e^{\varepsilon(1/t+1/s)}, \quad x \in \mathbb{R}^n,\ 0<t,s<1. \tag{5.23}$$

**Lemma 5.3 ([11,26]).** *The Gauss transform $U(x,t) = (u * E)(x,t)$ of $u \in \mathscr{F}'(\mathbb{R}^n)$ is a smooth solution of the heat equation $(\Delta - \partial/\partial_t)U = 0$ satisfying :*

(i) *For every $\varepsilon > 0$ there exists a positive constant $C_\varepsilon$ such that*

$$|U(x,t)| \le C_\varepsilon \exp(\varepsilon(1/t+|x|)) \quad x \in \mathbb{R}^n,\ t \in (0,\delta). \tag{5.24}$$

(ii) *$U(x,t) \to u$ as $t \to 0^+$ in the sense that for every $\varphi \in \mathscr{D}_{L^1}$,*

$$\langle u, \varphi\rangle = \lim_{t\to 0^+} \int U(x,t)\varphi(x)\,\mathrm{d}x.$$

*Conversely, every smooth solution $U(x,t)$ of the heat equation satisfying the estimate (5.24) can be uniquely expressed as $U(x,t) = (u * E)(x,t)$ for some $u \in \mathscr{F}'(\mathbb{R}^n)$.*

Similarly, we can represent bounded distributions and bounded hyperfunctions as initial values of solutions of the heat equation. In these cases, only the estimate (5.24) is replaced by the followings, respectively:
there existconstants $C > 0$ and $N \ge 0$ such that

$$|U(x,t)| \le Ct^{-N}, \quad x \in \mathbb{R}^n,\ t \in (0,\delta); \tag{5.25}$$

for every $\varepsilon > 0$ there exists a positive constant $C_\varepsilon$ such that

$$|U(x,t)| \le C_\varepsilon \exp(\varepsilon/t), \quad x \in \mathbb{R}^n,\ t \in (0,\delta). \tag{5.26}$$

Now we state and prove the main theorems.

**Theorem 5.3.** *Let $u, v, w \in \mathscr{F}'(\mathbb{R}^n)$. Then*

$$u \circ A - v \circ P_1 - w \circ P_2 \in \mathscr{D}'_{L^\infty}(\mathbb{R}^{2n})\ [\textit{resp. } \mathscr{A}'_{L^\infty}(\mathbb{R}^{2n})] \tag{5.27}$$

*if and only if*

$$u = c\cdot x + u_0, \quad v = c\cdot x + v_0, \quad w = c\cdot x + w_0, \tag{5.28}$$

*where $c \in \mathbb{C}^n$ and $u_0,\ v_0,\ w_0 \in \mathscr{D}'_{L^\infty}(\mathbb{R}^n)$ [resp. $\mathscr{A}'_{L^\infty}(\mathbb{R}^n)$].*

*Proof.* Convolving in (3.18) the tensor product $E_t(x)E_s(y)$ of n-dimensional heat kernels we have

$$\begin{aligned}[(u\circ A)*(E_t(\xi)E_s(\eta))](x,y) &= \left\langle u_\xi, \int E_t(x-\xi+\eta)E_s(y-\eta)\,d\eta\right\rangle\\ &= \langle u_\xi, (E_t*E_s)(x+y-\xi)\rangle\\ &= \langle u_\xi, (E_{t+s})(x+y-\xi)\rangle\\ &= U(x+y,t+s).\end{aligned}$$

Similarly, we have

$$\begin{aligned}[(v\circ P_1)*(E_t(\xi)E_s(\eta))](x,y) &= V(x,t),\\ [(w\circ P_1)*(E_t(\xi)E_s(\eta))](x,y) &= W(x,t),\end{aligned}$$

where $U(x,t), V(x,t)$ are the Gauss transforms of $u$, $v$, respectively. Thus, in view of Lemma 5.2, we have the following stability problem

$$|U(x+y,t+s)-V(x,t)-W(y,s)| \le \psi(t,s), \tag{5.29}$$

for all $x,y\in\mathbb{R}^n$, $t,s>0$, where $\psi(t,s)=C(1/t+1/s)^N$ [resp. $C_\varepsilon e^{\varepsilon(1/t+1/s)}$]. Now we apply Theorem 5.2. Since $U$, $V$, $W$ are continuous functions we have $A(x)=cx$ for some $c\in\mathbb{C}^n$. Thus we have

$$|U(x,t)-cx| \le \Psi_1(t), \tag{5.30}$$

$$|V(x,t)-cx| \le \Psi_2(t), \tag{5.31}$$

$$|W(x,t)-cx| \le \Psi_3(t), \tag{5.32}$$

for all $(x,t)\in G\times S$, where

$$\begin{aligned}\Psi_1(t) &= 3\psi\left(\tfrac{t}{2},\tfrac{t}{2}\right)+\left|V\left(0,\tfrac{t}{2}\right)\right|+\left|W\left(0,\tfrac{t}{2}\right)\right|,\\ \Psi_2(t) &= 4\psi(t,t)+|V(0,t)|,\\ \Psi_3(t) &= 4\psi(t,t)+|W(0,t)|.\end{aligned}$$

Now we consider the growth of $\Psi_j(t)$, $j=1,2,3$, as $t\to 0^+$. Letting $x=y=0$, $s=1$ in (5.29) and using the triangle inequality we have for $0<t<1$

$$\begin{aligned}|V(0,t)| &\le \psi(t,1)+|U(0,t+1)|+W(0,1)|,\\ &\le \psi(t,1)+M_1\end{aligned}$$

for some $M_1 > 0$, since $U$ is continuous. Similarly, we have

$$\begin{aligned}|W(0,t)| &\le \psi(1,t) + |U(0,t+1)| + V(0,1)|,\\ &\le \psi(1,t) + M_2\end{aligned}$$

for some $M_2 > 0$.

Thus, for the case when $u \circ A - v \circ P_1 - w \circ P_2 \in \mathscr{D}'_{L^\infty}(\mathbb{R}^{2n})$; there exist $C, N > 0$ such that

$$\Psi_j(t) \le Ct^{-N}, \qquad 0 < t < 1,\ j = 1,2,3,$$

and for the case when $u \circ A - v \circ P_1 - w \circ P_2 \in \mathscr{A}'_{L^\infty}(\mathbb{R}^{2n})$; for every $\varepsilon > 0$ there exists $C_\varepsilon > 0$ such that

$$\Psi_j(t) \le C_\varepsilon \exp(\varepsilon/t), \qquad 0 < t < 1,\ j = 1,2,3.$$

Note that $U(x,t) - c \cdot x$, $V(x,t) - c \cdot x$, $W(x,t) - c \cdot x$ are the Gauss transforms of $u - c \cdot x$, $v - c \cdot x$, $w - c \cdot x$, respectively. Now applying Lemma 5.3 for the inequalities (5.30)–(5.32) we have $u - c \cdot x$, $v - c \cdot x$, $w - c \cdot x \in \mathscr{D}'_{L^\infty}(\mathbb{R}^{2n})$ [resp. $\mathscr{A}'_{L^\infty}(\mathbb{R}^{2n})$]. □

As a direct consequence of Theorem 5.3 we have the following.

**Corollary 5.1.** *Let $u \in \mathscr{F}'(\mathbb{R}^n)$. Then*

$$u \circ A - u \circ P_1 - u \circ P_2 \in \mathscr{D}'_{L^\infty}(\mathbb{R}^{2n}) \ [\textit{resp.}\ \mathscr{A}'_{L^\infty}(\mathbb{R}^{2n})] \tag{5.33}$$

*if and only if*

$$u = c \cdot x + u_0, \tag{5.34}$$

*where $c \in \mathbb{C}^n$ and $u_0 \in \mathscr{D}'_{L^\infty}(\mathbb{R}^n)$ [resp. $\mathscr{A}'_{L^\infty}(\mathbb{R}^n)$].*

**Theorem 5.4.** *Let $u, v, w \in \mathscr{F}'(\mathbb{R}^n)$. Then*

$$2u \circ \frac{A}{2} - v \circ P_1 - w \circ P_2 \in \mathscr{D}'_{L^\infty}(\mathbb{R}^{2n}) \ [\textit{resp.}\ \mathscr{A}'_{L^\infty}(\mathbb{R}^{2n})] \tag{5.35}$$

*if and only if*

$$u = c \cdot x + u_0, \quad v = c \cdot x + v_0, \quad w = c \cdot x + w_0, \tag{5.36}$$

*where $c \in \mathbb{C}^n$ and $u_0$, $v_0$, $w_0 \in \mathscr{D}'_{L^\infty}(\mathbb{R}^n)$ [resp. $\mathscr{A}'_{L^\infty}(\mathbb{R}^n)$].*

*Proof.* Convolving in (5.35) the tensor product $E_t(x)E_s(y)$ of n-dimensional heat kernels we have following stability problem

$$\left| 2U\left(\frac{x+y}{2}, \frac{t+s}{4}\right) - V(x,t) - W(y,s) \right| \le \psi(t,s), \tag{5.37}$$

for all $x,y \in \mathbb{R}^n, t,s > 0$, where $U,V,W$ are the Gauss transforms of $u,v,w$, respectively. Letting $U_1(x,t) = U(x/2,t/4)$ and applying the proof of Theorem 5.3 we have

$$|U(x,t) - c \cdot x| \le \tfrac{1}{2}\Psi_1(4t), \tag{5.38}$$

$$|V(x,t) - c \cdot x| \le \Psi_2(t), \tag{5.39}$$

$$|W(x,t) - c \cdot x| \le \Psi_3(t), \tag{5.40}$$

for all $(x,t) \in G \times S$, where $\Psi_j(t)$, $j = 1,2,3$ are as in Theorem 5.3. Following the same approach as in Theorem 5.3 we have the result. □

As a direct consequence of Theorem 5.4 we have the following.

**Corollary 5.2.** *Let $u \in \mathscr{F}'(\mathbb{R}^n)$. Then*

$$2u \circ \frac{A}{2} - u \circ P_1 - u \circ P_2 \in \mathscr{D}'_{L^\infty}(\mathbb{R}^{2n}) \ [resp.\ \mathscr{A}'_{L^\infty}(\mathbb{R}^{2n})] \tag{5.41}$$

*if and only if*

$$u = c \cdot x + u_0, \tag{5.42}$$

*where $c \in \mathbb{C}^n$ and $u_0 \in \mathscr{D}'_{L^\infty}(\mathbb{R}^n)$ [resp. $\mathscr{A}'_{L^\infty}(\mathbb{R}^n)$].*

## 5.4 Hyers–Ulam–Rassias Stability of Quadratic Functional Equation in Fourier Hyperfunctions

In this section, we consider the following Hyers–Ulam–Rassias stability of the quadratic functional equation

$$\|u \circ A + u \circ B - 2u \circ P_1 - 2u \circ P_2\| \le \varepsilon(|x|^p + |y|^p) \tag{5.43}$$

in the space of Fourier hyperfunctions.

It is well known that the *weak semigroup property* of the heat kernel

$$(E_t * E_s)(x) = E_{t+s}(x)$$

holds for convolution. This semigroup property will be very useful later. Throughout this paper we denote by $\psi(x,y)$ or $\psi(z)$ a nonnegative continuous, homogeneous function defined on $\mathbb{R}^{2n}$ of degree $p \ge 0$, $p \ne 2$, that is, $\psi(rz) = r^p \psi(z)$ for all $r \ge 0$. It is easy to see that

$$\widetilde{\psi}(z,t) := \int \psi(\xi) E_t(z - \xi) d\xi$$

is well defined and $\widetilde{\psi}(z,t) \to \psi(z)$ locally uniformly as $t \to 0^+$. Also, $\widetilde{\psi}(z,t)$ satisfies the *weak homogeneity property*

$$\widetilde{\psi}(rz, r^2 t) = r^p \widetilde{\psi}(z,t) \tag{5.44}$$

for all $z \in \mathbb{R}^{2n}$, $t > 0$, $r \geq 0$.

We first consider the inequality (4.1) for $0 \leq p < 2$ or $p > 4$. For this case, we denote by $\Psi_p$,

$$\Psi_p(z,t) := \Sigma_{k=0}^{\infty} 4^{-k-1} \widetilde{\psi}(2^k z, 2^k t), \quad 0 \leq p < 2, \tag{5.45}$$

$$\Psi_p(z,t) := \Sigma_{k=1}^{\infty} 4^{k-1} \widetilde{\psi}(2^{-k} z, 2^{-k} t), \quad p > 4. \tag{5.46}$$

**Lemma 5.4.** *The summations (5.45) and (5.46) converge and $\Psi_p(z,t)$ has at most polynomial growth*

$$|\Psi_p(z,t)| \leq a|z|^p + bt^{p/2} \tag{5.47}$$

*for some positive constant $a, b > 0$ depending on p. Furthermore,*

$$\Psi_p(z,t) \to \frac{1}{|4-2^p|} \psi(z) \tag{5.48}$$

*locally uniformly as $t \to 0^+$.*

*Proof.* The inequality (5.47) follows from the weakhomogeneity property of $\widetilde{\psi}$. We just prove the convergence (5.48). For $0 \leq p < 2$ we write

$$|\Psi_p(z,t) - \frac{1}{4-2^p} \psi(z)| \leq \sum_{k=0}^{\infty} 4^{-k-1} |\widetilde{\psi}(2^k z,\, 2^k t) - \psi(2^k z)|$$
$$= \sum_{k=0}^{\infty} 2^{(p-2)k-2} |\widetilde{\psi}(z,\, 2^{-k} t) - \psi(z)|.$$

Now the convergence (5.48) for $0 \leq p < 2$ follows immediately from the above inequality since $\widetilde{\psi}(z,t) \to \psi(z)$ locally uniformly as $t \to 0^+$.

For the case $p > 4$ we write

$$|\Psi_p(z,t) - \frac{1}{2^p - 4} \psi(z)| \leq \sum_{k=1}^{\infty} 4^{k-1} |\widetilde{\psi}(2^{-k} z,\, 2^{-k} t) - \psi(2^{-k} z)|$$
$$= \sum_{k=1}^{\infty} 2^{(2-p/2)k-2} |\widetilde{\psi}(2^{-\frac{k}{2}} z,\, t) - \psi(2^{-\frac{k}{2}} z)|.$$

From the above inequality the convergence (5.48) for $p > 4$ also follows. □

Now we prove the Hyers–Ulam–Rassias type stability of *quadratic-additive* functional equation:

$$|f(x+y,t+s)+f(x-y,t+s)-2f(x,t)-2f(y,s)| \le \Phi(x,y,t,s) \tag{5.49}$$

for $x,\, y \in \mathbb{R}^n$, $t, s > 0$, where

$$\Phi(x,y,t,s) = [\psi(\xi,\eta) * (E_t(\xi)E_s(\eta))](x,y).$$

**Lemma 5.5.** *Let $f : \mathbb{R}^n \times (0,\infty) \to \mathbb{C}$ satisfy the inequality (4.7). Then there exists a unique function $Q(x,t)$ satisfying the quadratic-additive functional equation*

$$Q(x+y,t+s)+Q(x-y,t+s)-2Q(x,t)-2Q(y,s) = 0 \tag{5.50}$$

*such that*

$$\|f(x,t)-Q(x,t)\| \le \Psi_p^*(x,t), \tag{5.51}$$

*where if $0 \le p < 2$,*

$$\Psi_p^*(x,t) = \Psi_p(x,x,t) + 2^{p/2}[(4-2^{p/2+1})(4-2^{p/2})]^{-1}\widetilde{\psi}(0,\,0,\,t),$$

*and if $p > 4$,*

$$\Psi_p^*(x,t) = \Psi_p(x,x,t) + \Psi_p(0,0,t) + (2^{p/2+1}-4)^{-1}\,\widetilde{\psi}(0,\,0,\,t).$$

*Proof.* We first prove for $0 \le p < 2$. Putting $x = y = 0$ and $s = t$ in (5.49) and dividing the result by 4 we have

$$|2^{-1}f(0,2t) - f(0,t)| \le 4^{-1}\widetilde{\psi}(0,0,t).$$

By the induction argument we have

$$\begin{aligned} |2^{-n}f(0,2^n t) - f(0,t)| &\le \sum_{k=0}^{n} 2^{-k-2}\widetilde{\psi}(0,\,0,\,2^k t) \qquad (5.52)\\ &= \sum_{k=0}^{n} 2^{(p/2-1)k-2}\widetilde{\psi}(0,\,0,\,t)\\ &\le (4-2^{p/2+1})^{-1}\widetilde{\psi}(0,\,0,\,t) \end{aligned}$$

for all $n \in \mathbb{N}$, $t > 0$. Replacing $t$ by $2^m t$, respectively, in (5.52) and dividing the result by $2^m$ we have

$$|2^{-n-m}f(0,2^{n+m}t) - 2^{-m}f(0,2^m t)| \le 2^{(p/2-1)m}(4-2^{p/2+1})^{-1}\widetilde{\psi}(0,\,0,\,t). \tag{5.53}$$

The inequality (5.53) implies $h(t) := \lim_{m\to\infty} 2^{-m}f(0,\,2^m t)$ converges locally uniformly and it is easy to see that $h$ is the unique function satisfying

$$h(t+s) = h(t) + h(s), \tag{5.54}$$

$$|f(0,t) - h(t)| \le (4 - 2^{p/2+1})^{-1} \widetilde{\psi}(0,0,t) \tag{5.55}$$

for all $t, s > 0$. Putting $y = x$ and $s = t$ in (5.49), dividing the result by 4 and using the induction argument we have

$$|f(x,t) - 4^{-n} f(2^n x, 2^n t) - \sum_{k=1}^{n} 4^{-k} f(0, 2^k t)| \le \Psi_p(x,x,t).$$

It follows from (5.55) that

$$\left| \sum_{k=1}^{n} 4^{-k} f(0,2^k t) - (1 - 2^{-n}) h(t) \right| \le \sum_{k=1}^{n} 4^{-k} |f(0,2^k t) - h(2^k t)|$$

$$\le C_p \, \widetilde{\psi}(0,0,t).$$

where $C_p = 2^{p/2}[(4 - 2^{p/2+1})(4 - 2^{p/2})]^{-1}$. Thus, letting $F(x,t) = f(x,t) + h(t)$ we have

$$|F(x,t) - 4^{-n} F(2^n x, 2^n t)| \le \Psi_p(x,x,t) + C_p \, \widetilde{\psi}(0,0,t). \tag{5.56}$$

Replacing $x$ by $2^m x$ and $t$ by $2^m t$ in (5.56) and dividing by $4^m$ we have

$$|4^{-m} F(2^m x, 2^m t) - 4^{-n-m} F(2^{n+m} x, 2^{n+m} t)| \le 2^{(p-2)m-2} \Psi_p(x,x,2^{-m} t)$$

$$+ 2^{(p/2-2)m} C_p \, \widetilde{\psi}(0,0,t).$$

Since the right hand side of above inequality tends to 0 locally uniformly as $m \to \infty$,

$$g_m(x,t) := 4^{-m} F(2^m x, 2^m t)$$

forms a Cauchy sequence which converges locally uniformly.

Now we verify that $g(x,t) = \lim_{m \to \infty} g_m(x,t)$ satisfies

$$|F(x,t) - g(x,t)| \le \Psi_p(x,x,t) + C_p \, \widetilde{\psi}(0,0,t), \tag{5.57}$$

$$g(x+y,t+s) + g(x-y,t+s) - 2g(x,t) - 2g(y,s) = 0 \tag{5.58}$$

for all $x, y \in \mathbb{R}^n$, $t, s > 0$.

The inequality (5.57) follows immediately from (5.56). Now since $h(2^m t) = 2^m h(t)$ by (5.54) we have

$$\begin{aligned} g(x,t) : &= \lim_{m\to\infty} 4^{-m} F(2^m x, 2^m t) \\ &= \lim_{m\to\infty} 4^{-m} (f(2^m x, 2^m t) + h(2^m t)) \\ &= \lim_{m\to\infty} 4^{-m} f(2^m x, 2^m t). \end{aligned}$$

Now replacing $x$, $y$, $t$, $s$ by $2^m x$, $2^m y$, $2^m t$, $2^m s$ in (5.49), respectively, dividing the result by $4^m$ we have

$$\begin{aligned} &|g_m(x+y,t+s) + g_m(x-y,t+s) - 2g_m(x,t) - 2g_m(y,s)| \\ &\qquad \le 2^{(p-2)m} \widetilde{\psi}(x,y,2^{-m}t,2^{-m}s). \end{aligned}$$

Letting $m \to \infty$ in the above inequality we get (5.58). Now let $Q(x,t) = g(x,t) - h(t)$. Then $Q(x,t)$ satisfies (5.50) and (5.51).

Now we consider the case $p > 4$. Let $G(x,t) = f(x,t) - f(0,t)$. Then we get the inequality

$$\begin{aligned} &|G(x+y,t+s) + G(x-y,t+s) - 2G(x,t) - 2G(y,s)| \\ &\qquad \le \Phi(x,y,t,s) + \Phi(0,0,t,s). \end{aligned} \tag{5.59}$$

Replacing both $x$ and $y$ by $\frac{x}{2}$, $t$ and $s$ by $\frac{t}{2}$ in (5.59) we have

$$|G(x,t) - 4G(2^{-1}x, 2^{-1}t)| \le \widetilde{\psi}(2^{-1}x, 2^{-1}x, 2^{-1}t) + \widetilde{\psi}(0,0,2^{-1}t).$$

By the same method as in the case $0 \le p < 2$ (justreplacing $n$ by $-n$ in (5.56) we can verify that

$$g(x,t) := \lim_{m\to\infty} 4^m G(2^{-m}x, 2^{-m}t)$$

is the unique function satisfying the quadratic-additive functional equation (5.50) and the inequality

$$|F(x,t) - g(x,t)| \le \Psi_p(x,x,t) + \Psi_p(0,0,t), \tag{5.60}$$

for all $x, y \in \mathbb{R}^n$, $t, s > 0$.

On the other hand we can verify that

$$h(t) := \lim_{m\to\infty} 2^m f(0, 2^{-m}t)$$

is the unique function satisfying the Cauchy equation (4.12) and the inequality

$$|f(0,t)-h(t)| \le (2^{p/2+1}-4)^{-1}\,\widetilde{\psi}(0,0,t)$$

for all $t,s>0$.

Now $Q(x,t)=g(x,t)+h(t)$ is the function satisfying (5.50) and (5.51) with

$$\Psi_p^*(x,t)=\Psi_p(x,x,t)+\Psi_p(0,0,t)+(2^{p/2+1}-4)^{-1}\,\widetilde{\psi}(0,0,t).$$

Finally, we prove the uniqueness of $Q$. Let $Q_0(x,t)=Q(x,t)-Q(0,t)$. Then $Q_0(x,t)$ also satisfies the quadratic–additive functional equation

$$Q_0(x+y,t+s)+Q_0(x-y,t+s)-2Q_0(x,t)-2Q_0(y,s)=0. \tag{5.61}$$

Putting $y=0$ in (5.61) we have $Q_0(x,t+s)=Q_0(x,t)$ for all $x\in\mathbb{R}^n$, $t,s>0$. Thus, $Q_0(x,t)$ is independent of $t>0$ and we may write $Q_0(x,t):\equiv Q_0(x)$. Since $Q_0$ satisfies the quadratic functional equation

$$Q_0(x+y)+Q_0(x-y)-2Q_0(x)-2Q_0(y)=0,$$

and that

$$Q(rx,r^2t)=Q_0(rx)+Q(0,r^2t)=r^2Q(x,t). \tag{5.62}$$

for all rational numbers $r$.

Now suppose that $Q^*(x,t)$ also satisfies (5.50) and (5.51). Then we have

$$\begin{aligned}|Q(x,t)-Q^*(x,t)| &= r^{-2}|Q(rx,r^2t)-Q^*(rx,r^2t)|\\ &\le 2r^{-2}\Psi_p^*(rx,r^2t)=2r^{p-2}\Psi_p^*(x,t).\end{aligned}$$

Letting $r\to\infty$ if $0\le p<2$ and $r\to 0^+$ if $p>4$ we have $Q=Q^*$. □

Now we state and prove the main results of this section.

**Theorem 5.5.** *Let $u\in\mathcal{F}'$ satisfy the inequality*

$$\|u\circ A+u\circ B-2u\circ P_1-2u\circ P_2\|\le\psi(x,y). \tag{5.63}$$

*Then there exists a unique quadratic function*

$$q(x):=\sum_{1\le j\le k\le n} a_{jk}x_jx_k$$

*such that*

$$\|u-q(x)\|\le\frac{1}{|4-2^p|}\psi(x,x),\qquad 0<p<2\quad\text{or}\quad p>4, \tag{5.64}$$

$$\|u-q(x)\|\le\tfrac{1}{2}\psi(0,0),\qquad p=0. \tag{5.65}$$

*Proof.* Convolving in each side of the inequality (5.63) the tensor product $E_t(x)E_s(y)$ of n-dimensional heat kernels we have in view of the semigroup property of the heat kernel $(E_t * E_s)(x) = E_{t+s}(x)$,

$$[(u \circ A) * (E_t(\xi)E_s(\eta))](x, y) = \left\langle u_\xi, \int E_t(x - \xi + \eta)E_s(y - \eta)\,d\eta \right\rangle$$
$$= \langle u_\xi, (E_t * E_s)(x + y - \xi)\rangle = \tilde{u}(x + y, t + s).$$

Similarly, we have

$$[(u \circ B) * (E_t(\xi)E_s(\eta))](x, y) = \tilde{u}(x - y, t + s),$$
$$[(u \circ P_1) * (E_t(\xi)E_s(\eta))](x, y) = \tilde{u}(x, t),$$
$$[(u \circ P_2) * (E_t(\xi)E_s(\eta))](x, y) = \tilde{u}(y, s),$$

where $\tilde{u}(x,t)$ is the Gauss transform of $u$.

Thus, the inequality (5.63) is converted to the stability problem of quadratic-additive type functional equation:

$$|\tilde{u}(x + y, t + s) + \tilde{u}(x - y, t + s) - 2\tilde{u}(x, t) - 2\tilde{u}(y, s)| \le \Phi(x, y, t, s)$$

for $x, y \in \mathbb{R}^n$, $t, s > 0$, where $\Phi(x, y, t, s) = [\psi(\xi, \eta) * (E_t(\xi)E_s(\eta))](x, y)$.

By Lemma 5.1, there exists a unique function $Q(x,t)$ satisfying the quadratic-additive functional equation (5.50) such that

$$\|\tilde{u}(x,t) - Q(x,t)\| \le \Psi_p^*(x,t). \tag{5.66}$$

Since the Gauss transform $\tilde{u}$ a sooth function, $Q(x,t)$ is at least a continuous function as we see in the proof of Lemma 5.1. Thus, the solution $Q(x,t)$ has the form [6]

$$Q(x,t) = \sum_{1 \le i \le j \le n} a_{ij} x_i x_j + bt.$$

Thus, the left hand side of (5.66) tends to $\|u - q(x)\|$ as $t \to 0^+$. On the other hand, by Lemma 5.4 it is easy to see that if $p > 0$,

$$\Psi_p^*(x,t) \to \frac{1}{|4 - 2^p|}\psi(x,x)$$

in $\mathcal{F}'$, as $t \to 0^+$. Letting $t \to 0^+$ in (5.66) we get (5.64). Now if $p = 0$, then from the continuity and homogeneity, $\psi(x,y)$ must be a constant function. Thus, we have

$$\Psi_p^*(x,t) = \frac{1}{3}\psi(x,x) + \frac{1}{6}\psi(0,0) = \frac{1}{2}\psi(0,0). \qquad \square$$

As a direct consequence of the above result for $0 < p < 2$ or $p > 4$ we obtain the Hyers–Ulam–Rassias stability of quadratic functional equation.

**Theorem 5.6.** *Let $0 < p < 2$ or $p > 4$ and let $u \in \mathscr{F}'$ satisfy the inequality*

$$\|u \circ A + u \circ B - 2u \circ P_1 - 2u \circ P_2\| \le \varepsilon(|x|^p + |y|^p).$$

*Then there exists a unique quadratic function*

$$q(x) = \sum_{1 \le j \le k \le n} a_{jk} x_j x_k$$

*such that*

$$\|u - q(x)\| \le \frac{2\varepsilon}{|4 - 2^p|}|x|^p.$$

We also obtain the following stability theorem which improves the result in [3].

**Theorem 5.7.** *Let $u \in \mathscr{F}'$ satisfy the inequality*

$$\|u \circ A + u \circ B - 2u \circ P_1 - 2u \circ P_2\| \le \varepsilon.$$

*Then there exists a unique quadratic function*

$$q(x) = \sum_{1 \le j \le k \le n} a_{jk} x_j x_k$$

*such that*

$$\|u - q(x)\| \le \frac{1}{2}\varepsilon.$$

Now we consider the remaining case that $\psi(x,y)$ is of degree $2 < p \le 4$.

**Theorem 5.8.** *Let $u \in \mathscr{F}'$ satisfy the inequality*

$$\|u \circ A + u \circ B - 2u \circ P_1 - 2u \circ P_2\| \le \psi(x,y). \tag{5.67}$$

*where $\psi(x,y)$ is of degree $2 < p \le 4$. Then there exists a unique quadratic function*

$$q(x) := \sum_{1 \le j \le k \le n} a_{jk} x_j x_k$$

*such that*

$$\|u - q(x)\| \le \frac{1}{2^p - 4}(\psi(x,x) + 2^p \psi(x,0)).$$

*Proof.* As in Theorem 5.5, convolving in each side of the inequality (5.67) the tensor product $E_t(x)E_s(y)$ of $n$-dimensional heat kernels we have

$$|\tilde{u}(x+y,t+s) + \tilde{u}(x-y,t+s) - 2\tilde{u}(x,t) - 2\tilde{u}(y,s)| \le \Phi(x,y,t,s) \tag{5.68}$$

for $x, y \in \mathbb{R}^n$, $t,s > 0$, where $\Phi(x,y,t,s) = [\psi(\xi,\eta) * (E_t(\xi)E_s(\eta))](x,y)$.

It follows from the inequality (4.26) that

$$|\tilde{u}(x,t)| \le |\tilde{u}(x,t+s)+\tilde{u}(0,s)| + |\Phi(x,0,t,s)|.$$

Thus, it is easy to see that

$$U(x) := \limsup_{t\to 0^+} \tilde{u}(x,t)$$

exists since $\Phi(x,0,0^+,s) := \lim_{t\to 0^+}\Phi(x,0,t,s)$ exists. Letting $y=0$ and $t\to 0^+$ so that $\tilde{u}(x,t)\to U(x)$ in (5.68) we have

$$|\tilde{u}(x,s) - U(x) - \tilde{u}(0,s)| \le \frac{1}{2}\Phi(x,0,0^+,s). \tag{5.69}$$

From the inequality (5.68), (5.69) and the triangle inequality we have

$$\begin{aligned}|U(x+y) + U(x-y) - 2U(x) - 2U(y)| &\le \Phi(x,y,t,s)\\ &+\frac{1}{2}\Phi(x+y,0,0^+,t+s) + \frac{1}{2}\Phi(x-y,0,0^+,t+s)\\ &+\Phi(x,0,0^+,t) + \Phi(y,0,0^+,s) + \Phi(0,0,t,s)\end{aligned}$$

for all $x,y\in\mathbb{R}^n$, $t,s>0$. Letting $t,s\to 0^+$ in the above inequality we have

$$|U(x+y)+U(x-y)-2U(x)-2U(y)| \le \psi^*(x,y) \tag{5.70}$$

where $\psi^*(x,y) = \psi(x,y) + \frac{1}{2}\psi(x+y,0) + \frac{1}{2}\psi(x-y,0) + \psi(x,0) + \psi(y,0)$. Using the fact that $\psi^*(x,y)$ is a homogeneous function of degree $p>2$ and following the same method as in [8] we can verify that there exists a unique function $q(x)$ satisfying the quadratic functional equation

$$q(x+y)+q(x-y)-2q(x)-2q(y) = 0 \tag{5.71}$$

such that

$$|U(x)-q(x)| \le \frac{1}{2^p-4}\psi^*(x,x). \tag{5.72}$$

Since $q(x)$ is obtained by $q(x) := \lim_{n\to\infty} 4^n U(2^{-n}x)$ it is easy to see that $q(x)$ is continuous. Thus we have

$$q(x) = \sum_{1\le j\le k\le n} a_{jk}x_jx_k.$$

Now from (5.69) and (5.72) we have

$$|\tilde{u}(x,s) - q(x) - \tilde{u}(0,s)| \le \frac{1}{2}\Phi(x,0,0^+,s) + \frac{1}{2^p-4}\psi^*(x,x). \tag{5.73}$$

Letting $s \to 0^+$ in (5.73) we have

$$\|u - q(x)\| \le \frac{1}{2^p - 4}(\psi(x,x) + 0) + 2^p\psi(x,0)).$$

□

As a direct consequence of the above result we obtain the following Hyers–Ulam–Rassias stability of quadratic functional equation for $2 < p \le 4$.

**Theorem 5.9.** *Let* $2 < p \le 4$ *and let* $u \in \mathscr{F}'$ *satisfy the inequality*

$$\|u \circ A + u \circ B - 2u \circ P_1 - 2u \circ P_2\| \le \varepsilon(|x|^p + |y|^p).$$

*Then there exists a unique quadratic function*

$$q(x) = \sum_{1 \le j \le k \le n} a_{jk} x_j x_k$$

*such that*

$$\|u - q(x)\| \le \frac{\varepsilon(2^p + 2)}{2^p - 4}|x|^p.$$

## References


1. Aczél, J., Dhombres, J.: Functional equations inseveral variables. Cambridge University Press, New York-Sydney (1989)
2. Aoki, T.: On the stability of the linear transformation in Banach spaces. J. Math. Soc. Japan **2**, 64–66 (1950)
3. Baker, J.A.: On a functional equation of Aczél and Chung. Aequationes Math. **46**, 99–111 (1993)
4. Baker, J.A.: The stability of cosine functional equation. Proc. Amer. Math. Soc. **80**, 411–416 (1980)
5. Bourgin, D.G.: Class of transformations and bordering transformations. Bull. Amer. Math. Soc. **57**, 223–237 (1951)
6. Bourgin, D.G.: Multiplicative transformations. Proc. Nat. Academy Sci. U.S.A. **36**, 564–570 (1950)
7. Chung, J.: Distributional method for a class of functiuonal equations and their stabilities. Acta Math. Sinica **23**, 2017–2026 (2007)
8. Chung, J.: Stability of approximately quadratic Schwartz distributions. Nonlinear Anal. **67**, 175–186 (2007)
9. Chung, J.: A distributional version of functional equations and their stabilities. Nonlinear Anal. **62**, 1037–1051 (2005)
10. Chung, J., Chung, S.-Y., Kim, D.: A characterization for Fourier hyperfunctions. Publ. Res. Inst. Math. Sci. **30**, 203–208 (1994)
11. Chung, S.-Y., Kim, D., Lee, E.G.: Periodic hyperfunctions and Fourier series. Proc. Amer. math. Soc. **128**, 2421–2430 (1999)
12. Czerwik, S.: Stability of Functional Equations of Ulam-Hyers-Rassias Type. Hadronic Press, Inc., Palm Harbor (2003)

13. Deeba, E.Y., Koh, E.L.: The Pexider functional equations in distributions. Canad. J. Math. **42**, 304–314 (1990)
14. Deeba, E., Xie, S., Distributional analog of a functional equation. Appl. Math. Lett. **16**, 669–673 (2003)
15. Fenyö, I.: Über eine Losungsmethode gewisser Funktionalgleichungen. Acta Math. Acad. Sci. Hungar. **7**, 383–396 (1956)
16. Forti, G.L.: Hyer-Ulam stability of functional equation in several variables. Aeqationes Math. **50**, 143–190 (1995)
17. Forti, G.L.: The stability of homomorphisms and amenablity with applications to functional equations. Abh. Math. Sem. Univ. Hamburg **57**, 215–226 (1987)
18. Gelfand, I.M., Shilov, G.E.: Generalized Functions II. Academic Press, New York (1968)
19. Hörmander, L.: The Analysis of Linear Partial Differential Operator I. Springer–Verlag, Berlin–New York (1983)
20. Hyers, D.H.: On the stability of the linear functional equations. Proc. Nat. Acad. Sci. USA **27**, 222–224 (1941)
21. Hyers, D.H., Rassias, Th.M.: Approximate homomorphisms. Aequationes Math. **44**, 125–153 (1992)
22. Hyers, D.H., Isac, G., Rassias, Th.M.: Stability of Functional Equations in Several Variables. Birkhauser (1998)
23. Jung, S.M.: Hyers-Ulam-Rassias Stability of Functional Equations in Mathematical Analysis. Hadronic Press, Inc., Palm Harbor (2001)
24. Jun, K.W., Kim, H.M.: Stability problem for Jensen-type functional equations of cubic mappings, Acta Math. Sinica, English Series, **22(6)**, 1781–1788 (2006)
25. Kim, G.H., Lee, Y.H.: Boundedness of approximate trigonometric functional equations. Appl. Math. Lett. **31**, 439–443 (2009)
26. Matsuzawa, T.: A calculus approach to hyperfunctions III. Nagoya Math. J. **118**, 133–153 (1990)
27. Park, C.G.: Hyers-Ulam-Rassias stability of homomorphisms in quasi-Banach algabras. Bull. Sci. Math. **132**, 87–96 (2008)
28. Rassias, J.M., Rassias, M.J.: On the Ulam stability of Jensen and Jensen type mappings on restricted domains. J. Math. Anal. Appl. **281**, 516–524 (2003)
29. Rassias, J.M.: On Approximation of Approximately Linear Mappings by Linear Mappings. J. Funct. Anal. **46**, 126–130 (1982)
30. Rassias, Th.M.: On the stability of functional equations in Banach spaces. J. Math. Anal. Appl. **251**, 264–284 (2000)
31. Rassias, Th.M.: On the stability of linear mapping in Banach spaces. Proc. Amer. Math. Soc. **72**, 297–300 (1978)
32. Schwartz, L.: Théorie des Distributions. Hermann, Paris (1966)
33. Ulam, S.M.: A Collection of Mathematical Problems. Interscience Publ., New York (1960)
34. Widder, D.V.: The Heat Equation. Academic Press, New York (1975)

# Chapter 6
# Stability of Multi-Jensen Mappings in Non-Archimedean Normed Spaces

**Krzysztof Ciepliński**

*Dedicated to the memory of S.M. Ulam*
*on the 100th anniversary of his birth*

**Abstract** A function $f : V^n \longrightarrow W$, where $V$ and $W$ are normed spaces over a field of characteristic different from 2 and $n \geq 1$ is an integer, is called multi-Jensen if it satisfies Jensen's functional equation in each variable. In this note, we provide a proof of a generalized Hyers–Ulam stability of multi-Jensen mappings in non-Archimedean normed spaces, using the so-called direct method.

**Keywords** Generalized Hyers–Ulam stability • Non-Archimedean normed space • Multi-Jensen mapping • Multi-additive mapping • Direct method

**Mathematics Subject Classification (2000):** Primary 39B82, 46S10

## 6.1 Introduction

Throughout this paper we assume that $V$ and $W$ are normed spaces over a field of characteristic different from 2 and $n \geq 1$ is an integer. Moreover, $\mathbb{N}$ stands for the set of all positive integers and $\mathbb{Q}$ denotes the set of all rationals.

Let us recall that a function $f : V^n \longrightarrow W$ is called *multi-Jensen* if it satisfies Jensen's functional equation in each variable, that is

$$2f\left(x_1, \ldots, x_{i-1}, \frac{x_i + x_i'}{2}, x_{i+1}, \ldots, x_n\right) = f(x_1, \ldots, x_n) + f(x_1, \ldots, x_{i-1}, x_i', x_{i+1}, \ldots, x_n),$$

for all $i \in \{1, \ldots, n\}, x_1, \ldots, x_{i-1}, x_i, x_i', x_{i+1}, \ldots, x_n \in V$.

K. Ciepliński (✉)
Department of Mathematics, Pedagogical University, Podchorążych 2, 30-084 Kraków, Poland
e-mail: kc@up.krakow.pl

Th.M. Rassias and J. Brzdęk (eds.), *Functional Equations in Mathematical Analysis*, Springer Optimization and Its Applications 52, DOI 10.1007/978-1-4614-0055-4_6,

Similarly, $f$ is said to be *multi-additive* if it is additive (satisfies Cauchy's functional equation) in each variable. The notion of multi-Jensen mapping was introduced by Prager and Schwaiger in 2005 (see [20]) in connection with generalized polynomials.

In this note, we will provide a proof of a generalized Hyers–Ulam stability of the above system of $n$ Jensen's equations in non-Archimedean normed spaces, using the so-called direct (Hyers) method and some ideas from [3] and [17].

Regarding the stability of functional equations we study the problem posed by S.M. Ulam in 1940: "when is it true that the solution of an equation differing slightly from a given one, must of necessity be close to the solution of the given equation?". The first partial solution of Ulam's problem in the case of Cauchy's functional equation in the context of Banach spaces was given by Hyers (see [7]). During the last 70 years many extensions of Hyers's result for several types of functional equations have been investigated by a number of mathematicians worldwide (see, e.g., [6, 8, 11, 12, 15, 22]).

For instance, the stability of the Jensen functional equation

$$2f\left(\frac{x+y}{2}\right) = f(x) + f(y)$$

was a popular subject of research (cf. [2, 9, 10, 14, 17]). In particular, Prager and Schwaiger (see [21]) and the author (see [3,4]) studied different kinds of stability of the multi-Jensen mappings.

Let us recall that by a *non-Archimedean field* we mean a field $\mathbb{K}$ equipped with a function (*valuation*) $|\cdot| : \mathbb{K} \longrightarrow [0,\infty)$ such that

$$|r| = 0 \quad \text{if and only if} \quad r = 0,$$

$$|rs| = |r||s|, \qquad r, s \in \mathbb{K}$$

and

$$|r+s| \leq \max\{|r|, |s|\}, \qquad r, s \in \mathbb{K}.$$

In any non-Archimedean field we have

$$|1| = |-1| = 1$$

and

$$|n| \leq 1$$

for all $n \in \mathbb{N}$.

Let $X$ be a linear space over a field $\mathbb{K}$ with a non-Archimedean non-trivial valuation $|\cdot|$ (i.e., we additionally assume that there is an $r_0 \in \mathbb{K}$ such that $0 \neq |r_0| \neq 1$). A function $\|\cdot\| : X \longrightarrow [0,\infty)$ is said to be a *non-Archimedean norm* if it satisfies the following conditions:

$$\|x\| = 0 \quad \text{if and only if} \quad x = 0,$$

$$\|rx\| = |r|\|x\|, \qquad r \in \mathbb{K}, x \in X$$

and

$$\|x+y\| \leq \max\{\|x\|, \|y\|\}, \quad x, y \in X.$$

Then $(X, \|\cdot\|)$ is called a *non-Archimedean (normed) space*. In any such a space a sequence $(x_n)_{n\in\mathbb{N}}$ is Cauchy if and only if

$$\lim_{n\longrightarrow\infty} \|x_{n+1} - x_n\| = 0.$$

By a *complete non-Archimedean space* we mean the one in which every Cauchy sequence is convergent.

In 1899, Hensel discovered the $p$-adic numbers as a number theoretical analogue of power series in complex analysis. Fix a prime number $p$. For any non-zero rational number $x$, there exists a unique integer $n_x$ such that

$$x = \frac{a}{b} p^{n_x},$$

where $a$ and $b$ are integers not divisible by $p$. Then

$$|x|_p := p^{-n_x}$$

defines a non-Archimedean norm on $\mathbb{Q}$. The completion of $\mathbb{Q}$ with respect to the metric

$$d(x,y) = |x-y|_p$$

is called the *p-adic number field* (see for instance [23]). During the last three decades $p$-adic numbers have gained the interest of physicists for their research in particular in problems coming from quantum physics, $p$-adic strings and superstrings (see [13]).

In 2007, Moslehian and Rassias (see [18]) proved the generalized Hyers–Ulam stability of the Cauchy and quadratic functional equations in non-Archimedean normed spaces. After their results some papers (see for instance [5, 16, 17, 19]) on the stability of other equations in such spaces have been published.

## 6.2 Main Results

Our first result corresponds to some outcomes from [3] and [17].

**Theorem 6.1.** *Let $V$ and $W$ be non-Archimedean normed spaces over a non-Archimedean field of characteristic different from 2 with $0 < |3| < 1$. Assume also that the space $W$ is complete, $n \in \mathbb{N}$ and for every $i \in \{1, ..., n\}$, $\alpha_i, \beta_i \geq 0$ and $0 \leq p_i < 1$. If $f : V^n \longrightarrow W$ is a function satisfying*

$$\|f(x_1, ..., x_{i-1}, 0, x_{i+1}, ..., x_n)\| \leq \beta_i, \quad x_j \in V,$$
$$j \in \{1, ..., n\} \setminus \{i\}, \; i \in \{1, ..., n\} \quad (6.1)$$

*and*

$$\left\|2f\left(x_1, ..., x_{i-1}, \frac{x_i + x_i'}{2}, x_{i+1}, ..., x_n\right) - f(x_1, ..., x_n)\right.$$
$$\left.-f(x_1, ..., x_{i-1}, x_i', x_{i+1}, ..., x_n)\right\| \leq \alpha_i \max\{\|x_i\|^{p_i}, \|x_i'\|^{p_i}\},$$
$$x_i, x_i' \in V \setminus \{0\}, x_j \in V, j \in \{1, ..., n\} \setminus \{i\}, i \in \{1, ..., n\}, \quad (6.2)$$

*then for every* $i \in \{1, ..., n\}$ *there exists a multi-Jensen mapping* $F_i : V^n \longrightarrow W$ *for which*

$$\|f(x_1, ..., x_n) - F_i(x_1, ..., x_n)\| \leq \max\left\{\frac{\alpha_i}{|3|^{p_i}}\|x_i\|^{p_i}, |2|\beta_i\right\},$$
$$x_i \in V \setminus \{0\}, x_j \in V, j \in \{1, ..., n\} \setminus \{i\}. \quad (6.3)$$

*Proof.* Fix $i \in \{1, ..., n\}$, $x_i \in V \setminus \{0\}$ and $x_j \in V$ for $j \in \{1, ..., n\} \setminus \{i\}$. By (6.2) we get

$$\left\|2f\left(x_1, ..., x_{i-1}, \frac{x_i}{3}, x_{i+1}, ..., x_n\right)\right.$$
$$\left.-f(x_1, ..., x_n) - f\left(x_1, ..., x_{i-1}, -\frac{x_i}{3}, x_{i+1}, ..., x_n\right)\right\|$$
$$\leq \alpha_i \max\left\{\|x_i\|^{p_i}, \left\|-\frac{x_i}{3}\right\|^{p_i}\right\} = \frac{\alpha_i}{|3|^{p_i}}\|x_i\|^{p_i} \quad (6.4)$$

and

$$\left\|2f(x_1, ..., x_{i-1}, 0, x_{i+1}, ..., x_n) - f\left(x_1, ..., x_{i-1}, \frac{x_i}{3}, x_{i+1}, ..., x_n\right)\right.$$
$$\left.-f\left(x_1, ..., x_{i-1}, -\frac{x_i}{3}, x_{i+1}, ..., x_n\right)\right\|$$
$$\leq \alpha_i \max\left\{\left\|\frac{x_i}{3}\right\|^{p_i}, \left\|-\frac{x_i}{3}\right\|^{p_i}\right\} = \frac{\alpha_i}{|3|^{p_i}}\|x_i\|^{p_i}.$$

Hence and by (6.1) we obtain

$$\left\|f\left(x_1, ..., x_{i-1}, \frac{x_i}{3}, x_{i+1}, ..., x_n\right) + f\left(x_1, ..., x_{i-1}, -\frac{x_i}{3}, x_{i+1}, ..., x_n\right)\right\|$$
$$\leq \max\left\{\frac{\alpha_i}{|3|^{p_i}}\|x_i\|^{p_i}, |2|\beta_i\right\},$$

which together with (6.4) gives

$$\begin{aligned}
&\left\| 3\, f\left( x_1, \ldots, x_{i-1}, \frac{x_i}{3}, x_{i+1}, \ldots, x_n \right) - f(x_1, \ldots, x_n) \right\| \\
&\quad \le \max\left\{ \frac{\alpha_i}{|3|^{p_i}} \|x_i\|^{p_i}, \max\left\{ \frac{\alpha_i}{|3|^{p_i}} \|x_i\|^{p_i}, |2|\beta_i \right\} \right\} \\
&\quad = \max\left\{ \frac{\alpha_i}{|3|^{p_i}} \|x_i\|^{p_i}, |2|\beta_i \right\}. \qquad (6.5)
\end{aligned}$$

Fix an $x_i \in V$. From (6.5) it follows that for any $m \in \mathbb{N}$ we have

$$\begin{aligned}
&\left\| 3^{m+1}\, f\left( x_1, \ldots, x_{i-1}, \frac{x_i}{3^{m+1}}, x_{i+1}, \ldots, x_n \right) \right. \\
&\qquad \left. - 3^m f\left( x_1, \ldots, x_{i-1}, \frac{x_i}{3^m}, x_{i+1}, \ldots, x_n \right) \right\| \\
&\qquad \le \max\left\{ |3|^m \frac{\alpha_i}{|3|^{p_i}} \left\| \frac{x_i}{3^m} \right\|^{p_i}, |3|^m |2| \beta_i \right\} \\
&\qquad = \max\left\{ (|3|^{1-p_i})^m \frac{\alpha_i}{|3|^{p_i}} \|x_i\|^{p_i}, |3|^m |2| \beta_i \right\}. \qquad (6.6)
\end{aligned}$$

Therefore, the sequence

$$\left( 3^m f\left( x_1, \ldots, x_{i-1}, \frac{x_i}{3^m}, x_{i+1}, \ldots, x_n \right) \right)_{m \in \mathbb{N}}$$

is a Cauchy sequence. Since the space $W$ is complete, this sequence is convergent and we define $F_i : V^n \longrightarrow W$ by

$$F_i(x_1, \ldots, x_n) := \lim_{m \longrightarrow \infty} 3^m f\left( x_1, \ldots, x_{i-1}, \frac{x_i}{3^m}, x_{i+1}, \ldots, x_n \right), \quad (x_1, \ldots, x_n) \in V^n. \qquad (6.7)$$

Fix $x_i \in V \setminus \{0\}$ and $x_j \in V$ for $j \in \{1, \ldots, n\} \setminus \{i\}$. Using (6.5), (6.6) and induction one can show that for any $m \in \mathbb{N}$ we have

$$\begin{aligned}
&\left\| 3^m f\left( x_1, \ldots, x_{i-1}, \frac{x_i}{3^m}, x_{i+1}, \ldots, x_n \right) - f(x_1, \ldots, x_n) \right\| \\
&\qquad \le \max\left\{ \frac{\alpha_i}{|3|^{p_i}} \|x_i\|^{p_i}, |2|\beta_i \right\}.
\end{aligned}$$

Letting $m \longrightarrow \infty$ in this inequality we see that (6.3) holds true.

Next, fix $x_i' \in V \setminus \{0\}$, $m \in \mathbb{N}$ and note that according to (6.2) we have

$$\begin{aligned}\Bigg\| 2\cdot 3^m\, f\Big( x_1, ..., x_{i-1}, \frac{x_i + x_i'}{2\cdot 3^m}, x_{i+1}, ..., x_n\Big) \\ -3^m f\Big(x_1, ..., x_{i-1}, \frac{x_i}{3^m}, x_{i+1}, ..., x_n\Big) \\ -3^m f\Big(x_1, ..., x_{i-1}, \frac{x_i'}{3^m}, x_{i+1}, ..., x_n\Big)\Bigg\| \\ \leq |3|^m \frac{\alpha_i}{|3|^{mp_i}} \max\{\|x_i\|^{p_i}, \|x_i'\|^{p_i}\}.\end{aligned}$$

Finally, fix $k \in \{1, ..., n\} \setminus \{i\}$, $x_k, x_k' \in V \setminus \{0\}$ and assume that $k < i$ (the same arguments apply to the case where $k > i$). From (6.2) it follows that

$$\begin{aligned}\Bigg\| 2\cdot 3^m\, f\Big( x_1, ..., x_{k-1}, \frac{x_k + x_k'}{2}, x_{k+1}, ..., x_{i-1}, \frac{x_i}{3^m}, x_{i+1}, ..., x_n\Big) \\ -3^m f\Big(x_1, ..., x_{i-1}, \frac{x_i}{3^m}, x_{i+1}, ..., x_n\Big) \\ -3^m f\Big(x_1, ..., x_{k-1}, x_k', x_{k+1}, ..., x_{i-1}, \frac{x_i}{3^m}, x_{i+1}, ..., x_n\Big)\Bigg\| \\ \leq |3|^m \alpha_k \max\{\|x_k\|^{p_k}, \|x_k'\|^{p_k}\}.\end{aligned}$$

Letting $m \longrightarrow \infty$ in the above inequalities we see that the mapping $F_i$ is multi-Jensen.

□

**Theorem 6.2.** *Let $V$ and $W$ be non-Archimedean normed spaces over a non-Archimedean field of characteristic different from 2 with $0 < |3| < 1$. Assume also that the space $W$ is complete, $n \in \mathbb{N}$ and for every $i \in \{1, ..., n\}$, $\alpha_i \geq 0$ and $0 \leq p_i < 1$. If $f : V^n \longrightarrow W$ is a function satisfying (6.2) and*

$$f(x_1, ..., x_{i-1}, 0, x_{i+1}, ..., x_n) = 0, \quad x_j \in V,\ j \in \{1, ..., n\} \setminus \{i\}, \quad i \in \{1, ..., n\}, \tag{6.8}$$

*then for every $i \in \{1, ..., n\}$ there is a unique multi-additive mapping $F_i : V^n \longrightarrow W$ for which*

$$\|f(x_1, ..., x_n) - F_i(x_1, ..., x_n)\| \leq \frac{\alpha_i}{|3|^{p_i}} \|x_i\|^{p_i}, \quad (x_1, ..., x_n) \in V^n. \tag{6.9}$$

*Proof.* From Theorem 6.1 it follows that for every $i \in \{1, ..., n\}$ there exists a multi-Jensen mapping $F_i : V^n \longrightarrow W$ such that

$$\|f(x_1, ..., x_n) - F_i(x_1, ..., x_n)\| \leq \frac{\alpha_i}{|3|^{p_i}} \|x_i\|^{p_i},$$
$$x_i \in V \setminus \{0\},\ x_j \in V,\ j \in \{1, ..., n\} \setminus \{i\}. \tag{6.10}$$

Using (6.7) and (6.8) we also get

$$F_i(x_1, ..., x_{k-1}, 0, x_{k+1}, ..., x_n) = 0, \quad x_j \in V,\ j \in \{1, ..., n\} \setminus \{k\},$$
$$i, k \in \{1, ..., n\}.$$

Therefore (see for instance [1, Theorem 1.4]) every $F_i$ is multi-additive and, by (6.8) and (6.10), condition (6.9) holds true.

Now, fix an $i \in \{1, ..., n\}$ and assume that $F_i' : V^n \longrightarrow W$ is another multi-additive mapping satisfying (6.9). Then for any $(x_1, ..., x_n) \in V^n$ we have

$$\begin{aligned}
&\|F_i(x_1, ..., x_n) - F_i'(x_1, ..., x_n)\| \\
&= \lim_{m\longrightarrow\infty} |3|^m \left\| F_i\left(x_1, ..., x_{i-1}, \frac{x_i}{3^m}, x_{i+1}, ..., x_n\right) \right. \\
&\qquad \left. - F_i'\left(x_1, ..., x_{i-1}, \frac{x_i}{3^m}, x_{i+1}, ..., x_n\right) \right\| \\
&\leq \lim_{m\longrightarrow\infty} |3|^m \max\left\{ \left\| F_i\left(x_1, ..., x_{i-1}, \frac{x_i}{3^m}, x_{i+1}, ..., x_n\right) \right.\right. \\
&\qquad \left. - f\left(x_1, ..., x_{i-1}, \frac{x_i}{3^m}, x_{i+1}, ..., x_n\right) \right\|, \\
&\qquad \left\| f\left(x_1, ..., x_{i-1}, \frac{x_i}{3^m}, x_{i+1}, ..., x_n\right) \right. \\
&\qquad \left.\left. - F_i'\left(x_1, ..., x_{i-1}, \frac{x_i}{3^m}, x_{i+1}, ..., x_n\right) \right\| \right\} \\
&\leq \lim_{m\longrightarrow\infty} |3|^m \frac{\alpha_i}{|3|^{p_i}} \left\| \frac{x_i}{3^m} \right\|^{p_i} \\
&= \lim_{m\longrightarrow\infty} (|3|^{1-p_i})^m \frac{\alpha_i}{|3|^{p_i}} \|x_i\|^{p_i} = 0,
\end{aligned}$$

which proves that $F_i = F_i'$. □

## References


1. Brzdęk, J.: The Cauchy and Jensen diferences on semigroups. Publ. Math. Debrecen **48**, 117–136 (1996)
2. Cadariu L., Radu V.: Fixed points and the stability of Jensen's functional equation. JIPAM. J. Inequal. Pure Appl. Math. **4**, Article 4 (2003)
3. Ciepliński, K.: On multi-Jensen functions and Jensen difference. Bull. Korean Math. Soc. **45**, 729–737 (2008)
4. Ciepliński, K.: Stability of the multi-Jensen equation. J. Math. Anal. Appl. **363**, 249–254 (2010)
5. Ciepliński, K.: Stability of multi-additive mappings in non-Archimedean normed spaces. J. Math. Anal. Appl. **373**, 376–383 (2011)
6. Forti, G.L.: Hyers-Ulam stability of functional equations in several variables. Aequationes Math. **50**, 143–190 (1995)
7. Hyers, D.H.: On the stability of the linear functional equation. Proc. Nat. Acad. Sci. U. S. A. **27**, 222–224 (1941)
8. Hyers, D.H., Isac, G., Rassias, Th.M.: Stability of functional equations in several variables. Birkhuser Boston, Inc., Boston, MA (1998)
9. Jun, K.-W., Lee, Y.-H.: A generalization of the Hyers-Ulam-Rassias stability of Jensen's equation. J. Math. Anal. Appl. **238**, 305–315 (1999)
10. Jung, S.-M.: Hyers-Ulam-Rassias stability of Jensen's equation and its application. Proc. Amer. Math. Soc. **126**, 3137–3143 (1998)
11. Jung, S.-M.: Hyers-Ulam-Rassias stability of functional equations in mathematical analysis. Hadronic Press, Inc., Palm Harbor, FL (2001)
12. Kannappan, Pl.: Functional equations and inequalities with applications. Springer, New York (2009)
13. Khrennikov, A.: Non-Archimedean analysis: quantum paradoxes, dynamical systems and biological models. Kluwer Academic Publishers, Dordrecht (1997)
14. Kominek, Z.: On a local stability of the Jensen functional equation. Demonstratio Math. **22**, 499–507 (1989)
15. Maligranda, L.: A result of Tosio Aoki about a generalization of Hyers-Ulam stability of additive functions – a question of priority. Aequationes Math. **75**, 289–296 (2008)
16. Mirmostafaee, A.K.: Stability of quartic mappings in non-Archimedean normed spaces. Kyungpook Math. J. **49**, 289–297 (2009)
17. Moslehian, M.S.: The Jensen functional equation in non-Archimedean normed spaces. J. Funct. Spaces Appl. **7**, 13–24 (2009)
18. Moslehian, M.S., Rassias, Th.M.: Stability of functional equations in non-Archimedean spaces. Appl. Anal. Discrete Math. **1**, 325–334 (2007)
19. Moslehian, M.S., Sadeghi, G.: Stability of two types of cubic functional equations in non-Archimedean spaces. Real Anal. Exchange **33**, 375–383 (2008)
20. Prager W., Schwaiger, J.: Multi-affine and multi-Jensen functions and their connection with generalized polynomials. Aequationes Math. **69**, 41–57 (2005)
21. Prager W., Schwaiger, J.: Stability of the multi-Jensen equation. Bull. Korean Math. Soc. **45**, 133–142 (2008)
22. Rassias, Th.M.: On the stability of functional equations and a problem of Ulam. Acta Appl. Math. **62**, 23–130 (2000)
23. Robert, A.M.: A course in *p*-adic analysis. Springer-Verlag, New York (2000)

# Chapter 7
# On Stability of the Equation of Homogeneous Functions on Topological Spaces

**Stefan Czerwik**

**Abstract** Let $K$ be a cone of a linear space $X$ and $Y$ a sequentially complete locally convex linear topological Hausdorff space. Let $f\colon K \to Y$ and $g\colon K \to Y$ satisfy

$$\alpha^{-1} f(\alpha x) - g(x) \in U, \quad \alpha \in A,\, x \in K,$$

where $U$ is a bounded subset of $Y$ and $A \subset [1,\infty)$. Under some additional assumptions we prove that there exists exactly one positively homogeneous function $F\colon K \to Y$ such that the differences $F - f$ and $F - g$ are bounded on $K$, i.e. the equation of homogeneous functions is stable in the Ulam–Hyers sense.

**Keywords** Functional equations • Stability • Homogeneous functions

**Mathematics Subject Classification (2000):** Primary 39B82

## 7.1 Introduction

The concept of generalized Ulam–Hyers stability, called today Hyers–Ulam–Rassias stability, originated from the Rassias' stability Theorem proved in his paper: On the stability of the linear mapping in Banach spaces, Proc. Amer. Math. Soc. 72 (1978), 297–300.

First results about the Hyers–Ulam–Rassias stability of the equation of homogeneous functions have been obtained by the author in [2]. In the paper

S. Czerwik (✉)
Department of Mathematics, Silesian University of Technology, Kaszubska 23, 44-100 Gliwice, Poland
e-mail: stefan.czerwik@polsl.pl

Th.M. Rassias and J. Brzdęk (eds.), *Functional Equations in Mathematical Analysis*, Springer Optimization and Its Applications 52, DOI 10.1007/978-1-4614-0055-4_7,

[14], the authors have proved the result about the superstability of the homogenity condition. Some further generalizations can be found in [5, 8, 9, 11, 13].

For more information concerning the stability problems of functional equations the reader is referred to [3, 4, 6, 7, 10, 12].

## 7.2 Auxiliary Results

In the paper, $X$ will denote a real linear space and $Y$ a sequentially complete locally convex linear topological Hausdorff space. We recall that a set $K \subset X$ is said to be a cone if $sK \subset K$ for all $s > 0$.

Throughout this paper, the symbols $\mathbb{R}$ and $\mathbb{N}$ denote the sets of all real and positive integers, respectively.

Given sets $A, B \subset Y$ and a number $s \in \mathbb{R}$ we define, as usual, the sets $A+B := \{x \in Y\colon\ x = a+b,\ a \in A,\ b \in B\}$, and $sA := \{x \in Y\colon\ x = sa,\ a \in A\}$.

We start with the following.

**Lemma 7.1.** *Let $K \subset X$ be a cone and $f, g\colon K \to Y$ be given functions. Suppose that there exists an $A \subset [1, \infty)$, $1 \in A$ and a subset $U \subset Y$ such that*

$$\alpha^{-1} f(\alpha x) - g(x) \in U, \quad \alpha \in A,\ x \in K. \tag{7.1}$$

*Then*

$$\alpha^{-1} f(\alpha x) - f(x) \in U - U \tag{7.2}$$

*and*

$$\alpha^{-1} g(\alpha x) - g(x) \in U - \alpha^{-1} U \tag{7.3}$$

*for all $\alpha \in A$ and $x \in K$.*

*Proof.* We have from (7.1) for $\alpha \in A$, $x \in K$

$$\alpha^{-1} f(\alpha x) - f(x) = [\alpha^{-1} f(\alpha x) - g(x)] - [f(x) - g(x)] \in U - U.$$

Similarly

$$\begin{aligned}\alpha^{-1} g(\alpha x) - g(x) &= [\alpha^{-1} f(\alpha x) - g(x)] \\ &\quad - \alpha^{-1}[f(\alpha x) - g(\alpha x)] \in U - \alpha^{-1} U.\end{aligned}$$

□

**Lemma 7.2.** *Let the assumptions of Lemma 7.1 be satisfied. Then the following conditions hold true for all $\alpha \in A$, $x \in K$ and $n \in \mathbb{N}$*

$$\alpha^{-n}f(\alpha^n x) - f(x) \in \sum_{s=0}^{n-1} \alpha^{-s}(U - U), \tag{7.4}$$

$$\alpha^{-n}g(\alpha^n x) - g(x) \in \sum_{s=0}^{n-1} \alpha^{-s}(U - \alpha^{-1}U). \tag{7.5}$$

*Proof.* For $n = 1$ conditions (7.4) and (7.5) follow from Lemma 7.1. Now we have for $n \in \mathbb{N}$

$$\begin{aligned}
\alpha^{-(n+1)}f(\alpha^{n+1}x) - f(x) &= \alpha^{-(n+1)}f(\alpha^{n+1}x) \\
&\quad -\alpha^{-n}f(\alpha^n x) + \alpha^{-n}f(\alpha^n x) - f(x) \\
&= \alpha^{-n}[\alpha^{-1}f(\alpha \cdot \alpha^n x) - f(\alpha^n x)] + \alpha^{-n}f(\alpha^n x) \\
&\quad -f(x) \in \alpha^{-n}(U - U) + \sum_{s=0}^{n-1} \alpha^{-s}(U - U) \\
&= \sum_{s=0}^{n} \alpha^{-s}(U - U)
\end{aligned}$$

and by induction principle we get (7.4). Similarly, one can prove the formula (7.5). □

In the sequel, we will prove some technical properties of some operations on set. By *conv* $U$ we will denote the convex hull of a set $U$ and by *cl* $U$ the sequential closure of $U$.

**Lemma 7.3.** *If* $A \subset Y$ *and* $0 \le \alpha \le \beta$, *then* $\alpha A \subset \beta\ conv[A \cup \{0\}]$.

*Proof.* The case $\beta = 0$ is trivial. Let us take a $\beta > 0$, then we have

$$\alpha A \ni \alpha x = \beta\left(\frac{\alpha}{\beta}x\right) = \beta\left[\frac{\alpha}{\beta}x + \left(1 - \frac{\alpha}{\beta}\right)0\right] \in \beta\ conv[A \cup \{0\}].$$

□

*Remark 7.1 ([1]).* For every neighbourhood $W$ of zero in $Y$ there exists a convex neighbourhood $U$ of zero such that $U + U \subset W$.

*Remark 7.2.* For any subsets $A, B \subset Y$ and numbers $\alpha, \beta \in \mathbb{R}$ we have

$$\alpha(A + B) = \alpha A + \alpha B, \quad (\alpha + \beta)A \subset \alpha A + \beta A,$$

$$A \subset B \Rightarrow \alpha A \subset \beta A.$$

If, moreover, $A$ is a convex set and $\alpha_s \geq 0$, $s = 1,2,\ldots,n$, then

$$\sum_{s=1}^{n} \alpha_s A = \left( \sum_{s=1}^{n} \alpha_s \right) A.$$

The simple proof this Remark is left to the reader.

A set $A \subset Y$ is said to be bounded iff for every neighbourhood $V$ of zero there exists a number $s > 0$ such that $sA \subset V$.

Now we will prove the following.

**Lemma 7.4.** *If $A, B \subset Y$ are bounded, then*

$$A \cup B,\ A + B,\ conv\, A$$

*are bounded sets, too.*

*Proof.* Take a neighbourhood $W$ of zero. Because $Y$ is a locally convex space, we may assume that $W$ is a convex set. Since the sets $A$ and $B$ are bounded, there exist positive numbers $\alpha, \beta$ such that

$$\alpha A \subset W \text{ and } \beta B \subset W.$$

We may consider only the case: $\alpha \leq \beta$, $\beta > 0$. In view of Lemma 7.2 we have

$$\begin{aligned} \alpha(A \cup B) &= \alpha A \cup \alpha B \\ &\subset W \cup \beta\ conv[A \cup \{0\}] \\ &\subset W \cup \beta\ conv\left(\frac{1}{\beta} W\right) \\ &\subset W \cup W = W, \end{aligned}$$

which proves the boundedness of the set $A \cup B$.

To prove that the set $A + B$ is bounded, assume according to the Remark 7.1 that $U$ is the convex neighbourhood of zero such that

$$U + U \subset W.$$

We have for some $0 \leq \alpha \leq \beta$, $\beta > 0$ the conditions

$$\alpha A \subset U$$

and

$$\beta B \subset U.$$

Hence, in view of Lemma 7.3 and the convexity of $U$, we get

$$\begin{aligned}\alpha(A+B) &= \alpha A+\alpha B\\ &\subset U+\beta\, conv[B\cup\{0\}]\\ &\subset U\cup\beta\, conv\left(\frac{1}{\beta}U\right)\\ &= U+U\subset W,\end{aligned}$$

which means that $A+B$ is bounded.

Now let $\alpha \geq 0$ be such that $\alpha A\subset W$. Therefore

$$\alpha\, conv\, A = conv(\alpha A)\subset conv\, W = W,$$

i.e. *conv* $A$ is bounded set. □

## 7.3 Main Results

We are ready to prove the following.

**Theorem 7.1.** *Let the assumptions of Lemma 7.1 be satisfied. Assume that U is bounded. Then for every $\alpha \in A$, the sequences*

$$\{\alpha^{-n}f(\alpha^n x)\},\quad \{\alpha^{-n}g(\alpha^n x)\}$$

*are uniformly convergent on K. The functions*

$$\begin{aligned}F_\alpha(x) &:= \lim_{n\to\infty}\alpha^{-n}f(\alpha^n x),\quad x\in K,\\ G_\alpha(x) &:= \lim_{n\to\infty}\alpha^{-n}g(\alpha^n x),\quad x\in K\end{aligned}\tag{7.6}$$

*are $\alpha$ homogeneous, i.e.*

$$\begin{aligned}F_\alpha(\alpha x) &= \alpha F_\alpha(x),\quad x\in K,\\ G_\alpha(\alpha x) &= \alpha G_\alpha(x),\quad x\in K.\end{aligned}\tag{7.7}$$

*Moreover, for $x\in K$ and $\alpha\in A\backslash\{1\}$*

$$F_\alpha(x)-f(x)\in\alpha(\alpha-1)^{-1}cl\ conv(U-U),\tag{7.8}$$

$$G_\alpha(x)-g(x)\in\alpha(\alpha-1)^{-1}cl\ conv(U-\alpha^{-1}U).\tag{7.9}$$

*Proof.* Take an arbitrarily fixed $1 < \alpha \in A$ (case $\alpha = 1$ is trivial) and consider the Hyers sequence

$$F_\alpha^n(x) := \alpha^{-n} f(\alpha^n x), \quad x \in K,\ n \in \mathbb{N}.$$

We will verify that $\{F_\alpha^n(x)\}$ is a Cauchy sequence for every $x \in K$. Clearly, in view of (7.2) and Lemma 7.3, we have for all $m, n \in \mathbb{N}$, $n > m$ and $x \in K$

$$\begin{aligned}
F_\alpha^n(x) - F_\alpha^m(x) &= \alpha^{-m}[\alpha^{-(n-m)} f(\alpha^n x) - f(\alpha^m x)] \\
&= \alpha^{-m} \sum_{s=0}^{n-m-1} \alpha^{-s}[\alpha^{-1} f(\alpha^{m+s+1}x) - f(\alpha^{m+s}x)] \\
&\in \alpha^{-m} \sum_{s=0}^{n-m-1} \alpha^{-s}(U - U) \\
&\subset \alpha^{-m} \sum_{s=0}^{n-m-1} \alpha^{-s} conv(U - U) \\
&\subset \alpha^{1-m}(\alpha - 1)^{-1} conv(U - U).
\end{aligned}$$

Hence, it follows that $\{F_\alpha^n(x)\}$ is a Cauchy sequence uniformly convergent on $K$ (note that $conv(U - U)$ is a bounded set). Therefore, the function $F_\alpha$ is correctly defined. Similarly, applying (7.3) one can verify the same statement for the function $G_\alpha$. We have also

$$\begin{aligned}
F_\alpha(\alpha x) &= \lim_{n\to\infty} \alpha^{-n} f(\alpha^{n+1}x) \\
&= \alpha \lim_{n\to\infty} \alpha^{-(n+1)} f(\alpha^{n+1}x) = \alpha F_\alpha(x),
\end{aligned}$$

i.e. $F_\alpha$ (as well as $G_\alpha$) is $\alpha$ homogeneous.

We may write according to (7.1), Lemmas 7.2 and 7.3 for $\alpha > 1$

$$\begin{aligned}
F_\alpha^n(x) - f(x) &= \sum_{s=0}^{n-1} [\alpha^{-(s+1)} f(\alpha^{s+1}x) - \alpha^{-s} f(\alpha^s x)] \\
&\in \sum_{s=0}^{n-1} \alpha^{-s}(U - U) \\
&\subset \left(\sum_{s=0}^{n-1} \alpha^{-s}\right) conv(U - U) \\
&\subset \alpha(\alpha - 1)^{-1} conv(U - U).
\end{aligned}$$

Consequently, letting $n \to \infty$, we get for $\alpha > 1$

$$F_\alpha(x) - f(x) \in \alpha(\alpha - 1)^{-1} cl\ conv(U - U).$$

The condition (7.9) can be derived quite similarly. □

The next theorem is the main result of the paper.

**Theorem 7.2.** *Let the assumptions of Lemma 7.1 be satisfied. If A is bounded and int $A \neq 0$, then there exists exactly one positively homogeneous function $F : K \to Y$ such that*

$$F(x) - f(x) \in s(s-1)^{-1} cl\ conv(U - U), \quad x \in K \tag{7.10}$$

*and*

$$F(x) - g(x) \in s(s-1)^{-1} cl\ conv[(U-U) \cup U \cup (-U)], \quad x \in K, \tag{7.11}$$

*where* $s := \sup A < \infty$. *If* $s = \infty$, *then*

$$F(x) - f(x) \in cl\ conv(U - U), \quad x \in K \tag{7.12}$$

*and*

$$F(x) - g(x) \in cl\ conv[(U-U) \cup U \cup (-U)], \quad x \in K. \tag{7.13}$$

*Moreover*

$$F(x) = \lim_{n\to\infty} \alpha^{-n} f(\alpha^n x) = \lim_{n\to\infty} \alpha^{-n} g(\alpha^n x), \quad x \in K \tag{7.14}$$

*for* $\alpha \in A \backslash \{1\}$ *and the convergence is uniform on K.*

*Proof.* Take an arbitrary fixed $1 < \alpha \in A$. From Theorem 7.1 it follows that there exist the functions $F_\alpha$ and $G_\alpha$ satisfying the conditions (7.8) and (7.9).

Now we will prove that $F_\alpha = G_\alpha$ for $\alpha \in A \backslash \{1\}$. Indeed, according to (7.1) for $x \in K$, $y = \alpha^n x$, we get

$$\begin{aligned} F_\alpha^n(x) - G_\alpha^n(x) &= \alpha^{-n} f(\alpha^n x) - \alpha^{-n} g(\alpha^n x) \\ &= \alpha^{-n}[1^{-1} f(1 \cdot y) - g(y)] \in \alpha^{-n} U, \end{aligned}$$

whence since $U$ is the bounded set, for all $x \in K$

$$F_\alpha(x) - G_\alpha(x) = \lim_{n\to\infty} [F_\alpha^n(x) - G_\alpha^n(x)] = 0,$$

i.e. $F_\alpha = G_\alpha$.

Take any $\beta \in A \backslash \{1\}$. We claim that

$$F_\beta = F_\alpha, \quad x \in K, \tag{7.15}$$

which means that $F_\beta$ does not depend on $\beta$. Making sure of (7.8), we have for all $x \in K$

$$\beta^{-n}F_\alpha(\beta^n x) - \beta^{-n}f(\beta^n x) \in \beta^{-n}\alpha(\alpha-1)^{-1}cl\ conv(U-U),$$

which implies

$$\lim_{n\to\infty} \beta^{-n}F_\alpha(\beta^n x) = F_\beta(x), \quad x \in K.$$

Consequently, applying (7.8) and the homogeneity of $F_\alpha$, we obtain

$$\begin{aligned} F_\alpha(x) - F_\beta(x) &= \lim_{n\to\infty} [\alpha^{-n}F_\beta(\alpha^n x) - \beta^{-n}F_\alpha(\beta^n x)] \\ &= \lim_{n\to\infty} (\alpha\beta)^{-n}[\beta^n F_\beta(\alpha^n x) - \alpha^n F_\alpha(\beta^n x)] \\ &= \lim_{n\to\infty} (\alpha\beta)^{-n}[F_\beta(\alpha^n \beta^n x) - F_\alpha(\alpha^n\beta^n x)] \\ &= \lim_{n\to\infty} (\alpha\beta)^{-n}([F_\beta(\alpha^n\beta^n x) - f(\alpha^n\beta^n x)] \\ &\quad +[f(\alpha^n\beta^n x) - F_\alpha(\alpha^n\beta^n x)]) = 0. \end{aligned}$$

Putting $F := F_\alpha$, $\alpha \in A$, we have also

$$F(\alpha x) = F_\alpha(\alpha x) = \alpha F_\alpha(x) = \alpha F(x)$$

for all $x \in K$, $\alpha \in A$, i.e. $F$ is a homogeneous function on $A$.

To prove the homogenity for all positive real numbers, one can repeat the argument applied in the paper [6].

In fact, replacing $x$ by $\lambda^{-1}x$ in the condition

$$F(\lambda x) = \lambda F(x), \quad \lambda \in A,\ x \in K$$

one gets

$$F(\lambda^{-1}x) = \lambda^{-1}F(x), \quad \lambda \in A,\ x \in K.$$

Hence, by the induction

$$F\left(\alpha_1 \cdot \ldots \cdot \alpha_n \cdot \beta_1^{-1} \cdot \ldots \cdot \beta_m^{-1} x\right) = \alpha_1 \cdot \ldots \cdot \alpha_n \cdot \beta_1^{-1} \cdot \ldots \cdot \beta_m^{-1} F(x) \qquad (7.16)$$

for $\alpha_1, \ldots, \alpha_n, \beta_1^{-1}, \ldots, \beta_m^{-1} \in A$, $m, n \in \mathbb{N}$ and $x \in K$.

Because

$$\{\alpha_1 \cdot \ldots \cdot \alpha_n \cdot \beta_1^{-1} \cdot \ldots \cdot \beta_m^{-1} \colon \alpha_1, \ldots, \alpha_n, \beta_1^{-1}, \ldots, \beta_m^{-1} \in A,\ m,n \in \mathbb{N}\} = (0,\infty),$$

thus (7.16) proves that $F$ is a positively homogeneous function.

Taking into account the conditions (7.8), (7.9) and the definition of $F$, we obtain for $\alpha \in A \backslash \{1\}$ and $x \in K$

$$F(x) - f(x) \in \alpha(\alpha-1)^{-1}cl\ conv(U-U), \qquad (7.17)$$

$$F(x) - g(x) \in \alpha(\alpha-1)^{-1}cl\ conv(U-\alpha^{-1}U). \qquad (7.18)$$

Note that

$$\inf_{\alpha\in A\setminus\{1\}} \alpha(\alpha-1)^{-1} = \begin{cases} s(s-1)^{-1}, & s = \sup A < \infty, \\ 1, & s = \infty, \end{cases}$$

whence, on base of (7.17), we obtain (7.10) or (7.12), respectively.

To verify the conditions (7.11) and (7.13), we apply (7.5), Lemma 7.3 and Remark 7.2 to get for $\alpha \in A$

$$\begin{aligned}
\alpha^{-n}g(\alpha^n x) - g(x) &\in \sum_{s=0}^{n-1} \alpha^{-s}(U - \alpha^{-1}U) \\
&\subset \left(\sum_{s=0}^{n-1} \alpha^{-s}\right) [conv(U \cup \{0\}) - conv(U \cup \{0\})] \\
&\subset \alpha(\alpha-1)^{-1} conv[(U-U) \cup U \cup (-U)].
\end{aligned}$$

Thus, letting $n \to \infty$, we obtain

$$F(x) - g(x) \in \alpha(\alpha-1)^{-1} cl\ conv[(U-U) \cup U \cup (-U)],$$

hence, since the left hand side does not depend on $\alpha$

$$F(x) - g(x) \in s(s-1)^{-1} cl\ conv[(U-U) \cup U \cup (-U)]$$

for $\sup A = s < \infty$, or

$$F(x) - g(x) \in cl\ conv[(U-U) \cup U \cup (-U)]$$

for $\sup A = s = \infty$.

To prove the uniquennes part of the theorem, assume for the contrary that there exist two positively homogeneous functions $F_1, F_2 \colon K \to Y$ satisfying the condition

$$F_k(x) - f(x) \in \rho_k cl\ conv(U-U), \quad x \in K,$$

where $\rho$ and $k = 1, 2$ are constans. Then for $x \in K$, $n \in \mathbb{N}$ and $\alpha > 1$

$$\begin{aligned}
F_1(x) - F_2(x) &= \alpha^{-n}F_1(\alpha^n x) - \alpha^{-n}F_2(\alpha^n x) \\
&= \alpha^{-n}[F_1(\alpha^n x) - f(\alpha^n x)] - \alpha^{-n}[F_2(\alpha^n x) - f(\alpha^n x)] \\
&\in \alpha^{-n}\rho_1 cl\ conv(U-U) - \alpha^{-n}\rho_2 cl\ conv(U-U).
\end{aligned}$$

Consequently, by letting $n \to \infty$ we get $F_1(x) - F_2(x) = 0$ for $x \in K$. □

## References


1. Berge, C.: Topological Spaces, Including a Treatment of Multivalued Functions, Vector Spaces and Convexity. Oliver and Boyd, Edinburg–London (1963)
2. Czerwik, S.: On the stability of the homogeneous mapping. Math. Rep. Acad. Sci. Canada XIV **6**, 268–272 (1992)
3. Czerwik, S.: Functional Equations and Inequalities in Several Variables. World Scientific Publishing Company, New Jersey–London–Singapore–Hong Kong (2002)
4. Czerwik, S. (ed.): Stability of Functional Equations of Ulam-Hyers-Rassias Type. Hadronic Press, Inc., Palm Harbor (2003)
5. Czerwik, S.: Some results on the stability of the homogeneous functions. Bull. Korean Math. Soc. **42**, 29–37 (2005)
6. Hyers, D.H., Rassias, Th.M.: Approximate homomorphism. Aequationes Math. **44**, 125–153 (1992)
7. Hyers, D.H., Rassias, Th.M.: Stability of Functional Equations in Several Variables. Birkhauser, Boston–Basel–Berlin (1998)
8. Jabłoński, W.: On the stability of the homogeneous equation. Publ. Math. Debrecen **55**, 33–45 (1999)
9. Jung, S.-M.: Superstability of homogeneous functional equation. Kyungpook Math. J. **38**, 251–257 (1998)
10. Jung, S.-M.: Hyers–Ulam–Rassias Stability of Functional Equations in Mathematical Analisys. Hadronic Press, Inc., Palm Harbor (2003)
11. Kominek, Z., Matkowski, J.: On stability of the homogeneity condition. Result. Math. **27**, 373–380 (1995)
12. Rassias, Th.M.: On the stability of functional equations and a problem of Ulam. Acta Appl. Math. **62**, 23–130 (2000)
13. Schwaiger, J.: The Functional Equation of Homogeneity and its Stability Properties. Österreich. Akad. Wiss. Math.-Natur. Kl. Sitzungsber. II **205**, 3–12 (1996)
14. Tabor, J., Tabor, J.: Homogeneity is superstable. Publ. Math. Debrecen **45**, 123–130 (1994)

# Chapter 8
# Hyers–Ulam Stability of the Quadratic Functional Equation

**Elhoucien Elqorachi, Youssef Manar, and Themistocles M. Rassias**

*Dedicated to the memory of S.M. Ulam*
*on the 100th anniversary of his birth*

**Abstract** We prove a stability theorem for the quadratic functional equation

$$f(x+y)+f(x+\sigma(y))=2f(x)+2f(y), \quad x,y\in G,$$

where $G$ is an abelian group and $\sigma$ is an involution of $G$. We also prove that for functions $f$ from $G$ to an inner product space $E$, the inequality

$$\|2f(x)+2f(y)-f(x+\sigma(y))\| \leq \|f(x+y)\|, \quad x,y\in G.$$

implies that $f$ is a solution to the equation.

**Keywords** Hyers–Ulam stability • Quadratic functional equation • Group homomorphisms • Unbounded Cauchy difference • Abelian group

**Mathematics Subject Classification (2000):** Primary 39B52, 39B82

## 8.1 Introduction

The stability problem of functional equations originated from a question of Ulam [27] concerning the stability of group homomorphisms:

E. Elqorachi (✉) • Y. Manar
Faculty of Sciences, Department of Mathematics, University Ibn Zohr, Agadir, Morocco
e-mail: elqorachi@hotmail.com; manaryoussefr32@hotmail.fr

Th.M. Rassias
Department of Mathematics, National Technical University of Athens, Zografou Campus, 15780 Athens, Greece
e-mail: trassias@math.ntua.gr

Th.M. Rassias and J. Brzdęk (eds.), *Functional Equations in Mathematical Analysis*, Springer Optimization and Its Applications 52, DOI 10.1007/978-1-4614-0055-4_8,

*Given a group $G_1$ and a metric group $G_2$ with a metric $d(\cdot,\cdot)$ and $\varepsilon > 0$, does there exist a $\delta > 0$ such that if $f : G_1 \to G_2$ satisfies*

$$d(f(xy), f(x)f(y)) < \delta$$

*for all $x, y \in G_1$, then a homomorphism $g : G_1 \to G_2$ exists with*

$$d(f(x), g(x)) < \varepsilon$$

*for all $x \in G_1$?*

Hyers [8] considered the case of approximately additive mappings $f : E_1 \to E_2$, where $E_1$ and $E_2$ are Banach spaces and $f$ satisfies the inequality

$$\|f(x+y) - f(x) - f(y)\| \leq \varepsilon,$$

for all $x, y \in E_1$. He proved that the limit

$$a(x) = \lim_{n \to +\infty} \frac{f(2^n x)}{2^n}$$

exists for all $x \in E_1$ and that $a : E_1 \to E_2$ is the unique additive mapping satisfying

$$\|f(x) - a(x)\| \leq \varepsilon.$$

Aoki [1] and Rassias [16] provided a generalization of Hyers' theorem for additive and linear mappings, respectively, by allowing *the Cauchy difference to be unbounded.*

**Theorem 8.1 ([16]).** *Let $f : E_1 \longrightarrow E_2$ be a mapping from a normed vector space $E_1$ into a Banach spaces $E_2$ subject to the inequality*

$$\|f(x+y) - f(x) - f(y)\| \leq \varepsilon(\|x\|^p + \|y\|^p), \tag{8.1}$$

*for all $x, y \in E_1$, where $\varepsilon$ and $p$ are constants with $\varepsilon > 0$ and $p < 1$. Then the limit*

$$a(x) = \lim_{n \to +\infty} \frac{f(2^n x)}{2^n}$$

*exists for all $x \in E_1$ and $a : E_1 \longrightarrow E_2$ is the unique additive mapping which satisfies*

$$\|f(x) - a(x)\| \leq \frac{2\varepsilon}{2 - 2^p}\|x\|^p \tag{8.2}$$

*for all $x \in E_1$. If $p < 0$, then inequality (8.1) holds for $x, y \neq 0$ and (8.2) for $x \neq 0$.*

The paper of Rassias [16] has provided a lot of influence in the development of what we call generalized Hyers–Ulam stability of functional equations. A generalization of the Rassias theorem was obtained by Gavruta [5] by replacing *the unbounded Cauchy difference* by a general *control function* in the spirit of Rassias approach.

The generalized Hyers–Ulam stability problem for the quadratic functional equation

$$f(x+y)+f(x+\sigma(y))=2f(x)+2f(y), \quad x,y\in G, \tag{8.3}$$

where $\sigma$ is an involution of the abelian group $G$ was proved by Bouikhalene et al. [2]. Skof [26], Cholewa [3] and Czerwik [4] proved the generalized Hyers–Ulam stability of the quadratic functional equation (8.3) with $\sigma = -I$. The stability problems of several other functional equations have been extensively investigated by numerous authors (see for example [6, 9–13, 17–25] and the references therein).

In [14], Maksa and Volkmann proved that for functions $f: G \longrightarrow E$ from a group $G$ to inner product space $E$, the inequality

$$\|f(xy)\| \geq \|f(x)+f(y)\|, \quad x,y\in G,$$

implies

$$f(xy)=f(x)+f(y), \quad x,y\in G.$$

In [7], A. Gilanyi showed that if $G$ is a 2-divisible abelian group, then the functional inequality

$$\|2f(x)+2f(y)-f(x-y)\| \leq \|f(x+y)\|, \quad x,y\in G,$$

implies

$$f(x+y)+f(x-y)=2f(x)+2f(y), \quad x,y\in G.$$

The commutativity of $G$ may be replaced by the following Kannappan condition: $f(xyz)=f(xzy)$ for $x,y,z\in G$ (see [7]).

In [15], Rätz did not assume the condition of 2-divisibility upon the group G and studied variants of Gilanyi result.

In the present paper, the following results are proved: in Sect. 8.2, we provide a short proof of the main result in [2]; in Sect. 8.3, we give a proof of the fact that if $f: G\to E$ from a group $G$ to the inner product space $E$ satisfies the inequality

$$\|2f(x)+2f(y)-f(x+\sigma(y))\| \leq \|f(x+y)\|, \quad x,y\in G,$$

then $f$ satisfies the quadratic functional equation (8.3).

## 8.2 Hyers–Ulam Stability of (8.3)

In this section, we will provide a simple new proof of a stability theorem for (8.3) that was obtained in [2].

**Theorem 8.2.** *Let $G$ be an abelian group, $E$ be a Banach space and $f : G \to E$ a mapping which satisfies the inequality*

$$\|f(x+y)+f(x+\sigma(y))-2f(x)-2f(y)\| \leq \delta, \tag{8.4}$$

*for some $\delta > 0$ and for all $x, y \in G$. Then there exists a unique mapping $Q : G \to E$ satisfying (8.3) such that*

$$\|f(x)-Q(x)\| \leq \frac{3}{4}\delta, \quad x \in G.$$

*Proof.* For every element $x$ from the set

$$G_+ := \{z \in G \mid \sigma(z) = z\}$$

and $y \in G$, we get from (8.4) the inequality

$$\begin{aligned}\|f(x+y) &- f(x)-f(y)\| \\ &= \frac{1}{2}\|f(x+y)+f(y+\sigma(x))-2f(x)-2f(y)\| \\ &\leq \frac{\delta}{2}.\end{aligned}$$

Since $G_+$ is an abelian subgroup of $G$, then by applying [8], there exists a unique additive mapping $a : G_+ \longrightarrow E$ such that

$$\|f(x)-a(x)\| \leq \frac{\delta}{2}, \quad x \in G_+.$$

Now, for every $x$ from the set

$$G_- := \{z \in G \mid \sigma(z) = -z\}$$

and $y \in G$, we get from (8.4) that the following holds

$$\begin{aligned}\|f(x+y) &+ f(y+\sigma(x))-2f(y)-2f(x)\| \\ &= \|f(x+y)+f(y-x)-2f(y)-2f(x)\| \leq \delta.\end{aligned}$$

Since $G_-$ is a commutative subgroup of $G$, then from [3] there exists one mapping $q : G_- \longrightarrow E$ that satisfies the equation

$$q(x+y)+q(x-y)=2q(x)+2q(y), \quad x,y \in G_-,$$

and the inequality

$$\|f(x)-q(x)\| \le \frac{\delta}{2}, \quad x \in G_-. \tag{8.5}$$

On the other hand, by setting $y=x$ in (8.4) we get

$$\|4f(x)-f(x+\sigma(x))-f(2x)\| \le \delta. \tag{8.6}$$

By replacing $x$ by $x-\sigma(x)$ and $y$ by $x+\sigma(x)$ in (8.4), we obtain

$$\|f(2x)-f(x+\sigma(x))-f(x-\sigma(x))\| \le \frac{\delta}{2}. \tag{8.7}$$

By adding inequalities (8.6) and (8.7) by parts we obtain

$$\|4f(x)-2f(x+\sigma(x))-f(x-\sigma(x))\| \le \frac{3\delta}{2},$$

and consequently

$$\begin{aligned}
\|f(x) &- \frac{1}{2}a(x+\sigma(x)) - \frac{1}{4}q(x-\sigma(x))\| \\
&\le \|f(x)-\frac{1}{2}f(x+\sigma(x))-\frac{1}{4}f(x-\sigma(x))\| \\
&\quad +\|\frac{1}{2}f(x+\sigma(x))-\frac{1}{2}a(x+\sigma(x))\| \\
&\quad +\|\frac{1}{4}f(x-\sigma(x))-\frac{1}{4}q(x-\sigma(x))\| \\
&\le \frac{3\delta}{8}+\frac{\delta}{4}+\frac{\delta}{8}=\frac{3\delta}{4}.
\end{aligned}$$

Letting

$$Q(x)=\frac{1}{2}a(x+\sigma(x))+\frac{1}{4}q(x-\sigma(x)), \quad x \in G,$$

a simple computation shows that $Q$ is a solution of (8.3). For the uniqueness of $Q$, we apply the same argument as that in the proof used in [2]. □

## 8.3 Characterization of Quadratic Equation Having Values in an Inner Product Space

**Theorem 8.3.** *Let $G$ be an abelian group, $(E,\langle\cdot,\cdot\rangle)$ be an inner product space and the function $f: G \to E$ satisfies the functional inequality*

$$\|2f(x)+2f(y)-f(x+\sigma(y))\| \leq \|f(x+y)\|, \quad x,y \in G. \tag{8.8}$$

*Then $f$ satisfies the quadratic functional equation (8.3).*

*Proof.* For every element $y$ of the set

$$G_+ = \{z \in G \mid \sigma(z) = z\}$$

and $x \in G$, the inequality (8.8) implies that

$$\|2f(x)+2f(y)-f(y+x)\| \leq \|f(x+y)\|.$$

Then by using the triangle inequality, we obtain

$$\|f(x)+f(y)\| \leq \|f(x+y)\|,$$

for all $x,y \in G_+$, so in view of [14] we deduce that $f$: $G_+ \longrightarrow E$ satisfies the additive functional equation

$$f(x)+f(y) = f(x+y), \quad x,y \in G_+.$$

On the other hand, by (8.8) we get for any element $x$ of the set

$$G_- = \{z \in G \mid \sigma(z) = -z\}$$

and any element $y$ of $G$ the following holds

$$\|2f(x)+2f(y)-f(y-x)\| \leq \|f(x+y)\|.$$

Consequently, from [7, 15] we deduce that $f$: $G_- \longrightarrow E$ satisfies the quadratic functional equation:

$$f(x+y)+f(x-y) = 2f(x)+2f(y), \quad x,y \in G_-.$$

Finally, we conclude that the new functions $a$, $q: G \longrightarrow E$, defined by

$$x \longmapsto a(x) = f(x+\sigma(x)),$$
$$x \longmapsto q(x) = f(x-\sigma(x)),$$

satisfy the functional equations

$$a(x+y) = a(x) + a(y), \quad x,y \in G,$$

and

$$q(x+y) + q(x-y) = 2q(x) + 2q(y), \quad x,y \in G.$$

By setting $y = -x$ in (8.8), we get

$$2f(x) + 2f(-x) = f(x - \sigma(x)) = q(x).$$

By substituting in (8.8) $y$ by $y + \sigma(y)$, we get

$$\|f(x) + f(y+\sigma(y))\| \le \|f(x+y+\sigma(y))\|, \quad x,y \in G. \tag{8.9}$$

Replacing $x$ by $2x$ and $y$ by $-x$ in (8.9) we get

$$\begin{aligned}\|f(2x) - a(x)\|^2 &= \|f(2x)\|^2 + \|a(x)\|^2 \\ &\quad -2\mathrm{Re}\langle f(2x), a(x)\rangle \\ &\le \|q(x)\|^2.\end{aligned}$$

This inequality by replacing $x$ with $x/2$ implies

$$\|f(x)\|^2 + \frac{1}{4}\|a(x)\|^2 - \mathrm{Re}\langle f(x), a(x)\rangle \le \frac{1}{4}\|f(x) + f(-x)\|^2.$$

Using the parallelogram equation, we get

$$\begin{aligned}\|f(x)\|^2 &+ \frac{1}{4}\|a(x)\|^2 - \mathrm{Re}\langle f(x), a(x)\rangle \\ &\le \frac{1}{2}\|f(x)\|^2 + \frac{1}{2}\|f(-x)\|^2 \\ &\quad -\frac{1}{4}\|f(x) - f(-x)\|^2.\end{aligned} \tag{8.10}$$

Replacing $x$ by $-x$ in (8.10), we obtain

$$\begin{aligned}\|f(-x)\|^2 &+ \frac{1}{4}\|a(x)\|^2 + \mathrm{Re}\langle f(-x), a(x)\rangle \\ &\le \frac{1}{2}\|f(x)\|^2 + \frac{1}{2}\|f(-x)\|^2 \\ &\quad -\frac{1}{4}\|f(x) - f(-x)\|^2.\end{aligned} \tag{8.11}$$

By adding the result of (8.10) to the result (8.11) we get

$$\begin{aligned}\|f(x) - f(-x) - a(x)\|^2 &= \|a(x)\|^2 + \|f(x) - f(-x)\|^2 \\ &\quad -2\mathrm{Re}\langle f(x) - f(-x), a(x)\rangle \leq 0.\end{aligned}$$

Finally, we have

$$f(x) - f(-x) = a(x), \quad x \in G,$$

and

$$2f(x) + 2f(-x) = q(x), \quad x \in G.$$

So

$$f(x) = \frac{1}{2}a(x) + \frac{1}{4}q(x), \quad x \in G.$$

By simple computation we prove that $f$ is a solution of (8.3). □

*Remark 8.1.* The commutativity of $G$ used in Theorem 8.3 may be replaced by the Kannappan condition.

**Acknowledgements** Special thanks to Professor Janusz Brzdęk for reading a preliminary version of the paper.

## References

1. Aoki, T.: On the stability of the linear transformation in Banach spaces. J. Math. Soc. Japan **2**, 64–66 (1950)
2. Bouikhalene, B., Elqorachi, E., Rassias, Th.M.: On the generalized Hyers-Ulam stability of the quadratic functional equation with a general involution. Nonlinear Funct. Anal. Appl. **12**, 247–262 (2007)
3. Cholewa, P.W.: Remarks on the stability of functional equations. Aequationes Math. **27**, 76–86 (1984)
4. Czerwik, S.: On the stability of the quadratic mapping in normed spaces. Abh. Math. Sem. Univ. Hamburg. **62**, 59–64 (1992)
5. Găvruta, P.: A generalization of the Hyers-Ulam-Rassias stability of approximately additive mappings. J. Math. Anal. Appl. **184**, 431–436 (1994)
6. Gajda, Z.: On stability of additive mappings. Internat. J. Math. Math. Sci. **14**, 431–434 (1991)
7. Gilanyi, A.: Eine zur Parallelogrammgleichung aquivalente ungleichung. Aequationes Math. **62**, 303–309 (2001)
8. Hyers, D.H.: On the stability of the linear functional equation. Proc. Nat. Acad. Sci. U. S. A. **27**, 222–224 (1941)
9. Hyers, D.H., Rassias, Th.M.: Approximate homomorphisms. Aequationes Math. **44**, 125–153 (1992)
10. Hyers, D.H., Isac, G. Rassias, Th.M.: Stability of Functional Equations in Several Variables. Birkhäuser, Basel (1998)

11. Jung, S.-M.: Stability of the quadratic equation of Pexider type. Abh. Math. Sem. Univ. Hamburg **70**, 175–190 (2000)
12. Jung, S.-M., Sahoo, P.K.: Hyers–Ulam–Rassias stability of the quadratic equation of Pexider type. J. Korean Math. Soc. **38**, 645–656 (2001)
13. Lee, J.-R., An, J.-S., Park, C.: On the stability of quadratic functional equations. Abstr. Appl. Anal., doi: 10.1155/2008/628178
14. Maksa, Gy., Volkmann, P.: Characterization of group homomorphisms having values in an inner product space. Publ. Math. (Debrecen) **56**, 197–200 (2000)
15. Rätz, J.: On inequality associated with the Jordan-von Neumann functional equation. Aequationes Math. **54**, 191–200 (2003)
16. Rassias, Th.M.: On the stability of linear mapping in Banach spaces. Proc. Amer. Math. Soc. **72**, 297–300 (1978)
17. Rassias, Th.M.: On a modified Hyers-Ulam sequence. J. Math. Anal. Appl. **158**, 106–113 (1991)
18. Rassias, Th.M.: Functional Equations and Inequalities. Kluwer Academic Publishers, Dordrecht–Boston–London (2001)
19. Rassias, Th.M.: Functional Equations, Inequalities and Applications. Kluwer Academic Publishers, Dordrecht–Boston–London (2003)
20. Rassias, Th.M.: The problem of S.M. Ulam for approximately multiplicative mappings. J. Math. Anal. Appl. **246**, 352–378 (2000)
21. Rassias, Th.M.: On the stability of minimum points. Mathematica **45**, 93–104 (2003)
22. Rassias, Th.M.: On the stability of the functional equations and a problem of Ulam. Acta Appl. Math. **62**, 23–130 (2000)
23. Rassias, Th.M., Šemrl, P.: On the behavior of mappings which do not satisfy Hyers-Ulam stability. Proc. Amer. Math. Soc. **114**, 989–993 (1992)
24. Rassias, Th.M., Šemrl, P.: On the Hyers-Ulam stability of linear mappings. J. Math. Anal. Appl. **173**, 325–338 (1993)
25. Rassias, Th.M., Tabor, J. (eds.): Stability of Mappings of Hyers-Ulam Type. Hadronic Press, Inc., Palm Harbor (1994)
26. Skof, F.: Approssimazione di funzioni $\delta$-quadratic su dominio restretto. Atti. Accad. Sci. Torino Cl. Sci. Fis. Mat. Natur. **118**, 58–70 (1984)
27. Ulam, S.M.: A Collection of Mathematical Problems. Interscience Publ., New York (1961) (reprinted as: Problems in Modern Mathematics. Wiley, New York (1964))

# Chapter 9
# Intuitionistic Fuzzy Approximately Additive Mappings

**M. Eshaghi-Gordji, H. Khodaei, H. Baghani, and M. Ramezani**

*Dedicated to the memory of S.M. Ulam*
*on the 100th anniversary of his birth*

**Abstract** In this paper, we investigate the generalized Hyers–Ulam stability and the intuitionistic fuzzy continuity of the generalized additive functional equation

$$2^{n-1}\, a_1 f(x_1) = f\left(\sum_{i=1}^{n} a_i x_i\right)$$
$$+\sum_{k=2}^{n}\sum_{i_1=2}^{k}\sum_{i_2=i_1+1}^{k+1}\cdots\sum_{i_{n-k+1}=i_{n-k}+1}^{n} f\left(\sum_{\substack{i=1\\ i\neq i_1,\ldots,i_{n-k+1}}}^{n} a_i x_i - \sum_{r=1}^{n-k+1} a_{i_r} x_{i_r}\right)$$

in intuitionistic fuzzy Banach spaces, where $n \in \mathbb{N}\backslash\{1\}$ and $a_1,\ldots,a_n \in \mathbb{Z}\backslash\{0\}$ with $a_1 \neq \pm 1$.

**Keywords** Intuitionistic fuzzy normed space • Additive functional equation • Intuitionistic fuzzy stability • Intuitionistic fuzzy continuity

**Mathematics Subject Classification (2000):** Primary 39B82, 39B52

## 9.1 Introduction

In recent years, the fuzzy theory has emerged as the most active area of research in many branches of mathematics and engineering. This new theory was introduced by Zadeh [44], in 1965, and since then a large number of research papers have appeared that used the concept of fuzzy set/numbers and fuzzification of many

M. Eshaghi-Gordji (✉) • H. Khodaei • H. Baghani • M. Ramezani
Department of Mathematics, Semnan University, P.O. Box 35195-363, Semnan, Iran
e-mail: madjid.eshaghi@gmail.com; khodaei.hamid.math@gmail.com; h.baghani@gmail.com; ramezanimaryam873@gmail.com

Th.M. Rassias and J. Brzdęk (eds.), *Functional Equations in Mathematical Analysis*, Springer Optimization and Its Applications 52, DOI 10.1007/978-1-4614-0055-4_9,

classical theories has also been made. The notion of fuzziness has numerous very useful application in various fields; e.g., population dynamics [4], chaos control [10], computer programming [12], nonlinear dynamical systems [14], fuzzy physics [22], fuzzy topology [40], fuzzy stability [23–26, 33], nonlinear operators [28], statistical convergence [29, 30], etc.

The concept of intuitionistic fuzzy normed spaces has initially been introduced by Saadati and Park [38]. In [39], by modifying the separation condition and strengthening some conditions in the definition of Saadati and Park, Saadati et al. have obtained a modified case of intuitionistic fuzzy normed spaces. Many authors have considered the intuitionistic fuzzy normed linear spaces and intuitionistic fuzzy 2-normed space (see [2, 3, 9, 32]).

The stability problem of functional equations originated from a question of Ulam [42] in 1940, concerning the stability of group homomorphisms. Let $(G_1,.)$ be a group and let $(G_2,*)$ be a metric group with the metric $d(.,.)$. Given $\varepsilon > 0$, does there exist a $\delta > 0$ such that, if a mapping $h : G_1 \to G_2$ satisfies the inequality $d(h(x.y), h(x) * h(y)) < \delta$ for all $x, y \in G_1$, then there exists a homomorphism $H : G_1 \to G_2$ with $d(h(x), H(x)) < \varepsilon$ for all $x \in G_1$? In the other words, under what condition does there exists a homomorphism near an approximate homomorphism? The concept of stability for functional equation arises when we replace the functional equation by an inequality which acts as a perturbation of the equation. In 1941, Hyers [15] gave a first partial affirmative answer to the question of Ulam for Banach spaces. He proved that, for each mapping $f : X \to Y$ between Banach spaces such that

$$\|f(x+y) - f(x) - f(y)\| \leq \delta$$

for all $x, y \in X$ and some $\delta > 0$, there exists a unique additive mapping $A : X \to Y$ such that

$$\|f(x) - A(x)\| \leq \delta$$

for all $x \in X$. Aoki [1] and Rassias [34] provided a generalization of the Hyers theorem for additive and linear functions, respectively, by allowing the Cauchy difference to be unbounded.

**Theorem 9.1.** (*Rassias*). *Let $f : X \to Y$ be a function from a normed vector space $X$ into a Banach space $Y$ subject to the inequality*

$$\|f(x+y) - f(x) - f(y)\| \leq \varepsilon(\|x\|^p + \|y\|^p) \tag{9.1}$$

*for all $x, y \in X$, where $\varepsilon$ and $p$ are real constants with $\varepsilon > 0$ and $p < 1$. Then there exists a unique additive function $A : X \to Y$ satisfying*

$$\|f(x) - A(x)\| \leq \varepsilon\|x\|^p/(1 - 2^{p-1}) \tag{9.2}$$

*for all $x \in X$. If $p < 0$, then inequality (9.1) holds for $x, y \neq 0$ and (9.2) for $x \neq 0$. Also, if for each fixed $x \in X$ the function $t \mapsto f(tx)$ is continuous in $t \in \mathbb{R}$, then $A$ is linear.*

In 1991, Gajda [11] answered the question raised by Th.M. Rassias and concerning the case $p > 1$.

This new concept is known as generalized Hyers–Ulam stability or Hyers–Ulam–Rassias stability of functional equations. During the last decades several stability problems of functional equations have been investigated in the spirt of Hyers–Ulam–Rassias (see [5–8], [13], [16–20] and [35, 36] for more detailed information on stability of functional equations).

The functional equation

$$2^{n-1}a_1 f(x_1) = f\left(\sum_{i=1}^{n} a_i x_i\right) + \sum_{k=2}^{n}\sum_{i_1=2}^{k}\sum_{i_2=i_1+1}^{k+1}\cdots\sum_{i_{n-k+1}=i_{n-k}+1}^{n} f\left(\sum_{\substack{i=1\\ i\neq i_1,\ldots,i_{n-k+1}}}^{n} a_i x_i - \sum_{r=1}^{n-k+1} a_{i_r} x_{i_r}\right) \tag{9.3}$$

where $n \in \mathbb{N}\backslash\{1\}$ and $a_1,\ldots,a_n \in \mathbb{Z}\backslash\{0\}$ with $a_1 \neq \pm 1$, is called the generalized additive functional equation with $n$-variables, since the function $f(x) = ax$ is its solution. The stability problem for the generalized additive functional equation with $n$-variables was proved by Khodaei and Rassias [21]. As a special case, if $n = 2$ in (9.3), then the functional equation (9.3) reduces to

$$f(a_1x_1 - a_2x_2) + f(a_1x_1 + a_2x_2) = 2a_1 f(x_1);$$

also by putting $n = 3$ in (9.3), we obtain

$$\begin{aligned} 2^2 a_1 f(x_1) = & \ a_2x_2 + a_3x_3 + f(a_1x_1 - a_2x_2 - a_3x_3) \\ & + f(a_1x_1 - a_2x_2 + a_3x_3) + f(a_1x_1 + a_2x_2 - a_3x_3) + f(a_1x_1). \end{aligned}$$

Mursaleen and Mohiuddine investigated the intuitionistic fuzzy stability problems for the Jensen functional equation and the cubic functional equation in intuitionistic fuzzy Banach spaces. In this paper, we prove the generalized Hyers–Ulam stability and the intuitionistic fuzzy continuity of the functional equation (9.3) in intuitionistic fuzzy Banach spaces.

## 9.2 Preliminaries

We use the definition of intuitionistic fuzzy normed spaces given in [27, 31, 38] to investigate some stability results for the functional equation (9.3) in the intuitionistic fuzzy normed vector space setting.

**Definition 9.1.** ([41]) A binary operation $* : [0,1] \times [0,1] \to [0,1]$ is said to be a continuous $t$-norm if it satisfies the following conditions:

(a) $*$ is commutative and associative;
(b) $*$ is continuous;

(c) $a * 1 = a$ for all $a \in [0,1]$;
(d) $a * b \leq c * d$ whenever $a \leq c$ and $b \leq d$ for all $a,b,c,d \in [0,1]$.

**Definition 9.2 ([41]).** A binary operation $\star : [0,1] \times [0,1] \to [0,1]$ is said to be a continuous $t$-conorm if it satisfies the following conditions:

(a) $\star$ is commutative and associative;
(b) $\star$ is continuous;
(c) $a \star 0 = a$ for all $a \in [0,1]$;
(d) $a \star b \leq c \star d$ whenever $a \leq c$ and $b \leq d$ for all $a,b,c,d \in [0,1]$.

Using the continuous $t$-norm and $t$-conorm, Saadati and Park [38] have introduced the concept of intuitionistic fuzzy normed space.

**Definition 9.3 ([31, 38]).** The five-tuple $(X, \mu, \nu, *, \star)$ is said to be an intuitionistic fuzzy normed space (for short, IFNS) if $X$ is a vector space, $*$ is a continuous $t$-norm, $\star$ is a continuous $t$-conorm, and $\mu$, $\nu$ are fuzzy sets on $X \times (0, \infty)$ satisfying the following thirteen conditions:

$(IF_1)$ $\mu(x,t) + \nu(x,t) \leq 1$;
$(IF_2)$ $\mu(x,t) > 0$;
$(IF_3)$ $\mu(x,t) = 1$ if and only if $x = 0$;
$(IF_4)$ $\mu(\alpha x,t) = \mu(x, \frac{t}{|\alpha|})$ for each $\alpha \neq 0$;
$(IF_5)$ $\mu(x,t) * \mu(y,s) \leq \mu(x+y, t+s)$;
$(IF_6)$ $\mu(x,.) : (0,\infty) \to [0,1]$ is continuous;
$(IF_7)$ $\lim_{t\to\infty} \mu(x,t) = 1$ and $\lim_{t\to 0} \mu(x,t) = 0$;
$(IF_8)$ $\nu(x,t) < 1$;
$(IF_9)$ $\nu(x,t) = 0$ if and only if $x = 0$;
$(IF_{10})$ $\nu(\alpha x,t) = \nu(x, \frac{t}{|\alpha|})$ for each $\alpha \neq 0$;
$(IF_{11})$ $\nu(x,t) \star \nu(y,s) \geq \nu(x+y, t+s)$;
$(IF_{12})$ $\nu(x,.) : (0,\infty) \to [0,1]$ is continuous;
$(IF_{13})$ $\lim_{t\to\infty} \nu(x,t) = 0$ and $\lim_{t\to 0} \nu(x,t) = 1$,

for every $x, y \in X$ and $s, t > 0$.

The properties of IFNS, examples of intuitionistic fuzzy norms and the concepts of convergence and Cauchy sequences in an IFNS are given in [38].

**Definition 9.4.** Let $(X, \mu, \nu, *, \star)$ be an IFNS. Then a sequence $\{x_n\}$ is said to be convergent to $x \in X$ with respect to the intuitionistic fuzzy norm $(\mu, \nu)$ if, for every $\varepsilon > 0$ and $t > 0$, there exists $k \in \mathbb{N}$ such that $\mu(x_n - x, t) > 1 - \varepsilon$ and $\nu(x_n - x, t) < \varepsilon$ for all $n \geq k$. In this case we write $(\mu, \nu) - \lim x_n = x$.

**Definition 9.5.** Let $(X, \mu, \nu, *, \star)$ be an IFNS. Then $\{x_n\}$ is said to be Cauchy sequence with respect to the intuitionistic fuzzy norm $(\mu, \nu)$ if, for every $\varepsilon > 0$ and $t > 0$, there exists $k \in \mathbb{N}$ such that $\mu(x_n - x_m, t) > 1 - \varepsilon$ and $\nu(x_n - x_m, t) < \varepsilon$ for all $n, m \geq k$.

**Definition 9.6.** Let $(X,\mu,\nu,*,\star)$ be an IFNS. Then $(X,\mu,\nu,*,\star)$ is said to be complete if every intuitionistic fuzzy Cauchy sequence in $(X,\mu,\nu,*,\star)$ is intuitionistic fuzzy convergent in $(X,\mu,\nu,*,\star)$.

**Definition 9.7.** We say that a function $f: X \to Y$ between IFNS, $X$ and $Y$, is continuous at a point $x_0 \in X$ if for each sequence $\{x_n\}$ converging to $x_0$ in $X$, then the sequence $\{f(x_n)\}$ converges to $f(x_0)$. If $f: X \to Y$ is continuous at each $x \in X$, then $f: X \to Y$ is said to be continuous on $X$ (see [28, 31]).

In the rest of this paper, unless explicitly stated otherwise, we assume that $X$ is a linear space, $(Z,\mu',\nu')$ is an intuitionistic fuzzy normed space (IFNS) and $(Y,\mu,\nu)$ is an intuitionistic fuzzy Banach space (IFBS). For convenience, we use the following notion, for a given function $f: X \to Y$ and all $x_1,\dots,x_n \in X$,

$$D_f(x_1,..,x_n) := f\left(\sum_{i=1}^{n} a_i x_i\right) - 2^{n-1}a_1 f(x_1)$$

$$+\sum_{k=2}^{n}\sum_{i_1=2}^{k}\sum_{i_2=i_1+1}^{k+1}\cdots\sum_{i_{n-k+1}=i_{n-k}+1}^{n} f\left(\sum_{\substack{i=1\\ i\neq i_1,\dots,i_{n-k+1}}}^{n} a_i x_i - \sum_{r=1}^{n-k+1} a_{i_r} x_{i_r}\right),$$

where $n \in \mathbb{N}\backslash\{1\}$ and $a_1,\dots,a_n \in \mathbb{Z}\backslash\{0\}$ with $a_1 \neq \pm 1$.

## 9.3 Results in IFNS

In this section, we prove the generalized Hyers–Ulam stability of functional equation (9.3) in IFBS. Next, we will show that there exists a close relationship between the intuitionistic fuzzy continuity (IFC) behavior of an intuitionistic fuzzy almost additive function, control function and the unique additive function which approximates the almost additive function.

**Lemma 9.1 ([21]).** *Let $V_1$ and $V_2$ be real vector spaces. A function $f: V_1 \to V_2$ satisfies functional equation (9.3) if and only if $f: V_1 \to V_2$ is additive.*

**Theorem 9.2.** *Let $\ell \in \{-1,1\}$ be fixed and let $\varphi: X^n \to Z$ be a function such that*

$$\varphi(a_1x_1,\dots,a_1x_n) = \alpha\varphi(x_1,\dots,x_n) \tag{9.4}$$

*for all $x_1,\dots,x_n \in X$ and for some positive real number $\alpha$ with $\alpha\ell < a_1\ell$. Suppose that a function $f: X \to Y$, with $f(0) = 0$, satisfies the inequalities*

$$\begin{aligned}\mu\left(D_f(x_1,\dots,x_n),t\right) &\geq \mu'\left(\varphi(x_1,\dots,x_n),t\right),\\ \nu\left(D_f(x_1,\dots,x_n),t\right) &\leq \nu'\left(\varphi(x_1,\dots,x_n),t\right)\end{aligned} \tag{9.5}$$

*for all* $x_1, \ldots, x_n \in X$ *and* $t > 0$*. Then the limit*

$$A(x) = (\mu, \nu) - \lim_{k\to\infty} \frac{f(a_1^{\ell k}x)}{a_1^{\ell k}}$$

*exists for all* $x \in X$ *and* $A : X \to Y$ *is a unique additive function satisfying the following two inequalities*

$$\mu\,(A(x) - f(x), t) \geq \mu'\left(\varphi(x, \underbrace{0, \ldots, 0}_{n-1}), 2^{n-2}\ell(a_1 - \alpha)t\right),$$

$$\nu\,(A(x) - f(x), t) \leq \nu'\left(\varphi(x, \underbrace{0, \ldots, 0}_{n-1}), 2^{n-2}\ell(a_1 - \alpha)t\right) \tag{9.6}$$

*for all* $x \in X$ *and* $t > 0$*.*

*Proof.* Case 1: $\ell = 1$. Putting $x_1 = x$ and $x_i = 0 \ \ (i = 2, \ldots, n)$ in (9.5), we obtain

$$\begin{aligned}&\mu'(\varphi(x, 0, 0, \ldots, 0), t)\\ &\leq \mu\left(\left(\binom{n-1}{n-1} + \binom{n-1}{n-2} + \cdots + \binom{n-1}{1} + 1\right) f(a_1x) - 2^{n-1}a_1 f(x), t\right)\end{aligned}$$

and

$$\begin{aligned}&\nu'(\varphi(x, 0, 0, \ldots, 0), t)\\ &\geq \nu\left(\left(\binom{n-1}{n-1} + \binom{n-1}{n-2} + \cdots + \binom{n-1}{1} + 1\right) f(a_1x) - 2^{n-1}a_1 f(x), t\right)\end{aligned}$$

for all $x \in X$ and $t > 0$. Hence, the relation

$$1 + \sum_{\ell=1}^{n-1} \binom{n-1}{\ell} = 2^{n-1}$$

gives

$$\mu\,(2^{n-1}f(a_1x) - 2^{n-1}a_1 f(x), t) \geq \mu'(\varphi(x, 0, 0, \ldots, 0), t),$$

$$\nu\,(2^{n-1}f(a_1x) - 2^{n-1}a_1 f(x), t) \leq \nu'(\varphi(x, 0, 0, \ldots, 0), t) \tag{9.7}$$

for all $x \in X$ and $t > 0$. So

$$\mu\left(\frac{f(a_1x)}{a_1} - f(x), \frac{t}{2^{n-1}a_1}\right) \geq \mu'(\varphi(x,0,0,\ldots,0),t),$$
$$\nu\left(\frac{f(a_1x)}{a_1} - f(x), \frac{t}{2^{n-1}a_1}\right) \leq \nu'(\varphi(x,0,0,\ldots,0),t) \tag{9.8}$$

for all $x \in X$ and $t > 0$. Further, by our assumption,

$$\mu'(\varphi(a_1x,0,0,\ldots,0),t) = \mu'\left(\varphi(x,0,0,\ldots,0), \frac{t}{\alpha}\right),$$
$$\nu'(\varphi(a_1x,0,0,\ldots,0),t) = \nu'\left(\varphi(x,0,0,\ldots,0), \frac{t}{\alpha}\right) \tag{9.9}$$

for all $x \in X$ and $t > 0$. Replacing $x$ by $a_1^k x$ in (9.8) and using (9.9), we obtain

$$\begin{aligned}\mu\left(\frac{f(a_1^{k+1}x)}{a_1^{k+1}} - \frac{f(a_1^k x)}{a_1^k}, \frac{t}{2^{n-1}a_1^{k+1}}\right) &\geq \mu'(\varphi(a_1^k x,0,0,\ldots,0),t)\\ &= \mu'\left(\varphi(x,0,0,\ldots,0), \frac{t}{\alpha^k}\right),\\ \nu\left(\frac{f(a_1^{k+1}x)}{a_1^{k+1}} - \frac{f(a_1^k x)}{a_1^k}, \frac{t}{2^{n-1}a_1^{k+1}}\right) &\leq \nu'(\varphi(a_1^k x,0,0,\ldots,0),t)\\ &= \nu'\left(\varphi(x,0,0,\ldots,0), \frac{t}{\alpha^k}\right)\end{aligned} \tag{9.10}$$

for all $x \in X$, $t > 0$ and $k \geq 0$. Replacing $t$ by $\alpha^k t$ in (9.10), we see that

$$\mu\left(\frac{f(a_1^{k+1}x)}{a_1^{k+1}} - \frac{f(a_1^k x)}{a_1^k}, \frac{\alpha^k t}{2^{n-1}a_1^{k+1}}\right) \geq \mu'(\varphi(x,0,0,\ldots,0),t),$$
$$\nu\left(\frac{f(a_1^{k+1}x)}{a_1^{k+1}} - \frac{f(a_1^k x)}{a_1^k}, \frac{\alpha^k t}{2^{n-1}a_1^{k+1}}\right) \leq \nu'(\varphi(x,0,0,\ldots,0),t) \tag{9.11}$$

for all $x \in X$, $t > 0$ and $k > 0$. It follows from the equality

$$\frac{f(a_1^k x)}{a_1^k} - f(x) = \sum_{j=0}^{k-1}\left(\frac{f(a_1^{j+1}x)}{a_1^{j+1}} - \frac{f(a_1^j x)}{a_1^j}\right)$$

and (9.11) that

$$\begin{aligned}
&\mu\left(\frac{f(a_1^k x)}{a_1^k} - f(x), \sum_{j=0}^{k-1} \frac{\alpha^j t}{2^{n-1} a_1^{j+1}}\right) \\
&\qquad \geq \prod_{j=0}^{k-1} \mu\left(\frac{f(a_1^{j+1} x)}{a_1^{j+1}} - \frac{f(a_1^j x)}{a_1^j}, \frac{\alpha^j t}{2^{n-1} a_1^{j+1}}\right) \\
&\qquad \geq \mu'(\varphi(x,0,0,\ldots,0),t), \\
&\nu\left(\frac{f(a_1^k x)}{a_1^k} - f(x), \sum_{j=0}^{k-1} \frac{\alpha^j t}{2^{n-1} a_1^{j+1}}\right) \\
&\qquad \leq \widehat{\prod}_{j=0}^{k-1} \nu\left(\frac{f(a_1^{j+1} x)}{a_1^{j+1}} - \frac{f(a_1^j x)}{a_1^j}, \frac{\alpha^j t}{2^{n-1} a_1^{j+1}}\right) \\
&\qquad \leq \nu'(\varphi(x,0,0,\ldots,0),t)
\end{aligned} \tag{9.12}$$

for all $x \in X$, $t > 0$ and $k > 0$, where

$$\prod_{k=1}^{n} a_k = a_1 * a_2 * \ldots * a_n$$

and

$$\widehat{\prod}_{k=1}^{n} a_k = a_1 \star a_2 \star \cdots \star a_n.$$

Replacing $x$ by $a_1^m x$ in (9.12), we observe that

$$\begin{aligned}
\mu\left(\frac{f(a_1^{k+m} x)}{a_1^{k+m}} - \frac{f(a_1^m x)}{a_1^m}, \sum_{j=0}^{k-1} \frac{\alpha^j t}{2^{n-1} a_1^{j+m+1}}\right) &\geq \mu'(\varphi(a_1^m x,0,0,\ldots,0),t) \\
&= \mu'\left(\varphi(x,0,0,\ldots,0), \frac{t}{\alpha^m}\right), \\
\nu\left(\frac{f(a_1^{k+m} x)}{a_1^{k+m}} - \frac{f(a_1^m x)}{a_1^m}, \sum_{j=0}^{k-1} \frac{\alpha^j t}{2^{n-1} a_1^{j+m+1}}\right) &\leq \nu'(\varphi(a_1^m x,0,0,\ldots,0),t) \\
&= \nu'\left(\varphi(x,0,0,\ldots,0), \frac{t}{\alpha^m}\right)
\end{aligned}$$

for all $x \in X$, $t > 0$ and all $m \geq 0$, $k > 0$. Hence

$$\begin{aligned}
\mu\left(\frac{f(a_1^{k+m} x)}{a_1^{k+m}} - \frac{f(a_1^m x)}{a_1^m}, \sum_{j=m}^{k+m-1} \frac{\alpha^j t}{2^{n-1} a_1^{j+1}}\right) &\geq \mu'(\varphi(x,0,0,\ldots,0),t), \\
\nu\left(\frac{f(a_1^{k+m} x)}{a_1^{k+m}} - \frac{f(a_1^m x)}{a_1^m}, \sum_{j=m}^{k+m-1} \frac{\alpha^j t}{2^{n-1} a_1^{j+1}}\right) &\leq \nu'(\varphi(x,0,0,\ldots,0),t)
\end{aligned}$$

for all $x \in X$, $t > 0$ and all $m \geq 0$, $k > 0$. By the last inequality, we obtain

$$\mu\left(\frac{f(a_1^{k+m}x)}{a_1^{k+m}} - \frac{f(a_1^m x)}{a_1^m}, t\right) \geq \mu'\left(\varphi(x,0,0,\ldots,0), \frac{t}{\sum_{j=m}^{k+m-1} \frac{\alpha^j}{2^{n-1}a_1^{j+1}}}\right),$$

$$\nu\left(\frac{f(a_1^{k+m}x)}{a_1^{k+m}} - \frac{f(a_1^m x)}{a_1^m}, t\right) \leq \nu'\left(\varphi(x,0,0,\ldots,0), \frac{t}{\sum_{j=m}^{k+m-1} \frac{\alpha^j}{2^{n-1}a_1^{j+1}}}\right) \tag{9.13}$$

for all $x \in X$, $t > 0$ and all $m \geq 0$, $k > 0$. Since $0 < \alpha < a_1$ and

$$\sum_{j=0}^{\infty} \left(\frac{\alpha}{a_1}\right)^j < \infty,$$

the Cauchy criterion for convergence in IFNS shows that $\{f(a_1^k x)a_1^{-k}\}$ is a Cauchy sequence in $Y$. Since $Y$ is an IFBS, this sequence converges to some point $A(x) \in Y$. So one can define the function $A : X \to Y$ by

$$A(x) = (\mu,\nu) - \lim_{k\to\infty} \frac{f(a_1^k x)}{a_1^k} \tag{9.14}$$

for all $x \in X$. Fix $x \in X$ and put $m = 0$ in (9.13) to obtain

$$\mu\left(\frac{f(a_1^k x)}{a_1^k} - f(x), t\right) \geq \mu'\left(\varphi(x,0,0,\ldots,0), \frac{t}{\sum_{j=0}^{k-1} \frac{\alpha^j}{2^{n-1}a_1^{j+1}}}\right),$$

$$\nu\left(\frac{f(a_1^k x)}{a_1^k} - f(x), t\right) \leq \nu'\left(\varphi(x,0,0,\ldots,0), \frac{t}{\sum_{j=0}^{k-1} \frac{\alpha^j}{2^{n-1}a_1^{j+1}}}\right)$$

for all $x \in X$, $t > 0$ and all $k > 0$, from which we obtain

$$\begin{aligned}
\mu(A(x) - f(x), t) &\geq \mu\left(A(x) - \frac{f(a_1^k x)}{a_1^k}, \frac{t}{2}\right) * \mu\left(\frac{f(a_1^k x)}{a_1^k} - f(x), \frac{t}{2}\right) \\
&\geq \mu'\left(\varphi(x,0,\ldots,0), \frac{t}{\sum_{j=0}^{k-1} \frac{\alpha^j}{2^{n-2}a_1^{j+1}}}\right), \\
\nu(A(x) - f(x), t) &\leq \nu\left(A(x) - \frac{f(a_1^k x)}{a_1^k}, \frac{t}{2}\right) \star \mu\left(\frac{f(a_1^k x)}{a_1^k} - f(x), \frac{t}{2}\right) \\
&\leq \nu'\left(\varphi(x,0,\ldots,0), \frac{t}{\sum_{j=0}^{k-1} \frac{\alpha^j}{2^{n-2}a_1^{j+1}}}\right)
\end{aligned} \tag{9.15}$$

for $k$ large enough. Taking the limit as $k \to \infty$ in (9.15) and using the definition of IFNS, we obtain

$$\begin{aligned} \mu(A(x)-f(x),t) &\geq \mu'(\varphi(x,0,0,\ldots,0),2^{n-2}(a_1-\alpha)t), \\ \nu(A(x)-f(x),t) &\leq \nu'(\varphi(x,0,0,\ldots,0),2^{n-2}(a_1-\alpha)t) \end{aligned} \tag{9.16}$$

for all $x \in X$ and $t > 0$. It follows from (9.4) and (9.5) that

$$\begin{aligned} \mu\left(\frac{1}{a_1^k}D_f(a_1^k x_1,\ldots,a_1^k x_n),t\right) &\geq \mu'(\varphi(a_1^k x_1,\ldots,a_1^k x_n),a_1^k t) \\ &= \mu'\left(\varphi(x_1,\ldots,x_n),\frac{a_1^k t}{\alpha^k}\right), \\ \nu\left(\frac{1}{a_1^k}D_f(a_1^k x_1,\ldots,a_1^k x_n),t\right) &\leq \nu'(\varphi(a_1^k x_1,\ldots,a_1^k x_n),a_1^k t) \\ &= \nu'\left(\varphi(x_1,\ldots,x_n),\frac{a_1^k t}{\alpha^k}\right) \end{aligned} \tag{9.17}$$

for all $x_1,\ldots,x_n \in X$ and $t > 0$. Letting $k \to \infty$ in (9.17), we obtain

$$\mu(D_A(x_1,\ldots,x_n),t) = 1$$

for all $x_1,\ldots,x_n \in X$ and $t > 0$. Similarly, we obtain

$$\nu(D_A(x_1,\ldots,x_n),t) = 0$$

for all $x_1,\ldots,x_n \in X$ and $t > 0$. This means that $A$ satisfies (9.3). Thus, by Lemma 9.1, the function $A : X \to Y$ is additive.

Now, to prove the uniqueness property of $A$, let $A' : X \to Y$ be another additive function satisfying (9.6). Fix $x \in X$. Then clearly, $A(a_1^n x) = a_1^n A(x)$ and $A'(a_1^n x) = a_1^n A'(x)$ for all $n \in \mathbb{N}$. It follows from (9.6) and (9.9) that

$$\begin{aligned} \mu(A(x)-A'(x),t) &= \mu\left(\frac{A(a_1^k x)}{a_1^k}-\frac{A'(a_1^k x)}{a_1^k},t\right) \\ &\geq \mu\left(\frac{A(a_1^k x)}{a_1^k}-\frac{f(a_1^k x)}{a_1^k},\frac{t}{2}\right) * \mu\left(\frac{f(a_1^k x)}{a_1^k}-\frac{A'(a_1^k x)}{a_1^k},\frac{t}{2}\right) \\ &\geq \mu'\left(\varphi(a_1^k x,0,\ldots,0),\frac{a_1^k 2^{n-2}(a_1-\alpha)t}{2}\right) \\ &\quad * \mu'\left(\varphi(a_1^k x,0,\ldots,0),\frac{a_1^k 2^{n-2}(a_1-\alpha)t}{2}\right) \end{aligned}$$

$$= \mu'\left(\varphi(x,0,0,\ldots,0), \frac{a_1^k 2^{n-2}(a_1-\alpha)t}{2\alpha^k}\right)$$

$$* \mu'\left(\varphi(x,0,\ldots,0), \frac{a_1^k 2^{n-2}(a_1-\alpha)t}{2\alpha^k}\right),$$

$$\nu(A(x)-A'(x),t) \leq \nu'\left(\varphi(a_1^k x,0,\ldots,0), \frac{a_1^k 2^{n-2}(a_1-\alpha)t}{2}\right)$$

$$\star \nu'\left(\varphi(a_1^k x,0,\ldots,0), \frac{a_1^k 2^{n-2}(a_1-\alpha)t}{2}\right)$$

$$= \nu'\left(\varphi(x,0,\ldots,0), \frac{a_1^k 2^{n-2}(a_1-\alpha)t}{2\alpha^k}\right)$$

$$\star \nu'\left(\varphi(x,0,\ldots,0), \frac{a_1^k 2^{n-2}(a_1-\alpha)t}{2\alpha^k}\right)$$

for all $x \in X$ and $t > 0$. Since $\alpha < a_1$, we have

$$\lim_{k\to\infty} \mu'\left(\varphi(x,0,\ldots,0), \frac{a_1^k 2^{n-2}(a1-\alpha)t}{2\alpha^k}\right) = 1$$

and

$$\lim_{k\to\infty} \nu'\left(\varphi(x,0,\ldots,0), \frac{a_1^k 2^{n-2}(a_1-\alpha)t}{2\alpha^k}\right) = 0.$$

Thus, $\mu(A(x)-A'(x),t) = 1$ and $\nu(A(x)-A'(x),t) = 0$. Therefore $A(x) = A'(x)$.

*Case 2.* $\ell = -1$. We can state the proof in the same pattern as we did in the first case. Namely, replacing $x$ by $xa_1^{-1}$ in (9.7), we obtain

$$\mu\left(f(x) - a_1 f\left(\frac{x}{a_1}\right), \frac{t}{2^{n-1}}\right) \geq \mu'\left(\varphi\left(\frac{x}{a_1},0,\ldots,0\right),t\right),$$

$$\nu\left(f(x) - a_1 f\left(\frac{x}{a_1}\right), \frac{t}{2^{n-1}}\right) \leq \nu'\left(\varphi\left(\frac{x}{a_1},0,\ldots,0\right),t\right) \tag{9.18}$$

for all $x \in X$ and $t > 0$. Replacing $x$ and $t$ by $xa_1^{-k}$ and $ta_1^{-k}$ in (9.18), respectively, we obtain

$$\mu\left(a_1^k f\left(\frac{x}{a_1^k}\right) - a_1^{k+1} f\left(\frac{x}{a_1^{k+1}}\right), \frac{t}{2^{n-1}}\right) \geq \mu'\left(\varphi\left(\frac{x}{a_1^{k+1}},0,\ldots,0\right), \frac{t}{a_1^k}\right)$$

$$= \mu'\left(\varphi(x,0,\ldots,0), \left(\frac{\alpha}{a_1}\right)^k t\alpha\right),$$

$$\nu\left(a_1^k f\left(\frac{x}{a_1^k}\right) - a_1^{k+1} f\left(\frac{x}{a_1^{k+1}}\right), \frac{t}{2^{n-1}}\right) \leq \nu'\left(\varphi\left(\frac{x}{a_1^{k+1}}, 0, \ldots, 0\right), \frac{t}{a_1^k}\right)$$
$$= \nu'\left(\varphi(x, 0, \ldots, 0), \left(\frac{\alpha}{a_1}\right)^k t\alpha\right)$$

for all $x \in X$, $t > 0$ and all $k > 0$. One can deduce that

$$\mu\left(a_1^{k+m} f\left(\frac{x}{a_1^{k+m}}\right) - a_1^m f\left(\frac{x}{a_1^m}\right), t\right)$$
$$\geq \mu'\left(\varphi(x, 0, \ldots, 0), \frac{t}{\sum_{j=m+1}^{k+m} \frac{a_1^j}{2^{n-1} a_1 \alpha^j}}\right),$$
$$\nu\left(a_1^{k+m} f\left(\frac{x}{a_1^{k+m}}\right) - a_1^m f\left(\frac{x}{a_1^m}\right), t\right)$$
$$\leq \nu'\left(\varphi(x, 0, \ldots, 0), \frac{t}{\sum_{j=m+1}^{k+m} \frac{a_1^j}{2^{n-1} a_1 \alpha^j}}\right) \tag{9.19}$$

for all $x \in X$, $t > 0$ and all $m \geq 0$, $k \geq 0$, from which we conclude that $\{a_1^k f(x a_1^{-k})\}$ is a Cauchy sequence in the IFBS, $(Y, \mu, \nu)$. Therefore, there is a function $A : X \to Y$ defined by $A(x) := (\mu, \nu) - \lim_{k\to\infty} a_1^k f(x a_1^{-k})$. Employing (9.19) with $m = 0$, we obtain

$$\mu(A(x) - f(x), t) \geq \mu'(\varphi(x, 0, \ldots, 0), 2^{n-2}(\alpha - a_1)t),$$
$$\nu(A(x) - f(x), t) \leq \nu'(\varphi(x, 0, \ldots, 0), 2^{n-2}(\alpha - a_1)t)$$

for all $x \in X$ and $t > 0$. The proof of the uniqueness of $A$ for this case proceeds similarly to that in the previous case; hence it is omitted. □

From Theorem 9.2, we obtain the following corollary concerning the generalized Hyers–Ulam stability or Hyers–Ulam–Rassias stability [34] of additive functions satisfying (9.3), in normed spaces.

**Corollary 9.1.** *Let $X$ be a normed space and $Y$ be a Banach space. Let $\varepsilon, \lambda$ be non-negative real numbers such that $\lambda \neq 1$. Suppose that a function $f : X \to Y$, with $f(0) = 0$, satisfies*

$$\|D_f(x_1, \ldots, x_n)\| \leq \varepsilon \sum_{i=1}^{n} \|x_i\|^\lambda \tag{9.20}$$

*for all $x_1,\dots,x_n \in X$. Then the limit*

$$A(x) = \lim_{k\to\infty} \frac{f(a_1^{\ell k}x)}{a_1^{\ell k}}$$

*exists for all $x \in X$ and $A : X \to Y$ is a unique additive function satisfying*

$$\|f(x) - A(x)\| \le \frac{\varepsilon\|x\|^\lambda}{2^{n-2}\ell(a_1^\lambda - \alpha)} \tag{9.21}$$

*for all $x \in X$, where $\lambda\ell < \ell$.*

*Proof.* Define the functions $\mu$ and $\nu$ by

$$\mu(x,t) = \frac{t}{t + \|x\|}, \qquad \nu(x,t) = \frac{\|x\|}{t + \|x\|}.$$

It is easy to see that $(X,\mu,\nu)$ is an IFNS and $(Y,\mu,\nu)$ is an IFBS. Denote by $\varphi$ the function sending each $(x_1,\dots,x_n) \in X^n$ to $\varepsilon\sum_{i=1}^n \|x_i\|^\lambda \in \mathbb{R}$. Let $a * b = \min\{a,b\}$, $a \star b = \max\{a,b\}$ for all $a,b \in [0,1]$, and functions $\mu',\nu' : \mathbb{R}\times(0,\infty) \to [0,1]$ be given by

$$\mu'(x,t) = \frac{t}{t + |x|}, \qquad \nu'(x,t) = \frac{|x|}{t + |x|}.$$

Then $(\mu',\nu')$ is an intuitionistic fuzzy norm on $\mathbb{R}$ (see [43, Example 3.2]) and

$$\mu(D_f(x_1,\dots,x_n),t) \ge \mu'(\varphi(x_1,\dots,x_n),t),$$
$$\nu(D_f(x_1,\dots,x_n),t) \le \nu'(\varphi(x_1,\dots,x_n),t).$$

By Theorem 9.2, there exists a unique additive function $A : X \to Y$ that satisfies (9.3) and such that

$$\begin{aligned}
\frac{t}{t + \|f(x) - A(x)\|} &= \mu(f(x) - A(x),t) \ge \mu'(\varphi(x,0,\dots,0), 2^{n-2}\ell(a_1^\lambda - \alpha)t)\\
&= \mu'(\varepsilon\|x\|^\lambda, 2^{n-2}\ell(a_1^\lambda - \alpha)t) = \frac{2^{n-2}\ell(a_1^\lambda - \alpha)t}{2^{n-2}\ell(a_1^\lambda - \alpha)t + \varepsilon\|x\|^\lambda},\\
\frac{\|f(x) - A(x)\|}{t + \|f(x) - A(x)\|} &= \nu(f(x) - A(x),t) \le \nu'(\varphi(x,0,\dots,0), 2^{n-2}\ell(a_1^\lambda - \alpha)t)\\
&= \nu'(\varepsilon\|x\|^\lambda, 2^{n-2}\ell(a_1^\lambda - \alpha)t) = \frac{\varepsilon\|x\|^\lambda}{2^{n-2}\ell(a_1^\lambda - \alpha)t + \varepsilon\|x\|^\lambda},
\end{aligned}$$

which means that

$$\frac{t}{t+\|f(x)-A(x)\|} \geq \frac{2^{n-2}\ell(a_1^\lambda-\alpha)t}{2^{n-2}\ell(a_1^\lambda-\alpha)t+\varepsilon\|x\|^\lambda},$$

$$\frac{\|f(x)-A(x)\|}{t+\|f(x)-A(x)\|} \leq \frac{\varepsilon\|x\|^\lambda}{2^{n-2}\ell(a_1^\lambda-\alpha)t+\varepsilon\|x\|^\lambda}$$

and consequently

$$\|f(x)-A(x)\| \leq \frac{\varepsilon\|x\|^\lambda}{2^{n-2}\ell(a_1^\lambda-\alpha)}$$

for all $x \in X$. □

*Remark 9.1.* The generalized Hyers–Ulam stability problem for the case of $\lambda = 1$ was excluded in Corollary 9.1. In fact, functional equation (9.3) is not stable for $\lambda = 1$ in (9.20) (see [11,37]).

Now we examine some conditions under which the additive function found in Theorem 9.2 is continuous. In the following theorem, we investigate IFC of additive functions in IFNS. In fact, we will show that under some extra conditions in Theorem 9.2, the additive function $r \longmapsto A(rx)$ is IFC.

In the following result, we will use the terminologies of Theorem 9.2; we will also assume that all conditions of the theorem hold.

**Theorem 9.3.** *Let $(\mu_1, \nu_1)$ denote the intuitionistic fuzzy norm obtained in Corollary 9.1 on $\mathbb{R}$. Assume that, for all $x \in X$, the functions $r \longmapsto f(rx)$ (from $(\mathbb{R}, \mu_1, \nu_1)$ into $(Y, \mu, \nu)$) and $r \longmapsto \varphi(rx, 0, \ldots, 0)$ (from $(\mathbb{R}, \mu_1, \nu_1)$ into $(Z, \mu', \nu')$) are IFC. Then for all $x \in X$, the function $r \longmapsto A(rx)$ is IFC and $A(rx) = rA(x)$ for all $r \in \mathbb{R}$.*

*Proof.* First, consider the case: $\ell = 1$. Let $\{r_k\}$ be a sequence in $\mathbb{R}$ that converges to some $r \in \mathbb{R}$, and let $t > 0$. Let $\varepsilon > 0$ be given. Since $0 < \alpha < a_1$, so

$$\lim_{k\to\infty} \frac{2^{n-2}(a_1-\alpha)a_1^k t}{6\alpha^k} = \infty$$

and there is $m \in \mathbb{N}$ such that

$$\mu'\left(\varphi(rx,0,0,\ldots,0), \frac{2^{n-2}(a_1-\alpha)a_1^m t}{6\alpha^m}\right) > 1-\varepsilon,$$

$$\nu'\left(\varphi(rx,0,0,\ldots,0), \frac{2^{n-2}(a_1-\alpha)a_1^m t}{6\alpha^m}\right) < \varepsilon. \tag{9.22}$$

It follows from (9.16) and (9.22) that

$$\mu\left(\frac{f(a_1^m rx)}{a_1^m} - \frac{A(a_1^m rx)}{a_1^m}, \frac{t}{3}\right) > 1-\varepsilon, \quad \nu\left(\frac{f(a_1^m rx)}{a_1^m} - \frac{A(a_1^m rx)}{a_1^m}, \frac{t}{3}\right) < \varepsilon. \tag{9.23}$$

By the IFC of functions $r \longmapsto f(rx)$ and $r \longmapsto \varphi(rx,0,\ldots,0)$, we can find some $\iota \in \mathbb{N}$ such that for any $k \geq \iota$,

$$\mu\left(\frac{f(a_1^m r_k x)}{a_1^m} - \frac{f(a_1^m rx)}{a_1^m}, \frac{t}{3}\right) > 1-\varepsilon,$$
$$\nu\left(\frac{f(a_1^m r_k x)}{a_1^m} - \frac{f(a_1^m rx)}{a_1^m}, \frac{t}{3}\right) < \varepsilon \tag{9.24}$$

and

$$\mu'\left(\varphi(r_k x,0\ldots,0) - \varphi(rx,0,\ldots,0), \frac{2^{n-2}(a_1-\alpha)a_1^m t}{6\alpha^m}\right) > 1-\varepsilon,$$
$$\nu'\left(\varphi(r_k x,0\ldots,0) - \varphi(rx,0,\ldots,0), \frac{2^{n-2}(a_1-\alpha)a_1^m t}{6\alpha^m}\right) < \varepsilon. \tag{9.25}$$

Next, (9.22) and (9.25) yield the inequalities

$$\mu'\left(\varphi\left(r_k x,0,\ldots,0\right), \frac{2^{n-2}(a_1-\alpha)a_1^m t}{3\alpha^m}\right)\Bigg) > 1-\varepsilon,$$
$$\nu'\left(\varphi\left(r_k x,0,\ldots,0\right), \frac{2^{n-2}(a_1-\alpha)a_1^m t}{3\alpha^m}\right)\Bigg) < \varepsilon. \tag{9.26}$$

On the other hand,

$$\begin{aligned}
\mu\left(A(r_k x) - \frac{f(a_1^m r_k x)}{a_1^m}, \frac{t}{a_1^m}\right) &= \mu\left(\frac{A(a_1^m r_k x)}{a_1^m} - \frac{f(a_1^m r_k x)}{a_1^m}, \frac{t}{a_1^m}\right)\\
&\geq \mu'\left(\varphi(r_k x,0,\ldots,0), \frac{2^{n-2}(a_1-\alpha)t}{\alpha^m}\right),\\
\nu\left(A(r_k x) - \frac{f(a_1^m r_k x)}{a_1^m}, \frac{t}{a_1^m}\right) &= \nu\left(\frac{A(a_1^m r_k x)}{a_1^m} - \frac{f(a_1^m r_k x)}{a_1^m}, \frac{t}{a_1^m}\right)\\
&\leq \nu'\left(\varphi(r_k x,0,\ldots,0), \frac{2^{n-2}(a_1-\alpha)t}{\alpha^m}\right).
\end{aligned} \tag{9.27}$$

Inequalities (9.26) and (9.27) imply that

$$\mu\left(A(r_k x) - \frac{f(a_1^m r_k x)}{a_1^m}, \frac{t}{3}\right) > 1-\varepsilon, \quad \nu\left(A(r_k x) - \frac{f(a_1^m r_k x)}{a_1^m}, \frac{t}{3}\right) < \varepsilon. \tag{9.28}$$

So, it follows from (9.23), (9.24) and (9.28) that for any $k \geq \iota$,

$$\mu(A(r_k x) - A(rx), t) > 1-\varepsilon, \quad \nu(A(r_k x) - A(rx), t) < \varepsilon.$$

Therefore, for every choice of $x \in X$, $t > 0$ and $\varepsilon > 0$, we can find some $\iota \in \mathbb{N}$ such that $\mu(A(r_k x) - A(rx), t) > 1 - \varepsilon$ and $\nu(A(r_k x) - A(rx), t) < \varepsilon$ for every $k \geq \iota$. This shows that $A(r_k x) \to A(rx)$.

The proof for $\ell = -1$, proceeds similarly to that in the previous case.

It is not hard to see that $A(rx) = rA(x)$ for each rational number $r$. Let $r$ be a real number, then there exists a sequence $\{r_k\}$ of rational numbers such that $r_k \to r$. By the IFC of $A(x)$, for all $x \in X$,

$$A(rx) = \lim_{k \to \infty} A(r_k x) = \lim_{k \to \infty} r_k A(x) = rA(x)$$

for each $r \in \mathbb{R}$ (see [26, 31]). This completes the proof of the theorem. □

The following corollary concerns the Hyers–Ulam stability [15] of functional equation (9.3).

**Corollary 9.2.** *Let $X$ be a normed space and $Y$ be a Banach space. Let $\delta \geq 0$ be fixed. Suppose that a function $f : X \to Y$, with $f(0) = 0$, satisfies*

$$\|D_f(x_1, \ldots, x_n)\| \leq \delta \tag{9.29}$$

*for all $x_1, \ldots, x_n \in X$. Then there exists a unique $A : X \to Y$ such that*

$$\|f(x) - A(x)\| \leq \frac{\delta}{n2^{n-2}\ell(1-\alpha)} \tag{9.30}$$

*for all $x \in X$. Also, if the function $r \longmapsto f(rx)$ from $\mathbb{R}$ to $Y$ is continuous for all $x \in X$, then $A(rx) = rA(x)$ for all $r \in \mathbb{R}$.*

*Proof.* Take $\lambda = 0$, $\varepsilon = \frac{\delta}{n}$ and apply Corollary 9.1 and Theorem 9.3. □

## References


1. Aoki, T.: On the stability of the linear transformation in Banach spaces. J. Math. Soc. Japan **2**, 64–66 (1950)
2. Bag, T., Samanta, S.K.: Some fixed point theorems on fuzzy normed linear spaces. Inform. Sci. **177**, 3271–3289 (2007)
3. Bag, T., Samanta, S.K.: Fuzzy bounded linear operators. Fuzzy Sets and Systems **151**, 513–547 (2005)
4. Barros, L.C., Bassanezi, R.C., Tonelli, P.A.: Fuzzy modelling in population dynamics. Ecol. Model. **128**, 27–33 (2000)
5. Cholewa, P.W.: Remarks on the stability of functional equations. Aequationes Math. **27**, 76–86 (1984)
6. Czerwik, S.: On the stability of the quadratic mapping in normed spaces. Abh. Math. Sem. Univ. Hamburg **62**, 59–64 (1992)

7. Eshaghi Gordji, M., Khodaei, H.: Solution and stability of generalized mixed type cubic, quadratic and additive functional equation in quasi-Banach spaces. Nonlinear Anal. **71**, 5629–5643 (2009)
8. Faizev, V.A., Rassias, Th.M., Sahoo, P.K.: The space of $(\psi, \gamma)$–additive mappings on semigroups. Trans. Amer. Math. Soc. **354**, 4455–4472 (2002)
9. Fang, J.X.: On I-topology generated by fuzzy norm. Fuzzy Sets and Systems **157**, 2739–2750 (2006)
10. Fradkov, A.L., Evans, R.J.: Control of chaos: Methods and applications in engineering. Chaos Solitons Fractals **29**, 33–56 (2005)
11. Gajda, Z.: On stability of additive mappings. Int. J. Math. Math. Sci. **14**, 431–434 (1991)
12. Giles, R.: A computer program for fuzzy reasoning Fuzzy Sets and Systems **4**, 221–234 (1980)
13. Găvruta, P.: A generalization of the Hyers-Ulam-Rassias stability of approximately additive mappings. J. Math. Anal. Appl. **184**, 431–436 (1994)
14. Hong, L., Sun, J.Q.: Bifurcations of fuzzy nonlinear dynamical systems. Commun. Nonlinear Sci. Numer. Simul. **1**, 1–12 (2006)
15. Hyers, D.H.: On the stability of the linear functional equation. Proc. Natl. Acad. Sci. U.S.A. **27**, 222–224 (1941)
16. Hyers, D.H., Isac, G., Rassias, Th.M.: Stability of Functional Equations in Several Variables. Birkhauser, Basel (1998)
17. Hyers, D.H., Isac, G., Rassias, Th.M.: Topics in Nonlinear Analysis and Applications. World Scientific Publishing Company (1997)
18. Hyers, D.H., Rassias, Th.M.: Approximate homomorphisms. Aequationes Math. **44**, 125–153 (1992)
19. Isac, G., Rassias, Th.M.: Stability of $\psi$–additive mappings: Applications to nonlinear analysis. Int. J. Math. Math. Sci. **19**, 219–228 (1996)
20. Jung, S.M.: On the Hyers-Ulam-Rassias stability of a quadratic functional equation. J. Math. Anal. Appl. **232**, 384–393 (1999)
21. Khodaei, H., Rassias, Th.M.: Approximately generalized additive functions in several variables. Int. J. Nonlinear Anal. Appl. **1**, 22–41 (2010)
22. Madore, J.: Fuzzy physics. Ann. Phys. **219**, 187–198 (1992)
23. Miheţ, D.: The fixed point method for fuzzy stability of the Jensen functional equation. Fuzzy Sets and Systems **160**, 1663–1667 (2009)
24. Mirmostafaee, M., Mirzavaziri, M., Moslehian, M.S.: Fuzzy stability of the Jensen functional equation. Fuzzy Sets and Systems **159**, 730–738 (2008)
25. Mirmostafaee, A.K., Moslehian, M.S.: Fuzzy approximately cubic mappings. Inform. Sci. **178**, 3791–3798 (2008)
26. Mirmostafaee, A.K.: A fixed point approach to almost quartic mappings in quasi fuzzy normed spaces. Fuzzy Sets and Systems 160 (2009), 1653–1662.
27. Mohiuddine, S.A.: Stability of Jensen functional equation in intuitionistic fuzzy normed space. Chaos Solitons Fractals **42**, 2989–2996 (2009)
28. Mursaleen, M., Mohiuddine, S.A.: Nonlinear operators between intuitionistic fuzzy normed spaces and Frchet differentiation. Chaos Solitons Fractals **42**, 1010-1015 (2009)
29. Mursaleen, M., Mohiuddine, S.A.: Statistical convergence of double sequences in intuitionistic fuzzy normed spaces. Chaos Solitons Fractals **41**, 2414-2421 (2009)
30. Mursaleen, M., Mohiuddine, S.A., Edely, O.H.H.: On the ideal convergence of double sequences in intuitionistic fuzzy normed spaces. Comput. Math. Appl. **59**, 603–611 (2010)
31. Mursaleen, M., Mohiuddine, S.A.: On stability of a cubic functional equation in intuitionistic fuzzy normed spaces. Chaos Solitons Fractals **42**, 2997–3005 (2009)
32. Mursaleen, M., Lohani, Q.M.D.: Intuitionistic fuzzy 2-normed space and some related concepts. Chaos Solitons Fractals **42**, 224–234 (2009)
33. Park, C.: Fuzzy stability of a functional equation associated with inner product spaces. Fuzzy Sets and Systems **160**, 1632–1642 (2009)
34. Rassias, Th.M.: On the stability of the linear mapping in Banach spaces. Proc. Amer. Math. Soc. **72**, 297–300 (1978)

35. Rassias, Th.M.: New characterizations of inner product spaces. Bull. Sci. Math. **108**, 95–99 (1984)
36. Rassias, Th.M.: On the stability of functional equations and a problem of Ulam. Acta Appl. Math. **62**, 23–130 (2000)
37. Rassias, Th.M., Semrl, P.: On the behaviour of mappings which do not satisfy Hyers-Ulam stability. Proc. Amer. Math. Soc. **114**, 989–993 (1992)
38. Saadati, R., Park, J.H.: On the intuitionistic fuzzy topological spaces. Chaos Solitons Fractals **27**, 331–44 (2006)
39. Saadati, R., Sedghi, S., Shobe, N.: Modified intuitionistic fuzzy metric spaces and some fixed point theorems. Chaos Solitons Fractals, doi:10.1016/ j.chaos.2006.11.008
40. Saadati, R., Vaezpour, S.M., Cho, Y.J.: Quicksort algorithm: Application of a fixed point theorem in intuitionistic fuzzy quasi-metric spaces at a domain of words. J. Comput. Appl. Math. **228 (1)**, 219–225 (2009)
41. Schweizer, B., Sklar, A.: Satistical metric spaces. Pacific J. Math. **10**, 314-334 (1960)
42. Ulam, S.M.: Problems in Modern Mathematics. Chapter VI, Science Editions, Wiley, New York (1964)
43. Vijayabalaji, S., Thillaigovindan, N., Bae Jun, Y.: Intuitionistic Fuzzy n-normed linear space. Bull. Korean Math. Soc. **44**, 291–308 (2007)
44. Zadeh, L.A.: Fuzzy sets. Inform. Control **8**, 338–353 (1965)

# Chapter 10
# Generalized Hyers–Ulam Stability for General Quadratic Functional Equation in Quasi-Banach Spaces

**Jinmei Gao**

*Dedicated to the memory of S.M. Ulam*
*on the 100th anniversary of his birth*

**Abstract** In this paper, we investigate the generalized Hyers–Ulam stability of an $n$-dimensional quadratic functional equation

$$f\left(\sum_{i=1}^{n} x_i\right) + \sum_{1\leq i<j\leq n} f(x_i - x_j) = n\sum_{i=1}^{n} f(x_i) \qquad (n \geq 2)$$

in quasi-Banach spaces.

**Keywords** Stability • Functional equations • Quasi-Banach space • Quadratic function

**Mathematics Subject Classification(2000):** Primary 39B52, 39B72

## 10.1 Introduction

In 1940, Ulam [22, Chap. IV] gave a talk before a Mathematical Colloquium at the University of Wisconsin, in which he discussed a number of unsolved problems. Among those was the following question concerning the stability of homomorphism. Let $G_1$ be a group and $(G_2, d)$ a metric group with a metric $d$. Given a positive number $\varepsilon$, does there exist a $\delta > 0$ such that if $f : G_1 \to G_2$ satisfies $d(f(xy), f(x)f(y)) < \delta$ for all $x, y \in G_1$, then a homomorphism $T : G_1 \to G_2$ exists with $d(f(x), T(x)) < \varepsilon$ for all $x, y \in G_1$?

J. Gao (✉)
Department of Mathematics, Qingao University, Qingdao 266071, China

Department of Mathematics, Nankai University, Tianjin 300071, China
e-mail: gjinmei1979@163.com

Th.M. Rassias and J. Brzdęk (eds.), *Functional Equations in Mathematical Analysis*, Springer Optimization and Its Applications 52, DOI 10.1007/978-1-4614-0055-4_10,

In the case where the answer is affirmative, the functional equation for homomorphism will be called stable.

In 1942, Hyers [6] affirmatively answered the question of Ulam for the case of approximately additive mapping $f : E_1 \to E_2$, where $E_1, E_2$ are Banach spaces, i.e. $f$ satisfies the inequality

$$\|f(x+y) - f(x) - f(y)\| < \varepsilon , \qquad \forall x,y \in E_1 .$$

In 1978, Rassias [14] considered a generalized version of Hyers's theorem, which permitted the Cauchy difference to become unbounded. That is, he assumed that

$$\|f(x+y) - f(x) - f(y)\| < \varepsilon(\|x\|^p + \|y\|^p) , \qquad \forall x,y \in E_1,\ 0 \le p < 1 .$$

This result was later extended to all $p \neq 1$ and generalized by Gajda [5] and Rassias [15]. Since then several stability problems for various functional equations have been investigated by numerous mathematicians (see [3,9,12,16–19]). Eskandani [4], Rassias and Kim [13], Najati and Eskandani [11] investigated different functional equations in quasi-Banach spaces.

For a real constant $c$, the quadratic function $f(x) = cx^2$ satisfies the equation

$$f(x+y) + f(x-y) = 2f(x) + 2f(y) . \tag{10.1}$$

Hence, (10.1) is called a quadratic functional equation, and each solution of the quadratic functional equation (10.1) is called a quadratic function.

The Hyers–Ulam stability of the quadratic functional equation was first proved by Skof [21] for functions $f : E_1 \to E_2$, where $E_1$ is a normed space and $E_2$ a Banach space. Cholewa [2] demonstrated that Skof's theorem is also valid if $E_1$ is replaced by an Abelian group $G$. Bae and Jun [1] investigated the generalized Hyers–Ulam–Rassias stability of an $n$-dimensional quadratic functional equation

$$f\left(\sum_{i=1}^{n} x_i\right) + \sum_{1\le i<j\le n} f(x_i - x_j) = n\sum_{i=1}^{n} f(x_i) \qquad (n \ge 2) . \tag{10.2}$$

In this paper, we establish the Hyers–Ulam stability of (10.2) in quasi-Banach spaces.

We recall some basic facts concerning quasi-Banach spaces and some preliminary results.

**Definition 10.1.** Let $X$ by a real linear space. A quasi-norm is a real linear function on $X$ satisfying the following three conditions.

1. $\|x\| \ge 0$ for all $x \in X$ and $\|x\| = 0$ if and only if $x = 0$.
2. $\|\lambda x\| = |\lambda|\|x\|$ for all $x \in X, \lambda \in \mathbb{R}$.
3. There is a constant $K \ge 1$ such that $\|x+y\| \le K(\|x\| + \|y\|)$ for all $x, y \in X$.

The pair $(X, \|\cdot\|)$ is called a *quasi-normed space* if $\|\cdot\|$ is a quasi-norm on $X$. The smallest possible value of $K$ is called the modulus of concavity of $\|\cdot\|$. A *quasi-Banach space* is a complete quasi-normed space.

A quasi-norm $\|\cdot\|$ is called a $p$-norm $(0 < p < 1)$ if

$$\|x+y\|^p \leq \|x\|^p + \|y\|^p , \qquad \forall x,y \in X .$$

In this case, a quasi-Banach space is called a $p$-Banach space. By the Aoki–Rolewicz Theorem [20], each quasi-norm is equivalent to some $p$-norm. Since it is easier to work with $p$-norms than quasi-norms, henceforth we restrict our attention mainly to $p$-norms.

## 10.2 Main Results

Throughout this section, let $E$ and $F$ be a real vector space and a real p-Banach space, respectively. For a given mapping $f : E \to F$, we define

$$Df(x_1,x_2,...,x_n) = f\left(\sum_{i=1}^{n} x_i\right) + \sum_{1\leq i<j\leq n} f(x_i - x_j) - n\sum_{i=1}^{n} f(x_i) \tag{10.3}$$

for all $x_1, x_2, ..., x_n \in E$.

**Theorem 10.1.** *Let $\phi : E \to [0,\infty)$ be a function such that*

$$\Phi(x,x,...,x) = \sum_{k=1}^{\infty} \frac{1}{n^{2kp}} \phi(n^{k+1}x,...,n^{k+1}x)^p < \infty \tag{10.4}$$

*and*

$$\lim_{k\to\infty} \frac{1}{n^{2k}} \phi(n^k x_1,...,n^k x_n) = 0 \tag{10.5}$$

*for all $x, x_1, ..., x_n \in E$. If a function $f : E \to F$ satisfies the inequality*

$$\|Df(x_1,x_2,...,x_n)\| \leq \phi(x_1,x_2,...,x_n) \tag{10.6}$$

*for all $x_1, x_2, ..., x_n \in E$, then there exists a unique n-dimensional quadratic function $A : E \to F$ such that*

$$\|f(x) - A(x)\| \leq (\Phi(x,x,...,x))^{\frac{1}{p}} + \frac{Mn^2}{\sqrt[p]{n^{2p}-1}} \tag{10.7}$$

*for all $x$ in $E$. Moreover, if $f(tx)$ is continuous in $t$ for each fixed $x$ in $E$, then $A(rx) = r^2 A(x)$ for all $r \in \mathbb{R}, x \in E$.*

*Proof.* Putting $x_1 = x_2 = \cdots = x_n = x$ in inequality (10.4) yields

$$\left\| f(nx) + \frac{n(n-1)}{2} f(0) - n^2 f(x) \right\| \leq \phi(x,x,\ldots,x) \, ,$$

$$\|f(nx) - n^2 f(x)\|^p \leq \phi(x,x,\ldots,x)^p + \left( \frac{n(n-1)}{2} \|f(0)\| \right)^p .$$

Thus

$$\left\| f(x) - \frac{1}{n^2} f(nx) \right\|^p \leq \frac{1}{n^{2p}} \phi(x,x,\ldots,x)^p + \left( \frac{n-1}{2n} \|f(0)\| \right)^p \tag{10.8}$$

for all $x \in E$. Write $M = \frac{n-1}{2n} \|f(0)\|$. By induction on $n$ we show

$$\left\| f(x) - \frac{1}{n^{2k}} f(n^k x) \right\|^p \leq \sum_{i=1}^{k} \frac{1}{n^{2pi}} \phi(n^{i-1}x,\ldots,n^{i-1}x)^p + M \sum_{i=0}^{k-1} \frac{1}{n^{2pi}} \, . \tag{10.9}$$

The inequality (10.8) yields the validity of inequality (10.9) for $k = 1$. Assume now that inequality (10.9) holds for $k$. We shall prove it for $k+1$. Since the quasi-norm is p-subadditive, we get

$$\begin{aligned}
&\left\| f(x) - \frac{1}{n^{2k+2}} f(n^{k+1}x) \right\|^p \\
&= \left\| f(x) - \frac{1}{n^{2k}} f(n^k x) + \frac{1}{n^{2k}} f(n^k x) - \frac{1}{n^{2k+2}} f(n^{k+1}x) \right\|^p \\
&\leq \left\| f(x) - \frac{1}{n^{2k}} f(n^k x) \right\|^p + \left\| \frac{1}{n^{2k}} f(n^k x) - \frac{1}{n^{2k+2}} f(n^{k+1}x) \right\|^p \\
&\leq \sum_{i=1}^{k} \frac{\phi(n^{i-1}x,\ldots,n^{i-1}x)^p}{n^{2pi}} + \sum_{i=0}^{k-1} \frac{M^p}{n^{2pi}} + \frac{1}{n^{2pk}} \left[ \frac{\phi(n^k x,\ldots,n^k x)^p}{n^{2p}} + M^p \right] \\
&= \sum_{i=1}^{k+1} \frac{1}{n^{2pi}} \phi(n^{i-1}x,\ldots,n^{i-1}x)^p + M^p \sum_{i=0}^{k} \frac{1}{n^{2pi}} \, .
\end{aligned}$$

This completes the proof of inequality (10.9).

For $k > l > 0$ and $x \in E$, we have

$$\begin{aligned}
&\left\| \frac{1}{n^{2k}} f(n^k x) - \frac{1}{n^{2l}} f(n^l x) \right\|^p \\
&= \frac{1}{n^{2lp}} \left\| f(n^l x) - \frac{1}{n^{2(k-l)}} f(n^{k-l} \cdot n^l x) \right\|^p \\
&\le \frac{1}{n^{2lp}} \left( \sum_{i=1}^{k-l} \frac{1}{n^{2pi}} \phi(n^{i+l-1} x, \dots, n^{i+l-1} x)^p + M^p \sum_{i=0}^{k-l-1} \frac{1}{n^{2pi}} \right) \\
&= \sum_{i=l+1}^{k} \frac{1}{n^{2pi}} \phi(n^{i-1} x, \dots, n^{i-1} x)^p + M^p \sum_{i=l}^{k-l} \frac{1}{n^{2pi}} \,.
\end{aligned}$$

Consequently, it follows from inequality (10.4) that the sequence $\{n^{-2k} f(n^k x)\}$ is a Cauchy sequence in $F$ for all $x \in E$. Since $F$ is complete, the sequence converges in $F$ for all $x \in E$. So one can define the mapping $A : E \to F$ by

$$A(x) = \lim_{k \to \infty} \frac{1}{n^{2k}} f(n^k x), \qquad \forall x \in E \,.$$

Passing to the limit as $k \to \infty$ in inequality (10.9) we obtain inequality (10.7).

It follows from inequality (10.6) and equality (10.5) that

$$\begin{aligned}
\|DA(x_1, x_2, \dots, x_n)\| &= \lim_{k \to \infty} \frac{1}{n^{2k}} \|Df(n^k x_1, n^k x_2, \dots, n^k x_n)\| \\
&\le \lim_{k \to \infty} \frac{1}{n^{2k}} \phi(n^k x_1, n^k x_2, \dots, n^k x_n) = 0
\end{aligned}$$

for all $x \in E$. Hence, the mapping $A$ is an $n$-dimensional quadratic function. To show the uniqueness of $A$, let $C : E \to F$ be also $n$-dimensional quadratic function satisfying inequality (10.7). Then

$$\begin{aligned}
\|A(x) - C(x)\|^p &= \frac{1}{n^{2pk}} \|A(n^k x) - C(n^k x)\|^p \\
&\le \frac{1}{n^{2pk}} (\|A(n^k x) - f(n^k x)\|^p \\
&\quad + \|f(n^k x) - C(n^k x)\|^p) \\
&\le \frac{1}{n^{2pk}} 2\Phi(n^k x, n^k x, \dots, n^k x)
\end{aligned}$$

for all $x \in E$. Passing to the limit as $k \to \infty$, we obtain $A = C$.

It is obvious that $A(rx) = r^2A(x)$ holds for all $x \in E$ and all $r \in \mathbb{Q}$. To prove that $A$ is a homogeneous mapping of degree 2 for all real numbers as well, it suffices to prove that $A(tx)$ is continuous in $t$ for each fixed $x$ in $E$, which is ensured by the continuity of $f$ (see [7] for more details). □

**Theorem 10.2.** *Let $\phi : E \to [0,\infty)$ be a function such that $\phi(0,0,...,0) = 0$,*

$$\Phi(x,x,...,x) = \sum_{k=0}^{\infty} n^{2kp}\phi\left(\frac{x}{n^{k+1}},...,\frac{x}{n^{k+1}}\right)^p < \infty \tag{10.10}$$

*and*

$$\lim_{k\to\infty} n^{2k}\phi\left(\frac{x_1}{n^k},...,\frac{x_n}{n^k}\right) = 0 \tag{10.11}$$

*for all $x_1,x_2,...,x_n \in E$. If a function $f : E \to F$ satisfies the inequality*

$$\|Df(x_1,x_2,...,x_n)\| \le \phi(x_1,x_2,...,x_n) \tag{10.12}$$

*for all $x_1,x_2,...,x_n \in E$, then there exists a unique n-dimensional quadratic function $A : E \to F$ such that*

$$\|f(x) - A(x)\| \le (\Phi(x,x,...,x))^{\frac{1}{p}} \tag{10.13}$$

*for all $x$ in $E$. If, moreover, $f(tx)$ is continuous in $t$ for each fixed $x$ in $E$, then $A(rx) = r^2A(x)$ for all $r \in \mathbb{R}, x \in E$.*

*Proof.* It follows from $\phi(0,...,0) = 0$ that $f(0) = 0$. Similarly to the proof of Theorem 10.1, we have

$$\|f(nx) - n^2f(x)\| \le \phi(x,x,...,x)\,.$$

Replacing $x$ by $\frac{1}{n}x$, we get

$$\left\|f(x) - n^2f\left(\frac{x}{n}\right)\right\| \le \phi\left(\frac{x}{n},\frac{x}{n},...,\frac{x}{n}\right). \tag{10.14}$$

Replacing $x$ by $n^{-k}x$ in (10.14) and multiplying both sides by $n^{2k}$, we get

$$\left\|n^{2k}f\left(\frac{x}{n^k}\right) - n^{2k+2}f\left(\frac{x}{n^{k+1}}\right)\right\| \le n^{2k}\phi\left(\frac{x}{n^{k+1}},\frac{x}{n^{k+1}},...,\frac{x}{n^{k+1}}\right) \tag{10.15}$$

for all $x$ in $E$. For $k > l > 0$,

$$\begin{aligned}
&\left\| n^{2k} f\left(\frac{x}{n^k}\right) - n^{2l} f\left(\frac{x}{n^l}\right)\right\|^p \\
&\quad = \left\| \sum_{i=l}^{k-1} n^{2i} f\left(\frac{x}{n^i}\right) - n^{2(i+1)} f\left(\frac{x}{n^{i+l}}\right)\right\|^p \\
&\quad \leq \sum_{i=l}^{k-1} n^{2pi} \left\| f\left(\frac{x}{n^i}\right) - n^{2k+2} f\left(\frac{x}{n^{i+1}}\right)\right\|^p \\
&\quad \leq \sum_{i=l}^{k-1} n^{2pi} \phi\left(\frac{x}{n^{i+1}}, \ldots, \frac{x}{n^{i+1}}\right)^p . \qquad (10.16)
\end{aligned}$$

It follows from equality (10.10) that $\{n^{2k} f(n^{-k}x)\}$ is a Cauchy sequence in $F$. Because of the fact that $F$ is complete, the limit function

$$A(x) = \lim_{k\to\infty} n^{2k} f\left(\frac{x}{n^k}\right)$$

exists. Let $l = 0$ in inequality (10.16) and passing to the limit as $k \to \infty$, we get inequality (10.13). The rest of the proof is similar to that of Theorem 10.1. □

**Definition 10.2 ([13]).** Assume that $A$ and $(B, \leq)$ are closed under addition. We say that a function $\phi : A \to B$ is contractively subadditive if there exists a real constant $L$ with $0 < L < 1$ such that

$$\phi(x+y) \leq L(\phi(x) + \phi(y)) , \qquad \forall x, y \in A .$$

We say that $\phi : A \to (B, \geq)$ is expansively superadditive if

$$\phi(x+y) \geq \frac{1}{L}(\phi(x) + \phi(y)) , \qquad \forall x, y \in A .$$

If $\phi : A \to B$ is contractively subadditive, then $\phi$ satisfies $\phi(2x) \leq 2L\phi(x)$ and $\phi(2^k x) \leq (2L)^k \phi(x)$; thus, $\phi(\lambda x) \leq \lambda L \phi(x)$ and $\phi(\lambda^k x) \leq (\lambda L)^k \phi(x)$ for all $x \in A$, $k \in \mathbb{N}$ and positive integer $\lambda \geq 2$. Similarly, if $\phi : A \to B$ is expansively superadditive, then

$$\phi(\lambda x) \geq \frac{\lambda}{L} \phi(x)$$

and

$$\phi(\lambda^{-k} x) \leq \left(\frac{\lambda}{L}\right)^k \phi(x)$$

for all $x \in A$, $k \in \mathbb{N}$ and positive integer $\lambda \geq 2$.

**Corollary 10.1.** *Let $\phi : E \to [0,\infty)$ be a contractive subadditive function with constant L, where $L \leq n$. If a function $f : E \to F$ satisfies the inequality*

$$\|Df(x_1,x_2,...,x_n)\| \leq \phi(x_1,x_2,...,x_n)$$

*for all $x_1,x_2,...,x_n \in E$, then there exists a unique n-dimensional quadratic function $A : E \to F$. Such that*

$$\|f(x)-A(x)\| \leq \frac{1}{n}\phi(x,x,...,x)\frac{1}{\sqrt[p]{n^p-L^p}}+Mn^2\frac{1}{\sqrt[p]{n^{2p}-1}} \tag{10.17}$$

*for all x in E, where*

$$M = \frac{n-1}{2n}\|f(0)\|.$$

*Moreover, if $f(tx)$ is continuous in t for each fixed x in E, then*

$$A(rx) = r^2A(x)$$

*for all $r \in \mathbb{R}, x \in E$.*

*Proof.* It follows from (10.9) and contractive subadditivity of $\phi$ that

$$\begin{aligned}\left\|f(x)-\frac{1}{n^{2k}}f(n^kx)\right\|^p &\leq \sum_{i=1}^{k}\frac{1}{n^{2pi}}\phi(n^{i-1}x,...,n^{i-1}x)^p+M^p\sum_{i=0}^{k-1}\frac{1}{n^{2pi}}\\ &\leq \sum_{i=1}^{k}\frac{1}{n^{2pi}}(nL)^{(i-1)p}\phi(x,...,x)^p+M^p\sum_{i=0}^{k-1}\frac{1}{n^{2pi}}\\ &= \frac{1}{n^{2p}}\phi(x,...,x)^p\sum_{i=1}^{k}\left(\frac{L}{n}\right)^{(i-1)p}+M^p\sum_{i=0}^{k-1}\frac{1}{n^{2pi}}.\end{aligned}$$

For $k > l > 0$, we get

$$\left\|\frac{f(n^kx)}{n^{2k}}-\frac{f(n^lx)}{n^{2l}}\right\| \leq \frac{1}{n^{2p}}\phi(x,...,x)^p\sum_{i=l+1}^{k}\left(\frac{L}{n}\right)^{(i-1)p}+M^p\sum_{i=l}^{k-1}\frac{1}{n^{2pi}}.$$

Thus, $\{n^{-2k}f(n^kx)\}$ is Cauchy sequence in $F$. As $F$ is complete, the limit function

$$A(x) = \lim_{k\to\infty}\frac{f(n^kx)}{n^{2k}}$$

exists and satisfies inequality (10.17).

The rest of the proof is similar to that of Theorem 10.1. □

**Corollary 10.2.** *Let $\phi : E \to [0,\infty)$ be an expansive superadditive function with constant L satisfying $nL < 1$ and*

$$\phi(0,0,...,0) = 0.$$

*If a function $f : E \to F$ satisfies the inequality*

$$\|Df(x_1,x_2,...,x_n)\| \leq \phi(x_1,x_2,...,x_n)$$

*for all $x_1,x_2,...,x_n \in E$, then there exists a unique n-dimensional quadratic function $A : E \to F$, such that*

$$\|f(x) - A(x)\| \leq \frac{L}{n} \frac{1}{\sqrt[p]{1-(nL)^p}} \phi(x,x,...,x) \tag{10.18}$$

*for all x in E. If, moreover, $f(tx)$ is continuous in t for each fixed x in E, then*

$$A(rx) = r^2 A(x)$$

*for all $r \in \mathbb{R}, x \in E$.*

*Proof.* It follows from inequality (10.16) and expansive superadditivity of $\phi$ that

$$\begin{aligned}
\left\| n^{2k} f\left(\frac{x}{n^k}\right) - n^{2l} f\left(\frac{x}{n^l}\right) \right\|^p &\leq \sum_{i=l}^{k-1} n^{2pi} \phi\left(\frac{x}{n^{i+1}},...,\frac{x}{n^{i+1}}\right)^p \\
&\leq \sum_{i=l}^{k-1} n^{2pi} \left(\frac{L}{n}\right)^{(i+1)p} \phi(x,...,x)^p \\
&= \phi(x,...,x)^p L^{2p} \sum_{i=l}^{k-1} (nL)^{(i-1)p}.
\end{aligned}$$

Thus, $\{n^{2k} f(n^{-k}x)\}$ is Cauchy sequence in $F$. Because of the fact that $F$ is complete, the limit function

$$A = \lim_{k \to \infty} n^{2k} f\left(\frac{x}{n^k}\right)$$

exists. Let $l = 0$. Passing to the limit with $k \to \infty$, we get inequality (10.18).

The rest of the proof is similar as in the proof of Theorem 10.2. □

## 10.3 Stability Using Alternative Fixed Point

The first systematic study of fixed point theorems in nonlinear analysis is due to G.Isac and Th.M.Rassias [8]. In this section, we will investigate the stability of the given $n$-dimensional quadratic function (10.2) using the alternative fixed point.

**Theorem 10.3 ([10] the alternative of fixed point).** *Suppose that we are given a complete generalized metric space $(\Omega, d)$ and a strictly contractive mapping $T : \Omega \to \Omega$ with Lipschitz constant L. Then, for each given $x \in \Omega$, either*

$$d(T^m x, T^{n+1} x) = \infty, \qquad \forall n \in \mathbb{N},$$

*or there exists a natural number $n_0$ such that*

1. $d(T^m x, T^{n+1} x) = \infty$ *for all* $n \geq n_0$*;*
2. *the sequence* $\{T^n x\}$ *is convergent to a fixed point* $y^*$ *of* $T$*;*
3. $y^*$ *is the unique fixed point of* $T$ *in the set*

$$\Delta = \{y \in \Omega : d(T^{n_0} x, y) < \infty\} ;$$

4. $d(y, y^*) \leq \frac{1}{1-L} d(y, Ty)$ *for all* $y \in \Delta$.

Let $\phi : E^n \to [0, \infty)$ be a function such that

$$\lim_{k \to \infty} \frac{\phi(\lambda_i^k x_1, \ldots, \lambda_i^k x_n)}{\lambda_i^{2k}} = 0 \tag{10.19}$$

for all $x_1, x_2, \ldots, x_n \in E$, where $\lambda_i = n$ if $i = 0$ and $\lambda_i = \frac{1}{n}$ if $i = 1$.

**Theorem 10.4.** *Suppose that a function $f : E \to F$ satisfies*

$$\|Df(x_1, \ldots, x_n)\| \leq \phi(x_1, \ldots, x_n) \tag{10.20}$$

*for all $x_1, \ldots, x_n \in E$, and $f(0) = 0$. If there exists $L < 1$ such that*

$$\psi(x) = \phi\left(\frac{x}{n}, \ldots, \frac{x}{n}\right) \tag{10.21}$$

*has the property*

$$\psi(\lambda_i x) \leq L \cdot \lambda_i^2 \cdot \psi(x) \tag{10.22}$$

*for all $x \in E$, then there exists a unique n-dimensional quadratic function $A : E \to F$ such that*

$$\|f(x) - A(x)\| \leq \frac{L^{1-i}}{1-L} \psi(x)$$

*for all $x \in E$.*

*Proof.* Consider the set $\Omega = \{g : E \to E : \ g(0) = 0\}$ and introduce a generalized metric on $\Omega$ as follows:

$$d(f, g) = d(f, g)_\psi = \inf\{K \in [0, \infty) : \|f(x) - g(x) \leq K\psi(x)\} .$$

Then it is easy to see that $(\Omega, d)$ is a generalized complete metric space. We define a mapping $T : \Omega \to \Omega$ by: $Tf(x) = \lambda_i^{-2} f(\lambda_i x)$ for all $x \in E$. Note that, for all $f, g \in \Omega$ with $d(f,g) < K$, we get

$$\begin{aligned} \|Tf(x) - Tg(x)\| &= \left\| \frac{1}{\lambda_i^2} f(\lambda_i x) - \frac{1}{\lambda_i^2} g(\lambda_i x) \right\| \\ &= \frac{1}{\lambda_i^2} \|f(\lambda_i x) - g(\lambda_i x)\| \\ &\leq \frac{1}{\lambda_i^2} K \psi(\lambda_i x) \leq K\, L\, \psi(x)\,. \end{aligned}$$

Hence, we get $d(Tf, Tg) \leq K \cdot d(f,g)$ for all $f, g \in \Omega$. So $T$ is a strictly self-mapping of $\Omega$ with Lipschitz constant $L$.

Putting $x_1 = x_2 = \ldots = x_n = x$ and dividing both sides by $n^2$ in inequality (10.20) and using inequality (10.22) for $i = 0$ yields

$$\|f(x) - Tf(x)\| = \left\| f(x) - \frac{1}{n^2} f(nx) \right\| \leq \frac{1}{n^2}\, \psi(nx) \leq L\, \psi(x)$$

for all $x \in E$, that is $d(f, Tf) \leq L$.

Putting $x_1 = x_2 = \ldots = x_n = x$ and replacing $x$ by $\frac{1}{n} x$ in inequality (10.20) and using inequality (10.22) for $i = 1$ yields

$$\|f(x) - Tf(x)\| = \left\| f(x) - n^2 f\left(\frac{x}{n}\right) \right\| \leq \psi(x)$$

for all $x \in E$, that is $d(f, Tf) \leq L^0 = 1$.

Now from the fixed point alternative in both cases, there exists a fixed point $A$ of $T$ such that

$$A(x) = \lim \frac{1}{\lambda_i^2} f(\lambda_i x)$$

for all $x \in E$.

It follows from (10.20) that

$$\begin{aligned} \|DA(x_1, \ldots, x_n)\| &= \lim_{k \to \infty} \frac{1}{\lambda_i^2} \|Df(\lambda_i x_1, \ldots, \lambda_i x_n)\| \\ &\leq \lim_{k \to \infty} \frac{\phi(\lambda_i^k x_1, \ldots, \lambda_i^k x_n)}{\lambda_i^{2k}} = 0\,. \end{aligned}$$

Hence, $A$ is an $n$-dimensional quadratic function. According to the fixed point alternative, $A$ is the unique function such that

$$\|f(x) - A(x)\| \leq K \cdot \psi(x)$$

for some $K > 0$. Again using the fixed point alternative, we have

$$d(f,A) \le \frac{1}{1-L} d(f,Tf).$$

Hence, we conclude that

$$d(f,A) \le \frac{L^{1-i}}{1-L}.$$

□

**Corollary 10.3.** *Assume that* $\varepsilon \ge 0, \delta \ge 0$ *and* $p \ge 0$ *with* $p \ne 2$ *are given. Suppose a function* $f : E \to F$ *satisfies the inequality*

$$\|Df(x_1,\dots,x_n)\| \le \delta + \varepsilon \sum_{i=1}^{n} \|x_i\|^p$$

*for all* $x_1,\dots,x_n \in E$. *Furthermore, assume* $f(0) = 0$ *and* $\delta = 0$ *if* $p > 2$. *Then there exists a unique n-dimensional quadratic function* $A : E \to F$ *such that the inequality*

$$\|f(x) - A(x)\| \le \delta \frac{n^p}{n^2 - n^p} + \varepsilon \frac{\|x\|^p}{n - n^{p-1}}$$

*holds for all* $x \in E$ *and* $p < 2$, *or*

$$\|f(x) - A(x)\| \le \varepsilon \frac{\|x\|^p}{n^{p-1} - n}$$

*holds for all* $x \in E$ *and* $p > 2$.

*Proof.* Let

$$\phi(x_1,\dots,x_n) = \delta + \varepsilon \sum_{i=1}^{n} \|x_i\|^p$$

for all $x_1,\dots,x_n \in E$. Then

$$\lim_{k\to\infty} \frac{\phi(\lambda_i^k x_1,\dots,\lambda_i^k x_n)}{\lambda_i^{2k}} = \lim_{k\to\infty} \left( \frac{\delta}{\lambda_i^{2k}} + \lambda_i^{k(p-2)} \varepsilon \sum_{i=1}^{n} \|x_i\|^p \right) = 0$$

holds for all $x_1,\dots,x_n \in E$, where $p < 2$ if $i = 0$ and $p > 2, \delta = 0$ if $i = 1$.

It is easy to see that the inequality

$$\frac{1}{\lambda_i^2} \psi(\lambda_i x) = \frac{\delta}{\lambda_i^2} + \lambda_i^{p-2} \varepsilon \sum_{i=1}^{n} \|x_i\|^p \le \lambda_i^{p-2} \psi(x)$$

holds for all $x \in E$, where $p < 2$ if $i = 0$ and $p > 2$ if $i = 1$. Hence, the inequality (10.22) holds for $L = n^{p-2}$ when $p < 2$ and $L = n^{2-p}$ when $p > 2$. It follows from Theorem 10.4 that the conclusion in this corollary holds. □

**Corollary 10.4.** *Assume that $\theta \geq 0$ is given, if a function $f : E \to F$ satisfies the inequality*

$$\|Df(x_1,...,x_n)\| \leq \theta$$

*for all $x_1,...,x_n \in E$. Then there exists a unique n-dimensional quadratic function $A : E \to F$ such that*

$$\|f(x) - A(x)\| \leq \frac{\varepsilon}{n^2 - 1}$$

*for all $x \in E$.*

*Proof.* Putting $\delta = 0, p = 0$ and $\theta = \varepsilon/n$ in Corollary 3.3 yields the conclusion in this corollary. □

**Acknowledgements** I would like to express my sincere gratitude to Professor Ding Guanggui for his guidance and convey my heartfelt thanks to Professor Themistocles M.Rassias for his valuable comments.

The author was supported in part by Research Foundation for Doctor Programme (Grant No. 20060055010) and National Natural Science Foundation of China (Grant No. 10871101).

## References

1. Bae, J.H., Jun, K. W.: On the generalized Hyers-Ulam-Rassias stability of an n-dimensional quadratic functional equation. J. Math. Anal. Appl. **258**, 183–193 (2001)
2. Cholewa, P.W.: Remarks on the stability of functional equations. Aequationes Math. **27**, 76–86 (1984)
3. Czerwik, S.: Functional Equations and Inequalities in Several Variables. World Scientific Publishging Company, Singapore-New Jersey-London (2002)
4. Eskandani, G.Z.: On the Hyers-Ulam-Rassias stability of an addtive functional equation in quasi-Banach spaces. J. Math. Anal. Appl. **345**, 405–409 (2008)
5. Gajda, Z.: On stability of additive mappings. Int. J. Math. Math. Sci. **14**, 431–434 (1991)
6. Hyers, D.H.: On the stability of the linear functional equation. Proc. Nat. Acad. Sci. U.S.A. **27**, 222–224 (1941)
7. Hyers, D.H., Isac, G., Rassias, Th.M.: Stability of Functional Equations in Several Variables. Birkhäuser, Boston (1998)
8. Isac, G., Rassias, Th.M.: Stability of $\psi$-additive mappings: Applications to nonlinear analysis. Int. J. Math. Math. Sci. **19**, 219–228 (1996)
9. Jung, S.M.: Hyers-Ulam-Rassias Stability of Functional Equations in Mathematical Analysis. Hadronic Press Inc., Palm Harbor, Florida (2001)
10. Kang, D.S., Chu, H.Y.: Stability problem of Hyers-Ulam -Rassias for generalized forms of cubic functional equation. Acta. Math. Sin. **24**, 491–502 (2008)
11. Najiati, A., Eskandani, G.Z.: Stability of a mixed additive and cubic functional equation in quasi-Banach spaces. J. Math. Anal. Appl. **342**, 1318–1331 (2008)
12. Park, C.G., Rassias, Th.M.: Hyers-Ulam stability of a generalized Apollonius type quadratic mapping, J. Math. Anal. Appl. **322** (2006), 371–381
13. Rassias, J.M., Kim, H.M.: Generalized Hyers-Ulam stability for general additive functional equations in quasi-$\beta$-normed spaces. J. Math. Anal. Appl. **356**, 302–309 (2009)
14. Rassias, Th.M.: On the stability of the linear mapping in Banach spaces. Proc. Amer. Math. Soc. **72**, 297–300 (1978)

15. Rassias, Th.M.: Communication. 27th International Symposium on Functional Equations, Bielsko-Biała, Katowice, Kraków, Poland, 1989.
16. Rassias, Th.M.: On the stability of the quadratic functional equation and its applications. Studia Univ. Babes-Bolyai Math. **43(3)**, 89–124 (1998)
17. Rassias, Th.M.: On the stability of functional equations and a problem of Ulam. Acta Appl. Math. **62**, 23–130 (2000)
18. Rassias, Th.M.: Functional Equations and Inequalities. Kluwer Academic Publishers, Dordrecht (2000)
19. Rassias, Th.M.: Functional Equations, Inequalities and Applications Kluwer Academic Publishers, Dordrecht (2003)
20. Rolewicz, S.: Metric linear spaces. PWN-Polish Sci. Publ./Reidel, Warszawa-Dordrecht (1984)
21. Skof, F.: Proprieta locali e approssimazione di operatori. Rend. Sem. Mat. Fis. Milano. **53**, 113–129 (1983)
22. Ulam, S.M.: Problems in Modern Mathematics, Chapter VI. Science Editions, Wiley, New York (1960)

# Chapter 11
# Ulam Stability Problem for Frames

**Laura Găvruţa and Paşc Găvruţa**

*Dedicated to the memory of S.M. Ulam*
*on the 100th anniversary of his birth*

**Abstract** In this paper we give a solution to the Ulam stability problem for continuous Parseval frames, in finite dimensional Hilbert spaces. We prove that if $F$ is a nearly Parseval frame then there exists a Parseval frame near $F$. Also, we give generalizations of this result.

**Keywords** Hyers–Ulam-Rassias stability • Frames

**Mathematics Subject Classification (2010):** Primary 39B82, 42C15

## 11.1 Introduction

The notion of frames was first introduced by Duffin and Schaeffer [10] in 1952 to study some deep problems in nonharmonic Fourier series. After the fundamental paper [9] by Daubechies, Grossman and Meyer, frame theory began to be widely used, particularly in the more specialized context of wavelet frames and Gabor frames. Traditionally, frames have been used in signal processing, image processing, data compression and sampling in sampling theory. For other applications of frame theory see the references of paper [15].

A discrete frame is a countable family of elements in a separable Hilbert space which allows for a stable, not necessarily unique, decomposition of an arbitrary element into an expansion of the frame elements. The concept of a generalization of frames to a family indexed by some locally compact space endowed with a Radon measure was proposed by Kaiser [25] and independently by Ali, Antoine and Gazeau [1] (see also [26]). These frames are known as continuous frames.

L. Găvruţa (✉) • P. Găvruţa
Department of Mathematics, "Politehnica" University of Timişoara,
Piaţa Victoriei 2, 300006, Timişoara, Romania
e-mail: Laura.Gavruta@mat.upt.ro; pgavruta@yahoo.com

Th.M. Rassias and J. Brzdęk (eds.), *Functional Equations in Mathematical Analysis*, Springer Optimization and Its Applications 52, DOI 10.1007/978-1-4614-0055-4_11,

Gabardo and Han [12] called them *frames associated with measurable spaces*; in mathematical physics they are referred to as *coherent states* [1]. If in the definition of a continuous frame, the measure space $\Omega := \mathbb{N}$ and $\mu$ is the counting measure, the continuous frame will be a discrete frame. For details the reader could refer to[7].

**Definition 11.1.** Let $\mathcal{H}$ be a Hilbert space and $(\Omega,\mu)$ be a measure space with positive measure $\mu$. A mapping $F : \Omega \to \mathcal{H}$ is called *a continuous frame* with respect to $(\Omega,\mu)$ if

(*i*) $F$ is weakly measurable, i.e., for all $f \in \mathcal{H}$,

$$\omega \to \langle f, F(\omega)\rangle$$

is a measurable function on $\Omega$;

(*ii*) there exist constants $A, B > 0$ such that

$$A\|x\|^2 \le \int_\Omega |\langle x, F(\omega)\rangle|^2 \mathrm{d}\mu(\omega) \le B\|x\|^2, \quad x \in \mathcal{H}. \tag{11.1}$$

The constants $A$ and $B$ are called *continuous frame bounds*. If $A = B$, then F is called *tight continuous frame* and if $A = B = 1$ it is called a *Parseval frame*. The mapping $F$ is called *Bessel* if the second inequality in (11.1) holds. In this case, $B$ is called *Bessel constant*. If $\mu$ is a counting measure and $\Omega = \mathbb{N}$, $F$ is called a *discrete frame*.

Let $(\Omega,\mu)$ be a measure space and let $F$ be a Bessel mapping from $\Omega$ to $\mathcal{H}$. Then the operator

$$T : L^2(\Omega,\mu) \to \mathcal{H}$$

weakly defined by

$$\langle T\varphi, x\rangle = \int_\Omega \varphi(\omega)\langle F(\omega), x\rangle \mathrm{d}\mu(\omega), \quad x \in \mathcal{H}$$

is well defined, linear, bounded and its adjoint is given by

$$T^* : \mathcal{H} \to L^2(\Omega,\mu), \quad (T^*x)(\omega) = \langle x, F(\omega)\rangle, \quad \omega \in \Omega.$$

The operator $T$ is called *a pre-frame operator or synthesis operator* and $T^*$ is called *an analysis operator* of $F$. By composing $T$ with its adjoint $T^*$ we obtain the *frame operator*

$$S : \mathcal{H} \to \mathcal{H}, \quad Sx = \int_\Omega \langle x, F(\omega)\rangle F(\omega) \mathrm{d}\mu(\omega).$$

We know that $S$ is invertible and self-adjoint positive operator. It is easily shown that a $(\Omega,\mu)$-Bessel mapping $F$ is a $(\Omega,\mu)$-frame for $\mathcal{H}$ if and only if there exists a $(\Omega,\mu)$-Bessel mapping $G$ such that

$$\langle x, y\rangle = \int_\Omega \langle x, G(\omega)\rangle\langle F(\omega), y\rangle \mathrm{d}\mu(\omega), \quad x, y \in \mathcal{H}.$$

We call $G$ a *dual frame* for $F$ and $(F,G)$ a *dual pair*.

If $S$ is the frame operator for a $(\Omega,\mu)$-frame, then $S^{-1/2}F$ is a normalized tight $(\Omega,\mu)$-frame and $S^{-1}F$ is the dual of $F$. This dual is called the *standard dual.*

The stability problem as posed by Ulam [30, p. 63] reads as follows: given a group $G_1$, a metric group $G_2$ with metric $d$ and a positive number $\varepsilon$, find a positive number $\delta$ such that for every $f: G_1 \to G_2$ satisfying

$$d(f(xy), f(x)f(y)) \leq \delta, \quad x,y \in G_1$$

there exists a homomorphism $h: G_1 \to G_2$ with

$$d(f(x), h(x)) \leq \varepsilon, \quad x \in G_1.$$

In 1941, Hyers [21] gave an affirmative answer to the question of Ulam for additive Cauchy equation in Banach spaces.

Let $E_1$, $E_2$ be Banach spaces and let $f: E_1 \to E_2$ be a mapping satisfying:

$$\|f(x+y) - f(x) - f(y)\| \leq \delta.$$

for all $x,y \in E_1$ and $\delta > 0$. There exists a unique additive mapping $T: E_1 \to E_2$ which satisfies

$$\|f(x) - T(x)\| \leq \delta, \quad x \in E_1.$$

Hyers proved that the limit

$$T(x) = \lim_{n\to\infty} 2^{-n} f(2^n x)$$

exists for all $x \in E_1$. A generalized solution to Ulam's problem for approximately linear mappings was proved by Th.M. Rassias [28] in 1978. Th.M. Rassias considered a mapping $f: E_1 \to E_2$ such that $t \to f(tx)$ is continuous in $t$ for each fixed $x$. Assume that there exist $\theta \geq 0$ and $0 \leq p < 1$ such that

$$\|f(x+y) - f(x) - f(y)\| \leq \theta(\|x\|^p + \|y\|^p), \quad x,y \in E_1.$$

Then there exists a unique linear mapping $T: E_1 \to E_2$ such that

$$\|f(x) - T(x)\| \leq \frac{2\theta}{2-2^p}\|x\|^p, \quad x,y \in E_1.$$

Thus, the Hyers' Theorem follows as a special case of Th.M.Rassias' Theorem for $p = 0$. Th.M. Rassias' proof of his Theorem [28] applies as well for all real values of $p$ that are strictly less than zero.

In 1994, P. Găvruţa [14] provided a generalization of Th.M. Rassias' Theorem for the unbounded Cauchy difference and introduced the concept of generalized Hyers–Ulam–Rassias stability in the spirit of Th.M. Rassias approach.

**Theorem 11.1.** *Let G and E be an abelian group and a Banach space, respectively, and let $\varphi : G^2 \to [0,\infty)$ be a function satisfying*

$$\Phi(x,y) = \sum_{k=0}^{\infty} 2^{-k-1}\varphi(2^k x, 2^k y) < \infty$$

*for all $x, y \in G$. If a function $f : G \to E$ satisfies the inequality*

$$\|f(x+y) - f(x) - f(y)\| \leq \varphi(x,y)$$

*for any $x, y \in G$, then there exists a unique additive function $A : G \to E$ with*

$$\|f(x) - A(x)\| \leq \Phi(x,x)$$

*for all $x \in G$. If moreover G is a real normed space and $f(tx)$ is continuous in t for each fixed x in G, then A is a linear function.*

For more details see [3, 8, 18, 22, 23, 27, 28].

In this paper, we solve Ulam stability problem for continuous Parseval frames. For the case of wavelet transform, a solution of Ulam stability problem was posed by the second author in [17].

## 11.2 Continuous Frames in Finite Dimensional Hilbert Spaces

Let $F : \Omega \to \mathscr{H}$ be a weakly-measurable mapping with the property

$$\int_{\Omega} \|F(\omega)\|^2 \mathrm{d}\mu(\omega) < \infty. \tag{11.2}$$

**Proposition 11.1.** *If F satisfies (11.2), then F is a Bessel mapping.*

*Proof.* If

$$B = \int_{\Omega} \|F(\omega)\|^2 \mathrm{d}\mu(\omega) < \infty$$

using Cauchy–Schwarz inequality, we have

$$\int_{\Omega} |\langle x, F(\omega)\rangle|^2 \mathrm{d}\mu(\omega) \leq \int_{\Omega} \|x\|^2 \|F(\omega)\|^2 \mathrm{d}\mu(\omega) \leq B\|x\|^2.$$

□

In the following, we will prove that the converse of Proposition 11.1 holds if $\mathscr{H}$ is finite dimensional.

**Proposition 11.2.** *Let $\mathscr{H}$ be an n-dimensional Hilbert space and $F : \Omega \to \mathscr{H}$ be a Bessel mapping. Then F satisfies (11.2).*

*Proof.* Let

$$\{e_1, \ldots, e_n\}$$

be an orthonormal basis for $\mathscr{H}$. By Parseval identity for $\omega \in \Omega$ we have

$$\|F(\omega)\|^2 = \sum_{k=1}^{n} |\langle F(\omega), e_k \rangle|^2.$$

It follows

$$\begin{aligned}\int_\Omega \|F(\omega)\|^2 \mathrm{d}\mu(\omega) &= \sum_{k=1}^{n} \int_\Omega |\langle F(\omega), e_k \rangle|^2 \mathrm{d}\mu(\omega) \\ &\leq \sum_{k=1}^{n} B\|e_k\|^2 = Bn < \infty.\end{aligned}$$

□

We will derive a new identity for continuous frames in finite dimensional Hilbert spaces. First, we give a preliminary result.

**Lemma 11.1.** *Let $\mathscr{H}$ be an n-dimensional Hilbert space, $G, K : \Omega \to \mathscr{H}$ be continuous Parseval frames and $L \in \mathscr{L}(\mathscr{H})$ be self-adjoint. Then*

$$\int_\Omega \|LG(\omega)\|^2 \mathrm{d}\mu(\omega) = \int_\Omega \|LK(\omega)\|^2 \mathrm{d}\mu(\omega).$$

*Proof.* Using the Fubini theorem, we have

$$\begin{aligned}\int_\Omega \|LG(\omega)\|^2 \mathrm{d}\mu(\omega) &= \int_\Omega \int_\Omega |\langle LG(\omega), K(\sigma) \rangle|^2 \mathrm{d}\mu(\sigma) \mathrm{d}\mu(\omega) \\ &= \int_\Omega \int_\Omega |\langle LK(\sigma), G(\omega) \rangle|^2 \mathrm{d}\mu(\sigma) \mathrm{d}\mu(\omega) \\ &= \int_\Omega \int_\Omega |\langle LK(\sigma), G(\omega) \rangle|^2 \mathrm{d}\mu(\omega) \mathrm{d}\mu(\sigma) \\ &= \int_\Omega \|LK(\sigma)\|^2 \mathrm{d}\mu(\sigma).\end{aligned}$$

□

**Theorem 11.2.** *Let $F$ be a frame for $\mathscr{H}$ and $G$ a Parseval frame for $\mathscr{H}$, where $\mathscr{H}$ is a finite dimensional Hilbert space. Then it holds:*

$$\int_\Omega \|G(\omega)-F(\omega)\|^2 \mathrm{d}\mu(\omega) = \int_\Omega \|S^{-1/2}F(\omega)-F(\omega)\|^2 \mathrm{d}\mu(\omega) + \int_\Omega \|S^{1/4}G(\omega)-S^{-1/4}F(\omega)\|^2 \mathrm{d}\mu(\omega).$$

*Proof.* From the lemma above, with $K(\omega)=S^{-1/2}F(\omega)$, we have:

$$\int_\Omega \|G(\omega)\|^2 \mathrm{d}\mu(\omega) = \int_\Omega \|S^{-1/2}F(\omega)\|^2 \mathrm{d}\mu(\omega)$$

and

$$\int_\Omega \|S^{1/4}G(\omega)\|^2 \mathrm{d}\mu(\omega) = \int_\Omega \|S^{1/4}(S^{-1/2}F(\omega))\|^2 \mathrm{d}\mu(\omega) = \int_\Omega \|S^{-1/4}F(\omega)\|^2 \mathrm{d}\mu(\omega).$$

Thus

$$\begin{aligned}
&\int_\Omega \|G(\omega)-F(\omega)\|^2 \mathrm{d}\mu(\omega) - \int_\Omega \|S^{-1/2}F(\omega)-F(\omega)\|^2 \mathrm{d}\mu(\omega)\\
&= -2\mathrm{Re}\int_\Omega \langle G(\omega),F(\omega)\rangle \mathrm{d}\mu(\omega) + 2\int_\Omega \langle S^{-1/2}F(\omega),F(\omega)\rangle \mathrm{d}\mu(\omega)\\
&= -2\mathrm{Re}\int_\Omega \langle S^{1/4}G(\omega),S^{-1/4}F(\omega)\rangle \mathrm{d}\mu(\omega)\\
&\quad + \int_\Omega \|S^{-1/4}F(\omega)\|^2 \mathrm{d}\mu(\omega) + \int_\Omega \|S^{1/4}G(\omega)\|^2 \mathrm{d}\mu(\omega)\\
&= \int_\Omega \|S^{-1/4}F(\omega)-S^{1/4}G(\omega)\|^2 \mathrm{d}\mu(\omega).
\end{aligned}$$

□

**Corollary 11.1.** *Let $F$ be a frame for $\mathscr{H}$ finite dimensional Hilbert space, with frame operator $S$. For all Parseval frames $G$ of $\mathscr{H}$, the inequality:*

$$\int_\Omega \|G(\omega)-F(\omega)\|^2 \mathrm{d}\mu(\omega) \geq \int_\Omega \|S^{-1/2}F(\omega)-F(\omega)\|^2 \mathrm{d}\mu(\omega)$$

*takes place, and we have equality, if and only if $G(\omega)=S^{-1/2}F(\omega)$ a.e. $\omega\in\Omega$.*

*Proof.* Inequality follow identity and we have equality if and only if

$$S^{1/4}G(\omega) = S^{-1/4}F(\omega), \quad a.e.\ \omega \in \Omega$$

which is equivalent to

$$G(\omega) = S^{-1/2}F(\omega), \quad a.e.\ \omega \in \Omega.$$

□

For the discrete case one should see [13, 19] and references therein.

## 11.3 Ulam Stability of Parseval Frames

Let $\mathscr{H}$ be a Hilbert space and $F$ be a frame with frame bounds $A, B$. We denote by $S$ the frame operator

$$S : \mathscr{H} \to \mathscr{H}, \quad Sx = \int_{\Omega} \langle x, F(\omega)\rangle F(\omega) \mathrm{d}\mu(\omega).$$

The family $S^{-1/2}F$ is a Parseval frame, for $x \in \mathscr{H}$:

$$\begin{aligned}\int_{\Omega} |\langle x, S^{-1/2}F(\omega)\rangle|^2 \mathrm{d}\mu(\omega) &= \int_{\omega} |\langle S^{-1/2}x, F(\omega)\rangle|^2 \mathrm{d}\mu(\omega)\\ &= \langle S(S^{-1/2}x), S^{-1/2}x\rangle\\ &= \langle S^{-1/2}x, S^{-1/2}x\rangle = \|x\|^2.\end{aligned}$$

In the previous section we provided a proof of the fact that $S^{-1/2}F$ has a special property: it is the nearest by the given frame from all Parseval frames in $\mathscr{H}$. This fact was proved by Casazza and Kutyniok [4] for discrete case and it will be a corollary of a theorem that gives a new identity for frames.

In the following, we consider $0 \le \varepsilon < 1$ and $\mathscr{H}$ a n-dimensional Hilbert space. The frame $F$ is $\varepsilon$ *nearly Parseval* if, for all $x \in \mathscr{H}$,

$$(1-\varepsilon)\|x\|^2 \le \int_{\Omega} |\langle x, F(\omega)\rangle|^2 \mathrm{d}\mu(\omega) \le (1+\varepsilon)\|x\|^2.$$

**Theorem 11.3.** *If $F$ is an $\varepsilon$ near Parseval frame, then there exists a Parseval frame $G$ such that*

$$\int_{\Omega} \|G(\omega) - F(\omega)\|^2 \mathrm{d}\mu(\omega) \le n(1 - \sqrt{1-\varepsilon})^2.$$

*Moreover, the estimation is optimal.*

*Proof.* Let $S$ with eigenvalues $\{\lambda_k\}_{k=1}^n$ and a corresponding orthonormal set of eigenvectors $\{e_k\}_{k=1}^n$. It follows that

$$\|F(\omega)-S^{-1/2}F(\omega)\|^2 = \left\|\sum_{k=1}^n \langle F(\omega),e_k\rangle e_k - \sum_{k=1}^n \frac{1}{\sqrt{\lambda_k}}\langle F(\omega),e_k\rangle e_k\right\|^2$$
$$= \sum_{k=1}^n |\langle F(\omega),e_k\rangle|^2 \left|1-\frac{1}{\sqrt{\lambda_k}}\right|^2$$

and hence

$$\int_\Omega \|F(\omega)-S^{-1/2}F(\omega)\|^2 \mathrm{d}\mu(\omega) = \sum_{k=1}^n \left|1-\frac{1}{\sqrt{\lambda_k}}\right|^2 \int_\Omega |F(\omega),e_k|^2 \mathrm{d}\mu(\omega)$$
$$= \sum_{k=1}^n \left|1-\frac{1}{\sqrt{\lambda_k}}\right|^2 \langle Se_k,e_k\rangle$$
$$= \sum_{k=1}^n (\sqrt{\lambda_k}-1)^2.$$

We have

$$1-\varepsilon \le \lambda_k \le 1+\varepsilon, \quad k=1,2,\dots,n.$$

Since

$$\sqrt{1-\varepsilon}+\sqrt{1+\varepsilon} \le 2,$$

it follows

$$\sqrt{1-\varepsilon}-1 \le \sqrt{\lambda_k}-1 \le \sqrt{1+\varepsilon}-1 \le 1-\sqrt{1-\varepsilon}$$

and thus

$$\int_\Omega \|F(\omega)-S^{-1/2}F(\omega)\|^2 \mathrm{d}\mu(\omega) \le n(1-\sqrt{1-\varepsilon})^2.$$

We prove that the above estimation is optimal (i.e., the best possible). Consider the frame

$$\{f_j^\varepsilon\}_{j=1}^n,$$

where

$$f_j^\varepsilon = \sqrt{1-\varepsilon}u_k, \quad k=1,2,\dots,n.$$

Here, $\{u_k\}_{k=1}^n$ is an orthonormal basis of $\mathscr{H}$.

We have

$$\sum_{k=1}^n |\langle x,f_j^\varepsilon\rangle|^2 = \sum_{k=1}^n (1-\varepsilon)|\langle x,u_k\rangle|^2 = (1-\varepsilon)\|x\|^2$$

and the corresponding frame operator

$$Sx = \sum_{k=1}^{n} \langle x, f_j^{\varepsilon} \rangle f_j^{\varepsilon} = (1-\varepsilon) \sum_{k=1}^{n} \langle x, u_k \rangle u_k = (1-\varepsilon)x.$$

It follows

$$\begin{aligned}\sum_{k=1}^{n} \|f_j^{\varepsilon} - S^{-1/2} f_j^{\varepsilon}\|^2 &= \sum_{k=1}^{n} \|\sqrt{1-\varepsilon}u_k - u_k\|^2 \\ &= n(\sqrt{1-\varepsilon}-1)^2.\end{aligned}$$

□

**Corollary 11.2.** *Let $\{f_j\}_{j=1}^n$ be a basis for $\mathscr{H}$ such that*

$$(1-\varepsilon)\|x\|^2 \le \sum_{k=1}^{n} |\langle x, f_j \rangle|^2 \le (1+\varepsilon)\|x\|^2, \quad x \in \mathscr{H}.$$

*Then there exists an orthonormal basis $\{g_j\}_{j=1}^n$ of $\mathscr{H}$ such that*

$$\sum_{k=1}^{n} \|f_j - g_j\|^2 \le n(1-\sqrt{1-\varepsilon})^2.$$

*Proof.* Since $\{f_j\}_{j=1}^n$ is a basis, it follows that $\{S^{-\frac{1}{2}} f_j\}_{j=1}^n$ is an orthonormal family [5] and thus we can apply Theorem 11.2. □

**Theorem 11.4.** *If $F$ is an $\varepsilon$ near Parseval frame, then there exists an alternative dual $K$ of $F$ such that*

$$\int_{\Omega} \|K(\omega) - F(\omega)\|^2 \mathrm{d}\mu(\omega) \le n \frac{\varepsilon^2}{1-\varepsilon}.$$

*The estimation is optimal.*

*Proof.* We have

$$\begin{aligned}\|F(\omega) - S^{-1}F(\omega)\|^2 &= \left\| \sum_{k=1}^{n} \langle F(\omega), e_k \rangle e_k - \sum_{k=1}^{n} \frac{1}{\lambda_k} \langle F(\omega), e_k \rangle e_k \right\|^2 \\ &= \sum_{k=1}^{n} |\langle F(\omega), e_k \rangle|^2 \left| 1 - \frac{1}{\lambda_k} \right|^2\end{aligned}$$

and hence

$$\int_\Omega \|F(\omega)-S^{-1}F(\omega)\|^2 \mathrm{d}\mu(\omega) = \sum_{k=1}^n \left|1-\frac{1}{\lambda_k}\right|^2 \int_\Omega |\langle F(\omega), e_k\rangle|^2 \mathrm{d}\mu(\omega)$$

$$= \sum_{k=1}^n \left|1-\frac{1}{\lambda_k}\right|^2 \langle Se_k, e_k\rangle$$

$$= \sum_{k=1}^n \left|1-\frac{1}{\lambda_k}\right|^2 \lambda_k = \sum_{k=1}^n \frac{|\lambda_k-1|^2}{\lambda_k}.$$

From the hypothesis, we get

$$|\lambda_k - 1| \leq \varepsilon$$

and

$$\frac{1}{\lambda_k} \leq \frac{1}{1-\varepsilon}.$$

It follows

$$\int_\Omega \|F(\omega)-S^{-1}F(\omega)\|^2 \leq \frac{\varepsilon^2}{1-\varepsilon}.$$

We prove that the estimation is optimal.

Let $\{u_j\}_{j=1}^n$ be an orthonormal basis of $\mathscr{H}$ and take

$$f_j^\varepsilon = \sqrt{1-\varepsilon}u_j, \quad j = 1, 2, \dots, n.$$

Then

$$\|f_j^\varepsilon - S^{-1}f_j^\varepsilon\| = \|\sqrt{1-\varepsilon}u_j - \frac{1}{1-\varepsilon}\sqrt{1-\varepsilon}u_j\|$$

$$= \frac{\varepsilon}{\sqrt{1-\varepsilon}}$$

and therefore

$$\sum_{j=1}^n \|f_j^\varepsilon - S^{-1}f_j^\varepsilon\|^2 = n\frac{\varepsilon^2}{1-\varepsilon}.$$

□

In the following, we provide a generalization of the previous results.

**Theorem 11.5.** *Let $F$ be a frame for $\mathscr{H}$, $S$ be a frame operator and let $a \in \mathbb{R}$. Then the family*

$$S^{\frac{a-1}{2}} F(\omega)$$

*is a frame for $\mathscr{H}$ with frame operator $S^a$.*

*Proof.* For $x \in \mathscr{H}$, it follows

$$\begin{aligned}\int_{\Omega} \langle x, S^{\frac{a-1}{2}} F(\omega)\rangle S^{\frac{a-1}{2}} F(\omega) \mathrm{d}\mu(\omega) &= S^{\frac{a-1}{2}} \left( \int_{\Omega} \langle x, S^{\frac{a-1}{2}} F(\omega)\rangle F(\omega) \right) \mathrm{d}\mu(\omega) \\ &= S^{\frac{a-1}{2}} \left( \int_{\Omega} \langle S^{\frac{a-1}{2}} x, F(\omega)\rangle F(\omega) \right) \mathrm{d}\mu(\omega) \\ &= S^{\frac{a-1}{2}} S \left( S^{\frac{a-1}{2}} x \right) \\ &= S^a x.\end{aligned}$$

□

**Theorem 11.6.** *Let $a \leq 0$, $F$ be a frame for $\mathscr{H}$ and $S$ be a frame operator. Then we have:*

$$\int_{\Omega} \|F(\omega) - S^{\frac{a-1}{2}} F(\omega)\|^2 \mathrm{d}\mu(\omega) \leq n((\sqrt{1-\varepsilon})^a - \sqrt{1-\varepsilon})^2.$$

*The estimation is optimal.*

*Proof.* Let $\{\lambda_i\}_{i=1}^n$ be the eigenvalues of $S$ and $\{e_i\}_{i=1}^n$ the corresponding orthonormal set. We have

$$\begin{aligned}\|F(\omega) - S^{\frac{a-1}{2}} F(\omega)\|^2 &= \left\| \sum_{i=1}^n \langle F(\omega), e_i\rangle e_i - \sum_{i=1}^n \lambda_i^{\frac{a-1}{2}} \langle F(\omega), e_i\rangle e_i \right\|^2 \\ &= \left\| \sum_{i=1}^n \left(1 - \lambda_i^{\frac{a-1}{2}}\right) \langle F(\omega), e_i\rangle e_i \right\|^2 \\ &= \sum_{i=1}^n \left(1 - \lambda_i^{\frac{a-1}{2}}\right)^2 |\langle F(\omega), e_i\rangle|^2.\end{aligned}$$

So

$$\begin{aligned}\int_{\Omega} \|F(\omega) - S^{\frac{a-1}{2}} F(\omega)\|^2 \mathrm{d}\mu(\omega) &= \sum_{i=1}^n \left(1 - \lambda_i^{\frac{a-1}{2}}\right)^2 \int_{\Omega} |\langle F(\omega), e_i\rangle|^2 \mathrm{d}\mu(\omega) \\ &= \sum_{i=1}^n \left(1 - \lambda_i^{\frac{a-1}{2}}\right)^2 \langle S e_i, e_i\rangle \\ &= \sum_{i=1}^n \left(1 - \lambda_i^{\frac{a-1}{2}}\right)^2 \lambda_i \\ &= \sum_{i=1}^n (\sqrt{\lambda_i} - \sqrt{\lambda_i}^a)^2.\end{aligned}$$

The function

$$f(x) = x - x^a, \quad x > 0$$

has the derivative

$$f'(x) = 1 - ax^{a-1} \geq 0$$

and hence $f$ is nondecreasing . It follows that

$$\sqrt{1-\varepsilon} - (\sqrt{1-\varepsilon})^a \leq \sqrt{\lambda_i} - \sqrt{\lambda_i}^a \leq \sqrt{1+\varepsilon} - (\sqrt{1+\varepsilon})^a.$$

We have

$$\sqrt{1+\varepsilon} - (\sqrt{1+\varepsilon})^a \leq (\sqrt{1-\varepsilon})^a - \sqrt{1-\varepsilon},$$

which is equivalent with

$$\sqrt{1+\varepsilon} + \sqrt{1-\varepsilon} \leq (\sqrt{1+\varepsilon})^a + (\sqrt{1-\varepsilon})^a.$$

Indeed, if we denote by

$$g(a) = (\sqrt{1+\varepsilon})^a + (\sqrt{1-\varepsilon})^a,$$

then its derivative is

$$\begin{aligned} g'(a) &= (\sqrt{1+\varepsilon})^a \ln(\sqrt{1+\varepsilon}) + (\sqrt{1-\varepsilon})^a \ln(\sqrt{1-\varepsilon}) \\ &\leq (\sqrt{1-\varepsilon})^a \ln(\sqrt{1-\varepsilon^2}) \\ &\leq 0. \end{aligned}$$

It follows that $g$ is a non-increasing function.

If $a \leq 0$ it follows

$$g(a) \geq g(0) = 2 \geq \sqrt{1+\varepsilon} + \sqrt{1-\varepsilon}$$

and thus

$$\int_{\Omega} \|F(\omega) - S^{\frac{a-1}{2}} F(\omega)\|^2 \mathrm{d}\mu(\omega) \leq n((\sqrt{1-\varepsilon})^a - \sqrt{1-\varepsilon})^2.$$

We prove that the estimation is optimal.

We consider $\{u_j\}_{j=1}^n$ to be an orthonormal basis of $\mathscr{H}$ and we take:

$$f_j^{\varepsilon} = \sqrt{1-\varepsilon} u_j, \quad j = 1, 2, \ldots, n.$$

It holds

$$Sx = (1-\varepsilon)x$$

and

$$S^{-1/2} f_j^{\varepsilon} = u_j.$$

Therefore

$$S^{\frac{a-1}{2}} f_j^{\varepsilon} = S^{\frac{a}{2}} u_j = (\sqrt{1-\varepsilon})^a u_j$$

and thus

$$\sum_{j=1}^{n} \|f_j^{\varepsilon} - S^{\frac{a-1}{2}} f_j^{\varepsilon}\|^2 = n((\sqrt{1-\varepsilon})^a - \sqrt{1-\varepsilon})^2.$$

□

Independently, a recent paper [2] also computes the closest Parseval frame.

**Acknowledgements** The first author was supported by the Grant: POSDRU/88/1.5/S/49516, project: "Creşterea calităţii şi competivităţii cercetării doctorale prin acordarea de burse."

The authors thank Professors J. Brzdęk, P.G. Casazza, O. Christensen and Th.M. Rassias for carefully reading and useful comments.

## References

1. Ali, S.T., Antoine, J.P., Gazeau, J. P.: Continuous frames in Hilbert spaces. Ann. Physics **222**, 1–37 (1993)
2. Bodmann, B., Casazza, P.G.: The road to equal-norm Parseval frames. J. Funct. Anal. **258**, 397–420 (2010)
3. Brzdęk, J.: On a method of proving the Hyers–Ulam stability of functional equations on restricted domains. Aust. J. Math. Anal. Appl. **6**, Issue 1, Article 4, 1–10 (2009)
4. Casazza, P.G., Kutyniok, G.: A generalization of Gram–Schmidt orthogonalization generating all Parseval frames. Adv. Comput. Math. **27**, 65–78 (2007)
5. Casazza, P.G.: Local theory of frames and Schauder bases for Hilbert spaces. Illinois J. Math. **43**, 291–306 (1999)
6. Casazza, P.G.: Custom building finite frames. Contemp. Math. **345**, 61–86 (2004)
7. Christensen, O.: An Introduction to Frames and Riesz Bases. Birkhaüser, Boston (2003)
8. Czerwik, S.: Functional Equations and Inequalities in Several Variables. World Scientific, New Jersey (2002)
9. Daubechies, I., Grossmann, A., Meyer, Y.: Painless nonorthogonal expansions. J. Math. Phys. **27**, 1271–1283 (1986)
10. Duffin, R.J., Schaeffer, A.C.: A class of nonharmonic Fourier series. Trans. Amer. Math. Soc. **72**, 341–366 (1952)
11. Fornasier, M., Rauhut, H.: Continuous frames, function spaces, and the discretization problem. J. Fourier Anal. Appl. **11**, 245–287 (2005)
12. Gabardo, J.P., Han, D.: Frames associated with measurable space. Adv. Comput. Math. **18**, 127–147 (2003)
13. Găvruţa, L., Găvruţa, P., Zamani Eskandani, G.: Hyers–Ulam stability of frames in Hilbert spaces. Bul. Ştiinţ. Univ. Politeh. Timiş. Ser. Mat. Fiz. **55 (69) 2**, 60–67 (2010)
14. Găvruţa, P.: A generalization of the Hyers–Ulam–Rassias stability of approximately additive mappings. J. Math. Anal. Appl. **184**, 431–436 (1994)
15. Găvruţa, P.: On some identities and inequalities for frames in Hilbert spaces. J. Math. Anal. Appl. **321**, 469–478 (2006)
16. Găvruţa, P.: On the duality of fusion frames. J. Math. Anal. Appl. **333**, 871–879 (2007)

17. Găvruţa, P.: On the wavelet transform and chirps. In: Proceedings of the $7^{th}$ Symposium of Mathematics and its Applications, pp. 117–122, "Politehnica" University of Timisoara (1997)
18. Găvruţa, P., Găvruţa, L.: A new method for the generalized Hyers-Ulam-Rassias stability. Int. J. Nonlinear Anal. Appl. **1**, no. 2, 11–18 (2010)
19. Găvruţa, P., Ciurdariu, L., Găvruţa, L.: On the Hyers-Ulam stability of Parseval frames. Bul. Ştiinţ. Univ. Politeh. Timiş. Ser. Mat. Fiz. **51 (65) 2**, 12–17 (2006)
20. Han, D., Kornelson, K., Larson, D., Weber, E.: Frames for undergraduates. Amer. Math. Soc., SML/40, Providence, Rhode Island (2007)
21. Hyers, D.H.: On the stability of the linear functional equation. Proc. Natl. Acad. Soc. USA **27**, 222–224 (1941)
22. Hyers, D.H., Isac, G., Rassias, Th.M.: Stability of Functional Equations in Several Variables. Birkhäuser (1998)
23. Jung, S.-M.: Hyers–Ulam–Rassias Stability of Functional Equations in Mathematical Analysis. Hadronic Press, Palm Harbor, Florida (2001)
24. Kaiser, G.: Quantum Physics, Relativity and Complex Spacetime. North-Holland Mathematics Studies, Vol. **163**, Amsterdam (1990)
25. Kaiser, G.: A Friendly Guide to Wavelets. Birkhäuser, Boston (1994)
26. Rahimi, A., Najati, A., Dehgan, Y.N.: Continuous frame in Hilbert space. Methods Funct. Anal. Topology **12**, 170–182 (2006)
27. Rassias, J.M.: Solution of a problem of Ulam. J. Approx. Theory **57**, 268–273 (1989)
28. Rassias, Th.M.: On the stability of the linear mapping in Banach spaces. Proc. Amer. Math. Soc. **72**, 297–300 (1978)
29. Rassias, Th.M.: On the stability of functional equations and a problem of Ulam. Acta Appl. Math. **62**, 23–130 (2000)
30. Ulam, S.M.: A Collection of Mathematical Problems Interscience Publishers, Inc., New York (1968)

# Chapter 12
# Generalized Hyers–Ulam Stability of a Quadratic Functional Equation

**Kil-Woung Jun, Hark-Mahn Kim, and Jiae Son**

*Dedicated to the Memory of S.M. Ulam*
*on the 100th anniversary of his birth*

**Abstract** Let $a$ be a fixed integer with $a \neq -1, 0$. We obtain the general solution and the generalized Hyers–Ulam stability theorem for a quadratic functional equation

$$f(ax+y) + af(x-y) = (a+1)f(y) + a(a+1)f(x).$$

**Keywords** Generalized Hyers–Ulam stability • Quadratic mapping

**Mathematics Subject Classification (2000):** Primary 39B52, 39B82

## 12.1 Introduction

In 1940, Ulam [19] gave a talk before the Mathematics Club of the University of Wisconsin in which he discussed a number of unsolved problems. Among these was the following question concerning the stability of homomorphisms.

*Let $G$ be a group and let $G'$ be a metric group with metric $\rho(\cdot,\cdot)$. Given $\varepsilon > 0$, does there exist a $\delta > 0$ such that if $f : G \to G'$ satisfies $\rho(f(xy), f(x)f(y)) < \delta$ for all $x, y \in G$, then a homomorphism $h : G \to G'$ exists with $\rho(f(x), h(x)) < \varepsilon$ for all $x \in G$?*

In 1941, the first result concerning the stability of functional equations was presented by Hyers [8]. He has answered the question of Ulam for the case where $G_1$ and $G_2$ areBanach spaces.

K.-W. Jun (✉) • H.-M. Kim • J. Son
Department of Mathematics, Chungnam National University, 79 Daehangno, Yuseong-gu, Daejeon 305-764, Korea
e-mail: kwjun@math.cnu.ac.kr; hmkim@cnu.ac.kr; hand0627@cnu.ac.kr

Th.M. Rassias and J. Brzdęk (eds.), *Functional Equations in Mathematical Analysis*, Springer Optimization and Its Applications 52, DOI 10.1007/978-1-4614-0055-4_12,

Let $E_1$ and $E_2$ be real vector spaces. A function $f : E_1 \to E_2$ is called a quadratic function if and only if $f$ is a solution function of the quadratic functional equation

$$f(x+y)+f(x-y)=2f(x)+2f(y). \qquad (12.1)$$

It is well known that a function $f$ between real vector spaces is quadratic if and only if there exists a unique symmetric biadditive function $B$ such that $f(x)=B(x,x)$ for all $x$, where the mapping $B$ is given by

$$B(x,y)=\frac{1}{4}(f(x+y)-f(x-y)).$$

See [1,9] for the details.

The Hyers–Ulam stability of the quadratic functional equation (12.1) was first proved by Skof [18] for functions $f : E_1 \to E_2$, where $E_1$ is a normed space and $E_2$ is a Banach space. Cholewa [3] demonstrated that Skof's theorem is also valid if $E_1$ is replaced by an abelian group $G$. Assume that a function $f : G \to E$ satisfies the inequality

$$\|f(x+y)+f(x-y)-2f(x)-2f(y)\| \le \delta$$

for some $\delta \ge 0$ and for all $x,y \in G$, then there is a unique quadratic $Q : G \to E$ with

$$\|F(x)-Q(x)\| \le \frac{\delta}{2}$$

for all $x \in G$. S. Czerwik [4] proved the Hyers–Ulam–Rassias stability of quadratic functional equation (12.1). Let $E_1$ and $E_2$ be a real normed space and a real Banach space, respectively, and let $p \ne 2$ be a positive constant. If a function $f : E_1 \to E_2$ satisfies the inequality

$$\|f(x+y)+f(x-y)-2f(x)-2f(y)\| \le \varepsilon(\|x\|^p+\|y\|^p)$$

for some $\varepsilon > 0$ and for all $x,y \in E_1$, then there exists a unique quadratic function $q : E_1 \to E_2$ such that

$$\|f(x)-q(x)\| \le \frac{2\varepsilon}{|4-2^p|}\|x\|^p$$

for all $x \in E_1$. Furthermore, according to the theorem of Borelli and Forti [2], we know the following generalization of stability theorem for quadratic functional equation. Let $G$ be an abelian group and $E$ a Banach space, let $f : G \to E$ be a mapping with $f(0)=0$ satisfying the inequality

$$\|f(x+y)+f(x-y)-2f(x)-2f(y)\| \le \varphi(x,y)$$

for all $x, y \in G$. Assume that one of the series

$$\Phi(x,y) := \begin{cases} \sum_{k=0}^{\infty} \frac{1}{2^{2(k+1)}} \varphi(2^k x, 2^k y) < \infty; \\ \sum_{k=0}^{\infty} 2^{2k} \varphi\left(\frac{x}{2^{(k+1)}}, \frac{y}{2^{(k+1)}}\right) < \infty. \end{cases}$$

Then there exists a unique quadratic function $Q : G \to E$ such that

$$\|f(x) - Q(x)\| \leq \Phi(x,x)$$

for all $x \in G$. During the last three decades a number of papers and research monographs have been published on various generalizations and applications of the generalized Hyers–Ulam stability of several functional equations, and there are many interesting results concerning this problem [5, 6, 10, 12, 17]. In particular, J.M. Rassias investigated the stability of Ulam for the functional equation

$$f(ax+by) + f(bx-ay) = (a^2+b^2)[f(x)+f(y)]$$

in [13–15]. Now, we consider a new quadratic functional equation

$$f(ax+y) + af(x-y) = (a+1)f(y) + a(a+1)f(x) \tag{12.2}$$

for any fixed $a \in \mathbb{Z}$ with $a \neq 0, -1$, which is a modified and instrumental equation for the reference [11]. In this paper, we investigate the general solution of (12.2) and then prove the generalized Hyers–Ulam stability of (12.2).

## 12.2 Generalized Hyers–Ulam Stability of (12.2) for a $\neq -2$

First, we present the general solution of (12.2) in the class of all functions between vector spaces.

**Lemma 12.1.** *If vector spaces $X$ and $Y$ are common domain and range of the functions $f$ in both the functional equations (12.1) and (12.2), then the functional equation (12.2) is equivalent to the functional equation (12.1).*

*Proof.* Suppose that a function $f : X \to Y$ satisfies (12.2) for all $x, y \in X$. If we replace $x, y$ in (12.2) by 0, then we have $f(0) = 0$.

Setting $x, y$ by $0, -y$ in (12.2), respectively, yields

$$f(-y) + af(y) = (a+1)f(-y)$$

and so we conclude that $f$ is even.

Letting $y$ by $-y$ in (12.2), for all $x,y \in X$ we have

$$f(ax-y)+af(x+y) = (a+1)f(y)+a(a+1)f(x). \tag{12.3}$$

Replacing $y$ by $x-y$ in (12.2), one obtains

$$f((a+1)x-y)+af(y) = (a+1)f(x-y)+a(a+1)f(x) \tag{12.4}$$

for $x,y \in X$. Next, substituting $-ax+y$ for $y$ in (12.2) leads to

$$f(y)+af((a+1)x-y) = (a+1)f(ax-y)+a(a+1)f(x) \tag{12.5}$$

for all $x,y \in X$. Now, from (12.3) and (12.5), we get

$$\begin{aligned} af((a+1)x-y) &+ a(a+1)f(x+y) \\ &= a(a+2)f(y)+a(a+1)(a+2)f(x) \end{aligned} \tag{12.6}$$

for all $x,y \in X$. Finally, it is easily seen that (12.4) and (12.6) lead to (12.1).
The converse is obvious. □

Throughout this paper, let X be a vector space and let Y be a Banach space unless we give any specific reference. We will investigate the generalized Hyers–Ulam stability problem for the functional equation (12.2) for $a \neq -2$. Thus we find the condition that there exists a true quadratic function near an approximately quadratic function. For simplicity, we use the following abbreviation:

$$D_a f(x,y) := f(ax+y)+af(x-y)-(a+1)f(y)-a(a+1)f(x)$$

for all $x,y \in X$ and any fixed integer $a$ with $a \neq 0,-1,-2$; $D_a f$ is called the approximate remainder of (12.2) and acts as a perturbation of the equation.
Let $\varphi : X \times X \to [0,\infty)$ satisfy, for all $x,y \in X$, one of the conditions

$$\Phi_1(x,y) := \sum_{l=0}^{\infty} \frac{1}{(a+1)^{2(l+1)}} \varphi\left((a+1)^l x, (a+1)^l y\right) < \infty, \tag{12.7}$$

$$\Phi_2(x,y) := \sum_{l=0}^{\infty} (a+1)^{2l} \varphi\left(\frac{x}{(a+1)^{l+1}}, \frac{y}{(a+1)^{l+1}}\right) < \infty. \tag{12.8}$$

**Theorem 12.1.** *Assume that $\varphi$ satisfies (12.7) and a function $f : X \to Y$ satisfies*

$$\|D_a f(x,y)\| \leq \varphi(x,y) \tag{12.9}$$

*for all $x,y \in X$. Then there exists a unique quadratic function $Q : X \to Y$ satisfying*

$$\left\| f(x) - \frac{1}{a+2} f(0) - Q(x) \right\| \leq \Phi_1(x,x) \tag{12.10}$$

*for all $x \in X$, where $\|f(0)\| \le \varphi(0,0)/|a(a+1)|$. The function $Q$ is given by*

$$Q(x) = \lim_{k\to\infty} \frac{f((a+1)^k x)}{(a+1)^{2k}}.$$

*Proof.* Replacing $x, y$ by 0 in (12.9), we get

$$\|f(0)\| \le \frac{\varphi(0,0)}{|a(a+1)|}.$$

Replacing $y$ by $x$ in (12.9), we obtain

$$\|f((a+1)x) + af(0) - (a+1)^2 f(x)\| \le \varphi(x,x) \tag{12.11}$$

for $x \in X$. Dividing (12.11) by $(a+1)^2$, we get

$$\left\| \frac{1}{(a+1)^2}\bar{f}((a+1)x) - \bar{f}(x) \right\| \le \frac{1}{(a+1)^2}\varphi(x,x) \tag{12.12}$$

for $x \in X$, where

$$\bar{f}(x) = f(x) - \frac{f(0)}{a+2}$$

for $x \in X$. Thus using (12.12) and the triangle inequality we prove by induction that

$$\begin{aligned} &\left\| \frac{1}{(a+1)^{2k}}\ \bar{f}\ ((a+1)^k x) - \bar{f}(x) \right\| \\ &\qquad \le \sum_{l=0}^{k-1} \frac{1}{(a+1)^{2(l+1)}} \varphi\left((a+1)^l x, (a+1)^l x\right) \end{aligned} \tag{12.13}$$

for $x \in X$ and for all $k \in \mathbb{N}$. Therefore we deduce from the inequality (12.13) that for any integers $m, k$ with $m > k \ge 0$

$$\begin{aligned} &\left\| \frac{1}{(a+1)^{2m}}\ \bar{f}\ ((a+1)^m x) - \frac{1}{(a+1)^{2k}}\bar{f}((a+1)^k x) \right\| \\ &\quad = \frac{1}{(a+1)^{2k}} \sum_{l=0}^{m-k-1} \frac{1}{(a+1)^{2(l+1)}} \varphi((a+1)^{l+k}x, (a+1)^{l+k}x) \\ &\quad = \sum_{l=k}^{m-1} \frac{1}{(a+1)^{2(l+1)}} \varphi((a+1)^l x, (a+1)^l x). \end{aligned} \tag{12.14}$$

Since the right hand side of (12.14) tends to zero as $k \to \infty$, for all $x \in X$ the sequence

$$\left\{ \frac{1}{(a+1)^{2k}} \bar{f}((a+1)^k x) \right\}$$

is Cauchy and thus converges by the completeness of $Y$. Define $Q : X \to Y$ by

$$Q(x) = \lim_{k\to\infty} \frac{1}{(a+1)^{2k}} \left( f((a+1)^k x) - \frac{f(0)}{a+2} \right) = \lim_{k\to\infty} \frac{f((a+1)^k x)}{(a+1)^{2k}}$$

for all $x \in X$.

Replacing $x, y$ in (12.9) by $(a+1)^k x$, $(a+1)^k y$, respectively, dividing both sides by $(a+1)^{2k}$ and after then taking the limit in the resulting inequality, we have

$$Q(ax+y) + aQ(x-y) - (a+1)Q(y) - a(a+1)Q(x) = 0$$

so the function $Q$ is quadratic. Taking the limit in (12.13) as $k \to \infty$, we obtain that, for all $x \in X$,

$$\left\| f(x) - \frac{1}{a+2} f(0) - Q(x) \right\| \leq \Phi_1(x,x).$$

To prove the uniqueness of the quadratic function $Q$ subject to (12.10), let us assume that there exists a quadratic function $Q' : X \to Y$ which satisfies (12.2) and the inequality (12.10). Obviously, we have

$$Q(x) = (a+1)^{-2k} Q((a+1)^k x)$$

and

$$Q'(x) = (a+1)^{-2k} Q'((a+1)^k x)$$

for all $x \in X$. Hence it follows from (12.10) that

$$\begin{aligned} \left\| Q(x) - Q'(x) \right\| &= \frac{1}{(a+1)^{2k}} \left\| Q\left((a+1)^k x\right) - Q'\left((a+1)^k x\right) \right\| \\ &\leq \frac{2}{(a+1)^{2k}} \sum_{l=0}^{\infty} \frac{1}{(a+1)^{2(l+1)}} \varphi\left((a+1)^{l+k} x, (a+1)^{l+k} x\right) \\ &= 2 \sum_{l=k}^{\infty} \frac{1}{(a+1)^{2(l+1)}} \varphi\left((a+1)^l x, (a+1)^l x\right) \end{aligned}$$

for all $k \in \mathbb{N}$. Therefore, letting $k \to \infty$, one has $Q(x) - Q'(x) = 0$ for all $x \in X$, completing the proof of uniqueness. □

**Theorem 12.2.** *Assume that a function $f : X \to Y$ satisfies*

$$\|D_a f(x,y)\| \leq \varphi(x,y)$$

*for all $x, y \in X$ and $\varphi$ satisfies the condition (12.8). Then there exists a unique quadratic function $Q : X \to Y$ with*

$$\|f(x) - Q(x)\| \leq \Phi_2(x,x) \tag{12.15}$$

*for all* $x \in X$. *The function* $Q$ *is given by*

$$Q(x) = \lim_{k\to\infty} (a+1)^{2k} f\left(\frac{x}{(a+1)^k}\right)$$

*for all* $x \in X$.

*Proof.* In this case, $f(0) = 0$ since

$$\sum_{l=0}^{\infty} (a+1)^{2l} \varphi(0,0) < \infty$$

and so $\varphi(0,0) = 0$ by the assumption. Replacing $x$ by $x/(a+1)$ in (12.11), we obtain

$$\left\| f(x) - (a+1)^2 f\left(\frac{x}{a+1}\right) \right\| \le \varphi\left(\frac{x}{(a+1)}, \frac{x}{(a+1)}\right) \tag{12.16}$$

for $x \in X$. By induction and (12.16) we deduce that, for all $x \in X$, $k \in \mathbb{N}$,

$$\begin{aligned} &\left\| f(x) - (a+1)^{2k} f\left(\frac{x}{(a+1)^k}\right) \right\| \\ &\le \sum_{l=0}^{k-1} (a+1)^{2l} \varphi\left(\frac{x}{(a+1)^{l+1}}, \frac{x}{(a+1)^{l+1}}\right) \end{aligned} \tag{12.17}$$

The rest of the proof is similar to the proof of Theorem 12.1. □

In the following corollary, we have a stability result of (12.2) in the sense of Th.M. Rassias.

**Corollary 12.1.** *Let* $X$ *and* $Y$ *be a normed space and a Banach space, respectively. Let* $p, \varepsilon$ *be real numbers such that* $\varepsilon \ge 0$, $p \ne 2$. *Assume that a function* $f : X \to Y$ *satisfies the inequality*

$$\|D_a f(x,y)\| \le \varepsilon(\|x\|^p + \|y\|^p) \tag{12.18}$$

*for all* $x, y \in X$ *and* $X \setminus \{0\}$ *if* $p < 0$. *Then there exists a unique quadratic function* $Q : X \to Y$ *which satisfies the inequality*

$$\left\| f(x) - \frac{1}{a+2} f(0) - Q(x) \right\| \le \frac{2\varepsilon \|x\|^p}{|(a+1)^2 - |a+1|^p|}$$

*for all* $x \in X$ *and* $X \setminus \{0\}$ *if* $p < 0$. *The function* $Q$ *is given by*

$$Q(x) = \begin{cases} \lim_{k\to\infty} \dfrac{f((a+1)^k x)}{(a+1)^{2k}}, & \text{if } p < 2; \\ \lim_{k\to\infty} (a+1)^{2k} f\left(\dfrac{x}{(a+1)^k}\right), & \text{if } p > 2, \end{cases}$$

*for all* $x \in X$ *and* $X \setminus \{0\}$ *if* $p < 0$, *where* $f(0) = 0$ *if* $p > 0$.

*Proof.* If $p > 0$, we put $x = 0 = y$ in (12.18) and get $f(0) = 0$. Let

$$\varphi(x,y) := \varepsilon(\|x\|^p + \|y\|^p)$$

for all $x, y \in X$. Then applying Theorems 12.1 and 12.2 we obtain easily the statement. □

In the next corollary, we get a stability result of (12.2) in the sense of J.M. Rassias.

**Corollary 12.2.** *Let $X$ and $Y$ be a normed space and a Banach space, respectively. Let $\varepsilon, p_1, p_2$ be real numbers such that $p_1 p_2 \geq 0$, $\varepsilon \geq 0$, $P := p_1 + p_2 \neq 2$, and $f : X \to Y$ satisfy*

$$\|D_a f(x,y)\| \leq \varepsilon(\|x\|^{p_1} \|y\|^{p_2})$$

*for all $x, y \in X$ and $X \setminus \{0\}$ if $p_1, p_2 < 0$. Then there exists a unique quadratic function $Q : X \to Y$ which satisfies the inequality*

$$\left\| f(x) - \frac{1}{a+2} f(0) - Q(x) \right\| \leq \frac{\varepsilon \|x\|^P}{|(a+1)^2 - |a+1|^P|}$$

*for all $x \in X$ and $X \setminus \{0\}$ if $p_1, p_2 < 0$, where $f(0) = 0$ if $p_1, p_2 > 0$.*

*Proof.* We remark that

$$\varphi(x,y) = \varepsilon(\|x\|^{p_1} \|y\|^{p_2})$$

satisfies the condition (12.7) or (12.8) for all $x, y \in X$. By Theorems 12.1 and 12.2, we get the statement. □

As a consequence, we obtain the following Hyers–Ulam stability result of (12.2).

**Corollary 12.3.** *Assume that for some $\theta \geq 0$ a function $f : X \to Y$ satisfies*

$$\|D_a f(x,y)\| \leq \theta$$

*for all $x, y \in X$. Then there exists a unique quadratic function $Q : X \to Y$ which satisfies the inequality*

$$\left\| f(x) - \frac{1}{a+2} f(0) - Q(x) \right\| \leq \frac{\theta}{a(a+2)}$$

*for all $x \in X$.*

*Proof.* If we put $\varphi(x,y) := \theta$ , then $\varphi$ satisfies the condition (12.7), and so we get the desired result. □

## 12.3 Generalized Hyers–Ulam Stability of (12.2) for a = −2

We will investigate the generalized Hyers–Ulam stability problem for the functional equation (12.2) for the singular case

$$a = -2$$

excluded in Sect. 2. Thus, we find the condition that there exists a true quadratic function near an approximately quadratic function.

For simplicity, we write

$$D_{-2}f(x,y) := f(-2x+y) - 2f(x-y) + f(y) - 2f(x)$$

for all $x, y \in X$.

Let $\psi : X \times X \to [0,\infty)$ be a mapping satisfying one of the conditions

$$\Psi_1(x,y) := \sum_{l=0}^{\infty}\left(\frac{1}{2^{2l+1}}\psi\left(2^l x, 2^l y\right) + \frac{1}{4^{l+1}}\psi\left(2^l x, 2^{l+1} y\right)\right) < \infty, \quad (12.19)$$

$$\Psi_2(x,y) := \sum_{l=0}^{\infty}\left(2^{2l+1}\psi\left(\frac{x}{2^{l+1}}, \frac{y}{2^{l+1}}\right) + 4^l\psi\left(\frac{x}{2^{l+1}}, \frac{y}{2^l}\right)\right) < \infty \quad (12.20)$$

for all $x, y \in X$.

**Theorem 12.3.** *Assume that a function $f : X \to Y$ satisfies*

$$\|D_{-2}f(x,y)\| \le \psi(x,y) \quad (12.21)$$

*for all $x, y \in X$ and $\psi$ satisfies the condition (12.19). Then there exists a unique quadratic function $Q : X \to Y$ satisfying the estimation*

$$\|f(x) + f(0) - Q(x)\| \le \Psi_1(x,x)$$

*for all $x \in X$. The function $Q$ is given by*

$$Q(x) = \lim_{k\to\infty}\frac{f(2^k x)}{4^k}$$

*for all $x \in X$.*

*Proof.* Let $y := x$ in (12.21). Then we have

$$\|f(-x) - 2f(0) - f(x)\| \le \psi(x,x) \quad (12.22)$$

for all $x \in X$. Next, set $y := 2x$ in (12.21). Then one obtains

$$\|f(0) - 2f(-x) + f(2x) - 2f(x)\| \leq \psi(x, 2x) \tag{12.23}$$

for all $x \in X$. Thus it follows from (12.22) and (12.23) that

$$\|3f(0) + 4f(x) - f(2x)\| \leq 2\psi(x,x) + \psi(x, 2x) \tag{12.24}$$

for all $x \in X$. Defining $\tilde{f}(x) = f(x) + f(0)$ in (12.24), for all $x \in X$ we get

$$\left\| \frac{1}{4}\tilde{f}(2x) - \tilde{f}(x) \right\| \leq \frac{1}{2}\psi(x,x) + \frac{1}{4}\psi(x, 2x). \tag{12.25}$$

An induction argument together with (12.25) implies that, for all $x \in X$ and for all $k \in \mathbb{N}$,

$$\left\| \frac{1}{4^k}\tilde{f}(2^k x) - \tilde{f}(x) \right\| \leq \sum_{l=0}^{k-1} \left( \frac{1}{2^{2l+1}}\psi\left(2^l x, 2^l x\right) + \frac{1}{4^{l+1}}\psi\left(2^l x, 2^{l+1} x\right) \right).$$

Using the similar argument as for Theorem 12.1, we see that $Q : X \to Y$, given by

$$Q(x) = \lim_{n\to\infty} \frac{f(2^k x) + f(0)}{4^k} = \lim_{n\to\infty} \frac{f(2^k x)}{4^k}$$

for $x \in X$, is the unique quadratic mapping with $\|\tilde{f}(x) - Q(x)\| \leq \Psi_1(x,x)$ for $x \in X$. □

**Theorem 12.4.** *Assume that a function $f : X \to Y$ satisfies*

$$\|D_{-2}f(x,y)\| \leq \psi(x,y)$$

*for all $x, y \in X$ and $\psi$ satisfies the condition (12.20). Then there exists a unique quadratic function $Q : X \to Y$ satisfying the approximation*

$$\|f(x) - Q(x)\| \leq \Psi_2(x,x)$$

*for all $x \in X$. The function $Q$ is given by*

$$Q(x) = \lim_{k\to\infty} 4^k f\left(\frac{x}{2^k}\right)$$

*for all $x \in X$.*

*Proof.* The proof goes through by the similar argument to Theorem 12.3. □

The next corollary gives a stability result for (12.2) in the sense of Th.M. Rassias.

**Corollary 12.4.** *Let X and Y be a normed space and a Banach space, respectively. Let $p, \varepsilon$ be real numbers such that $\varepsilon \geq 0$, $p \neq 2$. Assume that a function $f : X \to Y$ satisfies and the inequality*

$$\|D_{-2}f(x,y)\| \leq \varepsilon(\|x\|^p + \|y\|^p) \tag{12.26}$$

*for all $x, y \in X$ and $X \setminus \{0\}$ if $p < 0$. Then there exists a unique quadratic function $Q : X \to Y$ which satisfies the inequality*

$$\|f(x) + f(0) - Q(x)\| \leq \frac{2^p + 5}{|2^p - 4|}\varepsilon\|x\|^p$$

*for all $x \in X$ and $X \setminus \{0\}$ if $p < 0$. The function $Q$ is given by*

$$Q(x) = \begin{cases} \lim_{k\to\infty} \dfrac{f(2^k x)}{4^k}, & if \quad p < 2; \\ \lim_{k\to\infty} 4^k f\left(\dfrac{x}{2^k}\right), & if \quad p > 2, \end{cases}$$

*for all $x \in X$ and $X \setminus \{0\}$ if $p < 0$, where $f(0) = 0$ if $p > 0$.*

*Proof.* Write $\psi(x,y) = \varepsilon(\|x\|^p + \|y\|^p)$ and apply Theorems 12.3 and 12.4. If $p > 0$, then $f(0) = 0$, since we get $\psi(x,y) = 0$ by putting $x, y = 0$ in (12.26). □

In the following corollary, we have a stability result of (12.2) in the sense of J.M. Rassias.

**Corollary 12.5.** *Let X and Y be a normed space and a Banach space, respectively. And let $\varepsilon, p_1, p_2$ be real numbers such that $p_1 p_2 \geq 0$, $\varepsilon \geq 0$ and $P := p_1 + p_2 \neq 2$. Assume that a function $f : X \to Y$ satisfies*

$$\|D_{-2}f(x,y)\| \leq \varepsilon(\|x\|^{p_1}\|y\|^{p_2})$$

*for all $x, y \in X$ and $X \setminus \{0\}$ if $p_1, p_2 < 0$. Then there exists a unique quadratic function $Q : X \to Y$ which satisfies the inequality*

$$\|f(x) + f(0) - Q(x)\| \leq \frac{2 + 2^{p_2}}{|4 - 2^P|}\varepsilon\|x\|^P$$

*for all $x \in X$ and $X \setminus \{0\}$ if $p_1, p_2 < 0$, where $f(0) = 0$ if $p_1, p_2 > 0$.*

*Proof.* If we put $\psi(x,y) := \varepsilon(\|x\|^{p_1}\|y\|^{p_2})$, then $\psi$ satisfies the condition (12.19) or (12.20) for all $x, y \in X$. And so we get the desired results. □

As a result, we obtain the following Hyers–Ulam stability result of (12.2).

**Corollary 12.6.** *Assume that for some $\theta > 0$, a function $f : X \to Y$ satisfies*

$$\|D_{-2}f(x,y)\| \leq \theta$$

*for all* $x, y \in X$. *Then there exists a unique quadratic function* $Q : X \to Y$, *which for all* $x \in X$ *satisfies the inequality*

$$\|f(x) + f(0) - Q(x)\| \leq \theta.$$

*Proof.* Since $\psi(x,y) = \theta$ satisfies the condition (12.19) for all $x, y \in X$, we get the desired result by Theorem 12.3. □

## References

1. J. Aczél, J., Dhombres, J.: Functional Equations in Several Variables. Cambridge University Press (1989)
2. Borelli, C., Forti, G.-L.: On a general Hyers-Ulm-stability result. Internat. J. Math. Math. Sci. **18**, 229–236 (1995)
3. Cholewa, P.W.: Remarks on the stability of functional equations. Aequationes Math. **27**, 76–86 (1984)
4. Czerwik, S.: On the stability of the mapping in normed spaces. Abh. Math. Sem. Univ. Hamburg **62**, 59–64 (1992)
5. Czerwik, S.: Functional Equations and Inequalities in Several Variables. World Scientific Publishing Company, New Jersey–London (2002)
6. Forti, G.-L.: Hyers-Ulam stability of functional equations in several variables. Aequationes Math. **50**, 143–190 (1995)
7. Găvruta, P.: A generalization of the Hyers–Ulam–Rassias stability of approximately additive mappings. J. Math. Anal. Appl. **184**, 431–436 (1994)
8. Hyers, D.H.: On the stability of the linear functional equation. Proc. Natl. Acad. Sci. U.S.A. **27**, 222–224 (1941)
9. Hyers, D.H., Rassias, Th.M.: Approximate homomorphisms. Aequationes Math. 44, 125-153 (1992)
10. Hyers, D.H., Isac, G., Rassias, Th.M.: Stability of Functional Equations in Several Variables. Birkhäuser, Boston–Basel–Berlin (1998)
11. Jun, K., Kim, H.: Ulam stability problem for generalized A-quadratic mappings. J. Math. Anal. Appl. **305**, 466–476 (2005)
12. Jung, S.-M.: Hyers–Ulam–Rassias Stability of Functional Equations in Mathematical Analysis. Hadronic Press Inc., Palm Harbor, Florida (2001)
13. Rassias, J.M.: On the stability of the Euler–Lagrange functional equation. Chinese J. Math. **20**, 185–190 (1992)
14. Rassias, J.M.: On the stability of the non-linear Euler-Lagrange functional equation in real normed linear spaces. J. Math. Phys. Sci. **28**, 231–235 (1994)
15. Rassias, J.M.: On the stability of the general Euler-Lagrange functional equation. Demonstratio Math. **29**, 755–766 (1996)
16. Rassias, Th.M.: On the stability of the linear mapping in Banach spaces. Proc. Amer. Math. Soc. **72**, 297–300 (1978)
17. Rassias, Th.M.: On the stability of functional equations and a problem of Ulam. Acta Appl. Math. **62**, 23–130 (2000)
18. Skof, F.: Local properties and approximations of operators. Rend. Sem. Mat. Fis. Milano **53**, 113–129 (1983)
19. Ulam, S.M.: A Collection of the Mathematical Problems. Interscience Publ., New York (1960)

# Chapter 13
# On the Hyers–Ulam–Rassias Stability of the Bi-Pexider Functional Equation

**Kil-Woung Jun and Yang-Hi Lee**

*Dedicated to the memory of S.M. Ulam*
*on the 100th anniversary of his birth*

**Abstract** In this paper, we obtain the Hyers–Ulam–Rassias stability of a bi-Pexider functional equation

$$f(x+y,z+w) = f_1(x,z) + f_2(x,w) + f_3(y,z) + f_4(y,w)$$

in the sense of Th.M. Rassias. Also, we establish the superstability of a bi-Jensen functional equation.

**Keywords** Solution • Stability • Bi-Jensen mapping • Functional equation

**Mathematics Subject Classification (2000):** Primary 39B52, 39B82

## 13.1 Introduction

In 1940, Ulam [7, p. 63] raised a question concerning the stability of homomorphisms: *Let $G_1$ be a group and let $G_2$ be a metric group with the metric $d(\cdot,\cdot)$. Given $\varepsilon > 0$, does there exists a $\delta > 0$ such that if a mapping $h : G_1 \to G_2$ satisfies the inequality*

$$d(h(xy), h(x)h(y)) < \delta$$

*for all $x, y \in G_1$, then there is a homomorphism $H : G_1 \to G_2$ with $d(h(x), H(x)) < \varepsilon$ forall $x \in G_1$?*

K.-W. Jun (✉)
Department of Mathematics, Chungnam National University, Taejon 305-764, Republic of Korea
e-mail: kwjun@cnu.ac.kr

Y.-H. Lee
Department of Mathematics Education, Kongju National University of Education, Kongju 314-711, Republic of Korea
e-mail: yanghi2@hanmail.net

Th.M. Rassias and J. Brzdęk (eds.), *Functional Equations in Mathematical Analysis*, Springer Optimization and Its Applications 52, DOI 10.1007/978-1-4614-0055-4_13,

The case of approximately additive mappings was solved by Hyers [4] under the assumption that $G_1$ and $G_2$ are Banach spaces. Aoki [1], Rassias [6], and Găvruta [3] gave successive generalizations of that result.

Throughout this paper, let $X$ be a normed space and $Y$ a Banach space. A mapping $g : X \to Y$ is called a Cauchy mapping (a Jensen mapping, respectively) if $g$ satisfies the functional equation $g(x+y) = g(x)+g(y)$ $(2g((x+y)/2) = g(x)+g(y)$, respectively). For given mappings $f, f_1, f_2, f_3, f_4 : X \times X \to Y$, we write

$$Df(x,y,z,w) := f(x+y,z+w) - f(x,z) - f(x,w) - f(y,z) - f(y,w),$$
$$Jf(x,y,z,w) := 4f\left(\frac{x+y}{2}, \frac{z+w}{2}\right) - f(x,z) - f(x,w) - f(y,z) - f(y,w),$$
$$P_4(x,y,z,w) := f(x+y,z+w) - f_1(x,z) - f_2(x,w) - f_3(y,z) - f_4(y,w)$$

for all $x,y,z,w \in X$. A mapping $f : X \times X \to Y$ is called a biadditive (bi-Jensen, bi-Pexider, respectively) mapping if $f$ satisfies the functional equations $Df = 0$ $(Jf = 0,\ P(f, f_1, f_2, f_3, f_4) := P_4 = 0$, respectively) and the functional equation $Df = 0$ $(Jf = 0,\ P(f, f_1, f_2, f_3, f_4) = 0$, respectively) is called a biadditive (bi-Jensen, bi-Pexider, respectively) functional equation. In 2006, Park and Bae [2] obtained the general solution and the Hyers–Ulam stability of a bi-Jensen functional equation. In this paper, we establish the Hyers–Ulam–Rassias stability of that equation.

## 13.2 Stability of a Bi-Pexider Functional Equation

We can easily check the following lemma.

**Lemma 13.1.** *Let $f, f_1, f_2, f_3, f_4 : X \times X \to Y$. Define $f', f'', f''' : X \times X \to Y$ by:*

$$f'(x,y) = f(x,y) - f(0,y), \quad f''(x,y) = f(x,y) - f(x,0),$$
$$f'''(x,y) = f(x,y) - f(x,0) - f(0,y) + f(0,0)$$

*for all $x,y \in X$. Then*

$$f'(x,y) - \frac{f'(2x,y)}{2} = \frac{1}{4}A_1\left(x,0,\frac{y}{2},\frac{y}{2}\right), \quad f''(x,y) - \frac{f''(x,2y)}{2} = \frac{1}{4}A_1\left(\frac{x}{2},\frac{x}{2},0,y\right),$$
$$f'''(x,y) - \frac{f'''(2x,2y)}{4} = -\frac{1}{4}A_2(x,x,y,y),$$
$$Jf(x,y,z,w) = A_1\left(\frac{x}{2},\frac{y}{2},\frac{z}{2},\frac{w}{2}\right), \quad Jf'''(x,y,z,w) = A_2\left(\frac{x}{2},\frac{y}{2},\frac{z}{2},\frac{w}{2}\right)$$

*for all $x,y,z,w \in X$, where*

$$\begin{aligned}A_1(x,y,z,w) &= P_4(x,y,z,w)+P_4(x,y,w,z)+P_4(y,x,z,w)+P_4(y,x,w,z)\\&\quad -P_4(x,x,z,z)-P_4(x,x,w,w)-P_4(y,y,z,z)-P_4(y,y,w,w),\\A_2(x,y,z,w) &= P_4(x,y,z,w)-P_4(x,y,0,0)-P_4(0,0,z,w)-P_4(x,0,z,0)\\&\quad -P_4(x,0,0,w)-P_4(0,y,z,0)-P_4(0,y,0,w)+2P_4(x,0,0,0)\\&\quad +2P_4(0,y,0,0)+2P_4(0,0,z,0)+2P_4(0,0,0,w)-3P_4(0,0,0,0).\end{aligned}$$

**Theorem 13.1.** *Let $0<p<1$, $0\le\varepsilon$ and let $f: X\times X\to Y$ be a mapping such that*

$$\|P(f,f_1,f_2,f_3,f_4)(x,y,z,w)\| \le \varepsilon(\|x\|^p+\|y\|^p+\|z\|^p+\|w\|^p) \tag{13.1}$$

*for all $x,y,z,w\in X$. Then there is a unique bi-Jensen mapping $F: X\times X\to Y$ with*

$$\|f(x,y)-F(x,y)\| \le \left(\frac{4}{|2-2^p|}+\frac{12}{|4-2^p|}\right)\varepsilon(\|x\|^p+\|y\|^p) \tag{13.2}$$

*for all $x,y\in X$ and $F(0,0)=f(0,0)$. The mapping $F: X\times X\to Y$ is given by*

$$F(x,y) := \lim_{j\to\infty}\frac{f(2^jx,2^jy)}{4^j}+\lim_{j\to\infty}\frac{f(2^jx,0)+f(0,2^jy)}{2^j}+f(0,0).$$

*Proof.* Let $f',f'',f''',A_1,A_2,A_3$ be as in the Lemma 13.1. By Lemma 13.1 and (13.1), we get

$$\left\|\frac{f'(2^lx,0)}{2^l}-\frac{f'(2^mx,0)}{2^m}\right\| = \left\|\sum_{j=l}^{m-1}\frac{A_1(2^jx,0,0,0)}{2^{j+2}}\right\| \le \sum_{j=l}^{m-1}\frac{2^{jp}\varepsilon}{2^{j-1}}\|x\|^p, \tag{13.3}$$

$$\left\|\frac{f''(0,2^ly)}{2^l}-\frac{f''(0,2^my)}{2^m}\right\| = \left\|\sum_{j=l}^{m-1}\frac{A_1(0,0,2^jy,0)}{2^{j+2}}\right\| \le \sum_{j=l}^{m-1}\frac{2^{jp}\varepsilon}{2^{j-1}}\|y\|^p, \tag{13.4}$$

$$\begin{aligned}\left\|\frac{f'''(2^lx,2^ly)}{4^l}-\frac{f'''(2^mx,2^my)}{4^m}\right\| &= \left\|\sum_{j=l}^{m-1}\frac{A_2(2^jx,2^jx,2^jy,2^jy)}{4^{j+1}}\right\|\\&\le 3\sum_{j=l}^{m-1}\frac{2^{jp}}{4^j}\varepsilon(\|x\|^p+\|y\|^p)\end{aligned} \tag{13.5}$$

for all $x,y\in X$ and for given integers $l,m$ $(0\le l<m)$. By the above inequalities, the sequences $\{2^{-n}(f(2^nx,0)-f(0,0))\}$, $\{2^{-n}(f(0,2^nx)-f(0,0))\}$, and $\{4^{-n}f'''(2^nx,2^ny)\}$ are Cauchy sequences for all $x,y\in X$. Since $Y$ is complete, those sequences converge for all $x,y\in X$ and we may define $F_1,F_2,F_3: X\times X\to Y$ by:

$$F_1(x,y) := \lim_{j\to\infty} \frac{f'(2^j x,0)}{2^j} = \lim_{j\to\infty} \frac{f(2^j x,0)}{2^j},$$

$$F_2(x,y) := \lim_{j\to\infty} \frac{f''(0,2^j y)}{2^j} = \lim_{j\to\infty} \frac{f(0,2^j y)}{2^j},$$

$$F_3(x,y) := \lim_{j\to\infty} \frac{f'''(2^j x,2^j y)}{4^j} = \lim_{j\to\infty} \frac{f(2^j x,2^j y)}{4^j}.$$

Putting $l = 0$ and taking $m \to \infty$ in (13.3), (13.4) and (13.5), one can obtain the inequalities

$$\|f(x,0) - f(0,0) - F_1(x,y)\| \le \frac{4}{2-2^p}\|x\|^p, \tag{13.6}$$

$$\|f(0,y) - f(0,0) - F_2(x,y)\| \le \frac{4}{2-2^p}\|y\|^p, \tag{13.7}$$

$$\|f'''(x,y) - F_3(x,y)\| \le \frac{12}{4-2^p}\|x\|^p + \frac{12}{4-2^p}\|y\|^p \tag{13.8}$$

for all $x,y \in X$. Since

$$JF_1(x,y,z,w) = \lim_{n\to\infty} \frac{A_1(2^n x, 2^n y, 0, 0)}{2^{n+1}} = 0,$$

$$JF_2(x,y,z,w) = \lim_{n\to\infty} \frac{A_1(0,0,2^n z, 2^n w)}{2^{n+1}} = 0,$$

$$JF_3(x,y,z,w) = \lim_{n\to\infty} \frac{A_2(2^n x, 2^n y, 2^n z, 2^n w)}{4^n} = 0$$

for all $x,y,z,w \in X$, $F$ is a bi-Jensen mapping satisfying (13.2), where

$$F(x,y) = F_1(x,y) + F_2(x,y) + F_3(x,y) + f(0,0)$$

for all $x,y \in X$. The uniqueness of such $F$ can be shown by the similar method as in the proof of [5, Theorem 2]. □

**Theorem 13.2.** *Let $p > 2$ and let $f : X \times X \to Y$ be a mapping satisfying (13.1) for all $x,y,z,w \in X$. Then there exists a unique bi-Jensen mapping $F : X \times X \to Y$ satisfying (13.2) for all $x,y \in X$. The mapping $F : X \times X \to Y$ is given by*

$$\begin{aligned} F(x,y) := & \lim_{j\to\infty} 4^j \left( f\left(\frac{x}{2^j}, \frac{y}{2^j}\right) - f\left(\frac{x}{2^j}, 0\right) - f\left(0, \frac{y}{2^j}\right) + f(0,0) \right) \\ & + \lim_{j\to\infty} 2^j \left( f\left(\frac{x}{2^j}, 0\right) - f(0,0) \right) \\ & + \lim_{j\to\infty} 2^j \left( f\left(0, \frac{y}{2^j}\right) - f(0,0) \right) + f(0,0). \end{aligned}$$

*Proof.* By the similar method as in the proof of Theorem 13.1, we can define the maps $F_1, F_2, F_3 : X \times X \to Y$ by

$$F_1(x,y) := \lim_{j\to\infty} 2^j \left(f\left(\frac{x}{2^j},0\right) - f(0,0)\right),$$

$$F_2(x,y) := \lim_{j\to\infty} 2^j \left(f\left(0,\frac{y}{2^j}\right) - f(0,0)\right),$$

$$F_3(x,y) := \lim_{j\to\infty} 4^j \left(f\left(\frac{x}{2^j},\frac{y}{2^j}\right) - f\left(\frac{x}{2^j},0\right) - f\left(0,\frac{y}{2^j}\right) + f(0,0)\right)$$

for all $x,y \in X$ and obtain the inequalities

$$\|f(x,0) - f(0,0) - F_1(x,y)\| \le \frac{4\varepsilon}{2^p - 2}\|x\|^p,$$

$$\|f(0,y) - f(0,0) - F_2(x,y)\| \le \frac{4\varepsilon}{2^p - 2}\|y\|^p,$$

$$\|f(x,y) - f(x,0) - f(0,y) + f(0,0) - F_3(x,y)\| \le \frac{12\varepsilon}{2^p - 4}(\|x\|^p + \|y\|^p)$$

for all $x,y \in X$. Since

$$JF_1(x,y) = \lim_{n\to\infty} 2^{n+1} A_1\left(\frac{x}{2^n},\frac{y}{2^n},0,0\right) = 0,$$

$$JF_2(x,y) = \lim_{n\to\infty} 2^{n+1} A_1\left(0,0,\frac{z}{2^n},\frac{w}{2^n}\right) = 0,$$

$$JF_3(x,y,z,w) = \lim_{n\to\infty} 4^n A_2\left(\frac{x}{2^n},\frac{y}{2^n},\frac{z}{2^n},\frac{w}{2^n}\right) = 0$$

for all $x,y \in X$, $F$ is a bi-Jensen mapping satisfying (13.2) where

$$F(x,y) = F_1(x,y) + F_2(x,y) + F_3(x,y) + f(0,0)$$

for all $x,y \in X$. The uniqueness of such $F$ can be shown by the similar method as in [5, Theorem 3]. □

**Theorem 13.3.** *Let $1 < p < 2$ and let $f : X \times X \to Y$ be a mapping satisfying (13.1) for all $x,y,z,w \in X$. Then there exists a unique bi-Jensen mapping $F : X \times X \to Y$ satisfying (13.2) for all $x,y \in X$. The mapping $F : X \times X \to Y$ is given by*

$$F(x,y) := \lim_{j\to\infty} \frac{1}{4^j}(f(2^j x, 2^j y) - f(2^j x, 0) - f(0, 2^j y)) + f(0,0)$$
$$+ \lim_{j\to\infty} 2^j \left(f\left(\frac{x}{2^j},0\right) - f(0,0)\right)$$
$$+ \lim_{j\to\infty} 2^j \left(f\left(0,\frac{y}{2^j}\right) - f(0,0)\right) + f(0,0).$$

*Proof.* Let $F_3$ be as in Theorem 13.1 and let $F_1$ and $F_2$ be as in Theorem 13.2. Then $F$ is a bi-Jensen mapping satisfying (13.2), where $F$ is defined by $F(x,y) = F_1(x,y)+F_2(x,y)+F_3(x,y)+f(0,0)$ for all $x,y \in X$. The uniqueness of such $F$ can be shown by the similar method in [5, Theorem 4]. □

*Remark 13.1.* Let $p, f, f_1, f_2, f_3, f_4$ be as in Theorem 13.1 (Theorem 13.2, Theorem 13.3, respectively). Then there exist a unique bi-Jensen mapping $F : X \times X \to Y$ and a unique bi-additive mapping $F_3 : X \times X \to Y$ such that, for all $x,y \in X$ and $i = 1,2,3,4$,

$$\|f(x,y) - F(x,y)\| \le \left(\frac{4}{|2(2^p-2)|} + \frac{12}{|2^p-4|}\right)(\|x\|^p + \|y\|^p),$$

$$\|f_i(x,y) - f_i(x,0) - f_i(0,y) + f_i(0,0) - F_3(x,y)\| \le \left(2 + \frac{12}{|4-2^p|}\right)(\|x\|^p + \|y\|^p).$$

*Proof.* Let $f''', F_3$ be as in the proof of Theorem 13.1 (Theorem 13.2, Theorem 13.3, respectively). Since $F_3$ is a bi-Jensen mapping, $F_3$ is a bi-additive mapping by the definition of $F_3$. Using (13.8) and the inequality

$$\begin{aligned}
&\|f'''(x,y) - (f_1(x,y) - f_1(x,0) - f_1(0,y) + f_1(0,0))\| \\
&\quad = \|P_4(x,0,y,0) - P_4(x,0,0,0) - P_4(0,0,y,0) + P_4(0,0,0,0)\| \\
&\quad \le 2\|x\|^p + 2\|y\|^p,
\end{aligned}$$

we get

$$\begin{aligned}
&\|f_1(x,y) - f_1(x,0) - f_1(0,y) + f_1(0,0) - F_3(x,y)\| \\
&\qquad \le \left(2 + \frac{12}{|4-2^p|}\right)(\|x\|^p + \|y\|^p)
\end{aligned}$$

for all $x,y \in X$. We can prove easily the other results by the similar method. □

**Theorem 13.4.** *Let $0 < p < 1, 0 \le \varepsilon$ and let $f : X \times X \to Y$ be a mapping such that*

$$\|P(f, f_1, f_2, f_3, f_4)(x,y,z,w)\| \le \varepsilon(\|x\|^p + \|y\|^p)(\|z\|^p + \|w\|^p) \tag{13.9}$$

*for all $x,y,z,w \in X$. Then there exists a unique bi-Jensen mapping $F : X \times X \to Y$ such that*

$$\|f(x,y) - F(x,y)\| \le \frac{8\varepsilon}{|4-4^p|}\|x\|^p\|y\|^p \tag{13.10}$$

*for all $x,y \in X$ and $F(0,0) = f(0,0)$. The mapping $F : X \times X \to Y$ is given by:*

$$F(x,y) := \lim_{j\to\infty} \frac{1}{4^j} f(2^j x, 2^j y) + f(x,0) + f(0,y) - f(0,0).$$

*Proof.* Let $f', f'', f''', A_1, A_2, A_3$ be as in the Lemma 13.1. Since

$$\left\| \frac{f'(2^l x, 0)}{2^l} - \frac{f'(2^m x, 0)}{2^m} \right\| = \left\| \sum_{j=l}^{m-1} \frac{A_1(2^j x, 2^j x, 0, 0)}{2^{j+1}} \right\| = 0,$$

$$\left\| \frac{f''(0, 2^l y)}{2^l} - \frac{f''(0, 2^m y)}{2^m} \right\| = \left\| \sum_{j=l}^{m-1} \frac{A_2(0, 0, 2^j y, 2^j y)}{2^{j+1}} \right\| = 0,$$

$$\left\| \frac{f'''(2^l x, 2^l y)}{4^l} - \frac{f'''(2^m x, 2^m y)}{4^m} \right\| = \left\| \sum_{j=l}^{m-1} \frac{A_3(2^j x, 2^j x, 2^j y, 2^j y)}{4^{j+1}} \right\|$$

$$\leq 2 \sum_{j=l}^{m-1} \frac{4^{jp}}{4^j} \varepsilon \|x\|^p \|y\|^p$$

for all $x, y \in X$, we can define $F_1, F_2, F_3 : X \times X \to Y$ by $F_1(x,y) := f(x,0) - f(0,0)$, $F_2(x,y) := f(0,y) - f(0,0)$, and $F_3(x,y) := \lim_{j\to\infty} 4^{-j} f(2^j x, 2^j y)$ for all $x, y \in X$. And we know that $F_3$ is a bi-Jensen mapping satisfying

$$\|f(x,y) + f(x,0) + f(0,y) - f(0,0) - F_3(x,y)\| \leq \frac{8\varepsilon}{4 - 4^p} \|x\|^p \|y\|^p$$

for all $x, y \in X$. Since

$$JF_1(x,y,z,w) = 2P_4\left(\frac{x}{2}, \frac{y}{2}, 0, 0\right) - P_4(x,0,0,0) - P_4(0,y,0,0) = 0,$$

$$JF_2(x,y,z,w) = 2P_4\left(0, 0, \frac{z}{2}, \frac{w}{2}\right) - P_4(0,0,z,0) - P_4(0,0,0,w) = 0$$

for all $x, y, z, w \in X$, $F$ is a bi-Jensen mapping satisfying (13.10), where $F(x,y) = F_3(x,y) + f(x,0) + f(0,y) - f(0,0)$ for all $x, y \in X$. The uniqueness of such $F$ can be shown by the similar method as in the proof of Theorem 13.1. □

**Theorem 13.5.** *Let $p > 1$ and let $f : X \times X \to Y$ be a mapping satisfying (13.9) for all $x, y, z, w \in X$. Then there exists a unique bi-Jensen mapping $F : X \times X \to Y$ satisfying (13.10) for all $x, y \in X$. The mapping $F : X \times X \to Y$ is given by*

$$F(x,y) := \lim_{j\to\infty} 4^j \left( f\left(\frac{x}{2^j}, \frac{y}{2^j}\right) - f\left(\frac{x}{2^j}, 0\right) - f\left(0, \frac{y}{2^j}\right) + f(0,0) \right)$$

$$+ f(x,0) + f(0,y) - f(0,0).$$

*Proof.* Since

$$\left\| f'''(x,y) - 4f'''\left(\frac{x}{2}, \frac{y}{2}\right) \right\| = \left\| A_3\left(\frac{x}{2}, \frac{x}{2}, \frac{y}{2}, \frac{y}{2}\right) \right\| \leq \frac{8\varepsilon}{4^p} \|x\|^p \|y\|^p,$$

for all $x,y \in X$, we get the bi-Jensen maps $F_1, F_2$ as in the proof of Theorem 13.4 and $F_3 : X \times X \to Y$ as in the proof of Theorem 13.2. Furthermore, we get the inequality

$$\|f(x,y) - f(x,0) - f(0,y) + f(0,0) - F_3(x,y)\| \le \frac{8\varepsilon}{4^p - 4}\|x\|^p\|y\|^p$$

for all $x,y \in X$. We show the uniqueness of $F$ analogously as for Theorem 13.3. □

**Theorem 13.6.** *Let $0 < p < 1, 0 \le \varepsilon$ and let $f : X \times X \to Y$ be a mapping satisfying (13.1) for all $x,y,z,w \in X$. Then there is a unique bi-Jensen mapping $F : X \times X \to Y$ with*

$$\|f(x,y) - F(x,y)\| \le \frac{4\varepsilon}{2-2^p}\|x\|^p + \left(\frac{4}{2-2^p} + \frac{8}{2^p}\right)\varepsilon\|y\|^p \tag{13.11}$$

*for all $x,y \in X$ and $F(0,0) = f(0,0)$. The mapping $F : X \times X \to Y$ is given by*

$$F(x,y) := \lim_{j\to\infty} \frac{1}{2^j}(f(2^j x,y) + f(0,2^j y)) + f(0,0).$$

*Proof.* Using Lemma 13.1 and (13.1), for given integers $l,m$ $(0 \le l < m)$ we get

$$\begin{aligned}\left\|\frac{f'(2^l x,y)}{2^l} - \frac{f'(2^m x,y)}{2^m}\right\| &= \left\|\sum_{j=l}^{m-1} \frac{1}{2^{j+2}} A_1\left(2^j x, 0, \frac{y}{2}, \frac{y}{2}\right)\right\| \\ &\le \sum_{j=l}^{m-1} \frac{2^{jp}\varepsilon}{2^{j-1}}\|x\|^p + \frac{4\varepsilon}{2^j\,2^p}\|y\|^p.\end{aligned}$$

By the similar method as in the proof of Theorem 13.1, we define $F_4 : X \times X \to Y$ by

$$F_4(x,y) := \lim_{j\to\infty} \frac{1}{2^j} f(2^j x,y)$$

for all $x,y \in X$ and obtain the inequality

$$\|f(x,y) - f(0,y) - F_4(x,y)\| \le \frac{4\varepsilon}{2-2^p}\|x\|^p + \frac{8\varepsilon}{2^p}\|y\|^p$$

for all $x,y \in X$. Since $\lim_{n\to\infty} 2^{-n} f(0,y) = 0$ for all $y \in X$, we have

$$JF_4(x,y,z,w) = \lim_{n\to\infty} \frac{1}{2^{n+1}} A_1\left(2^n x, 2^n y, \frac{z}{2}, \frac{w}{2}\right) = 0$$

for all $x,y,z,w \in X$. Let $F_2$ be as in the proof of Theorem 13.1. Then $F$ is a bi-Jensen mapping satisfying (13.11), where $F : X \times X \to Y$ is given by $F(x,y) = F_4(x,y) + f(0,y)$ for all $x,y \in X$. The uniqueness of such $F$ can be shown by the similar method as in the proof of [5, Theorem 5]. □

**Theorem 13.7.** *Let $0 < p < 1, 0 \leq \varepsilon$ and let $f : X \times X \to Y$ be a mapping satisfying (13.9) for all $x,y,z,w \in X$. Then there is a unique bi-Jensen mapping $F : X \times X \to Y$ with*

$$\|f(x,y) - F(x,y)\| \leq \frac{8\varepsilon}{(2-2^p)2^p}\|x\|^p\|y\|^p \tag{13.12}$$

*for all $x,y \in X$. Moreover $F(0,0) = f(0,0)$ and $F : X \times X \to Y$ is given by*

$$F(x,y) = \lim_{j\to\infty}\frac{f(2^jx,y)}{2^j} + f(0,y).$$

*Proof.* Using Lemma 13.1 and (13.1), for given integers $l,m$ $(0 \leq l < m)$ we get

$$\left\|\frac{f'(2^lx,y)}{2^l} - \frac{f'(2^mx,y)}{2^m}\right\| \leq \sum_{j=l}^{m-1}\frac{2^{(j-1)p}\varepsilon}{2^{j-2}}\|x\|^p\|y\|^p.$$

Analogously as in the proof of Theorem 13.2, we define $F_4 : X \times X \to Y$ by $F_4(x,y) = \lim_{j\to\infty} 2^{-j}f(2^jx,y)$ for all $x,y \in X$ and obtain the inequality

$$\|f(x,y) - f(0,y) - F_4(x,y)\| \leq \frac{8\varepsilon}{(2-2^p)2^p}\|x\|^p\|y\|^p$$

for all $x,y \in X$. Next, by Lemma 13.1, for all $x,y,z,w \in X$ we have

$$JF_4(x,y,z,w) = \lim_{n\to\infty}\frac{1}{2^{n+1}}A_1\left(2^nx, 2^ny, \frac{z}{2}, \frac{w}{2}\right) = 0.$$

Let $F_2$ be as in the proof of Theorem 13.4. Then $F$ is a bi-Jensen mapping satisfying (13.12), where $F : X \times X \to Y$ is given by $F(x,y) = F_4(x,y) + f(0,y)$ for all $x,y \in X$. The uniqueness of such $F$ can be shown as in the proof of [5, Theorem 5]. □

**Theorem 13.8.** *Let $p > 1, 0 \leq \varepsilon$ and let $f : X \times X \to Y$ be a mapping satisfying (13.9) for all $x,y,z,w \in X$. Then there exists a unique bi-Jensen mapping $F : X \times X \to Y$ satisfying (13.12) for all $x,y \in X$ with $F(0,0) = f(0,0)$. That mapping $F$ is given by*

$$F(x,y) := \lim_{j\to\infty} 2^j(f(2^{-j}x,y) - f(0,y)) + f(0,y).$$

*Proof.* Using Lemma 13.1 and (13.1), for given integers $l,m$ $(0 \leq l < m)$ we get

$$\left\|2^lf'\left(\frac{x}{2^l},y\right) - 2^mf'\left(\frac{x}{2^m},y\right)\right\| \leq \sum_{j=l}^{m-1}\frac{2^{j+3}\varepsilon}{2^{(j+2)p}}\|x\|^p\|y\|^p.$$

Next, similarly as in the proof of Theorem 13.7, we define $F_4 : X \times X \to Y$ by $F_4(x,y) := \lim_{j\to\infty} 2^j(f(2^{-j}x,y) - f(0,y))$ for all $x,y \in X$ and obtain the inequality

$$\|f(x,y) - f(0,y) - F_4(x,y)\| \leq \frac{8\varepsilon}{(2^p-2)2^p}\|x\|^p\|y\|^p$$

for all $x, y \in X$. Since $F_4(0,y) = 0$, we have

$$JF_4(x,y,z,w) = \lim_{n\to\infty} 2^n A_1\left(\frac{x}{2^{n+1}}, \frac{y}{2^{n+1}}, \frac{z}{2}, \frac{w}{2}\right) = 0$$

for all $x,y,z,w \in X$. Let $F_2$ be as in the proof of Theorem 13.4. Then $F$ is a bi-Jensen mapping satisfying (13.12), where $F : X \times X \to Y$ is given by $F(x,y) = F_4(x,y) + f(0,y)$ for all $x,y \in X$. The uniqueness of such $F$ can be shown as in [5, Theorem 5]. □

**Theorem 13.9.** *Let $p > 0, 0 \leq \varepsilon$ and let $f : X \times X \to Y$ be a mapping satisfying (13.9) for all $x,y,z,w \in X$. Then there exists a unique bi-Jensen mapping $F : X \times X \to Y$ satisfying (13.10) for all $x,y \in X$ with $F(0,0) = f(0,0)$. That mapping $F$ is given by*

$$F(x,y) = \begin{cases} \lim_{j\to\infty} 2^{-j} f(2^j x, y) + f(0,y) & \text{if } 0 < p < 1, \\ \lim_{j\to\infty} 2^j (f(2^{-j}x, y) - f(0,y)) + f(0,y) & \text{if } p > 1. \end{cases}$$

## 13.3 Superstability of a Bi-Pexider Functional Equation

The following two lemmas result from [5, Theorems 7 and 8].

**Lemma 13.2.** *Let $p < 0$ and $\varepsilon > 0$. Let $f : X \times X \to Y$ be a mapping such that $\|Jf(x,y,z,w))\| \leq \varepsilon(\|x\|^p + \|y\|^p + \|z\|^p + \|w\|^p)$ for all $x,y,z,w \in X\backslash\{0\}$. Then $f : X \times X \to Y$ is a bi-Jensen mapping.*

**Lemma 13.3.** *Let $p < 0$ and $\varepsilon > 0$. Let $f : X \times X \to Y$ be a mapping such that $\|Jf(x,y,z,w))\| \leq \varepsilon(\|x\|^p + \|y\|^p)(\|z\|^p + \|w\|^p)$ for all $x,y,z,w \in X\backslash\{0\}$.Then $f : X \times X \to Y$ is a bi-Jensen mapping.*

Finally, we show that we have superstability of a bi-Jensen mapping for $p < 0$. Namely, we have the following two theorems.

**Theorem 13.10.** *Let $p < 0$ and $f, f_1, f_2, f_3, f_4 : X \times X \to Y$ be mappings satisfying (13.1) for all $x,y,z,w \in X\backslash\{0\}$. Then $f$ is a bi-Jensen mapping.*

*Proof.* Since the inequality

$$\|Jf(x,y,z,w)\| = \left\|A_1\left(\frac{x}{2}, \frac{y}{2}, \frac{z}{2}, \frac{w}{2}\right)\right\| \leq \frac{8\varepsilon}{2^p}(\|x\|^p + \|y\|^p + \|z\|^p + \|w\|^p)$$

holds for all $x,y,z,w \in X\backslash\{0\}$, we can apply Lemma 13.2. □

**Theorem 13.11.** *Let $p < 0$ and $f, f_1, f_2, f_3, f_4 : X \times X \to Y$ be mappings satisfying (13.9) for all $x,y,z,w \in X\backslash\{0\}$. Then $f$ is a bi-Jensen mapping.*

*Proof.* It is enough to use Lemma 13.3, because for every $x,y,z,w \in X\backslash\{0\}$ we have

$$\|Jf(x,y,z,w)\| = \left\|A_1\left(\frac{x}{2},\frac{y}{2},\frac{z}{2},\frac{w}{2}\right)\right\| \leq \frac{8\varepsilon}{4^p}(\|x\|^p + \|y\|^p)(\|z\|^p + \|w\|^p).$$

□

## References


1. Aoki, T.: On the stability of the linear transformation in Banach spaces. J. Math. Soc. Japan **2**, 64–66 (1950)
2. Bae, J.-H., Park, W.-G.: On the solution of a bi-Jensen functional equation and its stability. Bull. Korean Math. Soc. **43**, 499–507 (2006)
3. Găvruta, P.: A generalization of the Hyers–Ulam–Rassias stability of approximately additive mappings. J. Math. Anal. Appl. **184**, 431–436 (1994)
4. Hyers, D.H.: On the stability of the linear functional equation. Proc. Natl. Acad. Sci. U.S.A. **27**, 222–224 (1941)
5. Jun, K.-W., Lee,Y.-H., Oh, J.-H.: On the Rassias stability of a bi-Jensen functional equation. J. Math. Ineq. **2**, 363–375 (2008)
6. Rassias, Th.M.: On the stability of the linear mapping in Banach spaces. Proc. Amer. Math. Soc. **72**, 297–300 (1978)
7. Ulam, S.M.: A Collection of Mathematical Problems. Interscience, New York (1968)

# Chapter 14
# Approximately Midconvex Functions

**Krzysztof Misztal, Jacek Tabor, and Józef Tabor**

*Dedicated to the memory of S.M. Ulam*
*on the 100th anniversary of his birth*

**Abstract** In this paper, we propose very general definition of approximate midconvexity. Let $\alpha\colon [0,\infty) \to \mathbb{R}$ be a given function. Let $X$ be a normed space and $V$ a convex subset of $X$. A function $f\colon V \to \mathbb{R}$ will be called $\alpha(\cdot)$-midconvex if

$$f\left(\frac{x+y}{2}\right) \le \frac{1}{2}f(x) + \frac{1}{2}f(y) + \alpha(\|x-y\|) \quad \text{for } x,y \in V.$$

The above definition simultaneously generalizes approximate and uniform midconvexities. We present several results concerning this notion.

**Keywords** Approximately convex function • Approximately midconvex function • Semiconvex function

**Mathematics Subject Classification (2001):** Primary 26A51, 26B25

## 14.1 Introduction

The paper deals with a generalization of the notion of approximately convex functions. The idea to consider convexity with the given accuracy is very natural. The term "approximate convexity" was introduced by Hyers and Ulam [7] in 1952.

---

K. Misztal (✉) • J. Tabor
Institute of Computer Science, Jagiellonian University, Łojasiewicza 6, 30-348 Kraków, Poland
e-mail: krzysztof.misztal@ii.uj.edu.pl; tabor@ii.uj.edu.pl

J. Tabor
Institute of Mathematics, University of Rzeszów, Rejtana 16A, 35-310 Rzeszów, Poland
e-mail: tabor@univ.rzeszow.pl

Th.M. Rassias and J. Brzdęk (eds.), *Functional Equations in Mathematical Analysis*, Springer Optimization and Its Applications 52, DOI 10.1007/978-1-4614-0055-4_14, 

Let $\delta \geq 0$ be a given number. We say that a real valued function $f : V \to \mathbb{R}$, where $V$ is convex, is *$\delta$-convex, $\delta$-midconvex respectively,* if

$$f(tx+(1-t)y) \leq tf(x)+(1-t)f(y)+\delta \quad \text{for } x,y \in V, t \in [0,1]$$
$$f\left(\frac{x+y}{2}\right) \leq \frac{f(x)+f(y)}{2}+\delta \quad \text{for } x,y \in V.$$

A function $f$ is called *approximately convex (approximately midconvex)* if it is $\delta$-convex ($\delta$-midconvex) with a certain $\delta \geq 0$. The paper of Hyers and Ulam gave a creative impulse to the theory of stability of convex functions. A survey of results concerning this topic can be found in the book [6].

We briefly describe the main idea of the paper. Let $\alpha : \mathbb{R}_+ \to \mathbb{R}$. We call a function $f$ *$\alpha(\cdot)$-midconvex* if

$$f\left(\frac{x+y}{2}\right) \leq \frac{f(x)+f(y)}{2}+\alpha(\|x-y\|) \quad \text{for } x,y \in V.$$

Up till now two separate versions of modified midconvexity were considered:

- One when $\alpha$ is negative, where we obtain the so-called uniformly midconvex functions (this is a strengthening of the midconvexity condition);
- The other when $\alpha$ is positive, and then we get approximately midconvex functions.

The basic novelty of our investigation relies on the resignation of the assumption that $\alpha$ has constant sign. This allows us to study simultaneously the two discussed cases in one generalized setting, consequently our results hold for approximately and uniformly midconvex functions.

We present a collection of results concerning this notion and its modifications or further generalizations. Our aim is to write the paper in such a way that it could be fully understand without previous knowledge of the subject. Therefore we place there also some already known results with their proofs.

In this paper (if not stated otherwise), $X$ denotes a normed space and $V$ a convex subset of $X$. By $\mathbb{R}_+$ we denote the set of nonnegative real numbers and by $\mathbb{N}_0$ the set of nonnegative integers. For a function $f : V \to \mathbb{R}$ we use the following notations

$$Cf(x,y,t) := f(tx+(1-t)y) - tf(x) - (1-t)f(y) \quad \text{for } x,y \in V,\ t \in [0,1];$$
$$Jf(x,y) := Cf\left(x,y,\frac{1}{2}\right) = f\left(\frac{x+y}{2}\right) - \frac{1}{2}f(x) - \frac{1}{2}f(y) \quad \text{for } x,y \in V,\ t \in [0,1].$$

The notions of convex and midconvex function were generalized and modified in many different directions. One should mention here so called $(M,N)$ – midpoint convex functions. With this direction the name of Matkowski [9] is strongly related. The notion of convex function on abelian group was defined and investigated by Jarczyk and Laczkovich [8]. Uniformly convex functions and strongly convex functions are very interesting classes of convex functions [3, 12]. But we will not consider the above mentioned generalizations and modifications of convexity.

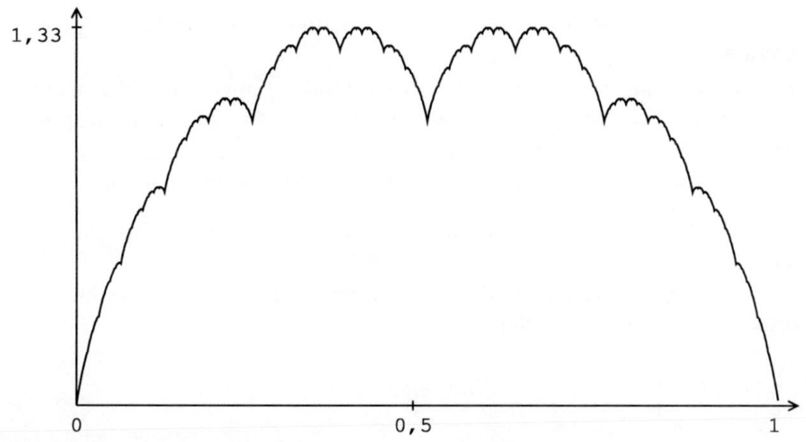


**Fig. 14.1** Graph of the Takagi function $T(x) = \sum_{k=0}^{\infty} 2^{-k+1}\operatorname{dist}(2^k x; \mathbb{Z})$ – an important example of an $\alpha(\cdot)$-midconvex function

Some important notions in the area of approximate convexity were defined by Rolewicz. He introduced in [13] the notions of paraconvex and strongly paraconvex functions. In the series of papers he investigated these notions (cf.[14]). There are also papers on this subject by another authors (cf. for instance [17]).

Let $\alpha\colon [0,\infty) \to [0,\infty)$ be a nondecreasing function with $\lim_{\rho\to 0^+} \rho^{-1}\alpha(\rho) = 0$. We say that a function $f\colon V \to \mathbb{R}$ is *$\alpha(\cdot)$-paraconvex* if there exists a $c > 0$ such that

$$Cf(x,y,t) \le c\alpha(\|x-y\|) \quad \text{for } x,y \in V, t \in [0,1],$$

and $f$ is called *strongly $\alpha(\cdot)$ – paraconvex* if there exists a $c > 0$ such that

$$Cf(x,y,t) \le c\min(t,1-t)\alpha(\|x-y\|) \quad \text{for } x,y \in V, t \in [0,1].$$

Investigations of approximate midconvexity in a similar spirit were initiated by Páles [11] and then continued by Házy, Z. Boros, Páles [2,5]. Simplifying a little bit one can say that they investigated functions satisfying the following condition

$$Jf(x,y) \le c\|x-y\|^p + \delta \quad \text{for } x,y \in V,$$

where $p$, $\delta$, $c$ are given positive constants. This way of understanding approximate midconvexity was generalized by the last two authors in [15], where the following definition was proposed.

Let $\alpha\colon [0,\infty) \to [0,\infty)$ be a given nondecreasing function. We say that a function $f\colon V \to \mathbb{R}$ is *$\alpha(\cdot)$-midconvex* if

$$Jf(x,y) \le \alpha(\|x-y\|) \quad \text{for } x,y \in V.$$

For a typical example of an $\alpha(\cdot)$-midconvex function see Fig. 14.1.

In this paper, we consider a more general definition which will be given in the next section.

There are also papers in the area of approximate convexity which use another definitions and terminology. The notion of semiconvex function (Jensen semiconvex function) should be mentioned here. In this case we relax simultaneously the inequality defining the convexity and condition concerning the domain of a function.

**Definition 14.1 ([4]).** Let $S$ be a subset of $\mathbb{R}^N$. By $[S]$ we denote the set of all pairs $(x,y) \in S \times S$ such that the line segment $[x,y]\colon = \{tx+(1-t)y\colon t \in [0,1]\}$ is contained in $S$. By $\mathscr{M}$ we denote the set of all nondecreasing upper semicontinuous functions $\omega\colon \mathbb{R}_+ \to \mathbb{R}_+$ such that

$$\lim_{\rho \to 0^+} \omega(\rho) = 0.$$

Let $\omega \in \mathscr{M}$. We say that a function $f\colon S \to \mathbb{R}$ is *$\omega$-semiconvex* if

$$Cf(x,y,t) \le t(1-t)\|x-y\|\omega(\|x-y\|) \quad \text{for } (x,y) \in [S], t \in [0,1].$$

We call $\omega$ a *modulus of semiconvexity of $f$*. If the above inequality holds for $t = \frac{1}{2}$, then we say that $f$ is *Jensen $\omega$-semiconvex*. We say that $f$ is *(Jensen) semiconvex* if it is (Jensen) $\omega$-semiconvex with a certain modulus of (Jensen) semiconvexity $\omega \in \mathscr{M}$.

Part of the Definition 14.1 was generalized in [16]. The main step relied on replacing $\mathbb{R}^n$ by a real topological vector space.

**Definition 14.2.** Let $X$ be a Hausdorff real topological vector space, $S$ an open subset of $X$, $\alpha\colon X \to [0,\infty)$ an even function locally bounded at zero and $t \in [0,1]$ a fixed number. A function $f\colon S \to \mathbb{R}$ is called *$(\alpha,t)$-preconvex* if

$$Cf(x,y,t) \le \alpha(x-y) \quad \text{for } x,y \in D \text{ such that } [x,y] \subset D.$$

In the case $t = 1/2$ we say that $f$ is *$\alpha$-premidconvex*.

At the end of the introduction we would like to mention that all the graphs presented in this paper were prepared by our computer program which is available on `http://www.ii.uj.edu.pl/~misztalk/index.php?page=convex`.

## 14.2 Direct Consequences of Approximate Midconvexity

As it is well known midconvexity implies $\mathbb{Q}$-convexity, that is each midconvex function $f\colon V \to \mathbb{R}$ satisfies the inequality

$$Cf(x,y,r) \le 0 \quad \text{for } x,y \in V,\ r \in [0,1] \cap \mathbb{Q}.$$

In this section, we will prove some generalizations of this result. We will try to answer the question what kind of approximate convexity does if follow from approximate midconvexity.

**Definition 14.3.** Let $\alpha\colon [0,\infty) \to \mathbb{R}$ be a given function bounded on each bounded set. A function $f\colon V \to \mathbb{R}$ will be called $\alpha(\cdot)$-midconvex if

$$Jf(x,y) \le \alpha(\|x-y\|) \quad \text{for } x,y \in V.$$

We would like to emphasize that the above definition is really very general. If we assume additionally that $\alpha$ takes its values in $\mathbb{R}_+$ we obtain the version of midconvexity considered in [15]. In case if $\alpha$ takes only nonpositive values we obtain midconvex (Jensen) version of uniform convexity [3]. In particular case if $\alpha(r) = -cr^2$ where $c > 0$ we obtain midconvex (Jensen) version of strong convexity ([12]).

We are going to generalize some results from [15]. The generalization rely on relaxing the assumption on $\alpha$. We resign from the assumption that $\alpha$ takes only nonnegative values. In spite of that, the results and the proofs remain the same. We "quote" both.

The following function plays a basic role in our investigations

$$d(x) := 2\mathrm{dist}(x,\mathbb{Z}) \quad \text{for } x \in \mathbb{R}.$$

**Proposition 14.1.** *Let $D \subset [0,1]$ be such that*

$$x \in [0,1/2] \cap D \Rightarrow 2x \in D,\ x \in [1/2,1] \cap D \Rightarrow 2x-1 \in D.$$

*Let $h : D \to \mathbb{R}$ be an upper bounded function such that*

$$h(x) \le h(2x)/2 + \alpha(2x) \quad \textit{for } x \in D \cap [0,1/2],$$
$$h(x) \le h(2x-1)/2 + \alpha(2-2x) \quad \textit{for } x \in D \cap [1/2,1].$$

*Then*

$$h(r) \le \sum_{k=0}^{\infty} \frac{1}{2^k}\alpha(d(2^k r)) \quad \textit{for } r \in D. \tag{14.1}$$

*Proof.* Notice that since $\alpha$ is bounded on bounded sets and $d(\mathbb{R}) = [0,1]$ the sum in (14.1) is well-defined and finite.

Let

$$C := \sup_{r\in D}\left( h(r) - \sum_{k=0}^{\infty} \frac{1}{2^k}\alpha(d(2^k r)) \right).$$

Since $h$ is upper bounded, we obtain that $C < \infty$. We are going to prove that $C \le 0$.

For an indirect proof, suppose that this is not the case. Choose an $r_0 \in D$ such that

$$h(r_0) - \sum_{k=0}^{\infty} \frac{1}{2^k} \alpha(d(2^k r_0)) > C/2.$$

Let us first consider the case when $r_0 \in [0, 1/2]$. Then we obtain

$$\begin{aligned} h(2r_0) &\geq 2h(r_0) - 2\alpha(2r_0) \\ &> 2\sum_{k=0}^{\infty} \frac{1}{2^k} \alpha(d(2^k r_0)) + 2C/2 - 2\alpha(d(r_0)) \\ &= \sum_{l=0}^{\infty} \frac{1}{2^l} \alpha(d(2^l(2r_0))) + C. \end{aligned}$$

Consequently

$$h(2r_0) - \sum_{l=0}^{\infty} \frac{1}{2^l} \alpha(d(2^l(2r_0))) > C,$$

a contradiction with the definition of $C$.

The case when $r_0 \in [1/2, 1]$ can be treated similarly (we begin with the inequality $h(2r_0 - 1) \geq 2h(r_0) - 2\alpha(2 - 2r_0)$ and proceed analogously). □

Directly from Proposition 14.1 we obtain the following result.

**Corollary 14.1.** *Let $h : [0,1] \to \mathbb{R}$ be an $\alpha(\cdot)$-midconvex function such that $h(0) = h(1) = 0$. Then*

$$h(r) \leq \sum_{k=0}^{\infty} \frac{1}{2^k} \alpha(d(2^k r)) \quad \textit{for } r \in [0,1] \cap \mathbb{Q}.$$

*Proof.* Let $r = k/n \in \mathbb{Q} \cap [0,1]$, where $k \in \mathbb{Z}$, $n \in \mathbb{N}$. We put

$$D = \{0, 1/n, \ldots, (n-1)/n, 1\}.$$

Then $D$ satisfies the assumptions of Proposition 14.1. Since $D$ is finite, $h|_D$ is bounded, and therefore Proposition 14.1 applied to $h|_D$ makes the proof complete. □

The main result in this direction reads as follows.

**Theorem 14.1.** *Let $f : V \to \mathbb{R}$ be an $\alpha(\cdot)$-midconvex function. Then*

$$f(rx + (1-r)y) \leq rf(x) + (1-r)f(y) + \sum_{k=0}^{\infty} \frac{1}{2^k} \alpha(d(2^k r)\|x - y\|) \tag{14.2}$$

*for $x, y \in V$, $r \in [0,1] \cap \mathbb{Q}$.*

*Proof.* Let us fix $x,y \in V$. We define the function $h : [0,1] \to \mathbb{R}$ by the formula

$$h(r) = f(rx+(1-r)y) - rf(x) - (1-r)f(y) \quad \text{for } r \in [0,1].$$

Then $h(0) = h(1) = 0$. By the $\alpha(\cdot)$-midconvexity of $f$ we obtain that

$$h\left(\frac{r+s}{2}\right) \le \frac{h(r)+h(s)}{2} + \alpha(|r-s|\|x-y\|) \quad \text{for } r,s \in [0,1].$$

It means that $h$ is $\theta(\cdot)$-midconvex, with $\theta(w) := \alpha(w\|x-y\|)$. By Corollary 14.1 we obtain that

$$h(q) \le \sum_{k=0}^{\infty} \frac{1}{2^k}\theta(d(2^k q)) = \sum_{k=0}^{\infty} \frac{1}{2^k}\alpha(d(2^k q)\|x-y\|) \quad \text{for } q \in [0,1] \cap \mathbb{Q},$$

that is

$$f(rx+(1-r)y) - rf(x) - (1-r)f(y) \le \sum_{k=0}^{\infty} \frac{1}{2^k}\alpha(d(2^k r)\|x-y\|)$$

for $r \in [0,1] \cap \mathbb{Q}$. □

We will prove later on that under certain additional assumptions on $\alpha$ the estimation of $Cf(x,y,r)$ given in Theorem 14.1 is optimal. However, in general, it is not optimal. Therefore, we will present another results in the similar spirit as in Theorem 14.1.

By $\mathcal{D}$ we denote the set of dyadic numbers in $\mathbb{Q}$, that is elements of the form $l/2^n$, where $l \in \mathbb{Z}$ and $n \in \mathbb{N}$. By the degree $\deg(q)$ of a dyadic number $q$ we understand the smallest nonnegative integer $n$ such that $2^n q \in \mathbb{Z}$.

**Proposition 14.2.** *Let $h : [0,1] \cap \mathcal{D} \to \mathbb{R}$, $h(0) = h(1) = 0$ be an $\alpha(\cdot)$-midconvex function. Then*

$$h(q) \le \sum_{k=0}^{\infty} \alpha(1/2^k) d(2^k q) \quad \textit{for } q \in [0,1] \cap \mathcal{D}. \tag{14.3}$$

*Proof.* Let us observe that for every $q \in \mathcal{D}$ the right hand side of (14.3) is finite, since the function $d$ is zero on integer numbers.

We are going to prove (14.3) by induction over the degree of $q$. Obviously, (14.3) is valid if $\deg(q) = 0$, that is if $q = 0$ or $q = 1$. Assume that for a certain $n \in \mathbb{N}_0$ (14.3) holds true for all $q \in [0,1] \cap \mathcal{D}$ such that $\deg(q) \le n$.

Let $q \in [0,1] \cap \mathcal{D}$, $\deg(q) = n+1$, that is $q = (2l+1)/2^{n+1}$ for a certain $l \in \{0,\dots,2^n - 1\}$. Then

$$q = (l/2^n + (l+1)/2^n)/2,$$

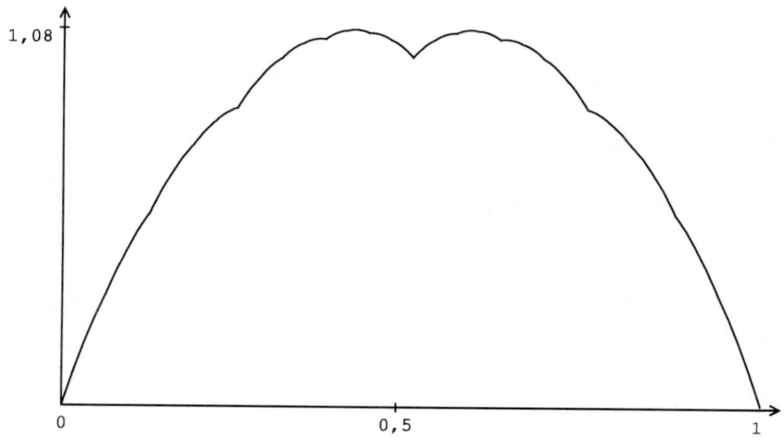


**Fig. 14.2** Graph of $\sum_{k=0}^{\infty} \alpha(2^{-k})d(2^k q)$ for $\alpha(x) = x^{1.5}$

and by the inductive assumption

$$h(q) \le [h(l/2^n) + h((l+1)/2^n)]/2 + \alpha(1/2^n) \tag{14.4}$$

$$\le \alpha(1/2^n) + \sum_{k=0}^{\infty} \alpha(1/2^k) \frac{d(l/2^{n-k}) + d((l+1)/2^{n-k})}{2}.$$

Obviously, $l/2^{n-k}, (l+1)/2^{n-k} \in \mathbb{Z}$ for all $k \ge n$. Let us observe that for every $k \in \{0, \dots, n-1\}$ there exists an $m_k \in \mathbb{Z}$ such that

$$l/2^{n-k}, (l+1)/2^{n-k} \in [m_k/2, (m_k+1)/2]. \tag{14.5}$$

Notice that $d(m) = 0$ and $d(m+1/2) = 1$ for all $m \in \mathbb{Z}$ and that $d$ is piecewise affine, more precisely,

$$d\left(\frac{x+y}{2}\right) = (d(x)+d(y))/2 \quad \text{for } x, y \in [m/2, (m+1)/2], m \in \mathbb{Z}.$$

Therefore, by applying (14.4) and (14.5) we obtain that

$$h(q) \le \alpha(1/2^n) + \sum_{k=0}^{n-1} \alpha(1/2^k) d((2l+1)/2^{n+1-k})$$

$$= \sum_{k=0}^{n} \alpha(1/2^k) d((2l+1)/2^{n+1-k}) = \sum_{k=0}^{\infty} \alpha(1/2^k) d(2^k q).$$

□

For a example of this estimation see Fig. 14.2.

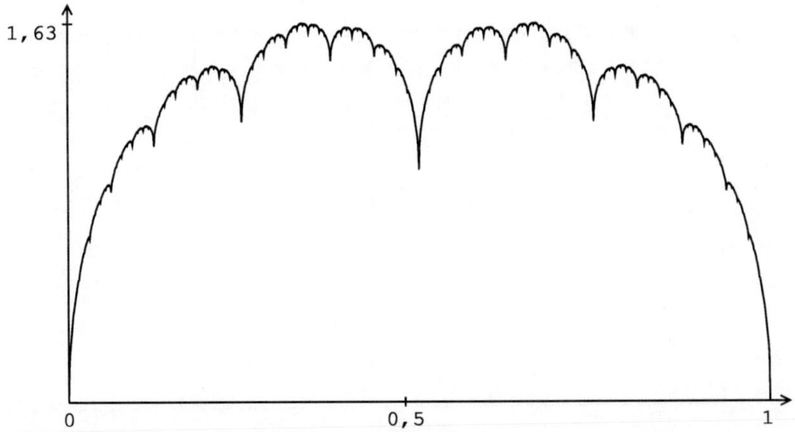


**Fig. 14.3** Graph of $\sum_{k=0}^{\infty} 2^{-k}\alpha(d(2^k q))$ for $\alpha(x) = x^{0.5}$

**Theorem 14.2.** *Let $f : V \to \mathbb{R}$ be an $\alpha(\cdot)$-midconvex function. Then*

$$f(rx + (1-r)y) - rf(x) - (1-r)f(y) \le \sum_{k=0}^{\infty} \alpha(\|x-y\|/2^k) d(2^k r)$$

*for all $x, y \in V$ and $r \in [0,1] \cap \mathscr{D}$.*

*Proof.* We fix fix arbitrarily $x, y \in V$ and define the function $h : [0,1] \to \mathbb{R}$ by the formula

$$h(r) = f(rx + (1-r)y) - rf(x) - (1-r)f(y) \quad \text{for } r \in [0,1].$$

Clearly, $h(0) = h(1) = 0$, and by the $\alpha(\cdot)$-midconvexity of $f$ we obtain that

$$h\left(\frac{s+t}{2}\right) \le \frac{h(s) + h(t)}{2} + \alpha(|s-t|\|x-y\|) \quad \text{for } s, t \in [0,1].$$

It means that $h$ is $\theta$-midconvex with $\theta(w) := \alpha(w\|x-y\|)$. By Proposition 14.2

$$h(r) \le \sum_{k=0}^{\infty} \theta(1/2^k) d(2^k r) = \sum_{k=0}^{\infty} \alpha(\|x-y\|/2^k) d(2^k r) \quad \text{for } r \in [0,1] \cap \mathscr{D},$$

and consequently

$$f(rx + (1-r)y) - rf(x) - (1-r)f(y) \le \sum_{k=0}^{\infty} \alpha(\|x-y\|/2^k) \cdot d(2^k r),$$

for $r \in [0,1] \cap \mathscr{D}$. □

For examples of this estimation see Figs. 14.3 and 14.4.

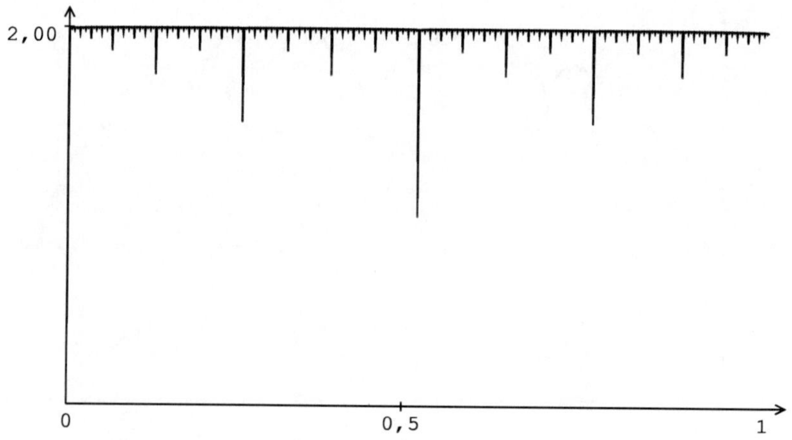


**Fig. 14.4** Graph of $\sum_{k=0}^{\infty} 2^{-k}\alpha(d(2^k q))$ for $\alpha(x) = x^{0.001}$ is very close to the example constructed in [10] which, as one can easily observe, is generated by $\alpha(x) = \mathrm{sign}(x)$

We conclude this section with a result concerning generalization of semiconvex functions (compare Definition 14.1).

**Corollary 14.2.** *Let $\omega\colon [0,\infty) \to \mathbb{R}$ be a given function and let $S$ be a subset of $X$. Let $f\colon S \to \mathbb{R}$ satisfy the following condition*

$$Jf(x,y) \le \frac{\|x-y\|}{4}\omega(\|x-y\|)$$

*for $x, y \in S$ such that the segment $[x,y]$ is included in $S$.*

*Then for every $x, y \in S$ such that $[x,y] \subset S$ we have*

$$Cf(x,y,r) \le \frac{1}{4}\sum_{k=0}^{\infty} \frac{\|x-y\|}{2^k}\omega\left(\frac{\|x-y\|}{2^k}\right) d(2^k r) \quad \text{for } r \in [0,1] \cap D.$$

## 14.3 Bernstein–Doetsch Type Results

The famous Bernstein–Doetsch theorem [1] states that if $V$ is open and convex subset of $\mathbb{R}^n$ and $f\colon V \to \mathbb{R}$ is midconvex and locally bounded at a point then it is continuous. As a consequence we obtain that under the above assumptions $f$ is convex. A similar result (and by similar method) can be obtained for approximately convex functions.

We begin with a lemma [6, Theorem 8.5 and Lemma 8.6].

**Lemma 14.1.** *Let $V$ be an open convex set, $\delta > 0$ and let $f \colon V \to \mathbb{R}$ be $\delta$-midconvex function locally bounded above at a point. Then $f$ is locally bounded at every point of $V$.*

**Theorem 14.3.** *Let $\alpha \colon [0,\infty) \to [0,\infty)$ be a given function bounded on each bounded subset of $\mathbb{R}_+$. Let $V$ be an open convex set and let $f \colon V \to \mathbb{R}$ be $\alpha(\cdot)$-midconvex and locally bounded above at a point. Then $f$ is locally bounded on each point of $V$.*

*If additionally*

$$\lim_{\rho \to 0^+} \alpha(\rho) = 0,$$

*then $f$ is continuous.*

*Proof.* Assume that $f$ is locally bounded above at $x_0 \in V$. Consider an arbitrary $r > 0$. We put

$$V_r \colon = V \cap B(x_0, r),$$

where $B(x_0, r)$ denotes the open ball centered at $x_0$ with radius $r$ and define

$$\delta \colon = \sup_{\rho \in (0,r]} \alpha(\rho).$$

Obviously, $V_r$ is open and convex and $\delta \in \mathbb{R}_+$. By Lemma 14.1 we obtain that $f$ is locally bounded at each point of $V_r$. Since $r$ was arbitrary it implies that $f$ is locally bounded at each point of $V$.

We assume now that

$$\lim_{\rho \to 0^+} \alpha(\rho) = 0. \tag{14.6}$$

By the first assertion we know that $f$ is locally bounded at each point of $V$. We fix arbitrarily $x \in V$ and put

$$m_f(x) \colon = \lim_{\rho \to 0^+} \inf_{B(x,\rho)} f,$$

$$M_f(x) \colon = \lim_{\rho \to 0^+} \sup_{B(x,\rho)} f.$$

It is obvious that

$$-\infty < m_f(x) \le f(x) \le M_f(x) < \infty. \tag{14.7}$$

We can find sequences $(x_n)_{n \in \mathbb{N}}, (y_n)_{n \in \mathbb{N}} \subset V$ such that

$$\lim_{n \to \infty} x_n = x, \qquad \lim_{n \to \infty} f(x_n) = m_f(x), \tag{14.8}$$

and

$$\lim_{n\to\infty} y_n = x, \qquad \lim_{n\to\infty} f(y_n) = M_f(x). \tag{14.9}$$

Let

$$z_n\colon = 2y_n - x_n \quad \text{for } n \in \mathbb{N}.$$

Without loss of generality we may assume that $(z_n)_{n\in\mathbb{N}} \subset V$.

We have

$$\begin{aligned} f(y_n) = f\left(\frac{x_n + z_n}{2}\right) &\le \frac{1}{2}f(x_n) + \frac{1}{2}f(y_n) + \alpha(\|z_n - x_n\|) \\ &= \frac{1}{2}f(x_n) + \frac{1}{2}f(y_n) + \alpha(2\|y_n - x_n\|) \end{aligned}$$

for $n \in \mathbb{N}$.

Making now use of (14.6), (14.8), (14.9) we get

$$\begin{aligned} M_f(x) &\le \frac{1}{2}m_f(x) + \frac{1}{2}\limsup_{n\to\infty} f(z_n) + \lim_{n\to\infty} \alpha(2\|y_n - x_n\|) \\ &\le \frac{1}{2}m_f(x) + \frac{1}{2}M_f(x). \end{aligned}$$

Whence we obtain that

$$M_f(x) \le m_f(x),$$

which together with (14.7) implies that $M_f(x) = m_f(x)$. But it means continuity of $f$ at the point $x$. □

Taking in Theorem 14.3 $\alpha(r) := \varepsilon r^p$, where $\varepsilon \ge 0$, $p > 0$ we obtain Theorems 1 and 3 from [5].

Theorem 14.3 can be modified in such a way that to cover the case of semiconvex functions (and even its generalization).

**Theorem 14.4.** *Let $\omega\colon [0,\infty) \to [0,\infty)$ be a given function, bounded on each bounded subset of $\mathbb{R}_+$. Let $V$ be an open and connected subset of $X$ and let $f\colon S \to \mathbb{R}$ satisfy the inequality*

$$Jf(x,y) \le \frac{\|x-y\|}{4}\omega(\|x-y\|) \quad \textit{for } x,y \in S \ \textit{such that: } [x,y] \subset S.$$

*If $f$ is locally bounded above at a point then $f$ is continuous.*

*Proof.* For $A \subset X$, $\delta > 0$ we denote

$$A_\delta := \{x \in X\colon \operatorname{dist}(x,A) < \delta\}.$$

Let $f$ be locally bounded above at $x_0 \in S$. Consider arbitrary point $x \in X$. Since $S$ is open and connected we can find in $S$ a broken line with vertices $x_0, x_1, \ldots, x_n = x$. Further we can find a $\delta_1 > 0$ such that $[x_0, x_1]_{\delta_1} \subset S$. Obviously $[x_0, x_1]_{\delta_1}$ is open and convex. We take

$$\alpha(r) := \frac{r}{4}\omega(r) \quad \text{for } r \in [0, \infty).$$

Then evidently

$$\lim_{r \to 0^+} \alpha(r) = 0.$$

By Theorem 14.3 we obtain that $f$ is continuous on $[x_0, x_1]_{\delta_1}$. It means, in particular, that $f$ is locally bounded at $x_1$. Continuing this procedure we obtain that $f$ is continuous at $x$. □

The results of Sect. 2, Theorems 14.3 and 14.4 can be formulated and proved also in a more general setting. Below we present versions of these results in a topological vector space.

Let $X$ be a Hausdorff real topological vector space, $S \subset X$ be an open set and $\alpha \colon X \to [0, \infty)$ be an even function locally bounded at zero.

**Definition 14.4.** We say that a function $f \colon S \to \mathbb{R}$ is $\alpha$-premidconvex if

$$Jf(x,y) \le \alpha(x-y) \quad \text{for } x, y \in S$$

such that $[x,y] \subset V$.

In [16], the following results have been proved.

**Theorem 14.5.** *Let $S$ be an open and connected subset of $X$ and let $f \colon S \to \mathbb{R}$ be an $\alpha$-premidconvex function locally bounded at a point. Then $f$ is locally bounded at every point.*

**Theorem 14.6.** *Let $S$ be an open and connected subset of $X$ and let $f \colon S \to \mathbb{R}$ be an $\alpha$-premidconvex function locally bounded above at a point. We assume additionally that $\alpha(0) = 0$ and that $\alpha$ is continuous at zero. Then $f$ is locally uniformly continuous.*

**Theorem 14.7.** *We assume that for each $x \in X$ the function $\mathbb{R}_+ \ni w \to \alpha(wx)$ is nondecreasing. Let $f \colon S \to \mathbb{R}$ be an $\alpha$-premidconvex function. Then*

$$Cf(x,y,r) \le \frac{1}{4} \sum_{k=0}^{\infty} \frac{1}{2^k} \alpha\left(d(2^k r)\frac{x-y}{2}\right)$$

*for all $x, y \in S$ such that $[x,y] \subset V$ and all $r \in [0,1] \cap \mathbb{Q}$.*

*If additionally $S$ is connected and $f$ is locally bounded at a point then the above inequality holds for all $x, y \in S$ such that $[x,y] \subset S$.*

The proofs of Theorems 14.5–14.7 are natural modifications of the corresponding their versions in normed space (we omit them to simplify and shorten the paper).

## References

1. Bernstein, F., Doetsch, G.: Zur Theorie der konvexen Funktionen. Math. Ann. **76**, 514–526 (1915)
2. Boros, Z.: An inequality for the Takagi functions. Math. Inequal. Appl. **11**, 757–765 (2008)
3. Butnariu, B., Iusem, A.N., Zălinescu, C.: On uniform convexity, total convexity and convergence of the proximal point and outer Bergman projection algorithms in Banach spaces. J. Convex Anal. **10**, 35–61 (2003)
4. Cannarsa, P., Sinestrari, C.: Semiconvex Functions, Hamiltonian-Jacobi Equations and Optimal Control. Progress in Nonlinear Differential Equations and Their Applications **58**, Birkhäuser, Boston (2004)
5. Házy, A., Páles, Zs.: On approximately midconvex functions. Bull. London Math. Soc. **36**, 339–350 (2004)
6. Hyers, D.H., Isac, G., Rassias, Th.M.: Stability of Functional Equations in Several Variables. Birkhäuser, Basel (1998)
7. Hyers, D.H., Ulam, S.M.: Approximately convex functions. Proc. Amer. Math. Soc. **3**, 821–828 (1952)
8. Jarczyk, W., Laczkovich, M.: Convexity on abelian groups. J. Convex Anal. **16**, 33–48 (2009)
9. Matkowski, J.: Generalized convex functions and a solution of a problem of Zs. Páles. Publ. Math. Debrecen **73**, 421–460 (2008)
10. Ng, C.T., Nikodem, K.: On approximately convex functions. Proc. Amer. Math. Soc. **118**, 103–108 (1993)
11. Páles, Zs.: On approximately convex functions. Proc. Amer. Math. Soc. **131**, 243–252 (2002)
12. Polovinkin, E.: Strongly convex analysis. Sb. Math. **187**, 259–286 (1996)
13. Rolewicz, S.: On $\alpha(\cdot)$-paraconvex and strongly $\alpha(\cdot)$-paraconvex functions. Control Cybernet. **29**, 367–377 (2000)
14. Rolewicz, S.: Paraconvex analysis. Control Cybernet. **34**, 951–965 (2005)
15. Tabor, J., Tabor, J.: Generalized approximate midconvexity. Control Cybernet. **38**, 655–669 (2009)
16. Tabor, J., Tabor, J., Żołdak, M.: Approximately convex functions on topological vector spaces. Publ. Math. Debrecen **77**, 115–123 (2010)
17. Zajiček, L.: A $C^1$ function which is nowhere strongly paraconvex and nowhere semiconcave. Control Cybernet. **36**, 803–810 (2007)

# Chapter 15
# The Hyers–Ulam and Ger Type Stabilities of the First Order Linear Differential Equations

**Takeshi Miura and Go Hirasawa**

*Dedicated to the memory of S.M. Ulam*
*on the 100th anniversary of his birth*

**Abstract** In this paper, we show a recent development of the Hyers–Ulam stability and Ger type stability of the first order linear differential equation $y' + py + q = 0$ for holomorphic functions on convex open sets of the complex number field and continuously differentiable functions on open intervals of the real number field.

**Keywords** Exponential functions • Hyers–Ulam stability • Ger type stability

**Mathematics Subject Classification (2000):** Primary 34K20; Secondary 26D10

## 15.1 Introduction

The stability problem of functional equations had been first raised by S. M. Ulam (cf. [26, Chap. VI]). "For what metric groups $G$ is it true that an $\varepsilon$-automorphism of $G$ is necessarily near to a strict automorphism? (An $\varepsilon$-automorphism of $G$ means, in this case, a transformation $f$ of $G$ into itself such that $\rho(f(x \cdot y), f(x) \cdot f(y)) < \varepsilon$ for all $x, y \in G$, where $\rho$ is the metric on $G$.)"

Hyers [8] gave an affirmative answer to the problem as follows:
*Let $E_1$ and $E_2$ be real Banach spaces, $\varepsilon \geq 0$, and $f\colon E_1 \to E_2$ satisfy the inequality $\|f(x+y) - f(x) - f(y)\| \leq \varepsilon$ for all $x, y \in E_1$. Then the limit $T(x) = \lim_{n\to\infty} 2^{-n} f(2^n x)$ exists for each $x \in E_1$, and $T\colon E_1 \to E_2$ is the unique additive mapping such that $\|f(x) - T(x)\| \leq \varepsilon$ for all $x \in E_1$. If, in addition, the mapping $\mathbb{R} \ni t \mapsto f(tx)$ is continuous for each fixed $x \in E_1$, then $T$ islinear.*

T. Miura (✉)
Department of Applied Mathematics and Physics, Yamagata University,
Yonezawa 992-8510, Japan
e-mail: miura@yz.yamagata-u.ac.jp

G. Hirasawa
Faculty of Engineering, Ibaraki University, Hitachi 316-8511, Japan
e-mail: gou@mx.ibaraki.ac.jp

Th.M. Rassias and J. Brzdęk (eds.), *Functional Equations in Mathematical Analysis*,
Springer Optimization and Its Applications 52, DOI 10.1007/978-1-4614-0055-4_15,

This result is called the *Hyers–Ulam stability* of the *additive* Cauchy equation $g(x+y) = g(x) + g(y)$. Aoki [2] introduced unbounded, additive Cauchy difference and generalized a result [8, Theorem 1] of Hyers. Th.M. Rassias [20], who independently introduced the unbounded Cauchy difference, was the first to prove the stability of the linear mapping between Banach spaces. The concept of the Hyers–Ulam–Rassias stability was originated from Rassias' paper [20] for the stability of the linear mapping. Rassias [20] generalized Hyers' Theorem as follows:

**Theorem 15.1.** *Suppose that $E_1$ and $E_2$ are two real Banach spaces and $f: E_1 \to E_2$ is a mapping. If there exist $\varepsilon \geq 0$ and $0 \leq p < 1$ such that*

$$\|f(x+y) - f(x) - f(y)\| \leq \varepsilon(\|x\|^p + \|y\|^p)$$

*for all $x, y \in E_1$, then there is a unique additive mapping $T: E_1 \to E_2$ such that*

$$\|f(x) - T(x)\| \leq \frac{2\varepsilon}{|2-2^p|} \|x\|^p$$

*for all $x \in E_1$. If, in addition, the mapping $\mathbb{R} \ni t \mapsto f(tx)$ is continuous for each fixed $x \in E_1$, then $T$ is linear.*

This result is, what is called, *the Hyers–Ulam–Rassias stability* of the additive Cauchy equation $g(x+y) = g(x) + g(y)$. The result of Hyers is just the case where $p = 0$. During the 27th International Symposium on Functional Equations, Rassias raised the problem whether a similar result holds for $1 \leq p$. Z. Gajda [5, Theorem 2] proved that Theorem B is valid for $1 < p$; In the same paper [5, Example], he also gave an example that a similar result does not hold for $p = 1$. Later, Th.M. Rassias and P. Šemrl [22, Theorem 2] gave another counter example for $p = 1$.

In connection with the stability of exponential functions, Alsina and Ger [1] remarked that the differential equation $y' = y$ has the Hyers–Ulam stability: more explicitly, if $I$ is an open interval, $\varepsilon > 0$ and $f: I \to \mathbb{R}$ is a differentiable function satisfying $|f'(t) - f(t)| \leq \varepsilon$ for all $t \in I$, then there exists a differentiable function $f_0: I \to \mathbb{R}$ such that $f_0'(t) = f_0(t)$ and $|f(t) - f_0(t)| \leq 3\varepsilon$ for all $t \in I$. Takahasi et al. [25] considered the Banach space valued differential equation $y' = \lambda y$, where $\lambda$ is a complex number. Then they proved the Hyers–Ulam stability of $y' = \lambda y$ under the condition that $\Re\lambda \neq 0$. This result is generalized by Miura, Miyajima and Takahasi [18]. They considered the Banach space valued $n$-th order linear differential equation with constant coefficients.

Let $\mathbb{C}$ be the complex number field. Taking the group structure of $\mathbb{C} \setminus \{0\}$ into account, R. Ger and P. Šemrl [6] considered the inequality

$$\left| \frac{f(x+y)}{f(x)f(y)} - 1 \right| \leq \varepsilon \qquad (\forall x, y \in S)$$

for a mapping $f\colon S \to \mathbb{C}\setminus\{0\}$, where $(S,+)$ is a semigroup. If $0 \le \varepsilon < 1$ and if $(S,+)$ is a cancellative abelian semigroup, then they proved that there is a unique function $f_0\colon S \to \mathbb{C}\setminus\{0\}$ such that $f_0(x+y) = f_0(x)f_0(y)$ for all $x,y \in S$ and that

$$\max\left\{\left|\frac{f(x)}{f_0(x)} - 1\right|, \left|\frac{f_0(x)}{f(x)} - 1\right|\right\} \le \sqrt{1 + \frac{1}{(1-\varepsilon)^2} - 2\sqrt{\frac{1+\varepsilon}{1-\varepsilon}}}$$

for all $x \in S$. The stability phenomena of this kind is called *Ger type stability*. Ger type stability of first order linear differential equation $y' = \lambda y$ for entire functions was studied in [14], where $\lambda \in \mathbb{C}\setminus\{0\}$.

Let $h\colon \mathbb{R} \to \mathbb{C}$ be a continuous function. We define

$$\tilde{h}(t) \stackrel{\text{def}}{=} \exp\int_0^t h(s)\mathrm{d}s$$

for every $t \in \mathbb{R}$. For such $h$ we set $C_h, D_h$ and $E_h$ as follows:

$$C_h \stackrel{\text{def}}{=} \sup_{t\in\mathbb{R}} \frac{1}{|\tilde{h}(t)|}\int_t^\infty |\tilde{h}(s)|ds\,, \qquad D_h \stackrel{\text{def}}{=} \sup_{t\in\mathbb{R}} \frac{1}{|\tilde{h}(t)|}\int_{-\infty}^t |\tilde{h}(s)|ds\,,$$

$$E_h \stackrel{\text{def}}{=} \sup_{t\in\mathbb{R}} \frac{1}{|\tilde{h}(t)|}\left|\int_0^t |\tilde{h}(s)|ds\right|.$$

We do not assume that each of $C_h, D_h$ and $E_h$ is finite. We say that the differential equation $y'(t) + h(t)y(t) = 0$ for $C^1(\mathbb{R},X)$ has Hyers–Ulam stability if and only if there exists a finite constant $K > 0$ with the following property: for each $\varepsilon \ge 0$ and $f \in C^1(\mathbb{R},X)$ satisfying

$$\|f'(t) + h(t)f(t)\| \le \varepsilon \qquad (\forall t \in \mathbb{R}),$$

there exists $g \in C^1(\mathbb{R},X)$ such that $g'(t) + h(t)g(t) = 0$ and $\|f(t) - g(t)\| \le K\varepsilon$ for all $t \in \mathbb{R}$. The Hyers–Ulam stability of the differential equation $y'(t) + h(t)y(t) = 0$ for $C^1(\mathbb{R},X)$ is characterized by the constants $C_h, D_h$ and $E_h$.

**Theorem 15.2 ([19, Corollary 2.3]).** *Let $h\colon \mathbb{R} \to \mathbb{C}$ be a continuous function. For each $f \in C^1(\mathbb{R},X)$ we define the linear operator $T_h\colon C^1(\mathbb{R},X) \to C(\mathbb{R},X)$ by*

$$(T_h u)(t) = u'(t) + h(t)u(t) \quad (\forall u \in C^1(\mathbb{R},X), \forall t \in \mathbb{R}).$$

*Then the operator $T_h$ has the Hyers–Ulam stability if and only if one of $C_h, D_h$ and $E_h$ is finite. More precisely, the following are true.*

1. *If $\inf_{t\in[0,\infty)} |\tilde{h}(t)| = 0$, then the operator $T_h$ has the Hyers–Ulam stability if and only if $C_h < \infty$.*
2. *If $\inf_{t\in(-\infty,0]} |\tilde{h}(t)| = 0$, then the operator $T_h$ has the Hyers–Ulam stability if and only if $D_h < \infty$.*
3. *If $\inf_{t\in\mathbb{R}} |\tilde{h}(t)| > 0$, then $T_h$ has the Hyers–Ulam stability if and only if $E_h < \infty$.*

It seems natural to consider the stability problem of the differential equation $y'(t) + h(t)y(t) = 0$ for another class of function-spaces, say holomorphic functions on a open set. In this paper, we show a recent development of such a problem.

## 15.2 Main Results

Let $\Omega$ be an open set of $\mathbb{C}$ and $X$ a complex Banach space. We denote by $H(\Omega, X)$ the set of all $X$-valued holomorphic functions defined on $\Omega$. First, we consider the Hyers–Ulam stability of the differential equation $y'(z) + p(z)y(z) + q(z) = 0$ for $H(\Omega, X)$. Then we give a sufficient condition for $p \in H(\Omega, \mathbb{C})$ in order that $y'(z) + p(z)y(z) + q(z) = 0$ have the Hyers–Ulam stability.

**Theorem 15.3 ([16, Theorem 2.2]).** *Let $\Omega$ be a convex open set of $\mathbb{C}$ with $0 \in \Omega$. Let $X$ be a complex Banach space and $q \in H(\Omega, X)$. Suppose that $p \in H(\Omega)$ satisfies*

$$C_p \stackrel{\text{def}}{=} \sup_{z\in\Omega} \frac{1}{|\tilde{p}(z)|} \left| \int_0^z |\tilde{p}(\zeta)|\, \mathrm{d}\zeta \right| < \infty.$$

*For each $\varepsilon \geq 0$ and $f \in H(\Omega, X)$ satisfying*

$$\|f'(z) + p(z)f(z) + q(z)\| \leq \varepsilon \qquad (\forall z \in \Omega), \tag{15.1}$$

*there exists $g \in H(\Omega, X)$ such that*

$$g'(z) + p(z)g(z) + q(z) = 0 \qquad (\forall z \in \Omega),$$
$$\|f(z) - g(z)\| \leq C_p \varepsilon \qquad (\forall z \in \Omega).$$

*Example 15.1 ([16, Example 2.1]).* Let $\Omega$ be a bounded convex open subset of $\mathbb{C}$, say $|z| \leq M$ for all $z \in \Omega$. If $0 \in \Omega$ and $p \in H(\Omega)$ is bounded, then we see that

$$C_p = \sup_{z\in\Omega} \frac{1}{|\tilde{p}(z)|} \left| \int_0^z |\tilde{p}(\zeta)|\, \mathrm{d}\zeta \right| < \infty.$$

So, $y'(z) + p(z)y(z) + q(z) = 0$ for $H(\Omega, X)$ has the Hyers–Ulam stability.

We give another sufficient condition in order that the equation $y'(z) + p(z)y(z) + q(z) = 0$ for $H(\Omega, X)$ have the Hyers–Ulam stability.

**Theorem 15.4 ([16, Theorem 2.3]).** *Let $\Omega$ be a convex open set of $\mathbb{C}$ with $0 \in \Omega$. Let $X$ be a complex Banach space, $q \in H(\Omega, X)$ and $p \in H(\Omega)$. Suppose that there exists $\lambda \in \partial\Omega$, the boundary of $\Omega$, such that*

$$D_p(\lambda) \stackrel{\text{def}}{=} \sup_{z\in\Omega} \frac{1}{|\tilde{p}(z)|} \left| \int_\lambda^z |\tilde{p}(\zeta)|\, \mathrm{d}\zeta \right| < \infty,$$

*where*

$$\int_{\lambda}^{z} |\tilde{p}(\zeta)|\,\mathrm{d}\zeta = \lim_{a \searrow 0} \int_{a}^{1} |\tilde{p}(\lambda + t(z-\lambda))|(z-\lambda)\,dt.$$

*For each* $\varepsilon \geq 0$ *and* $f \in H(\Omega,X)$ *satisfying* (15.1) *there is* $g \in H(\Omega,X)$ *such that*

$$g'(z) + p(z)g(z) + q(z) = 0\,, \qquad \|f(z) - g(z)\| \leq D_p(\lambda)\varepsilon \qquad (\forall z \in \Omega).$$

*Example 15.2 ([16, Example 2.2]).* Let $\Omega = \{z \in \mathbb{C} : |z| < 1\}$. We consider $p \in H(\Omega)$ defined by $p(z) = 1/(z+1)$ $(\forall z \in \Omega)$. Then we see that

$$D_p(-1) = \sup_{z\in\Omega} \frac{1}{|\tilde{p}(z)|} \left| \int_{-1}^{z} |\tilde{p}(\zeta)|\,\mathrm{d}\zeta \right| = \sup_{z\in\Omega} \frac{|z+1|}{2} = 1.$$

Here, we notice that

$$D_p(\lambda) = \sup_{z\in\Omega} \frac{1}{|\tilde{p}(z)|} \left| \int_{\lambda}^{z} |\tilde{p}(\zeta)|\,\mathrm{d}\zeta \right| = \infty$$

for each $\lambda \in \partial\Omega \setminus \{-1\}$, and that

$$C_p = \sup_{z\in\Omega} \frac{1}{|\tilde{p}(z)|} \left| \int_{0}^{z} |\tilde{p}(\zeta)|\,\mathrm{d}\zeta \right| = \infty.$$

This example shows that the finiteness of $C_p$ is not enough to characterize the stability of

$$y'(z) + p(z)y(z) + q(z) = 0$$

for $H(\Omega,X)$.

We can also consider the Ger type stability of the differential equation

$$y'(z) + p(z)y(z) + q(z) = 0$$

for $H(\Omega,\mathbb{C})$. Just for the sake of simplicity, we will write $H(\Omega)$ instead of $H(\Omega,\mathbb{C})$. By $H_0(\Omega)$, we denote the set of all $f \in H(\Omega)$ so that $0 \notin f(\Omega)$. We give two sufficient conditions for $p \in H_0(\Omega)$ in order that

$$y'(z) = p(z)y(z)$$

for $H_0(\Omega)$ have the Ger type stability.

**Theorem 15.5 ([13, Theorem 2.2]).** *Let* $\Omega$ *be a convex open subset of* $\mathbb{C}$ *with* $0 \in \Omega$. *Suppose that* $p \in H_0(\Omega)$ *satisfies*

$$M \stackrel{\text{def}}{=} \sup_{z\in\Omega} \int_{0}^{1} |zp(zt)|\,\mathrm{d}t < \infty.$$

*Then for every* $\varepsilon \geq 0$ *and* $f \in H_0(\Omega)$ *satisfying*

$$\left| \frac{f'(z)}{p(z)f(z)} - 1 \right| \leq \varepsilon \qquad (\forall z \in \Omega) \tag{15.2}$$

*there exists* $f_0 \in H_0(\Omega)$ *such that* $f_0{}'(z) = p(z)f_0(z)$ *and that*

$$\max\left\{ \left| \frac{f(z)}{f_0(z)} - 1 \right|, \left| \frac{f_0(z)}{f(z)} - 1 \right| \right\} \leq \mathrm{e}^{M\varepsilon} - 1 \qquad (\forall z \in \Omega).$$

**Theorem 15.6 ([13, Theorem 2.3]).** *Let* $\Omega$ *be a convex open subset of* $\mathbb{C}$ *with* $0 \in \Omega$ *and* $p \in H_0(\Omega)$. *Suppose that there exists* $\lambda \in \partial\Omega$, *the boundary of* $\Omega$, *such that*

$$L_\lambda \stackrel{\text{def}}{=} \sup_{z \in \Omega} \left\{ \lim_{a \searrow 0} \int_a^1 |(z-\lambda)p(\lambda + (z-\lambda)t)| \, \mathrm{d}t \right\} < \infty.$$

*Then for every* $\varepsilon \geq 0$ *and* $f \in H_0(\Omega)$ *satisfying* (15.2) *there exists* $f_0 \in H_0(\Omega)$ *such that*

$$f_0{}'(z) = p(z)f_0(z) \qquad (\forall z \in \Omega)$$

*and that*

$$\max\left\{ \left| \frac{f(z)}{f_0(z)} - 1 \right|, \left| \frac{f_0(z)}{f(z)} - 1 \right| \right\} \leq \mathrm{e}^{L_\lambda \varepsilon} - 1 \qquad (\forall z \in \Omega).$$

The idea of Theorems 15.5 and 15.6 is due to the following results on the Ger type stability of the differential equation $y'(t) = h(t)y(t)$ for $C(I, \mathbb{C}^\times)$, where $C(I, \mathbb{C}^\times)$ is the set of all continuous functions defined on an open interval $I \subset \mathbb{R}$ to the set of all non-zero complex numbers $\mathbb{C}^\times$. We denote by $C^1(I, \mathbb{C}^\times)$ the set of all continuously differentiable functions $f \in C(I, \mathbb{C}^\times)$.

**Theorem 15.7 ([17, Theorem 2.2]).** *Suppose that* $h \in C^1(I, \mathbb{C}^\times)$ *satisfies*

$$M \stackrel{\text{def}}{=} \sup_{t \in I} \left| \int_0^t |h(s)| \, ds \right| < \infty.$$

*Then to each* $\varepsilon \geq 0$ *and* $f \in C^1(I, \mathbb{C}^\times)$ *satisfying*

$$\left| \frac{f'(t)}{h(t)f(t)} - 1 \right| \leq \varepsilon \qquad (\forall t \in I)$$

*there exists* $f_0 \in C^1(I, \mathbb{C}^\times)$ *such that*

$$f_0{}'(t) = h(t)f_0(t) \qquad (\forall t \in I)$$

*and that*

$$\max\left\{\left|\frac{f(t)}{f_0(t)}-1\right|,\left|\frac{f_0(t)}{f(t)}-1\right|\right\}\leq e^{M\varepsilon}-1 \qquad (\forall t\in I).$$

*Example 15.3 ([17, Example 2.1]).* Set, for each $t\in\mathbb{R}$, $h(t)=2e^{-t^2}$. Then

$$\sup_{t\in\mathbb{R}}\left|\int_0^t |h(s)|\,ds\right| = \int_0^\infty h(s)\,ds=\sqrt{\pi}.$$

For each $\varepsilon$ with $\varepsilon>0$, we define

$$f(t)=\exp\int_0^t (1+\varepsilon)h(s)\,ds \qquad (\forall t\in\mathbb{R}).$$

It is easy to see that

$$\frac{f'(t)}{h(t)f(t)}-1=\varepsilon \qquad (\forall t\in\mathbb{R}).$$

By the proof of Theorem 15.7, we see that

$$f_0(t)=\exp\int_0^t h(s)\,ds$$

is a solution of the equation $y'=hy$ with

$$\max\left\{\left|\frac{f(t)}{f_0(t)}-1\right|,\left|\frac{f_0(t)}{f(t)}-1\right|\right\}\leq e^{\sqrt{\pi}\varepsilon}-1$$

for every $t\in\mathbb{R}$. By a simple calculation, we see that

$$\sup_{t\in\mathbb{R}}\left|\frac{f(t)}{f_0(t)}-1\right| = \sup_{t\in\mathbb{R}}\left|\exp\int_0^t \varepsilon h(s)\,ds-1\right| = e^{\sqrt{\pi}\varepsilon}-1.$$

Consequently, we see that the constant $e^{\sqrt{\pi}\varepsilon}-1$ is best possible. Therefore, the constant $e^{M\varepsilon}-1$ in Theorem 15.7 can not be improved in general.

In Theorem 15.7, we proved Ger type stability of the first order linear differential equation

$$y'(t)=h(t)y(t)$$

for $C^1(I,\mathbb{C}^\times)$ under the condition that

$$\sup_{t\in I}\left|\int_0^t |h(s)|\,ds\right|<\infty.$$

This condition is essential in the following sense.

**Theorem 15.8 ([17, Theorem 2.3]).** *Let $h \in C^1(I,\mathbb{C}^\times)$. Suppose that there exists a (finite) constant $K \geq 0$ with the following property: for each $f \in C^1(I,\mathbb{C}^\times)$ satisfying*

$$\left|\frac{f'(t)}{h(t)f(t)} - 1\right| \leq 1 \qquad (\forall t \in I)$$

*there exists $f_0 \in C^1(I,\mathbb{C}^\times)$ such that*

$$f_0{}'(t) = h(t)f_0(t) \qquad (\forall t \in I)$$

*and that*

$$\max\left\{\left|\frac{f(t)}{f_0(t)} - 1\right|, \left|\frac{f_0(t)}{f(t)} - 1\right|\right\} \leq K \qquad (\forall t \in I).$$

*Then we have*

$$\sup_{t\in I}\left|\int_0^t |h(s)|\,\mathrm{d}s\right| < \infty.$$

Unfortunately, we are not sure whether the Hyers–Ulam stability and Ger type stability of differential equations

$$y' + py + q = 0$$

for $H(\Omega,X)$ and $C^1(I,\mathbb{C}^\times)$ are characterized by a way similar to [19, Corollary 2.3].

**Acknowledgement** The first author was partly supported by the Grant-in-Aid for Scientific Research.

## References


1. Alsina, C., Ger, R.: On some inequalities and stability results related to the exponential function. J. Inequal. Appl. **2**, 373–380 (1998)
2. Aoki, T.: On the stability of the linear transformation in Banach spaces. J. Math. Soc. Japan **2**, 64–66 (1950)
3. Czerwik, S.: Functional Equations and Inequalitiesin Several Variables. World Scientific Publishing Co., Inc., River Edge, NJ (2002)
4. Czerwik, S. (ed.): Stability of Functional Equations of Ulam-Hyers-Rassias Type. Hadronic Press, Inc., Palm Harbor, FL (2003)
5. Gajda, Z.: On stability of additive mappings. Internat. J. Math. Math. Sci. **14**, 431–434 (1991)
6. Ger, R., Šemrl, P.: The stability of the exponential equation. Proc. Amer. Math. Soc. **124**, 779–787 (1996)
7. Hatori, O., Kobayashi, K., Miura, T., Takagi, H., Takahasi, S.-E.: On the best constant of Hyers–Ulam stability. J. Nonlinear Convex Anal. **5**, 387–393 (2004)
8. Hyers, D.H., On the stability of the linear functional equation. Proc. Nat. Acad. Sci. U.S.A., **27**, 222–224 (1941)

9. Hyers, D.H., Rassias, Th.M.: Approximate homomorphisms. Aequationes Math. **44**, 125–153 (1992)
10. Hyers, D.H., G. Isac, Rassias, Th.M.: Stability of Functional Equations in Several Variables. Birkhauser (1998)
11. Jung, S.-M.: Hyers–Ulam–Rassias Stability of Functional Equations in Mathematical Analysis. Hadronic Press, Inc., Palm Harbor, FL (2001)
12. Jung, S.-M., Rassias, Th.M.: Ulam's problem for approximate homomorphisms in connection with Bernoulli's differential equation. Appl. Math. Comput. **187**, 223–227 (2007)
13. Miura, T.: Ger type stability of thefirst order linear differential equation for holomorphic functions. Nonlinear Funct. Anal. Appl., to appear
14. Miura, T., Hirasawa, G., Takahasi, S.-E.: Ger type and Hyers–Ulam stabilities for the first order linear differential operators of entire functions. Int. J. Math. Math. Sci. **22**, 1151–1158 (2004)
15. Miura, T., Jung, S.-M., Takahasi, S.-E.: Hyers–Ulam–Rassias stability of the Banach space valued linear differential equations $y' = \lambda y$. J. Korean Math. Soc. **41**, 995–1005 (2004)
16. Miura, T., Oka, H., Takahasi, S.-E., Niwa, N.: Hyers–Ulam stability of the first order linear differential equation for Banach space-valued holomorphic mappings. J. Math. Inequal. **1**, 377–385 (2007)
17. Miura, T., Oka, H., Takahasi, S.-E., Niwa, N.: Ger type stability of the first order linear differential equation $y'(t) = h(t)y(t)$. Tamsui Oxf. J. Math. Sci. **24**, 445-456 (2008)
18. Miura, T., Takahasi, S.-E., Miyajima, S.: Hyers–Ulam stability of linear differential operator with constant coefficients. Math. Nachr. **258**, 90–96 (2003)
19. Miura, T., Takahasi, S.-E., Miyajima, S.: A characterization of Hyers–Ulam stability of first order linear differential operators. J. Math. Anal. Appl. **286**, 136–146 (2003)
20. Rassias, Th.M.: On the stability of the linear mapping in Banach spaces. Proc. Amer. Math. Soc. **72**, 297–300 (1978)
21. Rassias, Th.M.: On the stability of functional equations and a problem of Ulam. Acta Appl. Math. **62**, 23–130 (2000)
22. Rassias, Th.M., Šemrl, P.: On the behavior of mappings which do not satisfy Hyers–Ulam stability. Proc. Amer. Math. Soc. **114**, 989–993 (1992)
23. Takahasi, S.-E., Takagi, H., Miura, T., Miyajima, S.: The Hyers–Ulam stability constants of first order linear differential operators. J. Math. Anal. Appl. **296**, 403–409 (2004)
24. Takagi, H., Miura, T., Takahasi, S.-E.: Essential norm and stability constant of weighted composition operators on $C(X)$. Bull. Korean Math. Soc. **40**, 583–591 (2003)
25. Takahasi, S.-E., Miura, T., Miyajima, S.: On the Hyers–Ulam stability of the Banach space-valued differential equation $y' = \lambda y$. Bull. Korean Math. Soc. **39**, 309–315 (2002)
26. Ulam, S.M.: A Collection of Mathematical Problems. Interscience Tracts in Pure and Applied Mathematics, no. 8, Interscience Publishers, New York-London (1960)

# Chapter 16
# On the Butler–Rassias Functional Equation and its Generalized Hyers–Ulam Stability

**Takeshi Miura, Go Hirasawa, and Takahiro Hayata**

*Dedicated to the memory of S.M. Ulam*
*on the 100th anniversary of his birth*

**Abstract** In this paper, we present a recent development of the Hyers–Ulam stability of the Butler–Rassias functional equation $f(x+y)-f(x)f(y)=d\sin x\sin y$. We also show our most recent results on a Butler–Rassias type functional equation with complex variables.

**Keywords** Butler–Rassias functional equation • exponential functions • Hyers–Ulam stability

**Mathematics Subject Classification (2000):** Primary 34K20; Secondary 26D10

## 16.1 Butler–Rassias Functional Equation

In 2003, Butler [2] posed the following problem: show that for $d < -1$, there are exactly two solutions $f\colon \mathbb{R} \to \mathbb{R}$ of the functional equation

$$f(x+y)-f(x)f(y)=d\sin x\sin y \qquad (x,y\in\mathbb{R}). \tag{16.1}$$

T. Miura (✉)
Department of Applied Mathematics and Physics, Yamagata University,
Yonezawa 992-8510, Japan
e-mail: miura@yz.yamagata-u.ac.jp

G. Hirasawa
Faculty of Engineering, Ibaraki University, Hitachi 316-8511, Japan
e-mail: gou@mx.ibaraki.ac.jp

T. Hayata
Department of Informatics, Graduate School of Science and Engineering, Yamagata University,
Yonezawa 992-8510, Japan
e-mail: hayata@yz.yamagata-u.ac.jp

Th.M. Rassias and J. Brzdęk (eds.), *Functional Equations in Mathematical Analysis*,
Springer Optimization and Its Applications 52, DOI 10.1007/978-1-4614-0055-4_16,

The problem was excellently solved by Rassias [13] when he was a high school student. This functional equation we call Butler–Rassias functional equation.

**Theorem 16.1 (Rassias [13]).** *Let $d < -1$ be a constant. The solution of (16.1) is $f(x) = \pm c\sin x + \cos x$ for $x \in \mathbb{R}$, where $c = \sqrt{-d-1}$.*

By using the idea of Rassias [13], Jung [10] proved the stability of the Butler–Rassias functional equation in the sense of Hyers–Ulam.

**Theorem 16.2 (Jung [10]).** *Let $d < -1$. Then there is a constant $K = K(d) \geq 0$ such that, if $0 < \varepsilon < |d|$ and a function $f\colon \mathbb{R} \to \mathbb{R}$ satisfies the functional inequality*

$$|f(x+y) - f(x)f(y) - d\sin x\sin y| \leq \varepsilon \qquad (\forall x, y \in \mathbb{R}),$$

*then $|f(x) - f_0(x)| \leq K(\varepsilon + \sqrt{\varepsilon})$ for all $x \in \mathbb{R}$ and some solution $f_0$ of (16.1).*

## 16.2 Hyers–Ulam Stability

The stability phenomena of this kind are called *Hyers–Ulam stability*. Ulam (cf. [16, Chap. VI]) was the first to raise the stability problem of a functional equation. "For what metric groups $G$ is it true that an $\varepsilon$-automorphism of $G$ is necessarily near to a strict automorphism? (An $\varepsilon$-automorphism of $G$ means, in this case, a transformation $f$ of $G$ into itself such that $\rho(f(xy), f(x)f(y)) < \varepsilon$ for all $x, y \in G$, where $\rho$ is the metric on $G$.)" Hyers [6] gave an affirmative answer to the problem as follows.

**Theorem 16.3.** *Suppose that $E_1$ and $E_2$ are two real Banach spaces and $f\colon E_1 \to E_2$ is a mapping. If there exists $\varepsilon \geq 0$ such that*

$$\|f(x+y) - f(x) - f(y)\| \leq \varepsilon \qquad (\forall x, y \in E_1),$$

*then the limit*

$$T(x) = \lim_{n\to\infty} \frac{f(2^n x)}{2^n}$$

*exists for each $x \in E_1$ and $T\colon E_1 \to E_2$ is the unique additive mapping such that*

$$\|f(x) - T(x)\| \leq \varepsilon \qquad (\forall x \in E_1).$$

*If, in addition, the mapping $\mathbb{R} \ni t \mapsto f(tx)$ is continuous for each $x \in E_1$, then $T$ is linear.*

This result is called the *Hyers–Ulam stability* of the *additive* Cauchy equation. T. Aoki [1] introduced unbounded, additive Cauchy difference and generalized the result [6, Theorem 1] of Hyers. Rassias [14], who independently introduced the unbounded Cauchy difference, was the first to prove the stability of the linear mapping between Banach spaces. The concept of the Hyers–Ulam–Rassias stability was originated from Rassias' paper [14] for the stability of the linear mapping. Rassias [14] generalized Hyers' Theorem and completed Aoki's result as follows:

**Theorem 16.4.** *Suppose that $E_1$ and $E_2$ are two real Banach spaces and $f\colon E_1 \to E_2$ is a mapping. If there exist $\varepsilon \geq 0$ and $0 \leq p < 1$ such that*

$$\|f(x+y) - f(x) - f(y)\| \leq \varepsilon(\|x\|^p + \|y\|^p) \qquad (\forall x, y \in E_1),$$

*then there is a unique additive mapping $T\colon E_1 \to E_2$ such that*

$$\|f(x) - T(x)\| \leq \frac{2\varepsilon}{|2-2^p|}\|x\|^p \qquad (\forall x \in E_1).$$

*If, in addition, $\mathbb{R} \ni t \mapsto f(tx)$ is continuous for each $x \in E_1$, then $T$ is linear.*

This result is described by the term: *the Hyers–Ulam–Rassias stability* of the additive Cauchy functional equation. Since then the Hyers–Ulam–Rassias stability of various functional equations has been extensively studied (cf. [3–5, 7–9, 12]).

## 16.3 Recent Developments

In [15], with $\lambda, \rho, \alpha \in \mathbb{C}$, the authors solved the following Butler–Rassias type functional equation

$$f(x+y) + \lambda f(x)f(y) = \rho e^{\alpha(x+y)} \qquad (x, y \in \mathbb{R}). \tag{16.2}$$

**Proposition 16.1 (cf. [15]).** *Let $\lambda, \rho, \alpha \in \mathbb{C}$. If a function $f\colon \mathbb{R} \to \mathbb{C}$ satisfies (16.2), then*

$$f(x) = \frac{-1 \pm \sqrt{1+4\lambda\rho}}{2\lambda} e^{\alpha x} \qquad (x \in \mathbb{R}),$$

*where $\sqrt{z}$ denotes one of the square roots of the complex number $z$.*

One may replace the function $\rho e^{\alpha(x+y)}$ by some other functions. For example, $\phi(x+y)$ and $\psi(x)\psi(y)$ seem to be natural extensions. Unfortunately, such extensions do not change the set of solutions, as it is shown in the following result.

**Theorem 16.5 ([15], Theorems 1 and 2).** *Let $\lambda \in \mathbb{C} \setminus \{0\}$, $\phi, \psi\colon \mathbb{R} \to \mathbb{C}$ be continuous, and $\psi(0) \neq 0$. Let $f\colon \mathbb{R} \to \mathbb{C}$ satisfy one of the following two equations:*

$$f(x+y) + \lambda f(x)f(y) = \phi(x+y) \qquad (x, y \in \mathbb{R}),$$
$$f(x+y) + \lambda f(x)f(y) = \psi(x)\psi(y) \qquad (x, y \in \mathbb{R}).$$

*Then there exist* $\rho, \alpha \in \mathbb{C}$ *such that*

$$f(x) = \frac{-1 \pm \sqrt{1+4\lambda\rho}}{2\lambda} \mathrm{e}^{\alpha x} \qquad (x \in \mathbb{R}).$$

It seems natural to consider functions with "complex variable". In this setting, the following result has been obtained:

**Theorem 16.6 ([11]).** *Let* $\lambda \in \mathbb{C} \setminus \{0\}$ *and* $\phi, \psi \colon \mathbb{C} \to \mathbb{C}$ *be continuous functions with* $\psi(0) \neq 0$. *Let* $f \colon \mathbb{C} \to \mathbb{C}$ *satisfy one of the following two functional equations:*

$$f(z+w) + \lambda f(z)f(w) = \phi(z+w) \qquad (z, w \in \mathbb{C}),$$
$$f(z+w) + \lambda f(z)f(w) = \psi(z)\psi(w) \qquad (z, w \in \mathbb{C}).$$

*Then there exists* $\rho, \alpha, \beta \in \mathbb{C}$ *such that*

$$f(z) = \frac{-1 \pm \sqrt{1+4\lambda\rho}}{2\lambda} \mathrm{e}^{\alpha z + \beta \bar{z}} \qquad (z \in \mathbb{C}).$$

The next theorem concerns the Hyers–Ulam stability of a Butler–Rassias type functional equation.

**Theorem 16.7 ([15], Theorem 5).** *Let* $\lambda, \rho \in \mathbb{C} \setminus \{0\}$ *and* $\theta \in \mathbb{R}$. *There exists a constant* $K = K(\lambda, \rho)$ *with the following property: for any non-negative number* $\varepsilon < |\rho|$ *and any function* $g \colon \mathbb{R} \to \mathbb{C}$ *satisfying*

$$|g(x+y) + \lambda g(x)g(y) - \rho \mathrm{e}^{\mathrm{i}\theta(x+y)}| \leq \varepsilon \qquad (\forall x, y \in \mathbb{R}),$$

*there is a function* $f \colon \mathbb{R} \to \mathbb{C}$ *such that*

$$f(x+y) + \lambda f(x)f(y) = \rho \mathrm{e}^{\mathrm{i}\theta(x+y)} \qquad (\forall x, y \in \mathbb{R}),$$
$$|g(x) - f(x)| \leq K(\varepsilon + \sqrt{\varepsilon}) \qquad (\forall x \in \mathbb{R}).$$

Under an additional assumption, a quite similar result to Theorem 16.7 is obtained for a function $\psi(x)\psi(y)$ instead of $\rho \mathrm{e}^{\mathrm{i}\theta(x+y)}$.

**Theorem 16.8 ([15], Theorems 6).** *Let* $\lambda \in \mathbb{C} \setminus \{0\}$ *and* $\psi \colon \mathbb{R} \to \mathbb{C}$ *be a nonzero bounded function with* $\psi(0) = 0$, $\psi(c) \neq 0$ *and*

$$\psi(c)\psi(x+y) = \psi(x)\psi(y+c) + \psi(x+c)\psi(y) \qquad (\forall x, y \in \mathbb{R})$$

*for some* $c \in \mathbb{R}$. *There exists a constant* $K = K(\lambda, \psi, c)$ *with the following property: for any non-negative number* $\varepsilon < |\psi(c)|^2$ *and any function* $g \colon \mathbb{R} \to \mathbb{C}$ *satisfying*

$$|g(x+y) + \lambda g(x)g(y) - \psi(x)\psi(y)| \leq \varepsilon \qquad (\forall x, y \in \mathbb{R}),$$

*there is a function* $f : \mathbb{R} \to \mathbb{C}$ *such that*

$$f(x+y) + \lambda f(x)f(y) = \psi(x)\psi(y) \qquad (\forall x, y \in \mathbb{R}),$$
$$|g(x) - f(x)| \le K(\varepsilon + \sqrt{\varepsilon}) \qquad (\forall x \in \mathbb{R}).$$

If in Theorem 16.8, we take $\lambda = -1$ and $\psi(x) = \sqrt{-d}(\mathrm{e}^{\mathrm{i}x} - \mathrm{e}^{-\mathrm{i}x})/2$, with $d < -1$, then we obtain [10, Theorem 3.1].

It is a natural question whether the Hyers–Ulam stability of a Butler–Rassias type functional equation $f(z+w) - \lambda f(z)f(w) = \rho \mathrm{e}^{\alpha(z+w)}$ $(z, w \in \mathbb{C})$ holds or not. In fact, the following is true.

**Theorem 16.9 ([11]).** *Let* $\lambda, \rho, \alpha \in \mathbb{C} \setminus \{0\}$. *If a function* $g : \mathbb{C} \to \mathbb{C}$ *satisfies*

$$|g(z+w) + \lambda g(z)g(w) - \rho \mathrm{e}^{\alpha(z+w)}| \le \varepsilon \qquad (\forall z, w \in \mathbb{C})$$

*for some* $\varepsilon \ge 0$, *then one of the following two conditions holds:*

$$g(z) = \frac{-1 \pm \sqrt{1 + 4\lambda\rho}}{2\lambda} \mathrm{e}^{\alpha z} \qquad (\forall z \in \mathbb{C}),$$
$$\sup_{z \in \mathbb{C}} \left| g(z) - \frac{-1 \pm \sqrt{1 + 4\lambda\rho}}{2\lambda} \mathrm{e}^{\alpha z} \right| = \infty.$$

Theorem 16.9 states that an approximate solution $g : \mathbb{C} \to \mathbb{C}$ of a Butler–Rassias type functional equation $f(z+w) + \lambda f(z)f(w) = \rho \mathrm{e}^{\alpha(z+w)}$ is either a solution of the equation or it is far from exact solutions.

**Acknowledgement** The first author was partly supported by the Grant-in-Aid for Scientific Research.

## References

1. Aoki, T.: On the stability of the linear transformation in Banach spaces. J. Math. Soc. Japan **2**, 64–66 (1950)
2. Butler, S.:Problem no. 11030. Amer. Math. Monthly **110**, 637 (2003)
3. Czerwik, S.: Functional equations and inequalities in several variables. World Scientific Publishing Co., River Edge, NJ (2002)
4. Czerwik, S. (ed.): Stability of Functional Equations of Ulam–Hyers–Rassias Type. Hadronic Press, Palm Harbor, FL (2003)
5. Forti, G.L.: Hyers–Ulam stability of functional equations in several variables. Aequationes Math. **50**, 143–190 (1995)
6. Hyers, D.H.: On the stability of the linear functional equation. Proc. Nat. Acad. Sci. U.S.A. **27**, 222–224 (1941)
7. Hyers, D.H., Rassias, Th.M.: Approximate homomorphisms. Aequationes Math. **44**, 125–153 (1992)

8. Hyers, D.H., Isac, G., Rassias, Th.M.: Stability of Functional Equations in Several Variables. Birkhauser (1998)
9. Hyers, D.H., Ulam, S.M.: Approximate isometries of the space of continuous functions. Ann. Math. **48**, 285–289 (1947)
10. Jung, S.-M.: Hyers-Ulam stability of Butler–Rassias functional equation. J. Inequal. Appl. **2005**, 41–47 (2005)
11. Hirasawa, G., Hayata, T., Miura, T., Takahasi, S.-E., Takagi, H.: Stabilityof a Butler–Rassias type functional equation. In preparation
12. Miura, T., Miyajima, S., Takahasi, S.-E.: A characterization of Hyers-Ulam stability of first order linear differential operators. J. Math. Anal. Appl. **286**, 136–146 (2003)
13. Rassias, M.Th.: Solution of a functional equation problem of Steven Butler. Octogon Math. Mag. **12**, 152–153 (2004)
14. Rassias, Th.M.: On the stability of the linear mapping in Banach spaces. Proc. Amer. Math. Soc. **72**, 297–300 (1978)
15. Takahasi, S.-E., Miura, T., Takagi, H.: Exponential type functional equation and its Hyers-Ulam stability. J. Math. Anal. Appl. **329**, 1191–1203 (2007)
16. Ulam, S.M.: A Collection of Mathematical Problems. Interscience Tracts in Pure and Applied Mathematics, no. 8, Interscience Publishers, New York-London (1960)

# Chapter 17
# A Note on the Stability of an Integral Equation

**Takeshi Miura, Go Hirasawa, Sin-Ei Takahasi, and Takahiro Hayata**

*Dedicated to the memory of S.M. Ulam*
*on the 100th anniversary of his birth*

**Abstract** Let $p\colon \mathbb{R} \to \mathbb{C}$ be a continuous function. We give a sufficient condition in order that the integral equation $f(t) = f(0) + \int_0^t p(s)f(s)\,\mathrm{d}s$ have the Hyers–Ulam stability. We also prove that if $p$ has no zeros, then the sufficient condition is a necessary condition.

**Keywords** Exponential functions • Hyers–Ulam stability • Hyers–Ulam–Rassias stability

**Mathematics Subject Classification (2000):** Primary 34K20; Secondary 26D10

## 17.1 Introduction

The stability problem of functional equations has a long history. In fact, Ulam raised such a problem (cf. [15, Chapter VI]): "For what metric groups $G$ is it true that an $\varepsilon$-automorphism of $G$ is necessarily near to a strict automorphism?"

---

T. Miura (✉) • S.-E. Takahasi
Department of Applied Mathematics and Physics, Graduate School of Science and Engineering, Yamagata University, Yonezawa 992-8510, Japan
e-mail: miura@yz.yamagata-u.ac.jp; sin-ei@emperor.yz.yamagata-u.ac.jp

G. Hirasawa
Faculty of Engineering, Ibaraki University, Hitachi 316-8511, Japan
e-mail: gou@mx.ibaraki.ac.jp

T. Hayata
Department of Informatics, Graduate School of Science and Engineering, Yamagata University, Yonezawa 992-8510, Japan
e-mail: hayata@yz.yamagata-u.ac.jp

Th.M. Rassias and J. Brzdęk (eds.), *Functional Equations in Mathematical Analysis*, Springer Optimization and Its Applications 52, DOI 10.1007/978-1-4614-0055-4_17,

(An $\varepsilon$-automorphism of $G$ is a mapping $f: G \to G$ with $\rho(f(x \cdot y), f(x) \cdot f(y)) < \varepsilon$ for all $x, y \in G$.)

Hyers[5] gave an affirmative answer to the problem as follows. Suppose that $f: E_1 \to E_2$ is a mapping between two real Banach spaces, $\varepsilon \geq 0$, and

$$\|f(x+y) - f(x) - f(y)\| \leq \varepsilon \qquad (\forall x, y \in E_1).$$

Then the limit $T(x) = \lim_{n\to\infty} 2^{-n} f(2^n x)$ exists for each $x \in E_1$, and $T: E_1 \to E_2$ is the unique additive mapping such that

$$\|f(x) - T(x)\| \leq \varepsilon \qquad (\forall x \in E_1).$$

If, in addition, the mapping $\mathbb{R} \ni t \mapsto f(tx)$ is continuous for each $x \in E_1$, then $T$ is linear.

This result is called the *Hyers–Ulam stability* of the *additive* Cauchy equation $g(x+y) = g(x) + g(y)$. Here we note that Hyers [5] calls any solution of this equation a "linear" function or transformation. Hyers considered only *bounded* Cauchy difference $f(x+y) - f(x) - f(y)$. Aoki [2] and Rassias [12] introduced unbounded Cauchy difference

$$\|f(x+y) - f(x) - f(y)\| \leq \varepsilon(\|x\|^p + \|y\|^p)$$

independently, where $0 \leq p < 1$. They proved that there exists a unique additive mapping $T: E_1 \to E_2$ such that

$$\|f(x) - T(x)\| \leq \frac{2\varepsilon}{|2-2^p|} \|x\|^p$$

for all $x \in E_1$. Moreover, Rassias [12] proved that if the mapping $\mathbb{R} \ni t \mapsto f(tx)$ is continuous for each fixed $x \in E_1$, then $T$ is linear.

This result is, what is called, *the Hyers–Ulam–Rassias stability* of the additive Cauchy equation $g(x+y) = g(x) + g(y)$. The stability of various functional equations has been investigated [4, 6, 7, 13, 14].

Alsina and Ger [1] remarked that the Hyers–Ulam stability of the differential equation $y' = y$ holds. In fact, they proved that if $\varepsilon \geq 0$ and if $f$ is a differentiable function on an open interval $I$ into $\mathbb{R}$ with $|f'(t) - f(t)| \leq \varepsilon$ for all $t \in I$, then there exists a differentiable function $g: I \to \mathbb{R}$ such that $g'(t) = g(t)$ and $|f(t) - g(t)| \leq 3\varepsilon$ for all $t \in I$. Since then, the stability of several differential equations has been studied (cf. [3, 9–11]). Jung [8] studied the stability of a Volterra integral equation

$$y(t) = \int_c^t \phi(s, y(s))\, ds \tag{17.1}$$

defined on a *bounded interval*. He also remarked that the problem of the Hyers–Ulam stability of the Volterra integral equation defined on an *infinite interval* remains open. First, we give a negative answer to Jung's question.

*Example 17.1.* For each $\varepsilon > 0$, we set $f(t) = \varepsilon t e^{it}$ for every $t \in \mathbb{R}$. By a simple calculation, we have that

$$\left| f(t) - i\int_0^t f(s)\,ds \right| = \varepsilon |e^{it} - 1| \le 2\varepsilon \qquad (\forall t \in \mathbb{R}).$$

On the other hand, if a continuous function $g\colon \mathbb{R} \to \mathbb{C}$ satisfies $g(t) = i\int_0^t g(s)\,ds$ for every $t \in \mathbb{R}$, then we see that $g$ is identically 0. In fact, differentiating both sides with respect $t$, we get $g'(t) = ig(t)$ for every $t \in \mathbb{R}$. This implies that $g(t) = g(0)e^{it}$ for every $t \in \mathbb{R}$. Since $g(0) = 0$, we have that $g$ is identically 0 as claimed. We have

$$\sup_{t\in\mathbb{R}} |f(t) - g(t)| = \sup_{t\in\mathbb{R}} |f(t)| = \sup_{t\in\mathbb{R}} |\varepsilon t e^{it}| = \infty.$$

Consequently, the function $f(t) = \varepsilon t e^{it}$ is an approximate solution of the equation $y(t) = i\int_0^t y(s)\,ds$, while $f(t)$ is not near to the unique solution of $y(t) = i\int_0^t y(s)\,ds$. We thus conclude that the equation $y(t) = i\int_0^t y(s)\,ds$ is not stable in the sense of Hyers and Ulam.

Example 17.1 shows that the Volterra integral (17.1) defined on $\mathbb{R}$ has no Hyers–Ulam stability for $\phi(s, y(s)) = iy(s)$. In this paper, we consider the integral equation

$$y(t) - y(0) = \int_0^t p(s)y(s)\,ds \tag{17.2}$$

defined on $\mathbb{R}$, and investigate its Hyers–Ulam stability, where $p$ is a given continuous function.

## 17.2 Main Results

In this paper, we will denote by $X$ a complex Banach space with the norm $\|\cdot\|$. We write $C(\mathbb{R}, X)$ for the set of all continuous maps from $\mathbb{R}$ to $X$ and

$$\mathscr{P}(t) = \int_0^t p(\tau)\,d\tau \qquad (\forall t \in \mathbb{R})$$

for each continuous function $p\colon \mathbb{R} \to \mathbb{C}$. Then $\mathscr{P}'(t) = p(t)$ for all $t \in \mathbb{R}$. Next, for such $p$, we set:

$$C_p = \sup_{t\in\mathbb{R}} |e^{\mathscr{P}(t)}| \int_{-\infty}^t |p(s)e^{-\mathscr{P}(s)}|\,ds, \qquad D_p = \sup_{t\in\mathbb{R}} |e^{\mathscr{P}(t)}| \left| \int_0^t |p(s)e^{-\mathscr{P}(s)}|\,ds \right|,$$

$$E_p = \sup_{t\in\mathbb{R}} |e^{\mathscr{P}(t)}| \int_t^{\infty} |p(s)e^{-\mathscr{P}(s)}|\,ds.$$

Here, we do not assume that any of $C_p$, $D_p$ and $E_p$ is finite.

**Lemma 17.1.** *Let p be a complex-valued continuous function on* $\mathbb{R}$. *For each pair of maps* $f, u \in C(\mathbb{R}, X)$, *the following two conditions are equivalent.*

(i) $u(t) = f(t) - f(0) - \int_0^t p(s)f(s)\,\mathrm{d}s$ *for every* $t \in \mathbb{R}$.

(ii) $f(t) - f(0)\mathrm{e}^{\mathscr{P}(t)} = u(t) + \mathrm{e}^{\mathscr{P}(t)} \int_0^t p(s)\mathrm{e}^{-\mathscr{P}(s)}u(s)\,\mathrm{d}s$ *for every* $t \in \mathbb{R}$.

*In particular,*

$$f(t) - f(0) = \int_0^t p(s)f(s)\,\mathrm{d}s \qquad (\forall t \in \mathbb{R})$$

*if and only if* $f(t) = f(0)\mathrm{e}^{\mathscr{P}(t)}$ *for every* $t \in \mathbb{R}$.

*Proof.* (i) $\Rightarrow$ (ii) By assumption, we have

$$f(t) - u(t) = f(0) + \int_0^t p(s)f(s)\,\mathrm{d}s \qquad (\forall t \in \mathbb{R}).$$

Differentiating both sides with respect $t$, we have

$$(f(t) - u(t))' = p(t)f(t) = p(t)(f(t) - u(t)) + p(t)u(t).$$

The above equality, multiplied by $\mathrm{e}^{-\mathscr{P}(t)}$, shows that

$$(f(t) - u(t))'\mathrm{e}^{-\mathscr{P}(t)} = (f(t) - u(t))p(t)\mathrm{e}^{-\mathscr{P}(t)} + p(t)\mathrm{e}^{-\mathscr{P}(t)}u(t).$$

Since $\mathscr{P}'(t) = p(t)$, we have

$$(f(t) - u(t))p(t)\mathrm{e}^{-\mathscr{P}(t)} = -(f(t) - u(t))(\mathrm{e}^{-\mathscr{P}(t)})'.$$

Consequently

$$\begin{aligned}\{(f(t) - u(t))\mathrm{e}^{-\mathscr{P}(t)}\}' &= (f(t) - u(t))'\mathrm{e}^{-\mathscr{P}(t)} + (f(t) - u(t))(\mathrm{e}^{-\mathscr{P}(t)})' \\ &= p(t)\mathrm{e}^{-\mathscr{P}(t)}u(t)\end{aligned}$$

for every $t \in \mathbb{R}$. Here, we note that $\mathrm{e}^{-\mathscr{P}(0)} = 1$ and $u(0) = 0$. By integrating both sides of the above equality, we have

$$(f(t) - u(t))\mathrm{e}^{-\mathscr{P}(t)} - f(0) = \int_0^t p(s)\mathrm{e}^{-\mathscr{P}(s)}u(s)\,\mathrm{d}s,$$

which shows that

$$f(t) - f(0)\mathrm{e}^{\mathscr{P}(t)} = u(t) + \mathrm{e}^{\mathscr{P}(t)} \int_0^t p(s)\mathrm{e}^{-\mathscr{P}(s)}u(s)\,\mathrm{d}s$$

for every $t \in \mathbb{R}$.

(ii) $\Rightarrow$ (i) The above argument can be followed in the opposite direction, and so the proof is omitted. □

**Definition 17.1.** For $p \in C(\mathbb{R},\mathbb{C})$, define $T_p\colon C(\mathbb{R},X) \to C(\mathbb{R},X)$ by

$$T_p(f)(t) = f(t) - f(0) - \int_0^t p(s)f(s)\,\mathrm{d}s \qquad (\forall f \in C(\mathbb{R},X), t \in \mathbb{R}).$$

We say that $T_p$ has the Hyers–Ulam stability if and only if there exists a constant $K \geq 0$ with the following property: for every $\varepsilon \geq 0$ and $f \in C(\mathbb{R},X)$ satisfying

$$\|T_p(f)(t)\| \leq \varepsilon \qquad (\forall t \in \mathbb{R})$$

there exists $g \in C(\mathbb{R},X)$ such that

$$T_p(g)(t) = 0 \qquad \text{and} \qquad |f(t) - g(t)| \leq K\varepsilon \qquad (\forall t \in \mathbb{R}).$$

We call such $K$ a HUS constant for $T_p$. When the operator $T_p$ has the Hyers–Ulam stability, we simply say that (17.2) has the Hyers–Ulam stability.

**Theorem 17.1.** *Let $p \in C(\mathbb{R},\mathbb{C})$. If one of $C_p$, $D_p$ and $E_p$ is finite, then* (17.2) *has the Hyers–Ulam stability with a HUS constant $1 + C_p$, $1 + D_p$ and $1 + E_p$, respectively.*

*If, in addition, $\inf_{t\in\mathbb{R}} |\mathrm{e}^{-\mathscr{P}(t)}| = 0$, then for each $f \in C(\mathbb{R},X)$ with*

$$\sup_{t\in\mathbb{R}} \left\| f(t) - f(0) - \int_0^t p(s)f(s)\,\mathrm{d}s \right\| < \infty, \tag{17.3}$$

*there exists a unique $g \in C(\mathbb{R},X)$ such that, for each $t \in \mathbb{R}$,*

$$g(t) - g(0) = \int_0^t p(s)g(s)\,\mathrm{d}s \quad \text{and} \quad \sup_{t\in\mathbb{R}} |f(t) - g(t)| < \infty. \tag{17.4}$$

*Proof.* Let $\varepsilon \geq 0$ and $f \in C(\mathbb{R},X)$ satisfy

$$\left\| f(t) - f(0) - \int_0^t p(s)f(s)\,\mathrm{d}s \right\| \leq \varepsilon \qquad (\forall t \in \mathbb{R}).$$

Set

$$u(t) = f(t) - f(0) - \int_0^t p(s)f(s)\,\mathrm{d}s \qquad (\forall t \in \mathbb{R}).$$

Then $\|u(t)\| \leq \varepsilon$ for all $t \in \mathbb{R}$. By Lemma 17.1, we have

$$f(t) - f(0)\mathrm{e}^{\mathscr{P}(t)} = u(t) + \mathrm{e}^{\mathscr{P}(t)} \int_0^t p(s)\mathrm{e}^{-\mathscr{P}(s)}u(s)\,\mathrm{d}s \qquad (\forall t \in \mathbb{R}). \tag{17.5}$$

First, consider the case when $C_p < \infty$. Then $\int_{-\infty}^t |p(s)\mathrm{e}^{-\mathscr{P}(s)}|\,\mathrm{d}s$ exists for all $t \in \mathbb{R}$. Since $\|u(t)\| \leq \varepsilon$, we see that the integral $\int_{-\infty}^t p(s)\mathrm{e}^{-\mathscr{P}(s)}u(s)\,\mathrm{d}s \in X$ exists for every $t \in \mathbb{R}$. We have, for each $t \in \mathbb{R}$, that

$$\begin{aligned} f(t) - f(0)\mathrm{e}^{\mathscr{P}(t)} =& u(t) + \mathrm{e}^{\mathscr{P}(t)} \int_{-\infty}^t p(s)\mathrm{e}^{-\mathscr{P}(s)}u(s)\,\mathrm{d}s \\ & - \mathrm{e}^{\mathscr{P}(t)} \int_{-\infty}^0 p(s)\mathrm{e}^{-\mathscr{P}(s)}u(s)\,\mathrm{d}s. \end{aligned}$$

Let

$$g_1(t) := \mathrm{e}^{\mathscr{P}(t)} \left( f(0) - \int_{-\infty}^{0} p(s)\mathrm{e}^{-\mathscr{P}(s)} u(s)\,\mathrm{d}s \right) \qquad (\forall t \in \mathbb{R}).$$

Then, by Lemma 17.1, $g_1 \in C(\mathbb{R}, X)$ is a solution of (17.2) with

$$\begin{aligned} \|f(t) - g_1(t)\| &= \left\| u(t) + \mathrm{e}^{\mathscr{P}(t)} \int_{-\infty}^{t} p(s)\mathrm{e}^{-\mathscr{P}(s)} u(s)\,\mathrm{d}s \right\| \\ &\le \|u(t)\| + \left|\mathrm{e}^{\mathscr{P}(t)}\right| \int_{-\infty}^{t} \left\| p(s)\mathrm{e}^{-\mathscr{P}(s)} u(s) \right\| \mathrm{d}s \\ &\le \varepsilon + \varepsilon \left|\mathrm{e}^{\mathscr{P}(t)}\right| \int_{-\infty}^{t} \left| p(s)\mathrm{e}^{-\mathscr{P}(s)} \right| \mathrm{d}s \le (1 + C_p)\varepsilon \qquad (\forall t \in \mathbb{R}). \end{aligned}$$

In case where $E_p < \infty$, we prove that (17.2) has the Hyers–Ulam stability with a HUS constant $1 + E_p$ by an argument quite similar to the above.

We next consider the case when $D_p < \infty$. Let $g_2(t) = f(0)\mathrm{e}^{\mathscr{P}(t)}$ for $t \in \mathbb{R}$. Then $g_2 \in C(\mathbb{R}, X)$ is a solution of (17.2). According to (17.5), for each $t \in \mathbb{R}$,

$$\begin{aligned} |f(t) - g_2(t)| &= \left| u(t) + \mathrm{e}^{\mathscr{P}(t)} \int_{0}^{t} p(s)\mathrm{e}^{-\mathscr{P}(s)} u(s)\,\mathrm{d}s \right| \\ &\le |u(t)| + \left|\mathrm{e}^{\mathscr{P}(t)}\right| \left| \int_{0}^{t} \left| p(s)\mathrm{e}^{-\mathscr{P}(s)} u(s) \right| \mathrm{d}s \right| \\ &\le \varepsilon + \varepsilon \left|\mathrm{e}^{\mathscr{P}(t)}\right| \left| \int_{0}^{t} |p(s)\mathrm{e}^{-\mathscr{P}(s)}|\,\mathrm{d}s \right| \le (1 + D_p)\varepsilon. \end{aligned}$$

Finally, suppose that $\inf_{t\in\mathbb{R}} |\mathrm{e}^{-\mathscr{P}(t)}| = 0$. Let $f \in C(\mathbb{R}, X)$ and (17.3) hold. Suppose that $g_3, g_4 \in C(\mathbb{R}, X)$ satisfy

$$g_i(t) - g_i(0) = \int_{0}^{t} p(s) g_i(s)\,\mathrm{d}s \quad \text{and} \quad M_i = \sup_{t\in\mathbb{R}} |f(t) - g_i(t)| < \infty$$

for $i = 3, 4$. By Lemma 17.1, $g_i(t) = g_i(0)\mathrm{e}^{\mathscr{P}(t)}$ for $t \in \mathbb{R}$, $i = 3, 4$, which implies that

$$\begin{aligned} \|g_3(0) - g_4(0)\| &= \left|\mathrm{e}^{-\mathscr{P}(t)}\right| \|g_3(t) - g_4(t)\| \\ &\le \left|\mathrm{e}^{-\mathscr{P}(t)}\right| (\|g_3(t) - f(t)\| + \|f(t) - g_4(t)\|) \\ &\le \left|\mathrm{e}^{-\mathscr{P}(t)}\right| (M_3 + M_4) \qquad (\forall t \in \mathbb{R}). \end{aligned}$$

Since $\inf_{t\in\mathbb{R}} \left|\mathrm{e}^{-\mathscr{P}(t)}\right| = 0$, we have $g_3(0) = g_4(0)$. Thus, $g_3(t) = g_3(0)\mathrm{e}^{\mathscr{P}(t)} = g_4(t)$ for all $t \in \mathbb{R}$. We conclude that the element $g \in C(\mathbb{R}, X)$ is uniquely determined in the sense of (17.4). □

*Remark 17.1.* Let $p\colon \mathbb{R} \to \mathbb{C}$ be a continuous function so that (17.2) has the Hyers–Ulam stability. If $\inf_{t\in\mathbb{R}} |e^{-\mathscr{P}(t}| > 0$, then the uniqueness in the sense of Theorem 17.1 does not hold. To be more explicit, for each $f \in C(\mathbb{R}, X)$ with (17.3), there exist *infinitely many* $g \in C(\mathbb{R}, X)$ such that (17.4) holds for all $t \in \mathbb{R}$. In fact, let $f \in C(\mathbb{R}, X)$ satisfy (17.3). Set

$$\varepsilon = \sup_{t\in\mathbb{R}} \left\| f(t) - f(0) - \int_0^t p(s) f(s)\,\mathrm{d}s \right\| < \infty.$$

Since (17.2) have the Hyers–Ulam stability, there is a solution $g \in C(\mathbb{R}, X)$ of (17.2) with $\sup_{t\in\mathbb{R}} \|f(t) - g(t)\| \le K\varepsilon$, where $K$ is a HUS constant. According to Lemma 17.1, $g(t) = g(0)e^{\mathscr{P}(t)}$ for all $t \in \mathbb{R}$. Let $x \in X$ and $h(t) = xe^{\mathscr{P}(t)}$ for $t \in \mathbb{R}$. Then

$$\begin{aligned}\|f(t) - h(t)\| &\le \|f(t) - g(t)\| + \|g(t) - h(t)\| \\ &\le K\varepsilon + \|g(0) - x\|\,|e^{\mathscr{P}(t)}| \qquad (\forall t \in \mathbb{R}).\end{aligned}$$

Since $\sup_{t\in\mathbb{R}} |e^{\mathscr{P}(t)}| < \infty$, we conclude that $\sup_{t\in\mathbb{R}} \|f(t) - h(t)\| < \infty$, for any $x \in X$.

**Theorem 17.2.** *Let $p\colon \mathbb{R} \to \mathbb{R}$ be a continuous function.*

(i) *If $\lim_{t\to-\infty} \mathscr{P}(t) = \infty$, $\lim_{t\to\infty} \mathscr{P}(t) = -\infty$ and if $p(t) < 0$ on $\mathbb{R} \setminus [-\alpha_1, \alpha_1]$ for some $\alpha_1 > 0$, then $C_p < \infty$ and $D_p = E_p = \infty$.*

(ii) *If $\lim_{t\to\pm\infty} \mathscr{P}(t) = -\infty$, $p(t) > 0$ on $(-\infty, -\alpha_2)$ and if $p(t) < 0$ on $(\alpha_2, \infty)$ for some $\alpha_2 > 0$, then $D_p < \infty$ and $C_p = E_p = \infty$.*

(iii) *If $\lim_{t\to-\infty} \mathscr{P}(t) = -\infty$, $\lim_{t\to\infty} \mathscr{P}(t) = \infty$ and if $p(t) > 0$ on $\mathbb{R} \setminus [-\alpha_3, \alpha_3]$ for some $\alpha_3 > 0$, then $E_p < \infty$ and $C_p = D_p = \infty$.*

(iv) *If $\lim_{t\to\pm\infty} \mathscr{P}(t) = \infty$, $p(t) < 0$ on $(-\infty, \alpha_4)$ and if $p(t) > 0$ on $(\alpha_4, \infty)$ for some $\alpha_4 > 0$, then $C_p = D_p = E_p = \infty$.*

*Proof.* (i) Suppose that

$$\lim_{t\to-\infty} \mathscr{P}(t) = \infty, \qquad \lim_{t\to\infty} \mathscr{P}(t) = -\infty \tag{17.6}$$

and that $p(t) < 0$ on $\mathbb{R} \setminus [-\alpha_1, \alpha_1]$ for some $\alpha_1 > 0$. We see that $\int_{-\infty}^t |p(s)e^{-\mathscr{P}(s)}|\,\mathrm{d}s$ exists for every $t \in \mathbb{R}$. In fact, since $\mathscr{P}'(t) = p(t)$, for each $r < -\alpha_1$ we have

$$\begin{aligned}\int_r^{\alpha_1} |p(s)e^{-\mathscr{P}(s)}|\,\mathrm{d}s &= \int_r^{\alpha_1} (-p(s))e^{-\mathscr{P}(s)}\,\mathrm{d}s \\ &= \left[e^{-\mathscr{P}(s)}\right]_r^{\alpha_1} = e^{-\mathscr{P}(\alpha_1)} - e^{-\mathscr{P}(r)}.\end{aligned}$$

According to (17.6),

$$\int_{-\infty}^{\alpha_1} |p(s)e^{-\mathscr{P}(s)}|\,\mathrm{d}s = e^{-\mathscr{P}(\alpha_1)}.$$

Consequently, for every $t \in \mathbb{R}$ there exists

$$\int_{-\infty}^{t} |p(s)e^{-\mathscr{P}(s)}|\,ds = e^{-\mathscr{P}(\alpha_1)} + \int_{\alpha_1}^{t} |p(s)e^{-\mathscr{P}(s)}|\,ds.$$

Set $\psi_1(t) = \int_{-\infty}^{t} |p(s)e^{-\mathscr{P}(s)}|\,ds$ for $t \in \mathbb{R}$. Then $\lim_{t\to-\infty} \psi_1(t) = 0$. We see that $\lim_{t\to\infty} \psi(t) = \infty$. For if $t > \alpha_1$, we have

$$\int_{\alpha_1}^{t} |p(s)e^{-\mathscr{P}(s)}|\,ds = \int_{\alpha_1}^{t} (-p(s))e^{-\mathscr{P}(s)}\,ds = [e^{-\mathscr{P}(s)}]_{\alpha_1}^{t} = e^{-\mathscr{P}(t)} - e^{-\mathscr{P}(\alpha_1)}.$$

According to (17.6), we have $\lim_{t\to\infty} \int_{\alpha_1}^{t} |p(s)e^{-\mathscr{P}(s)}|\,ds = \infty$. Thus

$$\lim_{t\to\infty} \psi_1(t) = \psi_1(\alpha_1) + \lim_{t\to\infty} \int_{\alpha_1}^{t} |p(s)e^{-\mathscr{P}(s)}|\,ds = \infty$$

as claimed. Recall that $p(t) < 0$ for $t \in \mathbb{R} \setminus [-\alpha, \alpha]$. By the L'Hospital's rule

$$\lim_{t\to\pm\infty} \frac{\psi_1(t)}{e^{-\mathscr{P}(t)}} = \lim_{t\to\pm\infty} \frac{|p(t)e^{-\mathscr{P}(t)}|}{-p(t)e^{-\mathscr{P}(t)}} = 1.$$

So,

$$C_p = \sup_{t\in\mathbb{R}} |e^{\mathscr{P}(t)}| \int_{-\infty}^{t} |p(s)e^{-\mathscr{P}(s)}|\,ds = \sup_{t\in\mathbb{R}} \frac{\psi_1(t)}{e^{-\mathscr{P}(t)}} < \infty.$$

Note that

$$\lim_{t\to-\infty} \left| \int_{0}^{t} |p(s)e^{-\mathscr{P}(s)}|\,ds \right| = \psi_1(0) < \infty.$$

By (17.6), $D_p = \infty$. Since $\lim_{t\to\infty} \int_{\alpha_1}^{t} |p(s)e^{-\mathscr{P}(s)}|\,ds = \infty$, we have $E_p = \infty$.

(ii) Suppose that $\lim_{t\to\pm\infty} \mathscr{P}(t) = -\infty$, $p(t) > 0$ on $(-\infty, -\alpha_2)$ and $p(t) < 0$ on $(\alpha_2, \infty)$ for some $\alpha_2 > 0$. Set $\psi_2(t) = \int_0^t |p(s)e^{-\mathscr{P}(s)}|\,ds$ for $t \in \mathbb{R}$. Then

$$\begin{aligned}\psi_2(t) &= \int_0^{-\alpha_2} |p(s)e^{\mathscr{P}(s)}|\,ds + \int_{-\alpha_2}^{t} p(s)e^{-\mathscr{P}(s)}\,ds \\ &= \psi_2(-\alpha_2) + [-e^{-\mathscr{P}(s)}]_{-\alpha_2}^{t} = \psi_2(-\alpha_2) + e^{-\mathscr{P}(-\alpha_2)} - e^{-\mathscr{P}(t)}\end{aligned}$$

for each $t < -\alpha_2$. Thus, $\lim_{t\to-\infty} \psi_2(t) = -\infty$. Analogously, $\psi_2(t) = \psi(\alpha_2) + e^{-\mathscr{P}(t)} - e^{-\mathscr{P}(\alpha_2)}$ for $t > \alpha_2$, whence $\lim_{t\to\infty} \psi_2(t) = \infty$. By the L'Hospital's rule

$$\lim_{t\to-\infty} \frac{\psi_2(t)}{e^{-\mathscr{P}(t)}} = \lim_{t\to-\infty} \frac{|p(t)e^{-\mathscr{P}(t)}|}{-p(t)e^{-\mathscr{P}(t)}} = -1 \quad \text{and} \quad \lim_{t\to\infty} \frac{\psi_2(t)}{e^{-\mathscr{P}(t)}} = 1.$$

It follows that

$$D_p = \sup_{t\in\mathbb{R}} \left|e^{\mathscr{P}(t)}\right| \left|\int_0^t \left|p(s)e^{-\mathscr{P}(s)}\right| ds\right| = \sup_{t\in\mathbb{R}} \frac{|\psi_2(t)|}{e^{-\mathscr{P}(t)}} < \infty.$$

Since $\lim_{t\to\pm\infty} |\psi_2(t)| = \infty$, we have $C_p = E_p = \infty$.

(iii) Suppose that $\lim_{t\to-\infty} \mathscr{P}(t) = -\infty$, $\lim_{t\to\infty} \mathscr{P}(t) = \infty$ and that $p(t) > 0$ on $\mathbb{R}\setminus[-\alpha_3, \alpha_3]$ for some $\alpha_3 > 0$. We see that $\int_t^\infty \left|p(s)e^{-\mathscr{P}(s)}\right| ds$ exists for all $t \in \mathbb{R}$. In fact, for each $r > \alpha_3$

$$\begin{aligned}\int_{\alpha_3}^r \left|p(s)e^{-\mathscr{P}(s)}\right| ds &= \int_{\alpha_3}^r p(s)e^{-\mathscr{P}(s)}\, ds \\ &= \left[-e^{-\mathscr{P}(s)}\right]_{\alpha_3}^r = e^{-\mathscr{P}(\alpha_3)} - e^{-\mathscr{P}(r)}.\end{aligned}$$

Thus, $\int_{\alpha_3}^\infty \left|p(s)e^{-\mathscr{P}(s)}\right| ds = e^{-\mathscr{P}(\alpha_3)}$ and $\int_t^\infty \left|p(s)e^{-\mathscr{P}(s)}\right| ds$ exists for all $t \in \mathbb{R}$.

Let

$$\psi_3(t) = \int_t^\infty \left|p(s)e^{-\mathscr{P}(s)}\right| ds \qquad (\forall t \in \mathbb{R}).$$

Then $\lim_{t\to\infty} \psi_3(t) = 0$. Since $\lim_{t\to\infty} e^{\mathscr{P}(t)} = \infty$, by the L'Hospital's rule we get

$$\lim_{t\to\infty} \frac{\psi_3(t)}{e^{-\mathscr{P}(t)}} = \lim_{t\to\infty} \frac{-\left|p(t)e^{-\mathscr{P}(t)}\right|}{-p(t)e^{-\mathscr{P}(t)}} = 1. \tag{17.7}$$

Note that $\lim_{t\to-\infty} \psi_3(t) = \infty$. For, if $t < -\alpha_3$, then we have

$$\begin{aligned}\int_t^{-\alpha_3} \left|p(s)e^{-\mathscr{P}(s)}\right| ds &= \int_t^{-\alpha_3} p(s)e^{-\mathscr{P}(s)}\, ds \\ &= \left[-e^{-\mathscr{P}(s)}\right]_t^{-\alpha_3} = e^{-\mathscr{P}(t)} - e^{-\mathscr{P}(-\alpha_3)}.\end{aligned}$$

Therefore, $\lim_{t\to-\infty} \int_t^{-\alpha_3} \left|p(s)e^{-\mathscr{P}(s)}\right| ds = \infty$. It follows that

$$\lim_{t\to-\infty} \psi_3(t) = \lim_{t\to-\infty} \int_t^{-\alpha_3} \left|p(s)e^{-\mathscr{P}(s)}\right| ds + \int_{-\alpha_3}^\infty \left|p(s)e^{-\mathscr{P}(s)}\right| ds = \infty.$$

By the L'Hospital's rule, we have

$$\lim_{t\to-\infty} \frac{\psi_3(t)}{e^{-\mathscr{P}(t)}} = \lim_{t\to-\infty} \frac{-\left|p(t)e^{-\mathscr{P}(t)}\right|}{-p(t)e^{-\mathscr{P}(t)}} = 1.$$

According to (17.7), we conclude that

$$E_p = \sup_{t\in\mathbb{R}} \left|e^{\mathscr{P}(t)}\right| \int_t^\infty \left|p(s)e^{-\mathscr{P}(s)}\right| ds = \sup_{t\in\mathbb{R}} \frac{\psi_3(t)}{e^{-\mathscr{P}(t)}} < \infty.$$

Recall that

$$\lim_{t\to-\infty}\int_t^{-\alpha_3}\left|p(s)e^{-\mathscr{P}(s)}\right|ds=\infty$$

and therefore $C_p=\infty$. Since $\int_0^t\left|p(s)e^{-\mathscr{P}(s)}\right|ds\le\psi_3(0)$ for all $t>0$, we obtain that $D_p=\infty$.

(iv) Let $\lim_{t\to\pm\infty}\mathscr{P}(t)=\infty$, $p(t)<0$ on $(-\infty,\alpha_4)$ and $p(t)>0$ on $(\alpha_4,\infty)$ for some $\alpha_4>0$. Then

$$\int_{-\infty}^{-\alpha_4}\left|p(s)e^{-\mathscr{P}(s)}\right|ds=e^{-\mathscr{P}(-\alpha_4)}\quad\text{and}\quad\int_{\alpha_4}^{\infty}\left|p(s)e^{-\mathscr{P}(s)}\right|ds=e^{-\mathscr{P}(\alpha_4)}.$$

So, by $\lim_{t\to\pm\infty}\mathscr{P}(t)=\infty$, $C_p=D_p=\infty$. Finally, $E_p=\infty$, because, for each $t>0$,

$$\begin{aligned}\int_0^t\left|p(s)e^{-\mathscr{P}(s)}\right|ds&\le\int_0^{\alpha_4}\left|p(s)e^{-\mathscr{P}(s)}\right|ds+\int_{\alpha_4}^{\infty}\left|p(s)e^{-\mathscr{P}(s)}\right|ds\\&=\int_0^{\alpha_4}\left|p(s)e^{-\mathscr{P}(s)}\right|ds+e^{-\mathscr{P}(\alpha_4)}.\end{aligned}$$

□

*Example 17.2.* Let $p(t)=\sum_{k=0}^n a_kt^k$ be a polynomial where $n\in\mathbb{N}\cup\{0\}$, $a_k\in\mathbb{R}$ for $k=0,1,2,\ldots,n$ and $a_n\ne0$. By Theorem 17.2, the following four statements are true.

1. If $a_n<0$ and $n$ is even, then $C_p<\infty$ and $D_p=E_p=\infty$.
2. If $a_n<0$ and $n$ is odd, then $D_p<\infty$ and $C_p=E_p=\infty$.
3. If $a_n>0$ and $n$ is even, then $E_p<\infty$ and $C_p=D_p=\infty$.
4. If $a_n>0$ and $n$ is odd, then $C_p=D_p=E_p=\infty$.

So, Theorem 17.1 implies that (17.2) has the Hyers–Ulam stability unless $a_n>0$ and $n$ is odd. Clearly, one might ask whether (17.2) has the Hyers–Ulam stability in that last case. We will give an answer to this question.

**Theorem 17.3.** *Let $p\colon\mathbb{R}\to\mathbb{R}$ be a continuous function with*

$$\sup_{t\in(-\infty,0]}\mathscr{P}(t)=\sup_{t\in[0,\infty)}\mathscr{P}(t)=\infty.$$

*For every $\varepsilon>0$ there exists $f\in C(\mathbb{R},X)$ such that $\sup_{t\in\mathbb{R}}\|f(t)-g(t)\|=\infty$ for each solution $g\in C(\mathbb{R},X)$ of* (17.2) *and*

$$\left\|f(t)-f(0)-\int_0^t p(s)f(s)\,ds\right\|\le\varepsilon\qquad(\forall t\in\mathbb{R}).$$

*Proof.* For each $\varepsilon > 0$, take $x \in X$ with $\|x\| = \varepsilon$. Let $u\colon \mathbb{R} \to X$ be a continuous map defined by

$$u(t) = \begin{cases} 0 & (t < 0); \\ tx & (0 \le t \le 1); \\ x & (1 < t). \end{cases}$$

Set

$$f(t) = u(t) + \mathrm{e}^{\mathscr{P}(t)} \int_0^t p(s)\mathrm{e}^{-\mathscr{P}(s)} u(s)\,\mathrm{d}s \qquad (\forall t \in \mathbb{R}). \tag{17.8}$$

Since $f(0) = 0$, we have by Lemma 17.1 that

$$u(t) = f(t) - \int_0^t p(s)f(s)\,\mathrm{d}s \qquad (\forall t \in \mathbb{R}).$$

This implies that

$$\left\| f(t) - f(0) - \int_0^t p(s)f(s)\,\mathrm{d}s \right\| = \|u(t)\| \le \|x\| = \varepsilon \qquad (\forall t \in \mathbb{R}).$$

We show that

$$\int_0^t p(s)\mathrm{e}^{-\mathscr{P}(s)} u(s)\,\mathrm{d}s = \begin{cases} 0 & (t < 0); \\ -t\mathrm{e}^{-\mathscr{P}(t)}x + \displaystyle\int_0^t \mathrm{e}^{-\mathscr{P}(s)}x\,\mathrm{d}s & (0 \le t \le 1); \\ -\mathrm{e}^{-\mathscr{P}(t)}x + \displaystyle\int_0^1 \mathrm{e}^{-\mathscr{P}(s)}x\,\mathrm{d}s & (1 < t). \end{cases} \tag{17.9}$$

We only consider the case when $t \ge 0$. If $0 \le t \le 1$, then by the integration by parts we obtain

$$\begin{aligned} \int_0^t p(s)\mathrm{e}^{-\mathscr{P}(s)} u(s)\,\mathrm{d}s &= \int_0^t p(s)\mathrm{e}^{-\mathscr{P}(s)} sx\,\mathrm{d}s \\ &= [-\mathrm{e}^{-\mathscr{P}(s)}sx]_0^t + \int_0^t \mathrm{e}^{-\mathscr{P}(s)}x\,\mathrm{d}s = -t\mathrm{e}^{-\mathscr{P}(t)}x + \int_0^t \mathrm{e}^{-\mathscr{P}(s)}x\,\mathrm{d}s. \end{aligned}$$

If $1 < t$, then we have, by using the above equality

$$\begin{aligned} \int_0^t p(s)\mathrm{e}^{-\mathscr{P}(s)} u(s)\,\mathrm{d}s &= \int_0^1 p(s)\mathrm{e}^{-\mathscr{P}(s)} sx\,\mathrm{d}s + \int_1^t p(s)\mathrm{e}^{-\mathscr{P}(s)} x\,\mathrm{d}s \\ &= -\mathrm{e}^{-\mathscr{P}(1)}x + \int_0^1 \mathrm{e}^{-\mathscr{P}(s)}x\,\mathrm{d}s + [-\mathrm{e}^{-\mathscr{P}(s)}x]_1^t \\ &= -\mathrm{e}^{-\mathscr{P}(t)}x + \int_0^1 \mathrm{e}^{-\mathscr{P}(s)}x\,\mathrm{d}s. \end{aligned}$$

Consequently, we have (17.9), as claimed. It follows from (17.8) that

$$f(t) = \begin{cases} 0 & (t < 0); \\ e^{\mathscr{P}(t)} \int_0^t e^{-\mathscr{P}(s)} x \, ds & (0 \leq t \leq 1); \\ e^{\mathscr{P}(t)} \int_0^1 e^{-\mathscr{P}(s)} x \, ds & (1 < t). \end{cases}$$

Now suppose, on the contrary, that there exists a solution $g \in C(\mathbb{R}, X)$ of (17.2) such that $\sup_{t\in\mathbb{R}} \|f(t) - g(t)\| < \infty$. According to Lemma 17.1, we can write $g(t) = g(0)e^{\mathscr{P}(t)}$ for every $t \in \mathbb{R}$, and therefore $\sup_{t\in\mathbb{R}} \|f(t) - g(0)e^{\mathscr{P}(t)}\| \leq K$ for some $K \geq 0$. Since $f(t) = 0$ for $t < 0$, we have

$$\|g(0)\| e^{\mathscr{P}(t)} = \|g(0)e^{\mathscr{P}(t)}\| \leq K \qquad (\forall t < 0).$$

Since $\sup_{(-\infty,0]} \mathscr{P}(t) = \infty$, so $g(0) = 0$, whence $\|f(t)\| \leq K$ for every $t \in \mathbb{R}$. Consequently

$$\|f(t)\| = e^{\mathscr{P}(t)} \int_0^1 e^{-\mathscr{P}(s)} ds \|x\| = \left\| e^{\mathscr{P}(t)} \int_0^1 e^{-\mathscr{P}(s)} x \, ds \right\| \leq K \qquad (\forall t > 1).$$

Here, we notice that $\int_0^1 e^{-\mathscr{P}(s)} ds \|x\| \neq 0$. This is impossible because we have $\sup_{[0,\infty)} \mathscr{P}(t) = \infty$. □

*Example 17.3.* If $p\colon \mathbb{R} \to \mathbb{R}$ is a continuous function with $p(t) \geq 0$ for every $t \in \mathbb{R}$ and

$$\int_{-\infty}^{\infty} p(\tau) \, d\tau < \infty,$$

then we see that $C_p < \infty$, $D_p < \infty$ and $E_p < \infty$. In fact,

$$\int p(s) e^{-\mathscr{P}(s)} ds = -e^{-\mathscr{P}(s)} + c,$$

where $c$ is any constant. By the assumption, $-\infty < \lim_{t\to-\infty} \mathscr{P}(t) \leq 0$ and $0 \leq \lim_{t\to\infty} \mathscr{P}(t) < \infty$. It follows that

$$\int_{-\infty}^{t} p(s) e^{-\mathscr{P}(s)} ds < \infty \quad \text{and} \quad \int_{t}^{\infty} p(s) e^{-\mathscr{P}(s)} ds < \infty$$

for all $t \in \mathbb{R}$. Thus, $C_p < \infty$, $D_p < \infty$ and $E_p < \infty$. In particular, (17.2) has the Hyers–Ulam stability by Theorem 17.1.

**Theorem 17.4.** *Let $p\colon \mathbb{R} \to \mathbb{C} \setminus \{0\}$ be a continuous function. Suppose that for every $f \in C(\mathbb{R}, X)$ with*

$$\sup_{t\in\mathbb{R}} \left\| f(t) - f(0) - \int_0^t p(s) f(s) \, ds \right\| < \infty$$

*there exists a solution $g \in C(\mathbb{R},X)$ of* (17.2) *such that* $\sup_{t\in\mathbb{R}} |f(t)-g(t)| < \infty$. *Then one of $C_p$, $D_p$ and $E_p$ is finite.*

*In particular, if* (17.2) *has the Hyers–Ulam stability, then one of $C_p$, $D_p$ and $E_p$ is finite.*

*Proof.* Take $x \in X$ with $\|x\| = 1$. Set

$$u(t) = \left( \frac{|p(t)\mathrm{e}^{-\mathscr{P}(t)}|}{p(t)\mathrm{e}^{-\mathscr{P}(t)}} - \frac{|p(0)|}{p(0)} \right) x \qquad (\forall t \in \mathbb{R}).$$

Then $u(0) = 0$ with $\|u(t)\| \le 2$ for every $t \in \mathbb{R}$. Let us consider the map $f \in C(\mathbb{R},X)$ defined by

$$f(t) = u(t) + \mathrm{e}^{\mathscr{P}(t)} \int_0^t p(s)\mathrm{e}^{-\mathscr{P}(s)} u(s)\,\mathrm{d}s \qquad (\forall t \in \mathbb{R}).$$

Then $f(0) = u(0) = 0$. By Lemma 17.1, for each $t \in \mathbb{R}$ we have

$$u(t) = f(t) - \int_0^t p(s)f(s)\,\mathrm{d}s,$$

which implies that

$$\sup_{t\in\mathbb{R}} \left\| f(t) - f(0) - \int_0^t p(s)f(s)\,\mathrm{d}s \right\| = \sup_{t\in\mathbb{R}} \|u(t)\| \le 2.$$

By assumption, there exists $g \in C(\mathbb{R},X)$ such that

$$g(t) - g(0) = \int_0^t p(s)g(s)\,\mathrm{d}s \qquad (\forall t \in \mathbb{R})$$

and

$$\sup_{t\in\mathbb{R}} |f(t) - g(t)| < \infty.$$

By Lemma 17.1, $g(t) = g(0)\mathrm{e}^{\mathscr{P}(t)}$ for every $t \in \mathbb{R}$, and consequently

$$\sup_{t\in\mathbb{R}} \|f(t) - g(t)\| = \sup_{t\in\mathbb{R}} \left\| u(t) + \mathrm{e}^{\mathscr{P}(t)} \int_0^t p(s)\mathrm{e}^{-\mathscr{P}(s)} u(s)\,\mathrm{d}s - g(0)\mathrm{e}^{\mathscr{P}(t)} \right\| \le K$$

for some constant $K \ge 0$. Here, we note that

$$\int_0^t p(s)\mathrm{e}^{-\mathscr{P}(s)}\,\mathrm{d}s = \left[ -\mathrm{e}^{-\mathscr{P}(s)} \right]_0^t = 1 - \mathrm{e}^{-\mathscr{P}(t)}, \tag{17.10}$$

and therefore, by the definition of $u$ with (17.10), we have

$$\int_0^t p(s)\mathrm{e}^{-\mathscr{P}(s)} u(s)\,\mathrm{d}s = \int_0^t |p(s)\mathrm{e}^{-\mathscr{P}(s)}| x\,\mathrm{d}s - \frac{|p(0)|}{p(0)} (1 - \mathrm{e}^{-\mathscr{P}(t)}) x$$

for all $t \in \mathbb{R}$. This implies that

$$f(t)-g(t)=u(t)+\mathrm{e}^{\mathscr{P}(t)}\int_0^t \left|p(s)\mathrm{e}^{-\mathscr{P}(s)}\right|x\,\mathrm{d}s-\frac{|p(0)|}{p(0)}(\mathrm{e}^{\mathscr{P}(t)}-1)x-g(0)\mathrm{e}^{\mathscr{P}(t)}$$

for all $t \in \mathbb{R}$. Set $x_1 = |p(0)|x/p(0)+g(0)$. Then we have

$$\left|\mathrm{e}^{\mathscr{P}(t)}\right|\left\|\int_0^t \left|p(s)\mathrm{e}^{-\mathscr{P}(s)}\right|x\,\mathrm{d}s-x_1\right\| \le \|f(t)-g(t)\|+\left\|u(t)+\frac{|p(0)|}{p(0)}x\right\| \le K+1$$

and consequently

$$\left\|\int_0^t \left|p(s)\mathrm{e}^{-\mathscr{P}(s)}\right|x\,\mathrm{d}s-x_1\right\| \le (K+1)\left|\mathrm{e}^{-\mathscr{P}(t)}\right| \qquad (\forall t \in \mathbb{R}). \tag{17.11}$$

Now we have three possibilities:

(i) $\inf_{t\in(-\infty,0]}\left|\mathrm{e}^{-\mathscr{P}(t)}\right| = 0$. There exists a strictly monotone decreasing sequence $\{t_n\}_{n\in\mathbb{N}} \subset (-\infty,0]$ such that

$$t_n \searrow -\infty \quad (n\to\infty) \quad \text{and} \quad \left|\mathrm{e}^{-\mathscr{P}(t_n)}\right| < \frac{1}{n} \quad (\forall n \in \mathbb{N}).$$

By (17.11), we have

$$\left\|\int_0^{t_n} \left|p(s)\mathrm{e}^{-\mathscr{P}(s)}\right|x\,\mathrm{d}s-x_1\right\| < \frac{K+1}{n} \tag{17.12}$$

for every $n \in \mathbb{N}$. Since $\|x\|=1$,

$$\left\|\int_{t_n}^0 \left|p(s)\mathrm{e}^{-\mathscr{P}(s)}\right|\mathrm{d}s-\|x_1\|\right\| < \frac{K+1}{n} \qquad (\forall n \in \mathbb{N}). \tag{17.13}$$

We assert that $\int_{-\infty}^0 \left|p(s)\mathrm{e}^{-\mathscr{P}(s)}\right|\mathrm{d}s$ exists. To do this, we first show that

$$\sup_{t\in(-\infty,0]}\int_t^0 \left|p(s)\mathrm{e}^{-\mathscr{P}(s)}\right|\mathrm{d}s < \infty.$$

According to (17.13), there exists an $n_0 \in \mathbb{N}$ such that $n \ge n_0$ implies

$$\int_{t_n}^0 \left|p(s)\mathrm{e}^{-\mathscr{P}(s)}\right|\mathrm{d}s \le 1+\|x_1\|.$$

Since $t_n \searrow -\infty$, we see that

$$\sup_{t\in(-\infty,0]}\int_t^0 \left|p(s)\mathrm{e}^{-\mathscr{P}(s)}\right|\mathrm{d}s \le \max\left\{\int_{t_{n_0}}^0 \left|p(s)\mathrm{e}^{-\mathscr{P}(s)}\right|\mathrm{d}s,\, 1+\|x_1\|\right\}.$$

Therefore

$$\sup_{t\in(-\infty,0]}\int_t^0 \left|p(s)e^{-\mathscr{P}(s)}\right| ds < \infty.$$

Since the function

$$t \mapsto \int_t^0 \left|p(s)e^{-\mathscr{P}(s)}\right| ds$$

is monotone decreasing for $t \in (-\infty,0]$

$$\int_{-\infty}^0 \left|p(s)e^{-\mathscr{P}(s)}\right| ds = \sup_{t\in(-\infty,0]}\int_t^0 \left|p(s)e^{-\mathscr{P}(s)}\right| ds.$$

Thus, we have proved the existence of $\int_{-\infty}^0 \left|p(s)e^{-\mathscr{P}(s)}\right| ds$ as claimed. Next we show that

$$x_1 = -\int_{-\infty}^0 \left|p(s)e^{-\mathscr{P}(s)}\right| x\, ds. \tag{17.14}$$

In fact, by (17.12)

$$\begin{aligned}\left\|x_1+\int_{-\infty}^0 \left|p(s)e^{-\mathscr{P}(s)}\right| x\, ds\right\| &\le \left\|x_1-\int_0^{t_n} \left|p(s)e^{-\mathscr{P}(s)}\right| x\, ds\right\| \\ &\quad+\left\|\int_0^{t_n} \left|p(s)e^{-\mathscr{P}(s)}\right| x\, ds+\int_{-\infty}^0 \left|p(s)e^{-\mathscr{P}(s)}\right| x\, ds\right\| \\ &< \frac{K+1}{n}+\left|\int_0^{t_n} \left|p(s)e^{-\mathscr{P}(s)}\right| ds+\int_{-\infty}^0 \left|p(s)e^{-\mathscr{P}(s)}\right| ds\right| \to 0 \quad (n\to\infty),\end{aligned}$$

where we have used $|x| = 1$. Consequently, (17.14) holds as claimed. Thus, by (17.12)

$$\int_{-\infty}^t \left|p(s)e^{-\mathscr{P}(s)}\right| ds = \left\|\int_0^t \left|p(s)e^{-\mathscr{P}(s)}\right| x\, ds - x_1\right\| \le (K+1)\left|e^{-\mathscr{P}(t)}\right|$$

for every $t \in \mathbb{R}$. This proves that $C_p \le K+1 < \infty$.

(ii) $\inf_{t\in\mathbb{R}}\left|e^{-\mathscr{P}(t)}\right| > 0$. Set $L = \|x_1\|/\inf_{t\in\mathbb{R}}\left|e^{-\mathscr{P}(t)}\right|$. Then we have

$$\begin{aligned}\left\|\int_0^t \left|p(s)e^{-\mathscr{P}(s)}\right| x\, ds\right\| &\le (K+1)\left|e^{-\mathscr{P}(t)}\right| + \|x_1\| \\ &\le (K+1)\left|e^{-\mathscr{P}(t)}\right| + L\left|e^{-\mathscr{P}(t)}\right| \\ &= (K+L+1)\left|e^{-\mathscr{P}(t)}\right| \qquad (\forall t\in\mathbb{R}).\end{aligned}$$

With $\|x\| = 1$, this proves that $D_p < \infty$.

(iii) $\inf_{t\in[0,\infty)} \left|e^{-\mathscr{P}(t)}\right| = 0$. By a quite similar argument to (i), we can prove that

$$\int_0^\infty \left|p(s)e^{-\mathscr{P}(s)}\right| x\,ds = x_1,$$

and so

$$\int_t^\infty \left|p(s)e^{-\mathscr{P}(s)}\right| ds = \left\| x_1 - \int_0^t \left|p(s)e^{-\mathscr{P}(s)}\right| x\,ds \right\| \le (K+1)\left|e^{-\mathscr{P}(t)}\right|$$

for every $t \in \mathbb{R}$. This shows that $E_p \le K+1 < \infty$.

**Acknowledgement** The first and third authors were partly supported by the Grant-in-Aid for Scientific Research.

## References

1. Alsina, C., Ger, R.: On some inequalities and stability results related to the exponential function. J. Inequal. Appl. **2**, 373–380 (1998)
2. Aoki, T.: On the stability of the linear transformation in Banach spaces. J. Math. Soc. Japan **2** (1950), 64–66.
3. Byungbae, K., Jung, S.-M.: Bessel's differential equation and its Hyers–Ulam stability. J. Inequal. Appl. **2007** Art. ID 21640, 8 pp. (2007)
4. Ger, R., Šemrl, P.: The stability of the exponential equation. Proc. Amer. Math. Soc. **124**, 779–787 (1996)
5. Hyers, D.H.: On the stability of the linear functional equation. Proc. Nat. Acad. Sci. U.S.A., **27**, 222–224 (1941)
6. Hyers, D.H., Rassias, Th.M.: Approximate homomorphisms. Aequationes Math. **44**, 125–153 (1992)
7. Hyers, D.H., Isac, G., Rassias, Th.M.: Stability of Functional Equations in Several Variables. Birkhauser (1998)
8. Jung, S.-M.: A fixed point approach to the stability of a Volterra integral equation. Fixed Point Theory Appl. **2007**, Art ID 57064, 9 pp. (2007)
9. Jung, S.-M., Byungbae, K.: Chebyshev's differential equation and its Hyers–Ulam stability. Differ. Equ. Appl. **1**, 199–207 (2009)
10. Miura, T., Miyajima, S., Takahasi, S.-E.: Hyers–Ulam stability of linear differential operator with constant coefficients. Math. Nachr. **258**, 90–96 (2003)
11. Miura, T., Miyajima, S., Takahasi, S.-E.: A characterization of Hyers–Ulam stability of first order linear differential operators. J. Math. Anal. Appl. **286**, 136–146 (2003)
12. Rassias, Th.M.: On the stability of the linear mapping in Banach spaces. Proc. Amer. Math. Soc. **72**, 297–300 (1978)
13. Rassias, Th.M., Šemrl, P.: On the behavior of mappings which do not satisfy Hyers–Ulam stability. Proc. Amer. Math. Soc. **114**, 989–993 (1992)
14. Rassias, Th.M.: On the stability of functional equations in Banach spaces. J. Math. Anal. Appl. **251**, 264–284 (2000)
15. Ulam, S.M.: A Collection of Mathematical Problems. Interscience Tracts in Pure and Applied Mathematics, no. 8, Interscience Publishers, New York-London (1960)

# Chapter 18
# On the Stability of Polynomial Equations

**Abbas Najati and Themistocles M. Rassias**

*Dedicated to the memory of S.M. Ulam*
*on the 100th anniversary of his birth*

**Abstract** In this article, we prove the Hyers–Ulam type stability for the following two equations with real coefficients:

$$a_nx^n + a_{n-1}x^{n-1} + \cdots + a_1x + a_0 = 0 \quad \text{and} \quad e^x + \alpha x + \beta = 0$$

on a real interval $[a,b]$. More precisely, we show that if $x$ is an approximate solution of the equation $a_nx^n + a_{n-1}x^{n-1} + \cdots + a_1x + a_0 = 0$ (resp. $e^x + \alpha x + \beta = 0$), then there exists an exact solution of the equation near $x$.

**Keywords** Hyers-Ulam stability • Polynomial equation

**Mathematics Subject Classification (2000):** Primary 39B82; Secondary 34K20, 26D10

## 18.1 Introduction

The functional equation $(\xi)$ is stable if any function $g$ satisfying the equation $(\xi)$ *approximately* is near to 'a true' solution of $(\xi)$. The stability of functional equations was first introduced by Ulam [10, Chap. VI] in 1940. More precisely,

---

A. Najati (✉)
Faculty of Sciences, Department of Mathematics, University of Mohaghegh Ardabili, Ardabil, Iran
e-mail: a.nejati@yahoo.com

Th.M. Rassias
Department of Mathematics, National Technical University of Athens, Zografou Campus, 15780 Athens, Greece
e-mail: trassias@math.ntua.gr

Th.M. Rassias and J. Brzdęk (eds.), *Functional Equations in Mathematical Analysis*, Springer Optimization and Its Applications 52, DOI 10.1007/978-1-4614-0055-4_18,

Ulam proposed the following problem: Given a group $G_1$, a metric group $(G_2,d)$ and a positive number $\varepsilon$, does there exist a $\delta > 0$ such that if a function $f : G_1 \to G_2$ satisfies the inequality $d(f(xy), f(x)f(y)) < \delta$ for all $x, y \in G_1$, then there exists a group homomorphism $T : G_1 \to G_2$ such that $d(f(x), T(x)) < \varepsilon$ for all $x \in G_1$? As it is mentioned above, when this problem has a solution, we say that *the homomorphism equation is stable* in the class of functions from $G_1$ into $G_2$.

In 1941, Hyers [3] gave a partial solution of Ulam's problem, for the case of approximately additive mappings, under the assumption that $G_1$ and $G_2$ are Banach spaces. Aoki [1] and Rassias [8] provided a generalization of the Hyers' theorem for additive and linear mappings, respectively, by allowing the Cauchy difference to be unbounded. During the last decades various stability problems for a large variety of functional equations have been investigated by several mathematicians. A large list of references concerning the stability of functional equations can be found, e.g., in [2,4,5,9].

Recently, Li and Hua [6] investigated the Hyers–Ulam stability of the polynomial equation $x^n + \alpha x + \beta = 0$ on $[-1,1]$ and they posed the problem whether the polynomial equation $a_n x^n + a_{n-1}x^{n-1} + \cdots + a_1 x + a_0 = 0$ with real coefficients satisfies the Hyers–Ulam stability when it accepts some solutions in the real interval [a,b]?

In this paper, we prove the Hyers–Ulam stability for the following two equations

$$a_n x^n + a_{n-1}x^{n-1} + \cdots + a_1 x + a_0 = 0, \tag{18.1}$$

$$e^x + \alpha x + \beta = 0 \tag{18.2}$$

on $[-1,1]$.

We say that the equation $f(x) = 0$ has the Hyers–Ulam stability on the real interval $[a,b]$ if there exists a constant $K > 0$ with the following property: for every $\varepsilon > 0$ and $x \in [a,b]$ if $|f(x)| \le \varepsilon$, there exists some real number $z \in [a,b]$ such that $f(z) = 0$ and $|x - z| \le K\varepsilon$.

For an extensive information on the theory and applications of polynomial functions and their roots the reader is referred to [7].

## 18.2 Main Results

**Theorem 18.1.** *Let* $y \in [-1,1]$ *satisfy*

$$|a_n y^n + a_{n-1}y^{n-1} + \cdots + a_1 y + a_0| \le \varepsilon,$$

$$\sum_{m=2}^{n} m|a_m| < |a_1|,$$

*and*

$$|a_0| + \sum_{m=2}^{n} |a_m| < |a_1|.$$

*Then (18.1) has a unique solution in* $[-1,1]$*, namely, there exists a real number* $z \in [-1,1]$*, satisfying*

$$|y-z| \leq \frac{\varepsilon}{|a_1| - \sum_{m=2}^{n} m|a_m|}.$$

*Proof.* Consider the function $g : [-1,1] \to \mathbb{R}$ defined by

$$g(x) = -\frac{a_0}{a_1} - \frac{1}{a_1}\sum_{m=2}^{n} a_m x^m.$$

It is easy to see that $|g(x)| \leq 1$ for all $x \in [-1,1]$. Therefore, $g$ maps $[-1,1]$ into $[-1,1]$. For each $u, v \in [-1,1]$, we have

$$\begin{aligned}|g(u)-g(v)| &\leq \frac{1}{|a_1|}\sum_{m=2}^{n} |a_m u^m - a_m v^m| \\ &\leq \frac{1}{|a_1|}\sum_{m=2}^{n} |a_m||u-v||u^{m-1} + u^{m-2}v + \cdots + uv^{m-2} + v^{m-1}| \\ &\leq \frac{1}{|a_1|}\sum_{m=2}^{n} m|a_m||u-v|.\end{aligned}$$

Thus, $g$ is a contraction mapping and by Banach's contraction the mapping theorem,$g$ has a unique fixed point $z \in [-1,1]$, i.e., $g(z) = z$. Hence, $z$ is a solution of (18.1) and

$$\begin{aligned}|y-z| &\leq |y-g(y)| + |g(y)-g(z)| \\ &\leq \frac{1}{|a_1|}|a_n y^n + a_{n-1}y^{n-1} + \cdots + a_1 y + a_0| + \frac{1}{|a_1|}\sum_{m=2}^{n} m|a_m||y-z| \\ &\leq \frac{\varepsilon}{|a_1|} + \frac{1}{|a_1|}\sum_{m=2}^{n} m|a_m||y-z|.\end{aligned}$$

Therefore

$$|y-z| \leq \frac{\varepsilon}{|a_1| - \sum_{m=2}^{n} m|a_m|}.$$

□

**Theorem 18.2.** *Let* $|\alpha| > e + |\beta|$ *and* $y \in [-1,1]$ *satisfy*

$$|e^y + \alpha y + \beta| \leq \varepsilon.$$

*Then (18.2) has a unique solution in* $[-1,1]$*, namely* $z$*, satisfying*

$$|y-z| \leq \frac{\varepsilon}{|\alpha| - e}.$$

*Proof.* We define the function $g : [-1,1] \to \mathbb{R}$ by

$$g(x) = -\frac{1}{\alpha}(e^x + \beta).$$

It follows that $|g(x)| \leq 1$ for all $x \in [-1,1]$. Therefore, $g$ maps $[-1,1]$ into $[-1,1]$. For each $u, v \in [-1,1]$, we have

$$\begin{aligned}
|g(u) - g(v)| &\leq \frac{1}{|\alpha|}|e^u - e^v| \leq \frac{1}{|\alpha|}\sum_{n=1}^{\infty}\frac{|u^n - v^n|}{n!} \\
&\leq \frac{1}{|\alpha|}\sum_{n=1}^{\infty}\frac{1}{n!}|u-v||u^{n-1} + u^{n-2}v + \cdots + uv^{n-2} + v^{n-1}| \\
&\leq \frac{1}{|\alpha|}\sum_{n=0}^{\infty}\frac{1}{n!}|u-v| = \frac{e}{|\alpha|}|u-v|.
\end{aligned}$$

Thus, $g$ is a contraction mapping and by Banach's contraction mapping theorem, $g$ has a unique fixed point $z \in [-1,1]$, i.e., $g(z) = z$. Hence, $z$ is a solution of (18.2) and

$$\begin{aligned}
|y - z| &\leq |y - g(y)| + |g(y) - g(z)| \\
&\leq \frac{1}{|\alpha|}|e^y + \alpha y + \beta| + \frac{e}{|\alpha|}|y - z| \\
&\leq \frac{\varepsilon}{|\alpha|} + \frac{e}{|\alpha|}|y - z|.
\end{aligned}$$

Therefore

$$|y - z| \leq \frac{\varepsilon}{|\alpha| - e}.$$

□

*Remark 18.1.* Applying the same technique as in the proofs of Theorems 18.1 and 18.2, one can study the Hyers–Ulam stability of (18.1) and (18.2) defined on any finite interval $[a,b]$.

## References


1. Aoki, T.: On the stability of the linear transformation in Banach spaces. J. Math. Soc. Japan **2**, 64–66 (1950)
2. Czerwik, S.: Functional Equations and Inequalities in Several Variables. World Scientific, River Edge, NJ (2002)
3. Hyers, D.H.: On the stability of the linear functional equation. Proc. Nat. Acad. Sci. USA **27**, 222–224 (1941)

4. Hyers, D.H., Isac, G., Rassias, Th.M.: Stability of Functional Equations in Several Variables. Birkhäuser, Basel (1998)
5. Jung, S.-M.: Hyers–Ulam–Rassias Stability of Functional Equations in Mathematical Analysis. Hadronic Press, Palm Harbor (2001)
6. Li, Y., Hua, L.: Hyers–Ulam stability of a polynomial equation. Banach J. Math. Anal. **3**, 86–90 (2009)
7. Milovanovic, G.V., Mitrinovic, D.S., Rassias, Th.M.: Topics in Polynomials: Extremal Problems, Inequalities, Zeros. World Scientific Publishing Company, Singapore, New Jersey, London (1994)
8. Rassias, Th.M.: On the stability of the linear mapping in Banach spaces. Proc. Amer. Math. Soc. **72**, 297–300 (1978)
9. Rassias, Th.M.: On the stability of functional equations and a problem of Ulam. Acta Appl. Math. **62**, 23–130 (2000)
10. Ulam, S.M.: Problems in Modern Mathematics. Science Ed. Wiley, New York (1940)

# Chapter 19
# Isomorphisms and Derivations in Proper $JCQ^*$-Triples

**Choonkil Park and Madjid Eshaghi-Gordji**

*Dedicated to the memory of S.M. Ulam*
*on the 100th anniversary of his birth*

**Abstract** We investigate homomorphisms in proper $JCQ^*$-triples and derivations on proper $JCQ^*$-triples associated with the 3-variable Jensen functional equation

$$2f\left(\frac{x+y+z}{2}\right) = f(x) + f(y) + f(z),$$

which was introduced and investigated by Park, Cho and Han.

We moreover prove the Hyers–Ulam–Rassias stability of homomorphisms in proper $JCQ^*$-triples and of derivations on proper $JCQ^*$-triples. This is applied to investigate isomorphisms between proper $JCQ^*$-triples.

**Keywords** 3-variable Jensen functional equation • Hyers–Ulam–Rassias stability • Proper $JCQ^*$-triple homomorphism • Proper $JCQ^*$-triple derivation

**Mathematics Subject Classification (2010):** Primary 47Jxx, 39B52, 46B03, 17C65, 47B48, 47L60, 46L05

## 19.1 Introduction and Preliminaries

As it is extensively discussed in [54], the full description of a physical system $\mathscr{S}$ implies the knowledge of three basic ingredients: the sent of the observables, the set of the states and the dynamics that describes the time evolution of the system

---

C. Park
Department of Mathematics, Hanyang University, Seoul 133-791, Republic of Korea
e-mail: baak@hanyang.ac.kr

M. Eshaghi-Gordji (✉)
Department of Mathematics, Semnan University, P.O. Box 35195-363, Semnan, Iran
e-mail: madjid.eshaghi@gmail.com

Th.M. Rassias and J. Brzdęk (eds.), *Functional Equations in Mathematical Analysis*, Springer Optimization and Its Applications 52, DOI 10.1007/978-1-4614-0055-4_19,

by means of the time dependence of the expectation value of a given observable on a given state. Originally, the set of the observables was considered tobe a $C^*$-algebra [29]. In many applications, however, this was shown not to be the most convenient choice and the $C^*$-algebra was replaced by a von Neumann algebra, because the role of the representation turns out to be crucial mainly when long range interactions are involved (see [11] and references therein). Here we use a different algebraic structure, similar to the one considered in [23], which is suggested by the considerations above: because of the relevance of the unbounded operators in the description of $\mathscr{S}$, we will assume that the observables of the system belong to a quasi $*$-algebra $(A,A_0)$ (see [58] and references therein), while, in order to have a richer mathematical structure, we will use a slightly different algebraic structure: $(A,A_0)$ will be assumed to be a proper $CQ^*$-algebra, which has nicer topological properties. In particular, for instance, $A_0$ is a $C^*$-algebra.

Let $A$ be a linear space and $A_0$ is a $*$-algebra contained in $A$ as a subspace. We say that $A$ is a quasi $*$-algebra over $A_0$ if

1. The right and left multiplications of an element of $A$ and an element of $A_0$ are defined and linear;
2. $x_1(x_2a) = (x_1x_2)a$, $(ax_1)x_2 = a(x_1x_2)$ and $x_1(ax_2) = (x_1a)x_2$ for all $x_1,x_2 \in A_0$ and all $a \in A$;
3. An involution $*$, which extends the involution of $A_0$, is defined in $A$ with the property $(ab)^* = b^*a^*$ whenever the multiplication is defined.

A quasi $*$-algebra $(A,A_0)$ is said to be a locally convex quasi $*$-algebra if in $A$ a locally convex topology $\tau$ is defined such that

1. The involution is continuous and the multiplications are separately continuous;
2. $A_0$ is dense in $A[\tau]$.

Throughout this paper, we suppose that $(A[\tau],A_0)$ is a locally convex and complete quasi $*$-algebra. For an overview on partial $*$-algebra and related topics we refer to [2].

In a series of papers [7, 15, 17, 18], many authors have considered a special class of quasi $*$-algebras, called *proper $CQ^*$-algebras*, which arise as completions of $C^*$-algebras. They can be introduced in the following way:

Let $A$ be a right Banach module over the $C^*$-algebra $A_0$ with involution $*$ and $C^*$-norm $\|\cdot\|_0$ such that $A_0 \subset A$. We say that $(A,A_0)$ is a *proper $CQ^*$-algebra* if

1. $A_0$ is dense in $A$ with respect to its norm $\|\cdot\|$;
2. $(ab)^* = b^*a^*$ for all $a,b \in A_0$;
3. $\|y\|_0 = \sup_{a\in A,\|a\|\le 1} \|ay\|$ for all $y \in A_0$.

Several mathematicians have contributed works on these subjects (see [1, 4–10, 12–16, 19, 20, 24, 26, 33–35, 56, 57, 59, 60]).

Ulam [61] gave a talk before the Mathematics Club of the University of Wisconsin in which he discussed a number of unsolved problems. Among these was the following question concerning the stability of homomorphisms.

*We are given a group $G$ and a metric group $G'$ with metric $\rho(\cdot,\cdot)$. Given $\varepsilon > 0$, does there exist a $\delta > 0$ such that if $f : G \to G'$ satisfies $\rho(f(xy), f(x)f(y)) < \delta$ for all $x, y \in G$, then a homomorphism $h : G \to G'$ exists with $\rho(f(x), h(x)) < \varepsilon$ for all $x \in G$?*

By now an affirmative answer has been given in several cases, and some interesting variations of the problem have also been investigated.

Hyers [30] considered the case of approximately additive mappings $f : E \to E'$, where $E$ and $E'$ are Banach spaces and $f$ satisfies *Hyers inequality*

$$\|f(x+y) - f(x) - f(y)\| \leq \varepsilon$$

for all $x, y \in E$. It was shown that the limit

$$L(x) = \lim_{n\to\infty} \frac{f(2^n x)}{2^n}$$

exists for all $x \in E$ and that $L : E \to E'$ is the unique additive mapping satisfying

$$\|f(x) - L(x)\| \leq \varepsilon.$$

Rassias [47] provided the following generalization of Hyers' Theorem which allows the *Cauchy difference to be unbounded* (in this way he rediscovered a much earlier result of Aoki [3]).

**Theorem 19.1 (Rassias).** *Let $f : E \to E'$ be a mapping from a normed vector space $E$ into a Banach space $E'$ subject to the inequality*

$$\|f(x+y) - f(x) - f(y)\| \leq \varepsilon(\|x\|^p + \|y\|^p) \tag{19.1}$$

*for all $x, y \in E$, where $\varepsilon$ and $p$ are real constants with $\varepsilon > 0$ and $p < 1$. Then the limit*

$$L(x) = \lim_{n\to\infty} \frac{f(2^n x)}{2^n}$$

*exists for all $x \in E$ and $L : E \to E'$ is the unique additive mapping which satisfies*

$$\|f(x) - L(x)\| \leq \frac{2\varepsilon}{2 - 2^p}\|x\|^p \tag{19.2}$$

*for all $x \in E$. If $p < 0$ then inequality (19.1) holds for $x, y \neq 0$ and (19.2) for $x \neq 0$.*

Rassias [48] during the 27th International Symposium on Functional Equations asked the question whether such a theorem can also be proved for $p \geq 1$. Gajda [27] following the same approach as in Rassias [47], gave an affirmative solution to this question for $p > 1$. It was shown by Gajda [27], as well as by Rassias and Šemrl [53] that one cannot prove a Rassias' type theorem when $p = 1$. The counterexamples of Gajda [27], as well as of Rassias and Šemrl [53] have stimulated several mathematicians to invent new definitions of *approximately additive* or *approximately linear* mappings. The inequality (19.1) that was introduced for the first time by Aoki [3] and next by Rassias [47] provided a lot of influence in the development of a generalization of the Hyers–Ulam stability concept.This new concept is known as *Hyers–Ulam–Rassias stability* of functional equations (cf. the books of Czerwik [21, 22] and Hyers et al. [31]; see also Găvruta [28], who among others studied the Hyers–Ulam–Rassias stability of functional equations).

Beginning around the year 1980, the topic of approximate homomorphisms and their stability theory in the field of functional equations and inequalities was taken up by several mathematicians (cf. Hyers and Rassias [32], Rassias [51] and the references therein).

Rassias [43] following the spirit of the innovative approach of Rassias [47] for the unbounded Cauchy difference proved a similar stability theorem in which he replaced the factor $\|x\|^p + \|y\|^p$ by $\|x\|^p \cdot \|y\|^q$ for $p, q \in \mathbb{R}$ with $p + q \neq 1$ (see also [44] for a number of other new results). Several mathematician have contributed works on these subjects (see [36–39, 45, 46, 49–52, 55]).

A $C^*$-algebra $\mathscr{C}$, endowed with the Jordan product

$$x \circ y := \frac{xy + yx}{2}$$

on $\mathscr{C}$, is called a *Jordan $C^*$-algebra* (see [37, 38]). A $C^*$-algebra $A$, endowed with the Jordan triple product

$$\{z, x, w\} = \frac{1}{2}(zx^*w + wx^*z)$$

for all $z, x, w \in A$, is called a *$JC^*$-triple* (see [25]). Note that

$$\{z, x, w\} = (z \circ x^*) \circ w + (w \circ x^*) \circ z - (z \circ w) \circ x^*.$$

**Definition 19.1.** A proper $CQ^*$-algebra $(A, A_0)$, endowed with the Jordan triple product

$$\{z, x, w\} = \frac{1}{2}(zx^*w + wx^*z)$$

for all $x \in A$ and all $z, w \in A_0$, is called a *proper $JCQ^*$-triple*, and denoted by $(A, A_0, \{\cdot,\cdot,\cdot\})$.

**Definition 19.2.** Let $(A, A_0\{\cdot,\cdot,\cdot\})$ and $(B, B_0\{\cdot,\cdot,\cdot\})$ be proper $JCQ^*$-triples.

(i) A $\mathbb{C}$-linear mapping $H : A \to B$ is called a *proper $JCQ^*$-triple homomorphism* if $H(z) \in B_0$ and

$$H(\{z, x, w\}) = \{H(z), H(x), H(w)\}$$

for all $z, w \in A_0$ and all $x \in A$. If, in addition, the mapping $H : A \to B$ and the mapping $H|_{A_0} : A_0 \to B_0$ are bijective, then the mapping $H : A \to B$ is called a *proper $JCQ^*$-triple isomorphism*.

(ii) A $\mathbb{C}$-linear mapping $\delta : A_0 \to A$ is called a *proper $JCQ^*$-triple derivation* if

$$\delta(\{w_0, w_1, w_2\}) = \{\delta(w_0), w_1, w_2\} + \{w_0, \delta(w_1), w_2\} + \{w_0, w_1, \delta(w_2)\}$$

for all $w_0, w_1, w_2 \in A_0$.

This paper is organized as follows: In Sects. 2 and 3, we investigate homomorphisms and derivations in proper $JCQ^*$-triples for the 3-variable Jensen functional equation.

In Sects. 4 and 6, we prove the Hyers–Ulam–Rassias stability of homomorphisms and of derivations in proper $JCQ^*$-triples. In Sect. 5, we investigate isomorphisms between proper $JCQ^*$-triples.

## 19.2 Homomorphisms in Proper $JCQ^*$-Triples

Throughout this section, assume that $(A, A_0, \{\cdot,\cdot,\cdot\})$ is a proper $JCQ^*$-triple with $C^*$-norm $\|\cdot\|_{A_0}$ and norm $\|\cdot\|_A$, and that $(B, B_0, \{\cdot,\cdot,\cdot\})$ is a proper $JCQ^*$-triple with $C^*$-norm $\|\cdot\|_{B_0}$ and norm $\|\cdot\|_B$.

We investigate homomorphisms in proper $JCQ^*$-triples.

**Theorem 19.2.** *Let $r \neq 1$ and $\theta$ be nonnegative real numbers, and $f : A \to B$ a mapping satisfying $f(w) \in B_0$ for all $w \in A_0$ such that*

$$\|\mu f(x) + f(y) + f(z)\|_B \leq \left\| 2f\left(\frac{\mu x + y + z}{2}\right)\right\|_B, \tag{19.3}$$

$$\|f(\{w_0, x, w_1\}) - \{f(w_0), f(x), f(w_1)\}\|_B \leq \theta(\|w_0\|_A^{3r} + \|x\|_A^{3r} + \|w_1\|_A^{3r}) \tag{19.4}$$

*for all $\mu \in \mathbb{T}^1 := \{\lambda \in \mathbb{C}\colon |\lambda| = 1\}$, all $w_0, w_1 \in A_0$ and all $x, y, z \in A$. Then the mapping $f : A \to B$ is a proper $JCQ^*$-triple homomorphism.*

*Proof.* Let $\mu = 1$ in (19.3). By [42, Proposition 2.1], the mapping $f : A \to B$ is Cauchy additive.

Letting $z = 0$ and $y = -\mu x$ in (19.3), we get

$$\mu f(x) - f(\mu x) = \mu f(x) + f(-\mu x) = 0$$

for all $x \in A$. So $f(\mu x) = \mu f(x)$ for all $x \in A$. By the same reasoning as in the proof of [37, Theorem 2.1], the mapping $f : A \to B$ is $\mathbb{C}$-linear.

(i) Assume that $r < 1$. By (19.4),

$$\begin{aligned}\|f(\{w_0, x, w_1\}) - \{f(w_0), f(x), f(w_1)\}\|_B &= \lim_{n\to\infty} \frac{1}{8^n}\|f(8^n\{w_0, x, w_1\}) - \{f(2^n w_0), f(2^n x), f(2^n w_1)\}\|_B \\ &\leq \lim_{n\to\infty} \frac{8^{nr}}{8^n}\theta(\|w_0\|_A^{3r} + \|x\|_A^{3r} + \|w_1\|_A^{3r}) = 0\end{aligned}$$

for all $w_0, w_1 \in A_0$ and all $x \in A$. So

$$f(\{w_0, x, w_1\}) = \{f(w_0), f(x), f(w_1)\}$$

for all $w_0, w_1 \in A_0$ and all $x \in A$.

(ii) Assume that $r > 1$. By a similar method to the proof of the case (i), one can prove that the mapping $f : A \to B$ satisfies

$$f(\{w_0, x, w_1\}) = \{f(w_0), f(x), f(w_1)\}$$

for all $w_0, w_1 \in A_0$ and all $x \in A$.

Since $f(w) \in B_0$ for all $w \in A_0$, the mapping $f : A \to B$ is a proper $JCQ^*$-triple homomorphism, as desired. □

**Theorem 19.3.** *Let $r \neq 1$ and $\theta$ be nonnegative real numbers, and $f : A \to B$ a mapping satisfying* (19.3) *and $f(w) \in B_0$ for all $w \in A_0$ such that*

$$\|f(\{w_0, x, w_1\}) - \{f(w_0), f(x), f(w_1)\}\|_B \leq \theta \, \|w_0\|_A^r \, \|x\|_A^r \, \|w_1\|_A^r \qquad (19.5)$$

*for all $w_0, w_1 \in A_0$ and all $x \in A$. Then the mapping $f : A \to B$ is a proper $JCQ^*$-triple homomorphism.*

*Proof.* By the same reasoning as in the proof of Theorem 19.2, the mapping $f : A \to B$ is $\mathbb{C}$-linear.

(i) Assume that $r < 1$. By (19.5),

$$\begin{aligned} \|f(\{w_0, x, w_1\}) &- \{f(w_0), f(x), f(w_1)\}\|_B \\ &= \lim_{n\to\infty} \frac{1}{8^n} \|f(8^n\{w_0, x, w_1\}) - \{f(2^n w_0), f(2^n x), f(2^n w_1)\}\|_B \\ &\leq \lim_{n\to\infty} \frac{8^{nr}}{8^n} \theta \cdot \|w_0\|_A^r \cdot \|x\|_A^r \cdot \|w_1\|_A^r = 0 \end{aligned}$$

for all $w_0, w_1 \in A_0$ and all $x \in A$. So

$$f(\{w_0, x, w_1\}) = \{f(w_0), f(x), f(w_1)\}$$

for all $w_0, w_1 \in A_0$ and all $x \in A$.

(ii) Assume that $r > 1$. By a similar method to the proof of the case (i), one can prove that the mapping $f : A \to B$ satisfies

$$f(\{w_0, x, w_1\}) = \{f(w_0), f(x), f(w_1)\}$$

for all $w_0, w_1 \in A_0$ and all $x \in A$.

Therefore, the mapping $f : A \to B$ is a proper $JCQ^*$-triple homomorphism.

## 19.3 Derivations on Proper $JCQ^*$-Triples

Throughout this section, assume that $(A, A_0, \{\cdot,\cdot,\cdot\})$ is a proper $JCQ^*$-triple with $C^*$-norm $\|\cdot\|_{A_0}$ and norm $\|\cdot\|_A$.

We investigate derivations on proper $JCQ^*$-triples.

**Theorem 19.4.** *Let $r \neq 1$ and $\theta$ be nonnegative real numbers, and $f : A_0 \to A$ a mapping such that*

$$\|\mu f(x) + f(y) + f(z)\|_A \le \left\| 2f\left(\frac{\mu x + y + z}{2}\right)\right\|_A, \tag{19.6}$$

$$\begin{aligned} \|f(\{w_0,w_1,w_2\}) &- \{f(w_0),w_1,w_2\} - \{w_0,f(w_1),w_2\} \\ &- \{w_0,w_1,f(w_2)\}\|_A \le \theta(\|w_0\|_A^{3r} + \|w_1\|_A^{3r} + \|w_2\|_A^{3r}) \end{aligned} \tag{19.7}$$

*for all $\mu \in \mathbb{T}^1$ and all $w_0, w_1, w_2, x, y, z \in A_0$. Then the mapping $f : A_0 \to A$ is a proper $JCQ^*$-triple derivation.*

*Proof.* By the same reasoning as in the proof of Theorem 19.2, the mapping $f : A_0 \to A$ is $\mathbb{C}$-linear.

(i) Assume that $r < 1$. By (19.7),

$$\begin{aligned} \|f(\{w_0,w_1,w_2\}) &- \{f(w_0),w_1,w_2\} - \{w_0,f(w_1),w_2\} - \{w_0,w_1,f(w_2)\}\|_A \\ &= \lim_{n\to\infty} \frac{1}{8^n}\|f(8^n\{w_0,w_1,w_2\}) - \{f(2^n w_0),2^n w_1,2^n w_2\} \\ &\quad -\{2^n w_0, f(2^n w_1), 2^n w_2\} - \{2^n w_0, 2^n w_1, f(2^n w_2)\}\|_A \\ &\le \lim_{n\to\infty} \frac{8^{nr}}{8^n}\theta(\|w_0\|_A^{3r} + \|w_1\|_A^{3r} + \|w_2\|_A^{3r}) = 0 \end{aligned}$$

for all $w_0, w_1, w_2 \in A_0$. So

$$f(\{w_0,w_1,w_2\}) = \{f(w_0),w_1,w_2\} + \{w_0,f(w_1),w_2\} + \{w_0,w_1,f(w_2)\}$$

for all $w_0, w_1, w_2 \in A_0$.

(ii) Assume that $r > 1$. By a similar method to the proof of the case (i), one can prove that the mapping $f : A \to A$ satisfies

$$f(\{w_0,w_1,w_2\}) = \{f(w_0),w_1,w_2\} + \{w_0,f(w_1),w_2\} + \{w_0,w_1,f(w_2)\}$$

for all $w_0, w_1, w_2 \in A_0$.

Therefore, the mapping $f : A_0 \to A$ is a proper $JCQ^*$-triple derivation. □

**Theorem 19.5.** *Let $r \neq 1$ and $\theta$ be nonnegative real numbers, and $f : A_0 \to A$ a mapping satisfying* (19.6) *such that*

$$\|f(\{w_0, w_1, w_2\}) - \{f(w_0), w_1, w_2\} - \{w_0, f(w_1), w_2\} - \{w_0, w_1, f(w_2)\}\|_A \leq \theta \|w_0\|_A^r \|w_1\|_A^r \|w_2\|_A^r \quad (19.8)$$

*for all $w_0, w_1, w_2 \in A_0$. Then the mapping $f : A_0 \to A$ is a proper $JCQ^*$-triple derivation.*

*Proof.* The proof is similar to the proofs of Theorems 19.2 and 19.4. □

## 19.4 Stability of Homomorphisms in Proper $JCQ^*$-Triples

We prove the Hyers–Ulam–Rassias stability of homomorphisms in proper $JCQ^*$-triples.

**Theorem 19.6.** *Let $r < 1$ and $\theta$ be nonnegative real numbers, and let $f : A \to B$ be a mapping such that $f(w) \in B_0$ for all $w \in A_0$ and*

$$\left\|2f\left(\frac{\mu x + \mu y + \mu z}{2}\right) - \mu f(x) - \mu f(y) - \mu f(z)\right\|_B \leq \theta \|x\|_A^{r/3} \|y\|_A^{r/3} \|z\|_A^{r/3}, \quad (19.9)$$

$$\left\|2f\left(\frac{w_0 + w_1 + w_2}{2}\right) - f(w_0) - f(w_1) - f(w_2)\right\|_B \leq \theta \|w_0\|_{A_0}^{r/3} \|w_1\|_{A_0}^{r/3} \|w_2\|_{A_0}^{r/3}, \quad (19.10)$$

$$\|f(\{w_0, x, w_1\}) - \{f(w_0), f(x), f(w_1)\}\|_B \leq \theta \|w_0\|_A^r \|x\|_A^r \|w_1\|_A^r \quad (19.11)$$

*for all $\mu \in \mathbb{T}^1$, all $w_0, w_1, w_2 \in A_0$ and all $x, y, z \in A$. Then there exists a unique proper $JCQ^*$-triple homomorphism $H : A \to B$ such that*

$$\|f(x) - H(x)\|_B \leq \frac{2^r \theta}{3 \times 2^r - 2 \times 3^r} \|x\|_A^r \quad (19.12)$$

*for all $x \in A$.*

*Proof.* Let us assume $\mu = 1$ and $x = y = z$ in (19.9). Then we get

$$\left\|2f\left(\frac{3x}{2}\right) - 3f(x)\right\|_B \leq \theta \|x\|_A^r \quad (19.13)$$

for all $x \in A$. So

$$\left\| f(x) - \frac{2}{3} f\left(\frac{3x}{2}\right) \right\|_B \leq \frac{\theta}{3} \|x\|_A^r$$

for all $x \in A$. Hence

$$\begin{aligned} &\left\| \frac{2^l}{3^l} f\left(\frac{3^l x}{2^l}\right) - \frac{2^m}{3^m} f\left(\frac{3^m x}{2^m}\right) \right\|_B \\ &\qquad \leq \sum_{j=l}^{m-1} \left\| \frac{2^j}{3^j} f\left(\frac{3^j x}{2^j}\right) - \frac{2^{j+1}}{3^{j+1}} f\left(\frac{3^{j+1} x}{2^{j+1}}\right) \right\|_B \\ &\qquad \leq \frac{\theta}{3} \sum_{j=l}^{m-1} \frac{2^j 3^{rj}}{3^j 2^{rj}} \|x\|_A^r \end{aligned} \tag{19.14}$$

for all nonnegative integers $m$ and $l$ with $m > l$ and all $x \in A$. From this it follows that the sequence $\{2^n 3^{-n} f(3^n 2^{-n} x)\}$ is a Cauchy sequence for all $x \in A$. Since $B$ is complete, the sequence $\{2^n 3^{-n} f(3^n 2^{-n} x)\}$ converges. Thus, one can define the mapping $H : A \to B$ by

$$H(x) := \lim_{n\to\infty} \frac{2^n}{3^n} f\left(\frac{3^n x}{2^n}\right)$$

for all $x \in A$. Moreover, letting $l = 0$ and passing the limit $m \to \infty$ in (19.14), we get (19.12).

It follows from (19.9) that

$$\begin{aligned} &\left\| 2H\left(\frac{\mu(x+y+z)}{2}\right) - \mu H(x) - \mu H(y) - \mu H(z) \right\|_B \\ &\qquad = \lim_{n\to\infty} \frac{2^n}{3^n} \left\| 2f\left(\frac{3^n \mu(x+y+z)}{2^n}\right) - \mu f\left(\frac{3^n x}{2^n}\right) \right. \\ &\qquad\qquad \left. - \mu f\left(\frac{3^n y}{2^n}\right) - \mu f\left(\frac{3^n z}{2^n}\right) \right\|_B \\ &\qquad \leq \lim_{n\to\infty} \frac{2^n 3^{nr} \theta}{3^n 2^{nr}} \|x\|_A^{r/3} \|y\|_A^{r/3} \|z\|_A^{r/3} = 0 \end{aligned}$$

for all $\mu \in \mathbb{T}^1$ and all $x, y, z \in A$. So

$$2H\left(\frac{\mu x + \mu y + \mu z}{2}\right) = \mu H(x) + \mu H(y) + \mu H(z)$$

for all $\mu \in \mathbb{T}^1$ and all $x, y, z \in A$. By the same reasoning as in the proof of [37, Theorem 2.1], the mapping $H : A \to B$ is $\mathbb{C}$-linear.

Now, let $T : A \to B$ be another 3-variable Jensen mapping satisfying (19.12). Then we have

$$\begin{aligned}\|H(x) - T(x)\|_B &= \frac{2^n}{3^n}\left\|H\left(\frac{3^n x}{2^n}\right) - T\left(\frac{3^n x}{2^n}\right)\right\|_B \\ &\le \frac{2^n}{3^n}\left(\left\|H\left(\frac{3^n x}{2^n}\right) - f\left(\frac{3^n x}{2^n}\right)\right\|_B + \left\|T\left(\frac{3^n x}{2^n}\right) - f\left(\frac{3^n x}{2^n}\right)\right\|_B\right) \\ &\le \frac{2^{r+1}2^n 3^{nr}\theta}{3^n 2^{nr}(3\cdot 2^r - 2\cdot 3^r)}\|x\|_A^r,\end{aligned}$$

which tends to zero as $n \to \infty$ for all $x \in A$. So we can conclude that

$$H(x) = T(x)$$

for all $x \in A$. This proves the uniqueness of $H$.

It follows from (19.10) that

$$H(w) = \lim_{n\to\infty} \frac{2^n}{3^n} f\left(\frac{3^n w}{2^n}\right) \in B_0$$

for all $w \in A_0$. So it follows from (19.11) that

$$\begin{aligned}&\|H(\{w_0, x, w_1\}) - \{H(w_0), H(x), H(w_1)\}\|_B \\ &\quad = \lim_{n\to\infty} \frac{2^{3n}}{3^{3n}}\left\| f\left(\frac{3^{3n}\{w_0, x, w_1\}}{2^{3n}}\right)\right. \\ &\qquad \left. - \left\{f\left(\frac{3^n w_0}{2^n}\right), f\left(\frac{3^n x}{2^n}\right), f\left(\frac{3^n w_1}{2^n}\right)\right\}\right\|_B \\ &\quad \le \lim_{n\to\infty} \frac{2^{3n}3^{3nr}}{3^{3n}2^{3nr}}\,\theta\,\|w_0\|_A^r\,\|x\|_A^r\,\|w_1\|_A^r = 0\end{aligned}$$

for all $w_0, w_1 \in A_0$ and all $x \in A$. So

$$H(\{w_0, x, w_1\}) = \{H(w_0), H(x), H(w_1)\}$$

for all $w_0, w_1 \in A_0$ and all $x \in A$.

Thus, the mapping $H : A \to B$ is a unique proper $JCQ^*$-triple homomorphism satisfying (19.12), as desired. □

**Theorem 19.7.** *Let $r > 1$ and $\theta$ be nonnegative real numbers, and let $f : A \to B$ be a mapping satisfying* (19.9)–(19.11) *such that $f(w) \in B_0$ for all $w \in A_0$. Then there exists a unique proper $JCQ^*$-triple homomorphism $H : A \to B$ such that*

$$\|f(x) - H(x)\|_B \leq \frac{2^r\theta}{2 \times 3^r - 3 \times 2^r} \|x\|_A^r \tag{19.15}$$

*for all $x \in A$.*

*Proof.* It follows from (19.13) that

$$\left\| f(x) - \frac{3}{2} f\left(\frac{2x}{3}\right) \right\|_B \leq \frac{2^r\theta}{2 \times 3^r} \|x\|_A^r$$

for all $x \in A$. So

$$\begin{aligned}
&\left\| \frac{3^l}{2^l} f\left(\frac{2^l x}{3^l}\right) - \frac{3^m}{2^m} f\left(\frac{2^m x}{3^m}\right) \right\|_B \\
&\qquad \leq \sum_{j=l}^{m-1} \left\| \frac{3^j}{2^j} f\left(\frac{2^j x}{3^j}\right) - \frac{3^{j+1}}{2^{j+1}} f\left(\frac{2^{j+1} x}{3^{j+1}}\right) \right\|_B \\
&\qquad \leq \frac{2^r\theta}{2 \times 3^r} \sum_{j=l}^{m-1} \frac{3^j 2^{jr}}{2^j 3^{jr}} \|x\|_A^r
\end{aligned} \tag{19.16}$$

for all nonnegative integers $m$ and $l$ with $m > l$ and all $x \in A$. From this it follows that the sequence $\{3^n 2^{-n} f(2^n 3^{-n} x)\}$ is a Cauchy sequence for all $x \in A$. Since $B$ is complete, the sequence $\{3^n 2^{-n} f(2^n 3^{-n} x)\}$ converges. So one can define the mapping $H : A \to B$ by

$$H(x) := \lim_{n \to \infty} \frac{3^n}{2^n} f\left(\frac{2^n x}{3^n}\right)$$

for all $x \in A$. Moreover, letting $l = 0$ and passing the limit $m \to \infty$ in (19.16), we get (19.15).

The rest of the proof is similar to the proof of Theorem 19.6. □

## 19.5 Isomorphisms Between Proper $JCQ^*$-Triples

We investigate isomorphisms between proper $JCQ^*$-triples.

**Theorem 19.8.** *Let $r < 1$ and $\theta$ be nonnegative real numbers, and let $f : A \to B$ be a bijective mapping satisfying* (19.9) *and* (19.10) *such that*

$$f(\{w_0, x, w_1\}) = \{f(w_0), f(x), f(w_1)\} \tag{19.17}$$

*for all* $w_0, w_1 \in A_0$ *and all* $x \in A$. *If*

$$\lim_{n\to\infty} \frac{2^n}{3^n} f\left(\frac{3^n e}{2^n}\right) = e'$$

*and* $f|_{A_0} : A_0 \to B_0$ *is bijective, then the mapping* $f : A \to B$ *is a proper* $JCQ^*$*-triple isomorphism.*

*Proof.* By the same reasoning as in the proof of Theorem 19.6, there is a unique $\mathbb{C}$-linear mapping $H : A \to B$ satisfying (19.12). The mapping $H : A \to B$ is given by

$$H(x) := \lim_{n\to\infty} \frac{2^n}{3^n} f\left(\frac{3^n x}{2^n}\right)$$

for all $x \in A$.

Since $f(\{w_0, x, w_1\}) = \{f(w_0), f(x), f(w_1)\}$ for all $w_0, w_1 \in A_0$ and all $x \in A$,

$$\begin{aligned} H(\{w_0, x, w_1\}) &= \lim_{n\to\infty} \frac{2^{3n}}{3^{3n}} \left\{ f\left(\frac{3^n w_0}{2^n}\right), f\left(\frac{3^n x}{2^n}\right), f\left(\frac{3^n w_1}{2^n}\right) \right\} \\ &= \lim_{n\to\infty} \left\{ \frac{2^n}{3^n} f\left(\frac{3^n w_0}{2^n}\right), \frac{2^n}{3^n} f\left(\frac{3^n x}{2^n}\right), \frac{2^n}{3^n} f\left(\frac{3^n w_1}{2^n}\right) \right\} \\ &= \{H(w_0), H(x), H(w_1)\} \end{aligned}$$

for all $w_0, w_1 \in A_0$ and all $x \in A$.

It follows from (19.10) that

$$H(w) = \lim_{n\to\infty} (3k)^n f\left(\frac{w}{(3k)^n}\right) \in B_0$$

for all $w \in A_0$. So the mapping $H : A \to B$ is a proper $JCQ^*$-triple homomorphism.

By the assumption, for all $x \in A$,

$$\begin{aligned} H(x) &= H(ex) \\ &= \lim_{n\to\infty} \frac{2^n}{3^n} f\left(\frac{3^n ex}{2^n}\right) \\ &= \lim_{n\to\infty} \frac{2^n}{3^n} f\left(\frac{3^n e}{2^n} x\right) \\ &= \lim_{n\to\infty} \frac{2^n}{3^n} f\left(\frac{3^n e}{2^n}\right) f(x) \\ &= e' f(x) = f(x)\,. \end{aligned}$$

Hence, the bijective mapping $f : A \to B$ is a proper $JCQ^*$-triple isomorphism. □

**Theorem 19.9.** *Let $r > 1$ and $\theta$ be nonnegative real numbers, and let $f : A \to B$ be a bijective mapping satisfying (19.9), (19.10) and (19.17). If*

$$\lim_{n\to\infty} \frac{3^n}{2^n} f\left(\frac{2^n e}{3^n}\right) = e'$$

*and $f|_{A_0} : A_0 \to B_0$ is bijective, then the mapping $f : A \to B$ is a proper $JCQ^*$-triple isomorphism.*

*Proof.* By the same reasoning as in the proof of Theorem 19.7, there is a unique $\mathbb{C}$-linear mapping $H : A \to B$ satisfying (19.15). The mapping $H : A \to B$ is given by

$$H(x) := \lim_{n\to\infty} \frac{3^n}{2^n} f\left(\frac{2^n x}{3^n}\right)$$

for all $x \in A$.

The rest of the proof is similar to the proof of Theorem 19.8. □

## 19.6 Stability of Derivations on Proper $JCQ^*$-Triples

We prove the Hyers–Ulam–Rassias stability of derivations on proper $JCQ^*$-triples.

**Theorem 19.10.** *Let $r < 1$ and $\theta$ be nonnegative real numbers, and let $f : A_0 \to A$ be a mapping such that*

$$\left\| 2f\left(\frac{\mu(x+y+z)}{2}\right) - \mu f(x) - \mu f(y) - \mu f(z) \right\|_A \leq \theta \, \|x\|_{A_0}^{r/3} \, \|y\|_{A_0}^{r/3} \, \|z\|_{A_0}^{r/3}), \tag{19.18}$$

$$\|f(\{w_0, w_1, w_2\}) - \{f(w_0), w_1, w_2\} - \{w_0, f(w_1), w_2\} - \{w_0, w_1, f(w_2)\}\|_A \leq \theta \, \|w_0\|_A^{r/3} \, \|w_1\|_A^{r/3} \, \|w_2\|_A^{r/3} \tag{19.19}$$

*for all $\mu \in \mathbb{T}^1$ and all $w_0, w_1, w_2, x, y, z \in A_0$. Then there exists a unique proper $JCQ^*$-triple derivation $\delta : A_0 \to A$ such that*

$$\|f(w) - \delta(w)\|_A \leq \frac{2^r \theta}{3 \times 2^r - 2 \times 3^r} \|w\|_{A_0}^r \tag{19.20}$$

*for all $w \in A_0$.*

*Proof.* By the same reasoning as in the proof of Theorem 19.6, there exists a unique $\mathbb{C}$-linear mapping $\delta : A_0 \to A$ satisfying (19.20). The mapping $\delta : A_0 \to A$ is defined by

$$\delta(w) := \lim_{n\to\infty} \frac{2^n}{3^n} f\left(\frac{3^n w}{2^n}\right)$$

for all $w \in A_0$.

It follows from (19.19) that

$$\begin{aligned}
&\|\delta(\{w_0, w_1, w_2\}) - \{\delta(w_0), w_1, w_2\} - \{w_0, \delta(w_1), w_2\} - \{w_0, w_1, \delta(w_2)\}\|_A \\
&= \lim_{n\to\infty} \frac{2^{3n}}{3^{3n}} \left\| f\left(\frac{3^{3n}\{w_0, w_1, w_2\}}{2^{3n}}\right) - \left\{ f\left(\frac{3^n w_0}{2^n}\right), \frac{3^n w_1}{2^n}, \frac{3^n w_2}{2^n} \right\} \right. \\
&\quad \left. - \left\{ \frac{3^n w_0}{2^n}, f\left(\frac{3^n w_1}{2^n}\right), \frac{3^n w_2}{2^n} \right\} - \left\{ \frac{3^n w_0}{2^n}, \frac{3^n w_1}{2^n}, f\left(\frac{3^n w_2}{2^n}\right) \right\} \right\|_A \\
&\le \lim_{n\to\infty} \frac{2^{3n} 3^{3nr}}{3^{3n} 2^{3nr}} \theta \|w_0\|_A^r \|w_1\|_A^r \|w_2\|_A^r = 0
\end{aligned}$$

for all $w_0, w_1, w_2 \in A_0$. So

$$\delta(\{w_0, w_1, w_2\}) = \{\delta(w_0), w_1, w_2\} + \{w_0, \delta(w_1), w_2\} + \{w_0, w_1, \delta(w_2)\}$$

for all $w_0, w_1, w_2 \in A_0$.

Thus the mapping $\delta : A_0 \to A$ is a unique proper $JCQ^*$-triple derivation satisfying (19.20), as desired. □

**Theorem 19.11.** *Let $r > 1$ and $\theta$ be nonnegative real numbers, and let $f : A_0 \to A$ be a mapping satisfying* (19.18) *and* (19.19)*. Then there exists a unique proper $JCQ^*$-triple derivation $\delta : A_0 \to A$ such that*

$$\|f(w) - \delta(w)\|_A \le \frac{2^r \theta}{2 \times 3^r - 3 \times 2^r} \|w\|_{A_0}^r$$

*for all $w \in A_0$.*

*Proof.* The proof is similar to the proofs of Theorems 19.6, 19.7 and 19.10. □

## References


1. Alli, G., Sewell, G.L.: New methods and structures in the theory of the multi-mode Dicke laser model. J. Math. Phys. **36**, 5598–5626 (1995)
2. Antoine, J.P., Inoue, A., Trapani, C.: Partial ∗-Algebras and Their Operator Realizations. Kluwer, Dordrecht (2002)
3. Aoki, T.: On the stability of the linear transformation in Banach spaces. J. Math. Soc. Japan **2**, 64–66 (1950)

4. Bagarello, F.: Applications of topological $*$-algebras of unbounded operators. J. Math. Phys. **39**, 6091–6105 (1998)
5. Bagarello, F.: Fixed point results in topological $*$-algebras of unbounded operators. Publ. RIMS Kyoto Univ. **37**, 397–418 (2001)
6. Bagarello, F.: Applications of topological $*$-algebras of unbounded operators to modified quons. Nuovo Cimento B **117**, 593–611 (2002)
7. Bagarello, F., Inoue, A., Trapani, C.: Some classes of topological quasi $*$-algebras. Proc. Amer. Math. Soc. **129**, 2973–2980 (2001)
8. Bagarello, F., Inoue, A., Trapani, C.: $*$-derivations of quasi-$*$-algebras. Int. J. Math. Math. Sci. **21**, 1077–1096 (2004)
9. Bagarello, F., Inoue, A., Trapani, C.: Exponentiating derivations of quasi-$*$-algebras: possible approaches and applications. Int. J. Math. Math. Sci. **2005**, 2805–2820 (2005)
10. Bagarello, F., Karwowski, W.: Partial $*$-algebras of closed linear operators in Hilbert space. Publ. RIMS Kyoto Univ. **21**, 205–236 (1985); **22**, 507–511 (1986)
11. Bagarello, F., Morchio, G.: Dynamics of mean-field spin models from basic results in abstract differential equations. J. Stat. Phys. **66**, 849–866 (1992)
12. Bagarello, F., Sewell, G.L.: New structures in the theory of the laser model *II*: Microscopic dynamics and a non-equilibrium entropy principle. J. Math. Phys. **39**, 2730–2747 (1998)
13. Bagarello, F., Trapani, C.: Almost mean field Ising model: an algebraic approach. J. Stat. Phys. **65**, 469–482 (1991)
14. Bagarello, F., Trapani, C.: A note on the algebraic approach to the "almost" mean field Heisenberg model. Nuovo Cimento B **108**, 779–784 (1993)
15. Bagarello, F., Trapani, C.: States and representations of $CQ^*$-algebras. Ann. Inst. H. Poincaré **61**, 103–133 (1994)
16. Bagarello, F., Trapani, C.: The Heisenberg dynamics of spin systems: a quasi-$*$-algebras approach. J. Math. Phys. **37**, 4219–4234 (1996)
17. Bagarello, F., Trapani, C.: $CQ^*$-algebras: structure properties. Publ. RIMS Kyoto Univ. **32**, 85–116 (1996)
18. Bagarello, F., Trapani, C.: Morphisms of certain Banach $C^*$-modules. Publ. RIMS Kyoto Univ. **36**, 681–705 (2000)
19. Bagarello, F., Trapani, C.: Algebraic dynamics in $O^*$-algebras: a perturbative approach. J. Math. Phys. **43**, 3280–3292 (2002)
20. Bagarello, F., Trapani, C., Triolo, S.: Quasi $*$-algebras of measurable operators. Studia Math. **172**, 289–305 (2006)
21. Czerwik, S.: Functional Equations and Inequalities in Several Variables. World Scientific Publishing Company, New Jersey–London–Singapore–Hong Kong (2002)
22. Czerwik, S. (ed.): Stability of Functional Equations of Ulam-Hyers-Rassias Type. Hadronic Press, Palm Harbor (2003)
23. Ekhaguere, G.O.S.: Partial $W^*$-dynamical systems. In: Current Topics in Operator Algebras, Proceedings of the Satellite Conference of ICM-90, pp. 202–217, World Scientific, Singapore (1991)
24. G. Epifanio, G., Trapani, C.: Quasi-$*$-algebras valued quantized fields. Ann. Inst. H. Poincaré **46**, 175–185 (1987)
25. Fleming, R.J., Jamison, J.E.: Isometries on Banach Spaces: Function Spaces. Monographs and Surveys in Pure and Applied Mathematics vol. **129**, Chapman & Hall/CRC, Boca Raton–London–New York–Washington D.C. (2003)
26. Fredenhagen, K., Hertel, J.: Local algebras of observables and pointlike localized fields. Commun. Math. Phys. **80**, 555–561 (1981)
27. Gajda, Z.: On stability of additive mappings. Int. J. Math. Math. Sci. **14**, 431–434 (1991)
28. Găvruta, P.: A generalization of the Hyers-Ulam-Rassias stability of approximately additive mappings. J. Math. Anal. Appl. **184**, 431–436 (1994)
29. Haag, R., Kastler, D.: An algebraic approach to quantum field theory. J. Math. Phys. **5**, 848–861 (1964)

30. Hyers, D.H.: On the stability of the linear functional equation. Proc. Nat. Acad. Sci. U.S.A. **27**, 222–224 (1941)
31. Hyers, D.H., Isac, G., Rassias, Th.M.: Stability of Functional Equations in Several Variables. Birkh*ä*user, Basel (1998)
32. Hyers, D.H., Rassias, Th.M.: Approximate homomorphisms. Aequationes Math. **44**, 125–153 (1992)
33. Lassner, G.: Algebras of unbounded operators and quantum dynamics. Physica **124 A**, 471–480 (1984)
34. Morchio, G., Strocchi, F.: Mathematical structures for long range dynamics and symmetry breaking. J. Math. Phys. **28**, 622–635 (1987)
35. Pallu de la Barriére, R.: Algèbres unitaires et espaces d'Ambrose. Ann. Ecole Norm. Sup. **70**, 381–401 (1953)
36. Park, C.: Lie $*$-homomorphisms between Lie $C^*$-algebras and Lie $*$-derivations on Lie $C^*$-algebras. J. Math. Anal. Appl. **293**, 419–434 (2004)
37. Park, C.: Homomorphisms between Poisson $JC^*$-algebras. Bull. Braz. Math. Soc. **36**, 79–97 (2005)
38. Park, C.: Homomorphisms between Lie $JC^*$-algebras and Cauchy-Rassias stability of Lie $JC^*$-algebra derivations. J. Lie Theory **15**, 393–414 (2005)
39. Park, C.: Isomorphisms between unital $C^*$-algebras. J. Math. Anal. Appl. **307**, 753–762 (2005)
40. Park, C.: Approximate homomorphisms on $JB^*$-triples. J. Math. Anal. Appl. **306**, 375–381 (2005)
41. Park, C.: Isomorphisms between $C^*$-ternary algebras. J. Math. Phys. **47**, no. 10, Art. ID 103512, 12 pages (2006)
42. Park, C., Cho, Y., Han, M.: Functional inequalities associated with Jordan-von Neumann type additive functional equations. J. Inequal. Appl. **2007**, Art. ID 41820 (2007)
43. Rassias, J.M.: On approximation of approximately linear mappings by linear mappings. Bull. Sci. Math. **108**, 445–446 (1984)
44. Rassias, J.M.: Solution of a problem of Ulam. J. Approx. Theory **57**, 268–273 (1989)
45. Rassias, J.M.: Refined Hyers-Ulam approximation of approximately Jensen type mappings. Bull. Sci. Math. **131**, 89–98 (2007)
46. Rassias, J.M., Rassias, M.J.: Asymptotic behavior of alternative Jensen and Jensen type functional equations. Bull. Sci. Math. **129**, 545–558 (2005)
47. Rassias, Th.M.: On the stability of the linear mapping in Banach spaces. Proc. Amer. Math. Soc. **72**, 297–300 (1978)
48. Rassias, Th.M.: Problem 16; 2. In: Report of the 27th International Symp. on Functional Equations, p. 309, Aequationes Math. **39**, 292–293, (1990)
49. Rassias, Th.M.: The problem of S.M. Ulam for approximately multiplicative mappings. J. Math. Anal. Appl. **246**, 352–378 (2000)
50. Rassias, Th.M.: On the stability of functional equations in Banach spaces. J. Math. Anal. Appl. **251**, 264–284 (2000)
51. Rassias, Th.M.: On the stability of functional equations and a problem of Ulam. Acta Appl. Math. **62**, 23–130 (2000)
52. Rassias, Th.M.: Functional Equations, Inequalities and Applications. Kluwer Academic Publishers, Dordrecht–Boston–London (2003)
53. Rassias, Th.M., Šemrl, P.: On the Hyers-Ulam stability of linear mappings. J. Math. Anal. Appl. **173**, 325–338 (1993)
54. Sewell, G.L.: Quantum Mechanics and its Emergent Macrophysics. Princeton University Press, Princeton–Oxford (2002)
55. Skof, F.: Proprietà locali e approssimazione di operatori. Rend. Sem. Mat. Fis. Milano **53**, 113–129 (1983)
56. Streater, R.F., Wightman, A.S.: PCT, Spin and Statistics and All That. Benjamin Inc., New York (1964)
57. Thirring, W., Wehrl, A.: On the mathematical structure of the B.C.S.-model. Commun. Math. Phys. **4**, 303–314 (1967)

58. Trapani, C.: Quasi-$*$-algebras of operators and their applications. Rev. Math. Phys. **7**, 1303–1332 (1995)
59. Trapani, C.: Some seminorms on quasi-$*$-algebras. Studia Math. **158**, 99–115 (2003)
60. Trapani, C.: Bounded elements and spectrum in Banach quasi $*$-algebras. Studia Math. **172**, 249–273 (2006)
61. Ulam, S.M.: Problems in Modern Mathematics. Wiley, New York (1960)

# Chapter 20
# Fuzzy Stability of an Additive-Quartic Functional Equation: A Fixed Point Approach

**Choonkil Park and Themistocles M. Rassias**

*Dedicated to the memory of S.M. Ulam*
*on the 100th anniversary of his birth*

**Abstract** Mirmostafaee, Mirzavaziri and Moslehian have investigated the fuzzy stability problems for the Cauchy additive functional equation and the Jensen additive functional equation in fuzzy Banach spaces. Using the fixed point method, we prove the generalized Hyers–Ulam stability of the following additive-quartic functional equation

$$\begin{aligned} f(2x+y)+f(2x-y) = {} & 2f(x+y)+2f(-x-y)+2f(x-y)+2f(y-x) \\ & +14f(x)+10f(-x)-3f(y)-3f(-y) \end{aligned}$$

in fuzzy Banach spaces.

**Keywords** Fuzzy Banach space • Fixed point • Generalized Hyers–Ulam stability • Quartic mapping • Additive mapping

**Mathematics Subject Classification (2000):** Primary 46S40, 39B72; Secondary 39B52, 46S50, 26E50, 47H10

---

C. Park (✉)
Department of Mathematics, Hanyang University, Seoul 133-791, Republic of Korea
e-mail: baak@hanyang.ac.kr

Th.M. Rassias
Department of Mathematics, National Technical University of Athens, Zografou Campus, 15780 Athens, Greece
e-mail: trassias@math.ntua.gr

Th.M. Rassias and J. Brzdęk (eds.), *Functional Equations in Mathematical Analysis*, Springer Optimization and Its Applications 52, DOI 10.1007/978-1-4614-0055-4_20,

## 20.1 Introduction and Preliminaries

Katsaras [17] defined a fuzzy norm on a vector space to construct a fuzzy vector topological structure on the space. Some mathematicians have defined fuzzy norms on a vector space from various points of view [11, 19, 44]. In particular, Bag and Samanta [2], following Cheng and Mordeson [7], gave an idea of fuzzy norm in such a manner that the corresponding fuzzy metric is of Kramosil and Michalek type [18]. They established a decomposition theorem of a fuzzy norm into a family of crisp norms and investigated some properties of fuzzy normed spaces [3].

We use the definition of fuzzy normed spaces given in [2, 22, 23] to investigate a fuzzy version of the generalized Hyers–Ulam stability for the functional equation

$$\begin{aligned} f(2x+y)+f(2x-y) &= 2f(x+y)+2f(-x-y)+2f(x-y)+2f(y-x) \\ &\quad +14f(x)+10f(-x)-3f(y)-3f(-y) \end{aligned} \tag{20.1}$$

in the fuzzy normed vector space setting.

**Definition 20.1 ([2, 22–24]).** Let $X$ be a real vector space. A function $N : X \times \mathbb{R} \to [0,1]$ is called a *fuzzy norm* on $X$ if, for all $x, y \in X$ and all $s, t \in \mathbb{R}$,

$(N_1)$ $N(x,t) = 0$ for $t \leq 0$;
$(N_2)$ $x = 0$ if and only if $N(x,t) = 1$ for all $t > 0$;
$(N_3)$ $N(cx,t) = N\big(x, \frac{t}{|c|}\big)$ if $c \neq 0$;
$(N_4)$ $N(x+y, s+t) \geq \min\{N(x,s), N(y,t)\}$;
$(N_5)$ $N(x,\cdot)$ is a non-decreasing function of $\mathbb{R}$ and $\lim_{t\to\infty} N(x,t) = 1$;
$(N_6)$ for $x \neq 0$, $N(x,\cdot)$ is continuous on $\mathbb{R}$.

The pair $(X,N)$ is called a *fuzzy normed vector space*.

The properties of fuzzy normed vector spaces and examples of fuzzy norms are given in [22, 25].

**Definition 20.2 ([2, 22–24]).** Let $(X,N)$ be a fuzzy normed vector space. A sequence $\{x_n\}$ in $X$ is said to *be convergent* or *converge* if there exists an $x \in X$ such that $\lim_{n\to\infty} N(x_n - x, t) = 1$ for all $t > 0$. In this case, $x$ is called the *limit* of the sequence $\{x_n\}$ and we denote it by $N\text{-}\lim_{n\to\infty} x_n = x$.

**Definition 20.3 ([2, 22, 23]).** Let $(X,N)$ be a fuzzy normed vector space. A sequence $\{x_n\}$ in $X$ is called *Cauchy* if for each $\varepsilon > 0$ and each $t > 0$ there exists an $n_0 \in \mathbb{N}$ such that for all $n \geq n_0$ and all $p > 0$, we have $N(x_{n+p} - x_n, t) > 1 - \varepsilon$.

It is well-known that every convergent sequence in a fuzzy normed vector space is Cauchy. If each Cauchy sequence is convergent, then the fuzzy norm is said to be *complete* and the fuzzy normed vector space is called a *fuzzy Banach space*.

We say that a mapping $f : X \to Y$ between fuzzy normed vector spaces $X$ and $Y$ is continuous at a point $x_0 \in X$ if for each sequence $\{x_n\}$ converging to $x_0$ in $X$, then the sequence $\{f(x_n)\}$ converges to $f(x_0)$. If $f : X \to Y$ is continuous at each $x \in X$, then $f : X \to Y$ is said to be *continuous* on $X$ (see [3]).

The stability problem of functional equations originated from a question of Ulam [43] concerning the stability of group homomorphisms. Hyers [13] gave a first affirmative partial answer to the question of Ulam for Banach spaces. Hyers' Theorem was generalized by Aoki [1] for additive mappings and by Rassias [33] for linear mappings by considering an unbounded Cauchy difference. The paper of Rassias [33] has provided a lot of influence in the development of what we call *generalized Hyers–Ulam stability* or *Hyers–Ulam–Rassias stability* of functional equations. A generalization of the Th.M. Rassias theorem was obtained by Găvruta [12] by replacing the unbounded Cauchy difference by a general control function in the spirit of Th.M. Rassias' approach.

The functional equation

$$f(x+y)+f(x-y)=2f(x)+2f(y)$$

is called a *quadratic functional equation*. In particular, each solution of the quadratic functional equation is said to be a *quadratic function*. A generalized Hyers–Ulam stability problem for the quadratic functional equation was proved by Skof [42] for mappings $f : X \to Y$, where $X$ is a normed space and $Y$ is a Banach space. Cholewa [8] noticed that the theorem of Skof is still true if the relevant domain $X$ is replaced by an Abelian group. Czerwik [9] proved the generalized Hyers–Ulam stability of the quadratic functional equation.

In [20], Lee et al. considered the following quartic functional equation

$$f(2x+y)+f(2x-y)=4f(x+y)+4f(x-y)+24f(x)-6f(y). \tag{20.2}$$

It is easy to show that the function $f(x) = x^4$ satisfies the functional equation (20.2), which is called a *quartic functional equation* and every solution of the quartic functional equation is said to be a *quartic mapping*. The stability problems of several functional equations have been extensively investigated by a number of authors and there are many interesting results concerning this problem (see [14, 16, 28–31, 34–41]).

We recall a fundamental result in fixed point theory.

Let $X$ be a set. A function $d : X \times X \to [0,\infty]$ is called a *generalized metric* on $X$ if $d$ satisfies

(1) $d(x,y) = 0$ if and only if $x = y$;
(2) $d(x,y) = d(y,x)$ for all $x,y \in X$;
(3) $d(x,z) \le d(x,y) + d(y,z)$ for all $x,y,z \in X$.

**Theorem 20.1 ([4, 10]).** *Let $(X,d)$ be a complete generalized metric space and let a mapping $J : X \to X$ be strictly contractive with Lipschitz constant $L < 1$. Then, for each given element $x \in X$, either*

$$d\left(J^n x, J^{n+1} x\right) = \infty$$

*for all nonnegative integers $n$ or there exists a positive integer $n_0$ such that*

(1) $d(J^n x, J^{n+1} x) < \infty$ *for all* $n \ge n_0$;
(2) *the sequence* $\{J^n x\}$ *converges to a fixed point* $y^*$ *of* $J$;

(3) $y^*$ *is the unique fixed point of* $J$ *in the set* $Y = \{y \in X \mid d(J^{n_0}x, y) < \infty\}$;
(4) $d(y, y^*) \leq \frac{1}{1-L} d(y, Jy)$ *for all* $y \in Y$.

In 1996, Isac and Rassias [15] were the first to provide applications of stability theory of functional equations for the proof of new fixed point theorems with applications. By using fixed point methods, the stability problems of several functional equations have been extensively investigated by a number of authors (see [5, 6, 25–27, 32]).

This paper is organized as follows: In Sect. 2, we prove the generalized Hyers–Ulam stability of the additive-quartic functional equation (20.1) in fuzzy Banach spaces for an odd case. In Sect. 3, we prove the generalized Hyers–Ulam stability of the additive-quartic functional equation (20.1) in fuzzy Banach spaces for an even case.

Throughout this paper, assume that $X$ is a vector space and that $(Y, N)$ is a fuzzy Banach space.

## 20.2 Generalized Hyers–Ulam Stability of Functional Equation (20.1): An Odd Case

One can easily show that an even mapping $f : X \to Y$ satisfies (20.1) if and only if the even mapping $f : X \to Y$ is a quartic mapping, i.e.,

$$f(2x+y) + f(2x-y) = 4f(x+y) + 4f(x-y) + 24f(x) - 6f(y),$$

and that an odd mapping $f : X \to Y$ satisfies (20.1) if and only if the odd mapping mapping $f : X \to Y$ is an additive mapping, i.e.,

$$f(x+y) = f(x) + f(y).$$

In this section, we prove the generalized Hyers–Ulam stability of the additive-quartic functional equation (20.1) in fuzzy Banach spaces for an odd case.

For a given mapping $f : X \to Y$, we define

$$\begin{aligned} Cf(x,y) := {} & f(2x+y) + f(2x-y) - 2f(x+y) - 2f(-x-y) - 2f(x-y) \\ & -2f(y-x) - 14f(x) - 10f(-x) + 3f(y) + 3f(-y) \end{aligned}$$

for all $x, y \in X$.

Using the fixed point method, we prove the generalized Hyers–Ulam stability of the functional equation $Cf(x,y) = 0$ in fuzzy Banach spaces: an odd case.

**Theorem 20.2.** *Let* $\varphi : X^2 \to [0, \infty)$ *be a function such that there exists an* $L < 1$ *with*

$$\varphi(x,y) \leq \frac{L}{2} \varphi(2x, 2y)$$

*for all $x,y \in X$. Let $f : X \to Y$ be an odd mapping satisfying*

$$N(Cf(x,y),t) \geq \frac{t}{t+\varphi(x,y)} \tag{20.3}$$

*for all $x,y \in X$ and all $t > 0$. Then*

$$A(x) := N\text{-}\lim_{n\to\infty} 2^n f\left(\frac{x}{2^n}\right)$$

*exists for each $x \in X$ and defines an additive mapping $A : X \to Y$ such that*

$$N(f(x)-A(x),t) \geq \frac{(4-4L)t}{(4-4L)t+L\varphi(x,0)} \tag{20.4}$$

*for all $x \in X$ and all $t > 0$.*

*Proof.* Letting $y = 0$ in (20.3), we get

$$N(2f(2x)-4f(x),t) \geq \frac{t}{t+\varphi(x,0)} \tag{20.5}$$

for all $x \in X$.

Consider the set

$$S := \{g : X \to Y\}$$

and introduce a generalized metric on $S$:

$$d(g,h) = \inf\left\{\mu \in \mathbb{R}_+ : N(g(x)-h(x),\mu t) \geq \frac{t}{t+\varphi(x,0)}, \ \ \forall x \in X, \forall t > 0\right\},$$

where, as usual, $\inf\phi = +\infty$. It is easy to show that $(S,d)$ is complete (see the proof of [21, Lemma 2.1]).

Now we consider the linear mapping $J : S \to S$ such that

$$Jg(x) := 2g\left(\frac{x}{2}\right)$$

for all $x \in X$.

Let $g,h \in S$ be given such that $d(g,h) = \varepsilon$. Then

$$N(g(x)-h(x),\varepsilon t) \geq \frac{t}{t+\varphi(x,0)}$$

for all $x \in X$ and all $t > 0$. Hence

$$\begin{aligned} N(Jg(x)-Jh(x),L\varepsilon t) &= N\left(2g\left(\frac{x}{2}\right)-2h\left(\frac{x}{2}\right),L\varepsilon t\right) \\ &= N\left(g\left(\frac{x}{2}\right)-h\left(\frac{x}{2}\right),\frac{L}{2}\varepsilon t\right) \end{aligned}$$

$$\geq \frac{\frac{Lt}{2}}{\frac{Lt}{2}+\varphi\left(\frac{x}{2},0\right)} \geq \frac{\frac{Lt}{2}}{\frac{Lt}{2}+\frac{L}{2}\varphi(x,0)}$$
$$= \frac{t}{t+\varphi(x,0)}$$

for all $x \in X$ and all $t > 0$. So $d(g,h) = \varepsilon$ implies that $d(Jg,Jh) \leq L\varepsilon$. This means that

$$d(Jg,Jh) \leq Ld(g,h)$$

for all $g,h \in S$.

It follows from (20.5) that

$$N\left(f(x)-2f\left(\frac{x}{2}\right),\frac{L}{4}t\right) \geq \frac{t}{t+\varphi(x,0)}$$

for all $x \in X$ and all $t > 0$. So $d(f,Jf) \leq \frac{L}{4}$.

By Theorem 20.1, there exists a mapping $A : X \to Y$ satisfying the following:

(1) $A$ is a fixed point of $J$, i.e.,

$$A\left(\frac{x}{2}\right) = \frac{1}{2}A(x) \tag{20.6}$$

for all $x \in X$. Since $f : X \to Y$ is odd, $A : X \to Y$ is an odd mapping. The mapping $A$ is a unique fixed point of $J$ in the set

$$M = \{g \in S : d(f,g) < \infty\}.$$

This implies that $A$ is a unique mapping satisfying (20.6) such that there exists a $\mu \in (0,\infty)$ satisfying

$$N(f(x)-A(x),\mu t) \geq \frac{t}{t+\varphi(x,0)}$$

for all $x \in X$;

(2) $d(J^n f,A) \to 0$ as $n \to \infty$. This implies the equality

$$N\text{-}\lim_{n\to\infty} 2^n f\left(\frac{x}{2^n}\right) = A(x)$$

for all $x \in X$;

(3) $d(f,A) \leq \frac{1}{1-L}d(f,Jf)$, which implies the inequality

$$d(f,A) \leq \frac{L}{4-4L}.$$

This implies that inequality (20.4) holds.

By (20.3),

$$N\left(2^nDf\left(\frac{x}{2^n},\frac{y}{2^n}\right),2^nt\right) \geq \frac{t}{t+\varphi(2^{-n}x,2^{-n}y)}$$

for all $x,y \in X$, all $t > 0$ and all $n \in \mathbb{N}$. So

$$N\left(2^nDf\left(\frac{x}{2^n},\frac{y}{2^n}\right),t\right) \geq \frac{2^{-n}t}{2^{-n}t+2^{-n}L^n\varphi(x,y)}$$

for all $x,y \in X$, all $t > 0$ and all $n \in \mathbb{N}$. Since

$$\lim_{n\to\infty} \frac{2^{-n}t}{2^{-n}t+2^{-n}L^n\varphi(x,y)} = 1$$

for all $x,y \in X$ and all $t > 0$,

$$N(CA(x,y),t) = 1$$

for all $x,y \in X$ and all $t > 0$. Thus, the mapping $A : X \to Y$ is additive, as desired. □

**Corollary 20.1.** *Let $\theta \geq 0$ and let $p$ be a real number with $p > 1$. Let $X$ be a normed vector space with norm $\|\cdot\|$. Let $f : X \to Y$ be an odd mapping satisfying*

$$N(Cf(x,y),t) \geq \frac{t}{t+\theta(\|x\|^p+\|y\|^p)} \tag{20.7}$$

*for all $x,y \in X$ and all $t > 0$. Then $A(x) := N\text{-}\lim_{n\to\infty} 2^n f(2^{-n}x)$ exists for each $x \in X$ and defines an additive mapping $A : X \to Y$ such that*

$$N(f(x)-A(x),t) \geq \frac{2(2^p-2)t}{2(2^p-2)t+\theta\|x\|^p}$$

*for all $x \in X$ and all $t > 0$.*

*Proof.* The proof follows from Theorem 20.2 by taking

$$\varphi(x,y) := \theta(\|x\|^p+\|y\|^p)$$

for all $x,y \in X$. Then we can choose $L = 2^{1-p}$ and we get the desired result. □

**Theorem 20.3.** *Let $\varphi : X^2 \to [0,\infty)$ be a function such that there exists a real constant $L < 1$ with*

$$\varphi(x,y) \leq 2L\varphi\left(\frac{x}{2},\frac{y}{2}\right)$$

*for all $x, y \in X$. Let $f : X \to Y$ be an odd mapping satisfying (20.3). Then the limit*

$$A(x) := N\text{-}\lim_{n\to\infty} 2^{-n} f(2^n x)$$

*exists for each $x \in X$ and defines an additive mapping $A : X \to Y$ such that*

$$N(f(x) - A(x), t) \geq \frac{(4-4L)t}{(4-4L)t + \varphi(x,0)} \tag{20.8}$$

*for all $x \in X$ and all $t > 0$.*

*Proof.* Let $(S,d)$ be the generalized metric space defined in the proof of Theorem 20.2.

Consider the linear mapping $J : S \to S$ such that

$$Jg(x) := \frac{1}{2} g(2x)$$

for all $x \in X$.

Let $g, h \in S$ be given such that $d(g,h) = \varepsilon$. Then

$$N(g(x) - h(x), \varepsilon t) \geq \frac{t}{t + \varphi(x,0)}$$

for all $x \in X$ and all $t > 0$. Hence

$$\begin{aligned} N(Jg(x) - Jh(x), L\varepsilon t) &= N\left(\frac{1}{2} g(2x) - \frac{1}{2} h(2x), L\varepsilon t\right) \\ &= N(g(2x) - h(2x), 2L\varepsilon t) \\ &\geq \frac{2Lt}{2Lt + \varphi(2x,0)} \geq \frac{2Lt}{2Lt + 2L\varphi(x,0)} \\ &= \frac{t}{t + \varphi(x,0)} \end{aligned}$$

for all $x \in X$ and all $t > 0$. So $d(g,h) = \varepsilon$ implies that $d(Jg, Jh) \leq L\varepsilon$. This means that

$$d(Jg, Jh) \leq Ld(g,h)$$

for all $g, h \in S$.

It follows from (20.5) that

$$N\left(f(x) - \frac{1}{2} f(2x), \frac{1}{4} t\right) \geq \frac{t}{t + \varphi(x,0)}$$

for all $x \in X$ and all $t > 0$. So

$$d(f, Jf) \leq \frac{1}{4}.$$

By Theorem 20.1, there exists a mapping $A : X \to Y$ satisfying the following:

1. $A$ is a fixed point of $J$, i.e.,

$$A(2x) = 2A(x) \tag{20.9}$$

for all $x \in X$. Since $f : X \to Y$ is odd, $A : X \to Y$ is an odd mapping. The mapping $A$ is a unique fixed point of $J$ in the set

$$M = \{g \in S : d(f,g) < \infty\}.$$

This implies that $A$ is a unique mapping satisfying (20.9) such that there exists a $\mu \in (0,\infty)$ satisfying

$$N(f(x) - A(x), \mu t) \geq \frac{t}{t + \varphi(x,0)}$$

for all $x \in X$;

2. $d(J^n f, A) \to 0$ as $n \to \infty$. This implies the equality

$$N\text{-}\lim_{n\to\infty} \frac{1}{2^n} f(2^n x) = A(x)$$

for all $x \in X$;

3. $d(f,A) \leq \frac{1}{1-L} d(f,Jf)$, which implies the inequality

$$d(f,A) \leq \frac{1}{4-4L}.$$

This implies that the inequality (20.8) holds.

The rest of the proof is similar to the proof of Theorem 20.2. □

**Corollary 20.2.** *Let $\theta \geq 0$ and let $p$ be a real number with $0 < p < 1$. Let $X$ be a normed vector space with norm $\|\cdot\|$. Let $f : X \to Y$ be an odd mapping satisfying (20.7). Then*

$$A(x) := N\text{-}\lim_{n\to\infty} 2^{-n} f(2^n x)$$

*exists for each $x \in X$ and defines an additive mapping $A : X \to Y$ such that*

$$N(f(x) - A(x), t) \geq \frac{2(2-2^p)t}{2(2-2^p)t + \theta\|x\|^p}$$

*for all $x \in X$ and all $t > 0$.*

*Proof.* The proof follows from Theorem 20.3 by taking

$$\varphi(x,y) := \theta(\|x\|^p + \|y\|^p)$$

for all $x, y \in X$. Then we can choose $L = 2^{p-1}$ and we get the desired result. □

## 20.3 Generalized Hyers–Ulam Stability of Functional Equation (20.1): An Even Case

In this section, we prove the generalized Hyers–Ulam stability of the additive-quartic functional equation (20.1) in fuzzy Banach spaces for an even case.

Using the fixed point method, we prove the generalized Hyers–Ulam stability of the functional equation

$$Cf(x,y)=0$$

in fuzzy Banach spaces: an even case.

**Theorem 20.4.** *Let $\varphi : X^2 \to [0,\infty)$ be a function such that there exists an $L<1$ with*

$$\varphi(x,y) \le \frac{L}{16}\varphi(2x,2y)$$

*for all $x,y \in X$. Let $f : X \to Y$ be an even mapping satisfying $f(0)=0$ and (20.3). Then $Q(x) := N\text{-}\lim_{n\to\infty} 16^n f(2^{-n}x)$ exists for each $x \in X$ and defines a quartic mapping $Q : X \to Y$ such that*

$$N(f(x)-Q(x),t) \ge \frac{(32-32L)t}{(32-32L)t+L\varphi(x,0)} \tag{20.10}$$

*for all $x \in X$ and all $t>0$.*

*Proof.* Taking $y=0$ in (20.3), we get

$$N(2f(2x)-32f(x),t) \ge \frac{t}{t+\varphi(x,0)} \tag{20.11}$$

for all $x \in X$.

Let $(S,d)$ be the generalized metric space defined in the proof of Theorem 20.2. Now we consider the linear mapping $J : S \to S$ such that

$$Jg(x) := 16g\left(\frac{x}{2}\right)$$

for all $x \in X$.

Let $g,h \in S$ be given such that $d(g,h)=\varepsilon$. Then

$$N(g(x)-h(x),\varepsilon t) \ge \frac{t}{t+\varphi(x,0)}$$

for all $x \in X$ and all $t>0$. Hence

$$\begin{aligned} N(Jg(x)-Jh(x),L\varepsilon t) &= N\left(16g\left(\frac{x}{2}\right)-16h\left(\frac{x}{2}\right),L\varepsilon t\right) \\ &= N\left(g\left(\frac{x}{2}\right)-h\left(\frac{x}{2}\right),\frac{L}{16}\varepsilon t\right) \end{aligned}$$

$$\geq \frac{\frac{Lt}{16}}{\frac{Lt}{16} + \varphi\left(\frac{x}{2},0\right)}$$

$$\geq \frac{\frac{Lt}{16}}{\frac{Lt}{16} + \frac{L}{16}\varphi(x,0)}$$

$$= \frac{t}{t + \varphi(x,0)}$$

for all $x \in X$ and all $t > 0$. So $d(g,h) = \varepsilon$ implies that $d(Jg,Jh) \leq L\varepsilon$. This means that

$$d(Jg,Jh) \leq Ld(g,h)$$

for all $g,h \in S$. Moreover, it follows from (20.11) that

$$N\left(f(x) - 16f\left(\frac{x}{2}\right), \frac{L}{32}t\right) \geq \frac{t}{t + \varphi(x,0)}$$

for all $x \in X$ and all $t > 0$. So $d(f,Jf) \leq L/32$.

By Theorem 20.1, there exists a mapping $Q : X \to Y$ satisfying the following:

1. $Q$ is a fixed point of $J$, i.e.,

$$Q\left(\frac{x}{2}\right) = \frac{1}{16}Q(x) \tag{20.12}$$

   for all $x \in X$. Since $f : X \to Y$ is even, $Q : X \to Y$ is an even mapping. The mapping $Q$ is a unique fixed point of $J$ in the set $M = \{g \in S : d(f,g) < \infty\}$. This implies that $Q$ is a unique mapping satisfying (20.12) such that there exists a $\mu \in (0,\infty)$ satisfying

$$N(f(x) - Q(x), \mu t) \geq \frac{t}{t + \varphi(x,0)}$$

   for all $x \in X$;
2. $d(J^n f, Q) \to 0$ as $n \to \infty$. This implies the equality

$$N\text{-}\lim_{n\to\infty} 16^n f\left(\frac{x}{2^n}\right) = Q(x)$$

   for all $x \in X$;
3. $d(f,Q) \leq \frac{1}{1-L}d(f,Jf)$, which implies the inequality

$$d(f,Q) \leq \frac{L}{32 - 32L}.$$

   This implies that the inequality (20.10) holds.

The rest of the proof is similar to the proof of Theorem 20.2. □

**Corollary 20.3.** *Let $\theta \geq 0$ and let $p$ be a real number with $p > 4$. Let $X$ be a normed vector space with norm $\|\cdot\|$. Let $f : X \to Y$ be an even mapping satisfying $f(0) = 0$ and (20.7). Then $Q(x) := N\text{-}\lim_{n\to\infty} 16^n f(2^{-n}x)$ exists for each $x \in X$ and defines a quartic mapping $Q : X \to Y$ such that*

$$N(f(x) - Q(x), t) \geq \frac{2(2^p - 16)t}{2(2^p - 16)t + \theta\|x\|^p}$$

*for all $x \in X$ and all $t > 0$.*

*Proof.* The proof follows from Theorem 20.4 by taking $\varphi(x,y) := \theta(\|x\|^p + \|y\|^p)$ for all $x, y \in X$. Then we can choose $L = 2^{4-p}$ and we get the desired result. □

Similarly, we can obtain the following. We will omit the proof.

**Theorem 20.5.** *Let $\varphi : X^2 \to [0,\infty)$ be a function such that there exists an $L < 1$ with*

$$\varphi(x,y) \leq 16L\varphi\left(\frac{x}{2}, \frac{y}{2}\right)$$

*for all $x, y \in X$. Let $f : X \to Y$ be an even mapping satisfying $f(0) = 0$ and (20.3). Then $Q(x) := N\text{-}\lim_{n\to\infty} 16^{-n} f(2^n x)$ exists for each $x \in X$ and defines a quartic mapping $Q : X \to Y$ such that*

$$N(f(x) - A(x), t) \geq \frac{(32 - 32L)t}{(32 - 32L)t + \varphi(x, 0)}$$

*for all $x \in X$ and all $t > 0$.*

**Corollary 20.4.** *Let $\theta \geq 0$ and let $p$ be a real number with $0 < p < 4$. Let $X$ be a normed vector space with norm $\|\cdot\|$. Let $f : X \to Y$ be an even mapping satisfying $f(0) = 0$ and (20.7). Then $Q(x) := N\text{-}\lim_{n\to\infty} 16^{-n} f(2^n x)$ exists for each $x \in X$ and defines a quartic mapping $Q : X \to Y$ such that*

$$N(f(x) - Q(x), t) \geq \frac{2(16 - 2^p)t}{2(16 - 2^p)t + \theta\|x\|^p}$$

*for all $x \in X$ and all $t > 0$.*

*Proof.* The proof follows from Theorem 20.5 by taking $\varphi(x,y) := \theta(\|x\|^p + \|y\|^p)$ for all $x, y \in X$. Then we can choose $L = 2^{p-4}$ and we get the desired result. □

## References


1. Aoki, T.: On the stability of the linear transformation in Banach spaces. J. Math. Soc. Japan **2**, 64–66 (1950)
2. Bag, T., Samanta, S.K.: Finite dimensional fuzzy normed linear spaces. J. Fuzzy Math. **11**, 687–705 (2003)
3. Bag, T., Samanta, S.K.: Fuzzy bounded linear operators. Fuzzy Sets and Systems **151**, 513–547 (2005)
4. Cădariu, L., Radu, V.: Fixed points and the stability of Jensen's functional equation. J. Inequal. Pure Appl. Math. **4**, no. 1, Art. ID 4 (2003)
5. Cădariu, L., Radu, V.: On the stability of the Cauchy functional equation: a fixed point approach. Grazer Math. Ber. **346**, 43–52 (2004)
6. Cădariu, L., Radu, V.: Fixed point methods for the generalized stability of functional equations in a single variable. Fixed Point Theory Appl. **2008**, Art. ID 749392 (2008)
7. Cheng, S.C., Mordeson, J.M.: Fuzzy linear operators and fuzzy normed linear spaces. Bull. Calcutta Math. Soc. **86**, 429–436 (1994)
8. Cholewa, P.W.: Remarks on the stability of functional equations. Aequationes Math. **27**, 76–86 (1984)
9. Czerwik, S.: On the stability of the quadratic mapping in normed spaces. Abh. Math. Sem. Univ. Hamburg **62**, 59–64 (1992)
10. Diaz, J., Margolis, B.: A fixed point theorem of the alternative for contractions on a generalized complete metric space. Bull. Amer. Math. Soc. **74**, 305–309 (1968)
11. Felbin, C.: Finite dimensional fuzzy normed linear spaces. Fuzzy Sets and Systems **48**, 239–248 (1992)
12. Găvruta, P.: A generalization of the Hyers-Ulam-Rassias stability of approximately additive mappings. J. Math. Anal. Appl. **184**, 431–436 (1994)
13. Hyers, D.H.: On the stability of the linear functional equation. Proc. Nat. Acad. Sci. U.S.A. **27**, 222–224 (1941)
14. Hyers, D.H., Isac G., Rassias, Th.M.: Stability of Functional Equations in Several Variables. Birkhäuser, Basel (1998)
15. Isac, G., Rassias, Th.M.: Stability of $\psi$-additive mappings: Appications to nonlinear analysis. Int. J. Math. Math. Sci. **19**, 219–228 (1996)
16. Jung, S.M.: Hyers-Ulam-Rassias Stability of Functional Equations in Mathematical Analysis. Hadronic Press lnc., Palm Harbor, Florida (2001)
17. Katsaras, A.K.: Fuzzy topological vector spaces II. Fuzzy Sets and Systems **12**, 143–154 (1984)
18. Kramosil, I., Michalek, J.: Fuzzy metric and statistical metric spaces. Kybernetika **11**, 326–334 (1975)
19. Krishna, S.V., Sarma, K.K.M.: Separation of fuzzy normed linear spaces. Fuzzy Sets and Systems **63**, 207–217 (1994)
20. Lee, S., Im S., Hwang, I.: Quartic functional equations. J. Math. Anal. Appl. **307**, 387–394 (2005)
21. Miheţ, D., Radu, V.: On the stability of the additive Cauchy functional equation in random normed spaces. J. Math. Anal. Appl. **343**, 567–572 (2008)
22. Mirmostafaee, A.K., Mirzavaziri, M., Moslehian, M.S.: Fuzzy stability of the Jensen functional equation. Fuzzy Sets and Systems **159**, 730–738 (2008)
23. Mirmostafaee, A.K., Moslehian, M.S.: Fuzzy versions of Hyers-Ulam-Rassias theorem. Fuzzy Sets and Systems **159** (2008), 720–729.
24. Mirmostafaee, A.K. Moslehian, M.S.: Fuzzy approximately cubic mappings. Inform. Sci. **178**, 3791–3798 (2008)
25. Mirzavaziri, M., Moslehian, M.S.: A fixed point approach to stability of a quadratic equation. Bull. Braz. Math. Soc. **37**, 361–376 (2006)

26. Park, C.: Fixed points and Hyers-Ulam-Rassias stability of Cauchy-Jensen functional equations in Banach algebras. Fixed Point Theory Appl. **2007**, Art. ID 50175 (2007)
27. Park, C.: Generalized Hyers-Ulam-Rassias stability of quadratic functional equations: a fixed point approach. Fixed Point Theory Appl. **2008**, Art. ID 493751 (2008)
28. Park, C., Cho, Y., Han, M.: Functional inequalities associated with Jordan-von Neumann type additive functional equations. J. Inequal. Appl. **2007**, Art. ID 41820 (2007)
29. Park, C., Cui, J.: Generalized stability of $C^*$-ternary quadratic mappings. Abstr. Appl. Anal. **2007**, Art. ID 23282 (2007)
30. Park, C., Najati, A.: Homomorphisms and derivations in $C^*$-algebras. Abstr. Appl. Anal. **2007**, Art. ID 80630 (2007)
31. Park, C., Park, W., Najati, A.: Functional equations related to inner product spaces. Abstr. Appl. Anal. **2009**, Art. ID 907121 (2009)
32. Radu, V.: The fixed point alternative and the stability of functional equations. Fixed Point Theory **4**, 91–96 (2003)
33. Rassias, Th.M.: On the stability of the linear mapping in Banach spaces. Proc. Amer. Math. Soc. **72**, 297–300 (1978)
34. Rassias, Th.M.: Problem 16; 2. In: Report of the 27th International Symposium on Functional Equations, Aequationes Math. **39**, 309 (1990)
35. Rassias, Th.M.: On the stability of the quadratic functional equation and its applications. Stud. Univ. Babeş-Bolyai Math. **43**, 89–124 (1998)
36. Rassias, Th.M.: The problem of S.M. Ulam for approximately multiplicative mappings. J. Math. Anal. Appl. **246**, 352–378 (2000)
37. Rassias, Th.M.: On the stability of functional equations in Banach spaces. J. Math. Anal. Appl. **251**, 264–284 (2000)
38. Rassias, Th.M.: On the stability of functional equations and a problem of Ulam, Acta Appl. Math. **62**, 23–130 (2000)
39. Rassias, Th.M., Šemrl, P.: On the behaviour of mappings which do not satisfy Hyers-Ulam stability. Proc. Amer. Math. Soc. **114**, 989–993 (1992)
40. Rassias, Th.M., Šemrl, P.: On the Hyers-Ulam stability of linear mappings. J. Math. Anal. Appl. **173**, 325–338 (1993)
41. Rassias, Th.M., Shibata, K.: Variational problem of some quadratic functionals in complex analysis. J. Math. Anal. Appl. **228**, 234–253 (1998)
42. Skof, F.: Proprietà locali e approssimazione di operatori. Rend. Sem. Mat. Fis. Milano **53**, 113–129 (1983)
43. Ulam, S.M.: A Collection of the Mathematical Problems. Interscience Publ., New York (1960)
44. Xiao, J.Z., Zhu, X.H.: Fuzzy normed spaces of operators and its completeness. Fuzzy Sets and Systems **133**, 389–399 (2003)

# Chapter 21
# Selections of Set-Valued Maps Satisfying Functional Inclusions on Square-Symmetric Grupoids

**Dorian Popa**

*Dedicated to the memory of S.M. Ulam*
*on the 100th anniversary of his birth*

**Abstract** We give some results on existence of selections for set-valued maps satisfying functional inclusions on square-symmetric grupoids in connection with Hyers–Ulam stability of functional equations.

**Keywords** Functional inclusions • Hyers–Ulam stability • Selection

**Mathematics Subject Classification (2000):** Primary 39B82, 54C65

## 21.1 Introduction

This work is a survey paper containing some results obtained by the author concerning a connection between Hyers–Ulam stability of a functional equation and the existence of a selection, which is a solution of this functional equation, for a set-valued map satisfying a functional inclusion on square-symmetric grupoids (see [8, 11–13]). The classical result of D.H. Hyers on the stability of Cauchy's equation is the following one (see [4, 5, 7]):

*Let $X$ be a linear normed space, $Y$ a Banach space and $\varepsilon > 0$. Then for every function $g : X \to Y$ satisfying the inequality*

$$\|g(x+y) - g(x) - g(y)\| \leq \varepsilon, \quad x, y \in X, \tag{21.1}$$

*there exists a unique additive function $f : X \to Y$ with the property*

$$\|f(x) - g(x)\| \leq \varepsilon, \quad x \in X. \tag{21.2}$$

D. Popa (✉)
Department of Mathematics, Technical University of Cluj-Napoca, Str. Memorandumului 28, 400114 Cluj-Napoca, Romania
e-mail: Popa.Dorian@math.utcluj.ro

Th.M. Rassias and J. Brzdęk (eds.), *Functional Equations in Mathematical Analysis*, Springer Optimization and Its Applications 52, DOI 10.1007/978-1-4614-0055-4_21,

Smajdor [17, 18] and Gajda and Ger [3] observed that if $g$ is a solution of (21.1), then the set-valued map $F : X \to \mathscr{P}_0(Y)$ ($\mathscr{P}_0(Y)$ denotes the family of all nonempty subsets of $Y$) given by

$$F(x) = g(x) + B(0, \varepsilon),$$

where $B(0,\varepsilon) = \{y \in Y \mid \|y\| \leq \varepsilon\}$ is the closed ball of center 0 and radius $\varepsilon$ in $Y$, satisfies the inclusion $F(x+y) \subseteq F(x) + F(y)$, $x, y \in X$ and $f$ is an *additive selection* of $F$ (i.e., $f(x+y) = f(x) + f(y)$ and $f(x) \in F(x)$ for every $x, y \in X$).

Now the following question is pertinent. Under what conditions a subadditive set-valued map admits an additive selection. Let us recall the result of Z. Gajda and R. Ger to that question.

**Theorem 21.1 ([3]).** *Let $(S,+)$ be a commutative semigroup with zero, $X$ a Banach space over $\mathbb{R}$ and $F : S \to \mathscr{P}_0(X)$ a set-valued map with convex and closed values such that:*

1) $F(x+y) \subseteq F(x) + F(y),\ x, y \in S;$
2) $\sup_{x \in S} \delta(F(x)) < \infty.$

*Then $F$ admits a unique additive selection.*

The result of Gajda and Ger was extended by Nikodem and Popa to set-valued maps satisfying general linear inclusions (see [8, 11]). We show in what follows that similar results hold for set-valued maps satisfying functional inclusions on square-symmetric grupoids.

Let us recall that Rätz [15] pointed out the role of square-symmetry for the stability of functional equations and Páles [9], Páles, Volkmann, Luce [10] considered the stability of the Cauchy functional equation on square-symmetric grupoids.

A grupoid $(X,*)$ is called *square-symmetric* if

$$(x*y)*(x*y) = (x*x)*(y*y)$$

for all $x, y \in X$. An operation $*$ on $X$ is square-symmetric if and only if the function $\sigma_* : X \to X$, given by

$$\sigma_*(x) = x*x, \quad x \in X,$$

is an endomorphism of $(X,*)$. The grupoid $(X,*)$ is called *divisible* if $\sigma_*$ is an automorphism of $(X,*)$. The triple $(Y,*,d)$ is called a *metric grupoid* if $(Y,*)$ is a grupoid, $(Y,d)$ is a metric space and $*$ is a continuous operation with respect to the topology of $(Y,d)$. For a nonempty set $Y$ we denote by $\mathscr{P}_0(Y)$ the collection of all nonempty subsets of $Y$. If $(Y,d)$ is a metric space by $cl(Y)$ we denote the collection of all nonempty closed subsets of $Y$. In a linear normed space $(Y, \|\cdot\|)$, we define the following families of sets:

$$c(Y) := \{A \mid A \in \mathscr{P}_0(Y),\ A \text{ is convex set}\},$$
$$ccl(Y) := \{A \mid A \in \mathscr{P}_0(Y),\ A \text{ is closed and convex set}\},$$

$$cc(Y) := \{A \mid A \in \mathscr{P}_0(Y),\ A \text{ is compact and convex set}\}.$$

By $\mathbb{N} = \{0, 1, 2, \ldots\}$ we denote the set of all nonnegative integers.
Let $(Y, d)$ be a metric space. The *diameter* of a set $A \in \mathscr{P}_0(Y)$ is defined by

$$\delta(A) := \sup\{d(x,y) \mid x, y \in A\}.$$

The *Lipschitz modulus* of a function $f : Y \to Y$ is the smallest real extended number $L$ with the property

$$d(f(x), f(y)) \leq Ld(x,y), \quad x, y \in Y.$$

The Lipschitz modulus of a function $f$ is denoted by $Lipf$. Finally, recall that a *selection* of a set-valued map $F : X \to \mathscr{P}_0(Y)$ is a single-valued map $f : X \to Y$ with the property $f(x) \in F(x)$ for all $x \in X$.

## 21.2 Main Results

Let $(Y, \diamond, d)$ be a metric grupoid. We define an operation $\Diamond$ on $\mathscr{P}_0(Y)$ by

$$A \Diamond B = \{x \mid x = a \diamond b,\ a \in A,\ b \in B\}, \quad A, B \in \mathscr{P}_0(Y). \tag{21.3}$$

Suppose in what follows that the defined operation $\Diamond$ satisfies the condition: for all $\varepsilon > 0$ there exists $\eta > 0$ such that if $\delta(A), \delta(B) < \eta$, $A, B \in \mathscr{P}_0(Y)$, then:

$$\delta(A \Diamond B) < \varepsilon. \tag{21.4}$$

We give some results analogous to Theorem 21.1 for set-valued maps $F : X \to \mathscr{P}_0(Y)$ satisfying functional inclusions of the form

$$F(x * y) \subseteq F(x) \Diamond F(y),$$

$$F(x) \Diamond F(y) \subseteq F(x * y),$$

where $(X, *)$ is a square-symmetric grupoid.

Let us remark that if $\diamond$ is square-symmetric on $Y$ then $\Diamond$, is not necessary square-symmetric on $\mathscr{P}_0(Y)$. For the square-symmetry of $\Diamond$ it suffices that the operation $\diamond$ satisfies the condition of *bisymmetry* introduced by Aczél (see [1]). Hence, the following lemma holds.

**Lemma 21.1 ([12]).** *Let $(Y, \diamond)$ be a grupoid satisfying the condition*

$$(x_1 \diamond y_1) \diamond (x_2 \diamond y_2) = (x_1 \diamond x_2) \diamond (y_1 \diamond y_2) \tag{21.5}$$

*for* $x_1, x_2, y_1, y_2 \in Y$. *Then* $\sigma_\Diamond$ *is an increasing endomorphism of* $(\mathscr{P}_0(Y), \Diamond, \subseteq)$.

*Proof.* Let $A, B \in \mathscr{P}_0(Y)$. We have to prove that

$$\sigma_\Diamond(A \Diamond B) = \sigma_\Diamond(A) \Diamond \sigma_\Diamond(B).$$

Let $x \in \sigma_\Diamond(A \Diamond B)$. Then there exist $a_1, a_2 \in A$, $b_1, b_2 \in A$ such that $x = (a_1 \diamond b_1) \diamond (a_2 \diamond b_2)$. Taking account of the condition (21.5) it follows $x = (a_1 \diamond a_2) \diamond (b_1 \diamond b_2) \in \sigma_\Diamond(A) \Diamond \sigma_\Diamond(B)$. Hence $\sigma_\Diamond(A \Diamond B) \subseteq \sigma_\Diamond(A) \Diamond \sigma_\Diamond(B)$.

The reverse inclusion can be proved analogously. Let $A, B \in \mathscr{P}_0(Y)$, $A \subseteq B$. We prove that

$$\sigma_\Diamond(A) \subseteq \sigma_\Diamond(B).$$

Let $x \in \sigma_\Diamond(A)$. Then there exist $a_1, a_2 \in A$ such that $x = a_1 \diamond a_2 \in B \Diamond B$, hence $\sigma_\Diamond(A) \subseteq \sigma_\Diamond(B)$. □

Now we can give the first stability result of this paper.

**Theorem 21.2 ([12]).** *Let* $(X, *)$ *be a square-symmetric divisible grupoid,* $(Y, \diamond, d)$ *a complete metric bisymmetric divisible grupoid and* $(\mathscr{A}, \Diamond)$ *a divisible subgrupoid of* $(\mathscr{P}_0(Y), \Diamond)$.

*Suppose that* $F : X \to \mathscr{A}$ *is a set-valued map that satisfies:*

$$F(x * y) \subseteq F(x) \Diamond F(y), \quad x, y \in X. \tag{21.6}$$

*If*

$$\sigma_\Diamond^{-n} \circ F \circ \sigma_*^n(x) \in cl(Y) \tag{21.7}$$

*for every* $x \in X$ *and every* $n \in \mathbb{N}$ *and*

$$\lim_{n \to \infty} \delta(F \circ \sigma_*^n(x)) Lip(\sigma_\diamond^{-n}) = 0 \tag{21.8}$$

*for every* $x \in X$, *then there exists a unique selection* $f : X \to Y$ *of* $F$ *that satisfies the relation*

$$f(x * y) = f(x) \diamond f(y), \quad x, y \in X. \tag{21.9}$$

*Proof. Existence.* For $y = x$ the relation (21.6) becomes

$$F(\sigma_*(x)) \subseteq \sigma_\Diamond(F(x)), \quad x \in X,$$

and replacing $x$ by $\sigma_*^n(x)$, $n \in \mathbb{N}$, we get

$$F \circ \sigma_*^{n+1}(x) \subseteq \sigma_\Diamond \circ F \circ \sigma_*^n(x). \tag{21.10}$$

Now taking into account that $\sigma_\Diamond$ is increasing, it follows that $\sigma_\Diamond^{-1}$ is increasing too, and by (21.10) we get

$$\sigma_\Diamond^{-n-1} \circ F \circ \sigma_*^{n+1}(x) \subseteq \sigma_\Diamond^{-n} \circ F \circ \sigma_*^n(x), \quad x \in X. \tag{21.11}$$

Let $x \in X$ be fixed. Define the sequence of sets $(F_n(x))_{n \in \mathbb{N}}$ by

$$F_n(x) = \sigma_\Diamond^{-n} \circ F \circ \sigma_*^n(x), \quad n \in \mathbb{N}. \tag{21.12}$$

The sequence $(F_n(x))_{n \geq 0}$ is decreasing, in view of relation (21.11). We prove that

$$\lim_{n \to \infty} \delta(F_n(x)) = 0. \tag{21.13}$$

We have

$$\delta(F_n(x)) = \sup\{d(u,v) : u,v \in F_n(x)\}, \quad n \in \mathbb{N}.$$

Take $n > 0$ and $u,v \in \sigma_\diamond^{-n} \circ F \circ \sigma_*^n(x)$. Then

$$\sigma_\diamond^n(u) \in F \circ \sigma_*^n(x) \quad \text{and} \quad \sigma_\diamond^n(v) \in F \circ \sigma_*^n(x).$$

Let $\sigma_\diamond^n(u) = s$, $\sigma_\diamond^n(v) = t$, $s,t \in F \circ \sigma_*^n(x)$. Clearly

$$d(u,v) = d(\sigma_\diamond^{-n}(s), \sigma_\diamond^{-n}(t)) \leq Lip\, \sigma_\diamond^{-n} d(s,t) \leq Lip\, \sigma_\diamond^{-n} \delta(F \circ \sigma_*^n(x))$$

and

$$\delta(F_n(x)) \leq Lip\, \sigma_\diamond^{-n} \delta(F \circ \sigma_*^n(x)). \tag{21.14}$$

By (21.8) and (21.14) we obtain

$$\lim_{n \to \infty} \delta(F_n(x)) = 0,$$

whence

$$\bigcap_{n=0}^{\infty} F_n(x) \tag{21.15}$$

is a singleton, in view of the Cantor theorem for decreasing sequences of sets. Denote by $f(x)$ the unique element of this intersection.

The function $f : X \to Y$ is a selection of $F$, since

$$f(x) \in F_0(x) = F(x), \quad x \in X.$$

Now we show that $f$ satisfies (21.9). Let us prove first that

$$F_n(x * y) \subseteq F_n(x) \Diamond F_n(y), \quad x,y \in X,\ n \in \mathbb{N}. \tag{21.16}$$

Replacing $x$ by $\sigma_*^n(x)$ and $y$ by $\sigma_*^n(y)$, $n \in \mathbb{N}$, in

$$F(x*y) \subseteq F(x)\Diamond F(y)$$

we get

$$F \circ \sigma_n^*(x*y) \subseteq (F \circ \sigma_n^*(x))\Diamond(F \circ \sigma_n^*(y)), \quad n \in \mathbb{N}, \tag{21.17}$$

in view of the square-symmetry of $*$. Since $\sigma_\Diamond^{-n}$ is an increasing automorphism, by (21.17), it follows

$$\sigma_\Diamond^{-n} \circ F \circ \sigma_n^*(x*y) \subseteq (\sigma_\Diamond^{-n} \circ F \circ \sigma_n^*(x))\Diamond(\sigma_\Diamond^{-n} \circ F * \sigma_n^n(y)),$$

hence, the relation (21.16) is proved.

By (21.16) we get

$$d(f(x*y), f(x) \diamond f(y)) \leq \delta(F_n(x)\Diamond F_n(y)), \quad n \in \mathbb{N}, \tag{21.18}$$

and taking account of the condition (21.4) we obtain

$$d(f(x*y), f(x) \diamond f(y)) = 0,$$

if $n$ tends to infinity in (21.18). Hence, $f$ satisfies (21.9).

*Uniqueness.* Suppose that there exist two selections $f, g$ of $F$ that satisfy the relations

$$f(x*y) = f(x) \diamond f(y), \quad g(x*y) = g(x) \diamond g(y), \quad x, y \in X. \tag{21.19}$$

By (21.19), we get

$$f \circ \sigma_*^n(x) = \sigma_\diamond^n \circ f(x), \quad g \circ \sigma_*^n(x) = \sigma_\diamond^n \circ g(x), \quad x \in X, \ n \in \mathbb{N}, \tag{21.20}$$

and, taking into account that $f, g$ are selections of $F$,

$$f \circ \sigma_*^n(x) \in F \circ \sigma_*^n(x), \quad g \circ \sigma_*^n(x) \in F \circ \sigma_*^n(x), \quad x \in X, \ n \in \mathbb{N}. \tag{21.21}$$

Let $x \in X$ be fixed. Then, for every $x \in X$

$$d(\sigma_\diamond^n \circ f(x), \sigma_\diamond^n \circ g(x)) = d(f \circ \sigma_*^n(x), g \circ \sigma_*^n(x)) \leq \delta(F \circ \sigma_*^n(x)).$$

Let $\sigma_\diamond^n \circ f(x) = s$, $\sigma_\diamond^n \circ g(x) = t$, $s, t \in F \circ F_*^n(x)$. We have $f(x) = \sigma_\diamond^{-n}(s)$, $g(x) = \sigma_\diamond^{-n}(t)$ and

$$\begin{aligned} d(f(x), g(x)) &= d(\sigma_\diamond^{-n}(s), \sigma_\diamond^{-n}(t)) \leq Lip\, \sigma_\diamond^{-n} d(s,t) \\ &\leq Lip\, \sigma_\diamond^{-n} \delta(F \circ \sigma_n^*(x)), \quad n \in \mathbb{N}. \end{aligned}$$

Taking account of (21.8) it follows

$$\lim_{n\to\infty} Lip\,\sigma_{\diamond}^{-n}\delta(F\circ\sigma_n^*(x)) = 0,$$

hence, $f(x) = g(x)$. □

**Theorem 21.3 ([12]).** *Let $(X,*)$ be a square-symmetric divisible grupoid and $(Y,\diamond,d)$ a metric bisymmetric semigroup. Suppose that $F : X \to \mathscr{P}_0(Y)$ is a set-valued map satisfying the relation*

$$F(x*y) \subseteq F(x)\Diamond F(y), \quad x,y\in X. \tag{21.22}$$

*If*

$$\lim_{n\to\infty} \delta(F\circ\sigma_*^{-n}(x))Lip\,(\sigma_{\diamond}^{n}) = 0 \tag{21.23}$$

*for every $x\in X$, then $F$ is single valued and*

$$F(x*y) = F(x)\diamond F(y), \quad x,y\in X. \tag{21.24}$$

*Proof.* By the relation (21.22) we get

$$F(\sigma_*(x)) \subseteq \sigma_{\Diamond}(F(x)), \quad x\in X,$$

and replacing $x$ by $\sigma_*^{-n-1}(x)$, $n\in\mathbb{N}$, we obtain

$$F\circ\sigma_*^{-n}(x) \subseteq \sigma_{\Diamond}\circ F\circ\sigma_*^{-n-1}(x), \quad x\in X,$$

and taking into account that $\sigma_{\diamond}$ is increasing

$$\sigma_{\Diamond}^{n}\circ F\circ\sigma_*^{-n}(x) \subseteq \sigma_{\Diamond}^{n+1}\circ F\circ\sigma_*^{-n-1}(x), \quad x\in X. \tag{21.25}$$

Let $x\in X$ be fixed. The sequence of sets $(F_n(x))_{n\geq 0}$ defined by

$$F_n(x) = \sigma_{\Diamond}^{n}\circ F\circ\sigma_*^{-n}(x), \quad n\geq 0,$$

is increasing. Then $(\delta(F_n(x)))_{n\geq 0}$ is an increasing sequence of nonnegative numbers. As in the proof of Theorem 21.2 we obtain

$$\delta(F_n(x)) \leq \delta(F\circ\sigma_*^{-n}(x))Lip\,\sigma_{\diamond}^{n}$$

and taking account of (21.23) it follows

$$\lim_{n\to\infty}\delta(F_n(x)) = 0.$$

Then $\delta(F_n(x)) = 0$ for every $n \in \mathbb{N}$, hence $F_n(x)$ is single valued for every $n \in \mathbb{N}$ and $F_0(x) = F(x)$ satisfies the relation

$$F(x*y) = F(x) \diamond F(y), \quad x, y \in X.$$

□

Analogous results follow for the converse functional inclusion. We give them without proof since it is analogous to the proof of Theorem 21.2 and Theorem 21.3.

**Theorem 21.4 ([13]).** *Let $(X,*)$ be a square-symmetric divisible grupoid, $(Y,\diamond,d)$ a complete metric bisymmetric divisible grupoid and $F: X \to \mathscr{P}_0(Y)$ a set-valued map with the property*

$$F(x) \Diamond F(y) \subseteq F(x*y), \quad x, y \in X. \tag{21.26}$$

*If*

$$\sigma_{\Diamond}^{n} \circ F \circ \sigma_{*}^{-n}(x) \in cl(Y), \quad x \in X,\ n \in \mathbb{N}, \tag{21.27}$$

*and*

$$\lim_{n\to\infty} \delta(F \circ \sigma_{*}^{-n}(x)) Lip(\sigma_{\diamond}^{n}) = 0, \quad x \in X, \tag{21.28}$$

*then there exists a unique selection $f: X \to Y$ of $F$ with the property*

$$f(x) \diamond f(y) = f(x*y), \quad x, y \in X. \tag{21.29}$$

**Theorem 21.5 ([13]).** *Let $(X,*)$ be a square-symmetric divisible grupoid, $(Y,\diamond,d)$ a metric bisymmetric divisible grupoid and $\mathscr{A}$ a divisible subgrupoid of $(\mathscr{P}_0(Y), \Diamond)$. Suppose that $F: X \to \mathscr{A}$ is a set-valued map with the property*

$$F(x) \Diamond F(y) \subseteq F(x*y), \quad x, y \in X. \tag{21.30}$$

*If*

$$\lim_{n\to\infty} \delta(F \circ \sigma_{*}^{n}(x)) Lip(\sigma_{\diamond}^{-n}) = 0 \tag{21.31}$$

*for every $x \in X$, then $F$ is single valued and*

$$F(x) \diamond F(y) = F(x*y), \quad x, y \in X. \tag{21.32}$$

The following results are consequences of the previous theorems concerning the existence of selections for general linear inclusions and Hyers–Ulam stability for the general linear equation.

Suppose that $Y$ is a Banach space over $\mathbb{R}$ and $\diamond$ is defined by

$$x \diamond y = px + qy, \quad x, y \in Y, \tag{21.33}$$

where $p, q \in \mathbb{R}$ are given numbers.

The triple $(Y, \diamond, \|\cdot\|)$ is obviously a metric grupoid with a bisymmetric operation. Then for every $U,V \in \mathscr{P}_0(Y)$ the operation $\Diamond$ is defined by

$$U \Diamond V = pU + qV, \tag{21.34}$$

where "+" from the right hand side of (21.34) denotes the usual sum of two sets in a linear space.

**Corollary 21.1 ([11]).** *Let $(X,*)$ be a square-symmetric divisible groupoid, $Y$ a Banach space over $\mathbb{R}$, $p,q \in \mathbb{R}$, $p+q \neq 1$. Suppose that $F: X \to c(Y)$ is a set-valued map such that*

$$F(x*y) \subseteq pF(x) + qF(y), \quad x,y \in X, \tag{21.35}$$

*and the following conditions are satisfied:*

*(i)* $F \circ \sigma_*^n(x) \in cl(Y)$, $x \in X$, $n \in \mathbb{N}$;
*(ii) there exists $M > 0$ such that*

$$\delta(F \circ \sigma_*^n(x)) \leq M, \quad x \in X, \; n \in \mathbb{N}.$$

*Then there exists a unique selection $f: X \to Y$ of $F$ such that*

$$f(x*y) = pf(x) + qf(y), \quad x,y \in X. \tag{21.36}$$

*Proof.* Let $\mathscr{A} = c(Y)$. Then $\sigma_\Diamond(U) = (p+q)U$ for every $U \in c(Y)$, $\sigma_\Diamond$ is an automorphism of $(c(Y), \Diamond)$ for $p+q \neq 0$ and $\sigma_\Diamond^n(x) = (p+q)^n x$, $x \in X$, $n \in \mathbb{Z}$, with

$$Lip\, \sigma_\Diamond^n = |p+q|^n, \quad n \in \mathbb{Z}.$$

1. If $|p+q| > 1$, then

$$\sigma_\Diamond^{-n} \circ F \circ \sigma_*^n(x) = \frac{1}{(p+q)^n} F \circ \sigma_*^n(x) \in cl(Y), \quad x \in X, \; n \in \mathbb{N},$$

and

$$\delta(F \circ \sigma_*^n(x)) Lip\, \sigma_\diamond^{-n} \leq \frac{M}{|p+q|^n}, \quad x \in X, \; n \in \mathbb{N}.$$

So, by Theorem 21.2 it follows that there is a unique selection $f$ of $F$ that satisfies (21.36).

2. If $|p+q| < 1$, then

$$Lip\,(\sigma_\Diamond^n)\delta(F \circ \sigma_*^{-n}(x)) \leq M|p+q|^n, \quad x \in X, \; n \in \mathbb{N},$$

hence, in view of Theorem 21.3, $F$ is a single valued and satisfies the equation

$$F(x*y) = pF(x) + qF(y), \quad x,y \in X,$$

which means that it is its own selection

□

**Corollary 21.2 ([12]).** *Let $(X,*)$ be a square-symmetric divisible grupoid, $Y$ a Banach space over $\mathbb{R}$ and $B \in ccl(Y)$. Assume that $p,q \in \mathbb{R}$, $p+q > 1$, and $b \in Y$. Let $g : X \to Y$ be a function such that*

$$g(x*y) - pg(x) - qg(y) - b \in B, \quad x,y \in X. \tag{21.37}$$

*Then there exists a uniquely determined function $h : X \to Y$ such that*

$$h(x*y) = h(x) + h(y) + b, \quad x,y \in X, \tag{21.38}$$

*and*

$$g(x) - h(x) \in \frac{1}{p+q-1}B. \tag{21.39}$$

*Proof.* Define a set valued map $F : X \to ccl(Y)$ by:

$$F(x) = g(x) + \frac{1}{p+q-1}A, \quad x \in X,$$

where $A := b + B \in ccl(Y)$. We have

$$\begin{aligned} F(x*y) &= g(x*y) + \frac{1}{p+q-1}A \subseteq pg(x) + qg(y) + A + \frac{1}{p+q-1}A \\ &= pg(x) + qg(y) + \frac{p+q}{p+q-1}A = pF(x) + qF(y), \quad x,y \in X. \end{aligned}$$

By Corollary 21.1 it follows that there is a uniquely determined selection $f : X \to Y$ of $F$ that satisfies (21.36). The function $h : X \to Y$ given by

$$h(x) = f(x) - \frac{1}{p+q-1}b, \quad x \in X,$$

satisfies (21.38) and (21.39). □

**Corollary 21.3 ([13]).** *Let $(X,*)$ be a square-symmetric divisible grupoid, $(Y, \|\cdot\|)$ a Banach space over $\mathbb{R}$, $p,q \in \mathbb{R}$, $p+q \neq 0$, $p+q \neq 1$, and $F : X \to c(Y)$ a set-valued map with the property*

$$pF(x) + qF(y) \subseteq F(x*y), \quad x,y \in X. \tag{21.40}$$

*Suppose that there exists $M > 0$ such that $\delta(F(x)) \leq M$, $x \in X$, and*

$$F \circ \sigma_*^{-n}(x) \in cl(Y), \quad x \in X,\ n \in \mathbb{N}.$$

*Then there exists a unique selection $f : X \to Y$ of $F$ such that*

$$pf(x) + qf(y) = f(x * y), \quad x, y \in X. \tag{21.41}$$

Corollary 21.4 leads to the following stability result for the general linear equation.

**Corollary 21.4 ([13]).** *Let $(X, *)$ be a square-symmetric divisible grupoid, $(Y, \|\cdot\|)$ a Banach space over $\mathbb{R}$, $p, q, \varepsilon > 0$, $p + q < 1$, and $b \in Y$. Suppose that $f : X \to Y$ is a function satisfying*

$$\|f(x * y) - pf(x) - qf(y) - b\| \leq \varepsilon, \quad x, y \in X. \tag{21.42}$$

*Then there exists a unique function $g : X \to Y$ satisfying*

$$g(x * y) = pg(x) + qg(y) + b, \quad x, y \in X, \tag{21.43}$$

*and*

$$\|f(x) - g(x)\| \leq \frac{\varepsilon}{1 - p - q}, \quad x \in X. \tag{21.44}$$

The results proved in Corollaries 21.1–21.4 are analogous to some results obtained in [2, 6, 9, 11, 14, 16].

**Acknowledgement** This research is supported by Project PN 2-Partenership No.110118 MoDef.

## References

1. Aczél, J.: Lectures on Functional Equations and Their Applications. Academic Press (1966)
2. J. Brzdęk, J., Pietrzyk, A.: A note on stability of the general linear equation. Aequationes Math. **75**, 267–270 (2008)
3. Gajda, Z., Ger, R.: Subadditive multifunctions and Hyers-Ulam stability. Numerical Mathematics **80**, 281–291 (1987)
4. Hyers, D.H.: On the stability of the linear functional equation. Proc. Nat. Acad. Sci. USA **27**, 222–224 (1941)
5. Hyers, D.H., Isac, G., Rassias, Th.M.: Stability of Functional Equations in Several Variables. Birkhäuser, Boston (1998)
6. Kim, G.H.: Addendum to "On the stability of functional equations on square-symmetric grupoid". Nonlinear Analysis **62**, 365–381 (2005)
7. Kuczma, M.: An Introduction to the Theory of Functional Equations and Inequalities. Cauchy's Equation and Jensen Inequality. Państwowe Wydawnictwo Naukowe & Uniwersytet Śląski, Warszawa-Kraków-Katowice (1985)

8. Nikodem, K., Popa, D.: On selections of general linear inclusions. Publ. Math. Debrecen **75**, 239–249 (2009)
9. Páles, Zs.: Hyers-Ulam stability of the Cauchy functional equation on square-symmetric grupoids. Publ. Math. Debrecen **58**, 651–666 (2001)
10. Páles, Zs., Volkmann, P., Luce, R.D.: Hyers-Ulam stability of functional equations with square-symmetric operation. Proc. Nat. Acad. Sci. USA **95**, 12772–12775 (1998)
11. Popa, D.: A stability result for a general linear inclusion. Nonlinear Funct. Anal. Appl. **3**, 405–414 (2004)
12. Popa, D.: Functional inclusions on square-symmetric grupoids and Hyers-Ulam stability. Math. Inequal. Appl. **3**, 419–428 (2004)
13. Popa, D.: A property of a functional inclusion connected with Hyers–Ulam stability. J. Math. Inequal. **4**, 591–598 (2009)
14. Rassias Th.M., Tabor, J.: What is left Hyers-Ulam stability. J. Nat. Geometry **1**, 65–69 (1992)
15. Rätz, J.: On approximately additive mappings. In: Beckenbach, E.F. (ed.) General Inequalities 2, pp. 233–251. International Series in Numerical Mathematics **47**, Birkhäuser, Basel-Boston (1980)
16. Smajdor, A.: Additive selections of superadditive set-valued functions. Aequationes Math. **39**, 121–128 (1990)
17. Smajdor, W.: Subadditive set-valued functions. Glas. Mat. **21**, 343–348 (1986)
18. Smajdor, W.: Superadditive set-valued functions and Banach-Steinhaus theorem. Radovi Mat. **3**, 203–214 (1987)

# Chapter 22
# On Stability of Isometries in Banach Spaces

**Vladimir Yu. Protasov**

*Dedicated to the memory of S.M. Ulam*
*on the 100th anniversary of his birth*

**Abstract** We analyze the problem of stability of linear isometries (SLI) of Banach spaces. Stability means the existence of a function $\sigma(\varepsilon)$ such that $\sigma(\varepsilon) \to 0$ as $\varepsilon \to 0$ and for any $\varepsilon$-isometry $A$ of the space $X$ (i.e., $(1-\varepsilon)\|x\| \le \|Ax\| \le (1+\varepsilon)\|x\|$ for all $x \in X$) there is an isometry $T$ such that $\|A - T\| \le \sigma(\varepsilon)$. It is known that all finite-dimensional spaces, Hilbert space, the spaces $C(K)$ and $L_p(\mu)$ possess the SLI property. We construct examples of Banach spaces $X$, which have an infinitely smooth norm and are arbitrarily close to the Hilbert space, but fail to possess SLI, even for surjective operators. We also show that there are spaces that have SLI only for surjective operators. To obtain this result we find the functions $\sigma(\varepsilon)$ for the spaces $l_1$ and $l_\infty$. Finally, we observe some relations between the conditional number of operators and their approximation by operators of similarity.

**Keywords** Linear operator • Isometry • Stability • Approximation • Dvoretzky theorem

**Mathematics Subject Classification (2000):** Primary 39A30, 41A65, 46B20

## 22.1 Introduction

We consider approximations of "almost" isometries of Banach spaces by isometries. Let $X$ be an arbitrary Banach space. A linear operator $A : X \to X$ is an $\varepsilon$-isometry of $X$ with some $\varepsilon \in (0,1)$ if $(1-\varepsilon)\|x\| \le \|Ax\| \le (1+\varepsilon)\|x\|$ for all $x \in X$.

---

V.Y. Protasov (✉)
Department of Mechanics and Mathematics, Moscow State University,
Vorobyovy Gory, Moscow, 119992, Russia
e-mail: v-protassov@yandex.ru

Th.M. Rassias and J. Brzdęk (eds.), *Functional Equations in Mathematical Analysis*, Springer Optimization and Its Applications 52, DOI 10.1007/978-1-4614-0055-4_22, 

We say that $X$ possesses the *stability of linear isometries* (SLI) property if there exists a function $\sigma$ such that $\sigma(\varepsilon) \to 0$ as $\varepsilon \to 0$, and for any $\varepsilon$-isometry $A$ there exists a linear isometry $T$ of $X$ (not necessarily surjective), for which $\|Ax - Tx\| \leq \sigma(\varepsilon)\,\|x\|$. The *stability of surjective linear isometries* (SSLI) property is that any *surjective* $\varepsilon$-isometry can be $\sigma(\varepsilon)$-approximated by an isometry, where $\sigma(\varepsilon) \to 0$ as $\varepsilon \to 0$. We focus on the case $X = Y$.

Since any $\varepsilon$-isometry is obviously injective, it follows that for finite-dimensional spaces the properties SLI and SSLI coincide. In fact, it can be easily shown by the compactness argument that any finite-dimensional space $X$ possesses SLI with some function $\sigma(\varepsilon)$ depending on $X$ [1]. The separable Hilbert space $H$ also possesses SLI. This is established using the polar factorization of operators [1]. The problem for other Banach spaces is much more difficult. In [3], it was shown that the space $C(K)$ of continuous functions on a compact metric space $K$ possesses SLI. In [6], the compactness assumption was slightly relaxed, but for the weaker SSLI property. It was shown that for any locally compact Hausdorff space $M$ the space $C_0(M)$ of continuous functions vanishing at infinity has SSLI, and the function $\sigma(\varepsilon)$ was estimated. In [4], it was proved that any space $L_p(\mu)$, $1 \leq p < \infty$ possesses SLI, although the corresponding functions $\sigma(\varepsilon)$ were not specified. The SLI property for arbitrary Banach spaces was studied in [1, 2, 5–9]. In [7, 8] the SLI property was analyzed in the framework of Hyers–Ulam stability. The notion of SLI was generalized for any parameter $p \geq 0$ (our case corresponds to $p = 1$). In [5], it was shown that the uniform smoothness assumption for the norm of $X$ plays a role for SLI. However, none of the classical functional spaces satisfies that condition.

A natural question arises, whether all Banach spaces possess SLI or, at least, SSLI property? Which conditions on the space would guarantee SLI or SSLI? These problems were formulated by J. Chmielinski in the problem session of 12th ICFEI conference [10]. Examples of Banach spaces without SLI have been constructed in [1, 6]. In [9], it was shown that for nonlinear maps there are two-dimensional spaces with non-stable isometries. For linear operators all finite-dimensional spaces have SLI. The question remains what properties of the spaces (Radon–Nicodym property, Krein–Milman property, reflexivity, smoothness of the norm, etc.,) guarantee SLI or SSLI? In Sect. 22.2, we show that none of those assumptions suffices. There are Banach spaces, which are separable, arbitrarily close to the Hilbert space, have smooth (infinitely Frechet differentiable at any nonzero point) norms, but do not have SSLI (Theorem 22.1). In Sect. 22.3, we prove that SLI and SSLI are not equivalent. Namely, for any $d \geq 2$ the space $\mathbb{R}^d \oplus l_1$ does not possess SLI, although all its *surjective* isometries are stable (Theorem 22.3). In the proof we establish as an auxiliary result that both $l_1$ and $l_\infty$ have SSLI with $\sigma(\varepsilon) = 3\varepsilon$. For these spaces this sharpens the general results of [3,4,6]. In Sect. 22.4, we analyze the relation between the condition number of matrices with the stability function $\sigma(\varepsilon)$. Finally, in Sect. 22.5, several open problems are formulated.

## 22.2 Banach Spaces with Smooth Norms and Without SSLI

Let $X$ be a Banach space and $H$ be a separable Hilbert space. The space $X$ is equivalent to $H$ if $C_1\|x\|_H \leq \|x\|_X \leq C_2\|x\|_H,\ x \in X$, where $C_1, C_2$ are positive constants. In this case, $X$ is separable and reflexive, and therefore possesses both the Radon–Nicodym and Krein–Milman properties. The norm of $X$ is infinitely smooth if the function $f(x) = \|x\|_X$ is infinitely Frechet differentiable at any point $x \neq 0$. Let us show that there are spaces $X$ with smooth norm, arbitrarily close to the Hilbert space, i.e., the constants $C_1$ and $C_2$ are arbitrarily close to one, that do not possess SSLI.

**Theorem 22.1.** *There are Banach spaces that are arbitrarily close to the Hilbert space, have infinitely smooth norms, and do not possess the SSLI property.*

To construct examples of such spaces we consider any sequence of positive numbers $\alpha = \{\alpha_k\}_{k\in\mathbb{N}}$ satisfying the following two assumptions:

(a) $a_k < 1$ for all $k$, and the sequence increases: $\alpha_{k+1} > \alpha_k,\ k \in \mathbb{N}$;
(b) $\sum\limits_{k\in\mathbb{N}} (1-\alpha_k^2) < 1$.

This means, in particular, that $\alpha_k$ converges monotone to 1. We take the Hilbert space $H = l_2$ with the norm

$$\|x\| = \left(\sum_{k\in\mathbb{N}} |x_k|^2\right)^{1/2}$$

and introduce a new norm:

$$\|x\|_\alpha = \sup\left\{\|x\|, \left|\frac{x_1}{\alpha_1}\right|, \ldots, \left|\frac{x_k}{\alpha_k}\right|, \ldots\right\} \tag{22.1}$$

and write $H_\alpha$ for the space $H$ equipped with this norm. We call a linear bounded operator $T$ in $H_\alpha$ *coordinate symmetry* if $Te_j = \delta_j e_j, \delta_j \in \{-1, 1\},\ j \in \mathbb{N}$. So, a coordinate symmetry changes the signs of some entries.

**Lemma 22.1.** *Any linear isometry (not necessarily surjective) of the space $H_\alpha$ is a coordinate symmetry.*

The proof is outlined in the Appendix.

*Proof of Theorem (22.1).* Since

$$\|x\| \leq \|x\|_\alpha \leq \frac{1}{\alpha_1}\|x\|$$

for any $x \in H$, it follows that the norm $\|\cdot\|_\alpha$ is equivalent to $\|\cdot\|$. Moreover, choosing $\alpha_1$ close to 1 we make these norms arbitrarily close. For any $k \in \mathbb{N}$ let $A_k$ be the operator of interchanging the $k$th and $(k+1)$st coordinates: $A_k e_k = e_{k+1}, A_k e_{k+1} = e_k$ and $A_j e_j = e_j$ for all $j \notin \{k, k+1\}$ ($\{e_k\}_{k\in\mathbb{N}}$ is the basis of $H$). Observe first that

$$\left| \left\|A_k x\right\|_\alpha - \left\|x\right\|_\alpha \right| \leq \left(\frac{1}{\alpha_k} - \frac{1}{\alpha_{k+1}}\right) \left\|x\right\|_\alpha \tag{22.2}$$

for any $x \in H_\alpha$. To see this we consider two elements of the space $l_\infty$:

$$v_1 = \left( \|x\|, \left|\frac{x_1}{\alpha_1}\right|, \ldots, \left|\frac{x_k}{\alpha_k}\right|, \left|\frac{x_{k+1}}{\alpha_{k+1}}\right|, \ldots \right),$$

$$v_2 = \left( \|x\|, \left|\frac{x_1}{\alpha_1}\right|, \ldots, \left|\frac{x_k}{\alpha_{k+1}}\right|, \left|\frac{x_{k+1}}{\alpha_k}\right|, \ldots \right)$$

(we have interchanged $\alpha_k$ and $\alpha_{k+1}$). Applying the triangle inequality in the $l_\infty$-norm, we get

$$\begin{aligned}
\left| \left\|A_k x\right\|_\alpha - \left\|x\right\|_\alpha \right| &= \left| \left\|v_1\right\|_\infty - \left\|v_2\right\|_\infty \right| \leq \left\|v_1 - v_2\right\|_\infty \\
&= \max\left\{ \left(\frac{1}{\alpha_k} - \frac{1}{\alpha_{k+1}}\right) |x_k|, \left(\frac{1}{\alpha_k} - \frac{1}{\alpha_{k+1}}\right) |x_{k+1}| \right\} \\
&\leq \left(\frac{1}{\alpha_k} - \frac{1}{\alpha_{k+1}}\right) \|x\| \leq \left(\frac{1}{\alpha_k} - \frac{1}{\alpha_{k+1}}\right) \|x\|_\alpha .
\end{aligned}$$

Therefore

$$\left| \left\|A_k x\right\|_\alpha - \left\|x\right\|_\alpha \right| \leq \varepsilon_k \left\|x\right\|_\alpha \qquad \text{for all } x \in H_\alpha,$$

where $\varepsilon_k \to \infty$ as $k \to \infty$. Let us now show that for any isometry $T$ of $H_\alpha$ one has

$$\left\|A_k - T\right\|_\alpha \geq 1 \qquad \text{for all} \quad k \in \mathbb{N}. \tag{22.3}$$

Applying now Lemma 22.2 we immediately obtain (22.3). Indeed, since we have $\|\alpha_{k+1} e_{k+1}\|_\alpha = 1$ for each $k$, it follows that for any coordinate symmetry $T$

$$\left\|A_k - T\right\|_\alpha \geq \left\|(A_k - T)\, \alpha_{k+1} e_{k+1}\right\|_\alpha = \alpha_{k+1} \left\|e_k \pm e_{k+1}\right\|_\alpha \geq \frac{\alpha_{k+1}}{\alpha_k} \geq 1.$$

Combining (22.2) and (22.3) and taking into account that $1/\alpha_k - 1/\alpha_{k+1} \to 0$ as $k \to \infty$, we conclude that $H_\alpha$ does not possess SSLI.

The norm $\|\cdot\|_\alpha$ satisfies all the assumptions except for the smoothness. This norm is not differentiable at any point $x$, for which $\alpha_i x_i = \|x\|$ for some $i$. To make a smooth norm we first take an arbitrary sequence $\{\beta_i\}$ satisfying (a), (b) and such that $\beta_i < \alpha_i$ for all $i \in \mathbb{N}$. Then by a standard argument we construct for every $i$ an infinitely smooth norm $\varphi_i(x)$ in $\mathbb{R}^2$ such that

$$\varphi_i(x) = \sqrt{x_1^2 + x_2^2}$$

if $x_1 \le \beta_i \|x\|$, and $\varphi_i(x) = 1/\alpha_i$ if $x_1 \ge \alpha_i \|x\|$. Now consider a norm $\|\cdot\|_{\alpha,\beta}$ in $H$ defined as follows: $\|x\|_{\alpha,\beta} = \|x\|$ if $x_i < \beta_i \|x\|$ for all $i$, and

$$\|x\|_{\alpha,\beta} = \varphi_j\left(x_j, \sqrt{\|x\|^2 - x_j^2}\right)$$

if $x_j \ge \beta_j \|x\|$ for some $j$ (property (b) yields that such $j$ is always unique, if it exists). Note that $\|x\|_{\alpha,\beta} = \|x\|_\alpha$, provided $x_i < \beta_i \|x\|$ for all $i$, or $x_j \ge \alpha_j \|x\|$ for some $j$. This is easily checked that $\|\cdot\|_{\alpha,\beta}$ is a norm equivalent to the standard norm in $H$. The proof of Lemma 22.1 and of inequality (22.3) for this norm is literally the same as for $\|\cdot\|_\alpha$. It remains to establish the smoothness of the function $g(x) = \|\cdot\|_{\alpha,\beta}$. Let us show that $g$ is infinitely Frechet differentiable at any point $x \ne 0$. If $x_i < \beta_i \|x\|$ for all $i$, then the same holds for any point of some neighborhood of $x$ (this follows from (b)). Hence, $g(x) = \|x\|$ in this neighborhood, and $g$ is infinitely differentiable at $x$. If $x_j \ge \beta_j \|x\|$ for some $j$, then such $j$ is unique, and

$$g(x) = \varphi_j\left(x_j, \sqrt{\|x\|^2 - x_j^2}\right)$$

at some neighborhood of $x$ hence, it is infinitely differentiable at $x$.

## 22.3 SLI Property vs. SSLI

Now we are going to see that SSLI property does not imply SLI. As an auxiliary result we prove SSLI for the spaces $l_1$ and $l_\infty$ with $\sigma(\varepsilon) = 3\varepsilon$.

The first observation is for surjective SLI property the approximating isometry is also surjective. If $A$ is a surjective $\varepsilon$-isometry and $T$ is an isometry such that $\|A - T\| \le \sigma(\varepsilon)$, then $T$ is surjective, whenever $\varepsilon + \sigma(\varepsilon) < 1$. Indeed, if $T$ is not surjective, then its image is closed (Lemma 22.2), and hence there is $y \in S^*$ such that $(y, Ax) = 0$ for all $x \in X$. Let us remember that we write $S$ and $S^*$ for the unit spheres in $X$ and $X^*$ respectively. Further, take $\bar{x} \in S$ such that $(\bar{x}, y) \ge 1 - \alpha$, where $\alpha > 0$ is small, and take $z \in X$, for which $Az = \bar{x}$. We have

$$\|z\| \le \frac{1}{1-\varepsilon} \quad \text{and} \quad (Az - Tz, y) = (Az, y) = (\bar{x}, y) \ge 1 - \alpha.$$

On the other hand,

$$(Az - Tz, y) \le \sigma(\varepsilon)\,\|z\| \le \frac{\sigma(\varepsilon)}{1-\varepsilon}.$$

Hence

$$\frac{\sigma(\varepsilon)}{1-\varepsilon} \ge 1 - \alpha.$$

Taking $\alpha \to 0$, we obtain $\varepsilon + \sigma(\varepsilon) \ge 1$. The inverse inequality implies contradiction, which means that $T$ is surjective.

**Proposition 22.1.** *If a Banach space possesses SSLI with a function $\sigma$, then its dual space does with the same function.*

*Proof.* Let us first show that if $A^*$ is a surjective $\varepsilon$-isometry of the dual space $X^*$, where $\varepsilon \in [0,1/2)$, then $A$ is that of the primal space $X$. Indeed, the upper bound $\|Ax\| \le (1+\varepsilon)\|x\|$ is simple, because $\|A\| = \|A^*\| \le 1+\varepsilon$. To prove the lower bound we consider an arbitrary $x \in S$, take $x^* \in S^*$ such that $(x^*,x) = 1$, and $z \in X^*$, for which $A^*z = x^*$. Since $A^*z \le (1+\varepsilon)\,\|z\|$, it follows that

$$\begin{aligned} \|Ax\| &= \sup_{y\in S^*}(y,Ax) = \sup_{y\in S^*}(A^*y,x) \\ &\ge \frac{1}{\|z\|}(Az,x) = \frac{1}{\|z\|}(x^*,x) = \frac{1}{\|z\|} \ge \frac{1}{1+\varepsilon} \ge 1-\varepsilon. \end{aligned}$$

Suppose now that $X$ possesses SSLI. If $A^*$ is an $\varepsilon$-isometry in $X^*$, then so is $A$, and there is a surjective isometry $T$ such that $\|A-T\| \le \sigma(\varepsilon)$. Hence, $T^*$ is also an isometry and $\|A^*-T^*\| = \|A-T\| \le \sigma(\varepsilon)$. □

**Theorem 22.2.** *The spaces $l_1$ and $l_\infty$ possess the SSLI property with $\sigma(\varepsilon) = 3\,\varepsilon$.*

*Proof.* Due to Proposition 22.1 it suffices to prove the theorem for $l_1$. Let $A$ be a surjective $\varepsilon$-isometry of $l_1$ with $\varepsilon < 1/3$. Let $e_j$ be a basis element of $l_1$ (the $j$th coordinate is 1, all others are zeros). There is an element $x = (x^1,x^2,\dots) \in l_1$ such that $\|x\| = 1$ and $Ax = \lambda\,e_j$. Since $A$ is an $\varepsilon$-isometry, we have $|\lambda| \ge 1-\varepsilon$. The supremum of the convex function $\varphi(z) = \big|(Az,e_j)\big|$ on the unit ball of the space $l_1$ equals to its supremum on the set of its extreme points, which are $\pm e_i\,, i\in\mathbb{N}$. Hence, there exists $i$, for which $\varphi(e_i) \ge \varphi(x) = |\lambda|$. Thus, for any $j\in\mathbb{N}$ there is $i = i(j)$ such that $|(Ae_i,e_j)| \ge 1-\varepsilon$. If there are several such indices $i$ we take any of them as $i(j)$. Let us show that the map $j \mapsto i(j)$ is actually one-to-one on the set $\mathbb{N}$. The injectivity is easy: for any $i$ one has $\|Ae_i\| \le 1+\varepsilon < 2(1-\varepsilon)$; on the other hand, if $i(j_1) = i(j_2)$ for some $j_1 \ne j_2$, then

$$\|Ae_i\| \ge |(Ae_i,e_{j_1})| + |(Ae_i,e_{j_2})| \ge 2(1-\varepsilon),$$

which is a contradiction.

To prove the surjectivity we observe that for any $j$ one can write $e_j = \delta_j A\,e_i + r_j$, where $i = i(j)$, $\delta_j \in \{1,-1\}$ is the sign of $(Ae_i,e_j)$, and $\|r_j\| \le 3\varepsilon$. Indeed, the $j$th entry of $Ae_i$ is at least $1-\varepsilon$ by modulus, while $\|Ae_i\| \le 1+\varepsilon$. Whence, $\|r_j\| = \big\|\,\delta_j Ae_i - e_j\,\big\| \le 3\varepsilon$. For an arbitrary $s\in\mathbb{N}$ we have

$$A\,e_s = \sum_{j\in\mathbb{N}} z^j\,e_j = \sum_{j\in\mathbb{N}} z^j\,(\,\delta_j A\,e_i - r_j\,) = A\left(\sum_{j\in\mathbb{N}} \delta_j z^j\,e_i\right) - \sum_{j\in\mathbb{N}} z^j\,r_j.$$

Since

$$\left\|\sum_{j\in\mathbb{N}} z^j\,r_j\right\| \le \|z\|\,3\varepsilon,$$

we get

$$A\left(\sum_{j\in\mathbb{N}}\delta_j z^j e_{i(j)} - e_s\right) \le 3\varepsilon\,\|z\|.$$

If the index $s$ does not belong to the set $\{i(j) \mid j\in\mathbb{N}\}$, then

$$\left\|\sum_{j\in\mathbb{N}}\delta_j z^j e_{i(j)} - e_s\right\| = \sum_{j\in\mathbb{N}}|z^j| + 1 = \|z\| + 1,$$

which implies

$$A\left(\sum_{j\in\mathbb{N}}\delta_j z^j e_{i(j)} - e_s\right) \ge \big(\|z\| + 1\big)\big(1 - \varepsilon\big).$$

Thus, we obtain the inequality $3\varepsilon\,\|z\| \ge \big(\|z\| + 1\big)\big(1 - \varepsilon\big)$, from which we conclude

$$\varepsilon > \frac{1}{4} \quad\text{and}\quad \|z\| \ge \frac{1-\varepsilon}{4\varepsilon-1}.$$

On the other hand, $\|z\| = \|Ae_s\| \le 1+\varepsilon$. Combining these inequalities, we get

$$1+\varepsilon \ge \frac{1-\varepsilon}{4\varepsilon-1},$$

which does not hold for $\varepsilon \in [0,1/3]$. Thus, any index $s$ belongs to the image of the map $i(j)$, which proves the surjectivity. Since the map $i(j)$ is one-to-one, we take an inverse map $j(i)$ and consider the operator $T : l_1 \to l_1\,, T(e_i) = \delta_{j(i)}\, e_{j(i)}\,,\ i \in \mathbb{N}$. This is an isometry, and for any $i$ one has $\|A\, e_i - T\, e_i\| \le 3\varepsilon$. Indeed, writing $j = j(i)$ we see that the $j$th entry of the element $A\, e_i - T\, e_i$ equals to $(Ae_i, e_j) - \delta_j$, which is at most $\varepsilon$ by modulus, because $|(Ae_i, e_j)| \ge 1-\varepsilon$. The sum of modules of all other entries is at most

$$\|A\, e_i\| - (1-\varepsilon) \le (1+\varepsilon) - (1-\varepsilon) = 2\varepsilon.$$

Thus, $\|(A-T)e_i\| \le 3\varepsilon$ for all $i$. Since the norm of any operator in $l_1$ can be computed over the basis vectors, it follows that $\|A-T\| \le 3\varepsilon$. □

*Remark 22.1.* The result of Theorem 22.2 holds for the space $\mathbb{R}^d$ equipped with $l_1$-norm or with $l_\infty$-norm. The proof is the same as Theorem 22.2.

*Remark 22.2.* The main result of [4] yields that for any $p \in [1,+\infty)$ the space $l_p$ possesses not only SSLI, but SLI. However, the function $\sigma(\varepsilon)$ has not been specified in that work. Theorem 4 in [6] establishes SSLI of the space $C_0(M)$ of continuous vanishing at infinity functions on a locally compact Hausdorff space $M$. A special case of that result is the SSLI property of the subspace $c_0 \subset l_\infty$ that consists of vanishing sequences. In [3], it was shown that for any space $C(K)$, where $K$ is a compact metric space, we have $\sigma(\varepsilon) = 3\varepsilon$. Theorem 22.2 extends this result to $K = \mathbb{Z}$, i.e., to the space $l_\infty$.

Now we can prove that SLI and SSLI are not equivalent.

**Theorem 22.3.** *For any $d \geq 2$ the space $\mathbb{R}^d \oplus l_1$ possesses SSLI, but not SLI.*

The proof is outlined in Appendix. The idea is to apply the Dvoretzky theorem [11, 12], according to which for any $\varepsilon > 0$ there is $N(\varepsilon, d)$ such that for any $n$-dimensional Banach space $Y$, where $n \geq N$, there is an $\varepsilon$-isometry from the Euclidean space $\mathbb{R}^d$ to $Y$. Therefore, the operator $A_1 + A_2$, where $A_1$ is an $\varepsilon$-isometry of $\mathbb{R}^d$ to the $n$-dimensional subspace of $l_1$ corresponding to the first $n$ coordinates, and $A_2$ is the right $n$-shift of $l_1$, is a (non-surjective) $\varepsilon$-isometry. The distance from $A$ to the closest isometry is at least 2, because any isometry leaves $\mathbb{R}^d$ invariant. That is why this space does not possess SLI. On the other hand, it has SSLI, which can be derived from Theorem 22.3.

## 22.4 Finite Dimensional Spaces. The Conditional Number of Matrices

We consider the space $\mathbb{R}^d$ with some norm $\|\cdot\|$, maybe not Euclidean. The condition number $\kappa(A)$ of an operator $A$ acting on this space, or of the matrix of this operator (if the basis in $\mathbb{R}^d$ is fixed) is $\|A\| \cdot \|A^{-1}\|$. The condition number of matrices plays a crucial role in numerical methods of linear algebra and PDE. For any operator $\kappa(A) \geq 1$, and the equality takes place only if $A$ is a similarity operator, i.e., is proportional to an isometry. It appears that there is a stability result for this property: $\kappa(A)$ is close to 1 if and only if $A$ can be approximated by a similarity operator.

If $A$ can be $\varepsilon$-approximated by an isometry, then

$$\kappa(A) \leq \frac{1+\varepsilon}{1-\varepsilon}.$$

Thus, any operator close to an isometry has the conditional number close to one. The converse is not true, because the conditional number does not depend on multiplication of $A$ by a positive scalar, while the distance to isometries does. Nevertheless, a converse statement can be formulated as follows: If $\kappa(A)$ is close to 1, then $A$ can be approximated by a similarity operator, i.e., by $\lambda T$, where $\lambda$ is some scalar and $T$ is an isometry.

**Proposition 22.2.** *For any operator $A$ there is an isometry $T$ and a constant $\lambda > 0$ such that*

$$\|A - \lambda T\| \quad \leq \quad \|A\| \, \frac{\kappa+1}{2\kappa} \, \sigma\left(\frac{\kappa-1}{\kappa+1}\right),$$

*where $\kappa = \kappa(A)$ and $\sigma(\varepsilon)$ is the SLI function corresponding to a given norm in $\mathbb{R}^d$.*

*Proof.* For $\varepsilon = (\kappa - 1)/(\kappa + 1)$ we have

$$\|A\|\|A^{-1}\| = \frac{1+\varepsilon}{1-\varepsilon}.$$

Therefore, for $\mu = (1+\varepsilon)\|A\|^{-1}$ the operator $\mu A$ satisfies $\|\mu A\| = 1+\varepsilon$ and $\|(\mu A)^{-1}\| = 1/(1-\varepsilon)$. Whence, $\mu A$ is an $\varepsilon$-isometry, therefore, there is an isometry $T$ such that $\|\mu A - T\| \leq \sigma(\varepsilon)$. Thus,

$$\|A - \mu^{-1}T\| \leq \frac{\|A\|}{1+\varepsilon}\sigma(\varepsilon),$$

from which the proposition follows. □

Thus, the condition number $\kappa(A)$ indicates how the operator $A$ can be approximated by similarity operators. If the norm in $\mathbb{R}^d$ is Euclidean, then $\sigma(\varepsilon) = \varepsilon$, and hence the distance from $A$ to the closest similarity operator does not exceed $\|A\|(\kappa - 1)/(2\kappa)$. For the $l_1$-norm and $l_\infty$-norm we have $\sigma(\varepsilon) = 3\varepsilon$ (Theorem 22.2 and Remark 22.1), and hence this distance does not exceed $\|A\|(\kappa - 1)/(2\kappa)$.

## 22.5 Open Problems

1. Find sufficient conditions for a Banach space to possess SLI or SSLI.
2. Estimate $\sigma(\varepsilon)$ for the classical functional spaces, such as $l_p, L_p, W_p^k, C^k$.
3. Estimate the functions $\sigma(\varepsilon)$ for the space $\mathbb{R}^d$ with various norms, in particular, for the $l_p$-norm, $p \in (1, +\infty)$, $p \neq 2$.

## 22.6 Appendix

In the proof of Lemma 22.1, we use the following simple and well-know fact:

**Lemma 22.2.** *If $X$ and $Y$ are Banach spaces and $T : X \to Y$ is a continuous operator such that $\|Tx\| \geq C\|x\|$ for all $x$, then its image is closed.*

*Proof of Lemma 22.1.* Let $B_\alpha$ and $S_\alpha$ be the unit ball and the unit sphere in $H_\alpha$, respectively. Let $E_k = \{x \in H_\alpha \mid x_k = 0\}$ be the $k$th coordinate hyperplane. Let also

$$P_k^+ = \{x \in H_\alpha \mid \|x\| \leq 1,\, x_k = \alpha_k\}, \quad P_k^- = \{x \in H_\alpha \mid \|x\| \leq 1,\, x_k = -\alpha_k\}$$

for $k \in \mathbb{N}$. These are Euclidean balls of codimension 1 laying on $S_\alpha$ at the (Euclidean) distance of $\alpha_k$ from the origin. Since $(1-\alpha_k^2) + (1-\alpha_j^2) < 1$ for all $k \neq j$ (assumption (b)), all these balls are pairwise disjoint. In the $\alpha$-norm, the diameter of the balls $P_1^{\pm}$ is $\sqrt{1-\alpha_1^2}/\alpha_2$, and for all $k \geq 2$ the diameter of $P_k^{\pm}$ is $\sqrt{1-\alpha_k^2}/\alpha_1$. One can easily deduce from assumptions (a) and (b) that these diameters decrease in $k$.

Thus, for any point $x \in S_\alpha$ either $\|x\| = 1$ or $x \in \mathrm{int} P_k^\delta$ (the interior is in the corresponding hyperplane) for some $k \in \mathbb{N}$ and $\delta \in \{+,-\}$. All non-extreme points of $B_\alpha$ (midpoints of segments contained in $B_\alpha$) on the unit sphere $S_\alpha$ are precisely points from $\mathrm{int} P_k^\delta$. Consider now an arbitrary linear isometry $T$ of $H_\alpha$. We have $TS_\alpha \subset S_\alpha$. Since $T$ is injective, it takes any non-extreme point of $S_\alpha$ to a non-extreme point. So, the image of the open ball $\mathrm{int} P_j^+$ is in the union of all the balls $P_k^\delta$, $k \in \mathbb{N}$, $\delta \in \{+,-\}$. The image is connected (since the ball $\mathrm{int} P_j^+$ is and $T$ is continuous), whereas the union is disjoint, therefore the image lies entirely in one of these balls. Thus, for any $j$ we have $T(P_j^+) \subset P_m^\delta$ for some $\delta \in \{+,-\}$, where $m = m(j)$ is some function from $\mathbb{N}$ to $\mathbb{N}$. Observe that $m(j) \leq j$ for all $j$, otherwise some image $T(P_j^+)$, whose diameter is equal to the diameter of $P_j^+$ is contained in the ball $P_m^\delta$ of a smaller diameter. On the other hand, the function $m(j)$ is injective. Indeed, if $m(j) = m(k)$ for some $j \neq k$, then $TE_j \subset E_m$ and $TE_k \subset E_m$. Since the hyperplanes $E_j$ and $E_k$ span the whole $H_\alpha$, it follows that $TH_\alpha \subset E_m$. This is impossible, because the balls $P_m^+, P_m^-$, which contain $TP_j^+$, does not intersect $E_m$. So, $m(j)$ is injective and $m(j) \leq j$ for all $j$, hence $m(j) = j$ for all $j \in \mathbb{N}$. Thus, $TP_j^+$ is contained either in $P_j^+$ or in $P_j^-$. Consequently, for any $j$ the modules of $j$th coordinate of the vector $Te_j$ is at least 1. This yields that $T$ is a surjection. Otherwise, its image is a proper subspace of $H_\alpha$, which is closed by Lemma 22.2. Whence, there is $a \in H$, $\|a\| = 1$ such that $(a, Te_j) = 0$ for all $j \in \mathbb{N}$. Writing $z = Te_1$, we have

$$a_1 \bar{z}_1 = -\sum_{k=2}^{\infty} a_k \bar{z}_k,$$

whence

$$|a_1 \bar{z}_1| = \left| \sum_{k=2}^{\infty} a_k \bar{z}_k \right| \leq \sqrt{1-a_1^2}\,\sqrt{\|z\|^2 - |z_1|^2}$$

(Cauchy–Schwarz inequality). Since $\|T\alpha_1 e_1\| \leq 1$, we have $\|z\| \leq 1/\alpha_1$. On the other hand, $|z_1| \geq 1$. Therefore

$$|a_1| \leq \sqrt{1-a_1^2}\sqrt{\frac{1}{\alpha_1^2} - |z_1|^2},$$

which after elementary simplification implies $|a_1| \leq \sqrt{1-\alpha_1^2}$. In the same way we show that

$$|a_j| \leq \sqrt{1-\alpha_j^2} \quad \text{for every} \quad j \in \mathbb{N}.$$

This yields

$$\|a\| \le \left(\sum_j (1-\alpha_j^2)\right)^{1/2} < 1$$

(assumption (b)), which contradicts the assumption. Thus, $T$ is surjective. So, it maps the ball $P_j^+$ to the whole ball $P_j^\delta$, $\delta \in \{+,-\}$, and hence it takes its center to the center of the ball $P_j^\delta$. Consequently, for each $j$ the element $Te_j$ equals to either $e_j$ or $-e_j$, and so $T$ is a coordinate symmetry. □

*Proof of Theorem 22.3.* Let $Z = X \oplus Y$, where $X$ and $Y$ are isomorphic to $\mathbb{R}^d$ and $l_1$ respectively. Any element $z \in Z$ is considered as a sequence $\{z_k\}_{k\in\mathbb{N}}$ with the norm

$$|z\| = \sqrt{|z_1|^2 + \cdots + |z_d|^2} + \sum_{k=d+1}^{\infty} |z_k|.$$

Thus, $z = x + y$ and $\|z\| = \|x\|_2 + \|y\|_1$ (in the sequel we omit the indices of norms), where $x = (z_1, \ldots, z_d) \in X$ and $y = (z_{d+1}, \ldots) \in Y$.

Let us first show that $Z$ has SSLI. Let $A$ be a surjective $\varepsilon$-isometry with some small $\varepsilon$. Denote $A^{-1}e_j = x^j + y^j$, where $e_j$ is a basis vector, $j \ge d+1$. If $\|x^j\| = q$, then $(q + \|y^j\|)(1+\varepsilon) \ge 1$, and hence $\|y^j\| \le 1/(1-\varepsilon) - q$. Take a vector $b \in X$ orthogonal to $x^j$, $\|b\| = q$. Denote $z = Ab$ and assume that $z_{d+1} \ge 0$ (the opposite case is considered in the same way). Then

$$\|z + e_j\| = 1 + \|z\| \ge 1 + (1-\varepsilon)q.$$

On the other hand,

$$\|x^j + y^j + b\| = \|x^j + b\| + \|y^j\| = \sqrt{2}q + \|y^j\|.$$

Since $A(x^j + y^j + b) = z + e_j$, we have

$$\|z + e_j\| \le (\sqrt{2}q + \|y^j\|)(1+\varepsilon) \le \left(\frac{1}{1-\varepsilon} - q + \sqrt{2}q\right)(1+\varepsilon).$$

Thus

$$\left(\frac{1}{1-\varepsilon} - q + \sqrt{2}q\right) \quad \ge \quad 1 + (1-\varepsilon)q\,,$$

which implies that $q \le \left(2+\sqrt{2}\right)\varepsilon + o(\varepsilon)$ as $\varepsilon \to 0$. So, if $\varepsilon$ is small enough, then $\|x^j\| \le 4\varepsilon$ for every $j \ge 3$. For any $y \in Y$ we denote $A^{-1}y = \tilde{x} + \tilde{y}$, where $\tilde{x} \in X, \tilde{y} \in Y$, and consider the operator $B: Y \to Y$ defined as $By = \tilde{y}$. Clearly, $B$ is a linear surjective operator in $l_1$. We see that $\|(A^{-1} - B)e_j\| \le 4\varepsilon$ for all basis vectors $e_j$ of the space $Y$. Since those vectors are only extreme points of the unit ball of $Y$, it follows that $\|A^{-1} - B\|_Y \le 4\varepsilon$. This yields that $B$ is an $O(\varepsilon)$-isometry of $Y$, therefore it is invertible, and the operator $A_2 = B^{-1}$ is also an $O(\varepsilon)$-isometry. Hence, $\|A - A_2\|_Y \leq O(\varepsilon)$. Furthermore, for $e \in X, \|e\| = 1$, we denote $Ae = x + y$, where $x \in X, y \in Y$, and write $A_1$ for the operator on $X$ taking $e$ to $x$. Clearly

$$\|x\| + \|y\| \leq 1 + \varepsilon$$

and

$$A^{-1}y \leq \frac{\|y\|}{1-\varepsilon}.$$

Writing $A^{-1}y = \tilde{x} + \tilde{y}$, where $\tilde{x} \in X, \tilde{y} \in Y$, we get

$$\|\tilde{x}\| \leq 4\varepsilon \|A^{-1}y\| \leq \frac{4\varepsilon\|y\|}{1-\varepsilon}.$$

On the other hand,

$$1 = \|e\| \leq \|\tilde{x}\| + \|A^{-1}x\| \leq \frac{4\varepsilon\|y\|}{1-\varepsilon} + \frac{\|x\|}{1-\varepsilon}.$$

Substituting $\|x\| \leq 1 + \varepsilon - \|y\|$, we finally get $1 + \varepsilon - \|y\| + 4\varepsilon\|y\| \geq 1 - \varepsilon$, and therefore $\|y\| \leq 2\varepsilon/(1 - 4\varepsilon)$. Hence, $\|A - A_1\|_X \leq 2\varepsilon + o(\varepsilon)$. Thus, we approximate $A$ by the sum of two operators $A_1 + A_2$, $A_1 : X \to X, A_2 : Y \to Y$, both of them are surjective $O(\varepsilon)$-isometries. Since the spaces $\mathbb{R}^d$ and $l_1$ possess SSLI (Theorem 22.2), there are approximative isometries $\tilde{A}_1, \tilde{A}_2$. The operator $\tilde{A} = \tilde{A}_1 + \tilde{A}_2$ is an isometry of the space $Z$ and $\|A - \tilde{A}\| \leq O(\varepsilon)$, which concludes the proof. Thus, $Z$ possesses SSLI.

Let us now show that $Z$ does not possess SLI property. Writing $Y_n = \{y \in Y \,|\, y_j = 0, j \geq n+1\}$ for the $n$-dimensional subspace of $Y$ corresponding to the first $n$ coordinates, we apply the Dvoretzky theorem [11, 12] and obtain that for any $\varepsilon > 0$ there is an $\varepsilon$-isometry $B_n : X \to Y_n$, whenever $n$ is large enough. Furthermore, let $C_n : Y \to Y$ be the operator of shift by $n$, i.e., $C_n(y_1, y_2, \ldots) = (0, \ldots, 0, y_1, y_2, \ldots)$ ($n$ zeros). Then the operator $A_n = B_n + C_n$ is an $\varepsilon$-isometry of $Z$. However, the distance from $A_n$ to any isometry of $Z$ is at least 2. To show this it suffices to prove that every isometry respects the subspace $X$. Take any isometry of $Z$ and denote by $T$ its restriction to $X$. We have $T = T_x + T_y$, where $\mathrm{Im}\, T_x \subset X$ and $\mathrm{Im}\, T_y \subset Y$.

If the rank $k$ of the operator $T_x$ (i.e., the dimension of $\mathrm{Im}\, T_x$) is smaller than $d$, then in case $k \geq 1$ there is a subspace $L \subset X$ of dimension $d - k + 1$ and a vector $a \in X \setminus \{0\}$ such that $\|T_x u\|_2 = |(a, u)|$ for any $u \in L$. Hence, the operator $P : L \to l_1$, $Pu = \big((a, u), (T_y u)_1, (T_y u)_2, \ldots\big)$ is isometric, which is impossible, because there is no isometry from a Euclidean space of dimension $\geq 2$ to $l_1$. This is well-known, see, for instance [13]. If $k = 0$, then $T_x = 0$, and the operator $T = T_y$ isometrically maps $\mathbb{R}^d$ to $l_1$, which is again impossible. Thus, $T_x$ is nondegenerate. In this case the function $u \mapsto \|T_x u\|_2$ is differentiable at all points $u \neq 0$. Consequently, the function $g(u) = \|T_y u\|_1$ is differentiable for $u \neq 0$ as well.

Indeed, $\|T_y u\|_1 = \|Tu\| - \|T_x u\|_2 = \|u\|_2 - \|T_x u\|_2$ (the latter equality is because $T$ is an isometry). Thus, $g$ is a difference of two differentiable functions. On the other hand, for any nontrivial linear operator $A : \mathbb{R}^d \to l_1$, $d \geq 2$ the function $u \mapsto \|Au\|_1$ is not Gâteaux differentiable at some point $u \neq 0$ [13, p. 253]. Therefore, $T_y = 0$, and hence $TX \subset X$. □

**Acknowledgement** The research is supported by the grants RFBR 11-01-00329 and RFBR 10-01-00293, and by the grant of Dynasty foundation.

## References

1. Ding, G.G.: The approximation problem of almost isometric operators by isometric operators. Acta Math. Sci. **8**, 361–372 (1988)
2. Huang, S.Z.: Constructing operators which cannot be isometricaly approximated. Acta Math. Sci. **6**, 195–200 (1986)
3. Benyamini, Y.: Small into-isomorphisms between spaces of continuous functions. Proc. Amer. Math. Soc. **83**, 479–485 (1981)
4. Alspach, D.E.: Small into isomorphisms on $L_p$ spaces. Illinois J. Math. **27**, 300–314 (1983)
5. Xiang, S.: Small into isomorphisms on uniformly smooth spaces. J. Math. Anal. Appl. **290**, 310–315 (2004)
6. Jarosz, K.: Small isomorphisms of C(X,E) spaces. Pac. J. Math. **138**, 295–315 (1989)
7. Rassias, Th.M., Šemrl, P.: On the Hyers-Ulam stability of linear mappings. J. Math. Anal. Appl. **173**, 325 – 338 (1993)
8. Park, C., Rassias, Th.M: Isometric additive mappings of quasy-Banach spaces. Nonlinear Func. Anal. Appl. **12**, 377–385 (2007)
9. Dolinar, G.: Generalized stability of isometries. J. Math. Anal. Appl. 242, 39–56 (2000)
10. Report on the 12th ICFEI. Ann. Acad. Pedagog. Crac. Stud. Math. **7**, 125–159 (2008)
11. Dvoretzky, A.: Some results on convex bodies and Banach spaces. In: Proc. Internat. Sympos. Linear Spaces (Jerusalem, 1960), 123–160
12. Milman, V.D.: A new proof of the theorem of A. Dvoretzky on intersections of convex bodies. Funct. Anal. Appl. **5**, 288–295 (1971)
13. Fabian, M., Habala, P., Hajek, P., Montesinos Santalucia, V., Pelant, J., Zizler, V.: Functional Analysis and Infinite-Dimensional Geometry. CMS Books in Mathematics. Springer-Verlag, New York (2001)

# Chapter 23
# Ulam Stability of the Operatorial Equations

**Ioan A. Rus**

*Dedicated to the memory of S.M. Ulam*
*on the 100th anniversary of his birth*

**Abstract** Let $(E,+,\mathbb{R},\leq,\to)$ be an ordered linear $L$-space, $E_+ := \{e \in E \mid e \geq 0\}$, $(X,d)$ and $(Y,\rho)$ be two generalized metric spaces with $d(x,y)$, $\rho(x,y) \in E_+$, and $f,g : X \to Y$ be two operators. In this paper we present for the coincidence equation

$$f(x) = g(x)$$

four types of Ulam stability: Ulam–Hyers stability, generalized Ulam–Hyers stability, Ulam–Hyers–Rassias stability and generalized Ulam–Hyers–Rassias stability. Some illustrative examples are given, the relations of Ulam stability with the weakly Picard operator are studied and two research directions are also presented.

**Keywords** Generalized metric space • Operatorial equation • Ulam–Hyers stability • Ulam–Hyers–Rassias stability • Weakly Picard operator • Fixed point structure • Data dependence

**Mathematics Subject Classification (2010):** Primary 47H10, 54H25, 45N05, 39A11

## 23.1 Introduction

The basic hypostasies of data dependence in the theory of operatorial equations are the following (see, e.g., [5, 6, 14, 31, 43, 44, 46, 47, 50, 55–57]): monotony w.r.t. data, continuity w.r.t. data, differentiability w.r.t. parameters, Liapunov stability, asymptotic behavior, asymptotic equivalence, structural stability, analiticity of

---

I.A. Rus (✉)
Department of Applied Mathematics, Babeş–Bolyai University, Kogălniceanu Street, No. 1, 400084, Cluj-Napoca, Romania
e-mail: iarus@math.ubbcluj.ro

Th.M. Rassias and J. Brzdęk (eds.), *Functional Equations in Mathematical Analysis*, Springer Optimization and Its Applications 52, DOI 10.1007/978-1-4614-0055-4_23,

solutions, regularity of solutions, $G$-convergence, etc. On the other hand, in the theory of functional equations there are some special kind of data dependence (Ulam (1940; [59]), Hyers (1941; [15]), Aoki (1950; [2]), Bourgin (1951; [7]), Rassias (1978; [37]), Hyers (1983; [16]), Găvruţă (1994; [13]), Radu (2003; [36]), Jung [22], Hyers et al. [17], Cădariu [10], Brockner and Trif [8], Alsina and Ger [1], Hyers et al. [18], Páles [33], Rassias and Šemrl [42], Hyers and Rassias [19], Isac and Rassias [20], Isac and Rassias [21], Jung and Rassias [28], Prastaro and Rassias [35], Rassias [38, 39], Rassias (ed.) [40, 41], Trif [58], Wang [60], etc.).

With these considerations (notions, results, examples, counterexamples, etc.) in mind we introduced in [54] four types of Ulam stability for the operatorial equations in a metric space: Ulam–Hyers stability, generalized Ulam–Hyers stability, Ulam–Hyers–Rassias stability and generalized Ulam–Hyers–Rassias stability.

The aim and scope of this paper is to illustrate the complexity of Ulam stability of the operatorial equations in a generalized metric space. The outline of the paper is the following:

1. Generalized metric spaces
2. Ulam–Hyers stability on a generalized metric space
3. Ulam–Hyers and Ulam–Hyers–Rassias stability in the case of equations on a space of functions
4. Illustrative examples
5. Ulam stability of fixed point equations, via weakly Picard operators
6. Ulam stability of a fixed point structure on a metric space
7. Ulam stability of the equations with set-to-point operators

Throughout this paper, we use the terminology and the notations in [47] (see also [46, 56]). For the convenience of the reader, we recall that $\mathbb{N}$, $\mathbb{N}^*$, $\mathbb{Z}$, $\mathbb{R}$, $\mathbb{C}$ denote the sets of nonnegative integers, positive integers, integers, reals and complex numbers, respectively. Further, if $X$ is a nonempty set and $f : X \to X$ is an operator, then: $P(X) := \{Y \subset X \mid Y \neq \emptyset\}$, $f^0 := 1_X$, $f^1 := f$, $f^{n+1} := f \circ f^n$, $n \in \mathbb{N}$, are the iterate operators of the operator $f$, $F_f := \{x \in X \mid f(x) = x\}$ is the fixed point set of $f$.

## 23.2 Generalized Metric Spaces

Let $E$ be a nonempty set. Let

$$s(E) := \{(e_n)_{n\in\mathbb{N}^*} \mid e_n \in E,\ n \in \mathbb{N}^*\}$$

be the set of all sequences with the elements in $E$. Let $c(E) \subset s(E)$ be a subset of $s(E)$ and $Lim : c(E) \to E$ be an operator. By Fréchet (see [47, 56]), the triple $(E, c(E), Lim)$ is called an $L$-space if the following conditions are satisfied:

1. if $e_n = e$, $n \in \mathbb{N}^*$, then $(e_n)_{n\in\mathbb{N}^*} \in c(E)$ and $Lim(e_n)_{n\in\mathbb{N}^*} = e$;
2. if $(e_n)_{n\in\mathbb{N}^*} \in c(E)$ and $Lim(e_n)_{n\in\mathbb{N}^*} = e$, then all subsequences $(x_{n_i})_{i\in\mathbb{N}^*}$ of $(e_n)_{n\in\mathbb{N}^*}$ are in $c(E)$ and $Lim(x_{n_i})_{i\in\mathbb{N}^*} = e$.

By definition, an element of $c(E)$ is convergent and $e = Lim(e_n)_{n\in\mathbb{N}^*}$ is the limit of this sequence and we write

$$x_n \to e \quad \text{as} \quad n \to \infty.$$

We denote an $L$-space by $(E,\to)$. For some examples of $L$-spaces see [47,56,57].

In a standard way we define an ordered linear $L$-space (see [11,62]).

Let $(E,+,\mathbb{R},\leq,\to)$ be an ordered (real) linear $L$-space. Let

$$E_+ := \{e \in E \mid e \geq 0\}$$

and

$$E_+^* := \{e \in E \mid e \geq 0 \text{ and } e \neq 0\}.$$

Let $X$ be a nonempty set and $d : X \times X \to E_+$ be a generalized metric on $X$, i.e. $d$ satisfies the Fréchet axioms:

1. $d(x,y) = 0 \Leftrightarrow x = y$;
2. $d(x,y) = d(y,x),\ \forall x,y \in X$;
3. $d(x,y) \leq d(x,z) + d(z,y),\ \forall x,y,z \in X$.

For examples and the basic notions in a such generalized metric space see Zabreiko [62], De Pascale et al. [11]. See also Rus et al. [56] (pp. 77–93).

*Remark 23.1.* There are several concepts of the generalized metric spaces (premetric, pseudometric, quasimetric, ultrametric, etc.). We call a generalized metric space with $d(x,y) \in E_+$, $E_+$-metric space. For the fixed point theory in generalized metric spaces see [30,49,56] and the references therein.

## 23.3 Ulam–Hyers Stability on a Generalized Metric Space

Following [54], in what follows we present two types of Ulam stability for an operatorial equation in a generalized metric space.

Let $(X,d)$ and $(Y,\rho)$, with $d(x,y), \rho(x,y) \in E_+$, be two generalized metric spaces and $f,g : X \to Y$ be two operators. Let us consider the coincidence equation

$$f(x) = g(x) \tag{23.1}$$

**Definition 23.1.** By definition, (23.1) is Ulam–Hyers stable if there exists a linear increasing operator $c_{f,g} : E \to E$ such that: for each $\varepsilon \in E_+^*$ and each solution $y_* \in X$ of the inequality

$$\rho(f(x), g(x)) \leq \varepsilon \tag{23.2}$$

there exists a solution $x_* \in X$ of (23.1) such that $d(y_*, x_*) \leq c_{f,g}(\varepsilon)$.

**Definition 23.2.** Equation (23.1) is generalized Ulam–Hyers stable if there exists an increasing operator $\psi : E_+ \to E_+$, continuous in 0 with $\psi(0) = 0$, such that: for each $\varepsilon \in E_+^*$ and for each solution $y_* \in X$ of (23.2) there exists a solution $x_* \in X$ of (23.1) such that

$$d(y_*, x_*) \leq \psi(\varepsilon).$$

*Remark 23.2.* If $Y := X$ and $g := 1_X$, then we have the notions of Ulam–Hyers and generalized Ulam–Hyers stability for a fixed point equation.

*Remark 23.3.* If $E := \mathbb{R}$ is endowed with the usual structures, then the Definition 23.1 and 23.2 take the following forms:

**Definition 23.1$_{\mathbf{a}}$.** Equation (23.1) is Ulam–Hyers stable if there exists a positive real number $c_{f,g} > 0$ such that: for each $\varepsilon \in \mathbb{R}_+^*$ and each solution $y_* \in X$ of (23.2) there exists a solution $x_* \in X$ of (23.1) such that

$$d(y_*, x_*) \leq c_{f,g}\varepsilon.$$

**Definition 23.2$_{\mathbf{a}}$.** Equation (23.1) is generalized Ulam–Hyers stable if there exists an increasing function $\psi : \mathbb{R}_+ \to \mathbb{R}_+$, continuous in 0 with $\psi(0) = 0$, such that: for each $\varepsilon \in \mathbb{R}_+^*$ and each solution $y_* \in X$ of (23.2) there exists a solution $x_* \in X$ of (23.1) such that

$$d(y_*, x_*) \leq \psi(\varepsilon).$$

*Remark 23.4.* The case $E := s(\mathbb{R})$ with the usual linear structure, ordered structure and with termwise convergence, $\xrightarrow{t}$. In this case, the Definition 23.1 and 23.2 take the following forms:

**Definition 23.1$_{\mathbf{b}}$.** Equation (23.1) is Ulam–Hyers stable if there exists an infinite matrix $C = (c_{ij})$, $c_{ij} \in \mathbb{R}_+$, $i, j \in \mathbb{N}^*$ such that:

(a) $Cu$ is defined for each $u \in s(\mathbb{R}_+)$;
(b) for each $\varepsilon \in s(\mathbb{R}_+^*)$ and each solution $y_* \in X$ of (23.2) there exists a solution $x_* \in X$ of (23.1) such that

$$d(y_*, x_*) \leq C\varepsilon.$$

**Definition 23.2$_{\mathbf{b}}$.** Equation (23.1) is generalized Ulam–Hyers stable if there exists an increasing operator $\psi : s(\mathbb{R}_+) \to s(\mathbb{R}_+)$, continuous in 0 with $\psi(0) = 0$, such that: for each $\varepsilon \in s(\mathbb{R}_+^*)$ and for each solution $y_* \in X$ of (23.2) there exists a solution $x_* \in X$ of (23.1) such that

$$d(y_*, x_*) \leq \psi(\varepsilon).$$

**Definition 23.3.** Equation (23.1) is locally Ulam–Hyers stable if there exist a linear increasing operator $c_{f,g} : E \to E$ and $\varepsilon_* \in E_+^*$ such that: for each $\varepsilon \in E_+^*$ with $\varepsilon \leq \varepsilon_*$ and each solution $y_* \in X$ of (23.2) there exists a solution $x_* \in X$ of (23.1) such that

$$d(y_*, x_*) \leq c_{f,g}(\varepsilon).$$

In a similar way, we define the locally generalized Ulam–Hyers stability.

## 23.4 Ulam–Hyers and Ulam–Hyers–Rassias Stability in the Case of Equations in a Space of Functions

Let $\mathbb{K}$ be $\mathbb{R}$ or $\mathbb{C}$. Let $\Omega \subset \mathbb{K}^m$ be a nonempty subset of $\mathbb{K}^m$, $X$ a set of functions $x : \Omega \to \mathbb{K}$ and $f, g : X \to X$ two operators. We consider on $X$ the generalized metric $d$ defined by

$$d(x,y)(t) := |x(t) - y(t)|, \qquad \forall t \in \Omega.$$

In this case, in addition to Definitions 23.1 and 23.2 we present the following more restrictive definitions:

**Definition 23.4.** Equation (23.1) is Ulam–Hyers stable if there exists a real number $c_{f,g} > 0$ such that: for each $\varepsilon \in \mathbb{R}_+^*$ and for each solution $y_*$ of (23.2) there exists a solution $x_*$ of (23.1) with

$$|y_*(t) - x_*(t)| \leq c_{f,g}\varepsilon, \qquad \forall t \in \Omega.$$

**Definition 23.5.** Equation (23.1) is generalized Ulam–Hyers stable if there exists an increasing function $\psi : \mathbb{R}_+ \to \mathbb{R}_+$, continuous in 0, with $\psi(0) = 0$, such that: for each $\varepsilon \in \mathbb{R}_+^*$ and for each solution $y_*$ of (23.2) there exists a solution $x_*$ of (23.1) such that

$$|y_*(t) - x_*(t)| \leq \psi(\varepsilon), \qquad \forall t \in \Omega.$$

In this case, where the solutions are in a space of functions, we have other two notions: Ulam–Hyers–Rassias stability and generalized Ulam–Hyers–Rassias stability.

**Definition 23.6.** Let $\varphi : \Omega \to \mathbb{R}_+$ be a function. Equation (23.1) is Ulam–Hyers–Rassias stable with respect to $\varphi$ iff there exists $c_{f,g,\varphi} > 0$ such that for each $\varepsilon > 0$ and for each solution $y_*$ of the inequality

$$|f(y)(t) - g(y)(t)| \leq \varepsilon\varphi(t), \qquad \forall t \in \Omega \tag{23.3}$$

there exists a solution $x_*$ of (23.1) with

$$|y_*(t) - x_*(t)| \leq c_{f,g,\varphi}\varepsilon\varphi(t), \qquad \forall t \in \Omega.$$

**Definition 23.7.** Equation (23.1) is generalized Ulam–Hyers–Rassias stable, with respect to $\varphi$ iff there exists $c_{f,g,\varphi} > 0$ such that for each solution $y_*$ of the inequality

$$|f(y)(t) - g(y)(t)| \leq \varphi(t), \qquad \forall t \in \Omega \tag{23.4}$$

there exists a solution $x_*$ of (23.1) with

$$|y_*(t) - x_*(t)| \leq c_{f,g,\varphi}\varphi(t), \qquad \forall t \in \Omega.$$

## 23.5 Illustrative Examples

*Example 23.1.* Let us consider on $\mathbb{R}$ a metric $d$ ($d(x,y) \in \mathbb{R}_+$, $E := \mathbb{R}$, $X := \mathbb{R}$), the equation

$$x = \frac{1}{2}\sin x \tag{23.5}$$

and the corresponding inequality

$$d\left(y, \frac{1}{2}\sin y\right) \leq \varepsilon. \tag{23.6}$$

First of all, we remark that (23.5) has a unique solution, $x_* = 0$. Let us take as $d$ the following metrics:

(a) $d_1(u,v) := |u-v|$, the usual metric on $\mathbb{R}$;
(b) $d_2(u,v) := \sqrt{|u-v|}$, $u,v \in \mathbb{R}$;
(c) $d_3(u,v) := |u-v|/(1+|u-v|)$, $u,v \in \mathbb{R}$.

Let us consider (23.5) on $(\mathbb{R}, d_1)$. Let $y_*$ be a solution of the inequality (23.6), i.e.,

$$\left|y - \frac{1}{2}\sin y\right| \leq \varepsilon.$$

This implies that

$$|y_* - x_*| = |y_*| \leq \left|y_* - \frac{1}{2}\sin y_*\right| + \left|\frac{1}{2}\sin y_*\right| \leq \varepsilon + \frac{1}{2}|y_*|.$$

Hence, $|y_* - x_*| \leq 2\varepsilon$. So (23.5) is Ulam–Hyers stable in $(\mathbb{R}, d_1)$.

Now we consider (23.5) on the metric space $(\mathbb{R}, d_2)$. Let $y_*$ be a solution of the inequality

$$\sqrt{|y - \sin y|} \leq \varepsilon.$$

This implies that $|y_* - x_*| = |y_*| \leq 2\varepsilon^2$. So, (23.5) is generalized Ulam–Hyers stable on $(\mathbb{R}, d_2)$.

Finally, we consider the same (23.5) on $(\mathbb{R}, d_3)$. Let $y_*$ be a solution of the inequality

$$\frac{|y - \frac{1}{2}\sin y|}{1 + |y - \frac{1}{2}\sin y|} \leq \varepsilon.$$

For $\varepsilon_0 < 1$ we have

$$|y_* - x_*| = |y_*| \leq \frac{\varepsilon}{1-\varepsilon}, \qquad \forall 0 < \varepsilon \leq \varepsilon_0.$$

So, (23.5) is locally generalized Ulam–Hyers stable.

In general, we have the following result (see [54]) concerning the fixed point equation

$$x = f(x). \tag{23.7}$$

**Theorem 23.1.** *Let $X$ be a nonempty set, $f : X \to X$ be an operator. Let $d$ and $\rho$ be two metric equivalent metrics ($d(x,y), \rho(x,y) \in \mathbb{R}_+$), i.e., there exists $c_1, c_2 > 0$ such that*

$$c_1 d(x,y) \leq \rho(x,y) \leq c_2 d(x,y), \qquad \forall x,y \in X.$$

*Then the following two statements are equivalent:*

*(i) Equation (23.7) is Ulam–Hyers stable in $(X,d)$;*
*(ii) Equation (23.7) is Ulam–Hyers stable in $(X,\rho)$.*

*Example 23.2.* Let $X := C[0,1]$ and $d : C[0,1] \times C[0,1] \to C([0,1], \mathbb{R}_+)$ be a generalized metric defined by

$$d(u,v)(t) := |u(t) - v(t)|, \qquad \forall t \in [0,1].$$

Consider on this generalized metric space the nonlinear Volterra integral equation

$$x(t) = \int_0^t \sin(x(s))^2 ds, \qquad \forall t \in [0,1] \tag{23.8}$$

and the corresponding inequality with respect to $d$

$$\left| y(t) - \int_0^t \sin(y(s))^2 ds \right| \leq \varepsilon, \qquad \forall t \in [0,1]. \tag{23.9}$$

First of all, remark that the unique solution in $C[0,1]$ of (23.8) is $x_*(t) = 0$, $\forall t \in [0,1]$.

Let $y_* \in C[0,1]$ be a solution of (23.9). We have

$$|y_*(t) - x_*(t)| = |y_*(t)| \leq \varepsilon + \int_0^t |y_*(s)|^2 ds, \qquad t \in [0,1].$$

This inequality implies that

$$|y_*(t) - x_*(t)| \leq \frac{\varepsilon}{1-\varepsilon}, \qquad \forall \varepsilon < 1.$$

So, (23.8) is locally Ulam–Hyers stable.

*Example 23.3.* Let $(X,d)$ be a complete metric space ($d(x,y) \in \mathbb{R}_+$) and $f : X \to X$ an $\alpha$-contraction, i.e.

$$d(f(x), f(y)) \leq \alpha d(x,y), \qquad \forall x,y \in X, 0 \leq \alpha < 1.$$

Then, the fixed point equation (23.7) is Ulam–Hyers stable. Indeed, by the contraction principle (23.7) has a unique solution $x_* \in X$.

Let $y_*$ be a solution of the inequality

$$d(y, f(y)) \leq \varepsilon. \tag{23.10}$$

We have $d(y_*, x_*) \leq d(y_*, f(y_*)) + d(f(y_*), f(x_*)) \leq \varepsilon + \alpha d(y_*, x_*)$. So,

$$d(y_*, x_*) \leq \frac{\varepsilon}{1-\alpha}.$$

Now some examples for this abstract model.

(a) Let $X = \mathbb{R}$ with the usual metric. We consider on $\mathbb{R}$ the Kepler equation

$$x = m \sin x + M \tag{23.11}$$

where $0 < m < 1$ and $M \geq 0$. In this case (see (23.7))

$$f(x) = m \sin x + M.$$

The function $f$ is a $m$-contraction. So, the Kepler equation is Ulam–Hyers stable and if $y_* \in \mathbb{R}$ is a solution of (see also (23.10))

$$|y - m \sin y - M| \leq \varepsilon, \tag{23.12}$$

then

$$|y_* - x_*| \leq \frac{\varepsilon}{1-m}.$$

(b) Let

$$X := \{u \in C([0,1],[0,1]) \mid |u(t_1) - u(t_2)| \leq |t_1 - t_2|,\ t_1, t_2 \in [0,1]\}$$

with the metric

$$d(u,v) := \|u - v\|_\infty := \max_{0 \leq t \leq 1} |x(t) - y(t)|.$$

Then $(X, d)$ is a complete metric space.

We consider on $X$ the iterative Volterra integral equation

$$x(t) = \frac{1}{4}\int_0^t x(x(s))ds, \qquad \forall t \in [0,1]. \tag{23.13}$$

In this case (see (23.7)), $f : X \to X$ is defined by

$$f(x)(t) := \frac{1}{4}\int_0^t x(x(s))ds, \qquad \forall t \in [0,1].$$

We remark that $f(X) \subset X$ and $f$ is $1/2$-contraction. So, (23.13) is Ulam–Hyers stable and if $y_* \in X$ is a solution of the inequality (see also (23.10))

$$\left| y(t) - \int_0^t y(y(s))\mathrm{d}s \right| \leq \varepsilon, \tag{23.14}$$

then $\|y_* - x_*\|_\infty \leq 2\varepsilon$.

In the next section of this paper, we shall extend the above results in a more general frame, the frame of weakly Picard Operators.

*Example 23.4.* Let $X := C^1[0,T]$, $Y := C[0,T]$,

$$\rho(x,y)(t) := |x(t) - y(t)|, \qquad \forall t \in [0,T]$$

and

$$d := \rho\big|_{X \times X}.$$

Let $f, g : X \to Y$ be defined by $f(x) = x'$ and

$$g(x)(t) := h(t, x(t)), \qquad \forall t \in [0,T],$$

where $h \in C([0,T] \times \mathbb{R})$. Let $\varphi : [0,T] \to \mathbb{R}_+$ be a continuous function. In this case, (23.1)–(23.4) take the following forms with respect to the generalized metrics $d$ on $X$ and $\rho$ on $Y$:

$$x'(t) = h(t, x(t)), \qquad \forall t \in [0,T], \tag{23.15}$$

$$|y'(t) - h(t, y(t))| \leq \varepsilon, \qquad \forall t \in [0,T], \tag{23.16}$$

$$|y'(t) - h(t, y(t))| \leq \varepsilon\varphi(t), \qquad \forall t \in [0,T], \tag{23.17}$$

$$|y'(t) - h(t, y(t))| \leq \varphi(t), \qquad \forall t \in [0,T]. \tag{23.18}$$

The following remarks are very useful in the study of Ulam stability of the differential equation (23.15) (see [52, 53]):

*Remark 23.5.* If $x \in C^1[0,T]$ is a solution of (23.15) then $x$ is a solution of the functional-integral equation

$$x(t) = x(0) + \int_0^t h(s, x(s))\mathrm{d}s, \qquad \forall t \in [0,T]. \tag{23.19}$$

If $x \in C[0,T]$ is a solution of (23.19), then $x \in C^1[0,T]$ and $x$ is a solution of (23.15).

*Remark 23.6.* If $y \in C^1[0,T]$ is a solution of (23.15), then $y$ is a solution of the inequality

$$\left| y(t) - y(0) - \int_0^t h(s, y(s))\mathrm{d}s \right| \leq \varepsilon t, \qquad \forall t \in [0,T]. \tag{23.20}$$

*Remark 23.7.* If $y \in C^1[0,T]$ is a solution of (23.17), then $y$ is a solution of the inequality

$$\left| y(t) - y(0) - \int_0^t h(s,y(s))\mathrm{d}s \right| \le \varepsilon \int_0^t \varphi(s)\mathrm{d}s, \qquad \forall t \in [0,T]. \tag{23.21}$$

*Remark 23.8.* If $y \in C^1[0,T]$ is a solution of (23.18), then $y$ is a solution of the inequality

$$\left| y(t) - y(0) - \int_0^t h(s,y(s))\mathrm{d}s \right| \le \int_0^t \varphi(s)\mathrm{d}s, \qquad \forall t \in [0,T]. \tag{23.22}$$

For example, we have the following results ([52]).

**Theorem 23.2.** *We suppose that:*

*(i)* $h \in C([0,T] \times \mathbb{R})$ *and* $\varphi \in C[0,T]$ *is an increasing function;*
*(ii) there exists* $l_f > 0$ *such that* $|h(t,u) - h(t,v)| \le l_f |u - v|$ *for* $u, v \in \mathbb{R}$, $t \in [0,T]$;
*(iii) there exists* $\lambda_\varphi > 0$ *such that*

$$\int_0^t \varphi(s)\mathrm{d}s \le \lambda_\varphi \varphi(t), \qquad \forall t \in [0,T].$$

*Then* (23.15) *is Ulam–Hyers–Rassias stable with respect to* $\varphi$.

*Remark 23.9.* For more considerations of Ulam stability for the differential equations see [52, 53] (see also, e.g., [1, 23], [25–29], [32, 54]). For the Ulam stability of the integral equations see, e.g., [24, 51, 54].

*Example 23.5.* Let $(X,d)$ be a metric space ($d(x,y) \in \mathbb{R}_+$), $f, g : X \to X$ be two operators. Let us consider the coincidence equations

$$f(x) = g(x) \tag{23.23}$$

and the corresponding inequality

$$d(f(x), g(x)) \le \varepsilon, \qquad \varepsilon > 0. \tag{23.24}$$

Let $S_0 := \{x \in X \mid f(x) = g(x)\}$ and $S_\varepsilon := \{x \in X \mid d(f(x), g(x)) \le \varepsilon\}$, $\varepsilon > 0$. We have the following.

**Theorem 23.3.** *We suppose that* $S_0 \neq \emptyset$. *Then:*

*(a) If* (23.23) *is Ulam–Hyers stable with a coefficient* $c_{f,g} \in \mathbb{R}_+^*$, *then*

$$H_d(S_0, S_\varepsilon) \le c_{f,g}\varepsilon,$$

*where* $H_d : P(X) \times P(x) \to \mathbb{R}_+ \cup \{+\infty\}$ *is the Pompeiu–Hausdorff functional.*

*(b) If there exists $c \in \mathbb{R}_+^*$ such that*

$$H_d(S_0, S_\varepsilon) \leq c\varepsilon, \qquad \forall \varepsilon \in \mathbb{R}_+^*,$$

*then (23.23) is Ulam–Hyers stable with a coefficient $c_{f,g} = c + \tau$, for each $\tau > 0$.*

*Proof.* As is well known, the Pompeiu–Hausdorff functional is defined by

$$H_d(A,B) := \max\left(\sup_{a\in A}\inf_{b\in B} d(a,b),\ \sup_{b\in B}\inf_{a\in A} d(a,b)\right).$$

(a) First we remark that $S_0 \subset S_\varepsilon$, for all $\varepsilon \in \mathbb{R}_+^*$. Let $y_* \in S_\varepsilon$. Then there exists $x_* \in S_0$ with $d(y_*, x_*) \leq c_{f,g}\varepsilon$. Now the statement follows from [46, Lemma 8.1.3, p. 76]. This lemma reads as follows:
Let $(X,d)$ be a metric space and $A, B \in P(X)$. Let $\eta > 0$ be such that:

(1) for each $a \in A$ there exists $b \in B$ such that $d(a,b) \leq \eta$;
(2) for each $b \in B$ there exists $a \in A$ such that $d(a,b) \leq \eta$.

Then, $H_d(A,B) \leq \eta$.

(b) Follows from the definition of the Pompeiu–Hausdorff functional.
For example, let $X := [-1,1] \subset (\mathbb{R}, |\cdot|)$,

$$f(x) := \begin{cases} x & \text{for } x \in [-1,0[; \\ 1 & \text{for } x \in [0,1], \end{cases} \quad \text{and} \quad g(x) := \begin{cases} x & \text{for } x \in [-1,0[; \\ 1/2 & \text{for } x \in [0,1]. \end{cases}$$

In this case $S_0 = [-1,0[$ and

$$S_\varepsilon := \begin{cases} [-1,0[ & \text{for } 0 < \varepsilon < \frac{1}{2}; \\ [-1,1] & \text{for } \varepsilon \geq \frac{1}{2}. \end{cases}$$

From these we have that $H_{|\cdot|}(S_0, S_\varepsilon) \leq 2\varepsilon$ for $\varepsilon > 0$.
If $\varepsilon = 1/2$ and $y_* = 1$, then $x_* \in S_0$ with $d(y_*, x_*) \leq 1$ does not exists. But (23.23) is Ulam–Hyers stable with a coefficient $c$ if $c > 2$. □

## 23.6 Ulam–Hyers Stability of a Fixed Point Equation via Weakly Picard Operators

Let $(X,d)$ be a generalized metric space with $d(x,y) \in E_+$. In what follows we call a such metric space an $E_+$-metric space.

Following [47] we present the basic notions of weakly Picard operators in a $E_+$-metric space.

**Definition 23.8.** An operator $f : X \to X$ is weakly Picard operator ($WPO$) if the sequence $(f^n(x))_{n\in\mathbb{N}}$, of successive approximations, converges for all $x \in X$ and the limit (which may depend of $x$) is a fixed point of $f$.

**Definition 23.9.** If $f : X \to X$ is a $WPO$, then we define the operator $f^\infty : X \to X$ by $f^\infty(x) := \lim_{n\to\infty} f^n(x)$.

**Definition 23.10.** Let $f : X \to X$ be an $WPO$ and $c : E \to E$ an increasing linear operator. The operator $f$ is c-WPO iff

$$d(x, f^\infty(x)) \le c(d(x, f(x)), \qquad \forall x \in X.$$

**Definition 23.11.** Let $f : X \to X$ be an $WPO$ and $\psi : E_+ \to E_+$ an increasing operator, continuous in 0 with $\psi(0) = 0$. The operator $f$ is $\psi$-WPO iff

$$d(x, f^\infty(x)) \le \psi(d(x, f(x))), \qquad \forall x \in X.$$

*Remark 23.10.* For the $WPO$s in an $L$-space and a $\mathbb{R}_+$-metric space see [47]. For the $WPO$s in a $s(\mathbb{R}_+)$-metric space see [50, 54]. For the $WPO$s in a $K$-metric space see [57] (here $K$ is a cone in an ordered Banach space).

We have the following.

**Theorem 23.4.** *If $f : X \to X$ is c-WPO, then the equation*

$$x = f(x) \tag{23.25}$$

*is Ulam–Hyers stable.*

*Proof.* Let $\varepsilon \in E_+^*$. Consider the inequality

$$d(y, f(y)) \le \varepsilon. \tag{23.26}$$

Let $y_*$ be a solution of (23.26). We take the solution of (23.25), $x_* := f^\infty(y_*)$. Then, since $f$ is c-WPO, we have $d(y_*, x_*) \le c(d(y_*, f(y_*))) \le c(\varepsilon)$. So, (23.25) is Ulam–Hyers stable. □

In a similar way we have

**Theorem 23.5.** *If $f : X \to X$ is $\psi$-WPO, then* (23.25) *is generalized Ulam–Hyers stable.*

Now we give some applications of the above abstract results.

*Example 23.6.* Let $\Omega \subset \mathbb{R}^p$ be a bounded domain, $K \in C(\overline{\Omega} \times \overline{\Omega} \times \mathbb{R}^m, \mathbb{R}^m)$ and $g \in C(\overline{\Omega}, \mathbb{R}^m)$. We consider the system of Fredholm integral equations

$$x(t) = \int_\Omega K(t, s, x(s))\mathrm{d}s + g(t), \qquad \forall t \in \overline{\Omega}. \tag{23.27}$$

Consider on $C(\overline{\Omega},\mathbb{R}^m)$ the $\mathbb{R}^m_+$-metric $d : C(\overline{\Omega},\mathbb{R}^m)\times C(\overline{\Omega},\mathbb{R}^m)\to\mathbb{R}^m_+$, defined by:

$$d(x,y) := \begin{pmatrix} \|x_1-y_1\|_\infty \\ \vdots \\ \|x_m-y_m\|_\infty \end{pmatrix}.$$

So, we have a complete generalized metric space, $(C(\overline{\Omega},\mathbb{R}^m),d)$. Now we consider the operator $f : C(\overline{\Omega},\mathbb{R}^m)\to C(\overline{\Omega},\mathbb{R}^m)$, defined by

$$f(x)(t) := \int_\Omega K(t,s,x(s))\mathrm{d}s + g(t), \qquad \forall t\in\overline{\Omega}.$$

**Theorem 23.6.** *Suppose that there exists a matrix $L_K\in\mathbb{R}^{m\times m}_+$ such that*

$$\begin{pmatrix} |K_1(t,s,u)-K_2(t,s,v)| \\ \vdots \\ |K_m(t,s,u)-K_m(t,s,v)| \end{pmatrix} \le L_K \begin{pmatrix} |u_1-v_1| \\ \vdots \\ |u_m-v_m| \end{pmatrix}, \qquad \forall t,s\in\overline{\Omega}, u,v\in\mathbb{R}^m,$$

*and the matrix $mes(\Omega)L_K$ is such that $(mes(\Omega)L_K)^n\to 0$ as $n\to\infty$. Then*

*(a) Equation* (23.27) *has a unique solution $x_*\in C(\overline{\Omega},\mathbb{R}^m)$;*
*(b) Equation* (23.27) *is Ulam–Hyers stable.*

*Proof.* (a) Follows from the Perov fixed point theorem (see, e.g., [45, pp. 36–41] and [56, pp. 82–83]).
(b) Since $d(f(x),f(y))\le mes(\Omega)L_K d(x,y)$ for $x,y\in C(\overline{\Omega},\mathbb{R}^m)$, it follows that the operator $f$ is $(I_m - mes(\Omega)L_K)^{-1} - WPO$. Now Theorem 23.4 ends the proof. □

*Example 23.7.* Let $(X,d)$ be a $s(\mathbb{R}_+)$-metric space. Let $f : X\to X$ an operator and $C := (c_{ij})$, $c_{ij}\in\mathbb{R}_+$, $i,j\in\mathbb{N}^*$ be an infinite matrix such that $Cu$ is defined for each $u\in s(\mathbb{R}_+)$. We consider in $(X,d)$ the fixed point equation

$$x = f(x). \tag{23.28}$$

From Theorem 23.4, we derive the following.

**Theorem 23.7.** *If the operator $f$ is $C - WPO$, then* (23.28) *is Ulam–Hyers stable. Moreover, if $y_*$ is a solution of the inequality ($\varepsilon\in s(\mathbb{R}^*_+)$)*

$$d(x,f(x)) \le \varepsilon \tag{23.29}$$

*and $x_* := f^\infty(y_*)$, which is a solution of* (23.28), *then $d(y_*,x_*)\le C\varepsilon$.*

*Remark 23.11.* There are some results for the Ulam stability of the difference equations such as: [9], [34], [54], [61], .... We can put such equations as fixed point equations. For example, let us have the 2-order difference equation

$$x_{n+2} = f_n(x_n, x_{n+1}),\ n \in \mathbb{N}^* \tag{23.30}$$

in $s(\mathbb{B})$, where $(\mathbb{B}, |\cdot|)$ is a Banach space and $f_n : \mathbb{B}^2 \to \mathbb{B}$, $n \in \mathbb{N}^*$.

Let us consider the operator $F : s(\mathbb{B}) \to s(\mathbb{B})$, defined by

$$F(x_1, \dots, x_n, \dots) = (x_1, x_2, f_1(x_1, x_2), \dots, f_n(x_n, x_{n+1}), \dots).$$

In the terms of the operator $F$, the 2-order difference equation (23.30) takes the following form

$$x = F(x). \tag{23.31}$$

Let $d$ be the $s(\mathbb{R}_+)$-metric on $s(\mathbb{B})$ defined by, $d(x,y) := (|x_n - y_n|)_{n\in\mathbb{N}}$. Then the inequality

$$|y_{n+2} - f_n(y_n, y_{n+1})| \le \varepsilon_{n+2}, \tag{23.32}$$

with $\varepsilon_n \in \mathbb{R}_+^*, n \in \mathbb{N}^*$, takes the form

$$d(x, F(x)) \le \varepsilon, \tag{23.33}$$

where $\varepsilon \in s(\mathbb{R}_+^*)$.

From Theorem 23.4 we obtain

**Theorem 23.8.** *If the operator F is C-WPO, then the* 2*-order difference equation* (23.30) *is Ulam–Hyers stable.*

In the next part of this paper, we shall present two new research directions in the Ulam stability.

## 23.7 Ulam Stability of a Fixed Point Structure on a Metric Space

In order to formulate some open problems, we need certain notations and notions from the fixed point structure theory (see [48]).

Let $X$ and $Y$ be two sets. We denote by $\mathbb{M}(X,Y)$ the set of all operators $f : X \to Y$. If $Y = X$, then $\mathbb{M}(X) := \mathbb{M}(X,X)$.

Let $(X.d)$ be a metric space. Then we shall use the following notations:

$$P_b(X) := \{Y \in P(X) \mid \delta_d(Y) < +\infty\}, \quad P_{cp}(X) := \{Y \in P(X) \mid Y \text{ is compact}\},$$

$$P_{cl}(X) := \{Y \in P(X) \mid Y = \overline{Y}\}.$$

If $X$ is a Banach space, then $P_{cv}(X) := \{Y \in P(X) \mid Y \text{ is convex }\}$ and $P_{cp,cv}(X) := \{Y \in P(X) \mid Y \text{ is compact and convex }\}$.

**Definition 23.12.** A triple $(X, S(X), M)$ is a fixed point structure $(f.p.s.)$ on $X$ if

(i) $X$ is a nonempty set, $S(X) \subset P(X)$, $S(X) \neq \emptyset$;
(ii) $M : P(X) \multimap \bigcup_{Y \in P(X)} \mathbb{M}(Y)$, $Y \multimap M(Y) \subset \mathbb{M}(Y)$ is a multivalued operator;
(iii) if $Y \in S(X)$ and $f \in M(Y)$, then $F_f \neq \emptyset$, i.e., every $Y \in S(X)$ has the fixed point property with respect to $M(Y)$.

*Example 23.8.* The $f.p.s.$ of contractions. $(X,d)$ is a complete metric space, $S(X) := P_{cl}(X)$ and $M(Y) := \{f : Y \to Y \mid f \text{ is a contraction }\}$. By the contraction principle, the triple $(X, S(X), M)$ is a $f.p.s.$

*Example 23.9.* The $f.p.s.$ of graphic contractions. $(X,d)$ is a complete metric space, $S(X) := P_{cl}(X)$ and $M(Y) := \{f : Y \to Y \mid f$ is with closed graphic and there exists $\alpha \in [0,1[$ such that, $d(f^2(x), f(x)) \leq \alpha d(x, f(x))$, $\forall x \in Y\}$. By the graphic contraction principle the triple $(X, S(X), M)$ is a $f.p.s.$

*Example 23.10.* The fixed point structure of Caristi-Browder. $(X,d)$ is a complete metric space, $S(X) := P_{cl}(X)$ and $M(Y) := \{f : Y \to Y \mid f$ is with closed graphic and there is $\varphi : \mathbb{R}_+ \to \mathbb{R}_+$ such that $d(x, f(x)) \leq \varphi(x) - \varphi(f(x))$, $\forall x \in Y\}$. By the fixed point theorem of Caristi-Browder the triple $(X, S(X), M)$ is a $f.p.s.$

*Example 23.11.* The fixed point structure of Schauder. $X$ is a Banach space, $S(X) := P_{cp,cv}(X)$ and $M(Y) := C(Y,Y) := \{f : Y \to Y \mid f \text{ is continuous }\}$. By the fixed point theorem of Schauder, the triple $(X, S(X), M)$ is a $f.p.s.$

For other examples of fixed point structures see [48].

**Definition 23.13.** A fixed point structure $(X, S(X), M)$ on $(X,d)$ is Ulam–Hyers (generalized Ulam–Hyers) stable if for each $Y \in S(X)$ and each $f \in M(Y)$ the fixed point equation

$$x = f(x) \tag{23.34}$$

is Ulam–Hyers (generalized Ulam–Hyers) stable.

Now we can formulate our problems.

**Problem 23.1.** Which $f.p.s.$ are Ulam–Hyers (generalized Ulam–Hyers) stable?

**Problem 23.2.** Let $(X, S(X), M)$ be a $f.p.s.$ which is not Ulam–Hyers (generalized Ulam–Hyers) stable. Let $Y \in S(X)$. For which $f \in M(Y)$, the fixed point equation (23.34) is Ulam–Hyers (generalized Ulam–Hyers) stable?

For the considerations in Example 23.3, we have that the $f.p.s.$ of contractions is Ulam–Hyers stable. In a similar way, one can prove that the $f.p.s.$ of graphic contractions is Ulam–Hyers stable, and the $f.p.s.$ of strict $\varphi$-contraction (see, e.g., [46, 48, 56]) is generalized Ulam–Hyers stable.

Moreover, from Theorem 23.4 we have

**Theorem 23.9.** *Let $(X,S(X),M)$ be a fixed point structure on a metric space $(X,d)$. If for each $f \in M(Y)$, $Y \in S(X)$, there exists $c_f > 0$ such that $f$ is a $c_f$-WPO, then the fixed point structure $(X,S(X),M)$ is Ulam–Hyers stable.*

In general, the $f.p.s.$ of Caristi-Browder is not Ulam–Hyers stable but if the function $\varphi$ is such that $\varphi(x) \leq c_f d(x,f(x))$ for some $c_f > 0$ where $Y \in S(X)$ and $f \in M(Y)$, then the fixed point equation (23.34) is Ulam–Hyers stable.

## 23.8 Ulam Stability in the Case of Equations with Set-to-Point Operators

Let $(X,d)$ and $(Y,\rho)$ be two $E_+$-metric spaces, $Z \subset P(X)$, $Z \neq \emptyset$, and $\mathscr{D} : Z \times Z \to E_+$ be an operator.

Let $F, G : Z \to Y$ be two set-to-point operators. We consider on $Z$ the equation

$$F(A) = G(A). \tag{23.35}$$

**Definition 23.14.** Equation (23.35) is Ulam–Hyers stable if there exists an increasing linear operator $c_{F,G} : E \to E$ such that: for each $\varepsilon \in E_+^*$ and each solution $B_* \in Z$ of the inequality

$$\rho(F(B), G(B)) \leq \varepsilon \tag{23.36}$$

there exists a solution $A_* \in Z$ of (23.35) such that

$$\mathscr{D}(B_*, A_*) \leq c_{F,G}(\varepsilon).$$

**Definition 23.15.** Equation (23.35) is generalized Ulam–Hyers stable if there exists an increasing operator $\psi : E_+ \to E_+$, continuous in 0 with $\psi(0) = 0$, such that: for each $\varepsilon > 0$ and for each solution $B_* \in Z$ of (23.36) there exists a solution $A_* \in Z$ of (23.35) such that

$$\mathscr{D}(B_*, A_*) \leq \psi(\varepsilon).$$

**Problem 23.3.** To study the Ulam–Hyers (generalized Ulam–Hyers) stability of (23.35).

An interesting particular case of the Problem 23.3 is the following.

**Problem 23.4.** Let $(X,d)$ be a metric space ($d(x,y) \in \mathbb{R}_+$, $F : P_b(X) \to \mathbb{R}_+$ a set-to-point functional and $r > 0$ a given positive real number. We consider an $P_b(X)$ the Pompeiu–Hausdorff functional $H_d$. The problem is to study the Ulam–Hyers (generalized Ulam–Hyers) stability of the equation

$$F(A) = r \tag{23.37}$$

The corresponding inequality is the following ($\varepsilon \in \mathbb{R}_+^*$)

$$|F(A) - r| \leq \varepsilon. \tag{23.38}$$

**Problem 23.5.** To study the Problem 23.4 if $F := \delta_d$, the diameter functional.

**Problem 23.6.** To study the Problem 23.4 if $F := \alpha$, an abstract measure of noncompactness ([48]; see also, e.g., [3–5, 12, 30]).

**Problem 23.7.** To study the Problem 23.4 if $F := \beta$, an abstract measure of nonconvexity (see [5, 48] and the references therein).

## References


1. Alsina, C., Ger, R.: On some inequalities and stability results related to the exponential function. J. Inequal. Appl. **2**, 373–380 (1998)
2. Aoki, T.: On the stability of the linear transformation in Banach spaces. J. Math. Soc. Japan, **2**, 64–66 (1950)
3. Appell, J.: Measure of noncompactness, condensing operators and fixed points: an application – oriented survey. Fixed Point Theory **6**, 157–229 (2005)
4. Ayerbe Toledano, J.M., Dominguez Benavides, T., López Acedo, G.: Measures of Noncompactness in Metric Fixed Point Theory. Birkhäuser, Basel (1997)
5. Ban, A.I., Gal, S.G.: Defects of Properties in Mathematics. World Scientific (2002)
6. Berinde, V.: Iterative Approximation of Fixed Points. Springer, Berlin (2007)
7. Bourgin, D.G.: Classes of transformations and bordering transformations. Bull. Amer. Math. Soc. **57**, 223–237 (1951)
8. Breckner, W.W., Trif, T.: Convex Functions and Related Functional Equations. Cluj Univ. Press, Cluj–Napoca (2008)
9. Brzdęk, J., Popa, D., Xu, B.: The Hyers–Ulam stability of nonlinear recurrences. J. Math. Anal. Appl. **335**, 443–449 (2007)
10. Cădariu, L.: Stabilitate Ulam–Hyers–Bourgin pentru ecuaţii funcţionale. Editura Univ. de Vest Timişoara (2007)
11. De Pascale, E., Marino, G., Pietramala, P.: The use of the $E$-metric spaces in the search for fixed points. Le Matematiche **48** II, 367–376 (1993)
12. Deimling, K.: Nonlinear Functional Analysis. Springer, Berlin (1985)
13. Găvruţă, P.: A generalization of the Hyers–Ulam–Rassias stability of approximately additive mappings. J. Math. Anal. Appl. 184, 431–436 (1994)
14. Granas, A., Dugundji, J.: Fixed Point Theory. Springer, New York (2003)
15. Hyers, D.H.: On the stability of the linear functional equation. Proc. Nat. Acad. Sci. USA **27**, 222–224 (1941)
16. Hyers, D.H.: The stability of homomorphism and related topics. In: Th.M. Rassias (ed.) Global Analysis – Analysis on Manifolds, pp. 140–153. Teubner, Leipzig (1983)
17. Hyers, D.H., Isac, G., Rassias, Th.M.: Stability of Functional Equations in Several Variables. Birkhäuser, Bassel (1998)
18. Hyers, D.H., Isac, G., Rassias, Th.M.: On the asymptoticity aspect of Hyers–Ulam stability of mapping. Proc. Amer. Math. Soc. **126**, 425–430 (1998)
19. Hyers, D.H., Rassias, Th.M.: Approximate homomorphisms. Aequationes Math. **44**, 125–153 (1992)
20. Isac, G., Rassias, Th.M.: On the Hyers–Ulam stability of $\psi$–additive mappings. J. Approx. Theory **72**, 131–137 (1993)

21. Isac, G., Rassias, Th.M.: Stability of additive mappings: applications to nonlinear analysis. Int. J. Math. Math. Sci. **19**, 219–228 (1996)
22. Jung, S.–M.: Hyers–Ulam–Rassias stability of functional equations in mathematical analysis. Hadronic Press, Palm Harbor (2001)
23. Jung, S.–M.: Hyers–Ulam stability of linear differential equations of first order, III. J. Math. Anal. Appl. **311**, 139–146 (2005)
24. Jung, S.–M.: A fixed point approach to the stability of a Volterra integral equation. Fixed Point Theory Appl. **2007**, ID 57064, 9 pages (2007)
25. Jung, S.–M.: Hyers–Ulam stability of linear partial differential equations of first order. Appl. Math. Lett. **22**, 70–74 (2009)
26. Jung, S.–M., Lee, K.–S.: Hyers–Ulam–Rassias stability of linear differential equations of second order. J. Comput. Math. Optim. **3**, 193–200 (2007)
27. Jung, S.–M., Lee, K.–S.: Hyers–Ulam stability of first order linear partial differential equations with constant coefficients. Math. Ineq. Appl. **10**, 261–266 (2007)
28. Jung, S.–M., Rassias, Th.M.: Ulam's problem for approximate homomorphisms in connection with Bernoulli's differential equation. Appl. Math. Comput. **187**, 223–227 (2007)
29. Jung, S.–M., Rassias, Th.M.: Generalized Hyers–Ulam stability of Riccati differential equation. Math. Ineq. Appl. **11**, 777–782 (2008)
30. Kirk, W.A., Sims, B. (eds.): Handbook of Metric Fixed Point Theory. Kluwer, Boston, (2001)
31. Maruster, St.: The stability of gradient – like methods. Appl. Math. Comput. **117**, 103–115 (2001)
32. Miura, T., Jung, S.–M., Takahasi, S.–E.: Hyers–Ulam stability of the Banach space valued linear differential equations $y' = \lambda y$. J. Korean Math. Soc. **41**, 995–1005 (2004)
33. Páles, Zs.: Hyers–Ulam stability of the Cauchy functional equation on square–symmetric grupoids. Publ. Math. Debrecen **58**, 651–666 (2001)
34. Popa, D.: Hyers–Ulam–Rassias stability of a linear recurrence. J. Math. Anal. Appl. **309**, 591–597 (2005)
35. Prastaro, A., Rassias, Th.M.: Ulam stability in geometry of PDE's. Nonlinear Funct. Anal. Appl. **8(2)**, 259–278 (2003)
36. Radu, V.: The fixed point alternative and the stability of functional equations. Fixed Point Theory **4**, 91–96 (2003)
37. Rassias, Th.M.: On the stability of the linear mapping in Banach spaces. Proc. Amer. Math. Soc. **72**, 297–300 (1978)
38. Rassias, Th.M.: On a modified Hyers–Ulam sequence. J. Math. Anal. Appl. **158**, 106–113 (1991)
39. Rassias, Th.M.: On the stability of functional equations and a problem of Ulam. Acta Appl. Math. **62(1)**, 23–130 (2000)
40. Rassias, Th.M. (ed.): Functional Equations and Inequalities. Kluwer Academic Publishers, Dordrecht (2000)
41. Rassias, Th.M. (ed.): Functional Equations, Inequalities and Applications. Kluwer Academic Publishers, Dordrecht (2003)
42. Rassias, Th.M., Šemrl, P.: On the Hyers–Ulam stability of linear mappings. J. Math. Anal. Appl. **173**, 325–338 (1993)
43. Rassias, Th.M., Tabor, J. (eds.): Stability of mappings of Hyers–Ulam type. Hadronic, Palm Harbor (1994)
44. Reich, S., Zaslavski, A.J.: A stability result in fixed point theory. Fixed Point Theory **6**, 113–118 (2005)
45. Rus, I.A.: Principii şi aplicaţii ale teoriei punctului fix. Editura Dacia, Cluj–Napoca (1979)
46. Rus, I.A.: Generalized Contractions and Applications. Cluj University Press, Cluj–Napoca (2001)
47. Rus, I.A.: Picard operators and applications. Sci. Math. Jpn. **58**, 191–219 (2003)
48. Rus, I.A.: Fixed Point Structure Theory. Cluj University Press, Cluj–Napoca (2006)
49. Rus, I.A.: Fixed point theory in partial metric spaces. An. Univ. Vest Timiş. Ser. Mat.–Inform. **46**, No. 2, 149–160 (2008)

50. Rus, I.A.: The theory of a metrical fixed point theorem: theoretical and applicative relevances. Fixed Point Theory **9**, 541–559 (2008)
51. Rus, I.A.: Gronwall lemma approach to the Hyers–Ulam–Rassias stability of an integral equation. In: Pardalos, P., Rassias, Th.M., Khan, A.A. (eds.) Nonlinear Analysis and Variational Problems, pp. 147–152. Springer (2009)
52. Rus, I.A.: Ulam stability of ordinary differential equations. Stud. Univ. Babeş-Bolyai Math. **54**, No. 4, 125–133 (2009)
53. Rus, I.A.: Ulam stability of ordinary differential equations in a Banach space. Carpathian J. Math. **26**, 103–109 (2010)
54. Rus, I.A.: Remarks on Ulam stability of the operatorial equations. Fixed Point Theory **10**, 305–320 (2009)
55. Rus, I.A., Mureşan, A.S., Mureşan, V.: Weakly Picard operators on a set with two metrics. Fixed Point Theory **6**, 323–331 (2005)
56. Rus, I.A., Petruşel, A., Petruşel, G.: Fixed Point Theory. Cluj University Press, Cluj–Napoca (2008)
57. Rus, I.A., Petruşel, A., Şerban, M.A.: Weakly Picard operators: equivalent definitions, applications and open problems. Fixed Point Theory **7**, 3–22 (2006)
58. Trif, T.: On the stability of a functional equation deriving from an inequality of Popoviciu for convex functions. J. Math. Anal. Appl. **272**, 604–616 (2002)
59. Ulam, S.M.: A Collection of Mathematical Problems. Interscience Publ., New York (1960)
60. Wang, J.: Some further generalization of the Ulam–Hyers–Rassias stability of functional equations. J. Math. Anal. Appl. **263**, 406–423 (2001)
61. Xu, M.: Hyers–Ulam–Rassias stability of a system of first order linear recurrences. Bull. Korean Math. Soc. **44**, 841–849 (2007)
62. Zabrejko, P.P.: $K$–metric and $K$–normed linear spaces: survey. Collect. Math. **48**, 825–859 (1997)

# Chapter 24
# Stability of the Pexiderized Cauchy Functional Equation in Non-Archimedean Spaces

**G. Zamani Eskandani and Paşc Găvruţa**

*Dedicated to the memory of S.M. Ulam*
*on the 100th anniversary of his birth*

**Abstract** In this paper, we investigate the Hyers–Ulam–Rassias stability of the Pexiderized Cauchy functional equation in non-Archimedean spaces.

**Keywords** Hyers–Ulam–Rassias stability • Pexiderized Cauchy functional equation • Non-Archimedean space • p-adic field

**Mathematics Subject Classification (2000):** Primary 39B22, 39B82, 46S10

## 24.1 Introduction and Preliminaries

In 1940, Ulam [33] asked the first question on the stability problem. In 1941, Hyers [13] gave a solution to the problem of Ulam. This result was generalized by Aoki [1] for additive mappings and independently by Rassias [26] for linear mappings by considering an *unbounded Cauchy difference*.

**Theorem 24.1 (Th.M. Rassias).** *Let $f : E \longrightarrow E'$ be a mapping from a normed vector space $E$ into a Banach space $E'$ subject to the inequality*

$$\|f(x+y) - f(x) - f(y)\| \leq \varepsilon(\|x\|^p + \|y\|^p), \quad x,y \in E, \tag{24.1}$$

G.Z. Eskandani (✉)
Faculty of Mathematical Sciences, University of Tabriz, Tabriz, Iran
e-mail: zamani@tabrizu.ac.ir

P. Găvruţa
Department of Mathematics, University Politechnica of Timisoara, Piata Victoriei, No. 2, 300006 Timisoara, Romania
e-mail: pgavruta@yahoo.com

Th.M. Rassias and J. Brzdęk (eds.), *Functional Equations in Mathematical Analysis*, Springer Optimization and Its Applications 52, DOI 10.1007/978-1-4614-0055-4_24,

*where $\varepsilon > 0$ and $0 \leq p < 1$ are constant. Then the limit*

$$L(x) = \lim_{n\to\infty} \frac{1}{2^n} f(2^n x)$$

*exists for all $x \in E$ and $L : E \longrightarrow E'$ is the unique additive mapping which satisfies*

$$\|f(x) - L(x)\| \leq \frac{2\varepsilon}{2-2^p}\|x\|^p, \quad x \in E.$$

*Also, if for each $x \in E$ the function $f(tx)$ is continuous in $t \in \mathbb{R}$, then $L$ is linear.*

The paper of Th.M. Rassias has provided a lot of influence in the development of what we now call Hyers–Ulam–Rassias stability of functional equations (cf. the books of Czerwik [4] and Hyers et al. [14]). Rassias [24] replaced the factor $\|x\|^p + \|y\|^p$ by $\|x\|^p\|y\|^p$ for $p, q \in \mathbb{R}$ with $p + q \neq 1$ (see also [23, 25]). In the case $p + q = 1$, we have not stability [8]. In 1990, Rassias [28] during the 27th International Symposium on Functional Equations asked the question whether such a theorem can also be proved for $p \geq 1$. In 1991, Gajda [6] gave an affirmative solution to this question for $p > 1$. It was shown by Gajda [6], as well as by Rassias and Šemrl [31] that one cannot prove a Rassias type theorem when $p = 1$. P. Găvruta [9] proved that the function

$$f(x) = x \ln|x|,$$

if $x \neq 0$ and $f(0) = 0$ satisfies (24.1) with $\varepsilon = p = 1$ but

$$\sup_{x\neq 0} \frac{|f(x) - A(x)|}{|x|} \geq \sup_{n\in\mathbb{N}} \frac{|n\ln n - A(n)|}{n} = \sup_{n\in\mathbb{N}} |\ln n - A(1)| = \infty$$

for any additive function $A : \mathbb{R} \to \mathbb{R}$. In 1994, a further generalization of Rassias' Theorem was obtained by Găvruta [7], in which he replaced the bound $\varepsilon(\|x\|^p + \|y\|^p)$ by a general control function $\varphi(x,y)$. Isac and Rassias [16] replaced the factor $\|x\|^p + \|y\|^p$ by $\|x\|^{p_1} + \|y\|^{p_2}$ in Theorem 1.1 and solved stability problem when $p_2 \leq p_1 < 1$ or $1 < p_2 \leq p_1$, also they asked the question whether such a theorem can be proved for $p_2 < 1 < p_1$. Găvruta [9] gave a negative answer to this question. Isac and Rassias [15] applied the Hyers–Ulam–Rassias stability theory to prove fixed point theorems and study some new applications in Nonlinear Analysis. During the last two decades, a number of papers and research monographs have been published on various generalizations and applications of the Hyers–Ulam–Rassias stability to a number of functional equations and mappings (see [2, 3, 5, 10, 11, 17–20, 22, 27, 29, 30]).

A non-Archimedean field is a field $K$ equipped with a function (valuation) $|.|$ from $K$ into $[0,\infty)$ such that

(a) $|r| = 0$ if and only if $r = 0$,
(b) $|rs| = |r||s|$ for all $r, s \in K$,
(c) $|r + s| \leq \max\{|r|, |s|\}$ for all $r, s \in K$.

Clearly

$$|1| = |-1| = 1$$

and

$$|n| \leq 1$$

for all $n \in \mathbb{N}$. An example of a non-Archimedean valuation is the mapping $|.|$ taking everything but 0 into 1 and $|0| = 0$. This valuation is called trivial.

In 1897, Hensel [12] discovered the p-adic numbers as a number theoretical analogue of power series in complex analysis. Fix a prime number $p$. For any nonzero rational number $x$, there exists a unique integer $n_x \in \mathbb{Z}$ such that

$$x = \frac{a}{b} p^{n_x},$$

where $a$ and $b$ are integers not divisible by $p$. Then

$$|x|_p := p^{-n_x}$$

defines a non-Archimedean norm on $\mathbb{Q}$. The completion of $\mathbb{Q}$ with respect to the metric

$$d(x,y) = |x-y|_p$$

is denoted by $\mathbb{Q}_p$ which is called the $p$-adic number field; see [32].

Let $X$ be a vector space over a scalar field $K$ with a non-Archimedean non-trivial valuation $|\,.\,|$. A function $\|\,.\,\|: X \longrightarrow \mathbb{R}$ is a non-Archimedean norm (valuation) if it satisfies the following conditions:

(i) $\| x \| = 0$ if and only if $x = 0$;
(ii) $\| rx \| = | r | \| x \| \;(r \in K, x \in X)$;
(iii) the strong triangle inequality: $\| x+y \| \leq \max\{\| x \|, \| y \|\} \;(x, y \in X)$.

Then $(X, \|\;\|)$ is called a non-Archimedean space. Due to the fact that

$$\| x_n - x_m \| \leq \max\{\| x_{j+1} - x_j \| : m \leq j \leq n-1\} \quad (n > m),$$

a sequence $\{x_n\}$ is Cauchy if and only if the sequence

$$\{x_{n+1} - x_n\}$$

converges to zero in a non-Archimedean space. By a complete non-Archimedean space we mean one in which every Cauchy sequence is convergent.

In [21], Moslehian and Rassias have solved the stability problem for Cauchy and quadratic functional equations in non-Archimedean normed spaces.

In this paper, we give a generalization of the results from [7] in non-Archimedean spaces, and we investigate the Hyers–Ulam–Rassias stability of the Pexiderized Cauchy functional equation in non-Archimedean space. In the sequel $\mathbb{N}_0$ stands for the set of non-negative integers and $k$ is a fixed integer greater than 1.

## 24.2 Stability of the Pexiderized Cauchy Functional Equation

Throughout this section, let $X$ be non-Archimedean normed space and $Y$ be a complete non-Archimedean normed space.

**Theorem 24.2.** *Let $\varphi : X \times X \to [0,\infty)$ and $f,g,h : X \to Y$ be mappings with*

$$f(0) = 0,$$

$$\lim_{n\to\infty} \frac{1}{|k|^n}\varphi(k^n x, k^n y) = 0, \tag{24.2}$$

$$\|f(x+y) - g(x) - h(y)\| \le \varphi(x,y), \tag{24.3}$$

*and*

$$\tilde{M}_k(x) := \sup_{j\in\mathbb{N}_0} |k|^{-j} M_k(k^j x) < \infty \tag{24.4}$$

*for all $x,y \in X$, where*

$$M_k(x) := \max_{1\le i<k} M(x, ix)$$

*and*

$$M(x,y) := \max\{\varphi(x,y), \varphi(x,0), \varphi(0,y)\}.$$

*Then there exists a unique additive mapping $A_k : X \to Y$ such that, for all $x \in X$,*

$$\|f(x) - A_k(x)\| \le \frac{1}{|k|}\tilde{M}_k(x), \tag{24.5}$$

$$\|g(x) - A_k(x) - g(0)\| \le \max\left\{\frac{1}{|k|}\tilde{M}_k(x), \varphi(x,0)\right\}, \tag{24.6}$$

$$\|h(x) - A_k(x) - h(0)\| \le \max\left\{\frac{1}{|k|}\tilde{M}_k(x), \varphi(0,x)\right\}. \tag{24.7}$$

*Proof.* Replacing $y = 0$ in (24.3), we get

$$\|f(x) - g(x) - h(0)\| \le \varphi(x,0) \tag{24.8}$$

for all $x \in X$. Replacing $x = 0$ in (24.3), we get

$$\|f(y) - h(y) - g(0)\| \le \varphi(0,y) \tag{24.9}$$

for all $y \in X$. By using (24.8) and (24.9), we get

$$\|f(x+y) - f(x) - f(y)\| \le M(x,y) \tag{24.10}$$

for all $x, y \in X$. By induction on $k$, we show that

$$\|f(kx) - kf(x)\| \leq M_k(x) \tag{24.11}$$

for all $x \in X$. Letting $y = x$ in (24.10), we get

$$\|f(2x) - 2f(x)\| \leq M(x,x) \tag{24.12}$$

for all $x \in X$. So we get (24.11) for $k = 2$.

Assume that (24.11) holds for $k$. Letting $y = kx$ in (24.10), we get

$$\|f((k+1)x) - f(x) - f(kx)\| \leq M(x,kx) \tag{24.13}$$

for all $x \in X$. It follows from (24.11) and (24.13) that

$$\begin{aligned}
&\|f((k+1)x) - (k+1)f(x)\| \\
&\qquad \leq \max\{\|f((k+1)x) - f(x) - f(kx)\|, \|f(kx) - kf(x)\|\} \\
&\qquad \leq \max\{M(x,ix) : 1 \leq i < k+1\}.
\end{aligned}$$

This completes the induction argument.

Replacing $x$ by $k^n x$ in (24.11) and dividing both sides of (24.11) by $|k|^{n+1}$, we get

$$\left\| \frac{1}{k^{n+1}} f(k^{n+1}x) - \frac{1}{k^n} f(k^n x) \right\| \leq \frac{1}{|k|^{n+1}} M_k(k^n x) \tag{24.14}$$

for all $x \in X$ and all non-negative integers $n$. Therefore, we conclude from (24.2) and (24.14) that the sequence $\{k^{-n} f(k^n x)\}$ is a Cauchy sequence in $Y$ for all $x \in X$. Since $Y$ is complete the sequence $\{k^{-n} f(k^n x)\}$ converges in $Y$ for all $x \in X$. So one can define the mapping $A_k : X \to Y$ by

$$A_k(x) := \lim_{n\to\infty} \frac{1}{k^n} f(k^n x) \tag{24.15}$$

for all $x \in X$. Using induction one can show that

$$\left\| \frac{f(k^n x)}{k^n} - f(x) \right\| \leq \frac{1}{|k|} \max\left\{ \frac{M_k(k^j x)}{|k|^j} : 0 \leq j < n \right\} \tag{24.16}$$

for all $n \in N$ and all $x \in X$. By taking $n$ to approach infinity in (24.16) and using (24.4), (24.8), (24.9) and (24.15) one obtains (24.5)–(24.7). Now, we show that $A_k : X \to Y$ is an additive mapping. It follows from (24.2), (24.10) and (24.15) that

$$\begin{aligned}
\|A_k(x+y) - A_k(x) - A_k(y)\| &= \lim_{n\to\infty} \frac{1}{|k|^n} \|f(k^n x + k^n y) - f(k^n x) - f(k^n y)\| \\
&\leq \lim_{n\to\infty} \frac{1}{|k|^n} M(k^n x, k^n y) = 0
\end{aligned}$$

for all $x, y \in X$. Hence, the mapping $A_k : X \to Y$ is additive.

To prove the uniqueness of $A_k$, let $T : X \to Y$ be another additive mapping satisfying (24.5). Then

$$\begin{aligned}\|A_k(x) - T(x)\| &= \lim_{m\to\infty} |k|^{-m} \|A_k(2^m x) - T(2^m x)\| \\ &\leq \lim_{m\to\infty} |k|^{-m} \max\{\|A_k(k^m x) - f(k^m x)\|, \|T(k^m x) - f(k^m x)\|\} \\ &\leq \frac{1}{|k|} \lim_{m\to\infty} \lim_{n\to\infty} \max\left\{\frac{M_k(k^j x)}{|k|^j} : m \leq j < n+m\right\} = 0\end{aligned}$$

for all $x \in X$. So $A_k = T$. □

**Theorem 24.3.** *Let $\Phi : X \times X \to [0,\infty)$ and $f,g,h : X \to Y$ be mappings with $\Phi(0,0) = 0$, $f(0) = 0$,*

$$\|f(x+y) - g(x) - h(y)\| \leq \Phi(x,y),$$

$$\lim_{n\to\infty} |k|^n \Phi\left(\frac{x}{k^n}, \frac{y}{k^n}\right) = 0, \tag{24.17}$$

*and*

$$\tilde{M}_k(x) := \sup_{j\in\mathbb{N}_0} |k|^j M_k\left(\frac{x}{k^{j+1}}\right) < \infty \tag{24.18}$$

*for all $x,y \in X$, where*

$$M(x,y) = \max\{\Phi(x,y), \Phi(x,0), \Phi(0,x)\}$$

*and*

$$M_k(x) = \max_{1\leq i<k} M(x,ix).$$

*Then there exists a unique additive mapping $A_k : X \to Y$ such that, for all $x \in X$,*

$$\|f(x) - A_k(x)\| \leq \frac{1}{|k|}\tilde{M}_k(x), \tag{24.19}$$

$$\|g(x) - A_k(x) - g(0)\| \leq \max\left\{\frac{1}{|k|}\tilde{M}_k(x), \Phi(x,0)\right\}, \tag{24.20}$$

$$\|h(x) - A_k(x) - h(0)\| \leq \max\left\{\frac{1}{|k|}\tilde{M}_k(x), \Phi(0,x)\right\}. \tag{24.21}$$

*Proof.* Analogously, as in the proof of Theorem 24.2, we have

$$\|f(kx) - kf(x)\| \leq M_k(x) \tag{24.22}$$

for all $x \in X$. Replacing $x$ by $k^{-n-1}x$ in (24.22) and multiplying both sides of (24.22) by $|k|^n$, we get

$$\left\| k^{n+1} f\left(\frac{x}{k^{n+1}}\right) - k^n f\left(\frac{x}{k^n}\right) \right\| \le |k|^n M_k\left(\frac{x}{k^{n+1}}\right) \tag{24.23}$$

for all $x \in X$ and all non-negative integers $n$. Therefore, we conclude from (24.17) and (24.23) that the sequence $\{k^n f(k^{-n}x)\}$ is a Cauchy sequence in $Y$ for all $x \in X$. Since $Y$ is complete the sequence $\{k^n f(k^{-n}x)\}$ converges in $Y$ for all $x \in X$. So one can define the mapping $A_k : X \to Y$ by

$$A_k(x) := \lim_{n\to\infty} k^n f\left(\frac{x}{k^n}\right) \tag{24.24}$$

for all $x \in X$. Using induction one can show that

$$\left\| k^n f\left(\frac{x}{k^n}\right) - f(x) \right\| \le \max\left\{ |k|^j M_k\left(\frac{x}{k^{j+1}}\right) : 0 \le j < n \right\} \tag{24.25}$$

for all $n \in N$ and all $x \in X$. By taking $n$ to approach infinity in (24.25) and using (24.18) and (24.24) one obtains (24.19)–(24.21). The rest of the proof is similar to the proof of Theorem 24.2. □

**Theorem 24.4.** *Let $\psi : [0,\infty) \to [0,\infty)$ be a function with $\psi(0) = 0$ and $r$ be a fixed real number in $[1,+\infty)$ such that*

$$\psi(|k|) < |k|^r$$

*and*

$$\psi(|k|s) \le \psi(|k|)\psi(s)$$

*for all $s \in [0,\infty)$. Suppose that $f,g,h : X \to Y$ are mappings with*

$$f(0) = 0$$

*and*

$$\|f(x+y) - g(x) - h(y)\| \le \theta\left(\psi(\|x\|) + \psi(\|y\|)\right)$$

*for all $x,y \in X$. Then there exists a unique additive mapping $A_k : X \to Y$ such that, for all $x \in X$,*

$$\|f(x) - A_k(x)\| \le \frac{2\theta}{|k|}\psi(\|x\|),$$

$$\|g(x) - A_k(x) - g(0)\| \le \frac{2\theta}{|k|}\psi(\|x\|),$$

*and*

$$\|h(x) - A_k(x) - h(0)\| \le \frac{2\theta}{|k|}\psi(\|x\|).$$

*Proof.* Let

$$\varphi(x,y) = \theta\left(\psi(\|x\|) + \psi(\|y\|)\right)$$

for all $x,y \in X$. For each $n \in \mathbb{N}_0$ we have

$$\psi(|k|^n) \le (\psi(|k|))^n$$

and

$$\varphi(k^n x, k^n y) \le \theta\,(\psi(|k|))^n\left(\psi(\|x\|) + \psi(\|y\|)\right)$$

for all $x,y \in X$. By using Theorem 24.2, we can get the statement. □

**Theorem 24.5.** *Let $\psi : [0,\infty) \to [0,\infty)$ be a function with $\psi(0) = 0$ and $r$ be a fixed real number in $[1,+\infty)$ such that*

$$\psi(|k|) < |k|^r \quad \textit{and} \quad \psi(|k|s) \le \psi(|k|)\psi(s)$$

*for all $s \in [0,\infty)$. Suppose that $f,g,h : X \to Y$ are mappings with $f(0) = 0$ and*

$$\|f(x+y) - g(x) - h(y)\| \le \theta\,\psi(\|x\|)\,\psi(\|y\|)$$

*for all $x,y \in X$. Then there exists a unique additive mapping $A_k : X \to Y$ such that, for all $x \in X$,*

$$\|f(x) - A_k(x)\| \le \frac{\theta}{|k|}\left(\psi(\|x\|)\right)^2,$$

$$\|g(x) - A_k(x) - g(0)\| \le \frac{\theta}{|k|}\left(\psi(\|x\|)\right)^2,$$

$$\|h(x) - A_k(x) - h(0)\| \le \frac{\theta}{|k|}\left(\psi(\|x\|)\right)^2.$$

*Proof.* Let

$$\varphi(x,y) = \theta\,\psi(\|x\|)\,\psi(\|y\|)$$

for all $x,y \in X$. We have

$$\lim_{n\to\infty} \frac{\varphi(k^n x, k^n y)}{|k|^n} \le \varphi(x,y) \lim_{n\to\infty} \frac{\psi(|k|^n)^2}{|k|^n} = 0.$$

By using Theorem 24.2, we can get the statement. □

**Theorem 24.6.** *Let $\psi : [0,\infty) \to [0,\infty)$ be a function with $\psi(0) = 0$ and $r$ be a fixed real number in $(-\infty,+1]$ such that*

$$\psi(ts) \le \psi(t)\psi(s) \quad \textit{and} \quad \psi(|k|^{-1}) < |k|^{-r}$$

*for all* $t,s \in [0,\infty)$. *Suppose that* $f,g,h : X \to Y$ *be mappings with* $f(0) = 0$ *and*

$$\|f(x+y) - g(x) - h(y)\| \le \theta\,(\psi(\|x\|) + \psi(\|y\|)) \tag{24.26}$$

*for all* $x,y \in X$. *Then there exists a unique additive mapping* $A_k : X \to Y$ *such that*

$$\|f(x) - A_k(x)\| \le \frac{\theta}{|k|^{r+1}}\sigma_k(\psi)\psi(\|x\|),$$

$$\|g(x) - A_k(x) - g(0)\| \le \frac{\theta}{|k|^{r+1}}\sigma_k(\psi)\psi(\|x\|),$$

$$\|h(x) - A_k(x) - h(0)\| \le \frac{\theta}{|k|^{r+1}}\sigma_k(\psi)\psi(\|x\|)$$

*for all* $x \in X$, *where* $\sigma_k(\psi) := \max\{1 + \psi(|i|) : 1 \le i < k\}$.

*Proof.* Let

$$\Phi(x,y) = \theta\,(\psi(\|x\|) + \psi(\|y\|))$$

for all $x,y \in X$. We have

$$\lim_{n\to\infty} |k|^n \Phi\left(\frac{x}{k^n}, \frac{y}{k^n}\right) \le \Phi(x,y) \lim_{n\to\infty} |k|^n \psi(|k|^{-n}) = 0.$$

By using Theorem 24.3, one can obtain the statement. □

**Theorem 24.7.** *Let* $\psi : [0,\infty) \to [0,\infty)$ *be a function with* $\psi(0) = 0$ *and* $r$ *be a fixed real number in* $(-\infty, 1/2]$ *such that*

$$\psi(ts) \le \psi(t)\psi(s) \quad \text{and} \quad \psi(|k|^{-1}) < |k|^{-r}$$

*for all* $t,s \in [0,\infty)$. *Suppose that* $f,g,h : X \to Y$ *are mappings with* $f(0) = 0$ *and*

$$\|f(x+y) - g(x) - h(y)\| \le \theta\psi(\|x\|)\psi(\|y\|) \tag{24.27}$$

*for all* $x,y \in X$. *Then there exists a unique additive mapping* $A_k : X \to Y$ *such that*

$$\|f(x) - A_k(x)\| \le \frac{\theta}{|k|^{r+1}}\sigma_k(\psi)\psi(\|x\|)^2,$$

$$\|g(x) - A_k(x) - g(0)\| \le \frac{\theta}{|k|^{r+1}}\sigma_k(\psi)\psi(\|x\|)^2,$$

$$\|h(x) - A_k(x) - h(0)\| \le \frac{\theta}{|k|^{r+1}}\sigma_k(\psi)\psi(\|x\|)^2$$

*for all* $x \in X$, *where* $\sigma_k(\psi) := \max\{\psi(|i|) : 1 \le i < k\}$.

*Proof.* Let $\Phi(x,y) = \theta\psi(\|x\|)\psi(\|y\|)$ for all $x,y \in X$. By using Theorem 24.3, one can obtain the statement. □

**Theorem 24.8.** *Let $q$ be a real number with $q \neq 1$, $H : [0,\infty) \times [0,\infty) \to [0,\infty)$ be a homogeneous function of degree $q$ and $|k| < 1$. Suppose that $f,g,h : X \to Y$ are mappings with $f(0) = 0$ and*

$$\|f(x+y) - g(x) - h(y)\| \leq H(\|x\|,\|y\|) \tag{24.28}$$

*for all $x,y \in X$. Then there exists a unique additive mapping $A_k : X \to Y$ satisfying*

$$\|f(x) - A_k(x)\| \leq \begin{cases} \dfrac{\sigma_k(H)\|x\|^q}{|k|}, & q > 1; \\ \dfrac{\sigma_k(H)\|x\|^q}{|k|^{q+1}}, & q < 1, \end{cases}$$

*for all $x \in X$, where $\sigma_k(H) := \max\{\max\{H(1,|i|),H(1,0),H(0,|i|)\} : 1 \leq i < k\}$.*

*Proof.* The statement follows from Theorems 24.2 and 24.3. □

For the particular cases $H(x,y) = \theta(x^q + y^q)$, $H(x,y) = x^r y^s$ $(r+s = q)$ and $H(x,y) = \min\{x^q, y^q\}$, we have the following two corollaries.

**Corollary 24.1.** *Let $\theta$ and $q$ be positive real numbers such that $q \neq 1$ and $|k| < 1$. Suppose that $f,g,h : X \to Y$ are mappings with $f(0) = 0$ and*

$$\|f(x+y) - g(x) - h(y)\| \leq \theta(\|x\|^q + \|y\|^q)$$

*for all $x,y \in X$. Then there exists a unique additive mapping $A_k : X \to Y$ such that, for all $x \in X$,*

$$\|f(x) - A_k(x)\| \leq \begin{cases} \dfrac{2\theta\|x\|^q}{|k|}, & q > 1; \\ \dfrac{2\theta\|x\|^q}{|k|^{q+1}}, & q < 1. \end{cases}$$

**Corollary 24.2.** *Let $r,s$ and $q$ be positive real numbers and $|k| < 1$. Suppose that $f,g,h : X \to Y$ are mappings with $f(0) = 0$ and*

$$\|f(x+y) - g(x) - h(y)\| \leq \begin{cases} \|x\|^r\|y\|^s, & q := r+s \neq 1; \\ \min\{\|x\|^q,\|y\|^q\}, & q \neq 1, \end{cases}$$

*for all $x, y \in X$. Then there exists a unique additive mapping $A_k : X \to Y$ such that*

$$\|f(x) - A_k(x)\| \leq \begin{cases} \dfrac{\|x\|^q}{|k|}, & q > 1; \\ \dfrac{\|x\|^q}{|k|^{p+1}}, & q < 1, \end{cases}$$

*for all $x \in X$. Moreover, for all $x \in X$,*

$$f(x) = g(x) + h(0) = h(x) + g(0).$$

## References


1. Aoki, T.: On the stability of the linear transformation in Banach spaces. J. Math. Soc. Japan **2**, 64–66 (1950)
2. Brzdęk, J.: On a method of proving the Hyers–Ulam stability of functional equations on restricted domains. Aust. J. Math. Anal. Appl. **6**, no. 1, Art. 4, 10 pp. (2009)
3. Brzdęk, J.: On the quotient stability of a family of functional equations. Nonlinear Anal. **71**, 4396–4404 (2009)
4. Czerwik, S.: Functional Equations and Inequalities in Several Variables. World Scientific, New Jersey–London–Singapore–Hong Kong (2002)
5. Eskandani, G.Z.: On the Hyers-Ulam-Rassias stability of an additive functional equation in quasi-Banach spaces. J. Math. Anal. Appl. **345**, 405–409 (2008)
6. Gajda, Z.: On stability of additive mappings. Int. J. Math. Math. Sci. **14**, 431–434 (1991)
7. Găvruta, P.: A generalization of the Hyers–Ulam–Rassias stability of approximately additive mappings. J. Math. Anal. Appl. **184**, 431–436 (1994)
8. Găvruta, P.: An answer to question of John M. Rassias concerning the stability of Cauchy equation. In: Rassias, J.M. (ed.) Advances in Equations and Inequalities, pp. 67–71. Hadronic Press, Palm Harbor (1999)
9. Găvruta, P.: On a problem of G. Isac and Th. M. Rassias concerning the stability of mappings. J. Math. Anal. Appl. **261**, 543–553 (2001)
10. Găvruta, P.: On the Hyers-Ulam-Rassias stability of mappings. In: Milovanovic, G.V. (ed.) Recent Progress in Inequalities, pp. 465–469. Kluwer (1998)
11. Găvruta, P., Hossu, M., Popescu D., Căprău, C.: On the stability of mappings and an anwser to a problem of Th.M Rassias. Ann. Math. Blaise Pascal **2**, 55–60 (1995)
12. Hensel, K.: Über eine neue Begründung der Theorie der algebraischen Zahlen. Jahresber. Deutsch. Math. Verein **6**, 83–88 (1897)
13. Hyers, D.H.: On the stability of the linear functional equation. Proc. Nat. Acad. Sci. **27**, 222–224 (1941)
14. Hyers, D.H., Isac G., Rassias, Th.M.: Stability of Functional Equations in Several Variables. Birkhäuser, Basel (1998)
15. Isac, G., Rassias, Th.M.: Stability of $\psi$-additive mappings: Applications to nonlinear analysis. Int. J. Math. Math. Sci. **19**, 219–228 (1996)
16. Isac, G., Rassias, Th.M.: Functional inequalities for approximately additive mappings. In: Rassias, Th.M., Tabor, J. (eds.) Stability of Mappings of Hyers–Ulam Type, pp. 117–125. Hadronic Press, Palm Harbour (1994)
17. Jun, K., Lee, Y.: On the Hyers–Ulam–Rassias stability of a Pexiderized quadratic inequality. Math. Inequal. Appl. **4**, 93–118 (2001)

18. Jung, S.-M.: Hyers–Ulam–Rassias Stability of Functional Equations in Mathimatical Analysis. Hadronic Press, Palm Harbor (2001)
19. Kannappan, PL.: Quadratic functional equation and inner product spaces. Results Math. **27**, 368–372 (1995)
20. Moradlou, F., Vaezi, H., Eskandani, G.Z.: Hyers–Ulam–Rassias stability of a quadratic and additive functional equation in quasi-Banach spaces. Mediterr. J. Math. **6**, 233–248 (2009)
21. Moslehian, M.S., Rassias, Th.M.: Stability of functional equation in non-Archimedean spaces. Appl. Anal. Discrete Math. **1**, 325–334 (2007)
22. Park, C., Moradlou, F.: Stability of Homomorphisms and Derivations in C*-ternary Rings. Taiwanese J. Math. **13**, 1985–1999 (2009)
23. Rassias, J.M.: On approximation of approximately linear mappings by linear mappings. J. Funct. Anal. **46** 126–130 (1982)
24. Rassias, J.M.: On approximation of approximately linear mappings by linear mappings. Bull. Sci. Math. **108**, 445–446 (1984)
25. Rassias, J.M.: Solution of a problem of Ulam. J. Approx. Theory **57**, 268–273 (1989)
26. Rassias, Th.M.: On the stability of the linear mapping in Banach spaces. Proc. Amer. Math. Soc. **72**, 297–300 (1978)
27. Rassias, Th.M.: On a modified Hyers–Ulam sequence. J. Math. Anal. Appl. **158**, 106–113 (1991)
28. Rassias, Th.M.: Problem 16. In: Report of the 27th International Symposium on Functional Equations, p. 309. Aequationes Math. **39**, 292–293 (1990)
29. Rassias, Th.M.: On the stability of the quadratic functional equation and its applications. Studia Univ. Babeş-Bolyai Math. **43**, 89–124 (1998)
30. Rassias, Th.M. (ed.): Functional Equations and Inequalities. Kluwer Academic Publishers, Dordrecht–Boston–London (2000)
31. Rassias, Th.M., Šemrl, P.: On the behaviour of mappings which do not satisfy Hyers-Ulam stability. Proc. Amer. Math. Soc. **114** 989–993 (1992)
32. Robert, A.M.: A Course in p-adic Analysis. Springer-Verlag, New York (2000)
33. Ulam, S.M.: A Collection of the Mathematical Problems. Interscience Publ., New York (1960)

# Chapter 25
# Stability of the Quadratic–Cubic Functional Equation in Quasi–Banach Spaces

**Wanxiong Zhang and Zhihua Wang**

*Dedicated to the memory of S.M. Ulam*
*on the 100th anniversary of his birth*

**Abstract** In quasi–Banach spaces, we investigate the generalized Hyers–Ulam–Rassias stability of the quadratic-cubic functional equation

$$6f(x+y)-6f(x-y)+4f(3y)=3f(x+2y)-3f(x-2y)+9f(2y).$$

**Keywords** Hyers–Ulam–Rassias stability • Quadratic–cubic functional equation • Quasi–Banach space

**Mathematics Subject Classification (2000):** Primary 39B72, 39B82; Secondary 46B99

## 25.1 Introduction and Preliminaries

In 1940, Ulam [21] posed the first stability problem concerning group homomorphisms. In the next year, Hyers [8] gave a partial affirmative answer to the question of Ulam. Hyers' result was generalized by Aoki [3] for additive mappings and by Rassias [18] for linear mappings by considering an unbounded Cauchy difference.

W. Zhang (✉)
College of Mathematics and Statistics, Chongqing University, Chongqing 401331, People's Republic of China
e-mail: cqumatzwx@163.com

Z. Wang
School of Science, Hubei University of Technology, Wuhan, Hubei 430068, People's Republic of China
e-mail: matwzh2000@126.com

Th.M. Rassias and J. Brzdęk (eds.), *Functional Equations in Mathematical Analysis*, Springer Optimization and Its Applications 52, DOI 10.1007/978-1-4614-0055-4_25,

During the last decades several stability problems for various functional equations have been investigated by many authors for mappings with more general domains and ranges [2, 12, 14, 15].

It is well-known that a function $f$ between real vector spaces satisfies the following quadratic functional equation

$$f(x+y)+f(x-y)=2f(x)+2f(y) \tag{25.1}$$

for all $x,y$ if and only if there is a unique symmetric bi-additive function $B$ such that $f(x)=B(x,x)$ for all $x$, where $B$ is given by [1]

$$B(x,y)=\frac{1}{4}[f(x+y)-f(x-y)].$$

A stability problem for the quadratic functional equation (25.1) was studied by many authors [6, 10, 11, 13, 20].

Let both $E_1$ and $E_2$ be real vector spaces. Jun and Kim [9] proved that a function $f: E_1 \to E_2$ satisfies the functional equation

$$f(2x+y)+f(2x-y)=2f(x+y)+2f(x-y)+12f(x) \tag{25.2}$$

if and only if there exists a function $B: E_1\times E_1\times E_1 \to E_2$ such that $f(x)=B(x,x,x)$ for all $x\in E_1$, and $B$ is symmetric for each fixed one variable and additive for each fixed two variables, where $B$ is given by

$$B(x,y,z)=\frac{1}{24}[f(x+y+z)+f(x-y-z)-f(x+y-z)-f(x-y+z)]$$

for all $x,y,z\in E_1$, and established the generalized Hyers–Ulam–Rassias stability for (25.2). It is easy to see that the function

$$f(x)=ax^3$$

is a solution of the functional equation (25.2). Thus, it is natural that (25.2) is called a cubic functional equation and every solution of the cubic functional equation (25.2) is said to be a cubic function.

In this paper, using some ideas from [16, 17], we consider the following functional equation deriving from quadratic and cubic functions [5]:

$$6f(x+y)-6f(x-y)+4f(3y)=3f(x+2y)-3f(x-2y)+9f(2y). \tag{25.3}$$

It is easy to see that the function

$$f(x)=ax^3+bx^2$$

is a solution of the functional equation (25.3). We prove some stability results for (25.3) for mappings from quasi-normed space into $p$-Banach spaces. It should be noted that our results generalize those in [5] to quasi–Banach spaces.

We recall some basic facts concerning quasi–Banach spaces and some preliminary results.

**Definition 25.1 (cf. [4,19]).** Let $X$ be a real linear space. A *quasi-norm* is a real-valued function on $X$ satisfying the following:

(1) $\|x\| \geq 0$, $\forall x \in X$ and $\|x\| = 0$ if and only if $x = 0$.
(2) $\|\lambda x\| = |\lambda|\|x\|$, $\forall \lambda \in \mathbb{R}$ and $\forall x \in X$.
(3) There is a constant $K \geq 1$ such that $\|x+y\| \leq K(\|x\|+\|y\|)$, $\forall x,y \in X$.

It follows from condition (3) that

$$\left\|\sum_{i=1}^{2n} x_i\right\| \leq K^n \sum_{i=1}^{2n} \|x_i\|$$

and

$$\left\|\sum_{i=1}^{2n+1} x_i\right\| \leq K^{n+1} \sum_{i=1}^{2n+1} \|x_i\|$$

for all integers $n \geq 1$ and all $x_1, x_2, \ldots, x_{2n+1} \in X$.

The pair $(X, \|\cdot\|)$ is called a *quasi-normed space* if $\|\cdot\|$ is a *quasi-norm* on $X$. The smallest possible $K$ is called the *modulus of concavity* of $\|\cdot\|$. A *quasi–Banach space* is a complete *quasi-normed space*. A *quasi-norm* $\|\cdot\|$ is called a *p-norm* $(0 < p \leq 1)$ if

$$\|x+y\|^p \leq \|x\|^p + \|y\|^p$$

for all $x,y \in X$. In this case, a *quasi–Banach space* is called a *p-Banach space*.

Given a $p$-norm, the formula

$$d(x,y) := \|x-y\|^p$$

gives us a translation invariant metric on $X$. By the Aoki-Rolewicz theorem [19] (see also [4]), each quasi-norm is equivalent to some $p$-norm. Since it is much easier to work with $p$-norms than quasi-norms, henceforth we restrict our attention mainly to $p$-norms.

## 25.2 Hyers–Ulam–Rassias Stability of (25.3)

Throughout this section, assume that $X$ is a quasi-normed space with quasi-norm $\|\cdot\|_X$ and that $Y$ is a $p$-Banach space with $p$-norm $\|\cdot\|_Y$. Let $K$ be the modulus of concavity of $\|\cdot\|_Y$. In this section, using an idea of Găvruta [7] we prove the stability of (25.3). For convenience, we use the following abbreviation for a given function $f : X \to Y$:

$$Df(x,y) = 6f(x+y) - 6f(x-y) + 4f(3y) - 3f(x+2y) + 3f(x-2y) - 9f(2y)$$

for all $x,y \in X$.

We use the following Lemma 25.1 in this section; for a proof of it see [16].

**Lemma 25.1 (cf. [16]).** *Let $0 < p \leq 1$ and let $x_1, x_2, \ldots, x_n$ be non-negative real numbers. Then*

$$\left(\sum_{i=1}^{n} x_i\right)^p \leq \sum_{i=1}^{n} x_i^p. \tag{25.4}$$

**Theorem 25.1.** *Let $\varphi_e : X \times X \to [0, \infty)$ be a function such that*

$$\lim_{n\to\infty} \frac{1}{4^n} \varphi_e(2^n x, 2^n y) = 0, \qquad \forall x, y \in X, \tag{25.5}$$

*and*

$$M_e(x,y) := \sum_{i=0}^{\infty} \frac{1}{4^{ip}} \varphi_e^p(2^i x, 2^i y) < \infty, \qquad \forall y \in X, x \in \{0, y\}. \tag{25.6}$$

*Suppose that an even function $f : X \to Y$ with $f(0) = 0$ satisfies the inequality*

$$\|Df(x,y)\|_Y \leq \varphi_e(x,y), \qquad \forall x, y \in X. \tag{25.7}$$

*Then the limit*

$$Q(x) := \lim_{n\to\infty} 4^{-n} f(2^n x)$$

*exists for all $x \in X$ and $Q : X \to Y$ is a unique quadratic function satisfying*

$$\|f(x) - Q(x)\|_Y \leq \frac{K}{12} [\widetilde{\varphi}_e(x)]^{1/p}, \qquad \forall x \in X, \tag{25.8}$$

*where $\widetilde{\varphi}_e(x) := M_e(0,x) + (4K)^p M_e(x,x)$.*

*Proof.* In (25.7), we set $x = 0$, then replace $y$ by $x$ and obtain

$$\|4f(3x) - 9f(2x)\|_Y \leq \varphi_e(0,x), \qquad \forall x \in X. \tag{25.9}$$

Put $y = x$ in (25.7) to obtain

$$\|f(3x) + 3f(x) - 3f(2x)\|_Y \leq \varphi_e(x,x), \qquad \forall x \in X. \tag{25.10}$$

It follows from (25.9) and (25.10) that

$$\|3f(2x) - 12f(x)\|_Y \leq K[\varphi_e(0,x) + 4K\varphi_e(x,x)], \qquad \forall x \in X. \tag{25.11}$$

So

$$\|f(2x) - 4f(x)\|_Y \leq \frac{K}{3}[\varphi_e(0,x) + 4K\varphi_e(x,x)], \qquad \forall x \in X. \tag{25.12}$$

Let $\psi_e(x) := \frac{K}{3}[\varphi_e(0,x) + 4K\varphi_e(x,x)]$ for all $x \in X$. Then we get

$$\|f(2x) - 4f(x)\|_Y \leq \psi_e(x), \qquad \forall x \in X. \tag{25.13}$$

By Lemma 25.1 and (25.6), we infer that

$$\sum_{i=0}^{\infty} \frac{1}{4^{ip}} \psi_e^p(2^i x) < \infty, \qquad \lim_{n\to\infty} \frac{1}{4^n} \psi_e(2^n x) = 0 \tag{25.14}$$

for all $x \in X$. Replacing $x$ by $2^n x$ in (25.13) and dividing both sides of (25.13) by $4^{n+1}$, we get

$$\left\| \frac{1}{4^{n+1}} f(2^{n+1}x) - \frac{1}{4^n} f(2^n x) \right\|_Y \leq \frac{1}{4^{n+1}} \psi_e(2^n x), \qquad \forall x \in X, \tag{25.15}$$

and all non-negative integers $n$. Since $Y$ is a $p$-Banach space, we have

$$\begin{aligned} \left\| \frac{1}{4^{n+1}} f(2^{n+1}x) - \frac{1}{4^m} f(2^m x) \right\|_Y^p & \\ &\leq \sum_{i=m}^{n} \left\| \frac{1}{4^{i+1}} f(2^{i+1}x) - \frac{1}{4^i} f(2^i x) \right\|_Y^p \\ &\leq \frac{1}{4^p} \sum_{i=m}^{n} \frac{1}{4^{ip}} \psi_e^p(2^i x) \end{aligned} \tag{25.16}$$

for all $x \in X$ and all non-negative $n$ and $m$ with $n \geq m$. Therefore, we conclude from (25.14) and (25.16) that the sequence $\{4^{-n} f(2^n x)\}$ is a Cauchy sequence in $Y$ for all $x \in X$. Since $Y$ is complete, the sequence $\{4^{-n} f(2^n x)\}$ converges in $Y$ for all $x \in X$. So one can define the function $Q : X \to Y$ by

$$Q(x) := \lim_{n\to\infty} \frac{f(2^n x)}{4^n}, \qquad \forall x \in X. \tag{25.17}$$

Letting $m = 0$ and $n \to \infty$ in (25.16) and applying Lemma 25.1, we get (25.8).

Now, we show that $Q$ is quadratic. It follows from (25.5), (25.6) and (25.17) that

$$\|DQ(x,y)\|_Y = \lim_{n\to\infty} \frac{1}{4^n} \|Df(2^n x, 2^n y)\|_Y \leq \lim_{n\to\infty} \frac{1}{4^n} \varphi_e(2^n x, 2^n y) = 0$$

for all $x, y \in X$. Therefore, the function $Q : X \to Y$ satisfies (25.3). Since $Q(0) = 0$, then by [5, Lemma 2.1] we get that the function $Q : X \to Y$ is quadratic.

To prove the uniqueness of $Q$, let $Q' : X \to Y$ be another quadratic function satisfying (25.8). Since

$$\lim_{n\to\infty} \frac{1}{4^{np}} M_e(2^n x, 2^n y) = \lim_{n\to\infty} \sum_{i=n}^{\infty} \frac{1}{4^{ip}} \varphi_e^p(2^n x, 2^n y) = 0$$

for all $y \in X$ and all $x \in \{0,y\}$. Hence, $\lim_{n\to\infty} 4^{-np}\widetilde{\varphi}_e(2^n x) = 0$ for all $x \in X$. So it follows from (25.8) and (25.17) that

$$\|Q(x) - Q'(x)\|_Y^p = \lim_{n\to\infty} \frac{1}{4^{np}} \|f(2^n x) - Q'(2^n x)\|_Y^p \le \frac{K^p}{12^p} \lim_{n\to\infty} \frac{1}{4^{np}} \widetilde{\varphi}_e(2^n x) = 0$$

for all $x \in X$. So $Q = Q'$. □

**Theorem 25.2.** *Let $\Phi_e : X \times X \to [0,\infty)$ be a function such that*

$$\lim_{n\to\infty} 4^n \Phi_e\left(\frac{x}{2^n}, \frac{y}{2^n}\right) = 0, \qquad \forall x,y \in X, \tag{25.18}$$

*and*

$$\mathbf{M}_e(x,y) := \sum_{i=1}^{\infty} 4^{ip} \Phi_e^p\left(\frac{x}{2^i}, \frac{y}{2^i}\right) < \infty \tag{25.19}$$

*for all $y \in X$ and all $x \in \{0,y\}$. Suppose that an even function $f : X \to Y$ with $f(0) = 0$ satisfies the inequality*

$$\|Df(x,y)\|_Y \le \Phi_e(x,y), \qquad \forall x,y \in X. \tag{25.20}$$

*Then the limit $Q(x) := \lim_{n\to\infty} 4^n f(\frac{x}{2^n})$ exists for all $x \in X$ and $Q : X \to Y$ is a unique quadratic function satisfying*

$$\|f(x) - Q(x)\|_Y \le \frac{K}{12}\left[\widetilde{\Phi}_e(x)\right]^{\frac{1}{p}}, \qquad \forall x \in X, \tag{25.21}$$

*where $\widetilde{\Phi}_e(x) := \mathbf{M}_e(0,x) + (4K)^p \mathbf{M}_e(x,x)$.*

*Proof.* Similar to the proof of Theorem 25.1, we have

$$\|f(2x) - 4f(x)\|_Y \le \Phi_e(x), \qquad \forall x \in X, \tag{25.22}$$

where

$$\Phi_e(x) := \frac{K}{3}[\Phi_e(0,x) + 4K\Phi_e(x,x)].$$

By Lemma 25.1 and (25.19), we infer that

$$\sum_{i=1}^{\infty} 4^{ip} \Phi_e^p\left(\frac{x}{2^i}\right) < \infty, \qquad \forall x \in X. \tag{25.23}$$

Replacing $x$ by $2^{-n-1}x$ in (25.22) and multiplying both sides of (25.22) to $4^n$, we get

$$\left\|4^{n+1} f\left(\frac{x}{2^{n+1}}\right) - 4^n f\left(\frac{x}{2^{n+1}}\right)\right\|_Y \le 4^n \Phi_e\left(\frac{x}{2^{n+1}}\right), \qquad \forall x \in X, \tag{25.24}$$

and all non-negative integers $n$. Since $Y$ is a $p$-Banach space, we have

$$\begin{aligned}
&\left\|4^{n+1} f\left(\frac{x}{2^{n+1}}\right) - 4^m f\left(\frac{x}{2^m}\right)\right\|_Y^p \\
&\qquad \le \sum_{i=m}^{n} \left\|4^{i+1} f\left(\frac{x}{2^{i+1}}\right) - 4^i f\left(\frac{x}{2^i}\right)\right\|_Y^p \\
&\qquad \le \sum_{i=m}^{n} 4^{ip} \Phi_e^p\left(\frac{x}{2^{i+1}}\right)
\end{aligned} \tag{25.25}$$

for all $x \in X$ and all non-negative $n$ and $m$ with $n \ge m$. Therefore, we conclude from (25.23) and (25.25) that the sequence $\{4^n f(2^{-n}x)\}$ is a Cauchy sequence in $Y$ for all $x \in X$. Since $Y$ is complete, the sequence $\{4^n f(2^{-n}x)\}$ converges in $Y$ for all $x \in X$. So one can define the function $Q : X \to Y$ by $Q(x) := \lim_{n\to\infty} 4^n f(2^{-n}x)$ for all $x \in X$. Letting $m = 0$ and $n \to \infty$ in (25.25) and applying Lemma 25.1, we get (25.21). The rest of the proof is similar to the proof Theorem 25.1. □

**Corollary 25.1.** *Let $\theta, r, s$ be non-negative real numbers with $r, s > 2$ or $0 \le r, s < 2$. Suppose that an even function $f : X \to Y$ with $f(0) = 0$ satisfies the inequality*

$$\|Df(x,y)\|_Y \le \begin{cases} \theta, & r = s = 0, \\ \theta\|x\|_X^r, & r > 0, s = 0, \\ \theta\|y\|_X^s, & r = 0, s > 0, \\ \theta(\|x\|_X^r + \|y\|_X^s), & r > 0, s > 0, \end{cases} \tag{25.26}$$

*for all $x, y \in X$. Then there exists a unique quadratic function $Q : X \to Y$ satisfying*

$$\|f(x) - Q(x)\|_Y \le \begin{cases} \delta_e, & r = s = 0, \\ \alpha_e(x), & r > 0, s = 0, \\ \beta_e(x), & r = 0, s > 0, \\ \gamma_e(x), & r > 0, s > 0, \end{cases}$$

*for all $x \in X$, where*

$$\alpha_e(x) = \frac{K\theta}{3}\left\{\frac{(4K)^p}{|4^p - 2^{pr}|}\right\}^{\frac{1}{p}} \|x\|_X^r, \qquad \beta_e(x) = \frac{K\theta}{3}\left\{\frac{1 + (4K)^p}{|4^p - 2^{ps}|}\right\}^{\frac{1}{p}} \|x\|_X^s,$$

$$\gamma_e(x) = \frac{K\theta}{3}\left\{\frac{(4K)^p}{|4^p - 2^{pr}|}\|x\|_X^{pr} + \frac{1 + (4K)^p}{|4^p - 2^{ps}|}\|x\|_X^{ps}\right\}^{\frac{1}{p}},$$

$$\delta_e = \frac{K\theta}{3}\left\{\frac{1 + (4K)^p}{4^p - 1}\right\}^{\frac{1}{p}}.$$

*Proof.* The result follows by Theorems 25.1 and 25.2. □

**Corollary 25.2.** *Let $\theta \geq 0$ and $r,s$ be non-negative real numbers with $\lambda := r+s \neq 2$. Suppose that an even function $f : X \to Y$ with $f(0) = 0$ satisfies the inequality*

$$\|Df(x,y)\|_Y \leq \theta \|x\|_X^r \|y\|_X^s, \qquad \forall x,y \in X. \tag{25.27}$$

*Then there exists a unique quadratic function $Q : X \to Y$ satisfying*

$$\|f(x) - Q(x)\|_Y \leq \frac{K\theta}{3} \left\{ \frac{(4K)^p}{|4^p - 2^{p\lambda}|} \right\}^{\frac{1}{p}} \|x\|_X^{\lambda}, \qquad \forall x,y \in X.$$

*Proof.* The result follows by Theorems 25.1 and 25.2. □

**Theorem 25.3.** *Let $\varphi_o : X \times X \to [0,\infty)$ be a function such that*

$$\lim_{n\to\infty} \frac{1}{8^n} \varphi_o(2^n x, 2^n y) = 0, \qquad \forall x,y \in X, \tag{25.28}$$

*and*

$$M_o(x,y) := \sum_{i=0}^{\infty} \frac{1}{8^{ip}} \varphi_o^p(2^i x, 2^i y) < \infty \tag{25.29}$$

*for all $y \in X$ and all $x \in \{0,y\}$. Suppose that an odd function $f : X \to Y$ satisfies the inequality*

$$\|Df(x,y)\|_Y \leq \varphi_o(x,y), \qquad \forall x,y \in X. \tag{25.30}$$

*Then the limit $C(x) := \lim_{n\to\infty} 8^{-n} f(2^n x)$ exists for all $x \in X$ and $C : X \to Y$ is a unique cubic function satisfying*

$$\|f(x) - C(x)\|_Y \leq \frac{K}{24} [\widetilde{\varphi}_o(x)]^{1/p}, \qquad \forall x \in X, \tag{25.31}$$

*where*

$$\widetilde{\varphi}_o(x) := M_o(0,x) + (4K)^p M_o(x,x).$$

*Proof.* In (25.30), we set $x = 0$, then replace $y$ by $x$, and obtain

$$\|12f(x) + 4f(3x) - 15f(2x)\|_Y \leq \varphi_o(0,x), \qquad \forall x \in X. \tag{25.32}$$

Put $y = x$ in (25.30) to obtain

$$\|3f(x) - f(3x) + 3f(2x)\|_Y \leq \varphi_o(x,x), \qquad \forall x \in X. \tag{25.33}$$

It follows from (25.32) and (25.33) that

$$\|24f(x) - 3f(2x)\|_Y \le K[\varphi_o(0,x) + 4K\varphi_o(x,x)], \qquad \forall x \in X. \tag{25.34}$$

So

$$\|f(2x) - 8f(x)\|_Y \le \frac{K}{3}[\varphi_o(0,x) + 4K\varphi_o(x,x)], \qquad \forall x \in X. \tag{25.35}$$

Let

$$\psi_o(x) := \frac{K}{3}[\varphi_o(0,x) + 4K\varphi_o(x,x)]$$

for all $x \in X$. Then we get

$$\|f(2x) - 8f(x)\|_Y \le \psi_o(x), \qquad \forall x \in X. \tag{25.36}$$

By Lemma 25.1 and (25.29), we infer that

$$\sum_{i=0}^{\infty} \frac{1}{8^{ip}} \psi_o^p(2^i x) < \infty, \qquad \lim_{n\to\infty} \frac{1}{8^n} \psi_o(2^n x) = 0 \tag{25.37}$$

for all $x \in X$. Replacing $x$ by $2^n x$ in (25.36) and dividing both sides of (25.36) by $8^{n+1}$, we get

$$\left\| \frac{1}{8^{n+1}} f(2^{n+1}x) - \frac{1}{8^n} f(2^n x) \right\|_Y \le \frac{1}{8^{n+1}} \psi_e(2^n x) \tag{25.38}$$

for all $x \in X$ and all non-negative integers $n$. Since $Y$ is a $p$-Banach space,

$$\begin{aligned} \left\| \frac{1}{8^{n+1}} f(2^{n+1}x) - \frac{1}{8^m} f(2^m x) \right\|_Y^p &\le \sum_{i=m}^{n} \left\| \frac{1}{8^{i+1}} f(2^{i+1}x) - \frac{1}{8^i} f(2^i x) \right\|_Y^p \\ &\le \frac{1}{8^p} \sum_{i=m}^{n} \frac{1}{ip} \psi_o^p(2^i x) \end{aligned} \tag{25.39}$$

for all $x \in X$ and all non-negative $n$ and $m$ with $n \ge m$. Therefore, we conclude from (25.37) and (25.39) that the sequence $\{8^{-n} f(2^n x)\}$ is a Cauchy sequence in $Y$ for all $x \in X$. Since $Y$ is complete, the sequence $\{8^{-n} f(2^n x)\}$ converges in $Y$ for all $x \in X$. So one can define the function $C : X \to Y$ by

$$C(x) := \lim_{n\to\infty} \frac{f(2^n x)}{8^n}, \qquad \forall x \in X. \tag{25.40}$$

Letting $m = 0$ and $n \to \infty$ in (25.37) and applying Lemma 25.1, we get (25.31).

Now, we show that $C$ is cubic. It follows from (25.28), (25.29) and (25.40) that

$$\|DC(x,y)\|_Y = \lim_{n\to\infty} \frac{1}{8^n}\|Df(2^n x, 2^n y)\|_Y \le \lim_{n\to\infty} \frac{1}{8^n}\varphi_o(2^n x, 2^n y) = 0$$

for all $x,y \in X$. Therefore, the function $C : X \to Y$ satisfies (25.3). Since $C(-x) = -C(x)$, then by [5, Lemma 2.2] we get that the function $C : X \to Y$ is cubic.

To prove the uniqueness of $C$, let $C' : X \to Y$ be another cubic function satisfying (25.31). Since

$$\lim_{n\to\infty} \frac{1}{8^{np}} M_o(2^n x, 2^n y) = \lim_{n\to\infty} \sum_{i=n}^{\infty} \frac{1}{8^{ip}} \varphi_o^p(2^n x, 2^n y) = 0$$

for all $y \in X$ and all $x \in \{0, y\}$. Hence

$$\lim_{n\to\infty} 8^{-np}\widetilde{\varphi}_o(2^n x) = 0$$

for all $x \in X$. So it follows from (25.31) and (25.40) that

$$\begin{aligned}\|C(x) - C'(x)\|_Y^p &= \lim_{n\to\infty} \frac{1}{8^{np}}\|f(2^n x) - C'(2^n x)\|_Y^p \\ &\le \frac{K^p}{24^p} \lim_{n\to\infty} \frac{1}{8^{np}} \widetilde{\varphi}_o(2^n x) = 0\end{aligned}$$

for all $x \in X$. So $C = C'$. □

**Theorem 25.4.** *Let $\Phi_o : X \times X \to [0, \infty)$ be a function such that*

$$\lim_{n\to\infty} 8^n \Phi_o\left(\frac{x}{2^n}, \frac{y}{2^n}\right) = 0, \qquad \forall x, y \in X, \tag{25.41}$$

*and*

$$\mathbf{M}_o(x,y) := \sum_{i=1}^{\infty} 8^{ip} \Phi_o^p\left(\frac{x}{2^i}, \frac{y}{2^i}\right) < \infty \tag{25.42}$$

*for all $y \in X$ and all $x \in \{0, y\}$. Suppose that an odd function $f : X \to Y$ satisfies the inequality*

$$\|Df(x,y)\|_Y \le \Phi_o(x,y), \qquad \forall x, y \in X. \tag{25.43}$$

*Then the limit*

$$C(x) := \lim_{n\to\infty} 8^n f\left(\frac{x}{2^n}\right)$$

*exists for all $x \in X$ and $C : X \to Y$ is a unique cubic function satisfying*

$$\|f(x) - Q(x)\|_Y \leq \frac{K}{24}[\widetilde{\Phi}_o(x)]^{\frac{1}{p}}, \qquad \forall x \in X, \tag{25.44}$$

*where*

$$\widetilde{\Phi}_o(x) := \mathbf{M}_0(0,x) + (4K)^p \mathbf{M}_o(x,x).$$

*Proof.* Similar to the proof of Theorem 25.3, we have

$$\|f(2x) - 8f(x)\|_Y \leq \Phi_o(x), \qquad \forall x \in X, \tag{25.45}$$

where

$$\Phi_o(x) := \frac{K}{3}[\Phi_o(0,x) + 4K\Phi_o(x,x)].$$

By Lemma 25.1 and (25.42), we infer that

$$\sum_{i=1}^{\infty} 8^{ip}\Phi_o^p\left(\frac{x}{2^i}\right) < \infty, \qquad \forall x \in X. \tag{25.46}$$

Replace $x$ by $2^{-n-1}x$ in (25.22) and multiplying both sides of (25.45) by $8^n$, we get

$$\left\|8^{n+1}f\left(\frac{x}{2^{n+1}}\right) - 8^n f\left(\frac{x}{2^{n+1}}\right)\right\|_Y \leq 8^n\Phi_o\left(\frac{x}{2^{n+1}}\right) \tag{25.47}$$

for all $x \in X$ and all non-negative integers $n$. Since $Y$ is a $p$-Banach space,

$$\begin{aligned} &\left\|8^{n+1}f\left(\frac{x}{2^{n+1}}\right) - 8^m f\left(\frac{x}{2^m}\right)\right\|_Y^p \\ &\qquad \leq \sum_{i=m}^{n} \|8^{i+1}f\left(\frac{x}{2^{i+1}}\right) - 8^i f\left(\frac{x}{2^i}\right)\|_Y^p \\ &\qquad \leq \sum_{i=m}^{n} 8^{ip}\Phi_o^p\left(\frac{x}{2^{i+1}}\right) \end{aligned} \tag{25.48}$$

for all $x \in X$ and all non-negative $n$ and $m$ with $n \geq m$. Therefore, we conclude from (25.46) and (25.48) that the sequence $\{8^n f(2^{-n}x)\}$ is a Cauchy sequence in $Y$ for all $x \in X$. Since $Y$ is complete, the sequence $\{8^n f(2^{-n}x)\}$ converges in $Y$ for all $x \in X$. So one can define the function $C : X \to Y$ by $C(x) := \lim_{n\to\infty} 8^n f(2^{-n}x)$ for $x \in X$. Letting $m = 0$ and $n \to \infty$ in (25.48) and applying Lemma 25.1, we get (25.44). The rest of the proof is similar to the proof Theorem 25.3. □

The next two corollaries easily follow from Theorems 25.3 and 25.4.

**Corollary 25.3.** *Let $\theta, r, s$ be non-negative real numbers with $r,s > 3$ or $0 \leq r,s < 3$. Suppose that an odd function $f : X \to Y$ satisfies the inequality (25.26) for all $x,y \in X$. Then there exists a unique cubic function $C : X \to Y$ such that, for all $x \in X$,*

$$\|f(x) - Q(x)\|_Y \leq \begin{cases} \delta_o, & r = s = 0, \\ \alpha_o(x), & r > 0, s = 0, \\ \beta_o(x), & r = 0, s > 0, \\ \gamma_o(x), & r > 0, s > 0, \end{cases}$$

*where*

$$\alpha_o(x) = \frac{K\theta}{3}\left\{\frac{(4K)^p}{|8^p - 2^{pr}|}\right\}^{\frac{1}{p}} \|x\|_X^r, \qquad \beta_o(x) = \frac{K\theta}{3}\left\{\frac{1+(4K)^p}{|8^p - 2^{ps}|}\right\}^{\frac{1}{p}} \|x\|_X^s,$$

$$\gamma_o(x) = \frac{K\theta}{3}\left\{\frac{(4K)^p}{|8^p - 2^{pr}|}\|x\|_X^{pr} + \frac{1+(4K)^p}{|8^p - 2^{ps}|}\|x\|_X^{ps}\right\}^{\frac{1}{p}},$$

$$\delta_o = \frac{K\theta}{3}\left\{\frac{1+(4K)^p}{8^p - 1}\right\}^{\frac{1}{p}}.$$

**Corollary 25.4.** *Let $\theta \geq 0$ and $r,s$ be non-negative real numbers with*

$$\lambda := r + s \neq 3.$$

*Suppose that an odd function $f : X \to Y$ satisfies the inequality (25.27) for all $x,y \in X$. Then there exists a unique cubic function $C : X \to Y$ satisfying*

$$\|f(x) - C(x)\|_Y \leq \frac{K\theta}{3}\left\{\frac{(4K)^p}{|8^p - 2^{p\lambda}|}\right\}^{\frac{1}{p}} \|x\|_X^{\lambda}, \qquad \forall x \in X.$$

**Theorem 25.5.** *Let $\varphi : X \times X \to [0,\infty)$ be such that, for all $x,y \in X$ and $z \in \{0,y\}$,*

$$\lim_{n\to\infty} \frac{1}{4^n}\varphi(2^n x, 2^n y) = 0$$

*and*

$$\sum_{i=0}^{\infty} \frac{1}{4^{ip}}\varphi^p(2^i z, 2^i y) < \infty.$$

*Suppose that a function $f : X \to Y$ with $f(0) = 0$ satisfies the inequality*

$$\|Df(x,y)\|_Y \leq \varphi(x,y), \qquad \forall x,y \in X. \tag{25.49}$$

*Then there exist a unique quadratic function $Q : X \to Y$ and a unique cubic function $C : X \to Y$ satisfying (25.3) and such that, for all $x \in X$,*

$$\|f(x) - Q(x) - C(x)\|_Y \leq \frac{K^3}{24}\left\{[\widetilde{\varphi}_e(x) + \widetilde{\varphi}_e(-x)]^{1/p} + \frac{1}{2}[\widetilde{\varphi}_o(x) + \widetilde{\varphi}_o(-x)]^{1/p}\right\},$$

*where $\widetilde{\varphi}_e$, $\widetilde{\varphi}_o$ are given by:*

$$\widetilde{\varphi}_e(x) := \sum_{i=0}^{\infty} 4^{-ip}\{\varphi^p(0, 2^i x) + (4K)^p \varphi^p(2^i x, 2^i x)\}$$

*and*

$$\widetilde{\varphi}_o(x) := \sum_{i=0}^{\infty} 8^{-ip}\{\varphi^p(0, 2^i x) + (4K)^p \varphi^p(2^i x, 2^i x)\}.$$

*Proof.* Let

$$f_e(x) = \frac{1}{2}(f(x) + f(-x))$$

for $x \in X$. Then

$$f_e(0) = 0, \qquad f_e(-x) = f_e(x)$$

and

$$\|Df_e(x,y)\|_Y \leq \frac{K}{2}[\varphi(x,y) + \varphi(-x,-y)], \qquad \forall x,y \in X.$$

Write

$$\Phi(x,y) := \frac{K}{2}[\varphi(x,y) + \varphi(-x,-y)]$$

for all $x,y \in X$. Then it is easily seen that

$$\lim_{n\to\infty} 4^{-n}\Phi(2^n x, 2^n y) = 0$$

for all $x,y \in X$. Since

$$\Phi^p(x,y) \leq \frac{K^p}{2^p}[\varphi^p(x,y) + \varphi^p(-x,-y)], \qquad \forall x,y \in X,$$

so

$$\sum_{i=0}^{\infty} \frac{1}{4^{ip}} \Phi^p(2^n x, 2^n y) < \infty$$

for all $y \in X$ and all $x \in \{0,y\}$. Hence, in view of Theorem 25.1, there exists a unique quadratic function $Q : X \to Y$ satisfying

$$\|f_e(x) - Q(x)\|_Y \le \frac{K}{12}[\widetilde{\phi}_e(x)]^{\frac{1}{p}}, \qquad \forall x \in X, \tag{25.50}$$

where

$$\widetilde{\phi}_e(x) := \sum_{i=0}^{\infty} \frac{1}{4^{ip}} \{\Phi^p(0,2^i x) + (4K)^p \Phi^p(2^i x, 2^i x)\}.$$

It is clear that

$$\widetilde{\phi}_e(x) \le \frac{K^p}{2^p}[\widetilde{\varphi}_e(x) + \widetilde{\varphi}_e(-x)]$$

for all $x \in X$. Hence, from (25.50) we obtain

$$\|f_e(x) - Q(x)\|_Y \le \frac{K^2}{24}[\widetilde{\varphi}_e(x) + \widetilde{\varphi}_e(-x)]^{\frac{1}{p}}, \qquad \forall x \in X. \tag{25.51}$$

Let

$$f_o(x) = \frac{1}{2}(f(x) - f(-x))$$

for all $x \in X$. Then $f_o(0) = 0$, $f_o(-x) = -f_o(x)$ and

$$\|Df_o(x,y)\|_Y \le \Phi(x,y)$$

for all $x,y \in X$. From Theorem 25.3, there exists a unique cubic function $C : X \to Y$ satisfying

$$\|f_o(x) - Q(x)\|_Y \le \frac{K}{24}[\widetilde{\phi}_o(x)]^{1/p}, \qquad \forall x \in X, \tag{25.52}$$

where

$$\widetilde{\phi}_o(x) := \sum_{i=0}^{\infty} \frac{1}{8^{ip}} \{\Phi^p(0,2^i x) + (4K)^p \Phi^p(2^i x, 2^i x)\}.$$

Since

$$\widetilde{\phi}_o(x) \le \frac{K^p}{2^p}[\widetilde{\varphi}_o(x) + \widetilde{\varphi}_o(-x)]$$

for all $x \in X$, from (25.52) we get

$$\|f_e(x) - Q(x)\|_Y \le \frac{K^2}{48}[\widetilde{\varphi}_e(x) + \widetilde{\varphi}_e(-x)]^{1/p}, \qquad \forall x \in X. \tag{25.53}$$

Hence, (25.52) follows from (25.51) and (25.53). □

Now, it is easily seen that we have the following theorem and two corollaries.

**Theorem 25.6.** *Let $\Phi : X \times X \to [0,\infty)$ be a function such that*

$$\lim_{n\to\infty} 8^n \Phi\left(\frac{x}{2^n}, \frac{y}{2^n}\right) = 0, \qquad \forall x,y \in X, \tag{25.54}$$

*and*

$$\sum_{i=1}^{\infty} 8^{ip} \Phi^p\left(\frac{x}{2^i}, \frac{y}{2^i}\right) < \infty \tag{25.55}$$

*for all $y \in X$ and all $x \in \{0,y\}$. Suppose that a function $f : X \to Y$ with $f(0) = 0$ satisfies the inequality*

$$\|Df(x,y)\|_Y \leq \Phi(x,y), \qquad \forall x,y \in X. \tag{25.56}$$

*Then there exist a unique quadratic function $Q : X \to Y$ and a unique cubic function $C : X \to Y$ satisfying (25.3) and such that, for all $x \in X$,*

$$\|f(x) - Q(x) - C(x)\|_Y \leq \frac{K^3}{24}\left\{[\widetilde{\Phi}_e(x) + \widetilde{\Phi}_e(-x)]^{1/p} + \frac{1}{2}[\widetilde{\Phi}_o(x) + \widetilde{\Phi}_o(-x)]^{1/p}\right\},$$

*where $\widetilde{\Phi}_e$, $\widetilde{\Phi}_o$ are defined by:*

$$\widetilde{\Phi}_e(x) := \sum_{i=1}^{\infty} 4^{ip}\{\Phi^p(0,2^i x) + (4K)^p \Phi^p(2^i x, 2^i x)\}$$

*and*

$$\widetilde{\Phi}_o(x) := \sum_{i=1}^{\infty} 8^{ip}\{\Phi^p(0,2^i x) + (4K)^p \Phi^p(2^i x, 2^i x)\}.$$

**Corollary 25.5.** *Let $\theta, r, s$ be non-negative real numbers with $r,s > 3$ or $0 \leq r,s < 2$. Suppose that a function $f : X \to Y$ with $f(0) = 0$ satisfies the inequality (25.26) for all $x,y \in X$. Then there exist a unique quadratic function $Q : X \to Y$ and a unique cubic function $C : X \to Y$ satisfying (25.3) and*

$$\|f(x) - Q(x) - C(x)\|_Y \leq \begin{cases} 2^{1/p-1}K^2[\delta_e + \delta_o], & r = s = 0, \\ 2^{1/p-1}K^2[\alpha_e(x) + \alpha_o(x)], & r > 0, s = 0, \\ 2^{1/p-1}K^2[\beta_e(x) + \beta_o(x)], & r = 0, s > 0, \\ 2^{1/p-1}K^2[\gamma_e(x) + \gamma_o(x)], & r > 0, s > 0, \end{cases}$$

*for all $x \in X$, where $\delta_e, \delta_o, \alpha_e(x), \alpha_o(x), \beta_e(x), \beta_o(x), \gamma_e(x)$ and $\gamma_o(x)$ are defined as in Corollaries 25.1 and 25.3.*

**Corollary 25.6.** *Assume that $\theta, r, s$ are non-negative real numbers such that $\lambda := r + s \in (0,2) \cup (3,+\infty]$. Suppose that a function $f : X \to Y$ with $f(0) = 0$ satisfies the inequality (25.27) for all $x,y \in X$. Then there is a unique quadratic function*

*$Q : X \to Y$ and a unique cubic function $C : X \to Y$ satisfying (25.3) and such that, for all $x, y \in X$,*

$$\|f(x) - Q(x) - C(x)\|_Y \leq \frac{K^3\theta}{6}\left\{\left[\frac{2(4K)^p}{|4^p - 2^{p\lambda}|}\right]^{\frac{1}{p}} + \left[\frac{2(4K)^p}{|8^p - 2^{p\lambda}|}\right]^{\frac{1}{p}}\right\}\|x\|_X^{\lambda}.$$

Analogously, as in the proof Theorem 25.6, Theorems 25.2 and 25.3 yield the next theorem.

**Theorem 25.7.** *Let $\varphi : X \times X \to [0, \infty)$ be a function such that*

$$\lim_{n\to\infty} 4^n \varphi(2^n x, 2^n y) = 0 = \lim_{n\to\infty} 8^{-n}\varphi(2^n x, 2^n y)$$

*for all $x, y \in X$ and*

$$\sum_{i=1}^{\infty} 4^{ip}\varphi^p(2^i x, 2^i y) < \infty, \qquad \sum_{i=0}^{\infty} \frac{1}{8^{ip}}\varphi^p(2^i x, 2^i y) < \infty, \qquad \forall y \in X, x \in \{0, y\}.$$

*Suppose that a function $f : X \to Y$ with $f(0) = 0$ satisfies $\|Df(x,y)\|_Y \leq \varphi(x,y)$ for all $x, y \in X$. Then there exist a unique quadratic function $Q : X \to Y$ and a unique cubic function $C : X \to Y$ satisfying (25.3) and such that*

$$\|f(x) - Q(x) - C(x)\|_Y \leq \frac{K^3}{24}\left\{[\widetilde{\varphi}_e(x) + \widetilde{\varphi}_e(-x)]^{1/p} + \frac{1}{2}[\widetilde{\varphi}_o(x) + \widetilde{\varphi}_o(-x)]^{1/p}\right\}$$

*for all $x \in X$, where*

$$\widetilde{\varphi}_e(x) := \sum_{i=1}^{\infty} 4^{ip}\{\varphi^p(0, 2^i x) + (4K)^p\varphi^p(2^i x, 2^i x)\}$$

*and*

$$\widetilde{\varphi}_o(x) := \sum_{i=0}^{\infty} 8^{-ip}\{\varphi^p(0, 2^i x) + (4K)^p\varphi^p(2^i x, 2^i x)\}.$$

**Corollary 25.7.** *Let $\theta, r, s \in \mathbb{R}$, $\theta, r, s \geq 0$, $2 < r, s < 3$ and $f : X \to Y$ with $f(0) = 0$ satisfy $\|Df(x,y)\|_Y \leq \theta(\|x\|_X^r + \|y\|_X^s)$ for $x, y \in X$. Then there are a unique quadratic $Q : X \to Y$ and a unique cubic $C : X \to Y$ such that (25.3) holds and, for all $x \in X$,*

$$\|f(x) - Q(x) - C(x)\|_Y \leq \frac{K^3\theta}{6}\left\{\left[\frac{2(4K)^p}{|2^{pr} - 4^p|}\|x\|_X^{pr} + \frac{2 + 2(4K)^p}{|2^{ps} - 4^p|}\|x\|_X^{ps}\right]^{\frac{1}{p}} + \left[\frac{2(4K)^p}{|8^p - 2^{pr}|}\|x\|_X^{pr} + \frac{2 + 2(4K)^p}{|8^p - 2^{ps}|}\|x\|_X^{ps}\right]^{\frac{1}{p}}\right\}.$$

**Corollary 25.8.** *Let $\theta, r, s$ be non-negative real numbers with $2 < \lambda := r+s < 3$. Suppose that a function $f : X \to Y$ with $f(0) = 0$ satisfies the inequality*

$$\|Df(x,y)\|_Y \leq \begin{cases} \theta\|x\|_X^r, & r>0, s=0, \\ \theta\|y\|_X^s, & r=0, s>0, \\ \theta\|x\|_X^r\|y\|_X^s, & r>0, s>0, \end{cases}$$

*for all $x,y \in X$. Then there exist a unique quadratic function $Q : X \to Y$ and a unique cubic function $C : X \to Y$ satisfying (25.3) and such that, for all $x \in X$,*

$$\|f(x)-Q(x)-C(x)\|_Y \leq \begin{cases} 2^{1/p-1}K^2[\alpha_e'(x)+\alpha_o'(x)], & r>0, s=0, \\ 2^{1/p-1}K^2[\beta_e'(x)+\beta_o'(x)], & r=0, s>0, \\ 2^{1/p-1}K^2[\gamma_e'(x)+\gamma_o'(x)], & r>0, s>0, \end{cases}$$

*where*

$$\alpha_e'(x) = \frac{K\theta}{3}\left\{\frac{2(4K)^p}{2^{pr}-4^p}\right\}^{\frac{1}{p}}\|x\|_X^r, \qquad \alpha_0'(x) = \frac{K\theta}{3}\left\{\frac{2(4K)^p}{8^p-2^{pr}}\right\}^{\frac{1}{p}}\|x\|_X^r,$$

$$\beta_e'(x) = \frac{K\theta}{3}\left\{\frac{2+2(4K)^p}{2^{ps}-4^p}\right\}^{\frac{1}{p}}\|x\|_X^s, \qquad \beta_o'(x) = \frac{K\theta}{3}\left\{\frac{2+2(4K)^p}{8^p-2^{ps}}\right\}^{\frac{1}{p}}\|x\|_X^s,$$

$$\gamma_e'(x) = \frac{K\theta}{3}\left\{\frac{2(4K)^p}{2^{p\lambda}-4^p}\right\}^{\frac{1}{p}}\|x\|_X^\lambda, \qquad \gamma_o'(x) = \frac{K\theta}{3}\left\{\frac{2(4K)^p}{8^p-2^{p\lambda}}\right\}^{\frac{1}{p}}\|x\|_X^\lambda.$$

**Acknowledgement** Project No.CDJZR10 10 00 08 supported by the Fundamental Research Funds for the Central Universities.

## References


1. Aczél, J., Dhombres, J.: Functional Equations in Several Variables. Cambridge Univ. Press (1989)
2. Amyari, M., Moslehian, M.S.: Approximately ternary semigroup homomorphisms. Lett. Math. Phys. **77**, 1–9 (2006)
3. T. Aoki, On the stability of the linear transformation in Banach spaces. J. Math. Soc. Japan **2**(1950), 64–66.
4. Benyamini, Y., Lindenstrauss, J.: Geometric Nonlinear Functional Analysis, vol. 1. Colloq. Publ., vol. 48, Amer. Math. Soc., Providence, RI (2000)
5. Chang, I.S., Jung, Y.S.: Stability of a functional equation deriving from cubic and quadratic functions. J. Math. Anal. Appl. **283**, 491–500 (2003)
6. Cholewa, P.W.: Remarks on the stability of functional equations. Aequationes Math. **27**, 76–86 (1984)

7. Găvruta, P.: A generalization of the Hyers–Ulam–Rassias stability of approximately additive mappings. J. Math. Anal. Appl. **184**, 431–436 (1994)
8. Hyers, D.H.: On the stability of the linear functional equation. Proc. Natl. Acad. Sci. USA **27**, 222–224 (1941)
9. Jun, K.W., Kim, H.M.: The generalized Hyers–Ulam–Rassias stability of a cubic functional equation. J. Math. Anal. Appl. **274**, 867–878 (2002)
10. Jun, K.W., Lee, Y.H.: On the Hyers–Ulam–Rassias stability of a pexiderized quadratic inequality. Math. Inequal. Appl. **4**, 93–118 (2001)
11. Jung, S.M.: On the Hyers–Ulam stability of the functional equations that have the quadratic property. J. Math. Anal. Appl. **222**, 126–137 (1998)
12. Jung, S.M.: Hyers–Ulam–Rassias Stability of Functional Equations in Mathematical Analysis. Hadronic Press, Palm Harbor (2001)
13. Jung, S.M., Sahoo, P.K.: Hyers–Ulam stability of the quadratic equation of Pexider type. J. Korean Math. Soc. **38**, 645–656 (2001)
14. Mirzavaziri, M., Moslehian, M.S.: A fixed point approach to stability of a quadratic equation. Bull. Braz. Math. Soc. **37(3)**, 361–376 (2006)
15. Moslehian, M.S.: Approximately vanishing of topological cohomology groups. J. Math. Anal. Appl. **318**, 758–771 (2006)
16. Najati, A., Moghimi, M.B.: Stability of a functional equation deriving from quadratic and additive functions in quasi–Banach spaces. J. Math. Anal. Appl. **337**, 399–415 (2008)
17. Najati, A., Eskandani, G.Z.: Stability of a mixed additive and cubic functional equation in quasi–Banach spaces. J. Math. Anal. Appl. **342**, 1318–1331 (2008)
18. Rassias, Th.M.: On the stability of the linear mapping in Banach spaces. Proc. Amer. Math. Soc. **72**, 297–300 (1978)
19. Rolewicz, S.: Metric Linear Spaces. PWN–Polish Sci. Publ./Reidel, Warszawa–Dordrecht (1984)
20. Skof, F.: Local properties and approximations of operators. Rend. Sem. Mat. Fis. Milano **53**, 113–129 (1983)
21. Ulam, S.M.: Problems in Modern Mathematics. Science Eds. Wiley, New York (1964)

# Chapter 26
# $\mu$-Trigonometric Functional Equations and Hyers–Ulam Stability Problem in Hypergroups

## Trigonometric Formulas on Hypergroups

**D. Zeglami, S. Kabbaj, A. Charifi, and A. Roukbi**

*Dedicated to the memory of S.M. Ulam*
*on the 100th anniversary of his birth*

**Abstract** Let $(X,*)$ be a hypergroup and $\mu$ be a complex bounded measure on $X$. We determine the continuous and bounded solutions of each of the following three functional equations

$$\langle \delta_x * \mu * \delta_y, f \rangle = f(x)g(y) \pm g(x)f(y),\ x,y \in X,$$
$$\langle \delta_x * \mu * \delta_y, g \rangle = g(x)g(y) + f(x)f(y),\ x,y \in X.$$

In addition, when $\mu = \delta_e$, Hyers–Ulam stability problems for these functional equations on hypergroups are considered. The results obtained in this paper are natural extensions of previous works done in groups especially by Stetkær, Elqorachi, Redouani, and Székelyhidi.

**Keywords** Hypergroup • Trigonometric functionl equation • Hyers–Ulam stability • Spherical function • Polynomial hypergroup

**Mathematics Subject Classification (2000):** Primary 39B42; Secondary 30D05

## 26.1 Introduction

Chung, Kannappan and Ng [8] solved, on the group $G$, the functional equation

$$f(xy) = f(x)g(y) + g(x)f(y) + h(x)h(y), \qquad x,y \in G.$$

---

D. Zeglami • S. Kabbaj • A. Charifi (✉) • A. Roukbi
Faculty of Sciences, Department of Mathematics, Ibn Tofail University
BP:14000, Kenitra, Morocco
e-mail: charifi2000@yahoo.fr; zeglamidriss@yahoo.fr; samkabbaj@yahoo.fr; rroukbi.a2000@gmail.com

Th.M. Rassias and J. Brzdęk (eds.), *Functional Equations in Mathematical Analysis*, Springer Optimization and Its Applications 52, DOI 10.1007/978-1-4614-0055-4_26,

Stetkær [30] solved each of the following functional equations

$$\int_K f(x+k\cdot y)\mathrm{d}k = f(x)g(y) \pm g(x)f(y)\,, \qquad x,y \in G,$$

$$\int_K g(x+k\cdot y)\mathrm{d}k = g(x)g(y) + f(x)f(y)\,, \qquad x,y \in G,$$

where $K$ is a compact group acting by automorphisms on the abelian group $(G,+)$ and d$k$ it's normalized Haar measure. Poulsen and Stetkær [21] found the complete set of solutions of the functional equations

$$f(x\sigma(y)) = f(x)g(y) \pm g(x)f(y), \qquad x,y \in G,$$
$$g(x\sigma(y)) = g(x)g(y) + f(x)f(y), \qquad x,y \in G,$$

in which $G$ is an arbitrary group and $\sigma : G \longrightarrow G$ satisfies

$$\sigma(xy) = \sigma(x)\sigma(y)$$

and

$$\sigma o \sigma(x) = x$$

for all $x,y \in G$.

In [10], Elqorachi and Redwani solved the functional equations

$$\int_G f(xt\sigma(y))\mathrm{d}\mu(t) = f(x)g(y) \pm g(x)f(y), \qquad x,y \in G,$$

$$\int_G g(xt\sigma(y))\mathrm{d}\mu(t) = g(x)g(y) + f(x)f(y), \qquad x,y \in G,$$

with the only assumption that $\mu$ is a $\sigma$-invariant complex bounded measure on the group $G$.

As a continuation of these investigations we determine the continuous and bounded solutions of the following functional equations

$$\langle \delta_x * \mu * \delta_y, f\rangle = f(x)g(y) \pm g(x)f(y)\,, \qquad x,y \in X, \tag{26.1}$$
$$\langle \delta_x * \mu * \delta_y, g\rangle = g(x)g(y) + f(x)f(y)\,, \qquad x,y \in X, \tag{26.2}$$

where $(X,*)$ is a topological hypergroup and $\mu$ is a complex bounded measure on $X$. The solutions of these functional equations on $(X,*)$ are expressed by means of $\mu$-spherical functions $\chi$ (i.e $\chi(x)\chi(y) = \langle \delta_x * \mu * \delta_y, \chi\rangle, x,y \in X$) on $(X,*)$ and solutions of the equation

$$\langle \delta_x * \mu * \delta_y, f\rangle = f(x)\chi(y) + \chi(x)f(y)\,, \qquad x,y \in X.$$

The study of functional equations on hypergroups started with some recent results. Székelyhidy [33, 34] and Orosz and Székelyhidi [19] describe moment functions, additive functions and multiplicative functions in special cases of hypergroups.

In [20], sine and cosine functional equations are considered and solved on arbitrary polynomial hypergroups in a single variable and the method of solution is based on spectral synthesis. Székelyhidy in [33], deals with the stability of exponential (i.e. multiplicative) functions on hypergroups. Precisely, he proved the following result ([33, Theorem 7.1]), which will be used later in this paper.

**Theorem 26.1.** *Let $K$ be a hypergroup and let $f,g : K \longrightarrow \mathbb{C}$ be continuous functions with the property that the function*

$$x \longmapsto \int_K f d(\delta_x * \delta_y) - f(x)g(y)$$

*is bounded for all $y$ in $K$. Then either $f$ is bounded, or $g$ is exponential (i.e. multiplicative function).*

In last section of this paper, employing the idea of Székelyhidy in [32], we consider the Hyers–Ulam stability of each of the following two functional equations

$$\langle \delta_x * \delta_y, f \rangle = f(x)g(y) + g(x)f(y), \qquad x,y \in X, \tag{26.3}$$

$$\langle \delta_x * \delta_y, g \rangle = g(x)g(y) + f(x)f(y), \qquad x,y \in X, \tag{26.4}$$

where $(X,*)$ is a topological hypergroup which need not be abelian and $f,g$ are continuous fonctions on $X$.

The papers [10, 30, 32] are the essential motivation for the present work.

The contents of the present paper are as follows:

In the second section, we give some preliminaries on hypergroups and we prove some propositions which will be used in the proof of our results.

In the third section, we describe the set of bounded and continuous solutions of the functional equations (26.1) and (26.2) on hypergroups.

In the fourth section, Hyers–Ulam stability problems for the functional equations (26.3) and (26.4) on special case of hypergroups are considered. On the stability problem, the interested reader should refer to [1, 2, 4–7, 9, 11, 12, 14, 15, 18, 22–27].

## 26.2 Preliminary Results

### 26.2.1 Hypergroups

We start with some notation: For a locally compact Hausdorff space $X$, let $M(X)$ denote the complex space of all bounded regular Borel measures on $X$, if $\mu \in M(X)$, $\text{supp}(\mu)$ is the support of $\mu$. The unit point mass concentrated at $x$ is indicated by $\delta_x$. Let $K(X)$ be the complex algebra of all continuous complex-valued functions on $X$ with compact support and $C(X)$ (resp. $C_b(X)$) the complex algebra of all continuous (resp. continuous and bounded) complex-valued functions on $X$. Now, recall some basic notions and notation from the hypergroup theory.

**Definition 26.1.** If $M(X)$ is a Banach algebra with a associative multiplication $*$ (called a convolution), then $(X,*)$ is a hypergroup if the following axioms are satisfied.

**X1.** If $\mu$ and $\nu$ are probability measures, then so is $\mu * \nu$.

**X2.** The mapping $(\mu,\nu) \longmapsto \mu * \nu$ is continuous from $M(X) \times M(X)$ into $M(X)$ where $M(X)$ is endowed with the weak topology with respect to $K(X)$.

**X3.** There is an element $e \in X$ such that $\delta_e * \mu = \mu * \delta_e = \mu$ for all $\mu \in M(X)$.

**X4.** There is a homeomorphic mapping $x \longmapsto \check{x}$ of $X$ into itself such that $\check{(\check{x})} = x$ and $e \in supp(\delta_x * \delta_y)$ if and only if $y = \check{x}$.

**X5.** For all $\mu, \nu \in M(X)$, $\check{(\mu * \nu)} = \check{\nu} * \check{\mu}$ where $\check{\mu}$ is defined by

$$\langle \check{\mu}, f\rangle = \langle \mu, \check{f}\rangle = \int_X f(\check{t})\mathrm{d}\mu(t), \qquad f \in C_b(X).$$

**X6.** The mapping $(x,y) \longmapsto \mathrm{supp}(\delta_x * \delta_y)$ is continuous from $X \times X$ into the space of compact subsets of $X$ with the topology described in [16, Sect. 2.5].

A hypergroup $(X,*)$ is called commutative if its convolution is commutative. The definitive set of axioms was given first by Jewett in his encyclopaedic article [16]. On the hypergroup theory, the interested reader should refer to [3, 13].

We review some notations. Let $f \in C_b(X)$, for all $x \in X$ and $\mu \in M(X)$, we put $\langle \delta_x, f\rangle = f(x)$ and $\langle \mu, f\rangle = \int_X \langle \delta_x, f\rangle\, \mathrm{d}\mu(x)$. If $\mu, \nu \in M(X)$ then

$$\langle \mu * \nu, f\rangle = \int_X \int_X \langle \delta_x * \delta_y, f\rangle\, \mathrm{d}\mu(x)\mathrm{d}\mu(y).$$

*Example 26.1.* (i) Of course every locally compact group is a hypergroup with the usual group convolution (with respect to a left Haar measure).

(ii) ([16, Chap. 8]) Let $(G,+)$ be a locally compact abelian group and $K$ a compact subgroup of $Aut(G)$ with the normalized Haar measure d$k$. Then the space

$$G^K = \{K \cdot x : x \in G\}$$

of $K$-orbits in $G$ is a locally compact Hausdorff space with the quotient topology and becomes a commutative hypergroup with the (natural) convolution

$$(\delta_{K\cdot x} * \delta_{K\cdot y})(f) = \int_K f(K\cdot(x+ky))\mathrm{d}k, \qquad f \in K(G^K).$$

$(G^K,*)$ is called an orbit hypergroup; its neutral element is $K\cdot 0 = 0$ and the involution is $(K\cdot x)^- = K\cdot(-x)$.

(iii) An extensively studied class of hermitian hypergroups is represented by the polynomial hypergroups in one variable (see [3, 19]).

**Definition 26.2 ([3]).** Let $(X,*)$ be a hypergroup and $\chi : X \longrightarrow \mathbb{C}$ be a function. Then we say that

(i) $\chi$ is a multiplicative function on $(X,*)$ if it has the property

$$\langle \delta_x * \delta_y, \chi \rangle = \chi(x)\chi(y), \qquad x,y \in X.$$

(ii) $\chi$ is a hermitian function if $\chi(\check{x}) = \overline{\chi(x)}$ for all $x \in X$.
(iii) $\chi$ is a hypergroup character on $(X,*)$ if it is bounded, continuous, multiplicative and hermitian function.

If $(X,*)$ is commutative hypergroup, the set of characters with the weak topology defines an object which is, in some sense, dual to $(X,*)$.

**Definition 26.3.** Let $(X,*)$ be a hypergroup, $\mu \in M(X)$ and $\chi \in C_b(X)$, we say that $\chi$ is a $\mu$-spherical function on $(X,*)$ if it has the property

$$\langle \delta_x * \mu * \delta_y, \chi \rangle = \chi(x)\chi(y), \qquad x,y \in X.$$

We will say that a function $f \in C_b(X)$ satisfying

$$\langle \delta_x * \mu * \delta_y, f \rangle = f(x)\chi(y) + \chi(x)f(y), \qquad x,y \in X,$$

is associated to the $\mu$-spherical function $\chi$. We also need the following definition:

**Definition 26.4.** We call the continuous complex valued function $a$ on $X$ $\mu$-additive, if it satisfies

$$\langle \delta_x * \mu * \delta_y, a \rangle = a(x) + a(y), \qquad x,y \in X.$$

The set of $\mu$-additive functions on $(X,*)$ will be denoted by $\mathscr{A}(X)$.

### *26.2.2 Auxiliary Results*

Throughout this section we fix a measure $\mu$ in $M(X)$ and keep in mind the preceding notation. In the next two propositions certain assumptions are equivalent.

**Proposition 26.1.** *Let $(X,*)$ be a hypergroup and $f \in C_b(X)$. Then the following statements are equivalent.*

*(i)* $\langle \delta_x * \mu * \delta_y, f \rangle = f(x)f(y)$ *for all* $x,y \in X$.
*(ii)* $\langle \nu * \mu * w, f \rangle = \langle \nu, f \rangle \langle w, f \rangle$ *for all* $\nu, w \in M(X)$.

*Proof.* Assume that

$$\langle \delta_x * \mu * \delta_y, f \rangle = f(x)f(y)$$

for all $x,y \in X$ (i.e. $f$ is a $\mu$ -spherical function on $(X,*)$). Then

$$\begin{aligned}\langle \nu * \mu * w, f\rangle &= \int_X \int_X \langle \delta_x * \mu * \delta_y, f\rangle \, \mathrm{d}\nu(x)\mathrm{d}w(y)\\ &= \int_X \int_X f(x)f(y)\mathrm{d}\nu(x)\mathrm{d}w(y)\\ &= \int_X \langle \delta_x, f\rangle \mathrm{d}\nu(x) \int_X \langle \delta_y, f\rangle \mathrm{d}w(y)\\ &= \langle \nu, f\rangle \langle w, f\rangle .\end{aligned}$$

The other implication is trivial. □

**Proposition 26.2.** *Let $(X,*)$ be a hypergroup and $f,g \in C_b(X)$ . Then:*

*1. The following two statements are equivalent.*

*(i)* $\langle \delta_x * \mu * \delta_y, f\rangle = f(x)g(y) \pm g(x)f(y)$ *for all* $x,y \in X$.
*(ii)* $\langle \nu * \mu * w, f\rangle = \langle \nu, f\rangle \langle w, g\rangle \pm \langle w, f\rangle \langle \nu, g\rangle$ *for all* $\nu, w \in M(X)$.

*2. The following two statements are equivalent.*

*(a)* $\langle \delta_x * \mu * \delta_y, g\rangle = g(x)g(y) + f(x)f(y)$ *for all* $x,y \in X$.
*(b)* $\langle \nu * \mu * w, g\rangle = \langle \nu, g\rangle \langle w, g\rangle + \langle \nu, f\rangle \langle w, f\rangle$ *for all* $\nu, w \in M(X)$.

*Proof.* 1. Assume that $\langle \delta_x * \mu * \delta_y, f\rangle = f(x)g(y) \pm g(x)f(y)$ for all $x,y \in X$, then, for all $\nu, w \in M(X)$, we have

$$\begin{aligned}\langle \nu * \mu * w, f\rangle &= \int_X \int_X \langle \delta_x * \mu * \delta_y, f\rangle \, \mathrm{d}\nu(x)\mathrm{d}w(y)\\ &= \int_X \int_X (f(x)g(y) \pm g(x)f(y))\mathrm{d}\nu(x)\mathrm{d}w(y)\\ &= \langle \nu, f\rangle \langle w, g\rangle \pm \langle w, f\rangle \langle \nu, g\rangle .\end{aligned}$$

The other implication is trivial.
2. By similar arguments as in (1). □

The following proposition will be used in the proof of Theorem 26.4.

**Proposition 26.3.** *Let $(X,*)$ be a hypergroup and $f,g \in C_b(X)$ be solutions of the functional equation*

$$\langle \delta_x * \mu * \delta_y, g\rangle = g(x)g(y) + f(x)f(y), \qquad x,y \in X. \tag{26.5}$$

*Then there exists $\alpha \in \mathbb{C}$ such that*

$$\langle \delta_x * \mu * \delta_y, f\rangle = f(x)g(y) + g(x)f(y) + 2\alpha f(x)f(y), \qquad x,y \in X.$$

*Proof.* Proposition 26.3. is true for $f=0$. We will assume that $f\neq 0$. Using the formula (26.5) and Proposition 26.2 we get

$$\begin{aligned}\langle(\delta_x*\mu*\delta_y)*\mu*\delta_z,g\rangle &= g(z)\langle\delta_x*\mu*\delta_y,g\rangle+f(z)\langle\delta_x*\mu*\delta_y,f\rangle\\ &= g(z)g(x)g(y)+g(z)f(x)f(y)+f(z)\langle\delta_x*\mu*\delta_y,f\rangle.\end{aligned}$$

Similarly, we have

$$\begin{aligned}\langle\delta_x*\mu*(\delta_y*\mu*\delta_z),g\rangle &= g(x)\langle\delta_y*\mu*\delta_z,g\rangle+f(x)\langle\delta_y*\mu*\delta_z,f\rangle\\ &= g(x)g(y)g(z)+g(x)f(y)f(z)+f(x)\langle\delta_y*\mu*\delta_z,f\rangle.\end{aligned}$$

Then

$$g(z)f(x)f(y)+f(z)\langle\delta_x*\mu*\delta_y,f\rangle=g(x)f(y)f(z)+f(x)\langle\delta_y*\mu*\delta_z,f\rangle.$$

By adding $-f(x)g(y)f(z)$ to each side of the previous equality and rearranging terms we get

$$\begin{aligned}&f(z)[\langle\delta_x*\mu*\delta_y,f\rangle-f(x)g(y)-g(x)f(y)]\\ &\qquad=f(x)[\langle\delta_y*\mu*\delta_z,f\rangle-f(y)g(z)-g(y)f(z)].\end{aligned}\tag{26.6}$$

This allows us to write for $z=a$ such that $f(a)\neq 0$,

$$\langle\delta_x*\mu*\delta_y,f\rangle-f(x)g(y)-g(x)f(y)=f(x)k(y),$$

where

$$k(y)=\frac{1}{f(a)}\left(\langle\delta_y*\mu*\delta_a,f\rangle-f(y)g(a)-g(y)f(a)\right).$$

Then (26.6) can be written as $f(z)f(x)k(y)=f(x)f(y)k(z)$, which implies, for $x=z=a$, that $k(y)=k(a)/f(a)f(y)$. If we put $2\alpha=k(a)/f(a)$, then we get

$$\langle\delta_x*\mu*\delta_y,f\rangle=f(x)g(y)+g(x)f(y)+2\alpha f(x)f(y).$$

□

Similarly, we can prove the following proposition which will be used in the proof of Theorem 26.3.

**Proposition 26.4.** *Let $f,g\in C_b(X)$ be solutions of the functional equation*

$$\langle\delta_x*\mu*\delta_y,f\rangle=f(x)g(y)+g(x)f(y),\qquad x,y\in X.$$

*Then there exists $\beta\in\mathbb{C}$ such that*

$$\langle\delta_x*\mu*\delta_y,g\rangle=g(x)g(y)+\beta^2f(x)f(y),\qquad x,y\in X.$$

**Proposition 26.5.** *Let $f,g \in C_b(X)$ be solutions of the functional equation*

$$\langle \delta_x * \mu * \delta_y, f\rangle = f(x)g(y) - g(x)f(y), \qquad x,y \in X. \tag{26.7}$$

*Then*

$$\langle \delta_x * \mu * \delta_y, g\rangle = -\langle \delta_y * \mu * \delta_x, g\rangle, \qquad x,y \in G, \tag{26.8}$$

*and there exists $\alpha \in \mathbb{C}$ such that*

$$\langle \delta_y * \mu * \delta_x, g\rangle = g(x)g(y) - 2\alpha f(x)g(y) + \alpha^2 f(x)f(y), \qquad x,y \in X. \tag{26.9}$$

*Proof.* Using (26.7) and Proposition 26.2 we get

$$\begin{aligned}\langle \delta_x * \mu * (\delta_y * \mu * \delta_z), f\rangle &= f(x)\langle \delta_y * \mu * \delta_z, g\rangle - g(x)\langle \delta_y * \mu * \delta_z, f\rangle \\ &= f(x)\langle \delta_y * \mu * \delta_z, g\rangle - g(x)f(y)g(z) + g(x)g(y)f(z).\end{aligned}$$

On the other hand

$$\begin{aligned}\langle (\delta_x * \mu * \delta_y) * \mu * \delta_z, f\rangle &= \langle \delta_x * \mu * \delta_y, f\rangle g(z) - f(z)\langle \delta_x * \mu * \delta_y, g\rangle \\ &= f(x)g(y)g(z) - f(y)g(x)g(z) - f(z)\langle \delta_x * \mu * \delta_y, g\rangle.\end{aligned}$$

Hence

$$f(x)\left\{\langle \delta_y * \mu * \delta_z, g\rangle - g(y)g(z)\right\} = -f(z)\left\{\langle \delta_x * \mu * \delta_y, g\rangle + g(x)g(y)\right\}. \tag{26.10}$$

By letting $z = x$ in (26.10), we obtain

$$f(x)\left\{\langle \delta_y * \mu * \delta_x, g\rangle - g(x)g(y)\right\} = -f(x)\left\{\langle \delta_x * \mu * \delta_y, g\rangle + g(x)g(y)\right\}. \tag{26.11}$$

So for $x = x_0$ such that $f(x_0) \neq 0$ we get $\langle \delta_{x_0} * \mu * \delta_y, g\rangle = -\langle \delta_y * \mu * \delta_{x_0}, g\rangle$. Now if $f(x_0) = 0$, in view of (26.10), we get that

$$\langle \delta_{x_0} * \mu * \delta_y, g\rangle = -g(x_0)g(y), \qquad y \in G.$$

Since, if we let $x = z_0$ (for some $z_0 \in G$ such that $f(z_0) \neq 0$) and $z = x_0$ in (26.10) we obtain

$$\langle \delta_y * \mu * \delta_{x_0}, g\rangle = g(x_0)g(y), \qquad y \in G.$$

This proves (26.8). From (26.10) and (26.8) we obtain a new equation

$$f(x)\left\{\langle \delta_y * \mu * \delta_z, g\rangle - g(z)g(y)\right\} = f(z)\left\{\langle \delta_y * \mu * \delta_x, g\rangle - g(x)g(y)\right\}, \tag{26.12}$$

and choosing $z = z_0 \in G$ such that $f(z_0) \neq 0$ we obtain

$$\langle \delta_y * \mu * \delta_x, g\rangle = g(x)g(y) + h(y)f(x), \qquad x,y \in G, \tag{26.13}$$

where

$$h(y) = \frac{1}{f(z_0)} \left\{ \langle \delta_y * \mu * \delta_{z_0}, g \rangle - g(z_0)g(y) \right\}.$$

By (26.8), (26.13) and the definition of $h$ we get

$$h(y) = -2\alpha g(y) + \alpha^2 f(y), \tag{26.14}$$

where $\alpha = g(z_0)/f(z_0)$. By using (26.14) in (26.13) we obtain (26.9). □

We end this section by the following proposition which extends [29, Propositions 26.3 and 6].

**Proposition 26.6.** *Let $(X,*)$ be a hypergroup and $\chi_1, \chi_2, ..., \chi_n : X \longrightarrow \mathbb{C}$ be different measurable functions satisfying the functional equation*

$$\langle \delta_x * \mu * \delta_y, \chi \rangle = \chi(x)\chi(y), \qquad x, y \in X, \tag{26.15}$$

*and let $c_1, c_2, ..., c_n$ be non-zero complex numbers. Then the following two statements are valid.*

*(i) $\chi_1, \chi_2, ..., \chi_n$ are linearly independent in the vector space of all complex-valued functions on $X$.*
*(ii) If $g = c_1\chi_1 + c_2\chi_2 + \cdots + c_n\chi_n$ is continuous then each $\chi_j$, $j = 1, ..., n$ is continuous.*

*Proof.* (i) By similar arguments as in the proof of [28, Proposition 3].
(ii) It is an adaptation of the proof of [28, Proposition 6] to the solutions of (26.15). □

## 26.3 Trigonometric Functional Equations

In this section, we solve each of the following three functional equations:

$$\langle \delta_x * \mu * \delta_y, f \rangle = f(x)g(y) \pm g(x)f(y), \qquad x, y \in X,$$
$$\langle \delta_x * \mu * \delta_y, g \rangle = g(x)g(y) + f(x)f(y), \qquad x, y \in X,$$

where $\mu \in M(X)$. First, we solve the $\mu$-sine subtraction functional equation (26.16).

**Theorem 26.2.** *The solutions $f, g \in C_b(X)$ of the functional equation*

$$\langle \delta_x * \mu * \delta_y, f \rangle = f(x)g(y) - g(x)f(y), \qquad x, y \in X, \tag{26.16}$$

*can be listed as follows.*

*(i) $f = 0$ and $g \in C(X)$ is arbitrary.*

*(ii) There is a complex constant $\alpha$ such that $g = \alpha f$ and $f$ satisfies the functional equation*

$$\langle \delta_x * \mu * \delta_y, f \rangle = 0, \qquad x, y \in X.$$

*Proof.* Let us assume that $(f,g)$ solves the functional equation (26.16) and that $f \neq 0$. If there is $\alpha \in \mathbb{C}$ such that $g = \alpha f$, then according to (26.16), we get $\langle \delta_x * \mu * \delta_y, f \rangle = 0$, $x, y \in X$, i.e. case (ii). From now we assume that $f$ and $g$ are linearly independent. Proposition 26.4 tell us that

$$\langle \delta_x * \mu * \delta_y, g \rangle = -\langle \delta_y * \mu * \delta_x, g \rangle, \qquad x, y \in G, \tag{26.17}$$

and there exists $\alpha \in \mathbb{C}$ such that

$$\langle \delta_y * \mu * \delta_x, g \rangle = g(x)g(y) - 2\alpha f(x)g(y) + \alpha^2 f(x)f(y), \qquad x, y \in X. \tag{26.18}$$

From (26.17) and (26.18) we get

$$g(x)g(y) - 2\alpha g(x)f(y) + \alpha^2 f(x)f(y) = -g(y)g(x) + 2\alpha g(y)f(x) - \alpha^2 f(y)f(x),$$

whence $[g(x) - \alpha f(x)]\,[g(y) - \alpha f(y)] = 0$. So, $g = \alpha f$, in contradiction with the assumption that $f, g$ are linearly independent. □

Now, we solve the $\mu$-sine addition functional equation (26.19).

**Theorem 26.3.** *The solutions $f, g \in C_b(X)$ of the functional equation*

$$\langle \delta_x * \mu * \delta_y, f \rangle = f(x)g(y) + g(x)f(y),\ x, y \in X, \tag{26.19}$$

*can be listed as follows, with $\mu$-spherical functions $\chi, \psi : X \longrightarrow \mathbb{C}$ and $c \in \mathbb{C} \setminus \{0\}$.*

*(a) $f = 0$ and $g \in C(X)$ is arbitrary.*
*(b) $g = \chi/2$ and $f = c\chi$.*
*(c) $g = (\chi + \psi)/2$ and $f = c(\chi - \psi)$.*
*(d) $g = \chi$ and $f$ is a solution of the functional equation*

$$\langle \delta_x * \mu * \delta_y, f \rangle = f(x)\chi(y) + \chi(x)f(y), \qquad x, y \in X.$$

*Proof.* Checking that the stated pairs of functions satisfy (26.19) is done by elementary calculations, that we leave out. It is left to show that each solution $f, g \in C_b(X)$ of (26.19) falls into one of the categories (a)–(d).

Since $f = 0$ is the case (a) and $g = 0$ falls into the case (d) we may in the remainder of the proof assume that $f \neq 0$ and $g \neq 0$.

If there exists a constant $\alpha \in \mathbb{C}$ such that $g = \alpha f$, then $\alpha \neq 0$. Using (26.19), we get

$$\langle \delta_x * \mu * \delta_y, f \rangle = 2\alpha f(x)f(y),$$

this says that $\chi := 2\alpha f$ is a $\mu$-spherical function of $X$ into $\mathbb{C}$. Now $f = \chi/(2\alpha)$ and $g = \alpha f = \chi/2$, so we have case (b).

We may thus assume that $f$ and $g$ are linearly independent. According to Proposition 26.4 there exists a constant $\beta \in \mathbb{C}$ such that

$$\langle \delta_x * \mu * \delta_y, g \rangle = g(x)g(y) + \beta^2 f(x)f(y), \qquad x,y \in X. \tag{26.20}$$

Combining this with (26.19) we get that

$$\langle \delta_x * \mu * \delta_y, g \pm \beta f \rangle = (g \pm \beta f)(x)(g \pm \beta f)(y),$$

so $g \pm \beta f$ are $\mu$-spherical functions of $X$ into $\mathbb{C}$. If $\beta \neq 0$ we have case (c) with $\chi = g + \beta f$ and $\psi = g - \beta f$.

If $\beta = 0$ then $\chi := g$ is according to (26.20) a $\mu$-spherical function of $X$ into $\mathbb{C}$ and $f$ is a solution of the functional equation

$$\langle \delta_x * \mu * \delta_y, f \rangle = f(x)\chi(y) + \chi(x)f(y)$$

for all $x,y \in X$ (i.e. case (d)). □

*Remark 26.1.* The pair of functions $(a,1)$ is a solution of (26.19) for each $a \in \mathscr{A}(X)$.

We close this section by solving the functional equation

$$\langle \delta_x * \mu * \delta_y, g \rangle = g(x)g(y) + f(x)f(y), \qquad x,y \in X,$$

which is called "the $\mu$-cosh addition and the $\mu$-cosine subtraction formulas".

**Theorem 26.4.** *The solutions $f,g \in C_b(X)$ of the functional equation*

$$\langle \delta_x * \mu * \delta_y, g \rangle = g(x)g(y) + f(x)f(y), \qquad x,y \in X, \tag{26.21}$$

*can be listed as follows, where $\chi, \psi : X \longrightarrow \mathbb{C}$ denote $\mu$-spherical functions.*

*(a) $f = \pm ig$ and $g$ satisfies the equation $\langle \delta_x * \mu * \delta_y, g \rangle = 0$ for all $x,y \in X$.*

*(b) $g = \frac{1}{1+c^2}\chi$ and $f = \frac{c}{1+c^2}\chi$ with some $c \in \mathbb{C} \setminus \{\pm i\}$.*

*(c) $g = \frac{\chi + c^2\psi}{1+c^2}$ and $f = c\frac{\chi - \psi}{1+c^2}$ with some $c \in \mathbb{C} \setminus \{0,-i,i\}$.*

*(d) $f$ is associated to $\chi$ and $g = \chi + if$ or $g = \chi - if$.*

*Proof.* Checking that the stated pairs of functions satisfy (26.21) is done by elementary calculations, that we leave out. It is left to show that each solution $f,g \in C_b(X)$ of (26.21) falls into one of the categories (a)–(d).

If there exists a constant $c \in \mathbb{C}$ such that $f = cg$, using (26.21), we get

$$\langle \delta_x * \mu * \delta_y, g \rangle = \left(1 + c^2\right) g(x)g(y), \qquad x,y \in X.$$

The first case (a) is obvious if $1+c^2=0$. Now we assume that $1+c^2\neq 0$. Then $\chi=(1+c^2)g$ is a $\mu$-spherical function of $X$ into $\mathbb{C}$. This is case (b). From now on we can assume that $f$ and $g$ are linearly independent. According to Proposition 26.3 there exists a constant $\alpha\in\mathbb{C}$ such that

$$\langle \delta_x*\mu*\delta_y, f\rangle = f(x)g(y)+g(x)f(y)+2\alpha f(x)f(y), \qquad x,y\in X. \tag{26.22}$$

Combining the identities (26.21) and (26.22) we find that for any $\lambda\in\mathbb{C}$

$$\langle \delta_x*\mu*\delta_y, g-\lambda f\rangle = (g-\lambda f)(x)(g-\lambda f)(y)-(\lambda^2+2\alpha\lambda-1)f(x)f(y). \tag{26.23}$$

Let $\lambda_1$ and $\lambda_2$ denote the two roots of polynomial $\lambda^2+2\alpha\lambda-1=0$. Then $\lambda_1\neq 0$, $\lambda_2=-\lambda_1^{-1}$, and so $\chi=g-\lambda_1 f$ and $\psi=g+\lambda_1^{-1}f$ are by (26.23) $\mu$-spherical functions of $X$ into $\mathbb{C}$. Note that $\chi$ and $\psi$ are non-zero functions, because $f$ and $g$ are linearly independent.

If $\lambda_1\neq\lambda_2$, then we can express $f$ and $g$ by $\chi$ and $\psi$. This is case (c). If $\lambda_1$ and $\lambda_2$ coincide, then $\lambda_1=\lambda_2=-\alpha=\pm\mathrm{i}$ so $g=\chi\pm \mathrm{i}f$. From (26.23) we get in each of the two cases $\alpha=\pm\mathrm{i}$ that $f$ satisfies the sine addition formula

$$\langle \delta_x*\mu*\delta_y, f\rangle = f(x)\chi(y)+\chi(x)f(y), \qquad x,y\in X. \tag{26.24}$$

and $g=\chi\pm\mathrm{i}f$. So we have the case (d). □

*Remark 26.2.* If $\mu$ is a bounded probability measure then each of the pairs $(a,1+ia)$ and $(a,1-ia)$ where $a\in\mathscr{A}(X)$ is a solution of (26.21).

The next Corollary is an important consequence of Theorem 26.3 (take $\mu=\delta_e$ in (26.19)).

**Corollary 26.1.** *The solutions $f,g\in C_b(X)$ of the functional equation*

$$\langle \delta_x*\delta_y, f\rangle = f(x)g(y)+g(x)f(y), \qquad x,y\in X, \tag{26.25}$$

*can be listed as follows, where $\chi,\psi: X\longrightarrow\mathbb{C}$ denote continuous multiplicative functions and $c\in\mathbb{C}\setminus\{0\}$.*

1. *$f=0$ and $g\in C(X)$ is arbitrary.*
2. *$g=\chi/2$ and $f=c\chi$.*
3. *$g=(\chi+\psi)/2$ and $f=c(\chi-\psi)$.*
4. *$g=\chi$ and $f$ is a solution of the equation $\langle \delta_x*\delta_y, f\rangle = f(x)\chi(y)+\chi(x)f(y)$ for all $x,y\in X$.*

As an application of Theorem 26.4, if we take $\mu=\delta_e$ in (26.21), we get the following corollary.

**Corollary 26.2.** *The solutions $f,g\in C_b(X)$ of the functional equation*

$$\langle \delta_x*\delta_y, g\rangle = g(x)g(y)+f(x)f(y), \qquad x,y\in X, \tag{26.26}$$

*can be listed as follows, where* $\chi, \psi : X \longrightarrow \mathbb{C}$ *denote continuous multiplicative functions:*

*(a)* $g = 0$ *and* $f = 0$.

*(b)* $g = \frac{1}{1+c^2}\chi$ *and* $f = \frac{c}{1+c^2}\chi$, *where* $c \in \mathbb{C} \setminus \{\pm \mathrm{i}\}$.

*(c)* $g = \frac{\chi + c^2\psi}{1+c^2}$ *and* $f = c\frac{\chi - \psi}{1+c^2}$, *where* $c \in \mathbb{C} \setminus \{0, -\mathrm{i}, \mathrm{i}\}$.

*(d)* $f$ *is associated to* $\chi$ *and* $g = \chi + \mathrm{i}f$ *or* $g = \chi - \mathrm{i}f$.

**Corollary 26.3.** *The solutions* $f, g \in C(X)$ *of the functional equation*

$$\langle \delta_x * \delta_y, f \rangle = f(x)g(y) - g(x)f(y), \qquad x, y \in X, \tag{26.27}$$

*are the pairs* $(f, g)$, *where* $f = 0$ *and* $g \in C(X)$ *is arbitrary.*

*Proof.* Follows from Theorem 26.2 by taking $\mu = \delta_e$ in (26.16). □

*Remark 26.3.* For a hypergroup $X$ in which $(\delta_x * \delta_y) * \delta_z$ is a measure with compact support for all $x, y, z \in X$ we can assume in Corollaries 26.1 and 26.2 that $f, g \in C(X)$ (see [17, Remark 1]).

## 26.4 Stability of the Trigonometric Formulas

Székelyhidy in [32], studied the stability property of two well known functional equations: The sine and cosine functional equations:

$$f(xy) = f(x)g(y) + g(x)f(y), \qquad x, y \in G \tag{26.28}$$

and

$$g(xy) = g(x)g(y) + f(x)f(y), \qquad x, y \in G, \tag{26.29}$$

where $f, g$ are complex-valued functions on an amenable group $G$. In this section, employing the idea of L. Székelyhidy in [32], we consider the Hyers–Ulam stability of each of the functional equations (1.3) and (1.4). Throughout this section $(X, *)$ is a hypergroup, where the measure $\delta_x * \delta_y * \delta_z$ has compact support for all $x, y, z \in X$. As examples of such hypergroups: groups and polynomial hypergroups [3].

### *26.4.1 Stability of (26.3)*

In this subsection, we investigate the stability properties of the functional equation (26.3) and obtain a generalization of the stability results proved by Székelyhidi in [32] for (26.28).

**Lemma 26.1.** *Let $f,g \in C(X)$ satisfying the inequality, there is a constant $M > 0$ with*

$$|\langle \delta_x * \delta_y, f\rangle - f(x)g(y) - g(x)f(y)| \leq M, \qquad x,y \in X. \tag{26.30}$$

*Then either there exist $\alpha, \beta \in \mathbb{C}$, not both zero, and $L > 0$ such that one of the following two conditions is valid:*

$$|\alpha f(x) - \beta g(x)| \leq L, \qquad x \in X, \tag{26.31}$$

$$\langle \delta_x * \delta_y, f\rangle = f(x)g(y) + g(x)f(y), \qquad x,y \in X. \tag{26.32}$$

*Proof.* We prove that the (26.32) satisfied if the condition (26.31) fails.

Assume that $|\alpha f(x) - \beta g(x)| \leq L$ for some $L > 0$ implies $\alpha = \beta = 0$. We define the mapping

$$F(x,y) = \langle \delta_x * \delta_y, f\rangle - f(x)g(y) - f(y)g(x)$$

for $x,y \in X$. Then

$$\langle \delta_x * \delta_y, f\rangle = f(x)g(y) + f(y)g(x) + F(x,y), \qquad x,y \in X. \tag{26.33}$$

According to (26.30) we see that $|F(x,y)| \leq M$ for all $x,y \in X$. We will prove that $F(x,y) = 0$ for all $x,y \in X$.

We can choose $a \in X$ satisfying $f(a) \neq 0$. It is easy to show that

$$g(x) = \lambda_0 f(x) + \lambda_1 \langle \delta_x * \delta_a, f\rangle - \lambda_1 F(x,a), \qquad x \in X, \tag{26.34}$$

where $\lambda_0 = -g(a)/f(a)$ and $\lambda_1 = 1/f(a)$. Using (26.33) and (26.34) we get

$$\begin{aligned}
\langle (\delta_x * \delta_y) * \delta_z, f\rangle &= \int_X \langle \delta_t * \delta_z, f\rangle \, d(\delta_x * \delta_y)(t) \\
&= \int_X [f(t)g(z) + f(z)g(t) + F(t,z)] \, d(\delta_x * \delta_y)(t) \\
&= \langle \delta_x * \delta_y, f\rangle g(z) + \int_X F(t,z) d(\delta_x * \delta_y)(t) \\
&\quad + f(z) \int_X [\lambda_0 f(t) + \lambda_1 \langle \delta_t * \delta_a, f\rangle - \lambda_1 F(t,a)] \, d(\delta_x * \delta_y)(t) \\
&= \{f(x)g(y) + g(x)f(y) + F(x,y)\} g(z) \\
&\quad + \{\lambda_0 \langle \delta_x * \delta_y, f\rangle + \lambda_1 \langle \delta_x * \delta_y * \delta_a, f\rangle\} f(z) \\
&\quad - f(z) \int_X \lambda_1 F(t,a) d(\delta_x * \delta_y)(t) + \int_X F(t,z) d(\delta_x * \delta_y)(t) \\
&= \{f(x)g(y) + g(x)f(y) + F(x,y)\} g(z) \\
&\quad + \{\lambda_0 (f(x)g(y) + g(x)f(y) + F(x,y)\} f(z)
\end{aligned}$$

$$
\begin{aligned}
&+\lambda_1 f(z)\int_X \langle \delta_x * \delta_t, f\rangle\, d(\delta_y * \delta_a)(t)\\
&-\lambda_1 f(z)\int_X F(t,a)d(\delta_x * \delta_y)(t) + \int_X F(t,z)d(\delta_x * \delta_y)(t)\\
&= \{f(x)g(y)+g(x)f(y)+F(x,y)\}\, g(z)\\
&+\{\lambda_0(f(x)g(y)+g(x)f(y)+F(x,y)\}\, f(z)\\
&+\lambda_1 f(z)\Big\{f(x)\left\langle \delta_y * \delta_a, g\right\rangle + \left\langle \delta_y * \delta_a, f\right\rangle g(x)\\
&+\int_X F(x,t)d(\delta_y * \delta_a)(t)\Big\}\\
&-\lambda_1 f(z)\int_X F(t,a)d(\delta_x * \delta_y)(t)\\
&+\int_X F(t,z)d(\delta_x * \delta_y)(t)
\end{aligned}
$$

and

$$
\begin{aligned}
\left\langle \delta_x * (\delta_y * \delta_z), f\right\rangle &= \int_X \langle \delta_x * \delta_t, f\rangle\, d(\delta_y * \delta_z)(t)\\
&= \int_X [f(x)g(t)+f(t)g(x)+F(x,t)]\, d(\delta_y * \delta_z)(t)\\
&= f(x)\left\langle \delta_y * \delta_z, g\right\rangle + g(x)\left\langle \delta_y * \delta_z, f\right\rangle\\
&\quad + \int_X F(x,t)d(\delta_y * \delta_z)(t).
\end{aligned}
$$

It follows from those equalities that

$$
\begin{aligned}
f(x)\{g(y)g(z) &+ \lambda_0 g(y)f(z) + \lambda_1 \left\langle \delta_y * \delta_a, g\right\rangle f(z) - \left\langle \delta_y * \delta_z, g\right\rangle\}\\
&+ g(x)\{f(y)g(z)+\lambda_0 f(y)f(z)\\
&+\lambda_1 \left\langle \delta_y * \delta_a, f\right\rangle f(z) - \left\langle \delta_y * \delta_z, f\right\rangle\}\\
&= -F(x,y)g(z) - \lambda_0 F(x,y)f(z)\\
&-\lambda_1 \int_X F(x,t)d(\delta_y * \delta_a)(t)f(z)\\
&+\lambda_1 \int_X F(t,a)d(\delta_x * \delta_y)(t)f(z)\\
&-\int_X F(t,z)d(\delta_x * \delta_y)(t) + \int_X F(x,t)d(\delta_y * \delta_z)(t).
\end{aligned}
$$

Since $|F(x,y)| \le M$ and $\left\|\delta_y * \delta_a\right\| = \left\|\delta_y * \delta_z\right\| = \left\|\delta_x * \delta_y\right\| = 1$ for all $x,y \in X$, if we fix $y,z$ the right hand side of the above equation is bounded function of $x$.

Thus, by the assumption that $|\alpha f(x) - \beta g(x)| \leq L$ for some $L > 0$ we obtain $\alpha = \beta = 0$. The both sides of the above equation become zero. Consequently we have

$$\Big| - F(x,y)g(z) - \{\lambda_0 F x,y) + \lambda \int_X F(x,t)d(\delta_y * \delta_a)(t)$$

$$- \lambda_1 \int_X F(t,a)d(\delta_x * \delta_y)(t)\Big\} f(z)\Big|$$

$$= \left| \int_X F(t,z)d(\delta_x * \delta_y)(t) - \int_X F(x,t)d(\delta_y * \delta_z)(t) \right| \leq 2M.$$

Since the right hand side is bounded as a function of $z$ for all fixed $x, y \in X$, again by the assumption, we have $F(x,y) = 0$ for all $x, y \in X$. It follows that

$$\langle \delta_x * \delta_y, f \rangle = f(x)g(y) + g(x)f(y), \qquad x,y \in X. \tag{26.35}$$

□

**Theorem 26.5.** *Let $M$ be a positive constant and $f, g \in C(X)$ satisfy*

$$|\langle \delta_x * \delta_y, f \rangle - f(x)g(y) - g(x)f(y)| \leq M, \qquad x,y \in X. \tag{26.36}$$

*Then $f, g$ satisfy one of the following statements.*

*(i) $f = 0$ and $g \in C(X)$ is arbitrary.*
*(ii) $f$ and $g$ are bounded.*
*(iii) $f$ is unbounded and $g$ is a multiplicative function.*
*(iv) There exist a multiplicative function $\chi$, a continuous bounded function $b$, and a complex constant $c$ such that $f = c(\chi - b)$ and $g = (\chi + b)/2$.*
*(v) The pair of functions $f, g$ satisfy (26.3).*

*Proof.* Assume that the inequality (26.36) holds. If $f = 0$, then obviously $g$ can be chosen arbitrary in $C(X)$. This is case (i). If $f \neq 0$ is bounded, then the function

$$X \ni x \longmapsto f(x)g(y) + g(x)f(y)$$

is bounded for all $y \in X$, so $g$ is bounded. This is case (ii). If $f$ is unbounded and $g$ is bounded then the function $X \ni x \longmapsto f(y)g(x)$ is bounded for all $y \in X$. Consequently the function

$$X \ni x \longmapsto \langle \delta_x * \delta_y, f \rangle - f(x)g(y)$$

is bounded on $X$ for all $y \in X$. In view of [33, Theorem 7.1], $g$ is a multiplicative function. This is case (iii). If $f, g$ are unbounded functions, we distinguish two cases.

**First case**. We assume that there exist $\alpha, \beta \in \mathbb{C} \setminus \{0\}$ such that $\alpha f + \beta g$ is bounded, then $g$ can be written as

$$g = \frac{f}{2c} + b,$$

where $b$ is a bounded function and $c \in \mathbb{C} \setminus \{0\}$. Consequently, the function

$$G \ni x \longmapsto \langle \delta_x * \delta_y, f \rangle - \left( \frac{f(y)}{c} + b(y) \right) f(x)$$

is bounded for all $y \in X$. Hence, since $f$ is unbounded, by [33, Theorem 7.1], it follows that

$$\chi = \frac{f}{c} + b$$

is a multiplicative function. This is case (iv).

**Second case.** For all $\alpha, \beta \in \mathbb{C} \setminus \{0\}$, $\alpha f + \beta g$ is an unbounded function on $X$. According to Lemma 1, $f, g$ are solutions of (26.3). This is case (v). □

The following theorem describes all multiplicative functions on polynomial hypergroups (see [3]).

**Theorem 26.6.** *Let $X = (\mathbb{N}, *)$ be the polynomial hypergroup associated with the sequence of polynomials $(P_n)_{n \in \mathbb{N}}$. The function $\chi : \mathbb{N} \to \mathbb{C}$ is a multiplicative function on $X$ if and only if there is a complex number $\lambda$ with*

$$\chi(n) = P_n(\lambda)$$

*for all $n$ in $\mathbb{N}$.*

Hence, any multiplicative function on a polynomial hypergroup arises from the evaluations

$$n \longmapsto P_n(\lambda)$$

associated with the generating polynomials.

**Corollary 26.4.** *Let $X = (\mathbb{N}, *)$ be the polynomial hypergroup associated with the sequence of polynomials $(P_n)_{n \in \mathbb{N}}$, $M$ be a positive constant and $f, g \in C(X)$ satisfy*

$$|\langle \delta_n * \delta_m, f \rangle - f(n)g(m) - g(n)f(m)| \leq M, \qquad n, m \in X. \tag{26.37}$$

*Then $f, g$ satisfy one of the following statements.*

*(i)* *$f = 0$ and $g \in C(X)$ is arbitrary.*
*(ii)* *$f, g$ are bounded functions on $X$.*
*(iii)* *There is a complex constant $\lambda$ with $g(n) = P_n(\lambda)$ for $n \in \mathbb{N}$ and $f$ is unbounded on $X$.*
*(iv)* *There are complex constants $\lambda$, $c$ and a continuous bounded function $b$ on $X$ with $f(n) = c(P_n(\lambda) - b(n))$ and $g(n) = c(P_n(\lambda) + b(n))/2$.*
*(v)* *Functions $f$ and $g$ satisfy the equation*

$$\langle \delta_n * \delta_m, f \rangle = f(n)g(m) + g(n)f(m)$$

*for $n, m \in X$.*

*Proof.* Follows from Theorems 26.5 and 26.6. □

### 26.4.2 Stability of (26.4)

Now we deal with the Hyers–Ulam stability of (26.4) on hypergroups.

**Lemma 26.2.** *Let $f,g \in C(X)$ satisfying the inequality, there exists a positive constant $M$ such that*

$$\left|\langle \delta_x * \delta_y, g\rangle - g(x)g(y) - f(x)f(y)\right| \leq M \tag{26.38}$$

*for all $x,y \in X$. Then either there exist $\alpha,\beta \in \mathbb{C}$, not both zero, and $L > 0$ such that*

$$|\alpha f(x) - \beta g(x)| \leq L, \qquad x,y \in X \tag{26.39}$$

*or*

$$\langle \delta_x * \delta_y, g\rangle = g(x)g(y) + f(x)f(y), \qquad x,y \in X. \tag{26.40}$$

*Proof.* Assume that (26.39) holds only for $\alpha = \beta = 0$. We show that then (26.40) is satisfied.

Let $F(x,y) = \langle \delta_x * \delta_y, g\rangle - g(x)g(y) - f(x)f(y)$ for $x,y \in X$. Then

$$\langle \delta_x * \delta_y, g\rangle = g(x)g(y) + f(x)f(y) + F(x,y), \qquad x,y \in X. \tag{26.41}$$

According to (26.38) we see that $|F(x,y)| \leq M$ for all $x,y \in X$. We can choose $a \in X$ satisfying $f(a) \neq 0$. It is easy to show that

$$f(x) = \lambda_0 g(x) + \lambda_1 \langle \delta_x * \delta_a, g\rangle - \lambda_1 F(x,a), \qquad x \in X, \tag{26.42}$$

where $\lambda_0 = -g(a)/f(a)$ and $\lambda_1 = 1/f(a)$.

Using (26.41) and (26.42), for all $x,y,z \in X$ we obtain the following two sequences of equalities:

$$\begin{aligned}
\langle (\delta_x * \delta_y) * \delta_z, g\rangle &= \int_X \langle \delta_t * \delta_z, g\rangle \, d(\delta_x * \delta_y)(t) \\
&= \int_X [g(t)g(z) + f(t)f(z) + F(t,z)] \, d(\delta_x * \delta_y)(t) \\
&= \langle \delta_x * \delta_y, g\rangle g(z) + \int_X [f(t)f(z) + F(t,z)] \, d(\delta_x * \delta_y)(t) \\
&= \{g(x)g(y) + f(x)f(y) + F(x,y)\} g(z) \\
&\quad + f(z) \int_X [\lambda_0 g(t) + \lambda_1 \langle \delta_t * \delta_a, g\rangle - \lambda_1 F(t,a)] \, d(\delta_x * \delta_y)(t) \\
&\quad + \int_X F(t,z) d(\delta_x * \delta_y)(t) \\
&= \{g(x)g(y) + f(x)f(y) + F(x,y)\} g(z)
\end{aligned}$$

$$
\begin{aligned}
&+\lambda_0 f(z)\{g(x)g(y)+f(x)f(y)+F(x,y)\}\\
&+\lambda_1 f(z)\int_X \langle \delta_x * \delta_t, g\rangle\, d(\delta_y * \delta_a)(t)\\
&-\lambda_1 f(z)\int_X F(t,a)d(\delta_x * \delta_y)(t)+\int_X F(t,z)d(\delta_x * \delta_y)(t)\\
=&\ \{g(x)g(y)+f(x)f(y)+F(x,y)\}\,g(z)\\
&+\lambda_0 f(z)\{g(x)g(y)+f(x)f(y)+F(x,y)\}\\
&+\lambda_1 f(z)\Big\{g(x)\langle \delta_y * \delta_a, g\rangle + f(x)\langle \delta_y * \delta_a, f\rangle\\
&+\int_X F(x,t)d(\delta_y * \delta_a)(t)\Big\}\\
&-\lambda_1 f(z)\int_X F(t,a)d(\delta_x * \delta_y)(t)\\
&+\int_X F(t,z)d(\delta_x * \delta_y)(t),
\end{aligned}
$$

$$
\begin{aligned}
\langle \delta_x * (\delta_y * \delta_z), g\rangle &= \int_X \langle \delta_x * \delta_t, g\rangle\, d(\delta_y * \delta_z)(t)\\
&= \int_X [g(x)g(t)+f(x)f(t)+F(x,t)]\,d(\delta_y * \delta_z)(t)\\
&= g(x)\langle \delta_y * \delta_z, g\rangle + f(x)\langle \delta_y * \delta_z, f\rangle\\
&\quad + \int_X F(x,t)d(\delta_y * \delta_z)(t).
\end{aligned}
$$

It follows from those equalities that for all $x,y,z \in X$ we have

$$
\begin{aligned}
&g(x)\{g(y)g(z)+\lambda_0 g(y)f(z)+\lambda_1\langle \delta_y * \delta_a, g\rangle f(z)-\langle \delta_y * \delta_z, g\rangle\}\\
&\quad + f(x)\{f(y)g(z)+\lambda_0 f(y)f(z)\\
&\quad + \lambda_1\langle \delta_y * \delta_a, f\rangle f(z)-\langle \delta_y * \delta_z, f\rangle\}\\
&= - F(x,y)g(z)-\lambda_0 F(x,y)f(z)-\lambda_1\int_X F(x,t)d(\delta_y * \delta_a)(t)f(z)\\
&\quad + \lambda_1\int_X F(t,a)d(\delta_x * \delta_y)(t)f(z)\\
&\quad - \int_X F(t,z)d(\delta_x * \delta_y)(t)+\int_X F(x,t)d(\delta_y * \delta_z)(t).
\end{aligned}
$$

Since $|F(x,y)| \le M$ and $\|\delta_y * \delta_a\| = \|\delta_y * \delta_z\| = \|\delta_x * \delta_y\| = 1$ for all $x,y \in X$ the right hand side of the above equality is bounded function of $x$ for all fixed $y,z \in X$. Thus, by the assumption that (26.39) holds only for $\alpha = \beta = 0$, the both sides of the above equation become zero. Consequently, we have

$$F(x,y)g(z)+\left\{\lambda_0 F(x,y)+\lambda_1\int_X F(x,t)d(\delta_y*\delta_a)(t)\right.$$
$$\left.-\lambda_1\int_X F(t,a)d(\delta_x*\delta_y)(t)\right\}f(z)$$
$$=-\int_X F(t,z)d(\delta_x*\delta_y)(t)+\int_X F(x,t)d(\delta_y*\delta_z)(t).$$

Since the right hand side is bounded as function of $z$ for all fixed $x,y\in X$, again by the assumption we have $F(x,y)=0$ for all $x,y\in X$. It follows that $\langle\delta_x*\delta_y,g\rangle=g(x)g(y)+f(x)f(y)$ for all $x,y\in X$. □

**Theorem 26.7.** *Let $f,g\in C(X)$ satisfy (26.38). Then one of the followings statements is valid.*

*(i) $f$ and $g$ are bounded functions.*
*(ii) $g$ is a multiplicative function and $f$ is bounded.*
*(iii) $f,g$ are unbounded, and $g+\mathrm{i}f$ or $g-\mathrm{i}f$ is a bounded multiplicative function.*
*(iv) There exist a multiplicative function $\chi$, a continuous bounded function $b$ on $X$, and $c\in\mathbb{C}\setminus\{\pm\mathrm{i}\}$ such that*

$$g=\frac{c^2\chi+b}{c^2+1}\qquad\textit{and}\qquad f=\frac{c}{c^2+1}(\chi-b).$$

*(v) $f,g$ satisfy the equation $\langle\delta_x*\delta_y,g\rangle=g(x)g(y)+f(x)f(y)$ for all $x,y\in X$.*

*Proof.* Assume that (26.38) holds. Observe that if $f$ is bounded, then the function $X\times X\ni(x,y)\longmapsto\langle\delta_x*\delta_y,g\rangle-g(x)g(y)$ is bounded. So, by [33, Theorem 7.1], either $g$ is bounded or $g$ is a multiplicative function. This is case (i) and (ii). If $f$ is unbounded, then $g$ is unbounded. We distinguish two cases.

**First case.** We assume that there exist $\alpha,\beta\in\mathbb{C}\setminus\{0\}$ such that $\alpha f+\beta g$ is a bounded function on $X$. Then $g$ can be written as $g=cf+b$, where $b$ is a bounded function on $X$ and $c\in\mathbb{C}\setminus\{0\}$. Consequently, for all $y\in X$, bounded is the function

$$X\ni x\longmapsto\langle\delta_x*\delta_y,f\rangle-\frac{(c^2+1)f(y)+cb(y)}{c}f(x).$$

Now, by [33, Theorem 7.1] we obtain that $\chi=\frac{1}{c}(c^2+1)f+b$ is a multiplicative function, because $f$ is unbounded. Thus, we get (iii) for $c^2=-1$ and (iv) for $c^2\neq-1$.

**Second case.** If $\alpha f+\beta g$ is unbounded on $X$ for all $\alpha,\beta\in\mathbb{C}\setminus\{0\}$, then according to Lemma 26.2, $f,g$ are solutions of (26.4). This gives (v). □

Finally, from Theorems 26.6 and 26.7 we can easily derive the following corollary.

**Corollary 26.5.** *Let $X = (\mathbb{N}, *)$ be the polynomial hypergroup associated with the sequence of polynomials $(P_n)_{n\in\mathbb{N}}$, $M > 0$ be a constant, and $f, g \in C(X)$ satisfy*

$$|\langle \delta_n * \delta_m, g\rangle - g(n)g(m) - f(n)f(m)| \leq M, \qquad n,m \in X. \tag{26.43}$$

*Then one of the followings five statements is valid.*

*(i)* $f$ *and* $g$ *are bounded functions on* $X$.
*(ii)* $f$ *is bounded on* $X$ *and there is* $\lambda \in \mathbb{C}$ *with* $g(n) = P_n(\lambda)$ *for* $n \in \mathbb{N}$.
*(iii)* $f, g$ *are unbounded and there are* $\lambda \in \mathbb{C}$ *and* $\sigma \in \{-1,1\}$ *such that the function* $m \longmapsto P_m(\lambda)$ *is bounded on* $X$ *and* $g(n) + \sigma \mathrm{i} f(n) = P_n(\lambda)$ *for* $n \in \mathbb{N}$.
*(iv) There are* $\lambda \in \mathbb{C}$, *a continuous bounded function* $b$ *on* $X$, *and* $c \in \mathbb{C}\backslash\{\pm \mathrm{i}\}$ *with*

$$g(n) = \frac{c^2 P_n(\lambda) + b(n)}{c^2+1} \quad \text{and} \quad f(n) = \frac{c}{c^2+1}(P_n(\lambda) - b(n)), \qquad n \in \mathbb{N}.$$

*(v)* $f, g$ *satisfy the equation*

$$\langle \delta_n * \delta_m, g\rangle = g(n)g(m) + f(n)f(m), \qquad n,m \in X.$$

## References

1. Aoki, T., On the stability of the linear transformation in Banach spaces. J. Math. Soc. Japan **2**, 64–66 (1950)
2. Baker, J.A.: The stability of the cosine equation. Proc . Amer. Math. Soc. **80**, 411–416 (1980)
3. Bloom, W. R. and Heyer, H.: Harmonic analysis of probability measures on hypergroups. Walter der Gruyter, Berlin-New York (1995)
4. Chang, J., and Chung, J.: Hyers–Ulam Stability of trigonometric functional equations. Commun. Korean Math. Soc. **23**, 567–575 (2008)
5. Charifi, A., Bouikhalene B., Elqorachi E.: Hyers–Ulam–Rassias stability of a generalized Pexider functional equation. Banach J. Math. Anal. **1**, 176–185 (2007)
6. Charifi, A., Bouikhalene B., Elqorachi E.: Hyers–Ulam–Rassias stability of a generalized Jensen functional equation. Aust. J. Math. Anal. Appl. **6**, Article 19, 1–16 (2009)
7. Charifi, A., Bouikhalene B., Kabbaj S., Rassias J. M.: On the stability of Pexiderized version of the Gołąb-Schinzel equation. Comput. Math. Appl. **59**, 3193–3202 (2010)
8. Chung, J. K., Kannappan, Pl., Ng, C.T.: A generalization of the cosine-sine functional equation on groups. Linear Algebra Appl. **66**, 259–277 (1985)
9. Czerwik, S.: Functional Equations and Inequalities in Several Variables. World Scientific, New Jersey-London-Singapore-Hong Kong (2002)
10. Elqorachi, E., Redouani A.: Trigonometric formulas and $\mu$-spherical functions. Aequationes Math. **72**, 60–77 (2006)
11. Gavruta, P., An answer to question of John M. Rassias concerning the stability of Cauchy equation. In: Rassias, J.M. (ed.) Advances in Equations and Inequalities, pp. 67–71. Hadronic Press, Palm Harbor (1999)
12. Gavruta, P., Hossu, M., Popescu D., Caprau, C.: On the stability of mappings and an anwser to a problem of Th.M Rassias. Ann. Math. Blaise Pascal **2**, 55–60 (1995)
13. Heyer, H., Kawakami, S.: Extensions of Pontryagin hypergroups. Probab. Math. Statist. **26**, 245–260 (2006)

14. Hyers, D.H., Isac G., Rassias, Th.M.: Stability of Functional Equations in Several Variables. Birkhauser, Basel (1998)
15. Isac, G., Rassias, Th.M.: Functional inequalities for approximately additive mappings. In: Rassias, Th.M., Tabor, J. (eds.) Stability of Mappings of Hyers–Ulam Type, pp. 117125. Hadronic Press, Palm Harbour (1994)
16. Jewett, R. I.: Spaces with an abstract convolution of measures. Adv. Math. **18**, 1–101 (1975)
17. Kabbaj, S., Roukbi, A.: Zeglami, D.: Trigonometric formulas on hypergroups. Submited to Georgian Mathematical Journal
18. Moslehian, M.S., Rassias, Th.M.: Stability of functional equation in non-Archimedean spaces. Appl. Anal. Discrete Math. **1**, 325–334 (2007)
19. Orosz, Á., Székelyhidi, L.: Moment functions on polynomial hypergroups in several variables. Publ. Math. Debrecen **65**, 429–438 (2004)
20. Orosz, Á.: Sine and cosine equation on discrete polynomial hypergroups. Aequationes Math. **72**, 225–233 (2006)
21. Poulsen, Th. A., Stetkær, H.: On the trigonometric subtraction and addition formulas. Aequationes Math. **59**, 84–92 (2000)
22. Rassias, Th. M.: On the stability of linear mapping in Banach spaces. Proc. Amer. Math. Soc. **72** (1978), 297–300
23. Rassias, Th.M.: Problem 16. In: Report of the 27th International Symposium on Functional Equations, p. 309. Aequationes Math. **39** (1990)
24. Rassias, Th.M.: Functional equations inequalities and applications. Kluwer Academic Publishers, Dordercht-Boston-London (2003)
25. Rassias, Th.M. (ed.): Functional Equations and Inequalities. Kluwer Academic Publishers, Dordrecht-Boston-London (2000)
26. Rassias, Th.M., Šemrl, P.: On the behaviour of mappings which do not satisfy Hyers–Ulam stability. Proc. Amer. Math. Soc. **114**, 989–993 (1992)
27. Redouani, A., Elqorachi E., Bouikhalene, B.: Hyers-Ulam stability of the generalised trigonometric formulas. J. Ineq. Pure Appl. Math. **7**, Issue 2, Article 74 (2006)
28. Roukbi, A., Zeglami, D.: D'Alembert.s functional equations on hypergroups, Adv. Pure Appl. Math. **2**, 147–166 (2011)
29. Spector, R.: Mesure invariantes sur les hypergroupes, Trans. Amer. Math. Soc. **239**, 147–165 (1978)
30. Stetkær, H.: Trigonometric Functional equations of rectangular type. Aequationes Math. **56**, 251–270 (1998)
31. Székelyhidi, L.: On a theorem of Baker, Lawrence, and Zorzito. Proc. Amer. Math. Soc. **84** (1982), 95–96.
32. Székelyhidi, L.: The stability of the sine and cosine functional equations. Proc. Amer. Math. Soc. **110**, 109–115 (1990)
33. Székelyhidi, L.:Functional equations on hypergroups. In: Rassias, Th.M. (ed.) Functional equations, Inequalities and Applications, pp. 167–181. Kluwer Academic Publishers,Boston–Dordrecht–London (2003)
34. Székelyhidi, L.: Functional equations on topological Sturm-Liouville hypergroups. Math. Pannon. **17**, 169–182 (2006)

# Part II
# Topics in Mathematical Analysis

# Chapter 27
# On Multivariate Ostrowski Type Inequalities

**Chang-Jian Zhao and Wing-Sum Cheung**

*Dedicated to the memory of S.M. Ulam*
*on the 100th anniversary of his birth*

**Abstract** In this paper several multivariate Ostrowski integral inequalities are established. These generalize some existing results of Pachpatte and provide new estimates to inequalities of this type.

**Keywords** Multivariate Ostrowski's integral inequality • Ostrowski type inequality • $n$-fold integral

**Mathematics Subject Classification (2000):** Primary 26D15

## 27.1 Introduction

The following inequality is well known in the literature as Ostrowski's integral inequality (see, e.g., [4, 7]).

**Theorem 27.1.** *Let $f : [a,b] \to \mathbb{R}$ be a mapping that is differentiable on $(a,b)$ with the derivative $f' : (a,b) \to \mathbb{R}$ bounded on $(a,b)$, i.e., $\|f'\|_\infty = \sup_{t\in(a,b)} |f'(t)| < \infty$. Then, for all $x \in [a,b]$,*

$$\left| f(x) - \frac{1}{b-a}\int_a^b f(t)\mathrm{d}t \right| \le \left[ \frac{1}{4} + \frac{(x - \frac{a+b}{2})^2}{(b-a)^2} \right] (b-a)\|f'\|_\infty. \tag{27.1}$$

C.-J. Zhao (✉)
Department of Information and Mathematics Sciences, College of Science, China Jiliang University, Hangzhou 310018, People's Republic of China
e-mail: chjzhao@yahoo.com.cn; chjzhao@163.com; chjzhao@cjlu.edu.cn

W.-S. Cheung
Department of Mathematics, The University of Hong Kong, Pokfulam Road, Hong Kong, People's Republic of China
e-mail: wscheung@hku.hk

Th.M. Rassias and J. Brzdęk (eds.), *Functional Equations in Mathematical Analysis*, Springer Optimization and Its Applications 52, DOI 10.1007/978-1-4614-0055-4_27,

Many generalizations, extensions and variants of this inequality have appeared in the literature, see [1–10] and the references given therein. The main purpose of the present paper is to establish several multivariate Ostrowski integral inequalities. These results generalize the work of Pachpatte [10] in 2002, and provide new estimates to multivariate Ostrowski type inequalities.

## 27.2 Main Results

In what follows, $\mathbb{R}$ denotes the set of real numbers, $\mathbb{R}^n$ the $n$-dimensional Euclidean space. Let

$$D = \{(x_1,\ldots,x_n) : a_i < x_i < b_i (i = 1,\ldots,n)\}$$

and $\bar{D}$ be the closure of $D$. For a function $u(x) : \mathbb{R}^n \to \mathbb{R}$, we denote the first order partial derivatives by $\frac{\partial u(x)}{\partial x_i} (i = 1,\ldots,n)$ and $\int_D u(x)\mathrm{d}x$ stands for the $n$-fold integral

$$\int_{a_1}^{b_1} \cdots \int_{a_n}^{b_n} u(x_1,\ldots,x_n)\mathrm{d}x_1 \cdots \mathrm{d}x_n.$$

Our main results are established in the following theorems.

**Theorem 27.2.** *Let $f,g,h : \mathbb{R}^n \to \mathbb{R}$ be continuous functions on $\bar{D}$ and differentiable in $D$ such that*

$$\left\| \frac{\sum_{i=1}^n \frac{\partial f}{\partial x_i}(x_i - y_i)}{\sum_{i=1}^n \frac{\partial h}{\partial x_i}(x_i - y_i)} \right\|_\infty = \sup_{x\in D} \left| \frac{\sum_{i=1}^n \frac{\partial f(x)}{\partial x_i}(x_i - y_i)}{\sum_{i=1}^n \frac{\partial h(x)}{\partial x_i}(x_i - y_i)} \right| < \infty,$$

*and*

$$\left\| \frac{\sum_{i=1}^n \frac{\partial g}{\partial x_i}(x_i - y_i)}{\sum_{i=1}^n \frac{\partial h}{\partial x_i}(x_i - y_i)} \right\|_\infty = \sup_{x\in D} \left| \frac{\sum_{i=1}^n \frac{\partial g(x)}{\partial x_i}(x_i - y_i)}{\sum_{i=1}^n \frac{\partial h(x)}{\partial x_i}(x_i - y_i)} \right| < \infty.$$

*Then for every $x \in \bar{D}$,*

$$\left| f(x)g(x) - \frac{1}{2M}g(x)\int_D f(y)\mathrm{d}y - \frac{1}{2M}f(x)\int_D g(y)\mathrm{d}y \right|$$
$$\leq \frac{N}{2M}\left( |g(x)| \left\| \frac{\sum_{i=1}^n \frac{\partial f}{\partial x_i}(x_i - y_i)}{\sum_{i=1}^n \frac{\partial h}{\partial x_i}(x_i - y_i)} \right\|_\infty \right.$$
$$\left. + |f(x)| \left\| \frac{\sum_{i=1}^n \frac{\partial g}{\partial x_i}(x_i - y_i)}{\sum_{i=1}^n \frac{\partial h}{\partial x_i}(x_i - y_i)} \right\|_\infty \right), \tag{27.2}$$

*where $M = \mathrm{mes} D := \prod_{i=1}^n (b_i - a_i)$, $N = \int_D |h(x) - h(y)|\mathrm{d}y$, $\mathrm{d}y = \mathrm{d}y_1 s \mathrm{d}y_n$, and $\sum_{i=1}^n \frac{\partial h(x)}{\partial x_i}(x_i - y_i) \neq 0$.*

*Proof.* Let $x = (x_1, \ldots, x_n)$ and $y = (y_1, \ldots, y_n)$ $(x \in D,\ y \in \bar{D})$. On account of the $n$-dimensional version of the Cauchy's mean value theorem, we have

$$f(x) - f(y) = \frac{\sum_{i=1}^{n} \frac{\partial f(c)}{\partial x_i}(x_i - y_i)}{\sum_{i=1}^{n} \frac{\partial h(c)}{\partial x_i}(x_i - y_i)}(h(x) - h(y)), \tag{27.3}$$

$$g(x) - g(y) = \frac{\sum_{i=1}^{n} \frac{\partial g(d)}{\partial x_i}(x_i - y_i)}{\sum_{i=1}^{n} \frac{\partial h(d)}{\partial x_i}(x_i - y_i)}(h(x) - h(y)), \tag{27.4}$$

where

$$c = (y_1 + \alpha(x_1 - y_1), \ldots, y_n + \alpha(x_n - y_n))$$

and

$$d = (y_1 + \beta(x_1 - y_1), \ldots, y_n + \beta(x_n - y_n))$$

with $0 < \alpha < 1, 0 < \beta < 1$. Multiplying both sides of (27.3) and (27.4) by $g(x)$ and $f(x)$, respectively, and adding, we get

$$\begin{aligned} 2f(x)g(x) \; - \; g(x)f(y) - f(x)g(y) \\ &= g(x)\frac{\sum_{i=1}^{n} \frac{\partial f(c)}{\partial x_i}(x_i - y_i)}{\sum_{i=1}^{n} \frac{\partial h(c)}{\partial x_i}(x_i - y_i)}(h(x) - h(y)) \\ &\quad + f(x)\frac{\sum_{i=1}^{n} \frac{\partial g(d)}{\partial x_i}(x_i - y_i)}{\sum_{i=1}^{n} \frac{\partial h(d)}{\partial x_i}(x_i - y_i)}(h(x) - h(y)). \end{aligned} \tag{27.5}$$

Integrating both sides of (27.5) with respect to $y$ over $D$, using the fact that

$$\operatorname{mes} D > 0$$

and rearranging terms, we have

$$\begin{aligned} f(x)g(x) \; - \; \frac{1}{2M}g(x)\int_D f(y)\mathrm{d}y - \frac{1}{2M}f(x)\int_D g(y)\mathrm{d}y \\ &= \frac{1}{2M}g(x)\int_D \frac{\sum_{i=1}^{n} \frac{\partial f(c)}{\partial x_i}(x_i - y_i)}{\sum_{i=1}^{n} \frac{\partial h(c)}{\partial x_i}(x_i - y_i)}(h(x) - h(y))\mathrm{d}y \\ &\quad + \frac{1}{2M}f(x)\int_D \frac{\sum_{i=1}^{n} \frac{\partial g(d)}{\partial x_i}(x_i - y_i)}{\sum_{i=1}^{n} \frac{\partial h(d)}{\partial x_i}(x_i - y_i)}(h(x) - h(y))\mathrm{d}y. \end{aligned} \tag{27.6}$$

From (27.6) and from the properties of modulus we have

$$\left|f(x)g(x) - \frac{1}{2M}g(x)\int_D f(y)\mathrm{d}y - \frac{1}{2M}f(x)\int_D g(y)\mathrm{d}y\right|$$

$$\le \frac{1}{2M}|g(x)|\int_D \left|\frac{\sum_{i=1}^n \frac{\partial f(c)}{\partial x_i}(x_i-y_i)}{\sum_{i=1}^n \frac{\partial h(c)}{\partial x_i}(x_i-y_i)}\right| |h(x)-h(y)|\mathrm{d}y$$

$$+\frac{1}{2M}|f(x)|\int_D \left|\frac{\sum_{i=1}^n \frac{\partial g(d)}{\partial x_i}(x_i-y_i)}{\sum_{i=1}^n \frac{\partial h(d)}{\partial x_i}(x_i-y_i)}\right| |h(x)-h(y)|\mathrm{d}y$$

$$\le \frac{N}{2M}\left(|g(x)|\left\|\frac{\sum_{i=1}^n \frac{\partial f}{\partial x_i}(x_i-y_i)}{\sum_{i=1}^n \frac{\partial h}{\partial x_i}(x_i-y_i)}\right\|_\infty\right.$$

$$\left.+|f(x)|\left\|\frac{\sum_{i=1}^n \frac{\partial g}{\partial x_i}(x_i-y_i)}{\sum_{i=1}^n \frac{\partial h}{\partial x_i}(x_i-y_i)}\right\|_\infty\right).$$

□

*Remark 27.1.* If we take $h(x) = x$ and $h(y) = y$, then (27.2) becomes

$$\left|f(x)g(x) - \frac{1}{2M}g(x)\int_D f(y)\mathrm{d}y - \frac{1}{2M}f(x)\int_D g(y)\mathrm{d}y\right|$$

$$\le \frac{1}{2M}\sum_{i=1}^n\left(|g(x)|\left\|\frac{\partial f}{\partial x_i}\right\|_\infty + |f(x)|\left\|\frac{\partial g}{\partial x_i}\right\|_\infty\right)E_i(x),$$

where $\mathrm{d}y = \mathrm{d}y_1 s \mathrm{d}y_n$, and

$$E_i(x) = \int_D |x_i - y_i|\mathrm{d}y, \qquad M = \mathrm{mes}\, D =: \prod_{i=1}^n (b_i - a_i).$$

This was also obtained by Pachpatte in [10, Theorem 2.1].

*Remark 27.2.* If we take $h(x) = x$, $h(y) = y$, $g(x) = 1$ and hence $\partial g/\partial x_i = 0$ in Theorem 27.2, then inequality (27.2) reduces to an inequality of Milovanović in [7, Theorem 2], which is in turn is a further generalization of the well known Ostrowski's inequality (27.1).

**Theorem 27.3.** *Under the assumptions of Theorem 27.2, let $\omega(x)$ be nonnegative and integrable in D with $\int_D \omega(y)\mathrm{d}y > 0$. Then for every $x \in \bar{D}$, we have*

$$\left| f(x)g(x) - \frac{g(x)\int_D \omega(y)f(y)\mathrm{d}y + f(x)\int_D \omega(y)g(y)\mathrm{d}y}{2\int_D \omega(y)\mathrm{d}y} \right|$$

$$\leq \frac{1}{2\int_D \omega(y)\mathrm{d}y}\int_D \omega(y)\left( |g(x)| \left\| \frac{\sum_{i=1}^n \frac{\partial f}{\partial x_i}(x_i - y_i)}{\sum_{i=1}^n \frac{\partial h}{\partial x_i}(x_i - y_i)} \right\|_\infty \right.$$

$$\left. + |f(x)| \left\| \frac{\sum_{i=1}^n \frac{\partial g}{\partial x_i}(x_i - y_i)}{\sum_{i=1}^n \frac{\partial h}{\partial x_i}(x_i - y_i)} \right\|_\infty \right) \mathrm{d}y. \tag{27.7}$$

Multiplying both sides of (27.5) by $\omega(y)$ and integrating the resulting identity with respect to $y$ on $D$ and following the proof of inequality (27.2), it is not hard to arrive at inequality (27.7). The details are omitted here.

*Remark 27.3.* With $h(x) = x$ and $h(y) = y$, (27.7) reduces to a new inequality established by Pachpatte in [10, Theorem 2.1].

*Remark 27.4.* Again, if we take $h(x) = x$, $h(y) = y$, $g(x) = 1$ and hence $\partial g/\partial x_i = 0$ in Theorem 27.3, then inequality (7) reduces to an inequality of Milovanović in [7, Theorem 3], which is in turn a further generalization of the well known Ostrowski's inequality.

**Theorem 27.4.** *Under the assumptions of Theorem 27.2, for every $x \in \bar{D}$, we have*

$$\left| f(x)g(x) - f(x)\left(\frac{1}{M}\int_D g(y)\mathrm{d}y\right) \right.$$

$$\left. - g(x)\left(\frac{1}{M}\int_D f(y)\mathrm{d}y\right) + \frac{1}{M}\int_D f(y)g(y)\mathrm{d}y \right|$$

$$\leq \frac{\bar{N}}{M} \left\| \frac{\sum_{i=1}^n \frac{\partial f}{\partial x_i}(x_i - y_i)}{\sum_{i=1}^n \frac{\partial h}{\partial x_i}(x_i - y_i)} \right\|_\infty \times \left\| \frac{\sum_{i=1}^n \frac{\partial g}{\partial x_i}(x_i - y_i)}{\sum_{i=1}^n \frac{\partial h}{\partial x_i}(x_i - y_i)} \right\|_\infty, \tag{27.8}$$

*where*

$$\bar{N} = \int_D (h(x) - h(y))^2 \mathrm{d}y$$

*and M is defined as in Theorem 27.2.*

*Proof.* Multiplying the identities (27.3) and (27.4) and using the hypotheses, we get

$$f(x)g(x)p - f(x)g(y) - f(y)g(x) + f(y)g(y)$$

$$= \frac{\sum_{i=1}^n \frac{\partial f(c)}{\partial x_i}(x_i - y_i)}{\sum_{i=1}^n \frac{\partial h(c)}{\partial x_i}(x_i - y_i)} \times \frac{\sum_{i=1}^n \frac{\partial g(d)}{\partial x_i}(x_i - y_i)}{\sum_{i=1}^n \frac{\partial h(d)}{\partial x_i}(x_i - y_i)}(h(x) - h(y))^2, \tag{27.9}$$

where $c$, $d$ are as defined in the proof of Theorem 27.2. Integrating both sides of (27.9) with respect to $y$ on $D$ and rearranging the terms, we have

$$f(x)g(x) - f(x)\left(\frac{1}{M}\int_D g(y)\mathrm{d}y\right) - g(x)\left(\frac{1}{M}\int_D f(y)\mathrm{d}y\right) + \frac{1}{M}\int_D f(y)g(y)\mathrm{d}y$$
$$= \frac{1}{M}\int_D \frac{\sum_{i=1}^n \frac{\partial f(c)}{\partial x_i}(x_i - y_i)}{\sum_{i=1}^n \frac{\partial h(c)}{\partial x_i}(x_i - y_i)} \times \frac{\sum_{i=1}^n \frac{\partial g(d)}{\partial x_i}(x_i - y_i)}{\sum_{i=1}^n \frac{\partial h(d)}{\partial x_i}(x_i - y_i)}(h(x) - h(y))^2\mathrm{d}y.$$

Consequently, by properties of the modulus, we have

$$\left|f(x)g(x) - f(x)\left(\frac{1}{M}\int_D g(y)\mathrm{d}y\right)\right.$$
$$\left.-g(x)\left(\frac{1}{M}\int_D f(y)\mathrm{d}y\right) + \frac{1}{M}\int_D f(y)g(y)\mathrm{d}y\right|$$
$$\leq \frac{1}{M}\int_D \left|\frac{\sum_{i=1}^n \frac{\partial f(c)}{\partial x_i}(x_i - y_i)}{\sum_{i=1}^n \frac{\partial h(c)}{\partial x_i}(x_i - y_i)}\right| \times \left|\frac{\sum_{i=1}^n \frac{\partial g(d)}{\partial x_i}(x_i - y_i)}{\sum_{i=1}^n \frac{\partial h(d)}{\partial x_i}(x_i - y_i)}\right|$$
$$\times (h(x) - h(y))^2\mathrm{d}y$$
$$\leq \frac{1}{M}\int_D \left\|\frac{\sum_{i=1}^n \frac{\partial f}{\partial x_i}(x_i - y_i)}{\sum_{i=1}^n \frac{\partial h}{\partial x_i}(x_i - y_i)}\right\|_\infty \times \left\|\frac{\sum_{i=1}^n \frac{\partial g}{\partial x_i}(x_i - y_i)}{\sum_{i=1}^n \frac{\partial h}{\partial x_i}(x_i - y_i)}\right\|_\infty$$
$$\times (h(x) - h(y))^2\mathrm{d}y$$
$$= \frac{\bar{N}}{M}\left\|\frac{\sum_{i=1}^n \frac{\partial f}{\partial x_i}(x_i - y_i)}{\sum_{i=1}^n \frac{\partial h}{\partial x_i}(x_i - y_i)}\right\|_\infty \times \left\|\frac{\sum_{i=1}^n \frac{\partial g}{\partial x_i}(x_i - y_i)}{\sum_{i=1}^n \frac{\partial h}{\partial x_i}(x_i - y_i)}\right\|_\infty,$$

where

$$\bar{N} = \int_D (h(x) - h(y))^2\mathrm{d}y\,.$$

This proves (27.8). □

*Remark 27.5.* With $h(x) = x$ and $h(y) = y$, (27.8) reduces to the following inequality established by Pachpatte in [10, Theorem 2.2]:

$$\left|f(x)g(x) - f(x)\left(\frac{1}{M}\int_D g(y)\mathrm{d}y\right)\right.$$
$$\left.-g(x)\left(\frac{1}{M}\int_D f(y)\mathrm{d}y\right) + \frac{1}{M}\int_D f(y)g(y)\mathrm{d}y\right|$$
$$\leq \frac{1}{M}\int_D \left\{\left[\sum_{i=1}^n \left\|\frac{\partial f}{\partial x_i}\right\|_\infty |x_i - y_i|\right] \times \left[\sum_{i=1}^n \left\|\frac{\partial g}{\partial x_i}\right\|_\infty |x_i - y_i|\right]\right\}\mathrm{d}y.$$

**Theorem 27.5.** *Under the assumptions of Theorem 27.3, for every $x \in \bar{D}$, we have*

$$\left| f(x)g(x) - f(x)\left(\frac{1}{M}\int_D g(y)\mathrm{d}y\right) - g(x)\left(\frac{1}{M}\int_D f(y)\mathrm{d}y\right) \right.$$
$$\left. + \frac{1}{M^2}\left(\int_D f(y)\mathrm{d}y\right)\left(\int_D g(y)\mathrm{d}y\right)\right|$$
$$\leq \frac{N^2}{M^2}\left\| \frac{\sum_{i=1}^n \frac{\partial f}{\partial x_i}(x_i - y_i)}{\sum_{i=1}^n \frac{\partial h}{\partial x_i}(x_i - y_i)} \right\|_\infty \times \left\| \frac{\sum_{i=1}^n \frac{\partial g}{\partial x_i}(x_i - y_i)}{\sum_{i=1}^n \frac{\partial h}{\partial x_i}(x_i - y_i)} \right\|_\infty, \tag{27.10}$$

*where*

$$N = \int_D |h(x) - h(y)|\mathrm{d}y.$$

*Proof.* Integrating both sides of (27.3) and (27.4) with respect to $y$ over $D$, we get

$$f(x) - \frac{1}{M}\int_D f(y)\mathrm{d}y = \frac{1}{M}\int_D \frac{\sum_{i=1}^n \frac{\partial f(c)}{\partial x_i}(x_i - y_i)}{\sum_{i=1}^n \frac{\partial h(c)}{\partial x_i}(x_i - y_i)}(h(x) - h(y))\mathrm{d}y, \tag{27.11}$$

$$g(x) - \frac{1}{M}\int_D g(y)\mathrm{d}y = \frac{1}{M}\int_D \frac{\sum_{i=1}^n \frac{\partial g(d)}{\partial x_i}(x_i - y_i)}{\sum_{i=1}^n \frac{\partial h(d)}{\partial x_i}(x_i - y_i)}(h(x) - h(y))\mathrm{d}y. \tag{27.12}$$

Multiplying (27.11) and (27.12), we have

$$f(x)g(x) - f(x)\left(\frac{1}{M}\int_D g(y)\mathrm{d}y\right) - g(x)\left(\frac{1}{M}\int_D f(y)\mathrm{d}y\right)$$
$$+ \frac{1}{M^2}\left(\int_D f(y)\mathrm{d}y\right)\left(\int_D g(y)\mathrm{d}y\right)$$
$$= \frac{1}{M^2}\left[\int_D \frac{\sum_{i=1}^n \frac{\partial f(c)}{\partial x_i}(x_i - y_i)}{\sum_{i=1}^n \frac{\partial h(c)}{\partial x_i}(x_i - y_i)}(h(x) - h(y))\mathrm{d}y\right]$$
$$\times \left[\int_D \frac{\sum_{i=1}^n \frac{\partial g(d)}{\partial x_i}(x_i - y_i)}{\sum_{i=1}^n \frac{\partial h(d)}{\partial x_i}(x_i - y_i)}(h(x) - h(y))\mathrm{d}y\right]. \tag{27.13}$$

Using (27.13) and properties of the modulus, we have

$$\left| f(x)g(x) - f(x)\left(\frac{1}{M}\int_D g(y)\mathrm{d}y\right) - g(x)\left(\frac{1}{M}\int_D f(y)\mathrm{d}y\right) \right.$$
$$\left. + \frac{1}{M^2}\left(\int_D f(y)\mathrm{d}y\right)\left(\int_D g(y)\mathrm{d}y\right)\right|$$

$$\leq \frac{1}{M^2}\left[\int_D \left|\frac{\sum_{i=1}^n \frac{\partial f(c)}{\partial x_i}(x_i-y_i)}{\sum_{i=1}^n \frac{\partial h(c)}{\partial x_i}(x_i-y_i)}\right| |h(x)-h(y)|\mathrm{d}y\right]$$

$$\times\left[\int_D \left|\frac{\sum_{i=1}^n \frac{\partial g(d)}{\partial x_i}(x_i-y_i)}{\sum_{i=1}^n \frac{\partial h(d)}{\partial x_i}(x_i-y_i)}\right| |h(x)-h(y)|\mathrm{d}y\right]$$

$$\leq \frac{N^2}{M^2}\left\|\frac{\sum_{i=1}^n \frac{\partial f}{\partial x_i}(x_i-y_i)}{\sum_{i=1}^n \frac{\partial h}{\partial x_i}(x_i-y_i)}\right\|_\infty \times \left\|\frac{\sum_{i=1}^n \frac{\partial g}{\partial x_i}(x_i-y_i)}{\sum_{i=1}^n \frac{\partial h}{\partial x_i}(x_i-y_i)}\right\|_\infty,$$

where

$$N=\int_D |h(x)-h(y)|\mathrm{d}y.$$

□

*Remark 27.6.* Note that, with $h(x)=x$ and $h(y)=y$, inequality (27.10) reduces to the following inequality:

$$\Bigg|f(x)g(x) - f(x)\left(\frac{1}{M}\int_D g(y)\mathrm{d}y\right)$$

$$-g(x)\left(\frac{1}{M}\int_D f(y)\mathrm{d}y\right)+\frac{1}{M^2}\int_D f(y)\mathrm{d}y\int_D g(y)\mathrm{d}y\Bigg|$$

$$\leq \frac{1}{M^2}\left[\sum_{i=1}^n \left\|\frac{\partial f}{\partial x_i}\right\|_\infty E_i(x)\right]\times\left[\sum_{i=1}^n \left\|\frac{\partial g}{\partial x_i}\right\|_\infty E_i(x)\right],$$

where $E_i(x), M, \mathrm{d}y$ are defined as in Theorem 27.2. This is also another inequality established by Pachpatte in [10, Theorem 2.2].

**Acknowledgement** Chang-Jian Zhao: the research has been supported by National Natural Science Foundation of China (10971205).

Wing-Sum Cheung: the research has been partially supported by the Research Grants Council of the Hong Kong SAR, China (Project No.: HKU7016/07P).

## References

1. Anastassiou, G.A.: Multivariate Ostrowski type inequalities. Acta Math. Hungar. **76**, 267–278 (1997)
2. Barnett, N.S., Dragomir, S.S.: An Ostrowski type inequality for double integrals and Applications for cubature formulae. RGMIA Res. Rep. Coll. **1**, 13–23 (1998)
3. Dragomir, S.S., Barnett, N.S., Cerone, P.: An n-dimensional version of Ostrowski's inequality for mappings of the Höder type. RGMIA Res. Rep. Coll. **2(2)**, 169–180 (1999)

4. Dragomir, S.S., Rassias, Th.M.: Ostrowski Type Inequalities and Applications in Numerical Integration, Kluwer Academic Publishers, Dorderecht-Boston-London (2002)
5. Dragomir, S.S., Wang, S.: An inequality of Ostrowski-Grüss' type and its applications to the estimation of error bounds for some special means and for some numerical quadrature rules. Computers Math. Applic. **33**, 15–20 (1997)
6. Dragomir, S.S., Wang, S.: Applications of Ostrowski's inequality to the estimation of error bounds for some special means and some numerical quadrature rules. Appl. Math. Lett. **11**, 105–109 (1998)
7. Mitrinović, S., Pečarić, J.E., Fink, A.M.: Inequalities for functions and their integrals and derivatives, Kluwer Academic Publishers, Dordrecht (1994)
8. Pachpatte, B.G.: On an inequality of Ostrowski type in three independent variables. J. Math. Anal. Appl. **249**, 583–591 (2000)
9. Pachpatte, B.G.: On a new Ostrowski type inequality in two independent variables. Tamkang J. Math. **32**, 45–49 (2001)
10. Pachpatte, B.G.: On multivariate Ostrowski inequalities. J. Ineq. Pure Appl. Math. **3**, 1–5 (2002)

# Chapter 28
# Ternary Semigroups and Ternary Algebras

**Antoni Chronowski**

*Dedicated to the memory of S.M. Ulam*
*on the 100th anniversary of his birth*

**Abstract** The ternary algebraic, topological, ordered structures are used in the modern theoretical and mathematical physics and in the theory of functional equations. The subject-matter of this paper focuses on ternary semigroups of mappings and ternary algebras of mappings. The main theorem states that every $n$-ary (ternary) semigroup is embeddable into an $n$-ary (ternary) semigroup of mappings. The analysis of the structure of ternary semigroups of mappings by means of the Green's relations shows that these algebraic structures are a natural generalization of (binary) semigroups of mappings. The ternary semigroups of mappings are used for constructing the natural examples of ternary algebras, which are the counterparts of binary algebras.

**Keywords** Ternary semigroups of mappings • Ternary linear algebras

**Mathematics Subject Classification (2000):** Primary 20N10, 20N15; Secondary 17A40

## 28.1 Introduction

Dörnte [24] was the first to study $n$-ary (polyadic) groups as abstract algebras. The theory was later developed at great length by E. Post [47]. The different results related to $n$-ary groups one can find in many papers (e.g. [8, 17, 25, 29, 38, 54]). Very extensive investigations on the $n$-ary quasigroups were published by Belousov [5]. The foundations of the $n$-ary semigroups one can find in the papers (e.g. [15,37,52]).

A. Chronowski (✉)
Department of Mathematics, Pedagogical University, Podchorążych 2, 30-084 Kraków, Poland
e-mail: chron@up.krakow.pl

Th.M. Rassias and J. Brzdęk (eds.), *Functional Equations in Mathematical Analysis*, Springer Optimization and Its Applications 52, DOI 10.1007/978-1-4614-0055-4_28,

The notion of ternary semigroups was introduced by Lehmer in 1932. There are many papers on ternary semigroups (e.g. [1, 13, 14, 16, 18–20, 22, 23, 28, 41, 51, 53]).

An important area of research are ternary algebras: ternary linear algebras, involutive ternary linear algebras, normed ternary linear algebras, topological ternary linear algebras, Banach ternary algebras, $C^*$-ternary algebras (e.g. [2–4, 21, 33, 42, 46]). It is known (e.g. [2, 33, 46]), that the ternary algebras and the ternary semigroups are interesting for their applications to the problems of modern mathematical physics (e.g. the Nambu mechanics).

The ternary ($n$-ary) algebraic structures are considered in the theory of functional equations (e.g. [7, 9–12, 29, 35, 43]), particularly in the stability of functional equations (e.g. [1, 3, 39, 42, 45, 46, 48]).

The inspiration for the considerations on the functional equations and their stability on the ternary structures comes from many articles and books about this subject (e.g. [26, 27, 31, 32, 44, 49, 50]).

The considerations included in this paper are intended above all to draw attention to the ternary ($n$-ary) structures, which may find more interesting applications in the theory of functional equations and in the mathematical physics.

## 28.2 Basic Concepts: *n*-ary Algebraic Structures

In this section, certain basic definitions and results are presented. References will be made to these definitions and theorems throughout the paper, first of all in the case when $n = 3$.

**Definition 28.1.** Let $A$ be a nonempty set. A mapping $f : A^n \longrightarrow A$ is called an $n$-ary operation on the set $A$.

If $n = 2$, then a 2-ary operation $f$ is called an *operation*. If $n = 3$, then a 3-ary operation $f$ is called a *ternary operation*.

In this section, we consider $n$-ary operations for $n \geq 2$.

Let $f$ be an $n$-ary operation on the set $A$. Let us denote:

$$f(x_1^n) = f(x_1, \ldots, x_n),$$

$$f(x_1^k, \overset{s}{x}, x_{k+s+1}^n) = f(x_1, \ldots, x_k, x_{k+1}, \ldots, x_{k+s}, x_{k+s+1}, \ldots, x_n),$$

whenever

$$x_{k+1} = x_{k+2} = \cdots = x_{k+s} = x.$$

If $j < i$, then $x_i^j$ is the empty symbol, also $\overset{0}{x}$ is the empty symbol.

If $X_1, \ldots, X_n \in 2^A$, then we put

$$f(X_1, \ldots, X_n) = \{f(x_1, \ldots, x_n) : x_1 \in X_1, \ldots, x_n \in X_n\}.$$

Let us denote:

$$f(X^k, x^l, Y^m) = f(X_1, \ldots, X_k, X_{k+1}, \ldots, X_{k+l}, X_{k+l+1}, \ldots, X_{k+l+m}),$$

whenever $k+l+m=n$ and

$$X_1 = \cdots = X_k = X,$$
$$X_{k+1} = \cdots = X_{k+l} = \{x\},$$
$$X_{k+l+1} = \cdots = X_{k+l+m} = Y.$$

**Definition 28.2.** Let $A$ be a nonempty set, $f$ – an $n$-ary operation on the set $A$. The pair $(A, f)$ is said to be an $n$-ary groupoid (or $n$-groupoid).

If $n=2$, then a 2-ary groupoid $(A, f)$ is called a *groupoid*. If $n=3$, then a 3-ary groupoid $(A, f)$ is called a *ternary groupoid*.

**Definition 28.3.** An element $e \in A$ is called an identity of the $n$-ary groupoid $(A, f)$ if

$$f(\mathrm{e}^{i-1}, x, \mathrm{e}^{n-i}) = x$$

for every $x \in A$, where $i = 1, \ldots, n$.

**Definition 28.4.** A mapping $F : A \longrightarrow B$ is called a homomorphism of an $n$-ary groupoid $(A, f)$ into an $n$-ary groupoid $(B, f)$ if

$$F(f(x_1, x_2, \ldots, x_n)) = f(F(x_1), F(x_2), \ldots, F(x_n))$$

for all $x_1, x_2, \ldots, x_n \in A$.

**Definition 28.5.** An $n$-ary groupoid $(A, f)$ is said to be an $n$-ary quasigroup (or $n$-quasigroup) if for an arbitrary $i \in \{1, \ldots, n\}$ the equation

$$f(a_1^{i-1}, x, a_{i+1}^n) = a$$

has a unique solution in $A$ for any elements $a, a_1, \ldots, a_n \in A$.

If $n=2$, then a 2-ary quasigroup $(A, f)$ is called a *quasigroup*. If $n=3$, then a 3-ary quasigroup $(A, f)$ is called a *ternary quasigroup*.

**Definition 28.6.** An $n$-ary groupoid $(A, f)$ is said to be an $n$-ary semigroup (or $n$-semigroup) if for arbitrary $i, j \in \{1, \ldots, n\}$ the following equality

$$f\left(x_1^{i-1}, f\left(x_i^{i+n-1}\right), x_{i+n}^{2n-1}\right) = f\left(x_1^{j-1}, f\left(x_j^{j+n-1}\right), x_{j+n}^{2n-1}\right) \tag{28.1}$$

holds for any elements $x_1, \ldots, x_{2n-1} \in A$.

The condition (28.1) is called the *associative law*.

If $n = 2$, then a 2-ary semigroup $(A, f)$ is called a *semigroup*. If $n = 3$, then a 3-ary semigroup $(A, f)$ is called a *ternary semigroup*.

**Definition 28.7.** Let $(A, f)$ be a ternary semigroup. The equivalence relation $\alpha$ on the set $A$ is called:

(a) a left congruence if

$$\forall a, b, x_1, x_2 \in A \, [(a,b) \in \alpha \Longrightarrow (f(x_1, x_2, a), f(x_1, x_2, b)) \in \alpha];$$

(b) a right congruence if

$$\forall a, b, x_1, x_2 \in A \, [(a,b) \in \alpha \Longrightarrow (f(a, x_1, x_2), f(b, x_1, x_2)) \in \alpha].$$

**Definition 28.8.** A ternary semigroup $(A, f)$ is said to be regular if

$$\forall a \in A \exists x, y \in A \, [f(a, x, a, y, a) = a].$$

**Definition 28.9.** If an $n$-ary semigroup $(A, f)$ is an $n$-ary quasigroup, then $(A, f)$ is called an $n$-ary group (or $n$-group).

If $n = 2$, then a 2-ary group $(A, f)$ is called a *group*. If $n = 3$, then a 3-ary group $(A, f)$ is called a *ternary group*.

To illustrate the relationship between the $n$-ary groups and the (binary) groups we shall quote the following theorem.

**Theorem 28.1 ([29, 54]).** *An $n$-groupoid $(A, f)$ is an $n$-group if and only if*

$$f(x_1^n) = x_1 \cdot \alpha(x_2) \cdot \alpha^2(x_3) \cdot \cdots \cdot \alpha^{n-2}(x_{n-1}) \cdot a \cdot x_n$$

*for all $x_1, \ldots, x_n \in A$, whenever the following conditions hold:*

(a) *$(A, \cdot)$ is a group;*
(b) *$a \in A$ is a fixed element;*
(c) *$\alpha \in \mathrm{Aut}(A, \cdot)$, $\alpha(a) = a$, $\alpha^{n-1}(x) = a \cdot x \cdot a^{-1}$ for any $x \in A$.*

## 28.3 An $n$-ary Semigroup of Mappings. The Embedding Theorems

We will consider some $n$-ary (ternary) semigroups of mappings. The idea and significance of $n$-ary (ternary) semigroups of mappings will be explained in the problem of representing an $n$-ary (a ternary) semigroup as an $n$-ary (a ternary) semigroup of mappings. The main results of this section are the embedding theorems.

Let $X$ and $Y$ be nonempty sets. By $T(X,Y)$ we denote the set of all mappings of the set $X$ into the set $Y$.

Assume that $X_1, X_2,\ldots,X_{n-1}$, where $n \geq 3$, are nonempty sets. Consider the set

$$T[X_1,X_2,\ldots,X_{n-1}] = T(X_1,X_2) \times T(X_2,X_3) \times \cdots \times T(X_{n-2},X_{n-1}) \times T(X_{n-1},X_1).$$

The symbol $p_{ij}$ denotes a mapping from the set $T(X_i,X_j)$ for $i,j \in \{1,\ldots,n-1\}$. Consider the case $n=3$. Then

$$T[X_1,X_2] = T(X_1,X_2) \times T(X_2,X_1).$$

Define the ternary operation $f : T[X_1,X_2]^3 \longrightarrow T[X_1,X_2]$ by

$$f((p^1_{12},p^1_{21}),(p^2_{12},p^2_{21}),(p^3_{12},p^3_{21})) = (p^1_{12} \circ p^2_{21} \circ p^3_{12},\ p^1_{21} \circ p^2_{12} \circ p^3_{21})$$

for all

$$(p^1_{12},p^1_{21}),(p^2_{12},p^2_{21}),(p^3_{12},p^3_{21}) \in T[X_1,X_2].$$

We shall prove that the ternary operation $f$ is associative:

$$\begin{aligned}
&f(f((p^1_{12},p^1_{21}),\ (p^2_{12},p^2_{21}),\ (p^3_{12},p^3_{21})),\ (p^4_{12},p^4_{21}),(p^5_{12},p^5_{21}))\\
&\quad = f((p^1_{12} \circ p^2_{21} \circ p^3_{12},\ p^1_{21} \circ p^2_{12} \circ p^3_{21}),\ (p^4_{12},p^4_{21}),(p^5_{12},p^5_{21}))\\
&\quad = (p^1_{12} \circ p^2_{21} \circ p^3_{12} \circ p^4_{21} \circ p^5_{12},\ p^1_{21} \circ p^2_{12} \circ p^3_{21} \circ p^4_{12} \circ p^5_{21});
\end{aligned}$$

$$\begin{aligned}
&f((p^1_{12},p^1_{21}),\ f((p^2_{12},p^2_{21}),\ (p^3_{12},p^3_{21}),\ (p^4_{12},p^4_{21})),(p^5_{12},p^5_{21}))\\
&\quad = f((p^1_{12},p^1_{21}),\ (p^2_{12} \circ p^3_{21} \circ p^4_{12},\ p^2_{21} \circ p^3_{12} \circ p^4_{21}),\ (p^5_{12},p^5_{21}))\\
&\quad = (p^1_{12} \circ p^2_{21} \circ p^3_{12} \circ p^4_{21} \circ p^5_{12},\ p^1_{21} \circ p^2_{12} \circ p^3_{21} \circ p^4_{12} \circ p^5_{21});\\
&f((p^1_{12},p^1_{21}),\ (p^2_{12},p^2_{21}),\ f((p^3_{12},p^3_{21}),\ (p^4_{12},p^4_{21}),(p^5_{12},p^5_{21})))\\
&\quad = f((p^1_{12},p^1_{21}),\ (p^2_{12},p^2_{21}),\ (p^3_{12} \circ p^4_{21} \circ p^5_{12},\ p^3_{21} \circ p^4_{12} \circ p^5_{21}))\\
&\quad = (p^1_{12} \circ p^2_{21} \circ p^3_{12} \circ p^4_{21} \circ p^5_{12},\ p^1_{21} \circ p^2_{12} \circ p^3_{21} \circ p^4_{12} \circ p^5_{21})
\end{aligned}$$

for all

$$(p^1_{12},p^1_{21}),\ (p^2_{12},p^2_{21}),\ (p^3_{12},p^3_{21}),\ (p^4_{12},p^4_{21}),(p^5_{12},p^5_{21}) \in T[X_1,X_2].$$

Thus, we have obtained the following

**Proposition 28.1.** *The groupoid* $(T[X_1,X_2],f)$ *is a ternary semigroup.*

Now, we consider the general case. Define the $n$-ary operation

$$f : T[X_1,X_2,\ldots,X_{n-2},X_{n-1}]^n \longrightarrow T[X_1,X_2,\ldots,X_{n-2},X_{n-1}]$$

by the formula:

$$\begin{aligned}
&f((p^1_{12},p^1_{23},\dots,p^1_{n-2,n-1},p^1_{n-1,1}),(p^2_{12},p^2_{23},\dots,p^2_{n-2,n-1},p^2_{n-1,1}),\dots,\\
&\quad(p^{n-1}_{12},p^{n-1}_{23},\dots,p^{n-1}_{n-2,n-1},p^{n-1}_{n-1,1}),(p^n_{12},p^n_{23},\dots,p^n_{n-2,n-1},p^n_{n-1,1}))\\
&\quad=(p^1_{12}\circ p^2_{n-1,1}\circ\dots\circ p^{n-1}_{23}\circ p^n_{12},\ p^1_{23}\circ p^2_{12}\circ\dots\circ p^{n-1}_{34}\circ p^n_{23},\dots,\\
&\quad p^1_{n-2,n-1}\circ p^2_{n-3,n-2}\circ\dots\circ p^{n-1}_{n-1,1}\circ p^n_{n-2,n-1},p^1_{n-1,1}\circ p^2_{n-2,n-1}\circ\dots\\
&\quad\circ p^{n-1}_{12}\circ p^n_{n-1,1})
\end{aligned}\tag{28.2}$$

for all $(p^1_{12},p^1_{23},\dots,p^1_{n-2,n-1},p^1_{n-1,1}),\dots,(p^n_{12},p^n_{23},\dots,p^n_{n-2,n-1},p^n_{n-1,1})$ from the set $T[X_1,X_2,\dots,X_{n-2},X_{n-1}]$.

The proof of associativity of this $n$-ary operation $f$ may be modelled after the above proof of associativity of the ternary operation $f$.

So we have:

$$\begin{aligned}
&f((p^1_{12},p^1_{23},\dots,p^1_{n-2,n-1},p^1_{n-1,1}),(p^2_{12},p^2_{23},\dots,p^2_{n-2,n-1},p^2_{n-1,1}),\dots,\\
&\quad(p^{i-1}_{12},p^{i-1}_{23},\dots,p^{i-1}_{n-2,n-1},p^{i-1}_{n-1,1}),f((p^i_{12},p^i_{23},\dots,p^i_{n-2,n-1},p^i_{n-1,1}),\dots,\\
&\quad(p^n_{12},p^n_{23},\dots,p^n_{n-2,n-1},p^n_{n-1,1}),\dots,(p^{i+n-1}_{12},p^{i+n-1}_{23},\dots,p^{i+n-1}_{n-2,n-1},p^{i+n-1}_{n-1,1})),\\
&\quad(p^{i+n}_{12},p^{i+n}_{23},\dots,p^{i+n}_{n-2,n-1},p^{i+n}_{n-1,1}),\dots,(p^{2n-2}_{12},p^{2n-2}_{23},\dots,p^{2n-2}_{n-2,n-1},p^{2n-2}_{n-1,1}),\\
&\quad(p^{2n-1}_{12},p^{2n-1}_{23},\dots,p^{2n-1}_{n-2,n-1},p^{2n-1}_{n-1,1}))\\
&=(p^1_{12}\circ p^2_{n-1,1}\circ\dots\circ p^{n-1}_{23}\circ p^n_{12}\circ p^{n+1}_{n-1,1}\circ\dots\circ p^{2n-2}_{23}\circ p^{2n-1}_{12},\ p^1_{23}\circ p^2_{12}\circ\dots\\
&\quad\circ p^{n-1}_{34}\circ p^n_{23}\circ p^{n+1}_{12}\circ\dots\circ p^{2n-2}_{34}\circ p^{2n-1}_{23},\dots,p^1_{n-1,1}\circ p^2_{n-2,n-1}\circ\dots\\
&\quad\circ p^{n-1}_{12}\circ p^n_{n-1,1}\circ p^{n+1}_{n-2,n-1}\circ\dots\circ p^{2n-2}_{12}\circ p^{2n-1}_{n-1,1})
\end{aligned}$$

for all

$$(p^1_{12},p^1_{23},\dots,p^1_{n-2,n-1},p^1_{n-1,1}),\dots,(p^{2n-1}_{12},p^{2n-1}_{23},\dots,p^{2n-1}_{n-2,n-1},p^{2n-1}_{n-1,1})$$

from the set $T[X_1,X_2,\dots,X_{n-2},X_{n-1}]$ and for an arbitrary $i\in\{1,\dots,n\}$.

Thus we have proved the following

**Theorem 28.2.** *Let $X_1,\dots,X_{n-1}$, where $n\geq 3$, be nonempty sets. The groupoid*

$$(T[X_1,\dots,X_{n-1}],f)$$

*endowed with the $n$-ary operation $f$ given by the formula (28.2) is an $n$-ary semigroup.*

**Definition 28.10.** The $n$-ary semigroup $(T[X_1,\dots,X_{n-1}],f)$ is called the $n$-ary semigroup of mappings of sets $X_1,\dots,X_{n-1}$.

If the sets $X_1,\dots,X_{n-1}$ are pairwise disjoint, then $(T[X_1,\dots,X_{n-1}],f)$ is called the disjoint $n$-ary semigroup of mappings of sets $X_1,\dots,X_{n-1}$.

Let $(A,\cdot)$ be a semigroup. We define an $n$-ary operation $f : A^n \longrightarrow A$ by the formula:

$$f(x_1,\dots,x_n) = x_1 \cdot \ldots \cdot x_n \tag{28.3}$$

for all $x_1,\dots,x_n \in A$. It is clear that $(A,f)$ is an $n$-ary semigroup.

The $n$-ary semigroup $(A,f)$ with the $n$-ary operation defined by (28.3) is called an *$n$-ary semigroup reduct* of the semigroup $(A,\cdot)$. If $n=3$, then we say that $(A,f)$ is a *ternary semigroup reduct* of the semigroup $(A,\cdot)$.

Let $A$ be a nonempty set. Denote by $F_A$ the set of all finite sequences $(a_1,\dots,a_m)$ for $a_1,\dots,a_m \in A$. The empty set $\emptyset$ is treated as a finite sequence $\emptyset \in F_A$ and it is called an *empty sequence*.

The number $m$ is said to be a *length* of the $m$-termed sequence $a=(a_1,\dots,a_m) \in F_A$ and we write $l(a)=m$. For the empty sequence $\emptyset$ we put $l(\emptyset)=0$. We identify $a$ with $(a)$ for $a \in A$.

Let us define a binary operation on the set $F_A$ by the formula:

$$ab = (a_1,\dots,a_m,b_1,\dots,b_k) \tag{28.4}$$

for $a=(a_1,\dots,a_m) \in F_A$, $b=(b_1,\dots,b_k) \in F_A$.

The binary operation defined by the formula (28.4) is called a *concatenation* or *juxtaposition* on the set $F_A$.

With respect to the concatenation $F_A$ is a semigroup, called the *free semigroup* on the set $A$. The empty sequence is the identity of the free semigroup $F_A$.

Let $(A,f)$ be an $n$-ary semigroup. Consider the free semigroup $F_A$. The $n$-ary operation $f$ has a natural extension to the set of all sequences that, with some $m>0$, have the length $m(n-1)+1$. Namely, by recursion with $m>0$,

$$f(x_1,\dots,x_{m(n-1)+1}) = f(\dots f(f(x_1,\dots,x_n),x_{n+1},\dots,x_{2n-1}),\dots,x_{m(n-1)+1})$$

for $(x_1,\dots,x_{m(n-1)+1}) \in F_A$.

The proof of the next theorem is due to Monk and Sioson [37].

**Theorem 28.3.** *Every $n$-ary semigroup $(A,f)$ is embeddable into an $n$-ary semigroup reduct $(B,f)$ with identity.*

*Proof.* Consider the free semigroup $F_A$. Define the relation $R$ on the set $F_A$ by the following condition:

$(w_1)$ $aRb \Longleftrightarrow \exists p,q,r \in F_A\, [a=pqr \wedge l(q)=n \wedge b=pf(q)r]$
for $a,b \in F_A$.

Notice that there are no sequences in the set $F_A$ which are related by $R$ to the empty set $\emptyset$.

Observe that

$(w_2)$ if $b=f(b_1,\dots,b_n)$, then $(b_1,\dots,b_n)Rb$.

Indeed, it is enough to take $p = r = \emptyset$ and $q = (b_1,\dots,b_n)$ in the condition $(w_1)$.

Let $S$ be the equivalence relation on the set $F_A$ genereted by the relation $R$. Thus, we get:

$(w_3)$ if $a,b \in F_A$, then $aSb$ if and only if either $a = b$, or for some $n \in \mathscr{N}$ there is a finite sequence $a = c_1,\dots,c_k = b$ in which for each $i \in \{1,\dots,k-1\}$, either $c_iRc_{i+1}$ or $c_{i+1}Rc_i$.

Observe that $\emptyset Sb$ if and only if $b = \emptyset$, and so $[\emptyset]_S = \{\emptyset\}$.

First, we shall show that $S$ is a congruence on the free semigroup $F_A$. Assume that $a,b,c \in F_A$ and $aRb$. Thus, there exist $p,q,r \in F_A$ such that $a = pqr$, $l(q) = n$, $b = pf(q)r$. Hence, $ac = (pqr)c$, $bc = (pf(q)r)c$, and so

$$ac = pq(rc), \quad bc = pf(q)(rc).$$

Consequently, $(ac)R(bc)$. Similarly, $(ca)R(cb)$. By the condition $(w_3)$, we infer that $S$ is a congruence on the free semigroup $F_A$.

Therefore, we can consider the $n$-ary semigroup reduct $(F_A/S, f)$ of the quotient semigroup $F_A/S$. Notice that the equivalence class $[\emptyset]_S$ is the identity of the quotient semigroup $F_A/S$, what implies that $[\emptyset]_S$ is the identity of the $n$-ary semigroup reduct $(F_A/S, f)$.

Observe that if $a,b \in F_A$ and $aRb$, then

$$l(a) = l(b) + n - 1.$$

Using $(w_3)$ we have:

$(w_4)$ if $a,b \in F_A$ and $aSb$, then $l(a) \equiv l(b) \pmod{n-1}$.

The following statement will be found useful in the proof:

$(w_5)$ if $x \in A$, $b \in F_A$, and $xSb$, then either $x = b$ or there exists $m > 0$ such that $l(b) = m(n-1)+1$ and $f(b) = x$.

Indeed, in view of $(w_3)$ it follows that $x = b$ or $x = c_1,\dots,c_k = b$ for some $c_1,\dots,c_k \in F_A$ such that either $c_iRc_{i+1}$ or $c_{i+1}Rc_i$ for each $i \in \{1,\dots,k-1\}$. By induction on $k$ we shall prove the condition $(w_5)$.

If $k = 1$, then $x = c_1 = b$, and so $(w_5)$ holds. Assume that the condition $(w_5)$ holds for $k-1$, where $k > 1$. Suppose that $x = c_1,\dots,c_{k-1},c_k = b$. Since $xSc_{k-1}$, by the inductive assumption, we have two possibilities:

(a) $x = c_{k-1}$ or

(b) there exists $m > 0$ such that $l(c_{k-1}) = m(n-1)+1$ and $f(c_{k-1}) = x$.

Assume that (a) holds. If $x = c_{k-1}$, then $c_kRc_{k-1}$, because the case $c_{k-1}Rc_k$ is impossible in view of the definition of $R$. Hence, $c_kRx$. By the definition of $R$, there exist $p,q,r \in F_A$ such that

$$c_k = pqr, \quad l(q) = n, \quad x = pf(q)r.$$

Since $x \in A$, it follows that $p = r = \emptyset$. Thus, $c_k = q$, and so $x = f(c_k) = f(b)$. In this case, $(w_5)$ holds.

Assume that (b) holds. We know that $c_{k-1}Rc_k$ or $c_kRc_{k-1}$. First, suppose that $c_{k-1}Rc_k$. Thus, there exist elements $p,q,r \in F_A$ such that

$$c_{k-1} = pqr, \quad l(q) = n, \quad c_k = pf(q)r.$$

By (b),

$$l(pqr) = l(c_{k-1}) = m(n-1)+1$$

for some $m > 0$.

If $m = 1$, then $l(pqr) = n$. Since $l(q) = n$, it follows that $p = r = \emptyset$. Hence

$$c_{k-1} = q, \quad c_k = f(q) = f(c_{k-1}) = x,$$

and so $b = c_k = x$.

If $m > 1$, then

$$l(pqr) = m(n-1)+1,$$
$$l(c_k) = l(pf(q)r) = m(n-1)+1-n+1 = (m-1)(n-1)+1,$$

where $m-1 > 0$. Therefore

$$f(b) = f(c_k) = f(pf(q)r) = f(pqr) = f(c_{k-1}) = x.$$

Next, assume that $c_kRc_{k-1}$. There exist elements $p,q,r \in F_A$ such that

$$c_k = pqr, \quad l(q) = n, \quad c_{k-1} = pf(q)r.$$

By (b),

$$l(pf(q)r) = l(c_{k-1}) = m(n-1)+1$$

for some $m > 0$. Thus

$$l(c_k) = l(pqr) = m(n-1)+1+(n-1) = (m+1)(n-1)+1.$$

Consequently

$$f(b) = f(c_k) = f(pqr) = f(pf(q)r) = f(c_{k-1}) = x.$$

Define the mapping $\varphi : A \longrightarrow F_A/S$ by

$$\varphi(x) = [x]_S$$

for $x \in A$. Notice that $\varphi$ is an injection. Indeed, assume that $\varphi(x_1) = \varphi(x_2)$, where $x_1, x_2 \in A$. Thus

$$[x_1]_S = [x_2]_S,$$

whence $x_1 S x_2$, and so $x_1 = x_2$ in view of the condition $(w_5)$.

Assume that $x_1,\dots,x_n \in A$. Using the condition $(w_2)$ we get:

$$\varphi(f(x_1,\dots,x_n)) = [f(x_1,\dots,x_n)]_S = [(x_1,\dots,x_n)]_S = [x_1]_S\dots[x_n]_S$$
$$= f([x_1]_S,\dots,[x_n]_S) = f(\varphi(x_1),\dots,\varphi(x_n)).$$

Therefore, $\varphi$ is a monomorphism of the $n$-ary semigroup $(A,f)$ into the $n$-ary semigroup reduct $(F_A/S,f)$ of the quotient semigroup $F_A/S$. Taking $B = F_A/S$, we have proved the theorem. □

By Theorem 28.3 we get

**Corollary 28.1.** *Every n-ary semigroup can be extended to an n-ary semigroup with identity.*

In the sequel, continuing the argument applied in the proof of Theorem 28.3, we shall prove another theorem.

Let $(A,f)$ be an $n$-ary semigroup with identity $e$. For an arbitrary $i \in \{1,\dots,n-1\}$ let us consider the following sets:

$$X_i = \{M \in F_A/S :\ \exists a \in M\,[l(a) = i]\}.$$

From $(w_4)$ we see that the sets $X_1,\dots,X_{n-1}$ are pairwise disjoint.

We shall show that

$$(w_6)\forall a \in F_A\, \exists b \in F_A\, [a\,S\,b \wedge l(b) < n].$$

If $l(a) < n$, it is enough to take $b = a$. Assume that $a \in F_A$ and $m = l(a) \geq n$. Put $a = (a_1,\dots,a_m)$. By $(w_2)$,

$$(a_1,\dots,a_n)Sf(a_1,\dots,a_n).$$

If $m = n$, then it is enough to put $b = f(a_1,\dots,a_n)$. Suppose that $m > n$. Since $S$ is a congruence on the free semigroup $F_A$ and

$$(a_1,\dots,a_n)Sf(a_1,\dots,a_n),$$

it follows that

$$a = (a_1,\dots,a_n,a_{n+1},\dots,a_m)S(f(a_1,\dots,a_n),a_{n+1},\dots,a_m).$$

Observe that

$$l(f(a_1,\dots,a_n),a_{n+1},\dots,a_m) = l(a) - n + 1.$$

Proceeding in this way, we get $(w_6)$. By $(w_6)$, it follows that

$$X_1 \cup \dots \cup X_{n-1} = (F_A/S) \setminus \{[\emptyset]_S\}.$$

Let $x \in A$ be an arbitrary fixed element. Define the mappings:

$$p^x_{i,i+1} : X_i \longrightarrow X_{i+1} \quad \text{for } i = 1,\dots,n-2,$$
$$p^x_{n-1,1} : X_{n-1} \longrightarrow X_1$$

by the following formulas:

$$p^x_{i,i+1}([b]_S) = [xb]_S \text{ for } [b]_S \in X_i,$$
$$p^x_{n-1,1}([b]_S) = [xb]_S \text{ for } [b]_S \in X_{n-1}.$$

We shall prove that the mappings $p^x_{i,i+1}$ for $i = 1,\dots,n-2$ and $p^x_{n-1,1}$ are well-defined. Assume that $[b]_S = [b']_S$. Since $S$ is a congruence and $bSb'$, it follows that $(xb)S(xb')$, hence

$$[xb]_S = [xb']_S.$$

Take an arbitrary $i \in \{1,\dots,n-2\}$. Suppose that $[b]_S \in X_i$. Thus, $[b]_S = [a]_S$, where $l(a) = i$. Hence

$$p^x_{i,i+1}([b]_S) = p^x_{i,i+1}([a]_S) = [xa]_S \in X_{i+1}.$$

Suppose that $[b]_S \in X_{n-1}$. Thus, $[b]_S = [a]_S$, where $l(a) = n-1$. From $(w_2)$ it follows that

$$f(xa)S(xa).$$

Clearly $l(f(xa)) = 1$, hence $[xa]_S \in X_1$. Thus

$$p^x_{n-1,1}([b]_S) = p^x_{n-1,1}([a]_S) = [xa]_S \in X_1.$$

Let us define a mapping

$$\mu : A \longrightarrow T[X_1,\dots,X_{n-1}]$$

by the formula:

$$\mu(x) = (p^x_{12}, p^x_{23}, \dots, p^x_{n-2,n-1}, p^x_{n-1,1})$$

for $x \in A$. We shall prove that $\mu$ is a monomorphism of the $n$-ary semigroup $(A,f)$ into the disjoint $n$-ary semigroup

$$(T[X_1,\dots,X_{n-1}], f)$$

of mappings of sets $X_1,\dots,X_{n-1}$.

Assume that $x_1,\dots,x_n \in A$. We shall show that

$$\mu(f(x_1,x_2,\dots,x_{n-1},x_n)) = f(\mu(x_1),\mu(x_2),\dots,\mu(x_{n-1}),\mu(x_n)). \tag{28.5}$$

We have:

$$\mu(f(x_1,x_2,\dots,x_{n-1},x_n)) = \Big(p_{12}^{f(x_1,x_2,\dots,x_{n-1},x_n)}, p_{23}^{f(x_1,x_2,\dots,x_{n-1},x_n)}, \dots, p_{n-2,n-1}^{f(x_1,x_2,\dots,x_{n-1},x_n)}, p_{n-1,1}^{f(x_1,x_2,\dots,x_{n-1},x_n)}\Big);$$

$$f(\mu(x_1),\mu(x_2),\dots,\mu(x_{n-1}),\mu(x_n))$$

$$=f\Big(\Big(p_{12}^{x_1},p_{23}^{x_1},\dots,p_{n-2,n-1}^{x_1},p_{n-1,1}^{x_1}\Big),\Big(p_{12}^{x_2},p_{23}^{x_2},\dots,p_{n-2,n-1}^{x_2},p_{n-1,1}^{x_2}\Big),$$

$$\dots,\Big(p_{12}^{x_{n-1}},p_{23}^{x_{n-1}},\dots,p_{n-2,n-1}^{x_{n-1}},p_{n-1,1}^{x_{n-1}}\Big),\Big(p_{12}^{x_n},p_{23}^{x_n},\dots,p_{n-2,n-1}^{x_n},p_{n-1,1}^{x_n}\Big)\Big)$$

$$=\Big(p_{12}^{x_1}\circ p_{n-1,1}^{x_2}\circ\dots\circ p_{23}^{x_{n-1}}\circ p_{12}^{x_n},\ p_{23}^{x_1}\circ p_{12}^{x_2}\circ\dots\circ p_{34}^{x_{n-1}}\circ p_{23}^{x_n},$$

$$\dots,p_{n-2,n-1}^{x_1}\circ p_{n-3,n-2}^{x_2}\circ\dots\circ p_{n-1,1}^{x_{n-1}}\circ p_{n-2,n-1}^{x_n},\ p_{n-1,1}^{x_1}\circ p_{n-2,n-1}^{x_2}\circ$$

$$\dots\circ p_{12}^{x_{n-1}}\circ p_{n-1,1}^{x_n}\Big).$$

First, we shall show that

$$p_{12}^{f(x_1,x_2,\dots,x_{n-1},x_n)}=p_{12}^{x_1}\circ p_{n-1,1}^{x_2}\circ\dots\circ p_{23}^{x_{n-1}}\circ p_{12}^{x_n}.$$

Assume that $[a]_S\in X_1$. Therefore, in view of $(w_2)$ we get:

$$\begin{aligned}p_{12}^{f(x_1,x_2,\dots,x_{n-1},x_n)}([a]_S)&=[f(x_1,x_2,\dots,x_{n-1},x_n)a]_S\\&=[(x_1,x_2,\dots,x_{n-1},x_n)a]_S\\&=[x_1x_2\dots x_{n-1}x_na]_S\\&=(p_{12}^{x_1}\circ p_{n-1,1}^{x_2}\circ\dots\circ p_{23}^{x_{n-1}}\circ p_{12}^{x_n})([a]_S).\end{aligned}$$

Similarly, we can prove that

$$p_{23}^{f(x_1,x_2,\dots,x_{n-1},x_n)}=p_{23}^{x_1}\circ p_{12}^{x_2}\circ\dots\circ p_{34}^{x_{n-1}}\circ p_{23}^{x_n},$$

..................................................................

$$p_{n-2,n-1}^{f(x_1,x_2,\dots,x_{n-1},x_n)}=p_{n-2,n-1}^{x_1}\circ p_{n-3,n-2}^{x_2}\circ\dots\circ p_{n-1,1}^{x_{n-1}}\circ p_{n-2,n-1}^{x_n},$$
$$p_{n-1,1}^{f(x_1,x_2,\dots,x_{n-1},x_n)}=p_{n-1,1}^{x_1}\circ p_{n-2,n-1}^{x_2}\circ\dots\circ p_{12}^{x_{n-1}}\circ p_{n-1,1}^{x_n}.$$

Hence, (28.5) holds. Thus, $\mu$ is a homomorphism of $(A,f)$ into $(T[X_1,\dots,X_{n-1}],f)$.

Next, we shall prove that $\mu$ is an injection. Assume that

$$\mu(x)=\mu(y), \tag{28.6}$$

where $x,y\in A$. Clearly

$$\mu(x)=(p_{12}^{x},\dots,p_{n-1,1}^{x}),\qquad \mu(y)=(p_{12}^{y},\dots,p_{n-1,1}^{y}). \tag{28.7}$$

Consider the identity $e$ of the $n$-ary semigroup $(A,f)$. Of course,

$$[e^{n-1}]_S\in X_{n-1}.$$

Next, by (28.6) and (28.7)

$$p^x_{n-1,1} = p^y_{n-1,1}.$$

Hence, using $(w_2)$, we get

$$\begin{aligned}[x]_S &= [f(x,\mathrm{e}^{n-1})]_S = [x,\mathrm{e}^{n-1}]_S = [x\mathrm{e}^{n-1}]_S \\ &= p^x_{n-1,1}([\mathrm{e}^{n-1}]_S) = p^y_{n-1,1}([\mathrm{e}^{n-1}]_S) \\ &= [y\mathrm{e}^{n-1}]_S = [y,\mathrm{e}^{n-1}]_S = [f(y,\mathrm{e}^{n-1})]_S = [y]_S.\end{aligned}$$

Since $[x]_S = [y]_S$, it follows that $x = y$ in view of $(w_5)$. Thus, $\mu$ is a monomorphism of $(A,f)$ into

$$(T[X_1,\ldots,X_{n-1}],f).$$

If the $n$-ary semigroup $(A,f)$ has no identity element, then by Corollary 28.1 the $n$-ary semigroup $(A,f)$ can be extended to an $n$-ary semigroup $(B,f)$ with identity. Thus, $(A,f)$ is an $n$-ary subsemigroup of the $n$-ary semigroup $(B,f)$. It follows from the foregoing that the $n$-ary semigroup $(B,f)$ can be embedded into a disjoint $n$-ary semigroup $(T[X_1,\ldots,X_{n-1}],f)$. Therefore, the $n$-ary semigroup $(A,f)$ can also be embedded into a disjoint $n$-ary semigroup $(T[X_1,\ldots,X_{n-1}],f)$.

On the basis of the above considerations we have obtained the following

**Theorem 28.4.** *Every $n$-ary semigroup $(A,f)$ is embeddable into a disjoint $n$-ary semigroup $(T[X_1,\ldots,X_{n-1}],f)$ of mappings of sets $X_1,\ldots,X_{n-1}$.*

Theorem 28.4 implies the following

**Corollary 28.2.** *Every ternary semigroup $(A,f)$ is embeddable into a disjoint ternary semigroup $(T[X,Y],f)$ of mappings of sets $X$ and $Y$.*

## 28.4 Ideals and Green's Relations on Ternary Semigroups

A generalization of ideals of binary semigroups [30] are ideals of ternary semigroups [53].

**Definition 28.11.** Let $(A,f)$ be a ternary semigroup. A nonempty subset $I \subseteq A$ is called:

(a) a left ideal if $f(A,A,I) \subseteq I$;
(b) a right ideal if $f(I,A,A) \subseteq I$;
(c) a lateral ideal if $f(A,I,A) \subseteq I$;
(d) a two-sided ideal if $I$ is both a left and right ideal;
(e) an ideal if $I$ is a left, right, and lateral ideal.

We denote by $\mathrm{Id}_l(A)$, $\mathrm{Id}_r(A)$, $\mathrm{Id}_c(A)$, $\mathrm{Id}_j(A)$, $\mathrm{Id}(A)$ the sets of all left ideals, right ideals, lateral ideals, two-sided ideals, ideals of the ternary semigroup $(A,f)$, respectively.

It is easy to prove the following

**Proposition 28.2.** *Let $(A,f)$ be a ternary semigroup. If $\{I_t\}_{t\in T}$ is a nonempty family of ideals [ left ideals, right ideals, lateral ideals, two-sided ideals] of the ternary semigroup $(A,f)$ indexed by $t\in T$ such that*

$$\bigcap_{t\in T} I_t \neq \emptyset,$$

*then*

$$\bigcap_{t\in T} I_t$$

*is an ideal [a left ideal, right ideal, lateral ideal, two-sided ideal] of the ternary semigroup $(A,f)$.*

In view of Proposition 28.2, we can define ideals generated by sets. Let $(A,f)$ be a ternary semigroup. Let $X\subseteq A$ be a nonempty set. Consider the following sets:

$$I_l(X) = \bigcap\{I\in \mathrm{Id}_l(A) : X\subseteq I\},$$
$$I_r(X) = \bigcap\{I\in \mathrm{Id}_r(A) : X\subseteq I\},$$
$$I_c(X) = \bigcap\{I\in \mathrm{Id}_c(A) : X\subseteq I\},$$
$$I_j(X) = \bigcap\{I\in \mathrm{Id}_j(A) : X\subseteq I\},$$
$$I(X) = \bigcap\{I\in \mathrm{Id}(A) : X\subseteq I\}.$$

Of course the sets

$$I_l(X), \quad I_r(X), \quad I_c(X), \quad I_j(X), \quad I(X)$$

are suitable ideals of $(A,f)$ called, respectively, *a left ideal, right ideal, lateral ideal, two-sided ideal and an ideal of the ternary semigroup $(A,f)$ generated by the set $X$.*

Let $(A,f)$ be a ternary semigroup and $a\in A$. Throughout this paper we shall often write $a$ instead of $\{a\}$.

The sets

$$I_l(a), \quad I_r(a), \quad I_c(a), \quad I_j(a), \quad I(a)$$

are called, respectively, *the principal left ideal, right ideal, lateral ideal, two-sided ideal, ideal of the ternary semigroup $(A,f)$ generated by $a$.*

A straightforward reasoning yields the following

**Proposition 28.3.** *Let $(A,f)$ be a ternary semigroup. For each element $a\in A$ we have:*

(a) $I_l(a) = a\cup f(A,A,a)$;
(b) $I_r(a) = a\cup f(a,A,A)$;

(c) $I_c(a) = a \cup f(A,a,A) \cup f(A^2,a,A^2)$;
(d) $I_j(a) = a \cup f(A,A,a) \cup f(a,A,A) \cup f(A^2,a,A^2)$;
(e) $I(a) = a \cup f(A,A,a) \cup f(a,A,A) \cup f(A,a,A) \cup f(A^2,a,A^2)$.

The notion of ideals lead naturally to the consideration of certain equivalence relations on ternary semigroups. These equivalences, first studied by Green (1951) on binary semigroups [30], are very strong tools in the theory of ternary semigroups.

**Definition 28.12.** Let $(A, f)$ be a ternary semigroup. We define the following relations on the set $A$:

(a) $a\,L\,b \iff I_l(a) = I_l(b)$;
(b) $a\,R\,b \iff I_r(a) = I_r(b)$;
(c) $a\,C\,b \iff I_c(a) = I_c(b)$;
(d) $a\,J\,b \iff I_j(a) = I_j(b)$;
(e) $a\,T\,b \iff I(a) = I(b)$;
(f) $H = L \cap R$;
(h) $D = L \circ R$.

An immediate consequence of Definition 28.12 is the following

**Corollary 28.3.** *Let $(A, f)$ be a ternary semigroup. The relations L, R, C, J, T, H are equivalence relations on the set A.*

**Proposition 28.4.** *Let $(A, f)$ be a ternary semigroup. Then*

$$aLb \iff (a = b \vee \exists x_1,x_2,y_1,y_2 \in A\,[b = f(x_1,x_2,a) \wedge a = f(y_1,y_2,b)])$$

*for $a,b \in A$.*

*Proof.* Assume that $a,b \in A$. From the definition of the relation $L$ it follows that

$$aLb \iff a \cup f(A,A,a) = b \cup f(A,A,b).$$

Suppose that $aLb$. Since $b \in a \cup f(A,A,a)$, it follows that either $b = a$ or $b = f(x_1,x_2,a)$ for some $x_1,x_2 \in A$. Similarly, since $a \in b \cup f(A,A,b)$, it follows that either $a = b$ or $a = f(y_1,y_2,b)$ for some $y_1,y_2 \in A$.

Conversely, assume that $a = b$ or $b = f(x_1,x_2,a)$ and $a = f(y_1,y_2,b)$ for some $x_1,x_2,y_1,y_2 \in A$. The case $a = b$ is evident. Consider the second case. Take $z \in a \cup f(A,A,a)$. If $z = a$, then

$$z = f(y_1,y_2,b) \in f(A,A,b).$$

If $z \in f(A,A,a)$, then $z = f(z_1,z_2,a)$ for some $z_1,z_2 \in A$. Therefore

$$z = f(z_1,z_2,f(y_1,y_2,b)) = f(f(z_1,z_2,y_1),y_2,b) \in f(A,A,b).$$

Thus

$$a \cup f(A,A,a) \subseteq b \cup f(A,A,b).$$

Similarly

$$b \cup f(A,A,b) \subseteq a \cup f(A,A,a).$$

□

**Proposition 28.5.** *The relation L is a right congruence on the ternary semigroup $(A,f)$.*

*Proof.* We have to show that

$$\forall a,b,z_1,z_2 \in A\,[aLb \Longrightarrow f(a,z_1,z_2)Lf(b,z_1,z_2)].$$

Assume that $aLb$, where $a,b \in A$. From Proposition 28.4 it follows that $a=b$ or $b=f(x_1,x_2,a)$ and $a=f(y_1,y_2,b)$ for some $x_1,x_2,y_1,y_2 \in A$. The case when $a=b$ is evident. In the second case we get

$$f(b,z_1,z_2) = f(f(x_1,x_2,a),z_1,z_2) = f(x_1,x_2,f(a,z_1,z_2)),$$
$$f(a,z_1,z_2) = f(f(y_1,y_2,b),z_1,z_2) = f(y_1,y_2,f(b,z_1,z_2)).$$

Hence

$$f(a,z_1,z_2)Lf(b,z_1,z_2)$$

by Proposition 28.4. □

A similar argument applied to the relation $R$ yields the following two assertions:

**Proposition 28.6.** *Let $(A,f)$ be a ternary semigroup. Then*

$$aRb \Longleftrightarrow (a=b \vee \exists x_1,x_2,y_1,y_2 \in A\,[b=f(a,x_1,x_2) \wedge a=f(b,y_1,y_2)])$$

*for $a,b \in A$.*

**Proposition 28.7.** *The relation R is a left congruence on the ternary semigroup $(A,f)$.*

**Proposition 28.8.** *Let $(A,f)$ be a ternary semigroup. Then*

$$\begin{aligned} aCb \Longleftrightarrow (a=b \vee \exists x_1,x_2,x_3,x_4,y_1,y_2,y_3,y_4 \in A \\ [(b=f(x_1,a,x_2) \wedge a=f(y_1,b,y_2)) \\ \vee\ (b=f(x_1,a,x_2) \wedge a=f(y_1,f(y_2,b,y_3),y_4)) \\ \vee\ (b=f(x_1,f(x_2,a,x_3),x_4) \wedge a=f(y_1,b,y_2)) \\ \vee\ (b=f(x_1,f(x_2,a,x_3),x_4) \wedge a=f(y_1,f(y_2,b,y_3),y_4))]) \end{aligned}$$

*for $a,b \in A$.*

A straightforward calculation yields the proof of this proposition.

**Theorem 28.5.** *Let $(A, f)$ be a ternary semigroup. Then*

$$L \circ R = R \circ L.$$

*Proof.* By the definition of the composition of relations we get:

$$(a,b) \in L \circ R \Longleftrightarrow \exists c \in A \, [aRc \wedge cLb]$$

for $a, b \in A$.

Assume that $(a,b) \in L \circ R$, where $a, b \in A$. There exists $c \in A$ such that $aRc$ and $cLb$. Thus, in view of Propositions 28.4 and 28.6 we obtain:

$$aRc \Longleftrightarrow (a = c \vee (\exists x_1, x_2, y_1, y_2 \in A \, [c = f(a, x_1, x_2) \wedge a = f(c, y_1, y_2)])),$$

$$cLb \Longleftrightarrow (c = b \vee (\exists u_1, u_2, w_1, w_2 \in A \, [b = f(u_1, u_2, c) \wedge c = f(w_1, w_2, b)])).$$

If $a = c$, then $aLb$. Since $bRb$, it follows that $(a,b) \in R \circ L$. If $c = b$, then $aRb$. Since $aLa$, it follows that $(a,b) \in R \circ L$. Put $d = f(u_1, u_2, a)$. Hence

$$d = f(u_1, u_2, f(c, y_1, y_2)) = f(f(u_1, u_2, c), y_1, y_2) = f(b, y_1, y_2),$$

and so $d = f(b, y_1, y_2)$. Notice that

$$f(w_1, w_2, d) = f(w_1, w_2, f(b, y_1, y_2)) = f(f(w_1, w_2, b), y_1, y_2) = f(c, y_1, y_2) = a,$$

and so $a = f(w_1, w_2, d)$. Hence, $aLd$. Furthermore

$$f(d, x_1, x_2) = f(f(u_1, u_2, a), x_1, x_2) = f(u_1, u_2, f(a, x_1, x_2)) = f(u_1, u_2, c) = b,$$

and so $b = f(d, x_1, x_2)$. Hence, $dRb$. Thus, $(a,b) \in R \circ L$. Then the inclusion

$$L \circ R \subseteq R \circ L$$

is satisfied. Similarly, we can prove the inclusion $R \circ L \subseteq L \circ R$. □

In view of Theorem 28.5 we get the following

**Corollary 28.4.** *Let $(A, f)$ be a ternary semigroup. The relation $D$ is an equivalence relation on the set $A$.*

A consequence of Corollary 28.3 is the following

**Corollary 28.5.** *Let $(A, f)$ be a ternary semigroup. Then $L \subseteq D$, $R \subseteq D$ and $H \subseteq D$.*

**Proposition 28.9.** *Let $(A, f)$ be a ternary semigroup. Then $L \subseteq J$, $R \subseteq J$ and $H \subseteq J$.*

*Proof.* First, we prove that $L \subseteq J$. Assume that $(a,b) \in L$. In view of Proposition 28.4, $a = b$ or $b = f(x_1,x_2,a)$ and $a = f(y_1,y_2,b)$ for some $x_1,x_2,y_1,y_2 \in A$. If $a = b$, then $(a,b) \in J$. Suppose that $b = f(x_1,x_2,a)$ and $a = f(y_1,y_2,b)$. We shall show that

$$a \cup f(A,A,a) \cup f(a,A,A) \cup f(A^2,a,A^2)$$
$$= b \cup f(A,A,b) \cup f(b,A,A) \cup f(A^2,b,A^2).$$

By Proposition 28.4,

$$a \cup f(A,A,a) = b \cup f(A,A,b).$$

If $z \in f(a,A,A)$, then $z = f(a,z_1,z_2)$ for some $z_1,z_2 \in A$. Hence

$$z = f(f(y_1,y_2,b),z_1,z_2) = f(y_1,y_2,b,z_1,z_2) \in f(A^2,b,A^2).$$

Consequently

$$f(a,A,A) \subseteq f(A^2,b,A^2).$$

If $z \in f(A^2,a,A^2)$, then

$$z = f(z_1,z_2,a,z_3,z_4)$$

for some $z_1,z_2,z_3,z_4 \in A$. Hence

$$z = f(z_1,z_2,f(y_1,y_2,b),z_3,z_4) = f(f(z_1,z_2,y_1),y_2,b,z_3,z_4) \in f(A^2,b,A^2).$$

Thus

$$f(A^2,a,A^2) \subseteq f(A^2,b,A^2).$$

Consequently

$$I_j(a) \subseteq I_j(b).$$

Similarly, we can prove that $I_j(b) \subseteq I_j(a)$. Therefore, $I_j(a) = I_j(b)$, whence $(a,b) \in J$. So it follows that $L \subseteq J$.

Similar argument leads to the inclusion $R \subseteq J$. Clearly, $H \subseteq J$. □

## 28.5 A Ternary Semigroup of Mappings

In this section, we shall deal with the ternary semigroup of mappings of two sets [13], which is a counterpart of the semigroup of mappings of a set. To this end, we shall use the Green's relations. Moreover, a certain clear characterization of the structure of inverses in the ternary semigroup of mappings will be presented.

The similar considerations can be applied to the ternary semigroups of linear mappings and matrices [19].

Let $X$ and $Y$ be nonempty sets, $T(X,Y)$ be the set of all mappings of $X$ into $Y$ and $T[X,Y] = T(X,Y) \times T(Y,X)$. Define the ternary operation $f : T[X,Y]^3 \longrightarrow T[X,Y]$ by

$$f((p_1,q_1),(p_2,q_2),(p_3,q_3)) = (p_1 \circ q_2 \circ p_3, q_1 \circ p_2 \circ q_3)$$

for all $(p_i,q_i) \in T[X,Y]$, where $i = 1,2,3$.

By Proposition 28.1, $(T[X,Y],f)$ is a ternary semigroup.

**Definition 28.13.** The ternary semigroup $(T[X,Y],f)$ is called the ternary semigroup of mappings of sets $X$ and $Y$.

Suppose that $(p,q),(p',q') \in T[X,Y]$. Let us put

$$\mathrm{Im}(p,q) = (\mathrm{Im}(p),\mathrm{Im}(q)).$$

We write $\mathrm{Im}(p,q) \subseteq \mathrm{Im}(p',q')$ if and only if $\mathrm{Im}(p) \subseteq \mathrm{Im}(p')$ and $\mathrm{Im}(q) \subseteq \mathrm{Im}(q')$.

**Theorem 28.6.** *Assume that $(p,q),(p',q') \in T[X,Y]$. Then $\mathrm{Im}(p,q) \subseteq \mathrm{Im}(p',q')$ if and only if there exist $(p_1,q_1),(p_2,q_2) \in T[X,Y]$ such that*

$$(p,q) = f((p',q'),(p_1,q_1),(p_2,q_2)). \tag{28.8}$$

*Proof.* If the condition (28.8) holds, then $p = p' \circ q_1 \circ p_2$ and $q = q' \circ p_1 \circ q_2$. Thus, $\mathrm{Im}(p) \subseteq \mathrm{Im}(p')$ and $\mathrm{Im}(q) \subseteq \mathrm{Im}(q')$, and therefore $\mathrm{Im}(p,q) \subseteq \mathrm{Im}(p',q')$.

Conversely, assume that $\mathrm{Im}(p,q) \subseteq \mathrm{Im}(p',q')$. First, we will prove that there exist $q_1 \in T(Y,X)$ and $p_2 \in T(X,Y)$ such that $p = p' \circ q_1 \circ p_2$. Put $\alpha = \mathrm{Ker}(p)$ and $\alpha' = \mathrm{Ker}(p')$. Since $\mathrm{Im}(p) \subseteq \mathrm{Im}(p')$, it follows that for every $x \in X$ there exists an $x' \in X$ such that $p'(x') = p(x)$. Therefore, we can define the mapping $h : X/\alpha \longrightarrow X/\alpha'$ by the formula:

$$h([x]_\alpha) = [x']_{\alpha'} \iff p(x) = p'(x')$$

for $[x]_\alpha \in X/\alpha$. Notice that for $x'_1 \in X$ such that $p'(x'_1) = p'(x')$ we have

$$[x'_1]_{\alpha'} = [x']_{\alpha'}.$$

If $x_1 \in [x]_\alpha$, then $p(x_1) = p(x)$. Thus, the mapping $h$ is well-defined. Consider the choice function $w : X/\alpha' \longrightarrow X$. Notice that

$$(p' \circ w \circ h)([x]_\alpha) = p(x)$$

for $x \in X$. Consider the bijection

$$g : X/\alpha \longrightarrow \mathrm{Im}(p)$$

defined by

$$g([x]_\alpha) = p(x)$$

for all $[x]_\alpha \in X/\alpha$. Let $g_1 : Y \longrightarrow X/\alpha$ be a mapping such that

$$g_1|_{\mathrm{Im}(p)} = g^{-1}.$$

Put $q_1 = w \circ h \circ g_1$. Thus, $p' \circ q_1 \circ p = p$. Taking $p_2 = p$ we get $p' \circ q_1 \circ p_2 = p$. Similarly, there exist $p_1 \in T(X,Y)$ and $q_2 \in T(Y,X)$ such that $q' \circ p_1 \circ q_2 = q$. Thus, (28.8) holds. □

Suppose that $(p,q),(p',q') \in T[X,Y]$. We set $\mathrm{Ker}(p,q) = (\mathrm{Ker}(p), \mathrm{Ker}(q))$ and write $\mathrm{Ker}(p,q) \subseteq \mathrm{Ker}(p',q')$ if and only if $\mathrm{Ker}(p) \subseteq \mathrm{Ker}(p')$ and $\mathrm{Ker}(q) \subseteq \mathrm{Ker}(q')$.

**Theorem 28.7.** *Assume that* $(p,q),(p',q') \in T[X,Y]$. *Then* $\mathrm{Ker}(p',q') \subseteq \mathrm{Ker}(p,q)$ *if and only if there exist* $(p_1,q_1),(p_2,q_2) \in T[X,Y]$ *such that*

$$(p,q) = f((p_1,q_1),(p_2,q_2),(p',q')). \tag{28.9}$$

*Proof.* If the condition (28.9) holds, then $p = p_1 \circ q_2 \circ p'$ and $q = q_1 \circ p_2 \circ q'$. Thus, $\mathrm{Ker}(p') \subseteq \mathrm{Ker}(p)$ and $\mathrm{Ker}(q') \subseteq \mathrm{Ker}(q)$, and therefore

$$\mathrm{Ker}(p',q') \subseteq \mathrm{Ker}(p,q).$$

Conversely, assume that $\mathrm{Ker}(p',q') \subseteq \mathrm{Ker}(p,q)$. First, we will prove that there exist $p_1 \in T(X,Y)$ and $q_2 \in T(Y,X)$ such that $p = p_1 \circ q_2 \circ p'$. Put $\alpha' = \mathrm{Ker}(p')$. Consider the bijection $g : X/\alpha' \longrightarrow \mathrm{Im}(p')$ defined by

$$g([x]_{\alpha'}) = p'(x)$$

for $[x]_{\alpha'} \in X/\alpha'$. Let us take the choice function $w : X/\alpha' \longrightarrow X$. Notice that for every $x' \in [x]_{\alpha'} \in X/\alpha'$ we have $p'(x') = p'(x)$, and so

$$(x',x) \in \mathrm{Ker}(p') \subseteq \mathrm{Ker}(p),$$

hence $p(x') = p(x)$. Thus

$$p(w([x]_{\alpha'})) = p(x)$$

for all $[x]_{\alpha'} \in X/\alpha'$.

Consider a mapping $g_1 : Y \longrightarrow X/\alpha'$ such that

$$g_1|_{\mathrm{Im}(p')} = g^{-1}.$$

Put $q_2 = w \circ g_1$. Then $p \circ q_2 \circ p' = p$. Setting $p_1 = p$ we get

$$p = p_1 \circ q_2 \circ p'.$$

Similarly, there exist $q_1 \in T(Y,X)$ and $p_2 \in T(X,Y)$ such that $q = q_1 \circ p_2 \circ q'$. Thus (28.9) holds. □

Suppose that $p \in T(X,Y)$. Put

$$r(p) = \text{card}(\text{Im}(p)).$$

Assume that $(p,q),(p',q') \in T[X,Y]$. Let us set $r(p,q) = (r(p),r(q))$. We write $r(p,q) \leq r(p',q')$ if and only if $r(p) \leq r(p')$ and $r(q) \leq r(q')$.

**Theorem 28.8.** *Assume that $(p,q),(p',q') \in T[X,Y]$. Then $r(p,q) \leq r(p',q')$ if and only if there exist $(p_i,q_i) \in T[X,Y] (i = 1,\dots,4)$ such that*

$$(p,q) = f((p_1,q_1),(p_2,q_2),(p',q'),(p_3,q_3),(p_4,q_4)). \tag{28.10}$$

*Proof.* If the condition (28.10) holds, then

$$p = p_1 \circ q_2 \circ p' \circ q_3 \circ p_4$$

and

$$q = q_1 \circ p_2 \circ q' \circ p_3 \circ q_4.$$

Thus, $r(p) \leq r(p')$ and $r(q) \leq r(q')$, hence

$$r(p,q) \leq r(p',q').$$

Conversely, suppose that $r(p,q) \leq r(p',q')$. First, we will prove that there are $p_1, p_4 \in T(X,Y)$, $q_2, q_3 \in T(Y,X)$ with $p = p_1 \circ q_2 \circ p' \circ q_3 \circ p_4$. Since $r(p) \leq r(p')$, there exists an injection

$$s : \text{Im}(p) \longrightarrow \text{Im}(p').$$

Put $h = s \circ p$. Notice that $\text{Ker}(h) = \text{Ker}(p)$. Applying an argument similar to that in the proof of Theorem 28.7 we infer that there exist $p_1 \in T(X,Y)$ and $q_2 \in T(Y,X)$ such that $p = p_1 \circ q_2 \circ h$. Since $\text{Im}(h) \subseteq \text{Im}(p')$, using a method similar to that in the proof of Theorem 28.6 we get that there exist $p_4 \in T(X,Y)$ and $q_3 \in T(Y,X)$ such that $h = p' \circ q_3 \circ p_4$. Hence

$$p = p_1 \circ q_2 \circ p' \circ q_3 \circ p_4.$$

Similarly, there exist $p_2, p_3 \in T(X,Y)$ and $q_1, q_4 \in T(Y,X)$ such that

$$q = q_1 \circ p_2 \circ q' \circ p_3 \circ q_4.$$

Therefore, we have obtained (28.10). □

Assume that $(p,q),(p',q') \in T[X,Y]$. Set $r(p,q) \leq^* r(p',q')$ if and only if $r(p) \leq r(q')$ and $r(q) \leq r(p')$.

**Theorem 28.9.** *Assume that* $(p,q),(p',q') \in T[X,Y]$. *Then* $r(p,q) \leq^* r(p',q')$ *if and only if there exist* $(p_1,q_1),(p_2,q_2) \in T[X,Y]$ *such that*

$$(p,q) = f((p_1,q_1),(p',q'),(p_2,q_2)). \tag{28.11}$$

*Proof.* If the condition (28.11) holds, then

$$p = p_1 \circ q' \circ p_2$$

and

$$q = q_1 \circ p' \circ q_2.$$

Thus, $r(p) \leq r(q')$ and $r(q) \leq r(p')$, whence

$$r(p,q) \leq^* r(p',q').$$

Conversely, assume that

$$r(p,q) \leq^* r(p',q').$$

First, we prove that there exist $p_1,p_2 \in T(X,Y)$ such that $p = p_1 \circ q' \circ p_2$. Put $\alpha = \mathrm{Ker}(p)$ and $\beta' = \mathrm{Ker}(q')$. Since $r(p) = \mathrm{card}(X/\alpha)$ and $r(q') = \mathrm{card}(Y/\beta')$, it follows that

$$\mathrm{card}(X/\alpha) \leq \mathrm{card}(Y/\beta').$$

We define a mapping $h : X \longrightarrow Y/\beta'$ such that $\mathrm{Ker}(h) \subseteq \mathrm{Ker}(p)$.

Consider the two cases. Assume that

$$\mathrm{card}(X) > \mathrm{card}(Y/\beta').$$

Let $X_0$ be a subset of $X$ which has one and only one element in common with each equivalence class belonging to $X/\alpha$. Hence $\mathrm{card}(X_0) \leq \mathrm{card}(Y/\beta')$. Therefore, there exists an injection $b : X_0 \longrightarrow Y/\beta'$. Define the mapping $h : X \longrightarrow Y/\beta'$ by

$$h(x) = b(x_0)$$

for $x \in X$, where $x_0 \in [x]_\alpha \cap X_0$. Hence, $\mathrm{Ker}(h) = \mathrm{Ker}(p)$.

Next, consider the case

$$\mathrm{card}(X) \leq \mathrm{card}(Y/\beta').$$

Then there exists an injection $h : X \longrightarrow Y/\beta'$. Hence $\mathrm{Ker}(h) \subseteq \mathrm{Ker}(p)$. Fix a choice function $w : Y/\beta' \longrightarrow Y$. Put

$$g = q' \circ w \circ h.$$

Then $g : X \longrightarrow \mathrm{Im}(q')$. Notice that

$$\mathrm{Ker}(g) \subseteq \mathrm{Ker}(p).$$

Indeed, assume that $(x_1,x_2) \in \mathrm{Ker}(g)$. Then $g(x_1) = g(x_2)$ for $x_1,x_2 \in X$. Consequently, $q'(w(h(x_1))) = q'(w(h(x_2)))$, and so

$$[w(h(x_1))]_{\beta'} = [w(h(x_1))]_{\beta'}.$$

Hence, $w(h(x_1)) = w(h(x_2))$, which implies $h(x_1) = h(x_2)$. Therefore, $(x_1,x_2) \in \mathrm{Ker}(h) \subseteq \mathrm{Ker}(p)$. Define the mapping $s_1 : \mathrm{Im}(g) \longrightarrow X/\alpha$ by

$$s_1(g(x)) = [x]_\alpha$$

for every $g(x) \in \mathrm{Im}(g)$. Since $\mathrm{Ker}(g) \subseteq \mathrm{Ker}(p) = \alpha$, it follows that $s_1$ is well–defined. It is obvious that the mapping $s_1$ is a surjection. Let $s : X \longrightarrow X/\alpha$ be a surjection such that $s|_{\mathrm{Im}(g)} = s_1$. Fix a choice function $w_1 : X/\alpha \longrightarrow X$. Notice that

$$(p \circ w_1 \circ s \circ q' \circ w \circ h)(x) = p(x)$$

for every $x \in X$. Setting

$$p_1 = p \circ w_1 \circ s$$

and

$$p_2 = w \circ h$$

we have

$$p = p_1 \circ q' \circ p_2.$$

Similarly, there exist $q_1,q_2 \in T(Y,X)$ such that $q = q_1 \circ p' \circ q_2$. Thus, (28.11) holds. □

In view of Theorem 28.6, Proposition 28.3, and Definition 28.12 we can formulate the following

**Corollary 28.6.** *Assume that* $(p,q),(p',q') \in T[X,Y]$. *Then*

(a) $(p',q') \in I_r(p,q) \Longleftrightarrow \mathrm{Im}(p',q') \subseteq \mathrm{Im}(p,q)$;
(b) $I_r(p,q) = I_r(p',q') \Longleftrightarrow \mathrm{Im}(p,q) = \mathrm{Im}(p',q')$;
(c) $(p,q)\, R\, (p',q') \Longleftrightarrow \mathrm{Im}(p,q) = \mathrm{Im}(p',q')$.

According to Theorem 28.7, Proposition 28.3, and Definition 28.12 we obtain the following

**Corollary 28.7.** *Assume that* $(p,q),(p',q') \in T[X,Y]$. *Then*

(a) $(p',q') \in I_l(p,q) \Longleftrightarrow \mathrm{Ker}(p,q) \subseteq \mathrm{Ker}(p',q')$;
(b) $I_l(p,q) = I_l(p',q') \Longleftrightarrow \mathrm{Ker}(p,q) = \mathrm{Ker}(p',q')$;
(c) $(p,q)\, L\, (p',q') \Longleftrightarrow \mathrm{Ker}(p,q) = \mathrm{Ker}(p',q')$.

Corollaries 28.6, 28.7, and Definition 28.12 imply the following

**Corollary 28.8.** *If* $(p,q),(p',q') \in T[X,Y]$, *then* $(p,q)\ H\ (p',q')$ *if and only if*

$$\mathrm{Ker}(p,q) = \mathrm{Ker}(p',q') \quad \textit{and} \quad \mathrm{Im}(p,q) = \mathrm{Im}(p',q').$$

**Theorem 28.10.** *If* $(p,q),(p',q') \in T[X,Y]$, *then* $(p',q') \in I_j(p,q)$ *if and only if*

$$r(p',q') \leq r(p,q).$$

*Proof.* Assume that $(p',q') \in I_j(p,q)$. According to Proposition 28.3(d) we consider the following four cases.

(a) If $(p',q') = (p,q)$, then $r(p',q') = r(p,q)$.

(b) Suppose that
$$(p',q') = f((p_1,q_1),(p_2,q_2),(p,q))$$
for some $(p_1,q_1),(p_2,q_2) \in T[X,Y]$. It follows from Theorem 28.7 that
$$\mathrm{Ker}(p) \subseteq \mathrm{Ker}(p') \quad \text{and} \quad \mathrm{Ker}(q) \subseteq \mathrm{Ker}(q').$$
Therefore
$$r(p') = \mathrm{card}(X/\mathrm{Ker}(p')) \leq \mathrm{card}(X/\mathrm{Ker}(p)) = r(p).$$
Similarly, $r(q') \leq r(q)$. Hence, $r(p',q') \leq r(p,q)$.

(c) Suppose that
$$(p',q') = f((p,q),(p_1,q_1),(p_2,q_2))$$
for some $(p_1,q_1),(p_2,q_2) \in T[X,Y]$. By Theorem 28.6, $\mathrm{Im}(p',q') \subseteq \mathrm{Im}(p,q)$, and therefore $r(p',q') \leq r(p,q)$.

(d) Suppose that
$$(p',q') = f((p_1,q_1),(p_2,q_2),(p,q),(p_3,q_3),(p_4,q_4))$$
for some $(p_i,q_i) \in T[X,Y]$, where $i = 1,\ldots,4$. By Theorem 28.8, $r(p',q') \leq r(p,q)$.

Conversely, assume that $r(p',q') \leq r(p,q)$. Then, in view of Theorem 28.8, we deduce that $(p',q') \in I_j(p,q)$. □

According to Theorem 28.10 and Definition 28.12(d) we have the following

**Corollary 28.9.** *Assume that* $(p,q),(p',q') \in T[X,Y]$. *Then*

(a) $I_j(p,q) = I_j(p',q') \Longleftrightarrow r(p,q) = r(p',q')$;

(b) $(p,q)\ J\ (p',q') \Longleftrightarrow r(p,q) = r(p',q')$.

Corollaries 28.6, 28.7, 28.9 imply the following

**Corollary 28.10.** *The relations* $L, R, H, D, J$ *in the ternary semigroup* $T[X,Y]$ *satisfy the following set-inclusions :* $L \subseteq J$, $R \subseteq J, H \subseteq J$, $D \subseteq J$.

**Proposition 28.10.** *The relations* $D$ *and* $J$ *in the ternary semigroup* $T[X,Y]$ *are identical.*

*Proof.* In view of Corollary 28.10, it is enough to prove that $J \subseteq D$. Suppose that $(p,q)\, J\, (p',q')$ for $(p,q),(p',q') \in T[X,Y]$. This means that $r(p,q) = r(p',q')$. Since $r(p) = r(p')$, there exists a bijection

$$b : \mathrm{Im}(p) \longrightarrow \mathrm{Im}(p').$$

Put $p_1 = b \circ p$. Notice that $\mathrm{Ker}(p_1) = \mathrm{Ker}(p)$ and $\mathrm{Im}(p_1) = \mathrm{Im}(p')$. Similarly, one can construct the mapping $q_1 \in T(Y,X)$ such that $\mathrm{Ker}(q_1) = \mathrm{Ker}(q)$ and $\mathrm{Im}(q_1) = \mathrm{Im}(q')$. Thus, $(p,q)\, L\, (p_1,q_1)$ and $(p_1,q_1)\, R\, (p',q')$, and so $(p,q)\, D\, (p',q')$. □

According to Proposition 28.10 and Corollary 28.9 we obtain the following

**Corollary 28.11.** *If* $(p,q),(p',q') \in T[X,Y]$, *then* $(p,q)\, D\, (p',q')$ *if and only if*

$$r(p,q) = r(p',q').$$

The next result is an immediate consequence of Proposition 28.3 and Theorems 28.8 and 28.9.

**Theorem 28.11.** *If* $(p,q),(p',q') \in T[X,Y]$, *then* $(p',q') \in I_c(p,q)$ *if and only if*

$$r(p',q') \leq r(p,q) \quad or \quad r(p',q') \leq^* r(p,q).$$

Assume that $(p,q),(p',q') \in T[X,Y]$. Notice that

$$r(p,q) \leq^* r(p',q') \quad \text{and} \quad r(p',q') \leq^* r(p,q)$$

if and only if

$$r(p) = r(q') \quad \text{and} \quad r(q) = r(p').$$

Therefore, we set

$$r(p,q) \stackrel{*}{=} r(p',q')$$

if and only if

$$r(p) = r(q') \quad \text{and} \quad r(q) = r(p').$$

**Lemma 28.1.** *If* $r(p,q) \leq r(p',q')$ *and* $r(p',q') \leq^* r(p,q)$, *then* $r(p,q) = r(p',q')$.

*Proof.* Since

$$r(p') \leq r(q) \leq r(q') \leq r(p) \quad \text{and} \quad r(q') \leq r(p) \leq r(p') \leq r(q),$$

it follows that $r(p',q') \leq r(p,q)$. Consequently, $r(p,q) = r(p',q')$. □

**Theorem 28.12.** *If $(p,q),(p',q') \in T[X,Y]$, then $I_c(p,q) = I_c(p',q')$ if and only if*

$$r(p,q) = r(p',q') \quad or \quad r(p,q) \stackrel{*}{=} r(p',q').$$

*Proof.* We have $I_c(p,q) = I_c(p',q')$ if and only if

$$(p,q) \in I_c(p',q') \quad \text{and} \quad (p',q') \in I_c(p,q).$$

In view of Theorem 28.11 and Lemma 28.1, applying a straightforward calculation we get the desired result. □

The following corollary follows from Definition 28.12(c) and Theorem 28.12.

**Corollary 28.12.** *If $(p,q),(p',q') \in T[X,Y]$, then $(p,q)\ C\ (p',q')$ if and only if*

$$r(p,q) = r(p',q') \quad or \quad r(p,q) \stackrel{*}{=} r(p',q').$$

By Proposition 28.3(e) and Theorem 28.9, applying an argument similar to that in the proof of Theorem 28.10, we get the following result.

**Theorem 28.13.** *If $(p,q),(p',q') \in T[X,Y]$, then $(p',q') \in I(p,q)$ if and only if*

$$r(p',q') \leq r(p,q) \quad or \quad r(p',q') \leq^* r(p,q).$$

**Proposition 28.11.** *If $(p,q) \in T[X,Y]$, then $I(p,q) = I_c(p,q)$.*

The proof of the proposition above follows from Theorems 28.11 and 28.13.
By Proposition 28.11, Theorem 28.12 and Definition 28.12(e) the following corollary is valid.

**Corollary 28.13.** *If $(p,q),(p',q') \in T[X,Y]$, then*

(a) $I(p,q) = I(p',q') \Longleftrightarrow [r(p,q) = r(p',q') \vee r(p,q) \stackrel{*}{=} r(p',q')]$;
(b) $(p,q)\ T\ (p',q') \Longleftrightarrow [r(p,q) = r(p',q') \vee r(p,q) \stackrel{*}{=} r(p',q')]$.

Corollaries 28.12 and 28.13 yield

**Corollary 28.14.** *The relations $C$ and $T$ in the ternary semigroup $T[X,Y]$ are identical.*

**Corollary 28.15.** *The relations $C$ and $D$ in the ternary semigroup $T[X,Y]$ satisfy the set–inclusion $D \subseteq C$.*

This statement follows from Corollaries 28.11 and 28.12.

**Theorem 28.14.** *If $(p,q) \in T[X,Y]$ and $r(p) = r(q)$, then $C(p,q) = D(p,q)$.*

*Proof.* By Corollary 28.15 we have $D(p,q) \subseteq C(p,q)$. Suppose that $(p',q') \in C(p,q)$. If $r(p',q') = r(p,q)$, then $(p',q') \in D(p,q)$. If $r(p',q') \stackrel{*}{=} r(p,q)$, then $r(p) = r(q) = r(p') = r(q')$, and so $r(p',q') = r(p,q)$. This means that $(p',q') \in D(p,q)$. □

**Lemma 28.2.** *Assume that* $(p,q) \in T[X,Y]$ *and* $r(p) \neq r(q)$. *Then there exists a pair of mappings* $(p',q') \in T[X,Y]$ *such that*

(a) $r(p,q) \neq r(p',q')$;
(b) $r(p,q) \stackrel{*}{=} r(p',q')$.

*Proof.* First, we will construct $p' \in T(X,Y)$ such that $r(p') = r(q)$. Put $\mathrm{Ker}(q) = \beta$. There exists a surjection $g : X \longrightarrow Y/\beta$. Fix a choice function $w : Y/\beta \longrightarrow Y$. Put $p' = w \circ g$. Notice that $\mathrm{Im}(p') = \mathrm{Im}(w)$. Therefore,

$$r(p') = \mathrm{card}(\mathrm{Im}(w)) = \mathrm{card}(Y/\beta) = r(q).$$

Hence, and by the assumption, $r(p') \neq r(p)$. Similarly, one can construct $q' \in T(Y,X)$ such that $r(q') = r(p)$. Therefore, the conditions (a) and (b) hold. □

**Theorem 28.15.** *Assume that* $(p,q) \in T[X,Y]$ *and* $r(p) \neq r(q)$. *Then the C-class* $C(p,q)$ *is the union of the two distinct D-classes* $D_1$ *and* $D_2$ *defined by the formulas:*

$$D_1 = \{(p',q') \in T[X,Y] : r(p',q') = r(p,q)\}, \tag{28.12}$$

$$D_2 = \{(p',q') \in T[X,Y] : r(p',q') \stackrel{*}{=} r(p,q)\}. \tag{28.13}$$

*Proof.* Since $r(p) \neq r(q)$, it follows from Lemma 28.2 and Corollary 28.15 that the C-class $C(p,q)$ contains at least two distinct D-classes. Suppose that the C-class $C(p,q)$ contains three pairwise distinct D-classes $D(p_1,q_1), D(p_2,q_2), D(p_3,q_3)$. Then

$$r(p_1,q_1) \stackrel{*}{=} r(p_2,q_2) \quad \text{and} \quad r(p_2,q_2) \stackrel{*}{=} r(p_3,q_3).$$

Consequently

$$r(p_1) = r(q_2), \quad r(q_1) = r(p_2), \quad r(p_2) = r(q_3), \quad r(q_2) = r(p_3),$$

and so $r(p_1) = r(p_3)$ and $r(q_1) = r(q_3)$. Hence, $D(p_1,q_1) = D(p_3,q_3)$. This contradicts our assumption. □

We can extend the notion of an inverse in a (binary) semigroup to the ternary semigroup $T[X,Y]$. A pair $(p',q') \in T[X,Y]$ is an *inverse* of a pair $(p,q) \in T[X,Y]$ if and only if

$$f((p,q),(p',q'),(p,q)) = (p,q) \quad \text{and} \quad f((p',q'),(p,q),(p',q')) = (p',q').$$

**Theorem 28.16.** *For every pair* $(p,q) \in T[X,Y]$ *there exists an inverse*

$$(p',q') \in T[X,Y].$$

*Proof.* Put $\alpha = \mathrm{Ker}(p)$ and $\beta = \mathrm{Ker}(q)$. Define the bijections

$$g_1 : X/\alpha \longrightarrow \mathrm{Im}(p) \quad \text{and} \quad g_2 : Y/\beta \longrightarrow \mathrm{Im}(q)$$

by the formulas:

$$g_1([x]_\alpha) = p(x) \quad \text{and} \quad g_2([y]_\beta) = q(y)$$

for all $[x]_\alpha \in X/\alpha$ and $[y]_\beta \in Y/\beta$. Fix the choice functions

$$w_1 : X/\alpha \longrightarrow X \quad \text{and} \quad w_2 : Y/\beta \longrightarrow Y.$$

Consider the mappings $p_1 = w_2 \circ g_2^{-1}$ and $q_1 = w_1 \circ g_1^{-1}$. Notice that

$$Dom(p_1) = \text{Im}(q) \quad \text{and} \quad Dom(q_1) = \text{Im}(p).$$

Let $p' \in T(X,Y)$ and $q' \in T(Y,X)$ be mappings such that

$$p'|_{\text{Im}(q)} = p_1, \quad \text{Im}(p') = \text{Im}(p_1), \quad q'|_{\text{Im}(p)} = q_1, \quad \text{Im}(q') = \text{Im}(q_1).$$

Since $p \circ q' \circ p = p$ and $q \circ p' \circ q = q$, it follows that $f((p,q),(p',q'),(p,q)) = (p,q)$. For every $x \in X$ there exists an $x_1 \in \text{Im}(q)$ such that $p'(x) = p_1(x_1)$. Hence, there exists an element $y_1 \in Y$ such that $q(y_1) = x_1$. Since

$$p' \circ q \circ p' = p' \quad \text{and} \quad q' \circ p \circ q' = q',$$

it follows that

$$f((p',q'),(p,q),(p',q')) = (p',q').$$

Therefore, $(p',q')$ is an inverse of $(p,q)$ in $T[X,Y]$. □

From Definition 28.8 and Theorem 28.16 we can derive the next result.

**Corollary 28.16.** *The ternary semigroup $T[X,Y]$ is regular.*

Also, observe that immediately from Theorem 28.9 results the subsequent

**Proposition 28.12.** *If $(p',q') \in T[X,Y]$ is an inverse of $(p,q) \in T[X,Y]$, then*

$$r(p,q) \stackrel{*}{=} r(p',q').$$

It is easily seen that Corollary 28.12 and Proposition 28.12 imply

**Corollary 28.17.** *If $(p',q') \in T[X,Y]$ is an inverse of $(p,q) \in T[X,Y]$, then*

$$(p,q)\, C\, (p',q').$$

Assume that

$$E = \{(p,q) \in T[X,Y] : r(p) = r(q)\}$$

and

$$E^* = T[X,Y] \setminus E.$$

From Corollary 28.11 it follows that $D(p,q) \subseteq E$ for every $(p,q) \in E$. Therefore

$$E = \bigcup \{D(p,q) : (p,q) \in E\}.$$

**Proposition 28.13.** *For every C-class $C_0 \subseteq T[X,Y]$ precisely one of the following two conditions holds:*

(a) $C_0 \subseteq E$;
(b) $C_0 \subseteq E^*$.

*Proof.* Suppose that there exists a C-class $C_0 \subseteq T[X,Y]$ such that $(p_1,q_1),(p_2,q_2) \in C_0$, $(p_1,q_1) \in E$, and $(p_2,q_2) \in E^*$ for some $(p_1,q_1),(p_2,q_2) \in T[X,Y]$. From the foregoing and Theorem 28.14 it follows that $C_0 = C(p_1,q_1) = D(p_1,q_1) \subseteq E$. We have obtained a contradiction. □

Summarizing, we get the following theorem.

**Theorem 28.17.** *Given the ternary semigroup $T[X,Y]$.*

(A) *Assume that a C-class $C_0 \subseteq E$. Then every inverse $(p',q')$ of $(p,q) \in C_0$ is an element of the C-class $C_0$ ($C_0$ is a D-class).*
(B) *Assume that a C-class $C_0 \subseteq E^*$. Then $C_0 = D_1 \cup D_2$, where the D-classes $D_1$ and $D_2$ are defined by the formulas (28.12) and (28.13). Every inverse $(p',q')$ of $(p,q) \in D_1$ is an element of the D-class $D_2$. Every inverse $(p',q')$ of $(p,q) \in D_2$ is an element of the D-class $D_1$.*

*Proof.* The assertion (A) is an immediate consequence of Corollary 28.17. Next, we will prove the assertion (B). Assume that $C_0 = C(p_0,q_0)$. Therefore,

$$D_1 = \{(p,q) \in T[X,Y] : r(p,q) = r(p_0,q_0)\}$$

and

$$D_2 = \{(p,q) \in T[X,Y] : r(p,q) \stackrel{*}{=} r(p_0,q_0)\}.$$

Suppose that $(p,q) \in D_1$ and $(p',q')$ is an inverse of $(p,q)$. In view of Proposition 28.12 we get

$$r(p,q) \stackrel{*}{=} r(p',q'),$$

and so

$$r(p_0,q_0) \stackrel{*}{=} r(p',q').$$

Consequently, $(p',q') \in D_2$.

Suppose that $(p,q) \in D_2$ and $(p',q')$ is an inverse of $(p,q)$. By Proposition 28.12, $r(p,q) \stackrel{*}{=} r(p',q')$, and so $r(p_0,q_0) = r(p',q')$. Consequently, $(p',q') \in D_1$. □

## 28.6 Ternary Linear Algebras

The notion of a ternary linear algebra is a counterpart of the notion of a linear algebra. In this section we consider ternary linear algebras of linear mappings and matrices, involutive ternary linear algebras of linear mappings and matrices, a normed ternary linear algebra of continuous linear mappings. On the basis of these examples we formulate the general definitions of a ternary linear algebra, an involutive ternary linear algebra, a normed ternary linear algebra, a topological ternary linear algebra [21].

### *28.6.1 A Ternary Linear Algebra of Linear Mappings*

Let $(X,K,+,\cdot)$ and $(Y,K,+,\cdot)$ be vector spaces over a field $K$. Let $L(X,Y)$ be the set of all linear mappings of the space $X$ into the space $Y$. Of course, $L(Y,X)$ is the set of all linear mappings of the space $Y$ into the space $X$. Let us put

$$L[X,Y] = L(X,Y) \times L(Y,X).$$

Define the ternary operation $f : L[X,Y]^3 \longrightarrow L[X,Y]$ by the formula:

$$f((p_1,q_1),(p_2,q_2),(p_3,q_3)) = (p_1 \circ q_2 \circ p_3,\ q_1 \circ p_2 \circ q_3)$$

for all $(p_i,q_i) \in L[X,Y]$, where $i = 1,2,3$.

The groupoid $(L[X,Y],f)$ is a ternary semigroup (Proposition 28.1).

**Definition 28.14.** The ternary semigroup $(L[X,Y],f)$ is called the ternary semigroup of linear mappings of vector spaces $X$ and $Y$ over a field $K$.

Let $(L(X,Y),K,+,\cdot)$ and $(L(Y,X),K,+,\cdot)$ be vector spaces of linear mappings. Put

$$L[X,Y] = L(X,Y) \times L(Y,X).$$

Consider the Cartesian product $(L[X,Y],K,+,\cdot)$ of the following two vector spaces: $(L(X,Y),K,+,\cdot)$ and $(L(Y,X),K,+,\cdot)$. Let $(L[X,Y],f)$ be the ternary semigroup of linear mappings of vector spaces $X$ and $Y$ over a field $K$. The ternary operation $f$ will be called a *ternary multiplication* of linear mappings.

Assume that $(p_1,q_1)$, $(p_1',q_1')$, $(p_1'',q_1'')$, $(p_2,q_2)$, $(p_2',q_2')$, $(p_2'',q_2'')$, $(p_3,q_3)$, $(p_3',q_3')$, $(p_3'',q_3'') \in L[X,Y]$. We prove that the following conditions are satisfied:

$$\begin{aligned} f((p_1',q_1') + (p_1'',q_1''),(p_2,q_2),(p_3,q_3)) \\ = f((p_1',q_1'),(p_2,q_2),(p_3,q_3)) \\ + f((p_1'',q_1''),(p_2,q_2),(p_3,q_3)), \end{aligned} \tag{28.14}$$

$$
\begin{aligned}
f((p_1,q_1),&(p_2',q_2')+(p_2'',q_2''),(p_3,q_3)) \\
&= f((p_1,q_1),(p_2',q_2'),(p_3,q_3)) \\
&\quad + f((p_1,q_1),(p_2'',q_2''),(p_3,q_3)),
\end{aligned} \tag{28.15}
$$

$$
\begin{aligned}
f((p_1,q_1),&(p_2,q_2),(p_3',q_3')+(p_3'',q_3'')) \\
&= f((p_1,q_1),(p_2,q_2),(p_3',q_3')) \\
&\quad + f((p_1,q_1),(p_2,q_2),(p_3'',q_3'')).
\end{aligned} \tag{28.16}
$$

First, we prove (28.14). Indeed,

$$
\begin{aligned}
f((p_1',q_1')&+(p_1'',q_1''),(p_2,q_2),(p_3,q_3)) \\
&= f((p_1'+p_1'',q_1'+q_1''),(p_2,q_2),(p_3,q_3)) \\
&= ((p_1'+p_1'')\circ q_2\circ p_3,(q_1'+q_1'')\circ p_2\circ q_3) \\
&= (p_1'\circ q_2\circ p_3+p_1''\circ q_2\circ p_3,\ q_1'\circ p_2\circ q_3+q_1''\circ p_2\circ q_3) \\
&= (p_1'\circ q_2\circ p_3,q_1'\circ p_2\circ q_3)+(p_1''\circ q_2\circ p_3,q_1''\circ p_2\circ q_3) \\
&= f((p_1',q_1'),(p_2,q_2),(p_3,q_3))+f((p_1'',q_1''),(p_2,q_2),(p_3,q_3)).
\end{aligned}
$$

We can similarly prove that the conditions (28.15) and (28.16) hold.

When the conditions (28.14)–(28.16) are satisfied, we say, that the ternary multiplication $f$ is *distributive over addition* $+$ of linear mappings in $(L[X,Y],K,+,\cdot)$.

Next, we shall show that the following equalities are fulfilled:

$$
\begin{aligned}
f(a\cdot(p_1,q_1),&(p_2,q_2),(p_3,q_3)) \\
&= f((p_1,q_1),a\cdot(p_2,q_2),(p_3,q_3)) \\
&= f((p_1,q_1),(p_2,q_2),a\cdot(p_3,q_3)) \\
&= a\cdot f((p_1,q_1),(p_2,q_2),(p_3,q_3))
\end{aligned} \tag{28.17}
$$

for all $a\in K$ and $(p_1,q_1),(p_2,q_2),(p_3,q_3)\in L[X,Y]$. Indeed,

$$
\begin{aligned}
f(a\cdot(p_1,q_1),&(p_2,q_2),(p_3,q_3)) \\
&= f((a\cdot p_1,a\cdot q_1),(p_2,q_2),(p_3,q_3)) \\
&= ((a\cdot p_1)\circ q_2\circ p_3,(a\cdot q_1)\circ p_2\circ q_3) \\
&= (a\cdot(p_1\circ q_2\circ p_3),\ a\cdot(q_1\circ p_2\circ q_3)) \\
&= a\cdot(p_1\circ q_2\circ p_3,\ q_1\circ p_2\circ q_3) \\
&= a\cdot f((p_1,q_1),(p_2,q_2),(p_3,q_3)).
\end{aligned}
$$

Similar arguments lead to the remaining equalities (28.17).

Thus, we have obtained the system $(L[X,Y],K,+,\cdot,f)$ fulfilling the following conditions:

(a) $(L[X,Y],K,+,\cdot)$ is a vector space over a field $K$;
(b) $(L[X,Y],f)$ is a ternary semigroup;
(c) the ternary multiplication $f$ is distributive over addition $+$ of linear mappings in $(L[X,Y],K,+,\cdot)$;
(d) the equalities (28.17) are satisfied.

The system $(L[X,Y],K,+,\cdot,f)$ is said to be a *ternary linear algebra* of linear mappings of vector spaces $X$ and $Y$ over a field $K$.

### *28.6.2 A Ternary Linear Algebra of Matrices*

Let $K$ be a field. Let $M(m,n)$ denote the set of all $m\times n$ matrices over $K$. Of course, $M(n,m)$ denotes the set of all $n\times m$ matrices over $K$. Put

$$M[m,n]=M(m,n)\times M(n,m).$$

Define the ternary operation $f: M[m,n]^3 \longrightarrow M[m,n]$ by the formula:

$$f((A_1,B_1),(A_2,B_2),(A_3,B_3))=(A_1B_2A_3,\ B_1A_2B_3)$$

for all $(A_i,B_i)\in M[m,n]$, where $i=1,2,3$.

It is easy to check that the groupoid $(M[m,n],f)$ is a ternary semigroup.

**Definition 28.15.** The ternary semigroup $(M[m,n],f)$ is called the ternary semigroup of $m\times n$ matrices over a field $K$.

Let $(M(m,n),K,+,\cdot)$ and $(M(n,m),K,+,\cdot)$ be vector spaces of matrices over a field $K$. Put $M[m,n]=M(m,n)\times M(n,m)$. Consider the Cartesian product

$$(M[m,n],K,+,\cdot)$$

of the vector spaces $(M(m,n),K,+,\cdot)$ and $(M(n,m),K,+,\cdot)$. Let $(M[m,n],f)$ be the ternary semigroup of matrices over a field $K$.

The ternary operation $f$ will be called a *ternary multiplication* of matrices.

Assume that $(A_1,B_1)$, $(A_1',B_1')$, $(A_1'',B_1'')$, $(A_2,B_2)$, $(A_2',B_2')$, $(A_2'',B_2'')$, $(A_3,B_3)$, $(A_3',B_3')$, $(A_3'',B_3'')\in M[m,n]$. The folowing conditions are satisfied:

$$\begin{aligned} f((A_1',B_1')+(A_1'',B_1''),(A_2,B_2),(A_3,B_3)) \\ =f((A_1',B_1'),(A_2,B_2),(A_3,B_3)) \\ +f((A_1'',B_1''),(A_2,B_2),(A_3,B_3)), \end{aligned} \qquad (28.18)$$

$$
\begin{aligned}
&f((A_1,B_1),(A_2',B_2')+(A_2'',B_2''),(A_3,B_3)) \\
&\qquad = f((A_1,B_1),(A_2',B_2'),(A_3,B_3)) \\
&\qquad\quad + f((A_1,B_1),(A_2'',B_2''),(A_3,B_3)),
\end{aligned} \tag{28.19}
$$

$$
\begin{aligned}
&f((A_1,B_1),(A_2,B_2),(A_3',B_3')+(A_3'',B_3'')) \\
&\qquad = f((A_1,B_1),(A_2,B_2),(A_3',B_3')) \\
&\qquad\quad + f((A_1,B_1),(A_2,B_2),(A_3'',B_3'')).
\end{aligned} \tag{28.20}
$$

The proof of the conditions (28.18)–(28.20) is similar to the proof of (28.14)–(28.16).

When the conditions (28.18)–(28.20) are satisfied, we say, that the ternary multiplication $f$ is *distributive over addition* $+$ of matrices in $(M[m,n],K,+,\cdot)$.

The following equalities are fulfilled:

$$
\begin{aligned}
&f(a\cdot(A_1,B_1),(A_2,B_2),(A_3,B_3)) \\
&\qquad = f((A_1,B_1),a\cdot(A_2,B_2),(A_3,B_3)) \\
&\qquad = f((A_1,B_1),(A_2,B_2),a\cdot(A_3,B_3)) \\
&\qquad = a\cdot f((A_1,B_1),(A_2,B_2),(A_3,B_3))
\end{aligned} \tag{28.21}
$$

for all $a\in K$ and $(A_1,B_1),(A_2,B_2),(A_3,B_3)\in M[m,n]$.

The proof of the equalities (28.21) is similar to the proof of the equalities (28.17).

Thus, we have obtained the system $(M[m,n],K,+,\cdot,f)$ fulfilling the following conditions:

(a) $(M[m,n],K,+,\cdot)$ is a vector space over a field $K$;
(b) $(M[m,n],f)$ is a ternary semigroup;
(c) the ternary multiplication $f$ is distributive over addition $+$ of matrices in $(M[m,n],K,+,\cdot)$;
(d) the equalities (28.21) are satisfied.

The system $(M[m,n],K,+,\cdot,f)$ is said to be a *ternary linear algebra* of matrices over a field $K$.

Let $X$ and $Y$ be vector spaces over a field $K$ such that $\dim X=n$ and $\dim Y=m$. Assume that

$$(x_1,\ldots,x_n) \quad \text{and} \quad (y_1,\ldots,y_m)$$

are the bases of the vector spaces $X$ and $Y$, respectively. If $p\in L(X,Y)$ and $q\in L(Y,X)$, then $M(p)$ and $M(q)$ denote the matrices of the linear mappings $p$ and $q$, respectively. Let $F_1$ be an isomorphism of the vector space $(L(X,Y),K,+,\cdot)$ onto the vector space $(M(m,n),K,+,\cdot)$ defined by $F_1(p)=M(p)$ for $p\in L(X,Y)$.

Let $F_2$ be an isomorphism of the vector space $(L(Y,X),K,+,\cdot)$ onto the vector space $(M(n,m),K,+,\cdot)$ defined by $F_2(q)=M(q)$ for $q \in L(Y,X)$. It is easy to check that the mapping $F$, defined by the formula

$$F(p,q) = (F_1(p),F_2(q))$$

for all $(p,q) \in L[X,Y]$, is an isomorphism of the vector space $L[X,Y]$ onto the vector space $M[m,n]$. Notice that

$$\begin{aligned} F(f((p_1,q_1),&(p_2,q_2),(p_3,q_3))) \\ &= F(p_1 \circ q_2 \circ p_3, q_1 \circ p_2 \circ q_3) \\ &= (M(p_1 \circ q_2 \circ p_3), M(q_1 \circ p_2 \circ q_3)) \\ &= (M(p_1)\cdot M(q_2)\cdot M(p_3), M(q_1)\cdot M(p_2)\cdot M(q_3)) \\ &= f((M(p_1),M(q_1)),(M(p_2),M(q_2)),(M(p_3),M(q_3))) \\ &= f((F_1(p_1),F_2(q_1)),(F_1(p_2),F_2(q_2)),(F_1(p_3),F_2(q_3))) \\ &= f(F(p_1,q_1),F(p_2,q_2),F(p_3,q_3)) \end{aligned}$$

for all $(p_i,q_i) \in L[X,Y]$, where $i = 1,2,3$.

Because of the above properties of the mapping $F$, we say, that the ternary linear algebra

$$(L[X,Y],K,+,\cdot,f)$$

is *isomorphic* to the ternary linear algebra

$$(M[m,n],K,+,\cdot,f).$$

We have obtained the following

**Theorem 28.18.** *Let $X$ and $Y$ be vector spaces over a field $K$ such that* $\dim X = n$ *and* $\dim Y = m$*. Then the ternary linear algebra $(L[X,Y],K,+,\cdot,f)$ of linear mappings of vector spaces $X$ and $Y$ over a field $K$ is isomorphic to the ternary linear algebra*

$$(M[m,n],K,+,\cdot,f)$$

*of matrices over a field $K$.*

### 28.6.3 A Ternary Linear Algebra

On the basis of the above examples we can formulate a general definition of a ternary linear algebra.

**Definition 28.16.** A system $(A, K, +, \cdot, f)$ is said to be a ternary linear algebra over a field $K$ if the following conditions are satisfied:

(a) $(A, K, +, \cdot)$ is a vector space over a field $K$;
(b) $(A, f)$ is a ternary semigroup;
(c) for every $x_1, x_1', x_1'', x_2, x_2', x_2'', x_3, x_3', x_3'' \in A$,

$$f(x_1' + x_1'', x_2, x_3) = f(x_1', x_2, x_3) + f(x_1'', x_2, x_3),$$
$$f(x_1, x_2' + x_2'', x_3) = f(x_1, x_2', x_3) + f(x_1, x_2'', x_3),$$
$$f(x_1, x_2, x_3' + x_3'') = f(x_1, x_2, x_3') + f(x_1, x_2, x_3'');$$

(d) for all $a \in K$ and $x_1, x_2, x_3 \in A$,

$$f(a \cdot x_1, x_2, x_3) = f(x_1, a \cdot x_2, x_3) = f(x_1, x_2, a \cdot x_3) = a \cdot f(x_1, x_2, x_3).$$

**Definition 28.17.** A mapping $F : A \longrightarrow B$ is called an isomorphism of a ternary linear algebra $(A, K, +, \cdot, f)$ onto a ternary linear algebra $(B, K, +, \cdot, f)$ if the following conditions are satisfied:

(a) $F$ is an isomorphism of the vector space $(A, K, +, \cdot)$ onto the vector space $(B, K, +, \cdot)$;
(b) $F$ is a homomorphism of the ternary semigroup $(A, f)$ onto the ternary semigroup $(B, f)$.

We can similarly formulate the definitions of a homomorphism, monomorphism, an epimorphism, endomorphism, automorphism of ternary linear algebras.

### 28.6.4 An Involutive Ternary Linear Algebra of Linear Mappings of Unitary Vector Spaces

The symbol $K$ denotes the field $\mathscr{R}$ of real numbers or the field $\mathscr{C}$ of complex numbers. If $z \in \mathscr{C}$, then $\bar{z}$ denotes the complex number conjugate to $z$. Let $A = [a_{ij}]$ for $i = 1, \ldots, m$, $j = 1, \ldots, n$ be a matrix over the field $K$. We put

$$\overline{A} = [\overline{a}_{ij}]$$

for $i = 1, \ldots, m$, $j = 1, \ldots, n$. The symbol $A^T$ denotes the transposed matrix of the matrix $A$.

In this subsection, we shall consider non-zero *finite-dimensional* unitary spaces over the field $K$. A scalar product will be denoted by $( \, , \, )$. The symbol $L(X, Y)$ denotes the set of all linear mappings of a unitary space $X$ into a unitary space $Y$.

**Definition 28.18.** Let $X$ and $Y$ be unitary spaces. A linear mapping $p^* \in L(Y,X)$ is called an adjoint linear mapping of a linear mapping $p \in L(X,Y)$ if

$$(p(x),y) = (x,p^*(y))$$

for all $x \in X$ and $y \in Y$.

The proofs of the next three theorems are analogous to the proofs of the suitable theorems concerninig the adjoint endomorphisms of the unitary space $X$ [40].

**Theorem 28.19.** *Let $X$ and $Y$ be unitary spaces. For every linear mapping $p \in L(X,Y)$ there exists a unique adjoint linear mapping $p^* \in L(Y,X)$.*

**Theorem 28.20.** *Let $\mathscr{E}$ and $\mathscr{G}$ be normal orthogonal bases of the unitary spaces $X$ and $Y$, respectively. If a linear mapping $p \in L(X,Y)$ has the matrix $A$ in the bases $\mathscr{E}$ and $\mathscr{G}$, then the adjoint linear mapping $p^* \in L(Y,X)$ has the matrix $\overline{A}^{T}$ in the bases $\mathscr{E}$ and $\mathscr{G}$.*

**Theorem 28.21.** *Let $X$ and $Y$ be unitary spaces. If $p, p_1 \in L(X,Y)$, $q \in L(Y,X)$ and $a \in K$, then:*

(a) $(p^*)^* = p$;
(b) $(p+p_1)^* = p^* + p_1^*$;
(c) $(ap)^* = \overline{a}p^*$;
(d) $(p_1 \circ q \circ p)^* = p^* \circ q^* \circ p_1^*$.

Consider the vector spaces $(L(X,Y),K,+,\cdot)$ and $(L(Y,X),K,+,\cdot)$. Put

$$L[X,Y] = L(X,Y) \times L(Y,X).$$

Let $(L[X,Y],K,+,\cdot)$ be the Cartesian product of the vector spaces $(L(X,Y),K,+,\cdot)$ and $(L(Y,X),K,+,\cdot)$. Consider the ternary semigroup $(L[X,Y],f)$ of linear mappings of unitary spaces $X$ and $Y$, and the ternary linear algebra $(L[X,Y],K,+,\cdot,f)$ of linear mappings of unitary spaces $X$ and $Y$.

Define a mapping $* : L[X,Y] \longrightarrow L[X,Y]$ by the formula

$$(p,q)^* = (q^*,p^*) \tag{28.22}$$

for $(p,q) \in L[X,Y]$, where $p^*$ and $q^*$ are the adjoint linear mappings of $p \in L(X,Y)$ and $q \in L(Y,X)$, respectively. Therefore, $p^* \in L(Y,X)$ and $q^* \in L(X,Y)$, and so $(q^*,p^*) \in L[X,Y]$.

We will prove that the mapping (28.22) satisfies the following conditions:

(a) $((p,q)^*)^* = (p,q)$;
(b) $((p_1,q_1)+(p_2,q_2))^* = (p_1,q_1)^* + (p_2,q_2)^*$;
(c) $(a \cdot (p,q))^* = \overline{a} \cdot (p,q)^*$;
(d) $(f((p_1,q_1),(p_2,q_2),(p_3,q_3)))^* = f((p_3,q_3)^*,(p_2,q_2)^*,(p_1,q_1)^*)$.

In fact, using the theorem 28.21, respectively, we obtain:

$$((p,q)^*)^* = (q^*,p^*)^* = ((p^*)^*,(q^*)^*) = (p,q),$$

$$\begin{aligned}((p_1,q_1)+(p_2,q_2))^* &= (p_1+p_2,q_1+q_2)^* \\ &= ((q_1+q_2)^*,(p_1+p_2)^*) = (q_1^*+q_2^*,p_1^*+p_2^*) \\ &= (q_1^*,p_1^*)+(q_2^*,p_2^*) = (p_1,q_1)^*+(p_2,q_2)^*,\end{aligned}$$

$$\begin{aligned}(a\cdot(p,q))^* &= (ap,aq)^* = ((aq)^*,(ap)^*) = (\overline{a}q^*,\overline{a}p^*) \\ &= \overline{a}\cdot(q^*,p^*) = \overline{a}\cdot(p,q)^*,\end{aligned}$$

$$\begin{aligned}(f((p_1,q_1),(p_2,q_2),(p_3,q_3)))^* &= (p_1\circ q_2\circ p_3,q_1\circ p_2\circ q_3)^* \\ &= ((q_1\circ p_2\circ q_3)^*,(p_1\circ q_2\circ p_3)^*) = (q_3^*\circ p_2^*\circ q_1^*,p_3^*\circ q_2^*\circ p_1^*) \\ &= f((q_3^*,p_3^*),(q_2^*,p_2^*),(q_1^*,p_1^*)) = f((p_3,q_3)^*,(p_2,q_2)^*,(p_1,q_1)^*)\end{aligned}$$

for all $(p,q),(p_1,q_1),(p_2,q_2),(p_3,q_3) \in L[X,Y]$ and $a \in K$.

The mapping

$$* : L[X,Y] \longrightarrow L[X,Y],$$

defined by the formula (28.22), is said to be an *involution* (or *adjoint operation*) on the ternary linear algebra

$$(L[X,Y],K,+,\cdot,f)$$

of linear mappings of unitary spaces $X$ and $Y$.

Consider the ternary linear algebra $(L[X,Y],K,+,\cdot,f)$ of linear mappings of unitary spaces $X$ and $Y$.

The system

$$(L[X,Y],K,+,\cdot,f,*),$$

where $*$ is the involution (28.22), is said to be an *involutive ternary linear algebra* (or **-ternary algebra*) of linear mappings of unitary spaces $X$ and $Y$.

### 28.6.5 An Involutive Ternary Linear Algebra of Matrices

The symbol $K$ denotes the field $\mathscr{R}$ of real numbers or the field $\mathscr{C}$ of complex numbers. Let $(M[m,n],K,+,\cdot,f)$ be a ternary linear algebra of matrices over a field $K$. Define a mapping $* : M[m,n] \longrightarrow M[m,n]$ by the formula

$$(A,B)^* = (\overline{B}^{\mathrm{T}},\overline{A}^{\mathrm{T}}) \tag{28.23}$$

for $(A,B) \in M[m,n]$.

It is easy to prove that the mapping (28.23) satisfies the following conditions:

(a) $((A,B)^*)^* = (A,B)$;
(b) $((A_1,B_1)+(A_2,B_2))^* = (A_1,B_1)^* + (A_2,B_2)^*$;
(c) $(a\cdot(A,B))^* = \overline{a}\cdot(A,B)^*$;
(d) $(f((A_1,B_1),(A_2,B_2),(A_3,B_3)))^* = f((A_3,B_3)^*,(A_2,B_2)^*,(A_1,B_1)^*)$

for all $(A,B),(A_1,B_1),(A_2,B_2),(A_3,B_3) \in M[m,n]$ and $a \in K$.

The mapping $* : M[m,n] \longrightarrow M[m,n]$ defined by the formula (28.23) is said to be an *involution* (or *adjoint operation*) on the ternary linear algebra $(M[m,n],K,+,\cdot,f)$ of matrices over the field $K$.

Consider the ternary linear algebra $(M[m,n],K,+,\cdot,f)$ of matrices over the field $K$. The system

$$(M[m,n],K,+,\cdot,f,*),$$

where $*$ is the involution (28.23), is said to be an *involutive ternary linear algebra* (or **-ternary algebra*) of matrices over the field $K$.

Let $X$ and $Y$ be unitary spaces such that $\dim X = n$ and $\dim Y = m$. Assume that $\mathscr{E}$ and $\mathscr{G}$ are normal orthogonal bases of the unitary spaces $X$ and $Y$, respectively. If $p \in L(X,Y)$ and $q \in L(Y,X)$, then $M(p)$ and $M(q)$ denote the matrices of the linear mappings $p$ and $q$, respectively. Let $F_1$ be an isomorphism of the vector space $(L(X,Y),K,+,\cdot)$ onto the vector space $(M(m,n),K,+,\cdot)$ defined by

$$F_1(p) = M(p)$$

for $p \in L(X,Y)$. Let $F_2$ be an isomorphism of the vector space $(L(Y,X),K,+,\cdot)$ onto the vector space $(M(n,m),K,+,\cdot)$ defined by

$$F_2(q) = M(q)$$

for $q \in L(Y,X)$. It is easy to check that the mapping $F$ defined by the formula

$$F(p,q) = (F_1(p),F_2(q))$$

for $(p,q) \in L[X,Y]$, is an isomorphism of the ternary linear algebra

$$(L[X,Y],K,+,\cdot,f)$$

of linear mappings of unitary spaces $X$ and $Y$ onto the ternary linear algebra

$$(M[m,n],K,+,\cdot,f)$$

of matrices over the field $K$.

Using Theorem 28.20 we have:

$$\begin{aligned} F((p,q)^*) &= F(q^*,p^*) = (M(q^*),M(p^*)) \\ &= (\overline{M(q)}^{\mathrm{T}},\overline{M(p)}^{\mathrm{T}}) = (M(p),M(q))^* = (F(p,q))^* \end{aligned}$$

for $(p,q) \in L[X,Y]$.

Because of the above properties of the mapping $F$, we say, that the involutive ternary linear algebra $(L[X,Y],K,+,\cdot,f,*)$ is *isomorphic* to the involutive ternary linear algebra $(M[m,n],K,+,\cdot,f,*)$.

We have obtained the following

**Theorem 28.22.** *Let X and Y be unitary spaces such that* $\dim X = n$ *and* $\dim Y = m$. *Then the involutive ternary linear algebra* $(L[X,Y],K,+,\cdot,f,*)$ *of linear mappings of unitary spaces X and Y is isomorphic to the involutive ternary linear algebra* $(M[m,n],K,+,\cdot,f)$ *of matrices over a field K.*

### 28.6.6 An Involutive Ternary Linear Algebra

On the basis of the above examples we can formulate a general definition of an involutive ternary linear algebra over the field $K$, where $K$ denotes the field $\mathscr{R}$ of real numbers or the field $\mathscr{C}$ of complex numbers.

**Definition 28.19.** A system $(A,K,+,\cdot,f,*)$ is said to be an involutive ternary linear algebra (or *-ternary algebra) over the field $K$ if the following conditions are fulfilled:

(a) $(A,K,+,\cdot,f)$ is the ternary linear algebra over the field $K$;
(b) $*: A \ni x \longrightarrow x^* \in A$ is a mapping, which is called an involution (or adjoint operation), such that for all $x,x_1,x_2,x_3 \in A$ and $a \in K$ we have:

(b$_1$) $(x^*)^* = x$;
(b$_2$) $(x_1+x_2)^* = x_1^* + x_2^*$;
(b$_3$) $(a\cdot x)^* = \overline{a}\cdot x^*$;
(b$_4$) $(f(x_1,x_2,x_3))^* = f(x_3^*,x_2^*,x_1^*)$.

The involutive ternary linear algebra is a counterpart of the involutive algebra ([6, p. 26]).

**Definition 28.20.** A mapping $F: A \longrightarrow B$ is called an isomorphism of an involutive ternary linear algebra $(A,K,+,\cdot,f,*)$ onto an involutive ternary linear algebra $(B,K,+,\cdot,f,*)$ if the following conditions are satisfied:

(a) $F$ is an isomorphism of the ternary linear algebra $(A,K,+,\cdot,f)$ onto the ternary linear algebra $(B,K,+,\cdot,f)$;
(b) $F(x^*) = F(x)^*$ for every $x \in A$.

We can similarly formulate the definitions of a homomorphism, monomorphism, an epimorphism, endomorphism, automorphism of involutive ternary linear algebras.

### 28.6.7 Normed and Topological Ternary Linear Algebras

First, we shall list some known facts concerning the normed vector spaces, which will be useful in the further considerations.

The symbol $K$ denotes the field $\mathscr{R}$ of real numbers or the field $\mathscr{C}$ of complex numbers. Let $(X, \|\cdot\|)$ and $(Y, \|\cdot\|)$ be normed vector spaces. The Cartesian product $X \times Y$ of the vector spaces $X$ and $Y$ equipped with the norm

$$\|(x,y)\| = \sqrt{\|x\|^2 + \|y\|^2} \tag{28.24}$$

for $(x,y) \in X \times Y$, is a normed vector space.

The normed vector space $(X \times Y, \|\cdot\|)$ with the norm defined by (28.24) is called a *Cartesian product* of normed vector spaces $(X, \|\cdot\|)$ and $(Y, \|\cdot\|)$.

The vector space $C(X,Y)$ of all continuous linear mappings of $X$ into $Y$ equipped with the norm

$$\|p\| = \sup_{\|x\| \leq 1} \|p(x)\|$$

for $p \in C(X,Y)$, is a normed vector space.

**Lemma 28.3 ([34, p. 112]).** *Let $(X, \|\cdot\|)$, $(Y, \|\cdot\|)$, $(Z, \|\cdot\|)$ be normed vector spaces. If $p \in C(X,Y)$ and $q \in C(Y,Z)$, then the norm of $q \circ p \in C(X,Z)$ satisfies the following inequality*

$$\|q \circ p\| \leq \|q\| \, \|p\|.$$

**Theorem 28.23.** *Let $(X, \|\cdot\|)$, $(Y, \|\cdot\|)$, $(Z, \|\cdot\|)$ be normed vector spaces. Assume that $p, p_n \in C(X,Y)$, $q, q_n \in C(Y,Z)$ for $n \in \mathscr{N}$ and*

$$\lim_{n \to \infty} p_n = p, \quad \lim_{n \to \infty} q_n = q.$$

*Then*

$$\lim_{n \to \infty} (q_n \circ p_n) = q \circ p. \tag{28.25}$$

*Proof.* Applying Lemma 28.3 we get:

$$\begin{aligned}
\|q_n \circ p_n - q \circ p\| &= \|q_n \circ p_n - q \circ p_n + q \circ p_n - q \circ p\| \\
&= \|(q_n - q) \circ p_n + q \circ (p_n - p)\| \\
&\leq \|(q_n - q) \circ p_n\| + \|q \circ (p_n - p)\| \\
&\leq \|q_n - q\| \, \|p_n\| + \|q\| \, \|p_n - p\|
\end{aligned}$$

for $n \in \mathscr{N}$. Since

$$\lim_{n \to \infty} \|p_n - p\| = \lim_{n \to \infty} \|q_n - q\| = 0,$$

it follows from the above argument that

$$\lim_{n\to\infty} \|q_n \circ p_n - q \circ p\| = 0,$$

and consequently (28.25) holds. □

Put

$$C[X,Y] = C(X,Y) \times C(Y,X).$$

Consider the Cartesian product $(C[X,Y], \|\cdot\|)$ of the normed spaces $(C(X,Y), \|\cdot\|)$ and $(C(Y,X), \|\cdot\|)$. Define the ternary operation

$$f : C[X,Y]^3 \longrightarrow C[X,Y]$$

by the formula

$$f((p_1,q_1),(p_2,q_2),(p_3,q_3)) = (p_1 \circ q_2 \circ p_3, q_1 \circ p_2 \circ q_3)$$

for all $(p_i,q_i) \in C[X,Y]$, where $i = 1,2,3$. The groupoid $(C[X,Y], f)$ is a ternary semigroup.

We shall prove that

$$\|f((p_1,q_1),(p_2,q_2),(p_3,q_3))\| \leq \|(p_1,q_1)\| \, \|(p_2,q_2)\| \, \|(p_3,q_3)\| \tag{28.26}$$

for all $(p_i,q_i) \in C[X,Y]$, where $i = 1,2,3$.

Indeed, applying Lemma 28.3 we get:

$$\begin{aligned}
\|f((p_1,q_1),(p_2,q_2),(p_3,q_3))\| &= \|(p_1 \circ q_2 \circ p_3, q_1 \circ p_2 \circ q_3)\| \\
&= (\|p_1 \circ q_2 \circ p_3\|^2 + \|q_1 \circ p_2 \circ q_3)\|^2)^{\frac{1}{2}} \\
&\leq (\|p_1\|^2 \, \|q_2\|^2 \, \|p_3\|^2 + \|q_1\|^2 \, \|p_2\|^2 \, \|q_3)\|^2)^{\frac{1}{2}} \\
&\leq (\|p_1\|^2 + \|q_1\|^2)^{\frac{1}{2}} \, (\|p_2\|^2 + \|q_2\|^2)^{\frac{1}{2}} \, (\|p_3\|^2 + \|q_3\|^2)^{\frac{1}{2}} \\
&= \|(p_1,q_1)\| \, \|(p_2,q_2)\| \, \|(p_3,q_3)\|
\end{aligned}$$

for all $(p_i,q_i) \in C[X,Y]$, where $i = 1,2,3$.

**Definition 28.21.** A ternary semigroup $(A, f)$ defined on a topological space $(A, \tau)$ is called a topological ternary semigroup if the ternary operation $f : A^3 \longrightarrow A$ is continuous, where $A^3$ is the Cartesian product of the topological space $(A, \tau)$.

**Theorem 28.24.** *If $(X, \|\cdot\|)$ and $(Y, \|\cdot\|)$ are normed vector spaces, then $(C[X,Y], f)$ is a topological ternary semigroup.*

*Proof.* We prove that $f$ is continuous. Let

$$(p_i, q_i), (p_{in}, q_{in}) \in C[X,Y]$$

for $i = 1,2,3$, $n \in \mathcal{N}$, and

$$\lim_{n \to \infty} (p_{in}, q_{in}) = (p_i, q_i).$$

Notice that

$$\lim_{n \to \infty} (p_{in}, q_{in}) = (p_i, q_i) \Longleftrightarrow (\lim_{n \to \infty} p_{in} = p_i \wedge \lim_{n \to \infty} q_{in} = q_i)$$

for $i = 1,2,3$. In view of Theorem 28.23 we have:

$$\begin{aligned}
&\lim_{n \to \infty} f((p_{1n}, q_{1n}), (p_{2n}, q_{2n}), (p_{3n}, q_{3n})) \\
&\quad = \lim_{n \to \infty} (p_{1n} \circ q_{2n} \circ p_{3n}, q_{1n} \circ p_{2n} \circ q_{3n}) \\
&\quad = (\lim_{n \to \infty} (p_{1n} \circ q_{2n} \circ p_{3n}), \lim_{n \to \infty} (q_{1n} \circ p_{2n} \circ q_{3n})) \\
&\quad = (p_1 \circ q_2 \circ p_3, q_1 \circ p_2 \circ q_3) \\
&\quad = f((p_1, q_1), (p_2, q_2), (p_3, q_3)).
\end{aligned}$$

Therefore, the mapping $f$ is continuous. □

On the basis of the above considerations we obtain the system

$$(C[X,Y], K, +, \cdot, f, \|\cdot\|)$$

fulfilling the following conditions:

(a) $(C[X,Y], K, +, \cdot, f)$ is a ternary linear algebra;
(b) $(C[X,Y], K, +, \cdot, \|\cdot\|)$ is a normed vector space;
(c) The inequality (28.26) is satisfied.

The the system

$$(C[X,Y], K, +, \cdot, f, \|\cdot\|)$$

is said to be a *normed ternary linear algebra* of all continuous linear mappings of normed vector spaces $X$ and $Y$ over the field $K$.

Now, we can formulate a general definition of a normed ternary linear algebra which is modelled after the definition of a normed linear algebra ([36, p. 44]; [6, p. 27]).

**Definition 28.22.** Let $K = \mathscr{R}$ or $K = \mathscr{C}$. A system $(A, K, +, \cdot, f, \|\cdot\|)$ is said to be a normed ternary linear algebra over the field $K$ if the following conditions are satisfied:

(a) $(A, K, +, \cdot, f)$ is a ternary linear algebra over the field $K$;
(b) $(A, K, +, \cdot, \|\cdot\|)$ is a normed vector space;
(c) $\|f(x_1, x_2, x_3)\| \leq \|x_1\| \cdot \|x_2\| \cdot \|x_3\|$ for all $x_1, x_2, x_3 \in A$.

The ternary operation $f$ is called the *ternary multiplication* in the normed ternary linear algebra

$$(A, K, +, \cdot, f, \|\cdot\|).$$

Now, we give the definition of a topological ternary linear algebra which is modelled after the definition of a topological linear algebra ([36, p. 19]).

**Definition 28.23.** Let $K = \mathscr{R}$ or $K = \mathscr{C}$. A system $(A, K, +, \cdot, f, \tau)$ is said to be a topological ternary linear algebra over the field $K$ if the following conditions are satisfied:

(a) $(A, K, +, \cdot, f)$ is a ternary linear algebra over the field $K$;
(b) $(A, K, +, \cdot, \tau)$ is a topological vector space with topology $\tau$;
(c) $(A, f, \tau)$ is a topological ternary semigroup with topology $\tau$.

The ternary operation $f$ is called the *ternary multiplication* in the topological ternary linear algebra $(A, K, +, \cdot, f, \tau)$.

**Theorem 28.25.** *Every normed ternary linear algebra*

$$(A, K, +, \cdot, f, \|\cdot\|)$$

*is a topological ternary linear algebra.*

*Proof.* By Definitions 28.22 and 28.23 it is enough to show that the ternary multiplication $f$ is continuous.

So, suppose that $x_1, x_{1n} \in A$, $x_2, x_{2n} \in A$, $x_3, x_{3n} \in A$ for $n \in \mathscr{N}$, and

$$\lim_{n \to \infty} x_{1n} = x_1, \quad \lim_{n \to \infty} x_{2n} = x_2, \quad \lim_{n \to \infty} x_{3n} = x_3.$$

Notice that

$$\begin{aligned}
&\|f(x_{1n}, x_{2n}, x_{3n}) - f(x_1, x_2, x_3)\| \\
&\quad = \|(f(x_{1n}, x_{2n}, x_{3n}) - f(x_1, x_{2n}, x_{3n})) + (f(x_1, x_{2n}, x_{3n}) \\
&\quad\quad - f(x_1, x_2, x_{3n})) + (f(x_1, x_2, x_{3n}) - f(x_1, x_2, x_3))\| \\
&\quad = \|f(x_{1n} - x_1, x_{2n}, x_{3n}) + f(x_1, x_{2n} - x_2, x_{3n}) + f(x_1, x_2, x_{3n} - x_3)\|
\end{aligned}$$

$$\leq \|f(x_{1n}-x_1,x_{2n},x_{3n})\| + \|f(x_1,x_{2n}-x_2,x_{3n})\|$$
$$+ \|f(x_1,x_2,x_{3n}-x_3)\| \leq \|x_{1n}-x_1\| \, \|x_{2n}\| \, \|x_{3n}\|$$
$$+ \|x_1\| \, \|x_{2n}-x_2\| \, \|x_{3n}\| + \|x_1\| \, \|x_2\| \, \|x_{3n}-x_3\|.$$

Since

$$\lim_{n\to\infty} \|x_{1n}-x_1\| = \lim_{n\to\infty} \|x_{2n}-x_2\| = \lim_{n\to\infty} \|x_{3n}-x_3\| = 0,$$

it follows that

$$\lim_{n\to\infty} \|f(x_{1n},x_{2n},x_{3n}) - f(x_1,x_2,x_3)\| = 0,$$

and consequently

$$\lim_{n\to\infty} f(x_{1n},x_{2n},x_{3n}) = f(x_1,x_2,x_3).$$

Therefore, the ternary multiplication $f$ is continuous. □

## References


1. Amyari, M., Moslehian, M.S.: Approximate homomorphisms of ternary semigroups. Lett. Math. Phys. **77**, 1–9 (2006)
2. Ataguema, H., Makhlouf, A.: Notes on cohomologies of ternary algebras of associative type. J. Gen. Lie Theory Appl. **3**, 157–174 (2009)
3. Bavand Savadkouhi, M., Eshaghi Gordji, M., Rassias, J.M., Ghobadipour, N.: Approximate ternary Jordan derivations on Banach ternary algebras. J. Math. Phys. **50**, 042303, 1–9 (2009)
4. Bazunova, N., Borowiec, A., Kerner R.: Universal differential calculus on ternary algebras. Lett. Math. Phys. **67**, 195–206 (2004)
5. Belousov, V.D.: $n$-ary quasigroups (in Russian). Štiinca, Kišiniev (1972)
6. Bratteli, O., Robinson, D.W.: Operator Algebras and Quantum Statistical Mechanics 1. Springer-Verlag, New York–Heidelberg–Berlin (1979)
7. Chronowski, A.: On the Pexider equation and the Cauchy equation on 3-adic groups. Wyż. Szkoła Ped. Kraków, Rocznik Nauk.-Dydakt. 115 Prace Mat. **12**, 173–176 (1987)
8. Chronowski, A.: Congruences and centrality in $n$-groups. Bul. Ştiinţ. Inst. Politehn. Cluj-Napoca Ser. Mat. Mec. Apl. Construc. Maş. **33**, 113–128 (1990)
9. Chronowski, A.: The Pexider equation on $n$-semigroups and $n$-groups. Publ. Math. Debrecen **37**, 121–130 (1990)
10. Chronowski, A.: Decompositions and extensions of homomorphisms of $n$-groups. Demonstratio Math. **24**, 47–53 (1991)
11. Chronowski, A.: On extensions of homomorphisms and homotopies of commutative $n$-adic groups. Arch. Math. (Brno) **27b**, 139–148 (1991)
12. Chronowski, A.: On the conditional Cauchy equation on 3-adic groups. Zb. Rad. Prirod.-Mat. Fak. Ser. Mat. **23**, 269–282 (1993)
13. Chronowski, A.: A ternary semigroup of mappings. Demonstratio Math. **27**, 781–791 (1994)
14. Chronowski, A.: On ternary semigroups of homomorphisms of ordered sets. Arch. Math. (Brno) **30**, 85–95 (1994)
15. Chronowski, A.: Some properties of $\alpha$-ideals and generalized $\alpha$-ideals, $n$-semigroups and $n$-groups. Czechoslovak Math. J. **44**, 39–46 (1994)
16. Chronowski, A.: On some relationships between affine spaces and ternary semigroups of affine mappings. Demonstratio Math. **28**, 315–322 (1995)

17. Chronowski, A.: The interpretation and equivalence of the variety of $n$-groups. Quasigroups Related Systems **2**, 120–131 (1995)
18. Chronowski, A.: On ternary semigroups of lattice homomorphisms. Quasigroups Related Systems **3**, 55–72 (1996)
19. Chronowski, A.: Ternary semigroups of linear mappings and matrices. Annales Academiae Paedagogicae Cracoviensis 4, Studia Mathematica **1**, 5–17 (2001)
20. Chronowski, A.: Congruences on ternary semigroups. Ukrainian Math. J. **56**, 544–559 (2004)
21. Chronowski, A.: Ternary linear algebras and topological ternary structures. Tr. Inst. Prikl. Mat. Mekh. **11**, 62–70 (2005)
22. Chronowski, A., Novotný, M.: Induced isomorphisms of certain ternary semigroups. Arch. Math. (Brno) **31**, 205–216 (1995)
23. Chronowski, A., Novotný, M.: Ternary semigroups of morphisms of objects in categories. Arch. Math. (Brno) **31**, 147–153 (1995)
24. Dörnte, W.: Untersuchungen über einen verallgemeinerten gruppenbegriff. Math. Zeit. **29**, 1–19 (1928)
25. Dudek, W.A., Gazek, K., Gleichgewicht, B.: A note on the axioms of $n$-groups. Colloq. Math. Soc. J. Bolyai **29.** Universal Algebra, Esztergom (Hungary), 195–202 (1977)
26. Faiziev, V., Rassias, Th.M., Sahoo, P.K.: The space of $(\psi,\gamma)$-additive mappings on semigroups. Trans. Amer. Math. Soc. **354**, 4455–4472 (2002)
27. Faiziev, V., Rassias, Th.M.: The space of $(\psi,\gamma)$-pseudocharacters on semigroups. Nonlinear Funct. Anal. Appl. **5**, 107–126 (2000)
28. Gasanov, A.M.: Ternary semigroups of topological mappings of bounded open subsets of a finite–dimensional Euclidean space (in Russian). In: Maksudov, F.K. et al. (eds.) Special Questions of Algebra and Topology, pp. 28–45. Elm, Baku (1980)
29. Hosszú, M.: On the explicit form of $n$-group operations. Publ. Math. Debrecen **10**, 88–92 (1963)
30. Howie, J.M.: An Introduction to Semigroup Theory. Academic Press, London-New York-San Francisco (1976)
31. Hyers, D.H., Isac, G., Rassias, Th.M.: Stability of Functional Equations in Several Variables. Birkhauser, Boston–Basel–Berlin (1998)
32. Hyers, D.H., Rassias, Th.M.: Approximate homomorphisms. Aequationes Math. **44**, 125–153 (1992)
33. Kerner R.: Ternary algebraic structures and their applications in physics. In: G. Pogosyan, L. Mardoyan, A. Sissakyan (eds.) Proceedings of the Conference ICGTMP "Group-23", Dubna, Russia. Publication OIAI (2002). Preprint: arXiv:math-ph/0011023v1 14 Nov (2000)
34. Kołodziej, W.: Mathematical Analysis (in Polish). Polish Scientific Publishers, Warszawa (1979)
35. Mach, A.: The translation equation on certain $n$-groups. Aequationes Math. **47**, 11–30 (1994)
36. Mlak, W.: Introduction to Hilbert Spaces (in Polish). Polish Scientific Publishers, Warszawa (1972)
37. Monk, D., Sioson, F.M.: m-Semigroups, semigroups, and function representations. Fund. Math. **59**, 233–241 (1966)
38. Monk, D., Sioson, F.M.: On the general theory of $m$-groups. Fund. Math. **72**, 233–244 (1971)
39. Moslehian, M.S., Rassias, Th.M.: Generalized Hyers–Ulam stability of mappings on normed Lie triple systems. Math. Inequal. Appl. **11**, 371–380 (2008)
40. Mostowski, A., Stark, M.: Linear Algebra (in Polish). Polish Scientific Publishers, Warszawa (1975)
41. Novotný, M.: Ternary structures and groupoids. Czech. Math. J. **41**(116), 90-98 (1991)
42. Park, C.: Generalized Hyers–Ulam stability of $C^*$-ternary algebra homomorphisms. Dyn. Contin. Discrete Impuls. Syst. Ser. A Math. Anal. **16**, 67–79 (2009)
43. Park, C.-G., Rassias, Th.M.: Homomorphisms in $C^*$-ternary algebras and $JB^*$-triples. J. Math. Anal. Appl. **337**, 13–20 (2008)
44. Park, C.-G., Rassias, Th.M.: Stability of homomorphisms in $JC^*$-algebras. Pac.-Asian J. Math. **1**, 1–17 (2007)

45. Park, C., Rassias, Th.M.: Homomorphisms and derivations in proper $JCQ^*$-triples. J. Math. Anal. Appl. **337**, 1404–1414 (2008)
46. Park, K.-H., Jung, Y.-S.: Perturbations of higher ternary derivations on Banach ternary algebras. Commun. Korean Math. Soc. **23**, 387–399 (2008)
47. Post, E.L.: Polyadic groups. Trans. Amer. Math. Soc. **48**, 208–350 (1940)
48. Rassias, J.M., Kim H.-M.: Approximate homomorphisms and derivations between $C^*$-ternary algebras. J. Math. Phys. **49**, 063507, 1–10 (2008)
49. Rassias, Th.M.: Functional Equations and Inequalities. Kluwer Academic Publishers, Dordrecht–Boston–London (2001)
50. Rassias, Th.M.: Functional Equations, Inequalities and Applications. Kluwer Academic Publishers, Dordrecht–Boston–London (2003)
51. Rosenberg, J.: The application of ternary semigroups to the study of $n$-valued Sheffer functions. Notre Dame J. Form. Log. **10**, 90–94 (1969)
52. Sioson, F.M.: Cyclic and homogeneous $m$-semigroups. Proc. Japan Acad. **39**, 444–449 (1963)
53. Sioson, F.M.: Ideal theory in ternary semigroups. Math. Japon. **10**, 63–84 (1965)
54. Sokolov, E.I.: On the Gluskin-Hossz*ú* theorem on the D*ö*rnte $n$-groups (in Russian). Mat. Issled. **39**, 187–189 (1976)

# Chapter 29
# Popoviciu Type Functional Equations on Groups

**Małgorzata Chudziak**

*Dedicated to the memory of S.M. Ulam*
*on the 100th anniversary of his birth*

**Abstract** Let $m,n,M,N$ be positive integers, $(H,+)$ and $(G,+)$ be commutative groups, and $G$ be uniquely divisible by $m$ and $n$. We give a description of solutions $f: G \to H$ of the functional equation

$$Mf\left(\frac{x+y+z}{m}\right)+f(x)+f(y)+f(z) = N\left[f\left(\frac{x+y}{n}\right)+f\left(\frac{x+z}{n}\right)+f\left(\frac{y+z}{n}\right)\right].$$

**Keywords** Popoviciu equation • Quadratic equation • Additive function

**Mathematics Subject Classification (2000):** Primary 39B22.

## 29.1 Introduction

The functional equation

$$3f\left(\frac{x+y+z}{3}\right)+f(x)+f(y)+f(z) = 2\left[f\left(\frac{x+y}{2}\right)+f\left(\frac{x+z}{2}\right)+f\left(\frac{y+z}{2}\right)\right] \tag{29.1}$$

M. Chudziak (✉)
Department of Mathematics, University of Rzeszów, Rejtana 16 C, 35-959 Rzeszów, Poland
e-mail: mchudziak@univ.rzeszow.pl

Th.M. Rassias and J. Brzdęk (eds.), *Functional Equations in Mathematical Analysis*, Springer Optimization and Its Applications 52, DOI 10.1007/978-1-4614-0055-4_29,

has been considered for the first time by Popoviciu [6] in a connection with some inequalities for convex functions. The solutions and stability of (29.1) have been studied by Trif [7] in the case where the unknown function $f$ is acting between two real linear spaces. In [9], Smajdor has considered solutions and stability of (29.1) for functions mapping a commutative semigroup uniquely divisible by 2 and 3 into an abstract cone. (Let us recall that a group $(X,+)$ is said to be *uniquely divisible* by a given positive integer $k$ provided, for every $x \in X$, there exists a unique $y \in X$ such that $x = ky$; such an element will be denoted in a sequel by $\frac{x}{k}$). Recently, Brzdęk [1] has proved stability of (29.1) in the class of functions mapping a commutative semigroup uniquely divisible by 2 and 3 into a commutative semigroup endowed with a metric satisfying some additional natural assumptions.

The following generalization of (29.1)

$$m^2 f\left(\frac{x+y+z}{m}\right) + f(x) + f(y) + f(z) = n^2\left[f\left(\frac{x+y}{n}\right) + f\left(\frac{x+z}{n}\right) + f\left(\frac{y+z}{n}\right)\right], \tag{29.2}$$

has been studied by Lee [5] under the assumption that $m,n$ are non-zero integers such that $m+1 = 2n$ and the unknown function $f$ maps a real linear space into another one. A particular case $m = 3, n = 2$ has been considered earlier by the same author in [4], and the case $m = n = 1$ has been studied by Kannappan [3]. For some further generalization of (29.1) we refer to [8].

In our paper [2], we have determined the general solution of the equation

$$Mf\left(\frac{x+y+z}{m}\right) + f(x) + f(y) + f(z) = N\left[f\left(\frac{x+y}{n}\right) + f\left(\frac{x+z}{n}\right) + f\left(\frac{y+z}{n}\right)\right] \tag{29.3}$$

in a more general setting. Namely, we have assumed that $m,n,M,N$ are positive integers, $(G,+)$ is a commutative group uniquely divisible by $m$ and $n$, $(H,+)$ is a commutative group uniquely divisible by 2, and $f : G \to H$ is an unknown function. Clearly, the assumption that $(G,+)$ is uniquely divisible by $m$ and $n$, is forced by the form of equation (29.3). However, the assumption that $(H,+)$ is uniquely divisible by 2 is strictly connected with a method used in the proof. In the present paper we deal with the solutions of (29.3) in a more general case, without that assumption of divisibility by 2 in $H$.

## 29.2 Results

To simplify some statements in the sequel we will use the following hypothesis.

($\mathscr{H}$) $(H,+)$ and $(G,+)$ are commutative groups, $m,n,M,N$ are positive integers, and $G$ is uniquely divisible by $m$ and $n$.

Let us yet recall that a function $A : G \to H$ is additive provided

$$A(x+y) = A(x) + A(y) \quad \text{for } x,y \in G$$

and a function $Q : G \to H$ is quadratic provided

$$Q(x+y) + Q(x-y) = 2Q(x) + 2Q(y) \quad \text{for } x,y \in G.$$

We begin with an auxiliary proposition.

**Proposition 29.1.** *Assume that ($\mathscr{H}$) holds, a function $f : G \to H$ satisfies equation (29.3) for all $x,y,z \in G$, and*

$$f_{\mathrm{e}}(x) := f(x) + f(-x) - 2f(0) \quad \textit{for } x \in G; \tag{29.4}$$

$$f_{\mathrm{o}}(x) := f(x) - f(-x) \quad \textit{for } x \in G. \tag{29.5}$$

*Then functions $f_{\mathrm{e}}$ and $f_{\mathrm{o}}$ satisfy (29.3), the function $Q := 2f_{\mathrm{e}}$ is quadratic with $Q(0) = 0$, the function $A := 2f_{\mathrm{o}}$ is additive, and*

$$(M+3-3N)f(0) = 0, \tag{29.6}$$

$$(N-n^2)Q(x) = (M-m^2)Q(x) = 0 \quad \textit{for } x \in G, \tag{29.7}$$

$$(Mn+mn-2mN)A(x) = 0 \quad \textit{for } x \in G. \tag{29.8}$$

*Proof.* Condition (29.6) follows from (29.3) with $x = y = z = 0$. Further, by (29.3) and (29.6), $f_{\mathrm{e}}$ and $f_{\mathrm{o}}$ are solutions of (29.3).

Clearly, $f_{\mathrm{e}}$ is even and $f_{\mathrm{e}}(0) = 0$. Therefore, for every $x,y \in G$, we have

$$\begin{aligned} &Mf_{\mathrm{e}}\left(\frac{y}{m}\right) + 2f_{\mathrm{e}}(x) + f_{\mathrm{e}}(y) \\ &\quad = Mf_{\mathrm{e}}\left(\frac{x+y-x}{m}\right) + f_{\mathrm{e}}(x) + f_{\mathrm{e}}(y) + f_{\mathrm{e}}(-x) \\ &\quad = N\left[f_{\mathrm{e}}\left(\frac{x+y}{n}\right) + f_{\mathrm{e}}\left(\frac{x-y}{n}\right) + f_{\mathrm{e}}(0)\right]. \end{aligned}$$

Hence

$$Mf_{\mathrm{e}}\left(\frac{y}{m}\right) + 2f_{\mathrm{e}}(x) + f_{\mathrm{e}}(y) = N\left[f_{\mathrm{e}}\left(\frac{x+y}{n}\right) + f_{\mathrm{e}}\left(\frac{x-y}{n}\right)\right] \tag{29.9}$$

for $x, y \in G$. Taking $y = 0$ in (29.9), we get

$$Q(x) = 2f_e(x) = 2Nf_e\left(\frac{x}{n}\right) = NQ\left(\frac{x}{n}\right) \quad \text{for } x \in G. \tag{29.10}$$

Moreover, putting $x = 0$ in (29.9), we obtain

$$Mf_e\left(\frac{y}{m}\right) + f_e(y) = NQ\left(\frac{y}{n}\right) \quad \text{for } y \in G,$$

which together with (29.10) gives

$$Q(y) = MQ\left(\frac{y}{m}\right) \quad \text{for } y \in G. \tag{29.11}$$

Now, from (29.9)–(29.11) we deduce that $Q$ is quadratic. Consequently, (29.7) follows from (29.10) and (29.11).

Obviously, $f_o$ is odd and $f_o(0) = 0$. Thus, as we have already observed, for every $x, y, z \in G$, we get

$$\begin{aligned} Mf_o\left(\frac{x+y+z}{m}\right) &+ f_o(x) + f_o(y) + f_o(z) \\ &= N\left[f_o\left(\frac{x+y}{n}\right) + f_o\left(\frac{x+z}{n}\right) + f_o\left(\frac{y+z}{n}\right)\right]. \end{aligned} \tag{29.12}$$

Taking in (29.12) first $z = 0$, and then $y = z = 0$, for every $x, y \in G$ we obtain the following two equalities:

$$Mf_o\left(\frac{x+y}{m}\right) + f_o(x) + f_o(y) = N\left[f_o\left(\frac{x+y}{n}\right) + f_o\left(\frac{x}{n}\right) + f_o\left(\frac{y}{n}\right)\right], \tag{29.13}$$

$$Mf_o\left(\frac{x}{m}\right) + f_o(x) = 2Nf_o\left(\frac{x}{n}\right). \tag{29.14}$$

Next, by (29.14), for every $x, y \in G$, we get

$$Mf_o\left(\frac{x+y}{m}\right) + f_o(x+y) = 2Nf_o\left(\frac{x+y}{n}\right),$$

which together with (29.13), gives

$$f_o(x+y) - f_o(x) - f_o(y) = N\left[f_o\left(\frac{x+y}{n}\right) - f_o\left(\frac{x}{n}\right) - f_o\left(\frac{y}{n}\right)\right].$$

On the other hand, using the oddness of $f_o$ and applying (29.12), for $x,y \in G$, we have

$$N\left[f_o\left(\frac{x+y}{n}\right) - f_o\left(\frac{x}{n}\right) - f_o\left(\frac{y}{n}\right)\right]$$
$$= N\left[f_o\left(\frac{x+y}{n}\right) + f_o\left(\frac{y-(x+y)}{n}\right) + f_o\left(\frac{x-(x+y)}{n}\right)\right]$$
$$= Mf_o\left(\frac{x+y-(x+y)}{m}\right) + f_o(x) + f_o(y) - f_o(x+y).$$

Consequently

$$f_o(x+y) - f_o(x) - f_o(y) = f_o(x) + f_o(y) - f_o(x+y) \quad \text{for } x,y \in G,$$

which means that $A$ is an additive function.

Finally, making use of (29.14), for every $x \in G$, we get

$$MA\left(\frac{x}{m}\right) + A(x) = 2NA\left(\frac{x}{n}\right),$$

which implies (29.8). □

**Theorem 29.1.** *Let hypothesis* ($\mathscr{H}$) *be valid. If a function* $f: G \to H$ *satisfies* (29.3) *for all* $x,y,z \in G$*, then there exist a quadratic function* $Q: G \to H$ *with* $Q(0) = 0$*, an additive function* $A: G \to H$ *and a* $B \in H$ *such that* (29.7) *and* (29.8) *hold,*

$$(M+3-3N)B = 0 \tag{29.15}$$

*and*

$$4f(x) = Q(x) + A(x) + 4B \quad \textit{for } x \in G. \tag{29.16}$$

*Proof.* Assume that $f: G \to H$ satisfies (29.3). Let $f_e$ and $f_o$ be given by (29.4) and (29.5), respectively. Furthermore, $Q := 2f_e$, $A := 2f_o$ and $B := f(0)$. Then, according to Proposition 29.1, $Q$ is quadratic with $Q(0) = 0$, $A$ is additive and (29.7), (29.8) and (29.15) hold. Moreover, in view of (29.4) and (29.5), we get

$$4f(x) = 2(f_e(x) + f_o(x)) + 4f(0) = Q(x) + A(x) + 4B \quad \text{for } x \in G.$$

□

From Theorem 29.1 one can easily derive the following result (cf. [2, Theorem 1]).

**Corollary 29.1.** *Let $(\mathscr{H})$ be valid and $(H,+)$ be uniquely divisible by 2. Then a function $f : G \to H$ satisfies* (29.3) *for all $x,y,z \in G$ if and only if there exist a quadratic function $Q : G \to H$, an additive function $A : G \to H$ and a $B \in H$ such that* (29.7), (29.8) *and* (29.15) *hold, and*

$$f(x) = Q(x) + A(x) + B \quad \textit{for} \quad x \in G. \tag{29.17}$$

Now we prove one more auxiliary proposition.

**Proposition 29.2.** *Let $(\mathscr{H})$ be valid. Assume that a function $F : G \to H$ satisfies* (29.3) *for all $x,y,z \in G$, $F(0) = 0$ and*

$$2F(x) = 0 \quad \textit{for} \quad x \in G. \tag{29.18}$$

*Then*

$$F(2^k x) = 0 \quad \textit{for} \quad x \in G, \tag{29.19}$$

*where*

$$k = \begin{cases} 0 & \textit{whenever } M \textit{ and } N \textit{ are even;} \\ 1 & \textit{whenever exactly one of } M \textit{ and } N \textit{ is even;} \\ 2 & \textit{whenever } M \textit{ and } N \textit{ are odd.} \end{cases} \tag{29.20}$$

*Proof.* Note that if $M$ and $N$ are even, then in view of (29.18), (29.3) takes the form

$$F(x) + F(y) + F(z) = 0 \quad \text{for} \quad x,y,z \in G.$$

Thus, taking $y = z$ and applying (29.18) again, we get (29.19) with $k = 0$.

Next, consider the case where $M$ is odd and $N$ is even. Then, by (29.18), from (29.3) we derive that

$$MF\left(\frac{x+y+z}{m}\right) + F(x) + F(y) + F(z) = 0 \quad \text{for} \quad x,y,z \in G.$$

Hence

$$F\left(\frac{x+y+z}{m}\right) + (M-1)F\left(\frac{x+y+z}{m}\right) + F(x) + F(y) + F(z) = 0$$

for $x,y,z \in G$, so making use of (29.18), we get

$$F\left(\frac{x+y+z}{m}\right) + F(x) + F(y) + F(z) = 0 \quad \text{for} \quad x,y,z \in G.$$

Therefore, taking $x = 0$ and $y = z$, we obtain that

$$F\left(\frac{2z}{m}\right) = 0 \quad \text{for} \quad z \in G,$$

which gives (29.19) with $k = 1$.

In the case where $M$ is even and $N$ is odd, (29.3) takes the form

$$F(x)+F(y)+F(z)=F\left(\frac{x+y}{n}\right)+F\left(\frac{x+z}{n}\right)+F\left(\frac{y+z}{n}\right) \quad \text{for } x,y,z\in G.$$

Thus, repeating the previous arguments, we conclude that

$$F\left(\frac{2z}{n}\right)=0 \quad \text{for } z\in G,$$

which again gives (29.19) with $k=1$.

Finally, assume that $M$ and $N$ are odd. Then, by (29.18), from (29.3) we derive that

$$F\left(\frac{x+y+z}{m}\right)+F(x)+F(y)+F(z)$$
$$=F\left(\frac{x+y}{n}\right)+F\left(\frac{x+z}{n}\right)+F\left(\frac{y+z}{n}\right)$$

for $x,y,z\in G$. Taking in the latter equality $y=z$, in view of (29.18), we get

$$F\left(\frac{x+2z}{m}\right)+F(x)=F\left(\frac{2z}{n}\right) \quad \text{for } x,z\in G. \tag{29.21}$$

Next, applying (29.21) with $x=0$ and then with $z=0$, we obtain

$$F\left(\frac{2z}{m}\right)=F\left(\frac{2z}{n}\right) \quad \text{for } z\in G \tag{29.22}$$

and

$$F(x)=-F\left(\frac{x}{m}\right) \quad \text{for } x\in G, \tag{29.23}$$

respectively. Consequently, making use of (29.18) and (29.21)–(29.23), we conclude that

$$F(4x)=-F\left(\frac{4x}{m}\right)=-F\left(\frac{2x+2x}{m}\right)$$
$$=-F\left(\frac{2x}{n}\right)+F(2x)=-F\left(\frac{2x}{m}\right)+F(2x)$$
$$=F(2x)+F(2x)=2F(2x)=0 \quad \text{for } x\in G.$$

Thus, (29.19) holds with $k=2$. □

Now we are in a position to prove the following.

**Theorem 29.2.** *Let* $(\mathscr{H})$ *be valid. If a function* $f : G \to H$ *satisfies* (29.3) *for all* $x, y, z \in G$*, then there exist a quadratic function* $Q : G \to H$*, an additive function* $A : G \to H$ *and a* $B \in H$ *such that* (29.7), (29.8) *and* (29.15) *hold, and*

$$f(4^{k+1}x) = 2Q(x) + A(x) + B \quad \text{for} \quad x \in G, \tag{29.24}$$

*where* $k$ *is given by* (29.20).

*Proof.* Assume that $f : G \to H$ satisfies (29.3). Then, according to Theorem 29.1, there exist a quadratic function $q : G \to H$ with $q(0) = 0$, an additive $a : G \to H$ and a $b \in H$ such that

$$(N - n^2)q(x) = (M - m^2)q(x) = 0 \quad \text{for} \quad x \in G, \tag{29.25}$$

$$(Mn + mn - 2mN)a(x) = 0 \quad \text{for} \quad x \in G \tag{29.26}$$

and

$$4f(x) = q(x) + a(x) + 4b \quad \text{for} \quad x \in G. \tag{29.27}$$

Making use of (29.27), for every $x \in G$, we obtain

$$4f(4x) = q(4x) + a(4x) + 4b = 16q(x) + 4a(x) + 4f(0).$$

Thus

$$4\,[f(4x) - 4q(x) - a(x) - f(0)] = 0 \quad \text{for} \quad x \in G. \tag{29.28}$$

Let

$$F(x) := f(4x) - 4q(x) - a(x) - f(0) \quad \text{for} \quad x \in G. \tag{29.29}$$

Since $f$ satisfies (29.3), according to Proposition 29.1, (29.6) holds. So, taking into account (29.25) and (29.26), one can easily check that $F$ satisfies (29.3). Furthermore, (29.29) and (29.28) imply that $F(0) = 0$ and $4F(x) = 0$ for $x \in G$, respectively. Thus, the function $2F$ satisfies the assumptions of Proposition 29.2. Hence, applying Proposition 29.2, we obtain that

$$2F(2^k x) = 0 \quad \text{for} \quad x \in G,$$

where $k$ is as in (29.20). Therefore, the function $\tilde{F} : G \to H$ given by $\tilde{F}(x) = F(2^k x)$ for $x \in G$, also fulfils the assumptions of Proposition 29.2. Thus, we get that $\tilde{F}(2^k x) = 0$ for $x \in G$. Consequently, $F(4^k x) = 0$ for $x \in G$, so taking into account (29.29), we obtain (29.24) with $Q := 2^{4k+1}q$, $A := 4^k a$ and $B := f(0)$. □

The next result is an immediate consequence of Theorem 29.2.

**Corollary 29.2.** *Let* $(\mathscr{H})$ *be valid and* $G$ *be uniquely divisible by 2. Then a function* $f : G \to H$ *satisfies* (29.3) *for all* $x, y, z \in G$ *if and only if there exist a quadratic function* $Q : G \to H$*, an additive function* $A : G \to H$ *and a* $B \in H$ *such that* (29.7), (29.8) *and* (29.15) *hold, and*

$$f(x) = 2Q(x) + A(x) + B \quad \text{for} \quad x \in G. \tag{29.30}$$

Since every group (uniquely) divisible by an even number, is (uniquely) divisible by 2, we have also the following result.

**Corollary 29.3.** *Let hypothesis* $(\mathscr{H})$ *be valid. If* $m$ *or* $n$ *is an even number, then a function* $f : G \to H$ *satisfies* (29.3) *for all* $x, y, z \in G$ *if and only if there exist a quadratic function* $Q : G \to H$*, an additive function* $A : G \to H$ *and a* $B \in H$ *such that* (29.7), (29.8), (29.15) *and* (29.30) *hold.*

The next two corollaries generalize to some extend [7, Theorem 2.1] and [5, Theorem 2.1].

**Corollary 29.4.** *Let hypothesis* $(\mathscr{H})$ *be valid. If m or n is an even number, then a function* $f : G \to H$ *satisfies the functional equation*

$$mf\left(\frac{x+y+z}{m}\right) + f(x) + f(y) + f(z) = n\left[f\left(\frac{x+y}{n}\right) + f\left(\frac{x+z}{n}\right) + f\left(\frac{y+z}{n}\right)\right]$$

*for* $x, y, z \in G$ *if and only if there exist a quadratic function* $Q : G \to H$*, an additive function* $A : G \to H$ *and a* $B \in H$ *such that* $(m-1)Q(x) = (n-1)Q(x) = 0$ *for* $x \in G$, $(m-3n+3)B = 0$*, and* $f$ *is of the form* (29.30).

**Corollary 29.5.** *Let hypothesis* $(\mathscr{H})$ *be valid. If m or n is an even number, then a function* $f : G \to H$ *satisfies* (29.2) *for all* $x, y, z \in G$ *if and only if there exist a quadratic function* $Q : G \to H$*, an additive function* $A : G \to H$ *and a* $B \in H$ *such that* $(m^2 - 3n^2 + 3)B = 0$, $(m + 1 - 2n)A(x) = 0$ *for* $x \in G$*, and* $f$ *is of the form* (29.30).

## References

1. Brzdęk, J.: A note on stability of the Popoviciu functional equation on restricted domain. Demonstratio Math. **43**, 635–641 (2010)
2. Chudziak, M.: On a generalization of the Popoviciu equation on groups. Ann. Univ. Pedagog. Crac. Stud. Math. **9**, 49–53 (2010)
3. Kannappan, P.: Quadratic functional equation and inner product spaces. Results Math. **27**,368–372 (1995)
4. Lee, Y.W.: On the stability on a quadratic Jensen type functional equation. J. Math. Anal. Appl. **270**, 590–601 (2002)
5. Lee, Y.W.: Stability of a generalized quadratic functional equation with Jensen type. Bull. Korean Math. Soc. **42**, 57–73 (2005)
6. Popoviciu, T.: Sur certaines inégalités qui caractérisent les fonctions convexes. An. Ştiinţt. Univ. "Al. I. Cuza" Iaşi Secţ. I a Mat. (N.S.) **11**, 155–164 (1965)

7. Trif, T.: Hyers-Ulam-Rassias stability of a Jensen type functional equation. J. Math. Anal. Appl. **250**, 579–588 (2000)
8. Trif, T.: On the stability of a functional equation deriving from an inequality of Popoviciu for convex functions. J. Math. Anal. Appl. **272**, 604–616 (2002)
9. Smajdor, W.: Note on a Jensen type functional equation. Publ. Math. Debrecen **63**, 703–714 (2003)

# Chapter 30
# Norm and Numerical Radius Inequalities for Two Linear Operators in Hilbert Spaces: A Survey of Recent Results

**Sever S. Dragomir**

*Dedicated to the memory of S.M. Ulam*
*on the 100th anniversary of his birth*

**Abstract** The main aim of this paper is to survey some recent norm and numerical radius inequalities obtained by the author for composite operators generated by a pair of operators $(A,B)$ in complex Hilbert spaces under various assumptions. Applications in connection with classical results are also provided.

**Keywords** Bounded linear operators • Operator norm • Numerical radius • Inequalities for norms and numerical radius

**Mathematics Subject Classification (2000):** Primary 47A63, 47A99

## 30.1 Introduction

Let $(H;\langle\cdot,\cdot\rangle)$ be a complex Hilbert space. The *numerical range* of an operator $T$ is the subset of the complex numbers $\mathbb{C}$ given by $W(T)=\{\langle Tx,x\rangle,\ x\in H,\ \|x\|=1\}$, see for instance [26, p. 1]. It is well known that (see [26]):

1. The numerical range of an operator is convex;
2. The spectrum of an operator is contained in the closure of its numerical range;
3. $T$ is self-adjoint if and only if $W(T)$ is real.

The *numerical radius* $w(T)$ of an operator $T$ on $H$ is defined by $w(T):=\sup\{|\lambda|,\lambda\in W(T)\}=\sup\{|\langle Tx,x\rangle|,\|x\|=1\}$, [26, p. 8]. It is well known that

S.S. Dragomir (✉)
Department of Mathematics, School of Enginering & Science, Victoria University, PO Box 14428, Melbourne City, Victoria 8001, Australia

School of Computational & Applied Mathematics, University of the Witwatersrand, Private Bag-3, Wits-2050, Johannesburg, South Africa
e-mail: sever.dragomir@vu.edu.au

Th.M. Rassias and J. Brzdęk (eds.), *Functional Equations in Mathematical Analysis*, Springer Optimization and Its Applications 52, DOI 10.1007/978-1-4614-0055-4_30,

$w(\cdot)$ is a norm on the Banach algebra $B(H)$ of all bounded linear operators acting on $H$ and the following inequality holds true:

$$w(T) \leq \|T\| \leq 2w(T). \tag{30.1}$$

We recall some classical results involving the numerical radius of two linear operators $A, B$.

The following general result for the product of two operators holds [26, p. 37]:

**Theorem 30.1.** *If $A, B$ are two bounded linear operators on the Hilbert space $(H, \langle\cdot,\cdot\rangle)$, then $w(AB) \leq 4w(A)w(B)$. In the case that $AB = BA$, then $w(AB) \leq 2w(A)w(B)$.*

The following results are also well known [26, p. 38].

**Theorem 30.2.** *If $A$ is a unitary operator that commutes with another operator $B$, then*

$$w(AB) \leq w(B). \tag{30.2}$$

*If $A$ is an isometry and $AB = BA$, then (30.2) also holds true.*

We say that $A$ and $B$ *double commute* if $AB = BA$ and $AB^* = B^*A$. The following result holds [26, p. 38].

**Theorem 30.3 (Double commute).** *If the operators $A$ and $B$ double commute, then*

$$w(AB) \leq w(B)\|A\|.$$

As a consequence of the above, we have [26, p. 39]:

**Corollary 30.1.** *Let $A$ be a normal operator commuting with $B$. Then*

$$w(AB) \leq w(A)w(B).$$

For other results and historical comments on the above see [26, p. 39–41]. For more results on the numerical radius, see [27].

The main aim of this paper is to survey some recent inequalities obtained by the author in [9, 18–22] for composite operators generated by a pair of operators $(A, B)$ under various assumptions. For related results on operator and vector inequalities in Hilbert spaces see, [1–23] and [25]. For recent results on norm and numerical radius inequalities, see [29, 35].

### *30.1.1 General Inequalities for the Norm and Numerical Radius*

### *30.1.2 Some Preliminary Results*

The following result may be stated:

**Theorem 30.4 (Dragomir, 2008 [19]).** *Let $A, B : H \to H$ be two bounded linear operators on the Hilbert space $(H, \langle\cdot,\cdot\rangle)$. If $r > 0$ and*

$$\|A - B\| \leq r, \tag{30.3}$$

*then*

$$\left\| \frac{A^*A + B^*B}{2} \right\| \le w(B^*A) + \frac{1}{2}r^2. \tag{30.4}$$

*Proof.* For any $x \in H$, $\|x\| = 1$, we have from (30.3) that

$$\|Ax\|^2 + \|Bx\|^2 \le 2\mathrm{Re}\langle Ax, Bx\rangle + r^2. \tag{30.5}$$

However

$$\|Ax\|^2 + \|Bx\|^2 = \langle (A^*A + B^*B)x, x\rangle$$

and by (30.5) we obtain

$$\langle (A^*A + B^*B)x, x\rangle \le 2|\langle (B^*A)x, x\rangle| + r^2 \tag{30.6}$$

for any $x \in H$, $\|x\| = 1$.

Taking the supremum over $x \in H$, $\|x\| = 1$ in (30.6) we get

$$w(A^*A + B^*B) \le 2w(B^*A) + r^2 \tag{30.7}$$

and since the operator $A^*A + B^*B$ is self-adjoint, hence

$$w(A^*A + B^*B) = \|A^*A + B^*B\|$$

and by (30.7) we deduce the desired inequality (30.4). □

*Remark 30.1.* We observe that, from the proof of the above theorem, we have the inequalities

$$0 \le \left\| \frac{A^*A + B^*B}{2} \right\| - w(B^*A) \le \frac{1}{2}\|A - B\|^2, \tag{30.8}$$

provided that $A, B$ are bounded linear operators in $H$.

The second inequality in (30.8) is obvious while the first inequality follows by the fact that

$$\langle (A^*A + B^*B)x, x\rangle = \|Ax\|^2 + \|Bx\|^2 \ge 2|\langle (B^*A)x, x\rangle|$$

for any $x \in H$.

The inequality (30.4) is obviously a rich source of particular inequalities of interest.

Indeed, if we assume, for $\lambda \in \mathbb{C}$ and a bounded linear operator $T$, that we have $\|T - \lambda T^*\| \le r$, for a given positive number $r$, then by (30.8) we deduce the inequality

$$0 \le \left\| \frac{T^*T + |\lambda|^2 TT^*}{2} \right\| - |\lambda| w(T^2) \le \frac{1}{2}r^2. \tag{30.9}$$

Now, if we assume that for $\lambda \in \mathbb{C}$ and a bounded linear operator $V$ we have that $\|V - \lambda I\| \le r$, where $I$ is the identity operator on $H$, then by (30.4) we deduce the inequality

$$0 \le \left\| \frac{V^*V + |\lambda|^2 I}{2} \right\| - |\lambda| w(V) \le \frac{1}{2} r^2.$$

As a dual approach, the following result may be noted as well:

**Theorem 30.5 (Dragomir, 2008 [19]).** *Let $A, B : H \to H$ be two bounded linear operators on the Hilbert space $H$. Then*

$$\left\| \frac{A+B}{2} \right\|^2 \le \frac{1}{2} \left[ \left\| \frac{A^*A + B^*B}{2} \right\| + w(B^*A) \right]. \tag{30.10}$$

*Proof.* We obviously have

$$\begin{aligned} \|Ax + Bx\|^2 &= \|Ax\|^2 + 2\mathrm{Re}\langle Ax, Bx\rangle + \|Bx\|^2 \\ &\le \langle (A^*A + B^*B)x, x\rangle + 2|\langle (B^*A)x, x\rangle| \end{aligned}$$

for any $x \in H$.

Taking the supremum over $x \in H$, $\|x\| = 1$, we get

$$\|A + B\|^2 \le \|A^*A + B^*B\| + 2w(B^*A),$$

from where we get the desired inequality (30.10). □

*Remark 30.2.* The inequality (30.10) can generate some interesting particular results such as the following inequality

$$\left\| \frac{T+T^*}{2} \right\|^2 \le \frac{1}{2} \left[ \left\| \frac{T^*T + TT^*}{2} \right\| + w\left(T^2\right) \right], \tag{30.11}$$

holding for each bounded linear operator $T : H \to H$.

The following result concerning powers may be stated as well.

**Theorem 30.6 (Dragomir, 2008 [19]).** *Let $A, B : H \to H$ be two bounded linear operators on the Hilbert space $H$ and $p \ge 2$. Then*

$$\left\| \frac{A^*A + B^*B}{2} \right\|^{\frac{p}{2}} \le \frac{1}{4} \left[ \|A - B\|^p + \|A + B\|^p \right]. \tag{30.12}$$

*Proof.* We use the following inequality for vectors in inner product spaces obtained by Dragomir and Sándor in [23]:

$$2\left(\|a\|^p + \|b\|^p\right) \le \|a+b\|^p + \|a-b\|^p \tag{30.13}$$

for any $a, b \in H$ and $p \ge 2$.

Utilising (30.13), for any $x \in H$ we may write

$$2\left(\|Ax\|^p + \|Bx\|^p\right) \le \|Ax + Bx\|^p + \|Ax - Bx\|^p . \tag{30.14}$$

Now, observe that

$$\|Ax\|^p + \|Bx\|^p = \left(\|Ax\|^2\right)^{\frac{p}{2}} + \left(\|Bx\|^2\right)^{\frac{p}{2}}$$

and, by the elementary inequality

$$\frac{\alpha^q + \beta^q}{2} \ge \left(\frac{\alpha + \beta}{2}\right)^q, \quad \alpha, \beta \ge 0,\ q \ge 1,$$

we have

$$\left(\|Ax\|^2\right)^{\frac{p}{2}} + \left(\|Bx\|^2\right)^{\frac{p}{2}} \ge 2^{1-\frac{p}{2}} \left[\langle (A^*A + B^*B)x, x\rangle\right]^{\frac{p}{2}} . \tag{30.15}$$

Combining (30.14) with (30.15) we get

$$\frac{1}{4}\left[\|Ax - Bx\|^p + \|Ax + Bx\|^p\right] \ge \left|\left\langle \left(\frac{A^*A + B^*B}{2}\right) x, x \right\rangle\right|^{\frac{p}{2}} \tag{30.16}$$

for any $x \in H$, $\|x\| = 1$. Taking the supremum over $x \in H$, $\|x\| = 1$, and taking into account that

$$w\left(\frac{A^*A + B^*B}{2}\right) = \left\|\frac{A^*A + B^*B}{2}\right\|,$$

we deduce the desired result (30.12). □

*Remark 30.3.* If $p = 2$, then we have the inequality:

$$\left\|\frac{A^*A + B^*B}{2}\right\| \le \left\|\frac{A - B}{2}\right\|^2 + \left\|\frac{A + B}{2}\right\|^2,$$

for any $A, B$ bounded linear operators. This result can be obtained directly on utilising the parallelogram identity as well. We also should observe that for $A = T$ and $B = T^*$, $T$ a normal operator, the inequality (30.12) becomes

$$\|T\|^p \le \frac{1}{4}\left[\|T - T^*\|^p + \|T + T^*\|^p\right],$$

where $p \ge 2$.

The following result may be stated as well.

**Theorem 30.7 (Dragomir, 2008 [19]).** *Let $A,B : H \to H$ be two bounded linear operators on the Hilbert space $H$ and $r \geq 1$. If $A^*A \geq B^*B$ in the operator order or, equivalently, $\|Ax\| \geq \|Bx\|$ for any $x \in H$, then:*

$$\left\|\frac{A^*A+B^*B}{2}\right\|^r \leq \|A\|^{r-1}\|B\|^{r-1} w(B^*A) + \frac{1}{2}r^2 \|A\|^{2r-2}\|A-B\|^2. \tag{30.17}$$

*Proof.* We use the following inequality for vectors in inner product spaces due to Goldstein et al. [24]:

$$\|a\|^{2r} + \|b\|^{2r} \leq 2\|a\|^{r-1}\|b\|^{r-1}\operatorname{Re}\langle a,b\rangle + r^2\|a\|^{2r-2}\|a-b\|^2, \tag{30.18}$$

where $r \geq 1$, $a,b \in H$ and $\|a\| \geq \|b\|$.

Utilising (30.18) we can state that:

$$\begin{aligned}\|Ax\|^{2r} + \|Bx\|^{2r} \leq\ & 2\|Ax\|^{r-1}\|Bx\|^{r-1}|\langle Ax,Bx\rangle| \\ & + r^2\|Ax\|^{2r-2}\|Ax-Bx\|^2,\end{aligned} \tag{30.19}$$

for any $x \in H$. As in the proof of Theorem 30.6, we also have

$$2^{1-r}[\langle (A^*A+B^*B)x,x\rangle]^r \leq \|Ax\|^{2r} + \|Bx\|^{2r}, \tag{30.20}$$

for any $x \in H$. Therefore, by (30.19) and (30.20) we deduce

$$\begin{aligned}\left[\left\langle\left(\frac{A^*A+B^*B}{2}\right)x,x\right\rangle\right]^r \leq\ & \|Ax\|^{r-1}\|Bx\|^{r-1}|\langle Ax,Bx\rangle| \\ & + \frac{1}{2}r^2\|A\|^{2r-2}\|Ax-Bx\|^2\end{aligned} \tag{30.21}$$

for any $x \in H$.

Taking the supremum in (30.21) we obtain the desired result (30.17). □

*Remark 30.4.* Following [26, p. 156], we recall that the bounded linear operator $V$ is hyponormal, if $\|V^*x\| \leq \|Vx\|$ for all $x \in H$. Now, if we choose in (30.17) $A = V$ and $B = V^*$, then, on taking into account that for hyponormal operators $w(V^2) = \|V\|^2$, we get the inequality

$$\left\|\frac{V^*V+VV^*}{2}\right\|^r \leq \|V\|^{2r-2}\left[\|V\|^2 + \frac{1}{2}r^2\|V-V^*\|^2\right], \tag{30.22}$$

holding for any hyponormal operator $V$ and any $r \geq 1$.

### 30.1.3 Further Inequalities for an Invertible Operator

In this section, we assume that $B : H \to H$ is an invertible bounded linear operator and let $B^{-1} : H \to H$ be its inverse. Then, obviously,

$$\|Bx\| \geq \frac{1}{\|B^{-1}\|}\|x\|, \quad x \in H, \tag{30.23}$$

where $\|B^{-1}\|$ denotes the norm of the inverse $B^{-1}$.

**Theorem 30.8 (Dragomir, 2008 [19]).** *Let $A, B : H \to H$ be two bounded linear operators on $H$ and $B$ is invertible such that, for a given $r > 0$,*

$$\|A - B\| \leq r. \tag{30.24}$$

*Then*

$$\|A\| \leq \|B^{-1}\| \left[w(B^*A) + \frac{1}{2}r^2\right]. \tag{30.25}$$

*Proof.* The condition (30.24) is obviously equivalent to:

$$\|Ax\|^2 + \|Bx\|^2 \leq 2\mathrm{Re}\langle (B^*A)x, x\rangle + r^2 \tag{30.26}$$

for any $x \in H$, $\|x\| = 1$. Since, by (30.23),

$$\|Bx\|^2 \geq \frac{1}{\|B^{-1}\|^2}\|x\|^2, \quad x \in H$$

and

$$\mathrm{Re}\langle (B^*A)x, x\rangle \leq |\langle (B^*A)x, x\rangle|,$$

hence by (30.26) we get

$$\|Ax\|^2 + \frac{\|x\|^2}{\|B^{-1}\|^2} \leq 2|\langle (B^*A)x, x\rangle| + r^2 \tag{30.27}$$

for any $x \in H$, $\|x\| = 1$. Taking the supremum over $x \in H$, $\|x\| = 1$ in (30.27), we have

$$\|A\|^2 + \frac{1}{\|B^{-1}\|^2} \leq 2w(B^*A) + r^2. \tag{30.28}$$

By the elementary inequality

$$\frac{2\|A\|}{\|B^{-1}\|} \leq \|A\|^2 + \frac{1}{\|B^{-1}\|^2} \tag{30.29}$$

and by (30.28) we then deduce the desired result (30.25). □

*Remark 30.5.* If we choose above $B = \lambda I$, $\lambda \neq 0$, then we get the inequality

$$(0 \le) \|A\| - w(A) \le \frac{1}{2|\lambda|} r^2, \tag{30.30}$$

provided $\|A - \lambda I\| \le r$. This result has been obtained in the earlier paper [8].

Also, if we assume that $B = \lambda A^*$, $A$ is invertible, then we obtain

$$\|A\| \le \|A^{-1}\| \left[ w\left(A^2\right) + \frac{1}{2|\lambda|} r^2 \right], \tag{30.31}$$

provided $\|A - \lambda A^*\| \le r$, $\lambda \neq 0$.

The following result may be stated as well:

**Theorem 30.9 (Dragomir, 2008 [19]).** *Let $A, B : H \to H$ be two bounded linear operators on $H$. If $B$ is invertible and for $r > 0$,*

$$\|A - B\| \le r, \tag{30.32}$$

*then*

$$(0 \le) \|A\| \|B\| - w(B^* A) \le \frac{1}{2} r^2 + \frac{\|B\|^2 \|B^{-1}\|^2 - 1}{2 \|B^{-1}\|^2}. \tag{30.33}$$

*Proof.* The condition (30.32) is equivalent to $\|Ax\|^2 + \|Bx\|^2 \le 2\mathrm{Re}\langle Ax, Bx\rangle + r^2$ for any $x \in H$, which is clearly equivalent to

$$\|Ax\|^2 + \|B\|^2 \le 2\mathrm{Re}\langle B^* Ax, x\rangle + r^2 + \|B\|^2 - \|Bx\|^2. \tag{30.34}$$

Since

$$\mathrm{Re}\langle B^* Ax, x\rangle \le |\langle B^* Ax, x\rangle|, \quad \|Bx\|^2 \ge \frac{1}{\|B^{-1}\|^2} \|x\|^2$$

and $\|Ax\|^2 + \|B\|^2 \ge 2\|B\| \|Ax\|$ for any $x \in H$, hence by (30.34) we get

$$2\|B\| \|Ax\| \le 2|\langle B^* Ax, x\rangle| + r^2 + \frac{\|B\|^2 \|B^{-1}\|^2 - 1}{\|B^{-1}\|^2} \tag{30.35}$$

for any $x \in H$, $\|x\| = 1$. Taking the supremum over $x \in H$, $\|x\| = 1$ we deduce the desired result (30.33). □

*Remark 30.6.* If we choose in Theorem 30.9, $B = \lambda A^*$, $\lambda \neq 0$, $A$ is invertible, then we get the inequality:

$$(0 \le) \|A\|^2 - w\left(A^2\right) \le \frac{1}{2|\lambda|} r^2 + |\lambda| \cdot \frac{\|A\|^2 \|A^{-1}\|^2 - 1}{2 \|A^{-1}\|^2} \tag{30.36}$$

provided $\|A - \lambda A^*\| \le r$.

The following result may be stated as well.

**Theorem 30.10 (Dragomir, 2008 [19]).** *Let $A, B : H \to H$ be two bounded linear operators on $H$. If $B$ is invertible and for $r > 0$ we have*

$$\|A - B\| \le r < \|B\|, \tag{30.37}$$

*then*

$$\|A\| \le \frac{1}{\sqrt{\|B\|^2 - r^2}} \left( w(B^*A) + \frac{\|B\|^2 \|B^{-1}\|^2 - 1}{2\|B^{-1}\|^2} \right). \tag{30.38}$$

*Proof.* The first part of condition (30.37) is obviously equivalent to $\|Ax\|^2 + \|Bx\|^2 \le 2\mathrm{Re}\langle Ax, Bx\rangle + r^2$ for any $x \in H$, which is clearly equivalent to

$$\|Ax\|^2 + \|B\|^2 - r^2 \le 2\mathrm{Re}\langle B^*Ax, x\rangle + \|B\|^2 - \|Bx\|^2. \tag{30.39}$$

Since

$$\mathrm{Re}\langle B^*Ax, x\rangle \le |\langle B^*Ax, x\rangle|, \|Bx\|^2 \ge \frac{1}{\|B^{-1}\|^2} \|x\|^2$$

and, by the second part of (30.37),

$$\|Ax\|^2 + \|B\|^2 - r^2 \ge 2\sqrt{\|B\|^2 - r^2}\, \|Ax\|,$$

for any $x \in H$, hence by (30.39) we get

$$2\|Ax\| \sqrt{\|B\|^2 - r^2} \le 2|\langle B^*Ax, x\rangle| + \frac{\|B\|^2 \|B^{-1}\|^2 - 1}{\|B^{-1}\|^2} \tag{30.40}$$

for any $x \in H$, $\|x\| = 1$. Taking the supremum over $x \in H$, $\|x\| = 1$ in (30.40), we deduce the desired inequality (30.38). □

*Remark 30.7.* The above Theorem 30.10 has some particular cases of interest. For instance, if we choose $B = \lambda I$, with $|\lambda| > r$, then (30.37) is obviously fulfilled and by (30.38) we get

$$\|A\| \le \frac{w(A)}{\sqrt{1 - \left(\frac{r}{|\lambda|}\right)^2}}, \tag{30.41}$$

provided $\|A - \lambda I\| \le r$. This result has been obtained in the earlier paper [8].

On the other hand, if in the above we choose $B = \lambda A^*$ with $\|A\| \ge \frac{r}{|\lambda|}$ $(\lambda \neq 0)$, then by (30.38) we get

$$\|A\| \le \frac{1}{\sqrt{\|A\|^2 - \left(\frac{r}{|\lambda|}\right)^2}} \left[ w(A^2) + |\lambda| \cdot \frac{\|A\|^2 \|A^{-1}\|^2 - 1}{2\|A^{-1}\|^2} \right], \tag{30.42}$$

provided $\|A - \lambda A^*\| \le r$.

The following result may be stated as well.

**Theorem 30.11 (Dragomir, 2008 [19]).** *Let $A, B$ and $r$ be as in Theorem 30.8. Moreover, if*

$$\left\|B^{-1}\right\| < \frac{1}{r}, \tag{30.43}$$

*then*

$$\|A\| \le \frac{\left\|B^{-1}\right\|}{\sqrt{1 - r^2 \left\|B^{-1}\right\|^2}} w\left(B^*A\right). \tag{30.44}$$

*Proof.* Observe that by (30.28) we have

$$\|A\|^2 + \frac{1 - r^2 \left\|B^{-1}\right\|^2}{\left\|B^{-1}\right\|^2} \le 2w\left(B^*A\right). \tag{30.45}$$

Utilising the elementary inequality

$$2\frac{\|A\|}{\left\|B^{-1}\right\|}\sqrt{1 - r^2 \left\|B^{-1}\right\|^2} \le \|A\|^2 + \frac{1 - r^2 \left\|B^{-1}\right\|^2}{\left\|B^{-1}\right\|^2}, \tag{30.46}$$

which can be stated since (30.43) is assumed to be true; hence, by (30.45) and (30.46) we deduce the desired result (30.44). □

*Remark 30.8.* If we assume that $B = \lambda A^*$ with $\lambda \neq 0$ and $A$ an invertible operator, then, by applying Theorem 30.11, we get the inequality:

$$\|A\| \le \frac{\left\|A^{-1}\right\| w\left(A^2\right)}{\sqrt{|\lambda|^2 - r^2 \left\|A^{-1}\right\|^2}}, \tag{30.47}$$

provided $\|A - \lambda A^*\| \le r$ and $\left\|A^{-1}\right\| \le \frac{|\lambda|}{r}$.

The following result may be stated as well.

**Theorem 30.12 (Dragomir, 2008 [19]).** *Let $A, B : H \to H$ be two bounded linear operators. If $r > 0$ and $B$ is invertible with the property that $\|A - B\| \le r$ and*

$$\frac{1}{\sqrt{r^2 + 1}} \le \left\|B^{-1}\right\| < \frac{1}{r}, \tag{30.48}$$

*then*

$$\|A\|^2 \le w^2\left(B^*A\right) + 2w\left(B^*A\right) \frac{\left\|B^{-1}\right\| - \sqrt{1 - r^2 \left\|B^{-1}\right\|^2}}{\left\|B^{-1}\right\|}. \tag{30.49}$$

*Proof.* Let $x \in H$, $\|x\| = 1$. Then by (30.27) we have

$$\|Ax\|^2 + \frac{1}{\left\|B^{-1}\right\|^2} \le 2\left|\langle B^*Ax, x\rangle\right| + r^2, \tag{30.50}$$

and since $\left\|B^{-1}\right\|^{-2} - r^2 > 0$, we can conclude that $|\langle B^*Ax,x\rangle| > 0$ for any $x \in H$, $\|x\| = 1$.

Dividing (30.50) throughout by $|\langle B^*Ax,x\rangle| > 0$, we obtain

$$\frac{\|Ax\|^2}{|\langle B^*Ax,x\rangle|} \le 2 + \frac{r^2}{|\langle B^*Ax,x\rangle|} - \frac{1}{\|B^{-1}\|^2 |\langle B^*Ax,x\rangle|}. \tag{30.51}$$

Subtracting $|\langle B^*Ax,x\rangle|$ from both sides of (30.51), we get

$$\begin{aligned} \frac{\|Ax\|^2}{|\langle B^*Ax,x\rangle|} - |\langle B^*Ax,x\rangle| &\le 2 - |\langle B^*Ax,x\rangle| - \frac{1 - r^2\left\|B^{-1}\right\|^2}{|\langle B^*Ax,x\rangle| \left\|B^{-1}\right\|^2} \\ &= 2 - \frac{2\sqrt{1 - r^2 \|B^{-1}\|^2}}{\|B^{-1}\|} \\ &\quad - \left(\sqrt{|\langle B^*Ax,x\rangle|} - \frac{\sqrt{1 - r^2 \|B^{-1}\|^2}}{\|B^{-1}\| \sqrt{|\langle B^*Ax,x\rangle|}}\right)^2 \\ &\le 2\left(\frac{\|B^{-1}\| - \sqrt{1 - r^2 \|B^{-1}\|^2}}{\|B^{-1}\|}\right), \end{aligned} \tag{30.52}$$

which gives:

$$\|Ax\|^2 \le |\langle B^*Ax,x\rangle|^2 + 2|\langle B^*Ax,x\rangle| \frac{\|B^{-1}\| - \sqrt{1 - r^2 \|B^{-1}\|^2}}{\|B^{-1}\|}. \tag{30.53}$$

We also remark that, by (30.48) the quantity

$$\|B^{-1}\| - \sqrt{1 - r^2 \|B^{-1}\|^2} \ge 0,$$

hence, on taking the supremum in (30.53) over $x \in H$, $\|x\| = 1$, we deduce the desired inequality. □

*Remark 30.9.* It is interesting to remark that if we assume $\lambda \in \mathbb{C}$ with $0 < r \le |\lambda| \le \sqrt{r^2+1}$ and $\|A - \lambda I\| \le r$, then by (30.24) we can state the following inequality:

$$\|A\|^2 \le |\lambda|^2 w\left(A^2\right) + 2|\lambda| \left(1 - \sqrt{|\lambda|^2 - r^2}\right) w(A). \tag{30.54}$$

Also, if $\|A - A^*\| \le r$, $A$ is invertible and

$$\frac{1}{\sqrt{r^2+1}} \le \|A^{-1}\| \le \frac{1}{r},$$

then, by (30.49) we also have

$$\|A\|^2 \le w^2\left(A^2\right) + 2w\left(A^2\right)\,\frac{\|A^{-1}\| - \sqrt{1-r^2\|A^{-1}\|^2}}{\|A^{-1}\|}. \tag{30.55}$$

One can also prove the following result.

**Theorem 30.13 (Dragomir, 2008 [19]).** *Let $A,B : H \to H$ be two bounded linear operators. If $r > 0$ and $B$ is invertible with the property that $\|A-B\| \le r$ and*

$$\|B^{-1}\| \le \frac{1}{r},$$

*then*

$$\begin{aligned}(0 \le)\,\|A\|^2\|B\|^2 \;&-\; w^2(B^*A)\\ &\le 2w(B^*A)\,\frac{\|B\|}{\|B^{-1}\|}\left(\|B\|\,\|B^{-1}\| - \sqrt{1-r^2\|B^{-1}\|^2}\right).\end{aligned} \tag{30.56}$$

*Proof.* We subtract the quantity

$$\frac{|\langle B^*Ax,x\rangle|}{\|B\|^2}$$

from both sides of (30.51) to obtain

$$\begin{aligned}0 \le \frac{\|Ax\|^2}{|\langle B^*Ax,x\rangle|} &- \frac{|\langle B^*Ax,x\rangle|}{\|B\|^2}\\ &\le 2 - 2\,\frac{\sqrt{1-r^2\|B^{-1}\|^2}}{\|B\|\,\|B^{-1}\|}\\ &\quad - \left(\frac{\sqrt{|\langle B^*Ax,x\rangle|}}{\|B\|} - \frac{\sqrt{1-r^2\|B^{-1}\|^2}}{\sqrt{|\langle B^*Ax,x\rangle|}\,\|B^{-1}\|}\right)^2\\ &\le 2\frac{\left(\|B\|\,\|B^{-1}\| - \sqrt{1-r^2\|B^{-1}\|^2}\right)}{\|B\|\,\|B^{-1}\|},\end{aligned} \tag{30.57}$$

which is equivalent with

$$\begin{aligned}(0 \le)\,\|Ax\|^2\|B\|^2 \;&-\; |\langle B^*Ax,x\rangle|^2\\ &\le 2\frac{\|B\|\,|\langle B^*Ax,x\rangle|}{\|B^{-1}\|}\left(\|B\|\,\|B^{-1}\| - \sqrt{1-r^2\|B^{-1}\|^2}\right)\end{aligned} \tag{30.58}$$

for any $x \in H$, $\|x\| = 1$.

The inequality (30.58) also shows that

$$\|B\| \left\|B^{-1}\right\| \geq \sqrt{1-r^2 \left\|B^{-1}\right\|^2}$$

and then, by (30.58), we get

$$\begin{aligned}\|Ax\|^2 \|B\|^2 &\leq |\langle B^* Ax, x\rangle|^2 \\ &\quad + 2\frac{\|B\|}{\|B^{-1}\|} |\langle B^* Ax, x\rangle| \left( \|B\| \left\|B^{-1}\right\| - \sqrt{1-r^2 \left\|B^{-1}\right\|^2} \right) \qquad (30.59)\end{aligned}$$

for any $x \in X$, $\|x\| = 1$. Taking the supremum in (30.59) we deduce the desired inequality (30.56). □

*Remark 30.10.* The above Theorem 30.13 has some particular instances of interest as follows. If, for instance, we choose $B = \lambda I$ with $|\lambda| \geq r > 0$ and $\|A - \lambda I\| \leq r$, then by (30.56) we obtain the inequality

$$(0 \leq) \|A\|^2 - w^2(A) \leq 2|\lambda| w(A) \left(1 - \sqrt{1 - \frac{r^2}{|\lambda|^2}}\right). \qquad (30.60)$$

Also, if $A$ is invertible, $\|A - \lambda A^*\| \leq r$ and $\left\|A^{-1}\right\| \leq \frac{|\lambda|}{r}$, then by (30.56) we can state:

$$\begin{aligned}(0 \leq) \|A\|^4 &- w^2(A^2) \\ &\leq 2|\lambda| w(A^2) \cdot \frac{\|A\|}{\|A^{-1}\|} \left( \|A\| \left\|A^{-1}\right\| - \sqrt{1 - \frac{r^2}{|\lambda|^2} \left\|A^{-1}\right\|^2} \right). \qquad (30.61)\end{aligned}$$

## 30.2 Other Inequalities for a Product of Two Linear Operators

### *30.2.1 Some Preliminary Results*

For the complex numbers $\alpha, \beta$ and the bounded linear operator $T$ we define the following transform (see [16]):

$$C_{\alpha,\beta}(T) := (T^* - \overline{\alpha} I)(\beta I - T), \qquad (30.62)$$

where by $T^*$ we denote the adjoint of $T$.

We list some properties of the transform $C_{\alpha,\beta}(\cdot)$ that are of interest:

(i) For any $\alpha,\beta \in \mathbb{C}$ and $T \in B(H)$ we have:

$$C_{\alpha,\beta}(I) = (1-\overline{\alpha})(\beta-1)I, \qquad C_{\alpha,\alpha}(T) = -(\alpha I - T)^*(\alpha I - T), \quad (30.63)$$

$$C_{\alpha,\beta}(\gamma T) = |\gamma|^2 C_{\frac{\alpha}{\gamma},\frac{\beta}{\gamma}}(T) \quad \text{for each } \gamma \in \mathbb{C}\backslash\{0\}, \quad (30.64)$$

$$\left[C_{\alpha,\beta}(T)\right]^* = C_{\beta,\alpha}(T) \quad (30.65)$$

and

$$C_{\overline{\beta},\overline{\alpha}}(T^*) - C_{\alpha,\beta}(T) = T^*T - TT^*. \quad (30.66)$$

(ii) The operator $T \in B(H)$ is normal if and only if $C_{\overline{\beta},\overline{\alpha}}(T^*) = C_{\alpha,\beta}(T)$ for each $\alpha,\beta \in \mathbb{C}$.

We recall that a bounded linear operator $T$ on the complex Hilbert space $(H,\langle\cdot,\cdot\rangle)$ is called *accretive* if $\operatorname{Re}\langle Ty,y\rangle \geq 0$ for any $y \in H$.

Utilizing the following identity

$$\begin{aligned}\operatorname{Re}\left\langle C_{\alpha,\beta}(T)x,x\right\rangle &= \operatorname{Re}\left\langle C_{\beta,\alpha}(T)x,x\right\rangle \\ &= \frac{1}{4}|\beta-\alpha|^2 - \left\|\left(T - \frac{\alpha+\beta}{2}I\right)x\right\|^2\end{aligned} \quad (30.67)$$

that holds for any scalars $\alpha,\beta$ and any vector $x \in H$ with $\|x\| = 1$ we can give a simple characterization result that is useful in the following:

**Lemma 30.1 (Dragomir, 2009 [21]).** *For $\alpha,\beta \in \mathbb{C}$ and $T \in B(H)$ the following statements are equivalent:*

*(i) The transform $C_{\alpha,\beta}(T)$ (or, equivalently, $C_{\beta,\alpha}(T)$) is accretive;*
*(ii) The transform $C_{\overline{\alpha},\overline{\beta}}(T^*)$ (or, equivalently, $C_{\overline{\beta},\bar{\alpha}}(T^*)$) is accretive;*
*(iii) We have the norm inequality*

$$\left\|T - \frac{\alpha+\beta}{2}I\right\| \leq \frac{1}{2}|\beta-\alpha| \quad (30.68)$$

*or, equivalently,*

$$\left\|T^* - \frac{\bar{\alpha}+\bar{\beta}}{2}I\right\| \leq \frac{1}{2}|\beta-\alpha|. \quad (30.69)$$

*Remark 30.11.* In order to give examples of operators $T \in B(H)$ and numbers $\alpha, \beta \in \mathbb{C}$ such that the transform $C_{\alpha,\beta}(T)$ is accretive, it suffices to select a bounded linear operator $S$ and the complex numbers $z, w$ with the property that $\|S - zI\| \le |w|$ and, by choosing $T = S$, $\alpha = \frac{1}{2}(z+w)$ and $\beta = \frac{1}{2}(z-w)$ we observe that $T$ satisfies (30.68), i.e., $C_{\alpha,\beta}(T)$ is accretive.

### 30.2.2 Other Norm and Numerical Radius Inequalities

In light of the above results it is then natural to compare the quantities $\|AB\|$ and $w(A)w(B) + w(A)\|B\| + \|A\|w(B)$ provided that some information about the transforms $C_{\alpha,\beta}(A)$ and $C_{\gamma,\delta}(B)$ are available, where $\alpha, \beta, \gamma, \delta \in \mathbb{K}$.

**Theorem 30.14 (Dragomir, 2008, [21]).** *Let $A, B \in B(H)$ and $\alpha, \beta, \gamma, \delta \in \mathbb{K}$ be such that the transforms $C_{\alpha,\beta}(A)$ and $C_{\gamma,\delta}(B)$ are accretive, then*

$$\|BA\| \le w(A)w(B) + w(A)\|B\| + \|A\|w(B) + \frac{1}{4}|\beta - \alpha||\gamma - \delta|. \tag{30.70}$$

*Proof.* Since $C_{\alpha,\beta}(A)$ and $C_{\gamma,\delta}(B)$ are accretive, then, by Lemma 30.1 we have that

$$\left\|Ax - \frac{\alpha+\beta}{2}x\right\| \le \frac{1}{2}|\beta - \alpha| \quad \text{and} \quad \left\|B^*x - \frac{\bar{\gamma}+\bar{\delta}}{2}x\right\| \le \frac{1}{2}|\bar{\gamma} - \bar{\delta}|,$$

for any $x \in H, \|x\| = 1$. Next, utilizing the Schwarz inequality we may write that

$$\begin{aligned} &|\langle Ax - \langle Ax, x\rangle x, B^*y - \langle B^*y, y\rangle y\rangle| \\ &\quad \le \|Ax - \langle Ax, x\rangle x\| \, \|B^*y - \langle B^*y, y\rangle y\|, \end{aligned} \tag{30.71}$$

for any $x, y \in H$, with $\|x\| = \|y\| = 1$.

Since for any vectors $u, f \in H$ with $\|f\| = 1$ we have

$$\|u - \langle u, f\rangle f\| = \inf_{\mu \in \mathbb{K}} \|u - \mu f\|,$$

then obviously

$$\|Ax - \langle Ax, x\rangle x\| \le \left\|Ax - \frac{\alpha+\beta}{2}x\right\| \le \frac{1}{2}|\beta - \alpha|$$

and

$$\|B^*y - \langle B^*y, y\rangle y\| \le \left\|B^*y - \frac{\bar{\gamma}+\bar{\delta}}{2}y\right\| \le \frac{1}{2}|\gamma - \delta|$$

producing the inequality

$$\|Ax - \langle Ax,x\rangle x\| \, \|B^*y - \langle B^*y,y\rangle y\| \le \frac{1}{4}|\beta - \alpha| \, |\gamma - \delta| . \tag{30.72}$$

Now, observe that

$$\begin{aligned}&\langle BAx,y\rangle + \langle Ax,x\rangle\langle By,y\rangle\langle x,y\rangle - \langle Ax,x\rangle\langle Bx,y\rangle - \langle Ax,y\rangle\langle By,y\rangle \\ &\quad = \langle Ax - \langle Ax,x\rangle x, B^*y - \langle B^*y,y\rangle y\rangle,\end{aligned}$$

for any $x,y \in H$, with $\|x\| = \|y\| = 1$. Taking the modulus in the equality and utilizing its properties we have successively

$$\begin{aligned}&|\langle Ax - \langle Ax,x\rangle x, B^*y - \langle B^*y,y\rangle y\rangle| \\ &\quad \ge |\langle BAx,y\rangle| - |\langle Ax,x\rangle \langle Bx,y\rangle + \langle Ax,y\rangle \langle By,y\rangle - \langle Ax,x\rangle \langle By,y\rangle \langle x,y\rangle| \\ &\quad \ge |\langle BAx,y\rangle| - |\langle Ax,x\rangle \langle Bx,y\rangle| - |\langle Ax,y\rangle \langle By,y\rangle| - |\langle Ax,x\rangle \langle By,y\rangle \langle x,y\rangle|\end{aligned}$$

which is equivalent with

$$\begin{aligned}|\langle BAx,y\rangle| \le & |\langle Ax - \langle Ax,x\rangle x, B^*y - \langle B^*y,y\rangle y\rangle| + |\langle Ax,x\rangle \langle Bx,y\rangle| \\ & + |\langle Ax,y\rangle \langle By,y\rangle| + |\langle Ax,x\rangle \langle By,y\rangle \langle x,y\rangle| ,\end{aligned} \tag{30.73}$$

for any $x,y \in H$, with $\|x\| = \|y\| = 1$.

Finally, on making use of the inequalities (30.71)–(30.73) we can state that

$$\begin{aligned}|\langle BAx,y\rangle| \le & \frac{1}{4}|\beta - \alpha| \, |\gamma - \delta| + |\langle Ax,x\rangle \langle Bx,y\rangle| \\ & + |\langle Ax,y\rangle \langle By,y\rangle| + |\langle Ax,x\rangle \langle By,y\rangle \langle x,y\rangle| ,\end{aligned} \tag{30.74}$$

for any $x,y \in H$, with $\|x\| = \|y\| = 1$.

Taking the supremum in (30.74) over $\|x\| = \|y\| = 1$ and noticing that

$$\sup_{\|x\|=1} |\langle Ax,x\rangle| = w(A), \qquad \sup_{\|x\|=\|y\|=1} |\langle Ax,y\rangle| = \|A\| ,$$

$$\sup_{\|y\|=1} |\langle By,y\rangle| = w(B), \qquad \sup_{\|x\|=\|y\|=1} |\langle Bx,y\rangle| = \|B\| ,$$

$$\sup_{\|x\|=\|y\|=1} |\langle BAx,y\rangle| = \|BA\| , \qquad \sup_{\|x\|=\|y\|=1} |\langle x,y\rangle| = 1,$$

we deduce the desired result (30.70). □

*Remark 30.12.* It is an open problem whether or not the constant $1/4$ is best possible in the inequality (30.70).

A different approach is considered in the following result:

**Theorem 30.15 (Dragomir, 2008, [21]).** *With the assumptions from Theorem 30.14 we have the inequality*

$$\|BA\| \le w(A)\|B\| + \frac{1}{4}|\beta-\alpha|(|\gamma+\delta|+|\gamma-\delta|). \tag{30.75}$$

*Proof.* By the Schwarz inequality and taking into account the assumptions for the operators $A$ and $B$ we may state that

$$\begin{aligned}\frac{1}{4}|\beta-\alpha||\gamma-\delta| &\ge \left\|Ax-\frac{\alpha+\beta}{2}x\right\|\left\|B^*y-\frac{\bar{\gamma}+\bar{\delta}}{2}y\right\| \\ &\ge \|Ax-\langle Ax,x\rangle x\|\left\|B^*y-\frac{\bar{\gamma}+\bar{\delta}}{2}y\right\| \\ &\ge \left|\left\langle Ax-\langle Ax,x\rangle x, B^*y-\frac{\bar{\gamma}+\bar{\delta}}{2}y\right\rangle\right|, \end{aligned} \tag{30.76}$$

for any $x,y\in H$, with $\|x\|=\|y\|=1$. Now, since

$$\begin{aligned}&\langle BAx,y\rangle - \langle Ax,x\rangle\langle Bx,y\rangle - \frac{\gamma+\delta}{2}\langle Ax-\langle Ax,x\rangle x,y\rangle \\ &= \left\langle Ax-\langle Ax,x\rangle x, B^*y-\frac{\bar{\gamma}+\bar{\delta}}{2}y\right\rangle,\end{aligned}$$

on taking the modulus in this equality we have

$$\begin{aligned}&|\langle BAx,y\rangle| - |\langle Ax,x\rangle\langle Bx,y\rangle| - \left|\frac{\gamma+\delta}{2}\right||\langle Ax-\langle Ax,x\rangle x,y\rangle| \\ &\ge \left|\left\langle Ax-\langle Ax,x\rangle x, B^*y-\frac{\bar{\gamma}+\bar{\delta}}{2}y\right\rangle\right|,\end{aligned} \tag{30.77}$$

for any $x,y\in H$, with $\|x\|=\|y\|=1$. On making use of (30.76) and (30.77) we get

$$\begin{aligned}|\langle BAx,y\rangle| &\le |\langle Ax,x\rangle\langle Bx,y\rangle| \\ &\quad + \left|\frac{\gamma+\delta}{2}\right||\langle Ax-\langle Ax,x\rangle x,y\rangle| + \frac{1}{4}|\beta-\alpha||\gamma-\delta| \\ &\le |\langle Ax,x\rangle\langle Bx,y\rangle| + \left|\frac{\gamma+\delta}{2}\right|\left\|Ax-\frac{\alpha+\beta}{2}x\right\| \\ &\quad + \frac{1}{4}|\beta-\alpha||\gamma-\delta| \\ &\le |\langle Ax,x\rangle\langle Bx,y\rangle| + \frac{1}{4}|\beta-\alpha|(|\gamma+\delta|+|\gamma-\delta|),\end{aligned} \tag{30.78}$$

for any $x,y\in H$, with $\|x\|=\|y\|=1$. Taking the supremum over $\|x\|=\|y\|=1$ in (30.78) we deduce the desired inequality (30.75). □

In a similar manner we can state the following results as well:

**Theorem 30.16 (Dragomir, 2008, [21]).** *With the assumptions from Theorem 30.14 we have the inequality*

$$\|BA\| \leq w(A)\|B\| + \frac{1}{2}|\gamma+\delta|(w(A)+\|A\|) + \frac{1}{4}|\beta-\alpha||\gamma-\delta|. \tag{30.79}$$

Indeed, we observe that

$$\begin{aligned}
&\langle BAx,y\rangle - \langle Ax,x\rangle\langle Bx,y\rangle - \frac{\gamma+\delta}{2}\langle Ax,y\rangle + \frac{\gamma+\delta}{2}\langle Ax,x\rangle\langle x,y\rangle \\
&= \left\langle Ax - \langle Ax,x\rangle x, B^*y - \frac{\bar{\gamma}+\bar{\delta}}{2}y\right\rangle,
\end{aligned}$$

which produces the inequality

$$\begin{aligned}
|\langle BAx,y\rangle| &\leq \left|\left\langle Ax - \langle Ax,x\rangle x, B^*y - \frac{\bar{\gamma}+\bar{\delta}}{2}y\right\rangle\right| + |\langle Ax,x\rangle\langle Bx,y\rangle| \\
&\quad + \left|\frac{\gamma+\delta}{2}\right||\langle Ax,y\rangle| + \left|\frac{\gamma+\delta}{2}\right||\langle Ax,x\rangle||\langle x,y\rangle|,
\end{aligned}$$

for any $x,y \in H$, with $\|x\| = \|y\| = 1$.

On utilizing the same argument as in the proof of the above theorem, we get the desired result (30.79). The details are omitted.

### 30.2.3 Related Results

The following result concerning an upper bound for the norm of the operator product may be stated.

**Theorem 30.17 (Dragomir, 2008, [21]).** *With the assumptions from Theorem 30.14 we have the inequality*

$$\begin{aligned}
\|BA\| &\leq \left\|\frac{\alpha+\beta}{2}B + \frac{\gamma+\delta}{2} + A\frac{\alpha+\beta}{2} + \frac{\gamma+\delta}{2}I\right\| + \frac{1}{4}|\beta-\alpha||\gamma-\delta| \\
&\leq \min\left\{\left|\frac{\alpha+\beta}{2}\right|\left(\|B\| + \frac{|\beta-\alpha|}{2}\right), \left|\frac{\gamma+\delta}{2}\right|\left(\|A\| + \frac{|\gamma-\delta|}{2}\right)\right\} \\
&\quad + \frac{1}{4}|\beta-\alpha||\gamma-\delta|.
\end{aligned} \tag{30.80}$$

*Proof.* By the Schwarz inequality and utilizing the assumptions about $A$ and $B$ we have

$$\left|\left\langle Ax-\frac{\alpha+\beta}{2}x, B^{*}y-\frac{\bar{\gamma}+\bar{\delta}}{2}y\right\rangle\right| \leq \left\|Ax-\frac{\alpha+\beta}{2}x\right\|\left\|B^{*}y-\frac{\bar{\gamma}+\bar{\delta}}{2}y\right\| \leq \frac{1}{4}|\beta-\alpha||\gamma-\delta|, \tag{30.81}$$

for any $x,y\in H$, with $\|x\|=\|y\|=1$.

Also, the following identity is of interest in itself

$$\left\langle Ax-\frac{\alpha+\beta}{2}x, B^{*}y-\frac{\bar{\gamma}+\bar{\delta}}{2}y\right\rangle = \langle BAx,y\rangle+\frac{\alpha+\beta}{2}\frac{\gamma+\delta}{2}\langle x,y\rangle-\frac{\alpha+\beta}{2}\langle Bx,y\rangle - \frac{\gamma+\delta}{2}\langle Ax,y\rangle, \tag{30.82}$$

for any $x,y\in H$, with $\|x\|=\|y\|=1$.

This identity gives

$$\langle BAx,y\rangle = \left\langle Ax-\frac{\alpha+\beta}{2}x, B^{*}y-\frac{\bar{\gamma}+\bar{\delta}}{2}y\right\rangle + \left\langle \frac{\alpha+\beta}{2}Bx+\frac{\gamma+\delta}{2}Ax-\frac{\alpha+\beta}{2}\frac{\gamma+\delta}{2}x, y\right\rangle,$$

for any $x,y\in H$, with $\|x\|=\|y\|=1$.

Taking the modulus and utilizing (30.81) we get

$$\begin{aligned}|\langle BAx,y\rangle| &\leq \left|\left\langle Ax-\frac{\alpha+\beta}{2}x, B^{*}y-\frac{\bar{\gamma}+\bar{\delta}}{2}y\right\rangle\right| + \left|\left\langle \frac{\alpha+\beta}{2}Bx+\frac{\gamma+\delta}{2}Ax-\frac{\alpha+\beta}{2}\frac{\gamma+\delta}{2}x, y\right\rangle\right| \\ &\leq \frac{1}{4}|\beta-\alpha||\gamma-\delta| + \left\|\frac{\alpha+\beta}{2}Bx+\frac{\gamma+\delta}{2}Ax-\frac{\alpha+\beta}{2}\frac{\gamma+\delta}{2}x\right\|,\end{aligned}$$

for any $x,y\in H$, with $\|x\|=\|y\|=1$. Finally, taking the supremum over $\|x\|=\|y\|=1$ we deduce the first part of the desired inequality (30.80). The second part is obvious by the triangle inequality and by the assumptions on $A$ and $B$. □

The following particular case also holds.

**Corollary 30.2.** *Let $A \in B(H)$ and $\alpha, \beta \in \mathbb{K}$ be such that the transforms $C_{\alpha,\beta}(A)$ is accretive. Then, respectively,*

$$\|A^2\| \leq \frac{1}{4}|\beta-\alpha|^2 + \left|\frac{\alpha+\beta}{2}\right| \left\|2A - \frac{\alpha+\beta}{2} I\right\|$$
$$\left(\leq \frac{1}{4}|\beta-\alpha|^2 + \left|\frac{\alpha+\beta}{2}\right| \left(\|A\| + \frac{1}{2}|\beta-\alpha|\right)\right) \quad (30.83)$$

*and*

$$\|A\|^2 \leq \frac{1}{4}|\beta-\alpha|^2 + \left\|\frac{\bar{\alpha}+\bar{\beta}}{2} A^* + \frac{\alpha+\beta}{2} A - \left|\frac{\alpha+\beta}{2}\right|^2 I\right\|$$
$$\left(\leq \frac{1}{4}|\beta-\alpha|^2 + \left|\frac{\alpha+\beta}{2}\right| \left(\|A\| + \frac{1}{2}|\beta-\alpha|\right)\right). \quad (30.84)$$

The following result provides an approximation for the operator product in terms of some simpler quantities:

**Theorem 30.18 (Dragomir, 2008, [21]).** *With the assumptions from Theorem 30.14 we have the inequality*

$$\left\|BA - \frac{\alpha+\beta}{2} B - \frac{\gamma+\delta}{2} A + \frac{\alpha+\beta}{2}\frac{\gamma+\delta}{2} I\right\| \leq \frac{1}{4}|\beta-\alpha||\gamma-\delta|. \quad (30.85)$$

*Proof.* The identity (30.82) can be written in an equivalent form as

$$\left\langle Ax - \frac{\alpha+\beta}{2} x, B^* y - \frac{\bar{\gamma}+\bar{\delta}}{2} y \right\rangle$$
$$= \left\langle \left(BA - \frac{\alpha+\beta}{2} B - \frac{\gamma+\delta}{2} A + \frac{\alpha+\beta}{2}\frac{\gamma+\delta}{2} I\right) x, y \right\rangle, \quad (30.86)$$

for any $x, y \in H$, with $\|x\| = \|y\| = 1$.

Taking the modulus and making use of the inequality (30.81) we get

$$\left|\left\langle \left(BA - \frac{\alpha+\beta}{2} B - \frac{\gamma+\delta}{2} A + \frac{\alpha+\beta}{2}\frac{\gamma+\delta}{2} I\right) x, y \right\rangle\right|$$
$$\leq \frac{1}{4}|\beta-\alpha||\gamma-\delta|,$$

for any $x, y \in H$, with $\|x\| = \|y\| = 1$, which implies the desired result (30.85). □

**Corollary 30.3.** *Let $A \in B(H)$ and $\alpha, \beta \in \mathbb{K}$ be such that the transform $C_{\alpha,\beta}(A)$ is accretive, then*

$$\left\| A^2 - (\alpha+\beta)\, A + \left(\frac{\alpha+\beta}{2}\right)^2 I \right\| \le \frac{1}{4}\,|\beta-\alpha|^2 \tag{30.87}$$

*and*

$$\left\| A^*A - \frac{\alpha+\beta}{2}\, A^* - \frac{\bar{\alpha}+\bar{\beta}}{2}\, A + \left|\frac{\alpha+\beta}{2}\right|^2 I \right\| \le \frac{1}{4}\,|\beta-\alpha|^2, \tag{30.88}$$

*respectively.*

*Remark 30.13.* It is an open problem whether or not the constant $1/4$ is best possible in either of the inequalities (30.85), (30.87) or (30.88) above.

The next theorem provides an approximation for the operator $\frac{1}{2}(U^*U+UU^*)$ when some information about the real or imaginary part of the operator $U$ are given.

We recall that $U = \mathrm{Re}\,(U) + \mathrm{i}\mathrm{Im}\,(U)$, i.e.,

$$\mathrm{Re}\,(U) = \frac{1}{2}(U+U^*) \quad \text{and} \quad \mathrm{Im}\,(U) = \frac{1}{2\mathrm{i}}(U-U^*).$$

For simplicity, we denote by $A$ the real part of $U$ and by $B$ its imaginary part.

**Theorem 30.19 (Dragomir, 2008, [21]).** *Suppose that $a,b,c,d \in \mathbb{R}$ are such that $C_{a,c}(A)$ and $C_{b,d}(B)$ are accretive. Denote $\alpha := a + \mathrm{i}b$ and $\beta := c + \mathrm{i}d \in \mathbb{C}$, then*

$$\left\| \frac{1}{2}(U^*U+UU^*) - \frac{\bar{\alpha}+\bar{\beta}}{2}\, U - \frac{\alpha+\beta}{2}\, U^* + \left|\frac{\alpha+\beta}{2}\right|^2 I \right\| \le \frac{1}{4}\,|\alpha-\beta|^2. \tag{30.89}$$

*Proof.* It is well known that for any operator $T$ with the Cartesian decomposition

$$T = C + \mathrm{i}D$$

we have

$$\frac{1}{2}(T^*T+TT^*) = C^2 + D^2. \tag{30.90}$$

For any $z \in \mathbb{C}$ we also have the identity

$$\begin{aligned}&\frac{1}{2}(U^*U + UU^*) - \bar{z}\,U - z\,U^* + |z|^2 I \\ &= \frac{1}{2}\left[(U-zI)(U^*-\bar{z}I) + (U^*-\bar{z}I)(U-zI)\right].\end{aligned} \tag{30.91}$$

For $z = (\alpha+\beta)/2$ we observe that

$$\operatorname{Re}(U - zI) = A - \frac{a+c}{2} I$$

and

$$\operatorname{Im}(U - zI) = B - \frac{b+d}{2} I$$

and utilizing the identities (30.90) and (30.91) we deduce

$$\begin{aligned}
&\left\| \frac{1}{2}(U^*U + UU^*) - \bar{z}U - zU^* + |z|^2 I \right\| \\
&= \left\| \left(A - \frac{a+c}{2} I\right)^2 + \left(B - \frac{b+d}{2} I\right)^2 \right\| \\
&\leq \left\| A - \frac{a+c}{2} I \right\|^2 + \left\| B - \frac{b+d}{2} I \right\|^2 \\
&\leq \frac{1}{4}\left[ (c-a)^2 + (d-b)^2 \right] = \frac{1}{4}|\alpha - \beta|^2,
\end{aligned}$$

where for the last inequality we have used the fact that $C_{a,c}(A)$ and $C_{b,d}(B)$ are accretive.

*Remark 30.14.* It is an open problem whether or not the constant $1/4$ is best possible in (30.89).

## 30.3 Power Inequalities for the Numerical Radius of a Product

### *30.3.1 Inequalities for a Product of Two Operators*

The following result for powers of operators holds

**Theorem 30.20 (Dragomir, 2009 [22]).** *For any $A, B \in B(H)$ and $r \geq 1$, we have the inequality:*

$$w^r(B^*A) \leq \frac{1}{2}\|(A^*A)^r + (B^*B)^r\|. \tag{30.92}$$

*The constant $1/2$ is best possible.*

*Proof.* By the Schwarz inequality in the Hilbert space $(H; \langle \cdot, \cdot \rangle)$ we have:

$$\begin{aligned}
|\langle B^*Ax, x\rangle| &= |\langle Ax, Bx\rangle| \leq \|Ax\| \cdot \|Bx\| \\
&= \langle A^*Ax, x\rangle^{\frac{1}{2}} \cdot \langle B^*Bx, x\rangle^{\frac{1}{2}}, \qquad x \in H.
\end{aligned} \tag{30.93}$$

Utilising the arithmetic mean – geometric mean inequality and then the convexity of the function $f(t) = t^r$, $r \geq 1$, for any $x \in H$ we have successively,

$$\langle A^*Ax, x\rangle^{\frac{1}{2}} \cdot \langle B^*Bx, x\rangle^{\frac{1}{2}} \leq \frac{\langle A^*Ax, x\rangle + \langle B^*Bx, x\rangle}{2} \leq \left(\frac{\langle A^*Ax, x\rangle^r + \langle B^*Bx, x\rangle^r}{2}\right)^{\frac{1}{r}} \tag{30.94}$$

It is known that if $P$ is a positive operator then for any $r \geq 1$ and $x \in H$ with $\|x\| = 1$ we have the inequality (see for instance [31])

$$\langle Px, x\rangle^r \leq \langle P^r x, x\rangle. \tag{30.95}$$

Applying this property to the positive operator $A^*A$ and $B^*B$, we deduce that

$$\left(\frac{\langle A^*Ax, x\rangle^r + \langle B^*Bx, x\rangle^r}{2}\right)^{\frac{1}{r}} \leq \left(\frac{\langle (A^*A)^r x, x\rangle + \langle (B^*B)^r x, x\rangle}{2}\right)^{\frac{1}{r}} = \left(\frac{\langle [(A^*A)^r + (B^*B)^r] x, x\rangle}{2}\right)^{\frac{1}{r}} \tag{30.96}$$

for any $x \in H$, $\|x\| = 1$.

Now, on making use of the inequalities (30.93), (30.94) and (30.96), we get the inequality:

$$|\langle (B^*A)^r x, x\rangle|^r \leq \frac{1}{2}\langle [(A^*A)^r + (B^*B)^r] x, x\rangle \tag{30.97}$$

for any $x \in H$, $\|x\| = 1$.

Taking the supremum over $x \in H$, $\|x\| = 1$ in (30.97) and since the operator $[(A^*A)^r + (B^*B)^r]$ is self-adjoint, we deduce the desired inequality (30.92).

For $r = 1$ and $B = A$, we get on both sides of (30.92) the same quantity $\|A\|^2$ which shows that the constant $1/2$ is best possible in general in the inequality (30.92).

**Corollary 30.4.** *For any $A \in B(H)$ and $r \geq 1$ we have the inequalities:*

$$w^r(A) \leq \frac{1}{2}\|(A^*A)^r + I\| \tag{30.98}$$

*and*

$$w^r(A^2) \leq \frac{1}{2}\|(A^*A)^r + (AA^*)^r\|, \tag{30.99}$$

*respectively.*

A different approach is considered in the following result:

**Theorem 30.21 (Dragomir, 2009 [22]).** *For any $A,B\in B(H)$ and any $\alpha\in(0,1)$ and $r\geq 1$, we have the inequality:*

$$w^{2r}(B^*A)\leq\left\|\alpha(A^*A)^{\frac{r}{\alpha}}+(1-\alpha)(B^*B)^{\frac{r}{1-\alpha}}\right\|. \tag{30.100}$$

*Proof.* By Schwarz's inequality, we have:

$$\begin{aligned}|\langle(B^*A)x,x\rangle|^2 &\leq\langle(A^*A)x,x\rangle\langle(B^*B)x,x\rangle\\ &=\left\langle\left[(A^*A)^{\frac{1}{\alpha}}\right]^{\alpha}x,x\right\rangle\left\langle\left[(B^*B)^{\frac{1}{1-\alpha}}\right]^{1-\alpha}x,x\right\rangle,\end{aligned} \tag{30.101}$$

for any $x\in H$.

It is well known that (see for instance [31]) if $P$ is a positive operator and $q\in(0,1]$ then for any $u\in H$, $\|u\|=1$, we have

$$\langle P^q u,u\rangle\leq\langle Pu,u\rangle^q. \tag{30.102}$$

Applying this property to the positive operators

$$(A^*A)^{\frac{1}{\alpha}}\quad\text{and}\quad(B^*B)^{\frac{1}{1-\alpha}},\qquad(\alpha\in(0,1)),$$

we have

$$\begin{aligned}&\left\langle\left[(A^*A)^{\frac{1}{\alpha}}\right]^{\alpha}x,x\right\rangle\left\langle\left[(B^*B)^{\frac{1}{1-\alpha}}\right]^{1-\alpha}x,x\right\rangle\\ &\qquad\leq\left\langle(A^*A)^{\frac{1}{\alpha}}x,x\right\rangle^{\alpha}\left\langle(B^*B)^{\frac{1}{1-\alpha}}x,x\right\rangle^{1-\alpha},\end{aligned} \tag{30.103}$$

for any $x\in H$, $\|x\|=1$.

Now, utilising the weighted arithmetic mean – geometric mean inequality, i.e.,

$$a^{\alpha}b^{1-\alpha}\leq\alpha a+(1-\alpha)b,\qquad\alpha\in(0,1),\ a,b\geq 0,$$

we get

$$\begin{aligned}&\alpha\left\langle(A^*A)^{\frac{1}{\alpha}}x,x\right\rangle+(1-\alpha)\left\langle(B^*B)^{\frac{1}{1-\alpha}}x,x\right\rangle\\ &\qquad\geq\left\langle(A^*A)^{\frac{1}{\alpha}}x,x\right\rangle^{\alpha}\left\langle(B^*B)^{\frac{1}{1-\alpha}}x,x\right\rangle^{1-\alpha}\end{aligned} \tag{30.104}$$

for any $x\in H$, $\|x\|=1$.

Moreover, by the elementary inequality following from the convexity of the function $f(t)=t^r$, $r\geq 1$, namely

$$\alpha a+(1-\alpha)b\geq(\alpha a^r+(1-\alpha)b^r)^{\frac{1}{r}},\qquad\alpha\in(0,1),\ a,b\geq 0,$$

we deduce that

$$\alpha\left\langle (A^*A)^{\frac{1}{\alpha}}x,x\right\rangle + (1-\alpha)\left\langle (B^*B)^{\frac{1}{1-\alpha}}x,x\right\rangle$$
$$\le \left[\alpha\left\langle (A^*A)^{\frac{1}{\alpha}}x,x\right\rangle^r + (1-\alpha)\left\langle (B^*B)^{\frac{1}{1-\alpha}}x,x\right\rangle^r\right]^{\frac{1}{r}}$$
$$\le \left[\alpha\left\langle (A^*A)^{\frac{r}{\alpha}}x,x\right\rangle + (1-\alpha)\left\langle (B^*B)^{\frac{r}{1-\alpha}}x,x\right\rangle\right]^{\frac{1}{r}}, \qquad (30.105)$$

for any $x \in H$, $\|x\| = 1$, where, for the last inequality we used the inequality (30.95) for the positive operators

$$(A^*A)^{\frac{1}{\alpha}}$$

and

$$(B^*B)^{\frac{1}{1-\alpha}}.$$

Now, on making use of the inequalities (30.101) and (30.103)–(30.105), we get

$$|\langle (B^*A)x,x\rangle|^{2r} \le \left\langle \left[\alpha (A^*A)^{\frac{r}{\alpha}} + (1-\alpha)(B^*B)^{\frac{r}{1-\alpha}}\right]x,x\right\rangle \qquad (30.106)$$

for any $x \in H$, $\|x\| = 1$. Taking the supremum over $x \in H$, $\|x\| = 1$ in (30.106) produces the desired inequality (30.100). □

*Remark 30.15.* The particular case $\alpha = 1/2$ produces the inequality

$$w^{2r}(B^*A) \le \frac{1}{2}\left\|(A^*A)^{2r} + (B^*B)^{2r}\right\|, \qquad (30.107)$$

for $r \ge 1$. Notice that $\frac{1}{2}$ is best possible in (30.107) since for $r = 1$ and $B = A$ we get on both sides of (30.107) the same quantity $\|A\|^4$.

**Corollary 30.5.** *For any $A \in B(H)$ and $\alpha \in (0,1)$, $r \ge 1$, we have the inequalities*

$$w^{2r}(A) \le \left\|\alpha (A^*A)^{\frac{r}{\alpha}} + (1-\alpha)I\right\| \qquad (30.108)$$

*and*

$$w^{2r}(A^2) \le \left\|\alpha (A^*A)^{\frac{r}{\alpha}} + (1-\alpha)(AA^*)^{\frac{r}{1-\alpha}}\right\|, \qquad (30.109)$$

*respectively.*

*Moreover, we have*

$$\|A\|^{4r} \le \left\|\alpha (A^*A)^{\frac{r}{\alpha}} + (1-\alpha)(A^*A)^{\frac{r}{1-\alpha}}\right\|. \qquad (30.110)$$

### 30.3.2 Inequalities for the Sum of Two Products

The following result may be stated:

**Theorem 30.22 (Dragomir, 2009 [22]).** *For any $A,B,C,D \in B(H)$ and $r,s \geq 1$ we have:*

$$\left\| \frac{B^*A + D^*C}{2} \right\|^2 \leq \left\| \frac{(A^*A)^r + (C^*C)^r}{2} \right\|^{\frac{1}{r}} \left\| \frac{(B^*B)^s + (D^*D)^s}{2} \right\|^{\frac{1}{s}}. \tag{30.111}$$

*Proof.* By the Schwarz inequality in the Hilbert space $(H;\langle \cdot,\cdot \rangle)$ we have:

$$\begin{aligned} \left[ \langle A^*Ax,x \rangle^{\frac{1}{2}} \langle B^*By,y \rangle^{\frac{1}{2}} + \langle C^*Cx,x \rangle^{\frac{1}{2}} \cdot \langle D^*Dy,y \rangle^{\frac{1}{2}} \right]^2 & \\ \geq [|\langle B^*Ax,y \rangle| + |\langle D^*Cx,y \rangle|]^2 & \\ \geq |\langle B^*Ax,y \rangle + \langle D^*Cx,y \rangle|^2 & \\ = |\langle (B^*A + D^*C)x,y \rangle|^2, & \end{aligned} \tag{30.112}$$

for any $x,y \in H$.

Now, on utilising the elementary inequality:

$$(ab+cd)^2 \leq \left(a^2+c^2\right)\left(b^2+d^2\right), \qquad a,b,c,d \in \mathbb{R},$$

we then conclude that:

$$\begin{aligned} \left[ \langle A^*Ax,x \rangle^{\frac{1}{2}} \langle B^*By,y \rangle^{\frac{1}{2}} + \langle C^*Cx,x \rangle^{\frac{1}{2}} \langle D^*Dy,y \rangle^{\frac{1}{2}} \right]^2 & \\ \leq (\langle A^*Ax,x \rangle + \langle C^*Cx,x \rangle)(\langle B^*By,y \rangle + \langle D^*Dy,y \rangle), & \end{aligned} \tag{30.113}$$

for any $x,y \in H$.

Now, on making use of a similar argument to the one in the proof of Theorem 30.20, we have for $r,s \geq 1$ that

$$\begin{aligned} & (\langle A^*Ax,x \rangle + \langle C^*Cx,x \rangle)(\langle B^*By,y \rangle + \langle D^*Dy,y \rangle) \\ & \leq 4 \left\langle \left[ \frac{(A^*A)^r + (C^*C)^r}{2} \right] x,x \right\rangle^{\frac{1}{r}} \left\langle \left[ \frac{(B^*B)^s + (D^*D)^s}{2} \right] y,y \right\rangle^{\frac{1}{s}} \end{aligned} \tag{30.114}$$

for any $x,y \in H$, $\|x\| = \|y\| = 1$.

Consequently, by (30.112) – (30.114) we have:

$$\begin{aligned} \left\langle \left[ \frac{(A^*A)^r + (C^*C)^r}{2} \right] x,x \right\rangle^{\frac{1}{r}} \left\langle \left[ \frac{(B^*B)^s + (D^*D)^s}{2} \right] y,y \right\rangle^{\frac{1}{s}} & \\ \geq \left| \left\langle \left[ \frac{B^*A + D^*C}{2} \right] x,y \right\rangle \right|^2, & \end{aligned} \tag{30.115}$$

for any $x,y \in H$, $\|x\| = \|y\| = 1$.

Taking the supremum over $x,y \in H$, $\|x\| = \|y\| = 1$ we deduce the desired inequality (30.111). □

*Remark 30.16.* If $s = r$, then the inequality (30.111) is equivalent with:

$$\left\|\frac{B^*A+D^*C}{2}\right\|^{2r} \leq \left\|\frac{(A^*A)^r+(C^*C)^r}{2}\right\|\left\|\frac{(B^*B)^r+(D^*D)^r}{2}\right\|. \tag{30.116}$$

**Corollary 30.6.** *For any $A,C \in B(H)$ we have:*

$$\left\|\frac{A+C}{2}\right\|^{2r} \leq \left\|\frac{(A^*A)^r+(C^*C)^r}{2}\right\|, \tag{30.117}$$

*where $r \geq 1$. Also, we have*

$$\left\|\frac{A^2+C^2}{2}\right\|^{2} \leq \left\|\frac{(A^*A)^r+(C^*C)^r}{2}\right\|^{\frac{1}{r}}\left\|\frac{(AA^*)^s+(CC^*)^s}{2}\right\|^{\frac{1}{s}} \tag{30.118}$$

*for all $r,s \geq 1$, and in particular*

$$\left\|\frac{A^2+C^2}{2}\right\|^{2r} \leq \left\|\frac{(A^*A)^r+(C^*C)^r}{2}\right\|\left\|\frac{(AA^*)^r+(CC^*)^r}{2}\right\| \tag{30.119}$$

*for $r \geq 1$.*

The inequality (30.117) follows from (30.111) for $B = D = I$, while the inequality (30.118) is obtained from the same inequality (30.111) for $B = A^*$ and $D = C^*$.

Another particular result of interest is the following one:

**Corollary 30.7.** *For any $A,B \in B(H)$ we have:*

$$\left\|\frac{B^*A+A^*B}{2}\right\|^{2} \leq \left\|\frac{(A^*A)^r+(B^*B)^r}{2}\right\|^{\frac{1}{r}}\left\|\frac{(A^*A)^s+(B^*B)^s}{2}\right\|^{\frac{1}{s}} \tag{30.120}$$

*for $r,s \geq 1$ and, in particular,*

$$\left\|\frac{B^*A+A^*B}{2}\right\|^{r} \leq \left\|\frac{(A^*A)^r+(B^*B)^r}{2}\right\| \tag{30.121}$$

*for any $r \geq 1$.*

The inequality (30.119) follows from (30.111) for $D = A$ and $C = B$.

Another particular case that might be of interest is the following one.

**Corollary 30.8.** *For any $A,D \in B(H)$ we have:*

$$\left\|\frac{A+D}{2}\right\|^{2} \leq \left\|\frac{(A^*A)^r+I}{2}\right\|^{\frac{1}{r}}\left\|\frac{(DD^*)^s+I}{2}\right\|^{\frac{1}{s}}, \tag{30.122}$$

*where* $r,s \geq 1$. *In particular*

$$\|A\|^2 \leq \left\| \frac{(A^*A)^r + I}{2} \right\|^{\frac{1}{r}} \left\| \frac{(AA^*)^s + I}{2} \right\|^{\frac{1}{s}}. \quad (30.123)$$

*Moreover, for any* $r \geq 1$ *we have*

$$\|A\|^{2r} \leq \left\| \frac{(A^*A)^r + I}{2} \right\| \left\| \frac{(AA^*)^r + I}{2} \right\|.$$

The proof is obvious by the inequality (30.111) on choosing $B = I$, $C = I$ and writing the inequality for $D^*$ instead of $D$.

*Remark 30.17.* If $T \in B(H)$ and $T = A + \mathrm{i}C$, i.e., $A$ and $C$ are its Cartesian decomposition, then we get from (30.117) that

$$\|T\|^{2r} \leq 2^{2r-1} \left\| A^{2r} + C^{2r} \right\|,$$

for any $r \geq 1$.

Also, since

$$A = \mathrm{Re}(T) = \frac{T + T^*}{2}, \qquad C = \mathrm{Im}(T) = \frac{T - T^*}{2\mathrm{i}},$$

from (30.117) we get the following inequalities as well:

$$\|\mathrm{Re}(T)\|^{2r} \leq \left\| \frac{(T^*T)^r + (TT^*)^r}{2} \right\|$$

and

$$\|\mathrm{Im}(T)\|^{2r} \leq \left\| \frac{(T^*T)^r + (TT^*)^r}{2} \right\|$$

for any $r \geq 1$.

In terms of the *Euclidean radius* of two operators $w_e(\cdot,\cdot)$, where, as in [9],

$$w_e(T,U) := \sup_{\|x\|=1} \left( |\langle Tx, x\rangle|^2 + |\langle Ux, x\rangle|^2 \right)^{\frac{1}{2}},$$

we have the following result as well.

**Theorem 30.23 (Dragomir, 2009 [22]).** *For any* $A,B,C,D \in B(H)$ *and* $p,q > 1$ *with*

$$\frac{1}{p} + \frac{1}{q} = 1,$$

*we have the inequality:*

$$w_e^2(B^*A, D^*C) \le \| (A^*A)^p + (C^*C)^p \|^{\frac{1}{p}} \| (B^*B)^q + (D^*D)^q \|^{\frac{1}{q}}. \tag{30.124}$$

*Proof.* For any $x \in H$, $\|x\| = 1$ we have the inequalities

$$\begin{aligned}
&|\langle B^*Ax, x\rangle|^2 + |\langle D^*Cx, x\rangle|^2 \\
&\quad \le \langle A^*Ax, x\rangle \langle B^*Bx, x\rangle + \langle C^*Cx, x\rangle \langle D^*Dx, x\rangle \\
&\quad \le \left( \langle A^*Ax, x\rangle^p + \langle C^*Cx, x\rangle^p \right)^{\frac{1}{p}} \left( \langle B^*Bx, x\rangle^q + \langle D^*Dx, x\rangle^q \right)^{\frac{1}{q}} \\
&\quad \le \left( \langle (A^*A)^p x, x\rangle + \langle (C^*C)^p x, x\rangle \right)^{\frac{1}{p}} \\
&\qquad \times \left( \langle (B^*B)^q x, x\rangle + \langle (D^*D)^q x, x\rangle \right)^{\frac{1}{q}} \\
&\quad = \left\langle [(A^*A)^p + (C^*C)^p] x, x \right\rangle^{\frac{1}{p}} \left\langle [(B^*B)^q + (D^*D)^q] x, x \right\rangle^{\frac{1}{q}}.
\end{aligned}$$

Taking the supremum over $x \in H$, $\|x\| = 1$ and noticing that the operators

$$(A^*A)^p + (C^*C)^p, \qquad (B^*B)^q + (D^*D)^q$$

are self-adjoint, we deduce the desired inequality (30.124).

The following particular case is of interest.

**Corollary 30.9.** *For any $A, C \in B(H)$ and $p, q > 1$ with $\frac{1}{p} + \frac{1}{q} = 1$, we have:*

$$w_e^2(A, C) \le 2^{\frac{1}{q}} \| (A^*A)^p + (C^*C)^p \|^{\frac{1}{p}}.$$

The proof follows from (30.124) for $B = D = I$.

**Corollary 30.10.** *For any $A, D \in B(H)$ and $p, q > 1$ with $\frac{1}{p} + \frac{1}{q} = 1$, we have:*

$$w_e^2(A, D) \le \| (A^*A)^p + I \|^{\frac{1}{p}} \| (D^*D)^q + I \|^{\frac{1}{q}}.$$

### *30.3.3 Vector Inequalities for the Commutator*

The commutator of two bounded linear operators $T$ and $U$ is the operator $TU - UT$. For the usual norm $\|\cdot\|$ and for any two operators $T$ and $U$, by using the triangle inequality and the submultiplicity of the norm, we can state the following inequality:

$$\|TU - UT\| \le 2 \|U\| \|T\|. \tag{30.125}$$

In [17], the following result has been obtained as well

$$\|TU-UT\| \le 2\min\{\|T\|,\|U\|\}\min\{\|T-U\|,\|T+U\|\}. \tag{30.126}$$

By utilising Theorem 30.22 we can state the following result for the numerical radius of the commutator.

**Proposition 30.1 (Dragomir, 2009 [22]).** *For any $T,U\in B(H)$ and $r,s\ge 1$ we have*

$$\|TU-UT\|^2 \le 2^{2-\frac{1}{r}-\frac{1}{s}}\left\|(T^*T)^r+(U^*U)^r\right\|^{\frac{1}{r}}\left\|(TT^*)^s+(UU^*)^s\right\|^{\frac{1}{s}}. \tag{30.127}$$

*Proof.* Follows by Theorem 30.22 on choosing $B=T^*, A=U, D=-U^*$ and $C=T$. □

*Remark 30.18.* In particular, for $r=s$ we get from (30.127) that

$$\|TU-UT\|^{2r} \le 2^{2r-2}\left\|(T^*T)^r+(U^*U)^r\right\|\left\|(TT^*)^r+(UU^*)^r\right\| \tag{30.128}$$

and for $r=1$ we get

$$\|TU-UT\|^2 \le \|T^*T+U^*U\|\,\|TT^*+UU^*\|. \tag{30.129}$$

For a bounded linear operator $T\in B(H)$, the self-commutator is the operator $T^*T-TT^*$. Observe that the operator $V:=-\mathrm{i}(T^*T-TT^*)$ is self-adjoint and $w(V)=\|V\|$, i.e.,

$$w(T^*T-TT^*)=\|T^*T-TT^*\|.$$

Now, utilising (30.127) for $U=T^*$ we can state the following corollary.

**Corollary 30.11.** *For any $T\in B(H)$ we have the inequality:*

$$\|T^*T-TT^*\|^2 \le 2^{2-\frac{1}{r}-\frac{1}{s}}\left\|(T^*T)^r+(TT^*)^r\right\|^{\frac{1}{r}}\left\|(T^*T)^s+(TT^*)^s\right\|^{\frac{1}{s}}. \tag{30.130}$$

*In particular, we have*

$$\|T^*T-TT^*\|^r \le 2^{r-1}\left\|(T^*T)^r+(TT^*)^r\right\|, \tag{30.131}$$

*for any $r\ge 1$.*

*Moreover, for $r=1$ we have*

$$\|T^*T-TT^*\| \le \|T^*T+TT^*\|. \tag{30.132}$$

## 30.4 A Functional Associated with Two Bounded Linear Operators

### 30.4.1 Some Preliminary Facts

For two bounded linear operators $A,B$ in the Hilbert space $(H,\langle\cdot,\cdot\rangle)$, we define the functional

$$\mu(A,B) := \sup_{\|x\|=1}\{\|Ax\|\,\|Bx\|\}\,(\geq 0). \tag{30.133}$$

It is obvious that $\mu$ is symmetric and sub-additive in each variable, $\mu(A,A) = \|A\|^2$, $\mu(A,I) = \|A\|$, where $I$ is the identity operator, $\mu(\alpha A,\beta B) = |\alpha\beta|\,\mu(A,B)$ and $\mu(A,B) \leq \|A\|\,\|B\|$. We also have the following inequalities

$$\mu(A,B) \geq w(B^*A) \tag{30.134}$$

and

$$\mu(A,B)\,\|A\|\,\|B\| \geq \mu(AB,BA). \tag{30.135}$$

The inequality (30.134) follows by the Schwarz inequality $\|Ax\|\,\|Bx\| \geq |\langle Ax,Bx\rangle|$, $x \in H$, while (30.135) can be obtained by multiplying the inequalities $\|ABx\| \leq \|A\|\,\|Bx\|$ and $\|BAx\| \leq \|B\|\,\|Ax\|$.

From (30.134) we also get

$$\|A\|^2 \geq \mu(A,A^*) \geq w(A^2) \tag{30.136}$$

for any $A$.

Motivated by the above results we establish in this section several inequalities for the functional $\mu(\cdot,\cdot)$ under various assumptions for the operators involved, including operators satisfying the uniform $(\alpha,\beta)$-property and operators for which the transform $C_{\alpha,\beta}(\cdot,\cdot)$ is accretive.

### 30.4.2 General Inequalities

The following result concerning some general power operator inequalities may be stated:

**Theorem 30.24 (Dragomir, 2008 [20]).** *For any $A,B \in B(H)$ and $r \geq 1$ we have the inequality*

$$\mu^r(A,B) \leq \frac{1}{2}\|(A^*A)^r + (B^*B)^r\|. \tag{30.137}$$

*The constant $1/2$ is best possible.*

*Proof.* Utilising the arithmetic mean – geometric mean inequality and the convexity of the function $f(t) = t^r$ for $r \geq 1$ and $t \geq 0$ we have successively

$$\begin{aligned} \|Ax\| \, \|Bx\| &\leq \frac{1}{2}\left[\langle A^*Ax, x\rangle + \langle B^*Bx, x\rangle\right] \\ &\leq \left[\frac{\langle A^*Ax, x\rangle^r + \langle B^*Bx, x\rangle^r}{2}\right]^{\frac{1}{r}} \end{aligned} \tag{30.138}$$

for any $x \in H$.

It is well known that if $P$ is a positive operator, then for any $r \geq 1$ and $x \in H$ with $\|x\| = 1$ we have the inequality (see for instance [30])

$$\langle Px, x\rangle^r \leq \langle P^r x, x\rangle . \tag{30.139}$$

Applying this inequality to the positive operators $A^*A$ and $B^*B$ we deduce that

$$\left[\frac{\langle A^*Ax, x\rangle^r + \langle B^*Bx, x\rangle^r}{2}\right]^{\frac{1}{r}} \leq \left\langle \frac{[(A^*A)^r + (B^*B)^r]x}{2}, x\right\rangle^{\frac{1}{r}} \tag{30.140}$$

for any $x \in H$ with $\|x\| = 1$.

Now, on making use of the inequalities (30.138) and (30.140) we get

$$\|Ax\| \, \|Bx\| \leq \left\langle \frac{[(A^*A)^r + (B^*B)^r]x}{2}, x\right\rangle^{\frac{1}{r}} \tag{30.141}$$

for any $x \in H$ with $\|x\| = 1$. Taking the supremum over $x \in H$ with $\|x\| = 1$ we obtain the desired result (30.137).

For $r = 1$ and $B = A$ we get on both sides of (30.137) the same quantity $\|A\|^2$ which shows that the constant $1/2$ is best possible in general in the inequality (30.137).

**Corollary 30.12.** *For any $A \in B(H)$ and $r \geq 1$ we have the inequalities*

$$\mu^r(A, A^*) \leq \frac{1}{2}\|(A^*A)^r + (AA^*)^r\| \tag{30.142}$$

*and*

$$\|A\|^r \leq \frac{1}{2}\|(A^*A)^r + I\|, \tag{30.143}$$

*respectively.*

The following similar result for powers of operators can be stated as well:

**Theorem 30.25 (Dragomir, 2008 [20]).** *For any $A,B \in B(H)$, any $\alpha \in (0,1)$ and $r \geq 1$ we have the inequality*

$$\mu^{2r}(A,B) \leq \left\| \alpha (A^*A)^{\frac{r}{\alpha}} + (1-\alpha)(B^*B)^{\frac{r}{1-\alpha}} \right\|. \tag{30.144}$$

*The inequality is sharp.*

*Proof.* Observe that, for any $\alpha \in (0,1)$ we have

$$\begin{aligned} \|Ax\|^2 \|Bx\|^2 &= \langle (A^*A)x,x\rangle \langle (B^*B)x,x\rangle \\ &= \left\langle \left[(A^*A)^{\frac{1}{\alpha}}\right]^{\alpha} x,x\right\rangle \left\langle \left[(B^*B)^{\frac{1}{1-\alpha}}\right]^{1-\alpha} x,x\right\rangle, \end{aligned} \tag{30.145}$$

where $x \in H$.

It is well known that (see for instance [30]), if $P$ is a positive operator and $q \in (0,1)$, then

$$\langle P^q x,x\rangle \leq \langle Px,x\rangle^q. \tag{30.146}$$

Applying this property to the positive operators $(A^*A)^{1/\alpha}$ and $(B^*B)^{1/(1-\alpha)}$, where $\alpha \in (0,1)$, we have

$$\begin{aligned} &\left\langle \left[(A^*A)^{\frac{1}{\alpha}}\right]^{\alpha} x,x\right\rangle \left\langle \left[(B^*B)^{\frac{1}{1-\alpha}}\right]^{1-\alpha} x,x\right\rangle \\ &\qquad \leq \left\langle (A^*A)^{\frac{1}{\alpha}} x,x\right\rangle^{\alpha} \left\langle (B^*B)^{\frac{1}{1-\alpha}} x,x\right\rangle^{1-\alpha} \end{aligned} \tag{30.147}$$

for any $x \in H$ with $\|x\| = 1$.

Now, on utilising the weighted arithmetic mean-geometric mean inequality, i.e.,

$$a^{\alpha} b^{1-\alpha} \leq \alpha a + (1-\alpha) b,$$

where $\alpha \in (0,1)$ and $a,b \geq 0$, we get

$$\begin{aligned} &\alpha \left\langle (A^*A)^{\frac{1}{\alpha}} x,x\right\rangle + (1-\alpha) \left\langle (B^*B)^{\frac{1}{1-\alpha}} x,x\right\rangle \\ &\qquad \geq \left\langle (A^*A)^{\frac{1}{\alpha}} x,x\right\rangle^{\alpha} \left\langle (B^*B)^{\frac{1}{1-\alpha}} x,x\right\rangle^{1-\alpha} \end{aligned} \tag{30.148}$$

for any $x \in H$ with $\|x\| = 1$.

Moreover, by the elementary inequality

$$\alpha a + (1-\alpha) b \leq (\alpha a^r + (1-\alpha) b^r)^{\frac{1}{r}},$$

where $\alpha \in (0,1)$ and $a,b \geq 0$, we have successively

$$\alpha \left\langle (A^*A)^{\frac{1}{\alpha}} x,x \right\rangle + (1-\alpha) \left\langle (B^*B)^{\frac{1}{1-\alpha}} x,x \right\rangle$$
$$\leq \left[ \alpha \left\langle (A^*A)^{\frac{1}{\alpha}} x,x \right\rangle^r + (1-\alpha) \left\langle (B^*B)^{\frac{1}{1-\alpha}} x,x \right\rangle^r \right]^{\frac{1}{r}}$$
$$\leq \left[ \alpha \left\langle (A^*A)^{\frac{r}{\alpha}} x,x \right\rangle + (1-\alpha) \left\langle (B^*B)^{\frac{r}{1-\alpha}} x,x \right\rangle \right]^{\frac{1}{r}}, \quad (30.149)$$

for any $x \in H$ with $\|x\| = 1$, where for the last inequality we have used the property (30.139) for the positive operators $(A^*A)^{1/\alpha}$ and $(B^*B)^{1/(1-\alpha)}$.

Now, on making use of the identity (30.145) and the inequalities (30.147)–(30.149) we get

$$\|Ax\|^2 \|Bx\|^2 \leq \left[ \left\langle \left[ \alpha (A^*A)^{\frac{r}{\alpha}} + (1-\alpha) (B^*B)^{\frac{r}{1-\alpha}} \right] x,x \right\rangle \right]^{\frac{1}{r}}$$

for any $x \in H$ with $\|x\| = 1$. Taking the supremum over $x \in H$ with $\|x\| = 1$ we deduce the desired result (30.144).

Notice that the inequality is sharp since for $r = 1$ and $B = A$ we get on both sides of (30.144) the same quantity $\|A\|^4$.

**Corollary 30.13.** *For any $A \in B(H)$, any $\alpha \in (0,1)$ and $r \geq 1$, we have the inequalities*

$$\mu^{2r}(A,A^*) \leq \left\| \alpha (A^*A)^{\frac{r}{\alpha}} + (1-\alpha) (AA^*)^{\frac{r}{1-\alpha}} \right\|,$$
$$\|A\|^{2r} \leq \left\| \alpha (A^*A)^{\frac{r}{\alpha}} + (1-\alpha) I \right\|$$

*and*

$$\|A\|^{4r} \leq \left\| \alpha (A^*A)^{\frac{r}{\alpha}} + (1-\alpha) (A^*A)^{\frac{r}{1-\alpha}} \right\|,$$

*respectively.*

The following reverse of the inequality (30.134) may be stated as well:

**Theorem 30.26 (Dragomir, 2008 [20]).** *For any $A,B \in B(H)$ we have the inequalities*

$$(0 \leq) \mu(A,B) - w(B^*A) \leq \frac{1}{2} \|A-B\|^2 \quad (30.150)$$

*and*

$$\mu\left(\frac{A+B}{2}, \frac{A-B}{2}\right) \leq \frac{1}{2} w(B^*A) + \frac{1}{4} \|A-B\|^2, \quad (30.151)$$

*respectively.*

*Proof.* We have

$$\begin{aligned}\|Ax - Bx\|^2 &= \|Ax\|^2 + \|Bx\|^2 - 2\mathrm{Re}\,\langle B^*Ax, x\rangle \\ &\geq 2\,\|Ax\|\,\|Bx\| - 2\,|\langle B^*Ax, x\rangle|\,,\end{aligned} \tag{30.152}$$

for any $x \in H, \|x\| = 1$, which gives the inequality

$$\|Ax\|\,\|Bx\| \leq |\langle B^*Ax, x\rangle| + \frac{1}{2}\,\|Ax - Bx\|^2\,,$$

for any $x \in H, \|x\| = 1$.

Taking the supremum over $\|x\| = 1$ we deduce the desired result (30.150).

By the parallelogram identity in the Hilbert space $H$ we also have

$$\|Ax\|^2 + \|Bx\|^2 = \frac{1}{2}\left(\|Ax + Bx\|^2 + \|Ax - Bx\|^2\right) \geq \|Ax + Bx\|\,\|Ax - Bx\|\,,$$

for any $x \in H$.

Combining this inequality with the first part of (30.152) we get

$$\|Ax + Bx\|\,\|Ax - Bx\| \leq \|Ax - Bx\|^2 + 2\,|\langle B^*Ax, x\rangle|\,,$$

for any $x \in H$. Taking the supremum in this inequality over $\|x\| = 1$ we deduce the desired result (30.151). □

**Corollary 30.14.** *Let $A \in B(H)$. If $\mathrm{Re}\,(A) := \frac{A+A^*}{2}$ and $\mathrm{Im}\,(A) := \frac{A-A^*}{2i}$ are the real and imaginary parts of $A$, then we have the inequalities*

$$(0 \leq)\,\mu\,(A, A^*) - w\left(A^2\right) \leq 2\,\|\mathrm{Im}\,(A)\|^2$$

*and*

$$\mu\,(\mathrm{Re}\,(A)\,, \mathrm{Im}\,(A)) \leq \frac{1}{2} w\left(A^2\right) + \|\mathrm{Im}\,(A)\|^2\,,$$

*respectively.*

*Moreover, we have*

$$(0 \leq)\,\mu\,(\mathrm{Re}\,(A)\,, \mathrm{Im}\,(A)) - w\,(\mathrm{Re}\,(A)\,\mathrm{Im}\,(A)) \leq \frac{1}{2}\,\|A\|^2\,.$$

**Corollary 30.15.** *For any $A \in B(H)$ and $\lambda \in \mathbb{C}$ with $\lambda \neq 0$ we have the inequality (see also [7])*

$$(0 \leq)\,\|A\| - w\,(A) \leq \frac{1}{2\,|\lambda|}\,\|A - \lambda I\|^2\,. \tag{30.153}$$

For a bounded linear operator $T$ consider the quantity

$$\ell(T) := \inf_{\|x\|=1} \|Tx\|.$$

We can state the following result as well.

**Theorem 30.27 (Dragomir, 2008 [20]).** *For any $A, B \in B(H)$ with $A \neq B$ and such that $\ell(B) \geq \|A - B\|$ we have*

$$(0 \leq) \mu^2(A,B) - w^2(B^*A) \leq \|A\|^2 \|A - B\|^2. \tag{30.154}$$

*Proof.* Denote $r := \|A - B\| > 0$. Then for any $x \in H$ with $\|x\| = 1$ we have $\|Bx\| \geq r$ and by the first part of (30.152) we can write that

$$\|Ax\|^2 + \left(\sqrt{\|Bx\|^2 - r^2}\right)^2 \leq 2|\langle B^*Ax, x\rangle| \tag{30.155}$$

for any $x \in H$ with $\|x\| = 1$.

On the other hand we have

$$\|Ax\|^2 + \left(\sqrt{\|Bx\|^2 - r^2}\right)^2 \geq 2\|Ax\|\sqrt{\|Bx\|^2 - r^2} \tag{30.156}$$

for any $x \in H$ with $\|x\| = 1$.

Combining (30.155) with (30.156) we deduce

$$\|Ax\|\sqrt{\|Bx\|^2 - r^2} \leq |\langle B^*Ax, x\rangle|,$$

which is clearly equivalent to

$$\|Ax\|^2 \|Bx\|^2 \leq |\langle B^*Ax, x\rangle|^2 + \|Ax\|^2 \|A - B\|^2 \tag{30.157}$$

for any $x \in H$ with $\|x\| = 1$. Taking the supremum in (30.157) over $x \in H$ with $\|x\| = 1$, we deduce the desired inequality (30.154).

**Corollary 30.16.** *For any $A \in B(H)$ a non-self-adjoint operator in $B(H)$ and such that*

$$\ell(A^*) \geq \|\mathrm{Im}(A)\|$$

*we have*

$$(0 \leq) \mu^2(A, A^*) - w^2(A^2) \leq 4\|A\|^2 \|\mathrm{Im}(A)\|^2. \tag{30.158}$$

**Corollary 30.17.** *For any $A \in B(H)$ and $\lambda \in \mathbb{C}$ with $\lambda \neq 0$ and*

$$|\lambda| \geq \|A - \lambda I\|$$

*we have the inequality (see also [7])*

$$(0 \leq) \|A\|^2 - w^2(A) \leq \frac{1}{|\lambda|^2} \|A\|^2 \|A - \lambda I\|^2$$

*or, equivalently,*

$$(0 \leq) \sqrt{1 - \frac{\|A - \lambda I\|^2}{|\lambda|^2}} \leq \frac{w(A)}{\|A\|} (\leq 1).$$

### 30.4.3 Inequalities for Operators Satisfying the Uniform $(\alpha, \beta)$-property

The following result that may be of interest in itself holds:

**Lemma 30.2.** *Let $T \in B(H)$ and $\alpha, \beta \in \mathbb{C}$ with $\alpha \neq \beta$. The following statements are equivalent:*

*(i) We have*

$$\operatorname{Re}\langle \beta y - Tx, Tx - \alpha y\rangle \geq 0 \qquad (30.159)$$

*for any $x, y \in H$ with $\|x\| = \|y\| = 1$;*

*(ii) We have*

$$\left\| Tx - \frac{\alpha + \beta}{2} y \right\| \leq \frac{1}{2} |\alpha - \beta| \qquad (30.160)$$

*for any $x, y \in H$ with $\|x\| = \|y\| = 1$.*

*Proof.* This follows by the following identity

$$\operatorname{Re}\langle \beta y - Tx, Tx - \alpha y\rangle = \frac{1}{4} |\alpha - \beta|^2 - \left\| Tx - \frac{\alpha + \beta}{2} y \right\|^2,$$

that holds for any $x, y \in H$ with $\|x\| = \|y\| = 1$. □

*Remark 30.19.* For any operator $T \in B(H)$ if we choose $\alpha = a\|T\|(1 + 2i)$ and $\beta = a\|T\|(1 - 2i)$ with $a \geq 1$, then

$$\frac{\alpha + \beta}{2} = a\|T\| \quad \text{and} \quad \frac{|\alpha - \beta|}{2} = 2a\|T\|$$

showing that

$$\left\|Tx-\frac{\alpha+\beta}{2}y\right\| \le \|Tx\|+\left|\frac{\alpha+\beta}{2}\right| \le \|T\|+a\|T\|$$
$$\le 2a\|T\| = \frac{1}{2}|\alpha-\beta|,$$

that holds for any $x,y \in H$ with $\|x\| = \|y\| = 1$, i.e., $T$ satisfies the condition (30.159) with the scalars $\alpha$ and $\beta$ given above.

**Definition 30.1.** For given $\alpha,\beta \in \mathbb{C}$ with $\alpha \ne \beta$ and $y \in H$ with $\|y\| = 1$, we say that the operator $T \in B(H)$ has the $(\alpha,\beta,y)$-property if either (30.159) or, equivalently, (30.160) holds true for any $x \in H$ with $\|x\| = 1$. Moreover, if $T$ has the $(\alpha,\beta,y)$-property for any $y \in H$ with $\|y\| = 1$, then we say that this operator has the uniform $(\alpha,\beta)$-property.

*Remark 30.20.* The above Remark 30.19 shows that any bounded linear operator has the uniform $(\alpha,\beta)$-property for infinitely many $(\alpha,\beta)$ appropriately chosen. For a given operator satisfying an $(\alpha,\beta)$-property, it is an open problem to find the lower bound for the nonzero quantity $|\alpha-\beta|$.

The following results may be stated:

**Theorem 30.28 (Dragomir, 2008 [20]).** *Let $A,B \in B(H)$ and $\alpha,\beta,\gamma,\delta \in \mathbb{K}$ with $\alpha \ne \beta$ and $\gamma \ne \delta$. For $y \in H$ with $\|y\| = 1$ assume that $A^*$ has the $(\alpha,\beta,y)$-property while $B^*$ has the $(\gamma,\delta,y)$-property. Then*

$$\left|\|Ay\|\,\|By\| - \|BA^*\|\right| \le \frac{1}{4}|\beta-\alpha|\,|\gamma-\delta|. \tag{30.161}$$

*Moreover, if $A^*$ has the uniform $(\alpha,\beta)$-property and $B^*$ has the uniform $(\gamma,\delta)$-property, then*

$$\left|\mu(A,B) - \|BA^*\|\right| \le \frac{1}{4}|\beta-\alpha|\,|\gamma-\delta|. \tag{30.162}$$

*Proof.* $A^*$ has the $(\alpha,\beta,y)$-property while $B^*$ has the $(\gamma,\delta,y)$-property, then on making use of Lemma 30.2 we have that

$$\left\|A^*x-\frac{\alpha+\beta}{2}y\right\| \le \frac{1}{2}|\beta-\alpha|$$

and

$$\left\|B^*y-\frac{\gamma+\delta}{2}y\right\| \le \frac{1}{2}|\gamma-\delta|$$

for any $x,y \in H$ with $\|x\| = \|y\| = 1$.

Now, we make use of the following Grüss type inequality for vectors in inner product spaces obtained by the author in [1] (see also [2] or [6, p. 43]):

Let $(H,\langle\cdot,\cdot\rangle)$ be an inner product space over the real or complex number field $\mathbb{K}$, $u,v,e \in H$, $\|e\| = 1$, and $\alpha,\beta,\gamma,\delta \in \mathbb{K}$ such that

$$\operatorname{Re}\langle\beta e - u, u - \alpha e\rangle \geq 0, \qquad \operatorname{Re}\langle\delta e - v, v - \gamma e\rangle \geq 0 \tag{30.163}$$

or, equivalently,

$$\left\| u - \frac{\alpha+\beta}{2}e \right\| \leq \frac{1}{2}|\beta - \alpha|, \qquad \left\| v - \frac{\gamma+\delta}{2}e \right\| \leq \frac{1}{2}|\delta - \gamma|. \tag{30.164}$$

Then

$$|\langle u,v\rangle - \langle u,e\rangle\langle e,v\rangle| \leq \frac{1}{4}|\beta - \alpha||\delta - \gamma|. \tag{30.165}$$

Applying (30.165) for $u = A^*x$, $v = B^*z$ and $e = y$ we deduce

$$|\langle BA^*x,z\rangle - \langle x,Ay\rangle\langle By,z\rangle| \leq \frac{1}{4}|\beta - \alpha||\delta - \gamma|, \tag{30.166}$$

for any $x,z \in H$, $\|x\| = \|z\| = 1$, which is an inequality of interest in itself.

Observing that

$$||\langle BA^*x,z\rangle| - |\langle x,Ay\rangle\langle z,By\rangle|| \leq |\langle BA^*x,z\rangle - \langle x,Ay\rangle\langle By,z\rangle|,$$

then by (30.166) we deduce the inequality

$$||\langle BA^*x,z\rangle| - |\langle x,Ay\rangle\langle z,By\rangle|| \leq \frac{1}{4}|\beta - \alpha||\delta - \gamma|$$

for any $x,z \in H$, $\|x\| = \|z\| = 1$. This is equivalent to the following two inequalities

$$|\langle BA^*x,z\rangle| \leq |\langle x,Ay\rangle\langle z,By\rangle| + \frac{1}{4}|\beta - \alpha||\delta - \gamma| \tag{30.167}$$

and

$$|\langle x,Ay\rangle\langle z,By\rangle| \leq |\langle BA^*x,z\rangle| + \frac{1}{4}|\beta - \alpha||\delta - \gamma| \tag{30.168}$$

for any $x,z \in H$, $\|x\| = \|z\| = 1$.

Taking the supremum over $x,z \in H$, $\|x\| = \|z\| = 1$, in (30.167) and (30.168) we get the inequalities

$$\|BA^*\| \leq \|Ay\|\,\|By\| + \frac{1}{4}|\beta - \alpha||\delta - \gamma| \tag{30.169}$$

and

$$\|Ay\|\,\|By\| \leq \|BA^*\| + \frac{1}{4}|\beta-\alpha|\,|\delta-\gamma|, \tag{30.170}$$

which are clearly equivalent to (30.161).

Now, if $A^*$ has the uniform $(\alpha,\beta)$-property and $B^*$ has the uniform $(\gamma,\delta)$-property, then the inequalities (30.169) and (30.170) hold for any $y \in H$ with $\|y\| = 1$. Taking the supremum over $y \in H$ with $\|y\| = 1$ in these inequalities we deduce

$$\|BA^*\| \leq \mu(A,B) + \frac{1}{4}|\beta-\alpha|\,|\delta-\gamma|$$

and

$$\mu(A,B) \leq \|BA^*\| + \frac{1}{4}|\beta-\alpha|\,|\delta-\gamma|,$$

which are equivalent to (30.162). □

**Corollary 30.18.** *Let $A \in B(H)$ and $\alpha,\beta,\gamma,\delta \in \mathbb{K}$ with $\alpha \neq \beta$ and $\gamma \neq \delta$. For $y \in H$ with $\|y\| = 1$ assume that $A$ has the $(\alpha,\beta,y)$-property while $A^*$ has the $(\gamma,\delta,y)$-property. Then*

$$\left|\|A^*y\|\,\|Ay\| - \|A^2\|\right| \leq \frac{1}{4}|\beta-\alpha|\,|\gamma-\delta|.$$

*Moreover, if $A$ has the uniform $(\alpha,\beta)$-property and $A^*$ has the uniform $(\gamma,\delta)$-property, then*

$$\left|\mu(A,A^*) - \|A^2\|\right| \leq \frac{1}{4}|\beta-\alpha|\,|\gamma-\delta|.$$

The following results may be stated as well:

**Theorem 30.29 (Dragomir, 2008 [20]).** *Let $A,B \in B(H)$ and $\alpha,\beta,\gamma,\delta \in \mathbb{K}$ with $\alpha+\beta \neq 0$ and $\gamma+\delta \neq 0$. For $y \in H$ with $\|y\| = 1$ assume that $A^*$ has the $(\alpha,\beta,y)$-property while $B^*$ has the $(\gamma,\delta,y)$-property. Then*

$$\begin{aligned} &\left|\,\|Ay\|\,\|By\| \;-\; \|BA^*\|\,\right| \\ &\quad \leq \frac{1}{4}\frac{|\beta-\alpha|\,|\delta-\gamma|}{\sqrt{|\beta+\alpha|\,|\delta+\gamma|}}\sqrt{(\|A\|+\|Ay\|)(\|B\|+\|By\|)}. \end{aligned} \tag{30.171}$$

*Moreover, if $A^*$ has the uniform $(\alpha,\beta)$-property and $B^*$ has the uniform $(\gamma,\delta)$-property, then*

$$\left|\mu(A,B) - \|BA^*\|\right| \leq \frac{1}{2}\frac{|\beta-\alpha|\,|\delta-\gamma|}{\sqrt{|\beta+\alpha|\,|\delta+\gamma|}}\sqrt{\|A\|\,\|B\|}. \tag{30.172}$$

*Proof.* We make use of the following inequality obtained by the author in [5] (see also [6, p. 65]):

Let $(H, \langle \cdot, \cdot \rangle)$ be an inner product space over the real or complex number field $\mathbb{K}$, $u, v, e \in H$, $\|e\| = 1$, and $\alpha, \beta, \gamma, \delta \in \mathbb{K}$ with $\alpha + \beta \neq 0$ and $\gamma + \delta \neq 0$ such that

$$\operatorname{Re}\langle \beta e - u, u - \alpha e \rangle \geq 0, \qquad \operatorname{Re}\langle \delta e - v, v - \gamma e \rangle \geq 0$$

or, equivalently,

$$\left\| u - \frac{\alpha + \beta}{2} e \right\| \leq \frac{1}{2} |\beta - \alpha|, \qquad \left\| v - \frac{\gamma + \delta}{2} e \right\| \leq \frac{1}{2} |\delta - \gamma|.$$

Then

$$\begin{aligned} &|\langle u, v \rangle - \langle u, e \rangle \langle e, v \rangle| \\ &\leq \frac{1}{4} \frac{|\beta - \alpha| |\delta - \gamma|}{\sqrt{|\beta + \alpha| |\delta + \gamma|}} \sqrt{(\|u\| + |\langle u, e \rangle|)(\|v\| + |\langle v, e \rangle|)}. \end{aligned} \tag{30.173}$$

Applying (30.173) for $u = A^* x$, $v = B^* z$ and $e = y$ we deduce

$$\begin{aligned} &|\langle BA^* x, z \rangle - \langle x, Ay \rangle \langle By, z \rangle| \\ &\leq \frac{1}{4} \frac{|\beta - \alpha| |\delta - \gamma|}{\sqrt{|\beta + \alpha| |\delta + \gamma|}} \sqrt{(\|A^* x\| + |\langle x, Ay \rangle|)(\|B^* z\| + |\langle z, By \rangle|)} \end{aligned}$$

for any $x, y, z \in H$, $\|x\| = \|y\| = \|z\| = 1$.

Now, on making use of a similar argument to the one from the proof of Theorem 30.28, we deduce the desired results (30.171) and (30.172). The details are omitted. □

**Corollary 30.19.** *Let $A \in B(H)$ and $\alpha, \beta, \gamma, \delta \in \mathbb{K}$ with $\alpha + \beta \neq 0$ and $\gamma + \delta \neq 0$. For $y \in H$ with $\|y\| = 1$ assume that $A$ has the $(\alpha, \beta, y)$-property while $A^*$ has the $(\gamma, \delta, y)$-property. Then*

$$\left| \|A^* y\| \|Ay\| - \|A^2\| \right| \leq \frac{1}{4} \frac{|\beta - \alpha| |\delta - \gamma|}{\sqrt{|\beta + \alpha| |\delta + \gamma|}} \sqrt{(\|A\| + \|A^* y\|)(\|A\| + \|Ay\|)}.$$

*Moreover, if $A$ has the uniform $(\alpha, \beta)$-property and $A^*$ has the uniform $(\gamma, \delta)$-property, then*

$$\left| \mu(A, A^*) - \|A^2\| \right| \leq \frac{1}{2} \frac{|\beta - \alpha| |\delta - \gamma|}{\sqrt{|\beta + \alpha| |\delta + \gamma|}} \|A\|.$$

### 30.4.4 The Transform $C_{\alpha,\beta}(\cdot,\cdot)$ and Other Inequalities

For two given operators $T,U \in B(H)$ and two given scalars $\alpha,\beta \in \mathbb{C}$ consider the transform

$$C_{\alpha,\beta}(T,U) = (T^* - \bar{\alpha}U^*)(\beta U - T).$$

This transform generalizes the transform

$$C_{\alpha,\beta}(T) := (T^* - \bar{\alpha}I)(\beta I - T) = C_{\alpha,\beta}(T,I),$$

where $I$ is the identity operator, which has been introduced in [16] in order to provide some generalizations of the well known Kantorovich inequality for operators in Hilbert spaces.

We recall that a bounded linear operator $T$ on the complex Hilbert space $(H,\langle\cdot,\cdot\rangle)$ is called *accretive* if $\operatorname{Re}\langle Ty,y\rangle \geq 0$ for any $y \in H$.

Utilizing the following identity

$$\begin{aligned}\operatorname{Re}\left\langle C_{\alpha,\beta}(T,U)x,x\right\rangle &= \operatorname{Re}\left\langle C_{\beta,\alpha}(T,U)x,x\right\rangle \\ &= \frac{1}{4}|\beta-\alpha|^2\|Ux\|^2 - \left\|Tx - \frac{\alpha+\beta}{2}\cdot Ux\right\|^2, \qquad (30.174)\end{aligned}$$

that holds for any scalars $\alpha,\beta$ and any vector $x \in H$, we can give a simple characterization result that is useful in the following:

**Lemma 30.3.** *For $\alpha,\beta \in \mathbb{C}$ and $T,U \in B(H)$ the following statements are equivalent:*

*(i) The transform $C_{\alpha,\beta}(T,U)$ (or, equivalently, $C_{\beta,\alpha}(T,U)$) is accretive;*
*(ii) We have the norm inequality*

$$\left\|Tx - \frac{\alpha+\beta}{2}\cdot Ux\right\| \leq \frac{1}{2}|\beta-\alpha|\,\|Ux\| \qquad (30.175)$$

*for any $x \in H$.*

As a consequence of the above lemma we can state

**Corollary 30.20.** *Let $\alpha,\beta \in \mathbb{C}$ and $T,U \in B(H)$. If $C_{\alpha,\beta}(T,U)$ is accretive, then*

$$\left\|T - \frac{\alpha+\beta}{2}U\right\| \leq \frac{1}{2}|\beta-\alpha|\,\|U\|. \qquad (30.176)$$

*Remark 30.21.* In order to give examples of linear operators $T,U \in B(H)$ and numbers $\alpha,\beta \in \mathbb{C}$ such that the transform $C_{\alpha,\beta}(T,U)$ is accretive, it suffices to select two bounded linear operators $S$ and $V$ and the complex numbers $z,w$ $(w \neq 0)$

with the property that $\|Sx - zVx\| \le |w| \, \|Vx\|$ for any $x \in H$, and, by choosing $T = S$, $U = V$,

$$\alpha = \frac{1}{2}(z+w)$$

and

$$\beta = \frac{1}{2}(z-w)$$

we observe that $T$ and $U$ satisfy (30.175), i.e., $C_{\alpha,\beta}(T,U)$ is accretive.

We are able now to give the following result concerning other reverse inequalities for the case when the involved operators satisfy the accretivity property described above.

**Theorem 30.30 (Dragomir, 2008 [20]).** *Let $\alpha, \beta \in \mathbb{C}$ and $A, B \in B(H)$. If $C_{\alpha,\beta}(A,B)$ is accretive, then*

$$(0 \le)\, \mu^2(A,B) - w^2(B^*A) \le \frac{1}{4}|\beta - \alpha|^2 \|B\|^4. \tag{30.177}$$

*Moreover, if $\alpha + \beta \neq 0$, then*

$$(0 \le)\, \mu(A,B) - w(B^*A) \le \frac{1}{4}\frac{|\beta - \alpha|^2}{|\beta + \alpha|}\|B\|^2. \tag{30.178}$$

*In addition, if $\mathrm{Re}\left(\alpha\bar{\beta}\right) > 0$ and $B^*A \neq 0$, then also*

$$(1 \le)\, \frac{\mu(A,B)}{w(B^*A)} \le \frac{1}{2}\frac{|\beta + \alpha|}{\sqrt{\mathrm{Re}\left(\alpha\bar{\beta}\right)}} \tag{30.179}$$

*and*

$$(0 \le)\, \mu^2(A,B) - w^2(B^*A) \le \left(|\beta + \alpha| - 2\sqrt{\mathrm{Re}\left(\alpha\bar{\beta}\right)}\right) w(B^*A)\|B\|^2, \tag{30.180}$$

*respectively.*

*Proof.* By Lemma 30.3, since $C_{\alpha,\beta}(A,B)$ is accretive, then

$$\left\| Ax - \frac{\alpha+\beta}{2}Bx \right\| \le \frac{1}{2}|\beta - \alpha| \, \|Bx\| \tag{30.181}$$

for any $x \in H$.

We utilize the following reverse of the Schwarz inequality in inner product spaces obtained by the author in [3] (see also [6, p. 4]):

If $\gamma, \Gamma \in \mathbb{K}$ ($\mathbb{K} = \mathbb{C}, \mathbb{R}$) and $u, v \in H$ are such that

$$\mathrm{Re}\langle \Gamma v - u, u - \gamma v \rangle \ge 0 \tag{30.182}$$

or, equivalently,

$$\left\| u - \frac{\gamma + \Gamma}{2} v \right\| \le \frac{1}{2} |\Gamma - \gamma| \, \|v\|, \tag{30.183}$$

then

$$0 \le \|u\|^2 \|v\|^2 - |\langle u, v\rangle|^2 \le \frac{1}{4} |\Gamma - \gamma|^2 \|v\|^4. \tag{30.184}$$

Now, on making use of (30.184) for $u = Ax$, $v = Bx$, $x \in H$, $\|x\| = 1$ and $\gamma = \alpha$, $\Gamma = \beta$ we can write the inequality

$$\|Ax\|^2 \|Bx\|^2 \le |\langle B^* Ax, x\rangle|^2 + \frac{1}{4} |\beta - \alpha|^2 \|Bx\|^4,$$

for any $x \in H$, $\|x\| = 1$. Taking the supremum over $\|x\| = 1$ in this inequality produces the desired result (30.177).

Now, by utilizing the result from [5] (see also [6, p. 29]) namely:

If $\gamma, \Gamma \in \mathbb{K}$ with $\gamma + \Gamma \neq 0$ and $u, v \in H$ are such that either (30.182) or, equivalently, (30.183) holds true, then

$$0 \le \|u\| \|v\| - |\langle u, v\rangle| \le \frac{1}{4} \frac{|\Gamma - \gamma|^2}{|\Gamma + \gamma|} \|v\|^2. \tag{30.185}$$

Now, on making use of (30.185) for $u = Ax$, $v = Bx$, $x \in H$, $\|x\| = 1$ and $\gamma = \alpha, \Gamma = \beta$ and using the same procedure outlined above, we deduce the second inequality (30.178).

The inequality (30.179) follows from the result presented below obtained in [4] (see also [6, p. 21]):

If $\gamma, \Gamma \in \mathbb{K}$ with

$$\operatorname{Re}(\Gamma \bar{\gamma}) > 0$$

and $u, v \in H$ are such that either (30.182) or, equivalently, (30.183) holds true, then

$$\|u\| \|v\| \le \frac{1}{2} \frac{|\Gamma + \gamma|}{\sqrt{\operatorname{Re}(\Gamma \bar{\gamma})}} |\langle u, v\rangle|, \tag{30.186}$$

by choosing $u = Ax$, $v = Bx$, $x \in H$, $\|x\| = 1$ and $\gamma = \alpha, \Gamma = \beta$ and taking the supremum over $\|x\| = 1$.

Finally, on making use of the inequality (see [7])

$$\|u\|^2 \|v\|^2 - |\langle u, v\rangle|^2 \le \left( |\Gamma + \gamma| - 2\sqrt{\operatorname{Re}(\Gamma \bar{\gamma})} \right) |\langle u, v\rangle| \, \|v\|^2 \tag{30.187}$$

that is valid provided $\gamma, \Gamma \in \mathbb{K}$ with $\operatorname{Re}(\Gamma \bar{\gamma}) > 0$ and $u, v \in H$ are such that either (30.182) or, equivalently, (30.183) holds true, we obtain the last inequality (30.180). The details are omitted.

*Remark 30.22.* Let $M,m>0$ and $A,B\in B(H)$. If $C_{m,M}(A,B)$ is accretive, then

$$(0\leq)\,\mu^2(A,B)-w^2(B^*A)\leq\frac{1}{4}(M-m)^2\|B\|^4,$$

$$(0\leq)\,\mu(A,B)-w(B^*A)\leq\frac{1}{4}\frac{(M-m)^2}{m+M}\|B\|^2,$$

$$(1\leq)\,\frac{\mu(A,B)}{w(B^*A)}\leq\frac{1}{2}\frac{m+M}{\sqrt{mM}}$$

and

$$(0\leq)\,\mu^2(A,B)-w^2(B^*A)\leq\left(\sqrt{M}-\sqrt{m}\right)^2 w(B^*A)\|B\|^2,$$

respectively.

**Corollary 30.21.** *Let $\alpha,\beta\in\mathbb{C}$ and $A\in B(H)$. If $C_{\alpha,\beta}(A,A^*)$ is accretive, then*

$$(0\leq)\,\mu^2(A,A^*)-w^2(A^2)\leq\frac{1}{4}|\beta-\alpha|^2\|A\|^4.$$

*Moreover, if $\alpha+\beta\neq 0$, then*

$$(0\leq)\,\mu(A,A^*)-w\left(A^2\right)\leq\frac{1}{4}\frac{|\beta-\alpha|^2}{|\beta+\alpha|}\|A\|^2.$$

*In addition, if $\mathrm{Re}\left(\alpha\bar{\beta}\right)>0$ and $A^2\neq 0$, then also*

$$(1\leq)\,\frac{\mu(A,A^*)}{w(A^2)}\leq\frac{1}{2}\frac{|\beta+\alpha|}{\sqrt{\mathrm{Re}\left(\alpha\bar{\beta}\right)}}$$

*and*

$$(0\leq)\,\mu^2(A,A^*)-w^2\left(A^2\right)\leq\left(|\beta+\alpha|-2\sqrt{\mathrm{Re}\left(\alpha\bar{\beta}\right)}\right)w\left(A^2\right)\|A\|^2,$$

*respectively.*

*Remark 30.23.* In a similar manner, if $N,n>0$, $A\in B(H)$ and $C_{n,N}(A,A^*)$ is accretive, then

$$(0\leq)\,\mu^2(A,A^*)-w^2\left(A^2\right)\leq\frac{1}{4}(N-n)^2\|A\|^4,$$

$$(0\leq)\,\mu(A,A^*)-w\left(A^2\right)\leq\frac{1}{4}\frac{(N-n)^2}{n+N}\|A\|^2,$$

$$(1\leq)\,\frac{\mu(A,A^*)}{w(A^2)}\leq\frac{1}{2}\frac{n+N}{\sqrt{nN}}\qquad(\text{for }A^2\neq 0)$$

and

$$(0 \leq)\, \mu^2(A, A^*) - w^2(A^2) \leq \left(\sqrt{N} - \sqrt{n}\right)^2 w(A^2)\, \|A\|^2,$$

respectively.

## 30.5 Some Inequalities of the Grüss Type for the Numerical Radius

### *30.5.1 Introduction*

Motivated by the natural questions that arise, in order to compare the quantity $w(AB)$ with other expressions comprising the norm or the numerical radius of the involved operators $A$ and $B$ (or certain expressions constructed with these operators), we establish in this section some natural inequalities of the form

$$w(BA) \leq w(A)\, w(B) + K_1 \quad \text{(additive Grüss' type inequality)}$$

or

$$\frac{w(BA)}{w(A)\, w(B)} \leq K_2 \quad \text{(multiplicative Grüss' type inequality)}$$

where $K_1$ and $K_2$ are specified and desirably simple constants (depending on the given operators $A$ and $B$).

Applications in providing upper bounds for the non negative quantities

$$\|A\|^2 - w^2(A) \quad \text{and} \quad w^2(A) - w(A^2)$$

and the *super unitary* quantities

$$\frac{\|A\|^2}{w^2(A)} \quad \text{and} \quad \frac{w^2(A)}{w(A^2)}$$

are also given.

### *30.5.2 Numerical Radius Inequalities of Grüss Type*

For the complex numbers $\alpha, \beta$ and the bounded linear operator $T$ we define the following transform

$$C_{\alpha,\beta}(T) := (T^* - \overline{\alpha} I)(\beta I - T), \tag{30.188}$$

where by $T^*$ we denote the adjoint of $T$.

We list some properties of the transform $C_{\alpha,\beta}(\cdot)$ that are useful in the following:

(i) For any $\alpha,\beta \in \mathbb{C}$ and $T \in B(H)$ we have:

$$C_{\alpha,\beta}(I) = (1-\overline{\alpha})(\beta-1)I, \qquad C_{\alpha,\alpha}(T) = -(\alpha I - T)^*(\alpha I - T), \tag{30.189}$$

$$C_{\alpha,\beta}(\gamma T) = |\gamma|^2 C_{\frac{\alpha}{\gamma},\frac{\beta}{\gamma}}(T), \quad \gamma \in \mathbb{C}\backslash\{0\}, \tag{30.190}$$

$$\left[C_{\alpha,\beta}(T)\right]^* = C_{\beta,\alpha}(T) \tag{30.191}$$

and

$$C_{\overline{\beta},\overline{\alpha}}(T^*) - C_{\alpha,\beta}(T) = T^*T - TT^*. \tag{30.192}$$

(ii) The operator $T \in B(H)$ is normal if and only if $C_{\overline{\beta},\overline{\alpha}}(T^*) = C_{\alpha,\beta}(T)$ for each $\alpha,\beta \in \mathbb{C}$.

We recall that a bounded linear operator $T$ on the complex Hilbert space $(H,\langle\cdot,\cdot\rangle)$ is called *accretive* if $\operatorname{Re}\langle Ty,y\rangle \geq 0$ for any $y \in H$.

Utilizing the following identity

$$\begin{aligned}\operatorname{Re}\left\langle C_{\alpha,\beta}(T)x,x\right\rangle &= \operatorname{Re}\left\langle C_{\beta,\alpha}(T)x,x\right\rangle \\ &= \frac{1}{4}|\beta-\alpha|^2 - \left\|\left(T - \frac{\alpha+\beta}{2}I\right)x\right\|^2\end{aligned} \tag{30.193}$$

that holds for any scalars $\alpha,\beta$ and any vector $x \in H$ with $\|x\| = 1$ we can give a simple characterization result that is useful in the following:

**Lemma 30.4.** *For $\alpha,\beta \in \mathbb{C}$ and $T \in B(H)$ the following statements are equivalent:*

*(i) The transform $C_{\alpha,\beta}(T)$ (or, equivalently $C_{\beta,\alpha}(T)$) is accretive;*
*(ii) The transform $C_{\overline{\alpha},\overline{\beta}}(T^*)$ (or, equivalently $C_{\overline{\beta},\overline{\alpha}}(T^*)$) is accretive;*
*(iii) We have the norm inequality*

$$\left\|T - \frac{\alpha+\beta}{2}\cdot I\right\| \leq \frac{1}{2}|\beta-\alpha| \tag{30.194}$$

*or, equivalently,*

$$\left\|T^* - \frac{\bar{\alpha}+\bar{\beta}}{2}\cdot I\right\| \leq \frac{1}{2}|\beta-\alpha|. \tag{30.195}$$

*Remark 30.24.* In order to give examples of operators $T \in B(H)$ and numbers $\alpha,\beta \in \mathbb{C}$ such that the transform $C_{\alpha,\beta}(T)$ is accretive, it suffices to select a bounded linear operator $S$ and the complex numbers $z,w$ with the property that $\|S - zI\| \leq |w|$ and, by choosing $T = S$, $\alpha = \frac{1}{2}(z+w)$ and $\beta = \frac{1}{2}(z-w)$. We observe that $T$ satisfies (30.194), i.e., $C_{\alpha,\beta}(T)$ is accretive.

The following results compare the quantities $w(AB)$ and $w(A)w(B)$ provided that some information about the transforms $C_{\alpha,\beta}(A)$ and $C_{\gamma,\delta}(B)$ are available, where $\alpha,\beta,\gamma,\delta \in \mathbb{K}$.

**Theorem 30.31 (Dragomir, 2008 [18]).** *Let $A,B \in B(H)$ and $\alpha,\beta,\gamma,\delta \in \mathbb{K}$ be such that the transforms $C_{\alpha,\beta}(A)$ and $C_{\gamma,\delta}(B)$ are accretive, then*

$$w(BA) \le w(A)w(B) + \frac{1}{4}|\beta-\alpha||\gamma-\delta|. \tag{30.196}$$

*Proof.* Since $C_{\alpha,\beta}(A)$ and $C_{\gamma,\delta}(B)$ are accretive, then, on making use of Lemma 30.4 we have that

$$\left\| Ax - \frac{\alpha+\beta}{2}x \right\| \le \frac{1}{2}|\beta-\alpha|$$

and

$$\left\| B^*x - \frac{\bar{\gamma}+\bar{\delta}}{2}x \right\| \le \frac{1}{2}|\bar{\gamma}-\bar{\delta}|$$

for any $x \in H, \|x\| = 1$.

Now, we make use of the following Grüss type inequality for vectors in inner product spaces obtained by the author in [1] (see also [2] or [6, p. 43]):

Let $(H,\langle\cdot,\cdot\rangle)$ be an inner product space over the real or complex number field $\mathbb{K}$, $u,v,e \in H$, $\|e\| = 1$, and $\alpha,\beta,\gamma,\delta \in \mathbb{K}$ such that

$$\operatorname{Re}\langle \beta e - u, u - \alpha e\rangle \ge 0, \qquad \operatorname{Re}\langle \delta e - v, v - \gamma e\rangle \ge 0 \tag{30.197}$$

or equivalently,

$$\left\| u - \frac{\alpha+\beta}{2}e \right\| \le \frac{1}{2}|\beta-\alpha|, \qquad \left\| v - \frac{\gamma+\delta}{2}e \right\| \le \frac{1}{2}|\delta-\gamma|, \tag{30.198}$$

then

$$|\langle u,v\rangle - \langle u,e\rangle\langle e,v\rangle| \le \frac{1}{4}|\beta-\alpha||\delta-\gamma|. \tag{30.199}$$

Applying (30.199) for $u = Ax, v = B^*x$ and $e = x$ we deduce

$$|\langle BAx,x\rangle - \langle Ax,x\rangle\langle Bx,x\rangle| \le \frac{1}{4}|\beta-\alpha||\delta-\gamma|, \tag{30.200}$$

for any $x \in H, \|x\| = 1$, which is an inequality of interest in itself.

Observing that

$$|\langle BAx,x\rangle| - |\langle Ax,x\rangle\langle Bx,x\rangle| \le |\langle BAx,x\rangle - \langle Ax,x\rangle\langle Bx,x\rangle|,$$

then by (30.199) we deduce the inequality

$$|\langle BAx,x\rangle| \leq |\langle Ax,x\rangle \langle Bx,x\rangle| + \frac{1}{4}|\beta-\alpha||\delta-\gamma|, \tag{30.201}$$

for any $x \in H$, $\|x\| = 1$. On taking the supremum over $\|x\| = 1$ in (30.201) we deduce the desired result (30.196).

The following particular case provides a upper bound for the nonnegative quantity $\|A\|^2 - w(A)^2$ when some information about the operator $A$ is available:

**Corollary 30.22.** *Let $A \in B(H)$ and $\alpha,\beta \in \mathbb{K}$ be such that the transform $C_{\alpha,\beta}(A)$ is accretive, then*

$$(0 \leq)\,\|A\|^2 - w^2(A) \leq \frac{1}{4}|\beta-\alpha|^2. \tag{30.202}$$

*Proof.* Follows on applying Theorem 30.31 above for the choice $B = A^*$, taking into account that $C_{\alpha,\beta}(A)$ is accretive implies that $C_{\overline{\alpha},\overline{\beta}}(A^*)$ is the same and $w(A^*A) = \|A\|^2$.

*Remark 30.25.* Let $A \in B(H)$ and $M > m > 0$ are such that the transform

$$C_{m,M}(A) = (A^* - mI)(MI - A)$$

is accretive. Then

$$(0 \leq)\,\|A\|^2 - w^2(A) \leq \frac{1}{4}(M-m)^2. \tag{30.203}$$

A sufficient simple condition for $C_{m,M}(A)$ to be accretive is that $A$ is a selfadjoint operator on $H$ and such that $MI \geq A \geq mI$ in the partial operator order of $B(H)$.

The following result may be stated as well:

**Theorem 30.32 (Dragomir, 2008 [18]).** *Let $A,B \in B(H)$ and $\alpha,\beta,\gamma,\delta \in \mathbb{K}$ be such that $\operatorname{Re}(\beta\overline{\alpha}) > 0, \operatorname{Re}(\delta\overline{\gamma}) > 0$ and the transforms $C_{\alpha,\beta}(A), C_{\gamma,\delta}(B)$ are accretive, then*

$$\frac{w(BA)}{w(A)\,w(B)} \leq 1 + \frac{1}{4}\frac{|\beta-\alpha||\delta-\gamma|}{[\operatorname{Re}(\beta\overline{\alpha})\operatorname{Re}(\delta\overline{\gamma})]^{\frac{1}{2}}} \tag{30.204}$$

*and*

$$\begin{aligned} w(BA) \leq{}& w(A)\,w(B) + \left[\left(|\alpha+\beta| - 2[\operatorname{Re}(\beta\overline{\alpha})]^{\frac{1}{2}}\right)\right.\\ &\times \left.\left(|\delta+\gamma| - 2[\operatorname{Re}(\delta\overline{\gamma})]^{\frac{1}{2}}\right)\right]^{\frac{1}{2}} \times [w(A)\,w(B)]^{\frac{1}{2}} \end{aligned} \tag{30.205}$$

*respectively.*

*Proof.* With the assumptions (30.197) (or, equivalently, (30.198) in the proof of Theorem 30.31) and if $\operatorname{Re}(\beta\overline{\alpha}) > 0, \operatorname{Re}(\delta\overline{\gamma}) > 0$ then

$$
\begin{aligned}
&|\langle u,v\rangle-\langle u,e\rangle\langle e,v\rangle| \\
&\le\begin{cases}\dfrac{1}{4}\dfrac{|\beta-\alpha||\delta-\gamma|}{[\operatorname{Re}(\beta\overline{\alpha})\operatorname{Re}(\delta\overline{\gamma})]^{\frac{1}{2}}}|\langle u,e\rangle\langle e,v\rangle|,\\ \left[\left(|\alpha+\beta|-2[\operatorname{Re}(\beta\overline{\alpha})]^{\frac{1}{2}}\right)\left(|\delta+\gamma|-2[\operatorname{Re}(\delta\overline{\gamma})]^{\frac{1}{2}}\right)\right]^{\frac{1}{2}}\\ \qquad\qquad\times[|\langle u,e\rangle\langle e,v\rangle|]^{\frac{1}{2}}.\end{cases}
\end{aligned} \tag{30.206}
$$

The first inequality has been established in [4] (see [6, p. 62]) while the second one can be obtained in a canonical manner from the reverse of the Schwarz inequality given in [7]. The details are omitted.

Applying (30.199) for $u=Ax, v=B^*x$ and $e=x$ we deduce

$$
\begin{aligned}
&\left|\langle BAx,x\rangle-\langle Ax,x\rangle\langle Bx,x\rangle\right| \\
&\le\begin{cases}\dfrac{1}{4}\dfrac{|\beta-\alpha||\delta-\gamma|}{[\operatorname{Re}(\beta\overline{\alpha})\operatorname{Re}(\delta\overline{\gamma})]^{\frac{1}{2}}}|\langle A,x\rangle\langle Bx,x\rangle|,\\ \left[\left(|\alpha+\beta|-2[\operatorname{Re}(\beta\overline{\alpha})]^{\frac{1}{2}}\right)\left(|\delta+\gamma|-2[\operatorname{Re}(\delta\overline{\gamma})]^{\frac{1}{2}}\right)\right]^{\frac{1}{2}}\\ \qquad\qquad\times[|\langle A,x\rangle\langle Bx,x\rangle|]^{\frac{1}{2}}\end{cases}
\end{aligned} \tag{30.207}
$$

for any $x\in H, \|x\|=1$, which are of interest in themselves.

A similar argument to that in the proof of Theorem 30.31 yields the desired inequalities (30.204) and (30.205). The details are omitted. □

**Corollary 30.23.** *Let $A\in B(H)$ and $\alpha,\beta\in\mathbb{K}$ be such that $\operatorname{Re}(\beta\overline{\alpha})>0$ and the transform $C_{\alpha,\beta}(A)$ is accretive, then*

$$
(1\le)\frac{\|A\|^2}{w^2(A)}\le 1+\frac{1}{4}\frac{|\beta-\alpha|^2}{\operatorname{Re}(\beta\overline{\alpha})} \tag{30.208}
$$

*and*

$$
(0\le)\|A\|^2-w^2(A)\le\left(|\alpha+\beta|-2[\operatorname{Re}(\beta\overline{\alpha})]^{\frac{1}{2}}\right)w(A) \tag{30.209}
$$

*respectively.*

The proof is obvious from Theorem 30.32 on choosing $B=A^*$ and the details are omitted.

*Remark 30.26.* Let $A\in B(H)$ and $M>m>0$ are such that the transform $C_{m,M}(A)=(A^*-mI)(MI-A)$ is accretive. Then, on making use of Corollary 30.23, we may state the following simpler results

$$
(1\le)\frac{\|A\|}{w(A)}\le\frac{1}{2}\frac{M+m}{\sqrt{Mm}} \tag{30.210}
$$

and

$$(0 \leq) \|A\|^2 - w^2(A) \leq \left(\sqrt{M} - \sqrt{m}\right)^2 w(A) \tag{30.211}$$

respectively. These two inequalities were obtained earlier by the author using a different approach, see [8].

**Problem 30.1.** Find general examples of bounded linear operators realizing the equality case in each of the inequalities (30.196), (30.204) and (30.205), respectively.

### *30.5.3 Some Particular Cases of Interest*

The following result is well known in the literature (see for instance [33]):

$$w(A^n) \leq w^n(A),$$

for each positive integer $n$ and any operator $A \in B(H)$.

The following reverse inequalities for $n = 2$, can be stated:

**Proposition 30.2 (Dragomir, 2008 [18]).** *Let $A \in B(H)$ and $\alpha, \beta \in \mathbb{K}$ be such that the transform $C_{\alpha,\beta}(A)$ is accretive, then*

$$(0 \leq) w^2(A) - w(A^2) \leq \frac{1}{4}|\beta - \alpha|^2. \tag{30.212}$$

*Proof.* On applying the inequality (30.200) from Theorem 30.31 for the choice $B = A$, we get the following inequality of interest in itself:

$$\left|\langle Ax, x\rangle^2 - \langle A^2x, x\rangle\right| \leq \frac{1}{4}|\beta - \alpha|^2, \tag{30.213}$$

for any $x \in H, \|x\| = 1$. Since obviously,

$$|\langle Ax, x\rangle|^2 - |\langle A^2x, x\rangle| \leq \left|\langle Ax, x\rangle^2 - \langle A^2x, x\rangle\right|,$$

then by (30.213) we get

$$|\langle Ax, x\rangle|^2 \leq |\langle A^2x, x\rangle| + \frac{1}{4}|\beta - \alpha|^2, \tag{30.214}$$

for any $x \in H, \|x\| = 1$. Taking the supremum over $\|x\| = 1$ in (30.214) we deduce the desired result (30.212). □

*Remark 30.27.* Let $A \in B(H)$ and $M > m > 0$ are such that the transform $C_{m,M}(A) = (A^* - mI)(MI - A)$ is accretive. Then

$$(0 \leq) w^2(A) - w(A^2) \leq \frac{1}{4}(M - m)^2. \tag{30.215}$$

If $MI \geq A \geq mI$ in the partial operator order of $B(H)$, then (30.215) is valid.

Finally, we also have

**Proposition 30.3 (Dragomir, 2008 [18]).** *Let $A \in B(H)$ and $\alpha, \beta \in \mathbb{K}$ be such that $\operatorname{Re}(\beta\overline{\alpha}) > 0$ and the transform $C_{\alpha,\beta}(A)$ is accretive, then*

$$(1 \leq) \frac{w^2(A)}{w(A^2)} \leq 1 + \frac{1}{4}\frac{|\beta - \alpha|^2}{\operatorname{Re}(\beta\overline{\alpha})} \tag{30.216}$$

*and*

$$(0 \leq) w^2(A) - w(A^2) \leq \left(|\alpha + \beta| - 2[\operatorname{Re}(\beta\overline{\alpha})]^{\frac{1}{2}}\right) w(A), \tag{30.217}$$

*respectively.*

*Proof.* On applying the inequality (30.207) from Theorem 30.32 for the choice $B = A$, we get the following inequality of interest in itself:

$$\left|\langle Ax,x\rangle^2 - \langle A^2x,x\rangle\right| \leq \begin{cases} \frac{1}{4}\frac{|\beta-\alpha|^2}{\operatorname{Re}(\beta\overline{\alpha})} |\langle A,x\rangle|^2, \\ \left(|\alpha+\beta| - 2[\operatorname{Re}(\beta\overline{\alpha})]^{\frac{1}{2}}\right) |\langle A,x\rangle|. \end{cases} \tag{30.218}$$

for any $x \in H, \|x\| = 1$.

Now, on making use of a similar argument to the one in the proof of Proposition 30.2 we deduce the desired results (30.216) and (30.217). The details are omitted.

*Remark 30.28.* Let $A \in B(H)$ and $M > m > 0$ are such that the transform $C_{m,M}(A) = (A^* - mI)(MI - A)$ is accretive. Then, on making use of Proposition 30.3 we may state the following simpler results

$$(1 \leq) \frac{w^2(A)}{w(A^2)} \leq \frac{1}{4}\frac{(M+m)^2}{Mm} \tag{30.219}$$

and

$$(0 \leq) w^2(A) - w(A^2) \leq \left(\sqrt{M} - \sqrt{m}\right)^2 w(A), \tag{30.220}$$

respectively.

## 30.6 Some Inequalities for the Euclidean Operator Radius

### 30.6.1 Some Preliminary Facts

Let $B(H)$ denote the $C^*$-algebra of all bounded linear operators on a complex Hilbert space $H$ with inner product $\langle \cdot,\cdot \rangle$. For $A \in B(H)$, let $w(A)$ and $\|A\|$ denote the numerical radius and the usual operator norm of $A$, respectively. It is well known that $w(\cdot)$ defines a norm on $B(H)$, and for every $A \in B(H)$,

$$\frac{1}{2}\|A\| \leq w(A) \leq \|A\|. \tag{30.221}$$

For other results concerning the numerical range and radius of bounded linear operators on a Hilbert space, see [26] and [28].

In [32], Kittaneh has improved (30.221) in the following manner:

$$\frac{1}{4}\|A^*A+AA^*\| \leq w^2(A) \leq \frac{1}{2}\|A^*A+AA^*\|, \tag{30.222}$$

with the constants $1/4$ and $1/2$ as best possible.

Following Popescu's work [34], we consider the *Euclidean operator radius* of a pair $(C,D)$ of bounded linear operators defined on a Hilbert space $(H;\langle\cdot,\cdot\rangle)$. Note that in [34], the author has introduced the concept for an $n$-tuple of operators and pointed out its main properties.

Let $(C,D)$ be a pair of bounded linear operators on $H$. The *Euclidean operator radius* is defined by:

$$w_e(C,D) := \sup_{\|x\|=1}\left(|\langle Cx,x\rangle|^2+|\langle Dx,x\rangle|^2\right)^{\frac{1}{2}}. \tag{30.223}$$

As pointed out in [34], $w_e : B^2(H) \to [0,\infty)$ is a norm and the following inequality holds:

$$\frac{\sqrt{2}}{4}\|C^*C+D^*D\|^{\frac{1}{2}} \leq w_e(C,D) \leq \|C^*C+D^*D\|^{\frac{1}{2}}, \tag{30.224}$$

where the constants $\sqrt{2}/4$ and 1 are best possible in (30.224).

We observe that, if $C$ and $D$ are self-adjoint operators, then (30.224) becomes

$$\frac{\sqrt{2}}{4}\|C^2+D^2\|^{\frac{1}{2}} \leq w_e(C,D) \leq \|C^2+D^2\|^{\frac{1}{2}}. \tag{30.225}$$

We observe also that if $A \in B(H)$ and

$$A = B + \mathrm{i}C$$

is the Cartesian decomposition of $A$, then

$$\begin{aligned} w_e^2(B,C) &= \sup_{\|x\|=1}\left[|\langle Bx,x\rangle|^2+|\langle Cx,x\rangle|^2\right] \\ &= \sup_{\|x\|=1}|\langle Ax,x\rangle|^2 = w^2(A). \end{aligned}$$

By the inequality (30.225) and since (see [32])

$$A^*A + AA^* = 2\left(B^2 + C^2\right), \tag{30.226}$$

then we have

$$\frac{1}{16}\|A^*A + AA^*\| \le w^2(A) \le \frac{1}{2}\|A^*A + AA^*\|. \tag{30.227}$$

We remark that the lower bound for $w^2(A)$ in (30.227) provided by Popescu's inequality (30.224) is not as good as the first inequality of Kittaneh from (30.222). However, the upper bounds for $w^2(A)$ are the same and have been proved using different arguments.

The main aim of this section is to extend Kittaneh's result to the Euclidean radius of two operators and investigate other particular instances of interest. Related results connecting the Euclidean operator radius, the usual numerical radius of a composite operator and the operator norm are also provided.

### 30.6.2 *Some Inequalities for the Euclidean Operator Radius*

The following result concerning a sharp lower bound for the Euclidean operator radius may be stated:

**Theorem 30.33 (Dragomir, 2008 [9]).** *Let $B, C : H \to H$ be two bounded linear operators on the Hilbert space $(H; \langle \cdot, \cdot \rangle)$. Then*

$$\frac{\sqrt{2}}{2}\left[w\left(B^2 + C^2\right)\right]^{\frac{1}{2}} \le w_e(B,C) \left(\le \|B^*B + C^*C\|^{\frac{1}{2}}\right). \tag{30.228}$$

*The constant $\sqrt{2}/2$ is best possible in the sense that it cannot be replaced by a larger constant.*

*Proof.* We follow a similar argument to the one from [32].

For any $x \in H$, $\|x\| = 1$, we have

$$|\langle Bx, x\rangle|^2 + |\langle Cx, x\rangle|^2 \ge \frac{1}{2}\left(|\langle Bx, x\rangle| + |\langle Cx, x\rangle|\right)^2 \ge \frac{1}{2}|\langle (B \pm C)x, x\rangle|^2. \tag{30.229}$$

Taking the supremum in (30.229), we deduce

$$w_e^2(B,C) \ge \frac{1}{2}w^2(B \pm C). \tag{30.230}$$

Utilising the inequality (30.230) and the properties of the numerical radius, we have successively:

$$
\begin{aligned}
2w_e^2(B,C) &\geq \frac{1}{2}\left[w^2(B+C)+w^2(B-C)\right] \\
&\geq \frac{1}{2}\left\{w\left[(B+C)^2\right]+w\left[(B-C)^2\right]\right\} \\
&\geq \frac{1}{2}\left\{w\left[(B+C)^2+(B-C)^2\right]\right\} \\
&= w\left(B^2+C^2\right),
\end{aligned}
$$

which gives the desired inequality (30.228).

The sharpness of the constant will be shown in a particular case, later on. □

**Corollary 30.24.** *For any two self-adjoint bounded linear operators $B,C$ on $H$, we have*

$$\frac{\sqrt{2}}{2}\left\|B^2+C^2\right\|^{\frac{1}{2}} \leq w_e(B,C)\left(\leq \left\|B^2+C^2\right\|^{\frac{1}{2}}\right). \tag{30.231}$$

*The constant $\frac{\sqrt{2}}{2}$ is sharp in (30.231).*

*Remark 30.29.* The inequality (30.231) is better than the first inequality in (30.225) which follows from Popescu's first inequality in (30.224). It also provides, for the case that $B,C$ are the self-adjoint operators in the Cartesian decomposition of $A$, exactly the lower bound obtained by Kittaneh in (30.222) for the numerical radius $w(A)$. Moreover, since $1/4$ is a sharp constant in Kittaneh's inequality (30.222), it follows that $\sqrt{2}/2$ is also the best possible constant in (30.231) and (30.228), respectively.

The following particular case may be of interest:

**Corollary 30.25.** *For any bounded linear operator $A : H \to H$ and $\alpha,\beta \in \mathbb{C}$ we have:*

$$
\begin{aligned}
\frac{1}{2}w\left[\alpha^2A^2+\beta^2(A^*)^2\right] &\leq \left(|\alpha|^2+|\beta|^2\right)w^2(A) \\
&\left(\leq \left\||\alpha|^2A^*A+|\beta|^2AA^*\right\|\right).
\end{aligned} \tag{30.232}
$$

*Proof.* If we choose in Theorem 30.33, $B=\alpha A$ and $C=\beta A^*$, we get

$$w_e^2(B,C)=\left(|\alpha|^2+|\beta|^2\right)w^2(A)$$

and

$$w\left(B^2+C^2\right)=w\left[\alpha^2A^2+\beta^2(A^*)^2\right],$$

which, by (30.228) implies the desired result (30.232). □

*Remark 30.30.* If we choose in (30.232) $\alpha=\beta\neq 0$, then we get the inequality

$$\frac{1}{4}\left\|A^2+(A^*)^2\right\|\leq w^2(A)\left(\leq\frac{1}{2}\|A^*A+AA^*\|\right), \tag{30.233}$$

for any bounded linear operator $A\in B(H)$.

If we choose in (30.232), $\alpha=1$, $\beta=i$, then we get

$$\frac{1}{4}w\left[A^2-(A^*)^2\right]\leq w^2(A), \tag{30.234}$$

for every bounded linear operator $A:H\to H$.

The following result may be stated as well.

**Theorem 30.34 (Dragomir, 2008 [9]).** *For any two bounded linear operators $B,C$ on $H$ we have:*

$$\frac{\sqrt{2}}{2}\left[w^2(B+C)+w^2(B-C)\right]^{\frac{1}{2}}$$
$$\geq w_e(B,C)\geq\frac{\sqrt{2}}{2}\max\{w(B+C),w(B-C)\}. \tag{30.235}$$

*The constant $\sqrt{2}/2$ is sharp in both inequalities.*

*Proof.* The first inequality follows from (30.230).

For the second inequality, we observe that

$$|\langle Cx,x\rangle\pm\langle Bx,x\rangle|^2\leq w^2(C\pm B) \tag{30.236}$$

for any $x\in H$, $\|x\|=1$.

The inequality (30.236) and the parallelogram identity for complex numbers give:

$$2\left[|\langle Bx,x\rangle|^2+|\langle Cx,x\rangle|^2\right]=|\langle Bx,x\rangle-\langle Cx,x\rangle|^2+|\langle Bx,x\rangle+\langle Cx,x\rangle|^2$$
$$\leq w^2(B+C)+w^2(B-C), \tag{30.237}$$

for any $x\in H$, $\|x\|=1$.

Taking the supremum in (30.236) we deduce the desired result (30.235).

The fact that $\sqrt{2}/2$ is the best possible constant follows from the fact that for $B=C\neq 0$ one would obtain the same quantity $\sqrt{2}w(B)$ in all terms of (30.235).

□

**Corollary 30.26.** *For any two self-adjoint operators $B,C$ on $H$ we have:*

$$\frac{\sqrt{2}}{2}\left[\|B+C\|^2+\|B-C\|^2\right]^{\frac{1}{2}} \geq w_e(B,C) \geq \frac{\sqrt{2}}{2}\max\{\|B+C\|,\|B-C\|\}. \tag{30.238}$$

*The constant $\sqrt{2}/2$ is best possible in both inequalities.*

**Corollary 30.27.** *Let $A$ be a bounded linear operator on $H$. Then*

$$\frac{\sqrt{2}}{2}\max\left\{\left\|\frac{(1-\mathrm{i})A+(1+\mathrm{i})A^*}{2}\right\|,\left\|\frac{(1+\mathrm{i})A+(1-\mathrm{i})A^*}{2}\right\|\right\} \leq w(A) \leq \frac{\sqrt{2}}{2}\left[\left\|\frac{(1-\mathrm{i})A+(1+\mathrm{i})A^*}{2}\right\|^2+\left\|\frac{(1+\mathrm{i})A+(1-\mathrm{i})A^*}{2}\right\|^2\right]^{\frac{1}{2}}. \tag{30.239}$$

*Proof.* Follows from (30.238) applied for the Cartesian decomposition of $A$. □

The following result may be stated as well:

**Corollary 30.28.** *For any $A$ a bounded linear operator on $H$ and $\alpha,\beta \in \mathbb{C}$, we have:*

$$\frac{\sqrt{2}}{2}\left[w^2(\alpha A+\beta A^*)+w^2(\alpha A-\beta A^*)\right]^{\frac{1}{2}} \geq \frac{\sqrt{2}}{2}\max\{w(\alpha A+\beta A^*),w(\alpha A-\beta A^*)\} \geq \left(|\alpha|^2+|\beta|^2\right)^{\frac{1}{2}} w(A). \tag{30.240}$$

*Remark 30.31.* The above inequality (30.240) contains some particular cases of interest. For instance, if $\alpha=\beta\neq 0$, then by (30.240) we get

$$\frac{1}{2}\max\{\|A+A^*\|,\|A-A^*\|\} \leq w(A) \leq \frac{1}{2}\left[\|A+A^*\|^2+\|A-A^*\|^2\right]^{\frac{1}{2}}, \tag{30.241}$$

since, obviously $w(A+A^*)=\|A+A^*\|$ and $w(A-A^*)=\|A-A^*\|$, $A-A^*$ being a normal operator.

Now, if we choose in (30.240), $\alpha=1$ and $\beta=\mathrm{i}$, and taking into account that $A+\mathrm{i}A^*$ and $A-\mathrm{i}A^*$ are normal operators, then we get

$$\frac{1}{2}\max\{\|A+\mathrm{i}A^*\|,\|A-\mathrm{i}A^*\|\} \leq w(A) \leq \frac{1}{2}\left[\|A+\mathrm{i}A^*\|^2+\|A-\mathrm{i}A^*\|^2\right]^{\frac{1}{2}}. \tag{30.242}$$

The constant $1/2$ is best possible in both inequalities (30.241) and (30.242).

The following simple result may be stated as well.

**Proposition 30.4 (Dragomir, 2008 [9]).** *For any two bounded linear operators $B$ and $C$ on $H$, we have the inequality:*

$$w_e(B,C) \leq \left[w^2(C-B) + 2w(B)\,w(C)\right]^{\frac{1}{2}}. \tag{30.243}$$

*Proof.* For any $x \in H$, $\|x\| = 1$, we have

$$\begin{aligned}|\langle Cx,x\rangle|^2 &- 2\mathrm{Re}\left[\langle Cx,x\rangle\,\overline{\langle Bx,x\rangle}\right] + |\langle Bx,x\rangle|^2 \\ &= |\langle Cx,x\rangle - \langle Bx,x\rangle|^2 \leq w^2(C-B),\end{aligned}$$

giving

$$\begin{aligned}|\langle Cx,x\rangle|^2 + |\langle Bx,x\rangle|^2 &\leq w^2(C-B) + 2\mathrm{Re}\left[\langle Cx,x\rangle\,\overline{\langle Bx,x\rangle}\right] \\ &\leq w^2(C-B) + 2|\langle Cx,x\rangle|\,|\langle Bx,x\rangle| \end{aligned} \tag{30.244}$$

for any $x \in H$, $\|x\| = 1$.

Taking the supremum in (30.244) over $\|x\| = 1$, we deduce the desired inequality (30.243). □

In particular, if $B$ and $C$ are self-adjoint operators, then

$$w_e(B,C) \leq \left(\|B-C\|^2 + 2\|B\|\,\|C\|\right)^{\frac{1}{2}}. \tag{30.245}$$

Now, if we apply the inequality (30.245) for $B = \frac{A+A^*}{2}$ and $C = \frac{A-A^*}{2\mathrm{i}}$, where $A \in B(H)$, then we deduce:

$$w(A) \leq \left[\left\|\frac{(1+\mathrm{i})A + (1-\mathrm{i})A^*}{2}\right\|^2 + 2\left\|\frac{A+A^*}{2}\right\|\left\|\frac{A-A^*}{2}\right\|\right]^{\frac{1}{2}}.$$

The following result provides a different upper bound for the Euclidean operator radius than (30.243).

**Proposition 30.5 (Dragomir, 2008 [9]).** *For any two bounded linear operators $B$ and $C$ on $H$, we have*

$$w_e(B,C) \leq \left[2\min\left\{w^2(B), w^2(C)\right\} + w(B-C)\,w(B+C)\right]^{\frac{1}{2}}. \tag{30.246}$$

*Proof.* Utilising the parallelogram identity (30.237), we have, by taking the supremum over $x \in H, \|x\| = 1$, that

$$2w_e^2(B,C) = w_e^2(B-C, B+C). \qquad (30.247)$$

Now, if we apply Proposition 30.4 for $B-C, B+C$ instead of $B$ and $C$, then we can state

$$w_e^2(B-C, B+C) \le 4w^2(C) + 2w(B-C)\,w(B+C)$$

giving

$$w_e^2(B,C) \le 2w^2(C) + w(B-C)\,w(B+C). \qquad (30.248)$$

Now, if in (30.248) we swap the $C$ with $B$ then we also have

$$w_e^2(B,C) \le 2w^2(B) + w(B-C)\,w(B+C). \qquad (30.249)$$

The conclusion follows now by (30.248) and (30.249). □

### 30.6.3 Other Results

A different upper bound for the Euclidean operator radius is incorporated in the following

**Theorem 30.35 (Dragomir, 2008 [9]).** *Let $(H; \langle\cdot,\cdot\rangle)$ be a Hilbert space and $B, C$ two bounded linear operators on $H$. Then*

$$w_e^2(B,C) \le \max\left\{\|B\|^2, \|C\|^2\right\} + w(C^*B). \qquad (30.250)$$

*The inequality (30.250) is sharp.*

*Proof.* Firstly, let us observe that for any $y, u, v \in H$ we have successively

$$\begin{aligned}
&\|\langle y,u\rangle u + \langle y,v\rangle v\|^2 \\
&\quad = |\langle y,u\rangle|^2\|u\|^2 + |\langle y,v\rangle|^2\|v\|^2 + 2\mathrm{Re}\left[\langle y,u\rangle\overline{\langle y,v\rangle}\langle u,v\rangle\right] \\
&\quad \le |\langle y,u\rangle|^2\|u\|^2 + |\langle y,v\rangle|^2\|v\|^2 + 2|\langle y,u\rangle||\langle y,v\rangle||\langle u,v\rangle| \\
&\quad \le |\langle y,u\rangle|^2\|u\|^2 + |\langle y,v\rangle|^2\|v\|^2 + \left(|\langle y,u\rangle|^2 + |\langle y,v\rangle|^2\right)|\langle u,v\rangle| \\
&\quad \le \left(|\langle y,u\rangle|^2 + |\langle y,v\rangle|^2\right)\left(\max\left\{\|u\|^2, \|v\|^2\right\} + |\langle u,v\rangle|\right). \qquad (30.251)
\end{aligned}$$

On the other hand,

$$\left(|\langle y,u\rangle|^2 + |\langle y,v\rangle|^2\right)^2 = [\langle y,u\rangle\langle u,y\rangle + \langle y,v\rangle\langle v,y\rangle]^2$$
$$- [\langle y,\langle y,u\rangle u + \langle y,v\rangle v\rangle]^2 \le \|y\|^2 \|\langle y,u\rangle u + \langle y,v\rangle v\|^2 \tag{30.252}$$

for any $y,u,v \in H$.

Making use of (30.251) and (30.252) we deduce that

$$|\langle y,u\rangle|^2 + |\langle y,v\rangle|^2 \le \|y\|^2 \left[\max\left\{\|u\|^2, \|v\|^2\right\} + |\langle u,v\rangle|\right] \tag{30.253}$$

for any $y,u,v \in H$, which is a vector inequality of interest in itself.

Now, if we apply the inequality (30.253) for $y = x$, $u = Bx$, $v = Cx$, $x \in H$, $\|x\| = 1$, then we can state that

$$|\langle Bx,x\rangle|^2 + |\langle Cx,x\rangle|^2 \le \max\left\{\|Bx\|^2, \|Cx\|^2\right\} + |\langle Bx,Cx\rangle| \tag{30.254}$$

for any $x \in H$, $\|x\| = 1$, which is of interest in itself.

Taking the supremum over $x \in H$, $\|x\| = 1$, we deduce the desired result (30.250).

To prove the sharpness of the inequality (30.250) we choose $C = B$, $B$ a self-adjoint operator on $H$. In this case, both sides of (30.250) become $2\|B\|^2$. □

If information about the sum and the difference of the operators $B$ and $C$ is available, then one may use the following result:

**Corollary 30.29.** *For any two operators $B,C \in B(H)$ we have*

$$w_e^2(B,C) \le \frac{1}{2}\left\{\max\left\{\|B-C\|^2, \|B+C\|^2\right\} + w[(B^* - C^*)(B+C)]\right\}. \tag{30.255}$$

*The constant $\frac{1}{2}$ is best possible in (30.255).*

*Proof.* Follows by the inequality (30.250) written for $B+C$ and $B-C$ instead of $B$ and $C$ and by utilising the identity (30.247).

The fact that $1/2$ is best possible in (30.255) follows by the fact that for $C = B$, $B$ a self-adjoint operator, we get in both sides of the inequality (30.255) the quantity $2\|B\|^2$. □

**Corollary 30.30.** *Let $A : H \to H$ be a bounded linear operator on the Hilbert space $H$. Then:*

$$w^2(A) \le \frac{1}{4}\left[\max\left\{\|A+A^*\|^2, \|A-A^*\|^2\right\} + w[(A^* - A)(A+A^*)]\right]. \tag{30.256}$$

*The constant $\frac{1}{4}$ is best possible.*

*Proof.* If

$$B = \frac{A + A^*}{2}, \qquad C = \frac{A - A^*}{2i}$$

is the Cartesian decomposition of $A$, then

$$w_e^2(B, C) = w^2(A)$$

and

$$w(C^* B) = \frac{1}{4} w[(A^* - A)(A + A^*)].$$

Utilising (30.250) we deduce (30.256). □

*Remark 30.32.* If we choose in (30.250), $B = A$ and $C = A^*, A \in B(H)$ then we can state that

$$w^2(A) \le \frac{1}{2}\left[\|A\|^2 + w\left(A^2\right)\right]. \qquad (30.257)$$

The constant $1/2$ is best possible in (30.257).

Note that this inequality has been obtained in [12] by the use of a different argument based on the Buzano's inequality.

Finally, the following upper bound for the Euclidean radius involving different composite operators also holds:

**Theorem 30.36 (Dragomir, 2008 [9]).** *With the assumptions of Theorem 30.35, we have*

$$w_e^2(B, C) \le \frac{1}{2}\left[\|B^* B + C^* C\| + \|B^* B - C^* C\|\right] + w(C^* B). \qquad (30.258)$$

*The inequality (30.258) is sharp.*

*Proof.* We use (30.254) to write that

$$|\langle Bx, x\rangle|^2 + |\langle Cx, x\rangle|^2 \le \frac{1}{2}\left[\|Bx\|^2 + \|Cx\|^2 + \left|\|Bx\|^2 - \|Cx\|^2\right|\right] + |\langle Bx, Cx\rangle| \qquad (30.259)$$

for any $x \in H$, $\|x\| = 1$.

Since

$$\|Bx\|^2 = \langle B^* Bx, x\rangle, \qquad \|Cx\|^2 = \langle C^* Cx, x\rangle,$$

then (30.259) can be written as

$$|\langle Bx, x\rangle|^2 + |\langle Cx, x\rangle|^2 \le \frac{1}{2}\left[\langle (B^* B + C^* C)x, x\rangle + |\langle (B^* B - C^* C)x, x\rangle|\right] + |\langle Bx, Cx\rangle| \qquad (30.260)$$

$x \in H$, $\|x\| = 1$.

Taking the supremum in (30.260) over $x \in H$, $\|x\| = 1$ and noting that the operators $B^*B \pm C^*C$ are self-adjoint, we deduce the desired result (30.258).

The sharpness of the constant will follow from that of (30.263) pointed out below. □

**Corollary 30.31.** *For any two operators $B, C \in B(H)$, we have*

$$2w_e^2(B,C) \leq \|B^*B + C^*C\| + \|B^*C + C^*B\| + w[(B^* - C^*)(B+C)]. \quad (30.261)$$

*The constant 2 is best possible.*

*Proof.* If we write (30.258) for $B+C, B-C$ instead of $B, C$ and perform the required calculations then we get

$$w_e^2(B+C, B-C) \leq \|B^*B + C^*C\| + \|B^*C + C^*B\| + w[(B^* - C^*)(B+C)],$$

which, by the identity (30.247) is clearly equivalent with (30.261).

Now, if we choose in (30.261) $B = C$, then we get the inequality $w(B) \leq \|B\|$, which is a sharp inequality. □

**Corollary 30.32.** *If $B, C$ are self-adjoint operators on $H$ then*

$$w_e^2(B,C) \leq \frac{1}{2}\left[\|B^2 + C^2\| + \|B^2 - C^2\|\right] + w(CB). \quad (30.262)$$

The proof is obvious from Theorem 30.36.

We observe that, if $B$ and $C$ are chosen to be the Cartesian decomposition for the bounded linear operator $A$, then we can get from (30.262) that

$$w^2(A) \leq \frac{1}{4}\left\{\|A^*A + AA^*\| + \left\|A^2 + (A^*)^2\right\| + w[(A^* - A)(A + A^*)]\right\}. \quad (30.263)$$

The constant $1/4$ is best possible. This follows by the fact that for $A$ a self-adjoint operator, we obtain on both sides of (30.263) the same quantity $\|A\|^2$.

Now, if we choose in (30.258) $B = A$ and $C = A^*$, $A \in B(H)$, then we get

$$w^2(A) \leq \frac{1}{4}\{\|A^*A + AA^*\| + \|A^*A - AA^*\|\} + \frac{1}{2}w(A^2). \quad (30.264)$$

This inequality is sharp. The equality holds if, for instance, we assume that $A$ is normal, i.e., $A^*A = AA^*$. In this case we get on both sides of (30.264) the quantity $\|A\|^2$, since for normal operators, $w(A^2) = w^2(A) = \|A\|^2$.

## References


1. Dragomir, S.S.: A generalisation of Grüss' inequality in inner product spaces and applications. J. Math. Anal. Appl. **237**, 74–82 (1999)
2. Dragomir, S.S.: Some Grüss type inequalities in inner product spaces. J. Inequal. Pure Appl. Math. **4(2)**, Article 42 (2003)
3. Dragomir, S.S.: A counterpart of Schwarz's inequality in inner product spaces. East Asian Math. J. **20(1)**, 1–10 (2004)
4. Dragomir, S.S.: Reverses of Schwarz, triangle and Bessel inequalities in inner product spaces. J. Inequal. Pure Appl. Math. **5(3)**, Article 76 (2004)
5. Dragomir, S.S.: New reverses of Schwarz, triangle and Bessel inequalities in inner product spaces. Australian J. Math. Anal. Appl. **1**, Issue 1, Article 1, 1–18 (2004)
6. Dragomir, S.S.: Advances in Inequalities of the Schwarz, Gruss and Bessel Type in Inner Product Spaces. Nova Science Publishers Inc., New York (2005)
7. Dragomir, S.S.: Reverses of the Schwarz inequality generalising a Klamkin-McLenaghan result. Bull. Austral. Math. Soc. **73**, 69–78 (2006)
8. Dragomir, S.S.: Reverse inequalities for the numerical radius of linear operators in Hilbert spaces. Bull. Austral. Math. Soc. **73**, 255–262 (2006)
9. Dragomir, S.S.: Some inequalities for the Euclidean operator radius of two operators in Hilbert spaces. Linear Algebra Appl. **419**, 256–264 (2006)
10. Dragomir, S.S.: A survey of some recent inequalities for the norm and numerical radius of operators in Hilbert spaces. Banach J. Math. Anal. **1**, 154–175 (2007)
11. Dragomir, S.S.: Inequalities for some functionals associated with bounded linear operators in Hilbert spaces. Publ. Res. Inst. Math. Sci. **43** 1095–1110 (2007)
12. Dragomir, S.S.: Inequalities for the norm and the numerical radius of linear operators in Hilbert spaces. Demonstratio Math. **40**, 411–417 (2007)
13. Dragomir, S.S.: Norm and numerical radius inequalities for sums of bounded linear operators in Hilbert spaces. Facta Univ. Ser. Math. Inform. **22**, 61–75 (2007)
14. Dragomir, S.S.: The hypo-Euclidean norm of an *n*-tuple of vectors in inner product spaces and applications. J. Inequal. Pure Appl. Math. **8**, no. 2, Article 52, 22 pages (2007)
15. Dragomir, S.S.: Inequalities for the numerical radius, the norm and the maximum of the real part of bounded linear operators in Hilbert spaces. Linear Algebra Appl. **428**, no. 11-12, 2980–2994 (2008)
16. Dragomir, S.S.: New inequalities of the Kantorovich type for bounded linear operators in Hilbert spaces. Linear Algebra Appl. **428**, no. 11-12, 2750–2760 (2008)
17. Dragomir, S.S.: Some inequalities for commutators of bounded linear operators in Hilbert spaces. Preprint: RGMIA Res. Rep. Coll. **11**, No. 1, Article 7 (2008), http://www.staff.vu.edu.au/rgmia/v11n1.asp
18. Dragomir, S.S.: Some inequalities of the Grüss type for the numerical radius of bounded linear operators in Hilbert spaces. J. Ineq. Appl. **2008**, Art. Id. 763102, 9 pages. Preprint: RGMIA Res. Rep. Coll. **11**, No. 1 (2008), http://rgmia.vu.edu.au/reports.html
19. Dragomir, S.S.: Inequalities for the norm and the numerical radius of composite operators in Hilbert spaces. In: International Series of Numerical Mathematics **157**, pp. 135–146. Birkhäuser Verlag, Basel (2008)
20. Dragomir, S.S.: A functional associated with two bounded linear operators in Hilbert spaces and related inequalities. Italian Journal of Pure and Applied Mathematics, to appear. Preprint: RGMIA Res. Rep. Coll. **11**, Issue 3, Article 8 (2008), http://ajmaa.org/RGMIA/v11n3.php
21. Dragomir, S.S.: Norm and numerical radius inequalities for a product of two linear operators in Hilbert spaces. J. Math. Ineq. **2**, 499–510 (2009)
22. Dragomir, S.S.: Power inequalities for the numerical radius of a product of two operators in Hilbert spaces. Sarajevo J. Math. **5**(18), 269–278 (2009)
23. Dragomir, S.S., Sándor, J.: Some inequalities in prehilbertian spaces. Studia Univ. "Babeş-Bolyai"- Mathematica **32**, 71–78 (1987)

24. Goldstein, A., Ryff, J.V., Clarke, L.E.: Problem 5473. Amer. Math. Monthly **75**(3), 309 (1968)
25. Greub, W., Rheinboldt, W.: On a generalization of an inequality of L. V. Kantorovich. Proc. Amer. Math. Soc. **10**, 407–415 (1959)
26. Gustafson, K.E., Rao, D.K.M.: Numerical Range. Springer-Verlag, New York (1997)
27. Halmos, P.R.: Introduction to Hilbert Space and the Theory of Spectral Multiplicity. Chelsea Pub. Comp, New York (1972)
28. Halmos, P.R.: A Hilbert Space Problem Book. Springer-Verlag, New York–Heidelberg–Berlin (1982)
29. El-Haddad, M., Kittaneh, F.: Numerical radius inequalities for Hilbert space operators. II. Studia Math. **182**, 133–140 (2007)
30. Kittaneh, F.: Notes on some inequalities for Hilbert space operators. Publ. Res. Inst. Math. Sci. **24**, 283–293 (1988)
31. Kittaneh, F.: A numerical radius inequality and an estimate for the numerical radius of the Frobenius companion matrix. Studia Math. **158**, 11–17 (2003)
32. Kittaneh, F.: Numerical radius inequalities for Hilbert space operators. Studia Math. **168**, 73–80 (2005)
33. Pearcy, C.: An elementary proof of the power inequality for the numerical radius. Michigan Math. J. **13**, 289–291 (1966)
34. Popescu, G.: Unitary invariants in multivariable operator theory. Preprint: Ar$\chi$iv.math.OA/0410492
35. Yamazaki, T.: On upper and lower bounds for the numerical radius and an equality condition. Studia Math. **178**, 83–89 (2007)

# Chapter 31
# Cauchy's Functional Equation and Nowhere Continuous / Everywhere Dense Costas Bijections in Euclidean Spaces

**Konstantinos Drakakis**

*Dedicated to the memory of S.M. Ulam*
*on the 100th anniversary of his birth*

**Abstract** The definition of the Costas property is abstracted in the context of a group. Accordingly, using discontinuous solutions of Cauchy's functional equation, nowhere continuous Costas bijections are studied that satisfy the Costas property, and possess, in addition, dense graphs.

**Keywords** Costas arrays • Costas bijections • Cauchy functional equation • Nowhere continuous • Everywhere dense • Euclidean spaces

**Mathematics Subject Classification (2000):** Primary 05B10, 26A15, 11B05; Secondary 03E75, 11B13

## 31.1 Introduction

Costas arrays [3–5] are square arrangements of 1s/dots and 0s/blanks such that there is exactly one dot per row and column (i.e. they are permutation arrays), and such that (a) no four dots form a parallelogram, and (b) no three dots that lie on the same straight line are equidistant. They arose in the 1960s in connection with the development of SONAR/RADAR frequency-hopped waveforms with ideal auto-correlation properties [3, 4], but have been the subject of increasingly intensive mathematical study ever since Golomb published in 1984 [13, 14] some algebraic construction techniques (still the only ones available today) based on finite fields.

---

K. Drakakis (✉)
Complex & Adaptive Systems Laboratory (UCD CASL), University College Dublin, Belfield, Dublin 4, Ireland

The School of Electronic, Electrical & Mechanical Engineering, University College Dublin, Ireland
e-mail: Konstantinos.Drakakis@ucd.ie

Th.M. Rassias and J. Brzdęk (eds.), *Functional Equations in Mathematical Analysis*, Springer Optimization and Its Applications 52, DOI 10.1007/978-1-4614-0055-4_31,

Mathematicians are mainly concerned with the study of properties of Costas arrays, but also with the settlement of the question of the existence of a $n \times n$ Costas array for all $n \in \mathbb{N}^*$, which, despite all efforts, still remains open. Known Costas arrays today broadly fall into three categories: (a) the algebraically constructed ones, available for infinitely many, but not all, sizes $n$; (b) those discovered by exhaustive computer search, which currently has covered all sizes $n \leq 29$ [8,10,11]; and (c) four more Costas arrays, namely two for $n = 29$, and one for each of $n = 36$ and $n = 42$, constructed by extending smaller algebraically constructed Costas arrays [19].

This work first presents the Costas property, states the observation that it actually requires surprisingly little, namely only an underlying algebraic group structure, and hence proceeds to abstract it as a group property. It subsequently combines previously published results on the generalization of the Costas property in the continuum [7] along with the concept of sets with the Costas property in higher dimensions [6] and the topology of Euclidean spaces, to produce nowhere continuous Costas bijections whose graphs are everywhere dense in the Euclidean spaces $\mathbb{R}^{2n}$ and $\mathbb{C}^{2n}$.

## 31.2 Basics

The following notation is observed throughout this work. $n$ stands for a positive integer. The set of the first $n$ positive integers $\{1,2,\ldots,n\}$ is denoted by $[n]$. Furthermore, for any subset $A$ of a group $G$ under operation $\otimes$ with identity element $e$, $A^*$ denotes $A\setminus\{e\}$, while, for any $g \in G$,

$$g \otimes A := \{g \otimes a : a \in A\}.$$

For a set $S$, $|S|$ denotes its cardinal number. Finally, a partial order of vectors in $\mathbb{R}^n$ is defined as follows: for $x,y \in \mathbb{R}^n$, $x \leq y$ iff $\forall i \in [n]$, $x_i \leq y_i$, and $x < y$ iff $x \leq y$ and also $\exists i \in [n] : x_i < y_i$.

### 31.2.1 *The Costas Property*

For the benefit of the readers not familiar with Costas arrays, the definition and the basic facts about them are given below. The emphasis is, however, on the abstraction of the Costas property in groups.

#### 31.2.1.1 Costas Arrays and Costas Permutations

Let us begin with the usual definition of a Costas array and a Costas permutation:

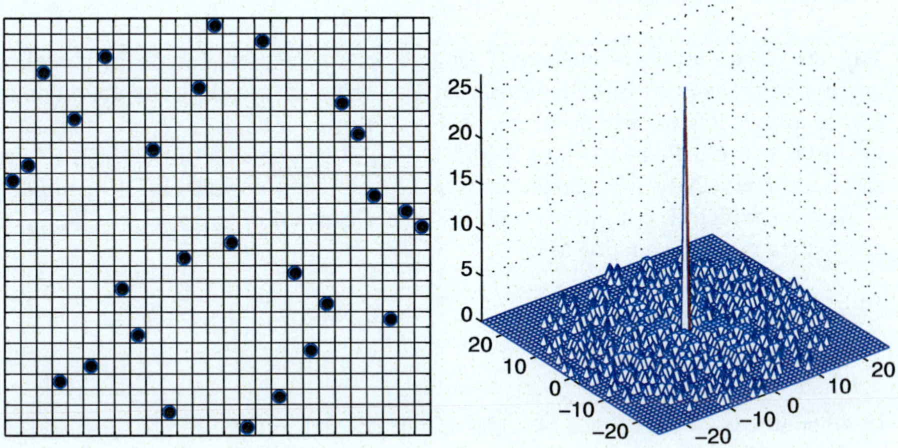


**Fig. 31.1** (*Left*) An example of a Costas array of order 27: note the haphazard positioning of its dots. (*Right*) Its auto-correlation

**Definition 31.1.** Let $f : [n] \to [n]$, $n \in \mathbb{N}^*$, be a bijection; then $f$ has the *Costas property* iff all vectors in the family $\{(f(i) - f(j), i - j) : 1 \le j < i \le n\}$, called the *distance vectors*,are distinct (in other words, no two of them are equal), or, equivalently, when

$$\forall i, j, k \in [n] : i + k, j + k \in [n], \quad f(i+k) - f(i) = f(j+k) - f(j) \Rightarrow i = j.$$

In this case, $f$ is called a *Costas permutation*.

The corresponding *Costas array* $A_f$ is the square $n \times n$ array where the elements at $(f(i), i)$, $i \in [n]$ are equal to 1 (dots), while all remaining elements are equal to 0 (blanks):

$$A_f = [a_{ij}] = \begin{cases} 1, & \text{if } i = f(j); \\ 0, & \text{otherwise,} \end{cases} \qquad j \in [n].$$

(In other words, for each $i$, $f(i)$ indicates the position of the dot in the $i$th column of the array.)

*Remark 31.1.* It is clear that the orbit of a Costas array under the symmetry group of the square (namely under horizontal/vertical flips and transposition) consists entirely of Costas arrays, hence out of a single Costas array $A$ eight can be constructed in total, or four if it happens to be symmetric ($A^T = A$).

Figure 31.1 (up) shows a Costas array of size 27, in fact the only Costas array of size 27 not obtainable by means of the algebraic constructions. Such Costas arrays, termed *sporadic*, appear to be growing fewer and fewer as the size $n$ increases.

#### 31.2.1.2 Motivation

Why are Costas arrays of interest? Though a detailed explanation of the modeling process, in the context of SONAR/RADAR systems, that leads to the definition of Costas arrays, lies outside the scope of this work (the interested reader is directed to Costas's original paper [4], or Sect. 3 of [9]), it can be said in brief that Costas arrays are mathematically interesting because their "auto-correlation" is "optimal".

More specifically, consider two binary $m \times n$ arrays $A = [a_{ij}]$ and $B = [b_{ij}]$, and define their cross-correlation to be

$$\Psi_{A,B}(u,v) = |\{(i,j) \in [m] \times [n] : \ (i+u, j+v) \in [m] \times [n], \ a_{ij} = b_{i+u,j+v} = 1\}|, \quad (u,v) \in \mathbb{Z}^2.$$

In other words, $A$ and $B$ are brought on top of each other, so that they overlap perfectly, and then $A$ gets slided by $u$ rows downwards and $v$ columns to the right: the number of pairs of overlapping dots at that position is taken to be the value of the cross-correlation of $A$ and $B$ at $u, v$.

Let now $A = B$ be a Costas array (so that $m = n$ as well), and consider the auto-correlation of $A$ $\Psi_{A,B}(u,v)$: clearly $\Psi_{A,A}(0,0) = n$, but, for $(u,v) \neq (0,0)$, necessarily $\Psi_{A,B}(u,v) \leq 1$, since there is no translation $(u,v)$ that can bring two dots of $A$ on two other dots of $A$; this would imply that $(u,v)$ is the distance vector between two pairs of dots, contradicting the fact that, in a Costas array, all distance vectors are distinct! For a random binary array, however, a translation may lead to the coincidence of two or more pairs of dots, yielding higher auto-correlation values between 1 and $n$. In this sense,Costas arrays have optimal auto-correlation. Figure 31.1 (up) shows the auto-correlation of the Costas array shown in the same figure (down).

#### 31.2.1.3 The Costas Property over an Arbitrary Group

The Costas property, as stated in Definition 31.1, can be abstracted over an arbitrary (possible non-abelian) group:

**Definition 31.2.** Let $G$ be a group under operation $\otimes$ and let $C$, $D$ be subsets of $G$. $C$ will be called a *left Costas set* in $G$ with respect to the *distance vectors subgroup* $D$, or simply a left Costas set if $D = G^*$, iff

$$\forall d \in D^*, \quad |(d \otimes C) \cap C| \leq 1,$$

or, equivalently, iff

$$\forall d \in D^*, \ \forall c_1, c_2 \in C, \quad d \otimes c_1, d \otimes c_2 \in C \Rightarrow c_1 = c_2.$$

A right Costas set is analogously defined. If $G$ is abelian the two definitions coincide.

*Remark 31.2.* Let $G = G_1 \times G_2$ be a Cartesian product of groups; then $C$ can be represented by points on the plane whose axes represent the points of $G_1$ and $G_2$, respectively, and its elements are ordered pairs of the form $c = (c_1, c_2), c_1 \in G_1, c_2 \in G_2$. Assuming that $C$ has either of the properties that each $g_1 \in G_1$ appears as the first coordinate of a $c \in C$ at most once, or that each $g_2 \in G_2$ appears as the second coordinate of a $c \in C$ at most once, $C$ represents the graph of a function with the Costas property (a Costas function, for short) from (a subset of) $G_1$ to $G_2$ or from (a subset of) $G_2$ to $G_1$, respectively. If both properties hold, the Costas function is injective, hence bijective by an appropriate redefinition of its domain and range.

*Remark 31.3.* The elements of $D$ can be considered as vectors (as opposed to scalars) when

$$G = G_1 \times \cdots \times G_k$$

is a tensor product of $k > 1$ groups. In that case $C$ can be interpreted geometrically to lie in a $k$-dimensional space, and $d \otimes C$, $d = \{d_1, \ldots, d_k\} \in D$ to be a shifted version of $C$ by $d_i$ in dimension $i$, $i \in [k]$. In particular, the Costas property forbids the existence of $c_1, c_2 \in C$, $c_1 \neq c_2$, such that, for some $d \in D$, $d \otimes c_1, d \otimes c_2 \in D$: no four points in $C$ can form a "parallelogram" (under $\otimes$), and no three points in $C$ that lie on a straight line can be equidistant.

*Remark 31.4.* Let $f : C_1 \subset G_1 \to G_2$ be a function; an equivalent definition of the Costas property with respect to $D$ is then that

$$\forall d \in D_1^*, \forall x, y \in C_1 : d \otimes x, d \otimes y \in C_1,$$
$$f(x \otimes d) - f(x) = f(y \otimes d) - f(y) \Rightarrow x = y,$$

where

$$D_1 = \{d_1 \in G_1 \mid \exists d_2 \in G_2 : (d_1, d_2) \in D\}.$$

*Remark 31.5.* Historically, the Costas property has been construed to include bijectivity: this is because the original SONAR/RADAR engineering application that introduced Costas arrays does not benefit any further from non-bijective Costas sets [4], namely Costas arrays with two or more dots on the same row or column, though such non-permutation Costas arrays can still be used successfully in other engineering applications, such as the predistribution of cryptographic keys in grid-based wireless sensor networks [2]. In this work, bijectivity/injectivity is no longer considered to be a prerequisite of (and is disassociated from) the Costas property itself.

*Remark 31.6.* A direct consequence of the definition is that $\forall d \in D$, $d \otimes C$ is also a Costas set with respect to $D$: the Costas property is translation invariant. In particular, Costas arrays can be shifted over an integer grid in either direction and still retain the Costas property.

## 31.2.2 Cauchy's Functional Equation

Cauchy's functional equation (CFE) will be useful below as an auxiliary building block. For the benefit of the readers who are not familiar with it, and for the sake of completeness, the basic facts about it are derived below. A detailed study of this equation can be found in [1].

### 31.2.2.1 Introduction: The Real Case

**Definition 31.3.** Let $f:\mathbb{R}\to\mathbb{R}$; it satisfies CFE iff

$$\forall x,y\in\mathbb{R},\quad f(x+y)=f(x)+f(y). \tag{31.1}$$

**Theorem 31.1.** *A solution $f$ of CFE (31.1) satisfies the following properties:*

1. $\forall q\in\mathbb{Q},\forall x\in\mathbb{R}\ f(qx)=qf(x)$.
2. *$f$ is continuous iff* $\exists c\in\mathbb{R}:\ \forall x\in\mathbb{R}\ f(cx)=cx$.
3. *$f$ is continuous everywhere iff it is continuous at a point.*
4. *$f$ is discontinuous iff its graph is everywhere dense on the real plane.*

*Proof.*

- Setting $x=y=0$ yields $f(0+0)=f(0)=f(0)+f(0)\Leftrightarrow f(0)=0$.
- Setting $y=-x$ yields $f(x-x)=f(0)=0=f(x)+f(-x)\Leftrightarrow f(-x)=-f(x)$.
- Setting $x=x_1,y=x_2+\cdots+x_n$ for $n\in\mathbb{N}^*$ yields $f(x_1+x_2+\cdots+x_n)=f(x_1)+f(x_2+\cdots+x_n)=\cdots=f(x_1)+f(x_2)+\cdots+f(x_n)$.
- Setting $x_1=x_2=\cdots=x_n=x$ yields $f(nx)=nf(x)$. Setting $y=nx$ yields

$$f\left(\frac{y}{n}\right)=\frac{1}{n}f(y).$$

Expressing the rational $q$ as an irreducible fraction of integers $m/n$, all of the above shows that, for any $x\in\mathbb{R}$:

$$f(qx)=f\left(\frac{m}{n}x\right)=\operatorname{sign}(m)f\left(\frac{|m|}{n}x\right)$$
$$=\operatorname{sign}(m)|m|f\left(\frac{1}{n}x\right)=\frac{m}{n}f(x)=qf(x).$$

Assume now $f$ is continuous: for every $x\in\mathbb{R}$ there exists a sequence $\{q_n\}$ of rationals such that $q_n\to x$. It follows that

$$f(x)=f(\lim q_n)=\lim f(q_n)=\lim q_nf(1)=xf(1)=cx,\quad c=f(1). \tag{31.2}$$

Conversely, every function of the form $f(x)=cx$ satisfies CFE (31.1) and is continuous.

Assume that $f$ is continuous at some point $x_0$; then

$$\lim_{y\to x} f(y) = f(y-x+x-x_0+x_0) = \lim_{y\to x} f(y-x+x_0)+f(x-x_0)$$
$$= \lim_{u\to x_0} f(u)+f(x-x_0) = f(x_0)+f(x-x_0) = f(x),$$

hence $f$ is continuous at (an arbitrary) $x$.

Assume now $f$ is not continuous: then, by what was just proved, it must be nowhere continuous and it cannot be linear (though it has to be linear over the rationals). This implies

$$\exists x_1, x_2 \in \mathbb{R}: \quad \frac{f(x_1)}{x_1} \neq \frac{f(x_2)}{x_2},$$

whence it follows that the two vectors $v_1 = (x_1, f(x_1))$, $v_2 = (x_2, f(x_2))$ are linearly independent and, consequently, span the entire real plane. The set of vectors, then, of the form $\{r_1v_1 + r_2v_2 : r_1, r_2 \in \mathbb{Q}\}$ are an everywhere dense subset of the real plane; but

$$r_1v_1 + r_2v_2 = r_1(x_1, f(x_1)) + r_2(x_2, f(x_2)) = (r_1x_1 + r_2x_2, f(r_1x_1 + r_2x_2)),$$

which means that the subset $\{f(x) : x = r_1x_1 + r_2x_2,\ r_1, r_2 \in \mathbb{Q}\}$ of the graph of $f$ is everywhere dense on the plane, whence the graph of $f$ itself is everywhere dense on the plane. Conversely, if the graph of $f$ is everywhere dense on the real plane, $f$ cannot possibly be continuous, or else it would be linear and thus would not possess an everywhere dense graph. This completes the proof. □

*Remark 31.7.* CFE, despite its simplicity, has been playing a prominent role in analysis: Hilbert's 5th problem [15] essentially proposes a generalization of this equation, while an important area of study is the existence of mappings in Banach spaces that satisfy CFE approximately, namely with an error whose norm is uniformly bounded by a fixed and small $\varepsilon > 0$, and may thus be considered to be "approximately additive" (this is known as the Hyers-Rassias-Ulam stability problem [16–18]).

*Remark 31.8.* All solutions of CFE are, by definition, additive functions in $\mathbb{R}$ and represent linear maps on the vector space $\mathbb{R}$ over the field $\mathbb{Q}$; continuous solutions are, furthermore, linear functions over $\mathbb{R}$; a discontinuous solution at a point is discontinuous everywhere and its graph is dense in $\mathbb{R}^2$. The reader is cautioned, therefore, that two distinct concepts of linearity have been considered above: linearity of maps over vector spaces and linearity of functions over $\mathbb{R}$. In particular, we reiterate that a discontinuous solution of CFE (whose graph is dense on the plane) is clearly not a linear function over $\mathbb{R}$ (namely, denoting the function by $f$, $f(x)/x$ is not constant and independent of $x$), while it is still a linear map on the vector space $\mathbb{R}$ over the field $\mathbb{Q}$!

#### 31.2.2.2 Bijective/Injective Solutions of CFE

**Theorem 31.2.** *There exist solutions of CFE (31.1) that are nowhere continuous bijections/injections, everywhere dense on the real plane.*

*Proof.* Consider $\mathbb{R}$ as a vector space over $\mathbb{Q}$. This vector space must necessarily have an uncountable basis, or else $\mathbb{R}$ itself would be countable: it follows by the Continuum Hypothesis that this basis can be indexed by the real numbers, and, therefore, that it can be described as $B = \{b_a : a \in \mathbb{R}\}$. By definition, any real number $x$ admits a (finite) linear expansion $x = q_1 b_{a_1} + \cdots + q_n b_{a_n}$ over this basis (where the rational $q$, the indices $a_i, i \in [n]$, and $n$ are obviously functions of $x$). Furthermore, a solution $f$ of CFE (31.1) can be considered as a linear map over this vector space, and thus can be written

$$f(x) = f(q_1 b_{a_1} + \cdots + q_n b_{a_n}) = q_1 f(b_{a_1}) + \cdots + q_n f(b_{a_n}),$$

which expresses the well-known result that a linear map over a vector space is unambiguously defined by its effect on the vector space basis $B$. Assuming now that $\exists c \in \mathbb{R} : \ \forall a \in \mathbb{R}, \ f(b_a) = cb_a$, it follows that $f(x) = cx$ for all $x$, namely that $f$ is linear. By Theorem 31.1, though, this will be the *only* case resulting to a linear $f$: in all other cases $f$ will be nowhere continuous and its graph everywhere dense. Choosing a bijection/injection $g$ over $\mathbb{R}$ (other than the identity) such that $\forall a \in \mathbb{R}, f(b_a) = b_{g(a)}$ results to an $f$ that is bijective/injective as well. □

#### 31.2.2.3 The Complex Case

Let us now study the properties of a function $f : \mathbb{C} \to \mathbb{C}$ such that

$$\forall u, v \in \mathbb{C} : \quad f(u+v) = f(u) + f(v). \tag{31.3}$$

Applying the real argument twice (on the real and the imaginary numbers) yields

$$\forall x, y \in \mathbb{Q}, \ z \in \mathbb{C} : \quad f((x+\mathrm{i}y)z) = xf(z) + yf(\mathrm{i}z). \tag{31.4}$$

Letting $z = 1$, this implies that $f(x+\mathrm{i}y) = xf(1) + yf(\mathrm{i})$, where $f(1)$ and $f(\mathrm{i})$ are, in general, complex values, and, therefore, their linear combinations over $\mathbb{Q}$ will form a dense subset of $\mathbb{C}$, unless $f(1)$ and $f(\mathrm{i})$ are linearly dependent over $\mathbb{R}$, namely

$$\exists a, b \in \mathbb{R}, \ |a| + |b| > 0 : \quad af(\mathrm{i}) + bf(1) = 0, \tag{31.5}$$

in which case $f(\mathbb{Q} + \mathrm{i}\mathbb{Q})$ is contained on a straight line of $\mathbb{C}$, and is dense therein. A verbatim repetition of the proof of the real case shows that $f$ is continuous on $\mathbb{C}$ iff it is continuous at a point: in this case, it follows that

$$\forall x, y \in \mathbb{R}, \quad f(x+\mathrm{i}y) = xf(1) + yf(\mathrm{i}). \tag{31.6}$$

It remains to be shown that $f$ is an injection/bijection, and that the graph of $f$, namely $\{(z, f(z)) : z \in \mathbb{C}\}$, is dense in $\mathbb{C}^2$. The corresponding proof in the real case relied heavily on the linearity of $f$ over rational multipliers, namely on the condition that $\forall q \in \mathbb{Q}, \forall x \in \mathbb{R}\ f(qx) = qf(x)$ (see Theorem 31.1), but this time the conditions stated in (31.3) and (31.4) are not linearity conditions, since they do not imply that $\forall q \in \mathbb{Q} + \mathrm{i}\mathbb{Q}, \forall z \in \mathbb{C}\ f(qz) = qf(z)$. Attention can still, however, be restricted to real linear solutions of (31.3) by imposing the compatibility condition

$$\forall x \in \mathbb{R}, \quad f(\mathrm{i}x) = \mathrm{i}f(x) : \tag{31.7}$$

this condition is equivalent to

$$\forall z \in \mathbb{C}, \quad f(\mathrm{i}z) = \mathrm{i}f(z),$$

because, letting $z = x + \mathrm{i}y,\ x, y \in \mathbb{R}$ yields

$$f(\mathrm{i}z) = f(\mathrm{i}x - y) = \mathrm{i}f(x) - f(y) = \mathrm{i}(f(x) + \mathrm{i}f(y)) = \mathrm{i}f(x + \mathrm{i}y) = \mathrm{i}f(z).$$

Applying this condition to (31.4) further yields

$$\forall x, y \in \mathbb{Q},\ z \in \mathbb{C}, \quad f((x + \mathrm{i}y)z) = (x + \mathrm{i}y)f(z), \tag{31.8}$$

while it is seen directly that

$$\forall x, y \in \mathbb{R}, \quad f(x + \mathrm{i}y) = f(x) + \mathrm{i}f(y).$$

In other words, every solution of (31.1) can be extended to a solution of (31.3) that obeys the linearity property (31.8) over the complex rationals $\mathbb{Q} + \mathrm{i}\mathbb{Q}$; such functions have the property that $f(\mathbb{R}) \subset \mathbb{R}$, which is equivalent to $f(\mathrm{i}\mathbb{R}) \subset \mathrm{i}\mathbb{R}$. As (31.5) and (31.7) are not compatible, $f(\mathbb{Q} + \mathrm{i}\mathbb{Q})$ is dense in $\mathbb{C}$.

Assuming $f$ is continuous, (31.6) shows that $\forall z \in \mathbb{C},\ f(z) = zf(1)$; assuming the function is nowhere continuous, hence nonlinear, there exist $z, w \in \mathbb{C}$ such that the vectors $(z, f(z))$ and $(w, f(w))$ are linearly independent. It follows that

$$\begin{aligned}\{p(z, f(z))\ +\ q(w, f(w)) :\ p, q \in \mathbb{Q} + \mathrm{i}\mathbb{Q}\}\\ = \{(pz + qw, f(pz + qw)) :\ p, q \in \mathbb{Q} + \mathrm{i}\mathbb{Q}\},\end{aligned}$$

namely a subset of the graph of $f$, is countably infinite and dense in $\mathbb{C}^2$.

It remains to be seen that such a discontinuous function can be constructed. The proof proceed as in Theorem 31.2: $\mathbb{C}^2$ is considered as a vector space over the field of complex rationals $\mathbb{Q} + \mathrm{i}\mathbb{Q}$ with an uncountable basis $B = \{b_a :\ a \in \mathbb{R}\}$. If $f : B \to B$ is a permutation other than the identity, $f : \mathbb{C} \to \mathbb{C}$ is a nowhere continuous bijection on $\mathbb{C}$ whose graph is dense in $\mathbb{C}^2$; if $f : B \to B$ is injective, $f : \mathbb{C} \to \mathbb{C}$ is a nowhere continuous injection on $\mathbb{C}$ whose graph is dense in $\mathbb{C}^2$.

### *31.2.3 Costas Functions and Bijections in the Continuum*

The following is a restatement of the definition of the Costas property (Definition 31.2), specifically in the context of functions whose domain and range are intervals of the real line:

**Definition 31.4.** Let $f: I_1 \to I_2 = f(I_1)$, where $I_1, I_2 \subset \mathbb{R}$ are intervals, and let $f$ be, in addition, continuously differentiable; then $f$ is said to have the Costas property iff

$$\forall x, y \in I_1,\ \forall d \in \mathbb{R}^* :\ x+d, y+d \in I_1,$$
$$f(x+d) - f(x) = f(y+d) - f(y) \Rightarrow x = y.$$

The following theorem is essentially a restatement of Theorem 5 in [7]. The assumptions are not the weakest possible, but it will be sufficient for our purposes:

**Theorem 31.3.** *Let $f: I_1 \to I_2$ be a strictly monotonic and continuously differentiable bijection, with a strictly monotonic derivative; then, $f$ has the Costas property.*

*Proof.* Let us choose two equidistant pairs of points in $I_1$, say $x$, $y$, $x+d$ and $y+d$ so that $y \geq x$ and $d > 0$; these may actually be three points if $y = x+d$. What needs to be shown is that

$$f(x) - f(x+d) = f(y) - f(y+d) \Rightarrow x = y.$$

If $y > x$, then exactly one of the two pairs of intervals

$$[x,y],\ [x+d, y+d] \quad \text{or} \quad [x, x+d],\ [y, y+d]$$

consists of intervals with disjoint interiors; without loss of generality, assume it is the second pair. Then, the Newton–Leibniz Theorem implies that

$$f(x+d) - f(x) = \int_x^{x+d} f'(u)\mathrm{d}u, \quad f(y+d) - f(y) = \int_y^{y+d} f'(u)\mathrm{d}u.$$

Now, since $f'$ is strictly monotonic, it is always the case that either

$$\forall u \in (x, x+d),\ v \in (y, y+d), \quad f'(u) < f'(v)$$

or

$$\forall u \in (x, x+d),\ v \in (y, y+d), \quad f'(u) > f'(v),$$

so that either $f(x) - f(x+d) < f(y) - f(y+d)$ or $f(x) - f(x+d) > f(y) - f(y+d)$, respectively. By contraposition, $f(x) - f(x+d) = f(y) - f(y+d) \Rightarrow x = y$. □

This work focuses exclusively on the case $I_1 = I_2 = \mathbb{R}$.

## 31.3 Results

### 31.3.1 Solutions of CFE in Higher Dimensions

It is simple to write bijective solutions of CFE in $\mathbb{R}^n$ using Cartesian products of bijective solutions in $\mathbb{R}$. Indeed, $f:\mathbb{R}^n \to \mathbb{R}^n$ can be defined as the Cartesian product

$$f(x_1,\ldots,x_n) = (f_1(x_1),\ldots,f_n(x_n)),$$

where $f_i:\mathbb{R}\to\mathbb{R}$, $i=1,\ldots,n$ are all solutions of CFE. Letting further $T_1$ and $T_2$ be invertible transformations in $\mathbb{R}^n$, new bijective solutions can be written using Cartesian product solutions $f$ as above as $T_2\circ f\circ T_1^{-1}$. What is more, assuming that, for each $i\in[n]$, $f_i$ is discontinuous, and hence that its graph is dense in $\mathbb{R}^2$, the graph of $f$ is dense in $\mathbb{R}^{2n}$: to see this, observe that, given an arbitrary $y=(y_1,\ldots,y_{2n})\in\mathbb{R}^{2n}$, for each $i\in[n]$ and for each $\varepsilon_i>0$, there exists an $x_i\in\mathbb{R}$ such that

$$|(x_i,f_i(x_i))-(y_i,y_{i+n})|<\varepsilon_i.$$

It further follows that $T_2\circ f\circ T_1^{-1}$, being a bijective linear transformation of $f$, is also dense in $\mathbb{R}^{2n}$.

For $n=2$, in particular, in view of the additive isomorphism (denoted by "$\leftrightarrow$" below) between $\mathbb{C}$ and $\mathbb{R}^2$, pairs of discontinuous solutions $f_1$ and $f_2$ in $\mathbb{R}$ can be combined to form nonlinear solutions $f=f_1+\mathrm{i}f_2\leftrightarrow(f_1,f_2)$ in $\mathbb{C}$: $f(x)=f_1(x_1)+\mathrm{i}f_2(x_2)$, $f_1,f_2:\mathbb{R}\to\mathbb{R}$, where $x=x_1+\mathrm{i}x_2\in\mathbb{C}\leftrightarrow(x_1,x_2)\in\mathbb{R}^2$. One, however, must keep in mind the compatibility condition: $f(\mathrm{i})=f_2(1)=\mathrm{i}f(1)=\mathrm{i}f_1(1)$. Repeated application of this isomorphism forms discontinuous solutions in $\mathbb{C}^n$, whose graphs are dense in $\mathbb{C}^{2n}$.

### 31.3.2 Costas Bijections in Higher Dimensions

Though it is tempting to use the Cartesian product construction of Sect. 31.3.1 to obtain Costas bijections in $\mathbb{R}^n$ out of Costas bijections in $\mathbb{R}$, it turns out that this is in fact not possible. To see this, let $f:\mathbb{R}^n\to\mathbb{R}^n$, $n\in\mathbb{N}^*$, be such that

$$f(x_1,\ldots,x_n) = (f_1(x_1),\ldots,f_n(x_n)),$$

where $f_i:\mathbb{R}\to\mathbb{R}$, $i\in[n]$ are Costas bijections in $\mathbb{R}$, and test whether $f$ is a Costas bijection in $\mathbb{R}^n$: for $x,y\in\mathbb{R}^n$ and $d\in(\mathbb{R}^n)^*$, it follows that

$$\begin{aligned} f(x+d)-f(x) &= f(y+d)-f(y) \\ &\Leftrightarrow \forall i\in[n],\quad f_i(x_i+d_i)-f(x_i)=f(y_i+d_i)-f(y_i) \\ &\Rightarrow \forall i\in[n]:\quad d_i\neq 0,\ x_i=y_i. \end{aligned}$$

Unfortunately, $d$ is allowed to have zero coordinates (as long as at least one coordinate is nonzero), and if $d_i = 0$ for some $i \in [n]$ one cannot conclude that $x_i = y_i$, and therefore that $x = y$. This conclusion actually holds if the condition that $d \in (\mathbb{R}^n)^*$ is substituted by the weaker condition that $d \in (\mathbb{R}^*)^n$; this, however, is not sufficient to ensure the validity of Definition 31.2, but rather leads to a weaker form of the Costas property.

It is, in fact, possible to obtain sufficient conditions for the existence of smooth Costas bijections in $\mathbb{R}^n$, using the same ideas as in Theorem 31.3, which impose a higher dimensional monotonicity condition on the partial derivatives of the bijection:

**Theorem 31.4.** *Let $f = (f_1,\dots,f_n) : \mathbb{R}^n \to \mathbb{R}^n$ be a bijection with continuous partial derivatives such that, for each $i \in [n]$, either $f_i$ or $-f_i$ satisfies the condition that, $\forall j \in [n]$ and $\forall s,t \in \mathbb{R}^n$ : $s \le t$, $\partial_j f_i(s) \le \partial_j f_i(t)$, and that $\partial_j f_i(s) = \partial_j f_i(t) \Rightarrow s = t$ (here, $\partial_j f_i$ stands for the partial derivative of $f_i$ with respect to the jth variable); then, $f$ is a Costas bijection.*

*Proof.* Using the above notation established above in this section, note that, for any $i \in [n]$, if $f_i$ has a continuous derivative everywhere, repeated application of the Newton-Leibniz theorem yields that:

$$\begin{aligned}
&f_i(x_1 + d_1, x_2 + d_2, \dots, x_n + d_n) - f_i(x_1, x_2, \dots, x_n) \\
&\quad = f_i(x_1 + d_1, x_2 + d_2, \dots, x_n + d_n) - f_i(x_1, x_2 + d_2, \dots, x_n + d_n) \\
&\quad + f_i(x_1, x_2 + d_2, \dots, x_n + d_n) - f_i(x_1, x_2, \dots, x_n) \\
&\quad = \int_{x_1}^{x_1+d_1} \partial_1 f_i(u_1, x_2 + d_2, \dots, x_n + d_n) du_1 \\
&\quad + f_i(x_1, x_2 + d_2, \dots, x_n + d_n) - f_i(x_1, x_2, \dots, x_n) \\
&\quad = \dots = \int_{x_1}^{x_1+d_1} \partial_1 f_i(u_1, x_2 + d_2, \dots, x_n + d_n) du_1 \\
&\quad + \int_{x_2}^{x_2+d_2} \partial_2 f_i(x_1, u_2, \dots, x_n + d_n) du_2 \\
&\quad + \dots + \int_{x_n}^{x_n+d_n} \partial_n f_i(x_1, x_2, \dots, u_n) du_n. \qquad (31.9)
\end{aligned}$$

Suppose now that

$$f(x+d) - f(x) = f(y+d) - f(y) \qquad (31.10)$$

holds for some $x \neq y$ and some $d$, where $x, y, d \in \mathbb{R}^n$. Without loss of generality, by swapping $x_i$ and $y_i$, and redefining $d_i$, if necessary, for some $i \in [n]$, it may be assumed that $y \ge x + d$ and $d > 0$, so that, $\forall i \in [n]$, the pairs of intervals $[x_i, x_i + d_i]$ and $[y_i, y_i + d_i]$ have disjoint interiors, and that the former lies always to the left of the latter. Suppose now further that, $\forall i,j \in [n]$, and $\forall s,t \in \mathbb{R}^n: \; s \leq t$, it holds true that $\partial_j f_i(s) \leq \partial_j f_i(t)$ and also that $\partial_j f_i(s) = \partial_j f_i(t) \Rightarrow s = t$; then, comparing

$$f_i(x_1+d_1,x_2+d_2,\ldots,x_n+d_n) - f_i(x_1,x_2,\ldots,x_n)$$

and

$$f_i(y_1+d_1,y_2+d_2,\ldots,y_n+d_n) - f_i(y_1,y_2,\ldots,y_n)$$

using (31.9), and assuming, without loss of generality, that $\forall i \in [n], f_i$ satisfies the condition of the statement, the conclusion follows that, for any $j \in [n]$ such that $d_j > 0$, and any $i \in [n]$,

$$\int_{x_j}^{x_j+d_j} \partial_j \, f_i(x_1,\ldots,x_{j-1},u_j,x_{j+1}+d_{j+1},\ldots,x_n+d_n)\mathrm{d}u_j$$
$$< \int_{y_j}^{y_j+d_j} \partial_j f_i(y_1,\ldots,y_{j-1},u_j,y_{j+1}+d_{j+1},\ldots,y_n+d_n)\mathrm{d}u_j$$

so that

$$f_i(x_1+d_1,x_2+d_2,\ldots,x_n+d_n) - f_i(x_1,x_2,\ldots,x_n)$$
$$< f_i(y_1+d_1,y_2+d_2,\ldots,y_n+d_n) - f_i(y_1,y_2,\ldots,y_n),$$

and (31.10) does not hold, a contradiction (if $-f_i$ instead of $f_i$ satisfies the condition of the statement, just swap the direction of the inequality). Therefore, (31.10) implies $x = y$, and this completes the proof. □

Generalizations of the Costas property in higher dimensions have been proposed in the literature before [6, 12], but considered exclusively discrete structures (in $\mathbb{Z}^n$).

### *31.3.3 Examples of Costas Bijections*

Some explicit constructions of Costas bijections in $\mathbb{R}^n$ are now provided. It is shown, in particular, that Costas bijections in $\mathbb{R}$ can be used to build higher-dimensional ones in $\mathbb{R}^n$, $n \geq 2$. In view of the additive isomorphism between $\mathbb{R}^2$ and $\mathbb{C}$ discussed in Sect. 31.3.1, attention will be focused on real Euclidean spaces, since the results obtained for $\mathbb{R}^n$ can be restated immediately in the context of $\mathbb{C}^m$ if $n = 2m$. This is possible because the Costas property is additive (or, more correctly, involves only one operation, which, without loss of generality, can be chosen to be addition, as it has been the case throughout this work).

### 31.3.3.1 Costas Bijections in $\mathbb{R}$

The purpose of this section is to provide examples of real functions satisfying the conditions of Theorem 31.3. Though there should be no doubt in the reader's mind about the existence of such functions, the author's experience was that one may not be able to think of a concrete example without spending some time and effort on it. Two examples are, therefore, provided below: the first is rather "ad hoc", while the second is more systematic.

*Example 31.1.* As a first example, consider the following case-defined function:

$$f(x) = \begin{cases} \ln(1+x), & x \geq 0; \\ -x^2 + x, & x < 0. \end{cases}$$

This function is bijective, as $f(\mathbb{R}_+) = \mathbb{R}_+$, $f(\mathbb{R}_-) = \mathbb{R}_-$, and strictly increasing in each of its two branches. It is also continuous everywhere, as it is continuous on each branch and also $f(0) = 0$, and continuously differentiable everywhere, as

$$f'(x) = \begin{cases} \frac{1}{1+x}, & x > 0; \\ 1, & x = 0; \\ -2x + 1, & x < 0. \end{cases}$$

Finally, the derivative is strictly decreasing on each branch; hence, everywhere by its continuity: in particular, $f'(x) \leq 1$, $x \geq 0$, while $f'(x) > 1$, $x < 0$.

*Example 31.2.* A more systematic construction leads to the desired Costas bijection through the specification of its derivative. Indeed, if $f : \mathbb{R} \to \mathbb{R}$ is such that its derivative $f'$ is positive, strictly increasing, and $0 < \alpha = \lim_{x \to -\infty} f'(x) < \beta = \lim_{x \to +\infty} f'(x) < +\infty$ (note that this is, in effect, the description of a sigmoid-like function, namely shaped like a slanted letter "s"), then the anti-derivative $f = \int f'$ will necessarily be strictly increasing itself, and, in addition, $f(\mathbb{R}) = \mathbb{R}$, because, asymptotically, $f(x) = \alpha x + o(x)$ for $x \ll -1$ and $f(x) = \beta x + o(x)$ for $x \gg +1$, whence $\lim_{x \to -\infty} f(x) = -\infty$, $\lim_{x \to +\infty} f(x) = +\infty$, and $f$ is continuous. Note that the conclusion holds even if $\beta = \infty$. Specific constructions include:

$$f'(x) = e^x + \alpha \Rightarrow f(x) = e^x + \alpha x + C, \quad C \in \mathbb{R};$$

$$f'(x) = \frac{e^x + \alpha}{e^x + 1}, \quad \alpha \in (0,1) \Rightarrow f(x) = \alpha x + (1-\alpha)\ln(1+e^x) + C, \quad C \in \mathbb{R};$$

$$f'(x) = \tan^{-1}(x) + \frac{\pi}{2} + \alpha \Rightarrow f(x) = x\tan^{-1}(x) + \left(\frac{\pi}{2} + \alpha\right)x$$
$$- \frac{1}{2}\ln(1+x^2) + C, \quad C \in \mathbb{R}.$$

Note that, in the last case, $\alpha = 0$ is allowed, since it still yields asymptotically that $f(x) \approx -\ln(|x|) \to -\infty$ as $x \to -\infty$.

#### 31.3.3.2 Costas Bijections in $\mathbb{R}^n$, $n \geq 2$

**Theorem 31.5.** *Let $f : \mathbb{R} \to \mathbb{R}$ satisfy the conditions of Theorem 31.3 and let $S = [s_{ij}]$ be an invertible $n \times n$ array, each row of which, denoted by $s_i$, $i \in [n]$, consists entirely of either positive or negative real numbers. Consider $F : \mathbb{R}^n \to \mathbb{R}^n$ such that*

$$F(x) = (f(s_1 \cdot x), \ldots, f(s_n \cdot x)),$$

*where "$\cdot$" stands for the ordinary dot product of two vectors. Then, $F$ is a Costas bijection in $\mathbb{R}^n$.*

*Proof.* It is first shown that $F$ is a bijection on $\mathbb{R}^n$, namely that, for arbitrary $w \in \mathbb{R}^n$, $F(x) = w$ has a unique solution $x$ (in the following derivation, $x$ is treated as a column vector):

$$\begin{aligned} F(x) = w &\Leftrightarrow (f(s_1 \cdot x), \ldots, f(s_n \cdot x)) = (w_1, \ldots, w_n) \\ &\Leftrightarrow s_1 \cdot x = f^{-1}(w_1), \ldots, s_n \cdot x = f^{-1}(w_n) \\ &\Leftrightarrow Sx = (f^{-1}(w_1), \ldots, f^{-1}(w_n))^T \\ &\Leftrightarrow x = S^{-1}(f^{-1}(w_1), \ldots, f^{-1}(w_n))^T, \end{aligned}$$

where the facts were used that $f$ is a bijection on $\mathbb{R}$, so that $x = f^{-1}(y)$ exists and is unique for any $y \in \mathbb{R}$, and that $S$ is invertible.

Furthermore, one observes that

$$\forall i, j \in [n], \quad \partial_j F_i(x) = s_{ij} f'(s_i \cdot x).$$

Since $f'$ is strictly monotonic and $s_i \cdot x$ is strictly monotonic with respect to all coordinates of $x$ as well (because all entries of $s_i$ have the same sign), it is concluded that $\partial_j F_i$ satisfies the conditions of Theorem 31.4 for all $i, j \in [n]$, and, therefore, that $F$ is a Costas bijection in $\mathbb{R}^n$. □

For example, if $f$ is the function considered in the previous example,

$$\forall i, j \in [n], \quad \partial_j F_i(x) = s_{ij}(ae^{s_i \cdot x} + b),$$

whence it is clear that, not only $\partial_j F_i$ is strictly monotonic for all $i, j \in [n]$, but also that, for each given $i$, the family of functions $\{\partial_j F_i : j \in [n]\}$ have the same type of strict monotonicity: either they are all strictly increasing or strictly decreasing. This is precisely the condition stated in Theorem 31.4.

### 31.3.4 A Conversion Theorem

Figure 31.1 suggests that the dots in a Costas array are scattered rather haphazardly, forming no visible pattern. This is, in fact, true for most Costas arrays (except perhaps for symmetric Costas arrays), and stands in stark contrast to the smooth Costas bijections in Euclidean spaces constructed by Theorems 31.3 and 31.4. Indeed, one might reasonably expect such bijections to exhibit no structure at all, and be discontinuous everywhere.

In this section, our two studies of Costas bijections and discontinuous solutions of CFE (presented separately till this point) get combined in order to produce the main result of this work, which allows the conversion of smooth Costas bijections into Costas bijections whose graphs are dense:

**Theorem 31.6.** *Let $c, f : \mathbb{R}^n \to \mathbb{R}^n$, so that $f$ is a smooth Costas bijection and $c$ is a discontinuous bijective solution of CFE; then, both $f \circ c$ and $c \circ f$ are Costas bijections whose graphs are dense in $\mathbb{R}^{2n}$.*

*Proof.* For $x, y \in \mathbb{R}^n$, $d \in (\mathbb{R}^n)^*$ consider

$$f \circ c(x+d) - f \circ c(x) = f \circ c(y+d) - f \circ c(y)$$
$$\Leftrightarrow f(c(x)+c(d)) - f(c(x)) = f(c(y)+c(d)) - f(c(y)).$$

Since $c$ is a bijection, $c(d) \neq 0$, and this implies, since $f$ is a Costas bijection, that

$$c(x) = c(y) \Leftrightarrow 0 = c(x) - c(y) = c(x-y) \Leftrightarrow x - y = 0 \Leftrightarrow x = y.$$

Similarly, by the additivity of $c$,

$$c \circ f(x+d) - c \circ f(x) = c \circ f(y+d) - c \circ f(y)$$
$$\Leftrightarrow c(f(x+d) - f(x) - (f(y+d) - f(y))) = 0.$$

Since $c$ is a bijection, this implies that $f(x+d) - f(x) = f(y+d) - f(y)$, and this, in turn, since $f$ is a Costas bijection, implies $x = y$.

The density of the graphs of these two functions follows from the fact that they both are compositions of a continuous bijective (hence strictly monotonic) function with a dense function. □

## 31.4 Summary and Conclusion

After a brief presentation of Costas arrays, the Costas property was abstracted as a property of groups. This allowed the consideration of smooth Costas bijections over the real line, as well as in higher dimensional Euclidean spaces: theorems with sufficient conditions for the construction of such bijections (essentially monotonicity conditions on the bijections and their derivatives) were stated and proved, and examples of explicit constructions were also given. In particular, higher dimensional constructions used one-dimensional ones as building blocks.

At the same time, Cauchy's functional equation (CFE) was considered in $\mathbb{R}$ and in $\mathbb{C}$. Its basic properties were reviewed, focusing on the facts that (a) although a solution of CFE can always be interpreted as a linear map over the vector space of $\mathbb{R}$ over $\mathbb{Q}$, or of $\mathbb{C}$ over $\mathbb{Q}+i\mathbb{Q}$, it is not always a linear function over $\mathbb{R}$ or $\mathbb{C}$, respectively; (b) the graph of a real or complex solution of CFE is dense in $\mathbb{R}^2$ or $\mathbb{C}^2$, respectively, iff it is a discontinuous (equivalently, nonlinear) function; and (c) that bijective solutions of CFE over $\mathbb{R}$ and $\mathbb{C}$ can be constructed. Solutions of CFE were subsequently considered in higher dimensions, again using one-dimensional solutions as building blocks.

These two independently considered subjects were finally combined to produce the main result of this work: the composition of a solution of CFE with a Costas bijection (in either order) is a new Costas bijection; and, if the solution of CFE used possesses a dense graph (which is, of course, the case of interest), so does the composition.

Throughout this work the case of real Euclidean spaces $\mathbb{R}^n$ was worked out in detail. Results about complex spaces $\mathbb{C}^n$ are readily obtainable through the additive isomorphism between $\mathbb{C}$ and $\mathbb{R}^2$.

**Acknowledgement** The author is indebted to Prof. Nigel Boston (Departments of Mathematics and Electrical & Computer Engineering, University of Wisconsin-Madison, US), Prof. Roderick Gow (School of Mathematical Sciences, University College Dublin), and Prof. Scott Rickard (School of Electrical, Electronic & Mechanical Engineering, University College Dublin) for the many useful discussions on the topic.

## References

1. Aczél, J., Dhombres, J.: Functional Equations in Several Variables. Cambridge University Press (1989)
2. Blackburn, S.R., Etzion, T., Martin, K.M., Paterson, M.B.: Efficient key predistribution for grid-based wireless sensor networks. In: Information Theoretic Security (Proceedings of the Third International Conference, ICITS 2008, Calgary, Canada, August 10–13, 2008), Lecture Notes in Computer Science **5155**, pp. 54–69. Springer (2008)
3. Costas, J.P.: Medium constraints on sonar design and performance. Technical Report Class **1** Rep. R65EMH33, GE Co. (1965)
4. Costas, J.P.: A study of detection waveforms having nearly ideal range-doppler ambiguity properties. Proceedings of the IEEE **72**, 996–1009 (1984)
5. Drakakis, K.: A review of Costas arrays, J. Appl. Math. **2006**
6. Drakakis, K.: On the generalization of the Costas property in higher dimensions. Adv. Math. Commun. **4**, 1–22 (2010)
7. Drakakis, K., Rickard, S.: On the generalization of the Costas property in the continuum. Adv. Math. Commun. **2**, 113–130 (2008)

8. Drakakis, K., Rickard, S., Beard, J., Caballero, R., Iorio, F., O'Brien, G., Walsh, J.: Results of the enumeration of Costas arrays of order 27. IEEE Trans. Inform. Theory **54**, 4684–4687 (2008)
9. Drakakis, K., Rickard, S. Gow, R.: Interlaced Costas arrays do not exist. Math. Probl. Eng. **2008**
10. Drakakis, K., Iorio, F., Rickard, S.: The enumeration of Costas arrays of order 28 and its consequences. Adv. Math. Commun. **5**(1), 69–86 (2011)
11. Drakakis, K., Iorio, F., Rickard, S., Walsh, J.: Results of the enumeration of Costas arrays of order 29. (accepted for publication in) Adv. Math. Commun.
12. Etzion, T.: Combinatorial designs with Costas arrays properties. Discrete Math. **93**, 143–154 (1991)
13. Golomb, S.: Algebraic constructions for Costas arrays. J. Combin. Theory Ser. A **37**, 13–21 (1984)
14. Golomb, S., Taylor, H.: Constructions and properties of Costas arrays. Proceedings of the IEEE **72**, 1143–1163 (1984)
15. Hilbert, D.: Mathematical problems. Lecture at the International Congress of Mathematicians (Paris, 1900); originally published in the Göttingen Nachrichten, 253–297 (1900). In English: Bull. Amer. Math. Soc. **8**, 437–479 (1902)
16. Hyers, D.H.: On the stability of the linear functional equation. Proc. Natl. Acad. Sci. USA **27**, 222–224 (1941)
17. Hyers, D.H., Rassias, Th.M.: Approximate homomorphisms. Aequationes Math. **44**, 125–153 (1992)
18. Th.M. Rassias: On the stability of linear mappings in Banach spaces. Proc. Amer. Math. Soc. **72**, 297–300 (1978)
19. Rickard, S.: Searching for Costas arrays using periodicity properties. IMA International Conference on Mathematics in Signal Processing, The Royal Agricultural College, Cirencester, December 2004.

# Chapter 32
# On Solutions of Some Generalizations of the Gołąb–Schinzel Equation

**Eliza Jabłońska**

*Dedicated to the memory of S.M. Ulam*
*on the 100th anniversary of his birth*

**Abstract** This paper is a survey devoted to the functional equation

$$f(x+M(f(x))y) = f(x)f(y),$$

which 'connects' the Gołąb–Schinzel equation with the exponential one, and to its generalization

$$f(x+M(f(x))y) = H(f(x), f(y)).$$

Our considerations refer to the paper: *Brzdęk, J.: The Gołąb–Schinzel equation and its generalizations. Aequationes Math.* ***70****, 14–24 (2005).*

**Keywords** Gołąb–Schinzel functional equation • Exponential equation

**Mathematics Subject Classification (2000):** Primary 39B52

## 32.1 Introduction

In 1959, Gołąb and Schinzel introduced in [32] the following functional equation

$$f(x+f(x)y) = f(x)f(y), \tag{32.1}$$

for the real continuous functions. The equation was obtained in connection with looking for subgroups of the centroaffine group of $\mathbb{R}^2$ (cf. [52]). It turned out that it is also useful when dealing with associative operations, subgroups of the

E. Jabłońska (✉)
Department of Mathematics, Rzeszów University of Technology, W. Pola 2, 35–959 Rzeszów, Poland
e-mail: elizapie@prz.edu.pl

Th.M. Rassias and J. Brzdęk (eds.), *Functional Equations in Mathematical Analysis*, Springer Optimization and Its Applications 52, DOI 10.1007/978-1-4614-0055-4_32,

semigroup of real affine mappings, classification of quasialgebras, subsemigroups of the group of affine mappings of a field, subgroups of the group $L_2^1$, differential equations arising in meteorology and fluid mechanics and classification of near-rings (informations on it can be find in [14]). In this connection, the equation (32.1) and also its generalizations have been studied by many authors in various classes of functions.

In 2005, Brzdęk published his survey paper [14] devoted to the Gołąb–Schinzel equation, its applications and further generalizations. At the end of the paper he formulated the following problem:

*To what extent can the results proved for equations*

$$f(x+f(x)^n y) = f(x)f(y) \tag{32.2}$$

*or*

$$f(x+f(x)^n y) = tf(x)f(y) \tag{32.3}$$

*be carried over to the case of the equation*

$$f(x+M(f(x))y) = H(f(x),f(y)), \tag{32.4}$$

*where* $n \in \mathbb{N}$, $X$ *is a linear space over a commutative field* $\mathbb{K}$, $t \in \mathbb{K}$, $f : X \to \mathbb{K}$, $M : \mathbb{K} \to \mathbb{K}$ *and* $H : \mathbb{K}^2 \to \mathbb{K}$ *?*

This problem has been motivated by the results of the paper [13], where for the first time equation (32.4) has appeared. Since that time numerous papers answering Brzdęk's question have been published.

## 32.2 Solutions of a Generalization of (32.2)

Let $X$ be a linear space over a commutative field $\mathbb{K}$. In this part of the paper, we give theorems characterizing solutions of the functional equation

$$f(x+M(f(x))y) = f(x)f(y), \tag{32.5}$$

where $f : X \to \mathbb{K}$ and $M : \mathbb{K} \to \mathbb{K}$ are both unknown functions. This equation 'connects' two classical functional equations, which seem to be of a quite different nature: the exponential equation

$$f(x+y) = f(x)f(y) \tag{32.6}$$

(for information on it we refer the reader to [46, Chap. XIII §1] or [2, pp. 25–33, 52–57]) and the Gołąb–Schinzel equation; i.e. (32.6) and (32.1) are particular cases of (32.5) with $M = 1$ and $M = \mathrm{id}_{\mathbb{K}}$, respectively. So, looking for solutions of (32.5) seems to be interesting.

Clearly, (32.5) also generalizes equation (32.2), where $M(z) = z^n$ for $z \in \mathbb{K}$.

### 32.2.1 Continuous Solutions

As we already have mentioned, for the first time the continuous real solutions of (32.1) were determined by S. Gołąb and A. Schinzel in [32]. Next, the problem of continuous solutions of (32.1) has been studied by many authors (see [3, 6, 27, 28, 31, 44, 49, 52, 53]). Consequently, we have the following result:

**Theorem 32.1 ([14, Theorem 1]).** *Let $X$ be a topological linear space over $\mathbb{K} \in \{\mathbb{R}, \mathbb{C}\}$ and $f : X \to \mathbb{K}$ be a continuous solution of* (32.1).

*(i) If $X = \mathbb{K} = \mathbb{R}$, then either $f = 0$ or $f$ is of one of the following forms:*

$$f(x) = cx + 1 \text{ for } x \in \mathbb{R}$$

*or*

$$f(x) = \max\{cx + 1, 0\} \text{ for } x \in \mathbb{R}$$

*with some $c \in \mathbb{R}$.*

*(ii) If $X = \mathbb{K} = \mathbb{C}$, then either $f = h \circ g$ for a continuous solution $h : \mathbb{R} \to \mathbb{R}$ of* (32.1) *and an $\mathbb{R}$-linear functional $g : \mathbb{C} \to \mathbb{R}$, or*

$$f(x) = cx + 1 \text{ for } x \in X$$

*with some $c \in \mathbb{C}$.*

*(iii) If $X \neq \mathbb{K}$, then there are a continuous solution $h : \mathbb{K} \to \mathbb{K}$ of* (32.1) *and a continuous linear functional $g : X \to \mathbb{K}$ such that $f = h \circ g$.*

In [13] J. Brzdęk obtained the following

**Theorem 32.2 ([13, Corollary 4]).** *Let $X$ be a real linear space and $J$ a nontrivial real interval. Let $f : X \to J$ be continuous on rays, $M : J \to \mathbb{R}$ be continuous and $M \circ f \neq 1$. Then $f$ and $M$ satisfy* (32.5) *if and only if one of the following two conditions holds:*

*(i) $J = \mathbb{R}$ and there are $c > 0$ and a nontrivial linear functional $g : X \to \mathbb{R}$ such that*

$$\begin{aligned} f(x) &= |g(x) + 1|^{\frac{1}{c}} \operatorname{sgn}(g(x) + 1) \text{ for } x \in X, \\ M(y) &= |y|^c \operatorname{sgn} y \text{ for } y \in \mathbb{R}\,; \end{aligned} \tag{32.7}$$

*(ii) $[0, \infty) \subset J$ and there are $c > 0$ and a nontrivial linear functional $g : X \to \mathbb{R}$ such that*

$$\begin{aligned} f(x) &= (\max\{0, g(x) + 1\})^{\frac{1}{c}} \text{ for } x \in X, \\ M(y) &= y^c \text{ for } y \in [0, \infty). \end{aligned} \tag{32.8}$$

Everyone can see some similarities between continuous solutions of (32.1) and (32.5).

**Problem 32.1.** What can we say on continuous solutions of (32.5) in the complex case?

### 32.2.2 General Solution

Let $X$ be a linear space over a commutative field $\mathbb{K}$. A description of the general solution of (32.1) in the class of functions mapping $X$ into $\mathbb{K}$ is due to Wołodźko [52]. His result has been modified by Javor [43]. In [8] Brzdęk gives the general solution of (32.3). Now, we present an analogous result for (32.5).

**Theorem 32.3 ([34, Proposition 1]).** *Let $X$ be a linear space over a commutative field $\mathbb{K}$, $f: X \to \mathbb{K}$, $f \neq 0$, $f \neq 1$ and $M: \mathbb{K} \to \mathbb{K}$. If functions $f$ and $M$ satisfy* (32.5), *then there exist a subgroup $W$ of $(\mathbb{K} \setminus \{0\}, \cdot)$, a subgroup $A$ of $(X, +)$, and a function $w: W \to X$ such that:*

$$\text{if } w(a) \in A, \text{ then } a = 1,$$

$$M(a)A = A \text{ for } a \in W,$$

$$w(ab) - M(a)w(b) - w(a) \in A \text{ for } a, b \in W,$$

$$f(x) = \begin{cases} a, & \text{if } x \in w(a) + A \text{ for some } a \in W; \\ 0, & \text{otherwise.} \end{cases}$$

*Remark 32.1.* The case $f = 1$ has to be excluded from Theorem 32.3 if $M(1) = 0$, because then $M(a)A = \{0\} \neq X = A$. If $f = 1$, $M$ satisfy (32.5) and $M(1) \neq 0$, then the statement of Theorem 32.3 holds for $A = X$ and $W = \{1\}$.

### 32.2.3 Algebraically Interior Point

In [7], J. Brzdęk considered (32.2), where $n \in \mathbb{N}$, $X$ is a linear space over $\mathbb{K} \in \{\mathbb{R}, \mathbb{C}\}$, $f: X \to \mathbb{K}$ and $\{x \in X : f(x) \neq 0\}$ has an algebraically interior point. Following the idea in [7], the author generalized his result in the following way:

**Theorem 32.4 ([37, Theorem 1]; cf. also [34,35]).** *Let $X$ be a linear space over the field $\mathbb{K} \in \{\mathbb{R}, \mathbb{C}\}$. Functions $f: X \to \mathbb{K}$ and $M: \mathbb{K} \to \mathbb{K}$ satisfy* (32.5) *and the set $\{x \in X : f(x) \neq 0\}$ has an algebraically interior point if and only if one of the following statements holds:*

*(i)* $f = 1$;
*(ii)* $f: X \to \mathbb{K} \setminus \{0\}$ *is a nontrivial exponential function and* $M \circ f = 1$;
*(iii)* $\mathbb{K} = \mathbb{R}$ *and there exist a multiplicative injection $H: \mathbb{R} \to \mathbb{R}$ and a nontrivial $\mathbb{R}$-linear functional $g: X \to \mathbb{R}$ such that either*

$$\begin{aligned} f(x) &= H(g(x) + 1) \text{ for } x \in X, \\ M(y) &= H^{-1}(y) \text{ for } y \in H(\mathbb{R}) \end{aligned} \tag{32.9}$$

*or*

$$f(x) = H(\max\{0, g(x)+1\}) \text{ for } x \in X,$$
$$M(y) = H^{-1}(y) \text{ for } y \in H([0,\infty))\,; \tag{32.10}$$

*(iv)* $\mathbb{K} = \mathbb{C}$ *and one of the following two conditions holds:*

1. *there exist a multiplicative injection* $H : \mathbb{C} \to \mathbb{C}$ *and a nontrivial* $\mathbb{C}$*-linear functional* $g : X \to \mathbb{C}$ *such that*

$$f(x) = H(g(x)+1) \text{ for } x \in X,$$
$$M(y) = H^{-1}(y) \text{ for } y \in H(\mathbb{C}); \tag{32.11}$$

2. *there exist a multiplicative function* $H : \mathbb{R} \to \mathbb{C}$ *and a nontrivial* $\mathbb{R}$*-linear functional* $g : X \to \mathbb{R}$ *such that either* $f$ *and* $M$ *are given by* (32.9) *and* $H$ *is injective, or* $f$ *and* $M$ *are given by* (32.10) *and* $H\big|_{[0,\infty)}$ *is injective.*

### *32.2.4 Measurable Solutions*

For the first time the Lebesgue measurable solutions of the Gołąb–Schinzel equation have been studied by Popa [50]. He proved that each Lebesgue measurable real function satisfying (32.1) is continuous or equal to 0 almost everywhere. On the other hand, it is well-known that each Lebesgue measurable real exponential function is continuous (see e.g. [2, p. 29, Theorem 5]). Those two results confirm that (32.1) and (32.6) have different nature.

Brzdęk [10, 11] has generalized the mentioned result in several directions; more exactly, he proved that every Christensen [Baire, respectively] measurable function $f$ mapping a separable $F$-space $X$ over $\mathbb{K} \in \{\mathbb{R}, \mathbb{C}\}$ into $\mathbb{K}$ and satisfying (32.5) is continuous or equal to 0 almost everywhere in the Christensen [Baire] sense. In [41] the author has proved that the result of Brzdęk for Baire measurable solution of (32.5) also holds even if $X$ is a linear topological space over $\mathbb{K}$. Here we present some generalization of those results.

**Theorem 32.5 ([39, Theorem 16], [40, Corolarry 4.3]).** *Let* $X$ *be a separable* $F$*-space [a linear topological space, respectively] over* $\mathbb{K} \in \{\mathbb{R}, \mathbb{C}\}$*. If a Christensen [Baire] measurable function* $f : X \to \mathbb{K}$ *and a continuous function* $M : \mathbb{K} \to \mathbb{K}$ *satisfy* (32.5)*, then either* $M \circ f$ *is continuous or* $f$ *is equal to* 0 *almost everywhere in the Christensen [Baire] sense.*

**Problem 32.2.** Are there any discontinuous function $f$ and continuous function $M$, satisfying the assumptions of Theorem 32.5, such that $M \circ f$ is continuous?

**Problem 32.3.** The proofs of the results of Theorem 32.5, presented in [39] and [40], run in two various ways for the Christensen and Baire measurable functions $f$. How can these two proofs be unified in an abstract way?

### 32.2.5 Bounded Solutions

Denote $|T| = \{|t| : t \in T\}$ for a set $T \subset \mathbb{C}$. In the previous subsection, the result from [11], concerning Baire measurable solutions of (32.2), has been mentioned. Actually that result in [11] is stronger; namely it is proved that each solution $f$ mapping a separable $F$-space $X$ over $\mathbb{K} \in \{\mathbb{R}, \mathbb{C}\}$ into $\mathbb{K}$ of (32.2) with $|f(A)| \subset (0,a)$ for a set $A \subset X$ of second category with the Baire property and for a number $a > 0$, is continuous or satisfies $|f(X)| \subset \{0,1\}$; and, moreover, in the case $\mathbb{K} = \mathbb{R}$, $f$ has to be continuous.

An analogous result for solutions $f : X \to \mathbb{R}$ of (32.5) bounded on a nonzero Christensen measurable set has been proved in [38] (cf. [33]), as an answer to a question of Brzdęk [14, Problem, p. 18] in the real case; in the complex case the problem seems still to be open.

With reference to two problems from [14] (see [14, pp. 18 and 21]), the author characterized in [40] the bounded solutions of (32.5) under the following hypothesis:

(H) $X$ is a linear topological space over $\mathbb{K} \in \{\mathbb{R}, \mathbb{C}\}$, $\mathfrak{M}$ is a $\sigma$-algebra of its subsets and there exists a nontrivial proper linearly invariant $\sigma$-ideal $\mathscr{I} \subset 2^X$ such that

(H1) $\operatorname{int}(A+B) \neq \emptyset$ for every $A \in \mathfrak{M} \setminus \mathscr{I}$ and $B \in 2^X \setminus \mathscr{I}$;
(H2) $\operatorname{int}[(g(A)+1)(g(A)+1)] \neq \emptyset$ for every $A \in \mathfrak{M} \setminus \mathscr{I}$ and $g \in X^* \setminus \{0\}$,

where $X^*$ is the space of all linear continuous functionals on $X$.

The hypothesis (H) holds in the following two well-known cases:

(a) $X$ is a linear topological space over $\mathbb{K} \in \{\mathbb{R}, \mathbb{C}\}$, $\mathfrak{M}$ denotes the $\sigma$-algebra of all subsets of $X$ having the Baire property and $\mathscr{I}$ denotes the $\sigma$-ideal of all subsets of $X$ of the first category;
(b) $X = \mathbb{K}^n$, where $\mathbb{K} \in \{\mathbb{R}, \mathbb{C}\}$ and $n \in \mathbb{N}$, $\mathfrak{M}$ denotes the $\sigma$-algebra of all Lebesgue measurable subsets of $\mathbb{K}^n$ and $\mathscr{I}$ denotes the $\sigma$-ideal of all Lebesgue zero subsets of the space.

**Theorem 32.6 ([40, Theorem 1]).** *Assume that (H) is valid. Let $D \in \mathfrak{M} \setminus \mathscr{I}$, $f : X \to \mathbb{K}$, $|f(D)| \subset (0,a)$ for some $a > 0$ and let $M : \mathbb{K} \to \mathbb{K}$ be continuous. If functions $f$ and $M$ satisfy* (32.5)*, then:*

*(i) in the case $\mathbb{K} = \mathbb{R}$, $f$ is continuous;*
*(ii) in the case $\mathbb{K} = \mathbb{C}$, either $M \circ f$ is continuous or*

$$|f(x)| \in \{0,1\} \text{ for } x \in X.$$

In the author's paper [42], the solutions of (32.5) bounded on a nonzero Christensen measurable set are determined. Here we present that result in details.

**Theorem 32.7 ([42, Theorem 1]).** *Let $X$ be a real separable $F$-space, $f: X \to \mathbb{R}$, $M: \mathbb{R} \to \mathbb{R}$ and $|f(D)| \subset (0,a)$ for a nonzero Christensen measurable set $D \subset X$ and a positive number $a$. If functions $f$ and $M$ satisfy* (32.5)*, then $M \circ f$ is continuous.*

**Problem 32.4.** If in Theorem 32.7 we also assume that $M$ is continuous, then $f$ has to be continuous. How does the thesis of Theorem 32.6 change after excepting the assumption about the continuity of $M$?

Moreover, in [37] solutions of (32.5) have been characterized under the assuption that $f$ is bounded on a set having an algebraically interior point.

**Theorem 32.8 ([37, Theorem 2]).** *Let $X$ be a linear space over the field $\mathbb{K} \in \{\mathbb{R}, \mathbb{C}\}$. Let $B \subset X$ be a set having an algebraically interior point, $f: X \to \mathbb{K}$, $M: \mathbb{K} \to \mathbb{K}$,*

$$|f(B)| \subset (0,a) \text{ for some } a > 0 \tag{32.12}$$

*and, moreover, in the case $\mathbb{K} = \mathbb{C}$,*

$$\left| \arg \frac{f(z_1)}{f(z_2)} \right| < \frac{2\pi}{3} \text{ for every } z_1, z_2 \in B. \tag{32.13}$$

*Functions $f$ and $M$ satisfy* (32.5) *if and only if one of the following four statements holds:*

*(i)* $f = 1$;
*(ii)* $M \circ f = 1$ *and there exists a nontrivial $\mathbb{R}$-linear functional $g: X \to \mathbb{K}$ such that $f(x) = \exp g(x)$ for each $x \in X$;*
*(iii)* $\mathbb{K} = \mathbb{R}$ *and there exist a multiplicative function $H: \mathbb{R} \to \mathbb{R}$, that is continuous on $\mathbb{R} \setminus \{0\}$, and a nontrivial $\mathbb{R}$-linear functional $g: X \to \mathbb{R}$ such that either* (32.9) *or* (32.10) *holds;*
*(iv)* $\mathbb{K} = \mathbb{C}$ *and one of the following conditions holds:*

1. *there are a multiplicative injection $H: \mathbb{C} \to \mathbb{C}$, that is continuous on $\mathbb{C} \setminus \{0\}$, and a nontrivial $\mathbb{C}$-linear functional $g: X \to \mathbb{C}$ such that* (32.11) *holds;*
2. *there are a multiplicative function $H: \mathbb{R} \to \mathbb{C}$, that is continuous on $\mathbb{R} \setminus \{0\}$, and a nontrivial $\mathbb{R}$-linear functional $g: X \to \mathbb{R}$ such that either* (32.9) *or* (32.10) *holds.*

*Remark 32.2.* If in Theorem 32.8, in the case where $\mathbb{K} = \mathbb{C}$ and $H: \mathbb{C} \to \mathbb{C}$, the function $f: X \to \mathbb{C}$ satisfies (32.12) but not (32.13), then the function $H$ need not be continuous on $\mathbb{C} \setminus \{0\}$ (see [37, Remark 1]).

*Remark 32.3.* There are discontinuous functions $f$ and $M$ satisfying one of conditions (ii)–(iv) of Theorem 32.8 (see [34, Example 2]).

### 32.2.6 Darboux Property

In [9], Brzdęk proved that each solution of (32.3), mapping a real linear topological space $X$ into $\mathbb{R}$ and having the Darboux property, is continuous. Here we present a more general result.

**Theorem 32.9 ([36, Corollary 1]).** *Let $X$ be a real linear topological space. If functions $f : X \to \mathbb{R}$ and $M : \mathbb{R} \to \mathbb{R}$, both having the Darboux property, satisfy* (32.5)*, then either $f$ is continuous or $f$ is a nontrivial exponential function.*

*Remark 32.4.* We know that a function $f : X \to \mathbb{R}$ having the Darboux property and satisfying the exponential equation need not be continuous (see [45], cf. [9, Remark 2]).

**Problem 32.5.** Let $X$ be a real linear topological space and $f : X \to \mathbb{R}$ have the Darboux property. What can we say on solutions $f$ and $M : \mathbb{R} \to \mathbb{R}$ of (32.5), without the assumption that $M$ has the Darboux property?

## 32.3 Solutions of a Generalization of (32.3)

Let $X$ be a linear space over a commutative field $\mathbb{K}$. In this section of the paper we recall theorems on solutions of (32.4), where $f : X \to \mathbb{K}$, $M : \mathbb{K} \to \mathbb{K}$ and $H : \mathbb{K}^2 \to \mathbb{K}$ are unknown functions. A particular case of (32.4) is the functional equation

$$f(x+y) = H(f(x), f(y)), \tag{32.14}$$

that is known as addition formulas or addition theorems (see e.g. [1, pp. 49–81]), as well as the Gołąb–Schinzel equation. Moreover, (32.4) also generalizes (32.2), (32.3) and (32.5).

### 32.3.1 Continuous Solutions

As we wrote in Introduction, the first paper concerning (32.4) has been written by J. Brzdęk. In [13], he proved the following

**Theorem 32.10 ([13, Theorem 2]).** *Let $X$ be a real linear space and $J$ a nontrivial real interval. Let $f : X \to J$ be continuous on rays, $M : J \to \mathbb{R}$ be continuous and $H : J^2 \to J$ be symmetric (i.e. $H(u,v) = H(v,u)$ for $u, v \in J$). Then $f$, $M$, $H$ satisfy* (32.4) *if and only if one of the following conditions holds:*

*(i) there is $c \in J$ such that $f(x) = c = H(c,c)$ for $x \in \mathbb{R}$;*
*(ii) $M$ is bijective, there is a nontrivial linear functional $g : X \to \mathbb{R}$ such that*

$$f(x) = M^{-1}(g(x) + 1) \text{ for } x \in X$$

*and*

$$H(z,w) = M^{-1}(M(z)M(w)) \text{ for } z,w \in J;$$

*(iii)* *there exist a a nontrivial linear functional* $g : X \to \mathbb{R}$ *and an injective continuous function* $h : [0,\infty) \to J$ *such that*

$$M(y) = h^{-1}(y) \text{ for } y \in h([0,\infty)),$$

$$f(x) = h(\max\{g(x)+1,0\}) \text{ for } x \in X,$$

$$H(z,w) = h(M(z)M(w)) \text{ for } z,w \in h([0,\infty));$$

*(iv)* *there exist a nontrivial linear functional* $g : X \to \mathbb{R}$ *and an injective continuous function* $h : \mathbb{R} \to J$ *such that* $M(h(\mathbb{R})) = \{1\}$, $f = h \circ g$ *and*

$$H(z,w) = h(h^{-1}(z) + h^{-1}(w)) \text{ for } z,w \in f(X).$$

In his next paper [15], Brzdęk continued considerations on continuous functions $f : \mathbb{R} \to J$ and $M : J \to \mathbb{R}$ satisfying (32.4) without the assumption of symmetry of $H$.

In [24], Chudziak studied (32.4) under the same assumptions as in Theorem 32.10, but he replaced the symmetry of $H$ with the associativity of $H$ on $f(X)^2$ (i.e. $H(H(u,v),w) = H(u,H(v,w))$ for $u,v,w \in f(X)$). He obtained the following result:

**Theorem 32.11 ([24, Theorem 1]).** *Let* $X$ *be a real linear space and* $J$ *a nontrivial real interval. Let* $f : X \to J$ *be continuous on rays,* $M : J \to \mathbb{R}$ *continuous and* $H : J^2 \to J$ *be associative on* $f(X)^2$*. Then* $f$, $M$, $H$ *satisfy* (32.4) *if and only if either* $H$ *is symmetric on the set* $f(X)^2$ *and one of conditions* (i)–(iv) *of Theorem 32.10 holds, or* $H$ *is not symmetric on* $f(X)^2$ *and one of the following two statements is valid:*

(a) $M(u) = 0$ *for* $u \in f(X)$ *and* $H(u,v) = u$ *for* $u,v \in f(X)$*;*
(b) *there exist a nontrivial linear functional* $g : X \to \mathbb{R}$*, an injective continuous function* $h : \mathbb{R} \to J$ *and a* $c > 0$ *such that*

$$M(u) = \begin{cases} h^{-1}(u) & \text{if } u \in h([0,\infty)); \\ 0 & \text{if } u \in h((-\infty,0)), \end{cases}$$

$$H(u,v) = \begin{cases} h(h^{-1}(u)h^{-1}(v)) & \text{if } u \in h([0,\infty)); \\ u & \text{if } u \in h((-\infty,0)) \end{cases}$$

*and*

$$f(x) = \begin{cases} h(g(x)+1) & \text{if } g(x)+1 \geq 0; \\ h(c(g(x)+1)) & \text{otherwise}. \end{cases}$$

### 32.3.2 General Solution

Following ideas of Wołodźko [52], P. Javor [43] and Brzdęk [8], Mureńko [48] has described the general solution of (32.4). Here we present her result.

**Theorem 32.12 ([34, Proposition 1]).** *Let $X$ be a linear space over a commutative field $\mathbb{K}$. Assume that $f : X \to \mathbb{K}$ is nonconstant, $M : \mathbb{K} \to \mathbb{K}$, $M(f(X)) \setminus \{1\} \neq \emptyset$, $M^{-1}(\{0\}) = \{0\}$, and $H : \mathbb{K}^2 \to \mathbb{K}$ is symmetric and associative. Let a binary operation $\circ : \mathbb{K}^2 \to \mathbb{K}$ be given by*

$$a \circ b = H(a,b) \text{ for } a,b \in \mathbb{K}.$$

*If $f$, $M$ and $H$ satisfy* (32.4)*, then there exist a subgroup $W \subset \mathbb{K} \setminus \{0\}$ of the semigroup $(\mathbb{K}, \circ)$ (with neutral element $e$), a subgroup $A$ of $(X,+)$ and a function $w : W \to X$ such that:*

$$\text{if } w(a) \in A, \text{ then } a = e,$$

$$M(a)A = A \text{ for } a \in W,$$

$$w(a \circ b) - M(a)w(b) - w(a) \in A \text{ for } a,b \in W,$$

$$f(x) = \begin{cases} a, & \text{if } x \in w(a) + A \text{ for some } a \in W; \\ 0, & \text{otherwise}, \end{cases}$$

$$u \circ 0 = 0 \text{ for } u \in W \cup \{0\}.$$

As we see, mainly because of the additional assumption: $M^{-1}(\{0\}) = \{0\}$, Theorem 32.12 does not fully generalize Theorem 32.3.

### 32.3.3 Measurable Solutions

With reference to papers [10, 11, 50], Mureńko [47] proved some result concerning Lebesgue and Baire measurable solutions of (32.4).

**Theorem 32.13 ([47, Theorems 12 and 13]).** *Assume that $f : \mathbb{R} \to \mathbb{R}$ is Lebesgue [Baire, respectively] measurable, $M : \mathbb{R} \to \mathbb{R}$, $M^{-1}(\{0\}) = \{0\}$, $H : \mathbb{R}^2 \to \mathbb{R}$ is symmetric and associative. If $f$, $M$ and $H$ satisfy* (32.4)*, then either $M \circ f$ is continuous or $f$ is constant almost everywhere in the Lebesgue [Baire] sense.*

## 32.4 A New Functional Equation

In 2006, Chudziak [23] published a paper concerning solutions of

$$f(x + g(x)y) = f(x) \circ f(y),$$

where $g : \mathbb{R} \to \mathbb{R}$ is continuous, $(S, \circ)$ is a semigroup and $f : \mathbb{R} \to S$. This equation generalizes (32.4) and (32.5). In this connection, the following problem seems to be of interest:

*To what extent can the results proved for equations* (32.4) *and* (32.5) *be carried over to the case of the equation*

$$f(x+g(x)y) = H(f(x), f(y)), \tag{32.15}$$

*where X is a linear space over a commutative field* $\mathbb{K}$, $f, g : X \to \mathbb{K}$ *and* $H : \mathbb{K}^2 \to \mathbb{K}$?

Equation (32.15) is a particular case of the functional equation

$$f(x+g(x)y) = H(h(x), k(y)),$$

considered for the first time by Vincze [51].

## 32.5 Final Remarks

In the final part of the survey paper [14], Brzdęk mentioned that the problem of the stability of equations of the Gołąb–Schinzel type, which has been formulated by R. Ger at the 38th International Symposium on Functional Equations (Noszvaj, Hungary, 2000) (see [29]), supplies a very large field for research. Indeed, in recent years stability of the equations of the Gołąb–Schinzel type has been extensively studied by Chudziak et al. (see [16, 17] and [20–26]). They solved the problem in some special classes of functions, so the subject is still worth of considerations. Further, in [19], for the first time the problem of the stability of the pexiderized Gołąb–Schinzel equation has been studied.

In recent years, some new results concerning conditional versions of the Gołąb–Schinzel equation have been proved (cf. [4, 5, 18]; see also [14] for the further references).

Moreover, for the first time, at the 44th International Symposium on Functional Equations (Louisville, Kentucky, USA, 2006), Ger presented a description of functions satisfying the Gołąb–Schinzel equation almost everywhere with respect to the planar Lebesgue measure, under the assumption of local integrability (see [30]). This result seems to open a new area of research.

## References

1. Aczél, J.: Lectures on Functional Equations and their Applications. Academic Press (1966)
2. Aczél, J., Dhombres, J.: Functional Equations in Several Variables. Encyclopedia of Mathematics and its Applications v. 31. Cambridge University Press (1989)

3. Baron, K.: On the continuous solutions of the Gołąb–Schinzel equation. Aequationes Math. **38**, 155–162 (1989)
4. Brillouët–Belluot, N., Brzdęk, J.: On continuous solutions and stability of a conditional Gołąb–Schinzel equation. Publ. Math. Debrecen **72**, 441–450 (2008)
5. Brillouët–Belluot, N., Brzdęk, J., Chudziak, J.: On continuous solutions of a class of conditional equations. Publ. Math. Debrecen **75**, 11–22 (2009)
6. Brillouët, N., Dhombres, J.: Équations fonctionnelles et recherche de sous–groupes. Aequationes Math. **31**, 253–293 (1986)
7. Brzdęk, J.: Subgroups of the group $Z_n$ and a generalization of the Gołąb–Schinzel functional equation. Aequationes Math. **43**, 59–71 (1992)
8. Brzdęk, J.: Some remarks on solutions of the functional equation $f(x+f(x)^n y) = tf(x)f(y)$. Publ. Math. Debrecen **43**, 147–160 (1993)
9. Brzdęk, J.: On solutions of the Gołąb–Schinzel functional equation. Publ. Math. Debrecen **44**, 235–241 (1994)
10. Brzdęk, J.: The Christensen measurable solutions of a generalization of the Gołąb–Schinzel functional equation. Ann. Polon. Math. **64**, 195–205 (1996)
11. Brzdęk, J.: Bounded solutions of the Gołąb–Schinzel equation. Aequationes Math. **59**, 248–254 (2000)
12. Brzdęk, J.: On the continuous solutions of a generalization of the Gołąb–Schinzel equation. Publ. Math. Debrecen **63**, 421–429 (2003)
13. Brzdęk, J.: A generalization of addition formulae. Acta Math. Hungar. **101**, 281–291 (2003)
14. Brzdęk, J.: The Gołąb–Schinzel equation and its generalizations. Aequationes Math. **70**, 14–24 (2005)
15. Brzdęk, J.: A remark on solutions of a generalization of the addition formulae. Aequationes Math. **71**, 288–293 (2006)
16. Brzdęk, J.: On the quotient stability of a family of functional equations. Nonlinear Anal. **71**, 4396–4404 (2009)
17. Brzdęk, J.: On stability of a family of functional equaions. Acta Math. Hungar. **128**, 139–149 (2010) DOI: 10.1007/s10474-010-9169-8
18. Brzdęk, J., Mureńko, A.: On a conditional Gołąb–Schinzel equation. Arch. Math. (Basel) **84**, 503–511 (2005)
19. Charifi, A., Bouikhalene, B., Kabbaj, S., Rassias, J.M.: On the stability of a pexiderized Gołąb–Schinzel equation. Comput. Math. Appl. **59**, 3193–3202 (2010)
20. Chudziak, J.: Approximate solutions of the genearalized Gołąb–Schinzel functional equation. J. Approx. Theory **136**, 21–25 (2005)
21. Chudziak, J.: On a functional inequality related to the stability problem for the Gołąb–Schinzel equation. Publ. Math. Debrecen **67**, 199–208 (2005)
22. Chudziak, J.: Stability of the generalized Gołąb–Schinzel equation. Acta Math. Hungar. **113**, 133–144 (2006)
23. Chudziak, J.: Semigroup–valued solutions of the Gołąb–Schinzel functional equation. Abh. Math. Sem. Univ. Hamburg **76**, 91–98 (2006)
24. Chudziak, J.: Associative solutions of a composite generalization of addition formulae. Acta Math. Hungar. **117**, 141–152 (2007)
25. Chudziak, J.: Stability problem for the Gołąb–Schinzel type functional equations. J. Math. Anal. Appl. **339**, 454–460 (2008)
26. Chudziak, J., Tabor, J.: On the stability of the Gołąb–Schinzel functional equation. J. Math. Anal. Appl. **302**, 196–200 (2005)
27. Daróczy, Z.: Az $f[x+yf(x)] = f(x)f(y)$ függvenyegyenlet folytones megoldasairol Hilbert – terekben. Matematikai Lapok **17**, 339–343 (1966)
28. Gebert, H.: A direct proof of a theorem of K. Baron. Ann. Math. Sil. **9**, 101–103 (1995)
29. Ger, R.: A collection of problems in stability theory. In: Report of Meeting (The Thirty–Eight International Symposium on Functional Equations, June 11–18, 2000, Noszvaj, Hungary), Aequationes Math. **61**, 284 (2001)
30. Ger, R.: Gołąb–Schinzel equation almost everywhere. In: Report of Meeting (The Forty–Forth International Symposium of Functional Equations, May 14–20, 2006, Louisville, Kentucky, USA), Aequationes Math. **73**, 178 (2007)

31. Gheorghiu, O.E., Gołąb, S.: Über ein System von Funktionalgleichungen. Ann. Polon. Math. **17**, 223–243 (1966)
32. Gołąb, S., Schinzel, A.: Sur l'équation fonctionnelle $f(x+f(x)y) = f(x)f(y)$. Publ. Math. Debrecen **6**, 113–125 (1959)
33. Jabłońska, E.: Continuity of Lebesgue measurable solutions of a generalized Gołąb–Schinzel equation. Demonstratio Math. **39**, 91–96 (2006)
34. Jabłońska, E.: On solutions of a generalization of the Gołąb–Schinzel equation. Aequationes Math. **71**, 269–279 (2006)
35. Jabłońska, E.: A short note concerning solutions of a generalization of the Gołąb–Schinzel equation. Aequationes Math. **74**, 318–320 (2007)
36. Jabłońska, E.: Functions having the Darboux property and satisfying some functional equation. Colloq. Math. **114**, 113–118 (2009)
37. Jabłońska, E.: Bounded solutions of a generalized Gołąb–Schinzel equation. Demonstratio Math. **42**, 533–547 (2009)
38. Jabłońska, E.: Solutions of some functional equation bounded on nonzero Christensen measurable sets. Acta Math. Hungar. **125**, 113–119 (2009)
39. Jabłońska, E.: Christensen measurable solutions of some functional equation. Nonlinear Anal. **75**, 2465–2473 (2010)
40. Jabłońska, E.: Solutions of a Gołąb–Schinzel–type functional equation bounded on 'big' sets in an abstract sense. Bull. Aust. Math. Soc. **81**, 430–441 (2010)
41. Jabłońska, E.: Baire measurable solutions of a generalized Gołąb–Schinzel equation. Commentationes Math. **50**, 69–72 (2010)
42. Jabłońska, E.: Christensen measurability and some functional equations. Aequationes Math. **81**, 155–165 (2011)
43. Javor, P.: On the general solution of the functional equation $f(x+yf(x)) = f(x)f(y)$. Aequationes Math. **1**, 235–238 (1968)
44. Javor, P.: Continuous solutions of the functional equation $f(x+yf(x)) = f(x)f(y)$. In: Proc. Internat. Sympos. on Topology and its Applications (Herceg–Novi, 1968) 206–209, Savez Drůstava Mat. Fiz. i Astronom., Belgrade (1969)
45. Jones, F.B.: Connected and disconnected plane sets and the functional equation $f(x+y) = f(x)+f(y)$. Bull. Amer. Math. Soc. **48**, 115–120 (1942)
46. Kuczma, M.: An Introduction to the Theory of Functional Equations and Inequalities. Cauchy's equation and Jensen's inequality. Państwowe Wydawnictwo Naukowe & Uniwersytet Śląski (1985)
47. Mureńko, A.: On solutions of a common generalization of the Gołąb–Schinzel equation and of the addition formulae. J. Math. Anal. Appl. **341**, 1236–1240 (2008)
48. Mureńko, A.: On the general solution of a generalization of the Gołąb–Schinzel equation. Aequationes Math. **77**, 107–118 (2009)
49. Plaumann, P., Strambach, S.: Zweidimensionale Quasialgebren mit Nullteilern. Aequationes Math. **15**, 249–264 (1977)
50. Popa, C.G.: Sur l'équation fonctionnelle $f(x+f(x)y) = f(x)f(y)$. Ann. Polon. Math. **17**, 193–198 (1965)
51. Vincze, E.:Über die Lösung der Funktionalgleichung $f(y+xg(y)) = L(h(x),k(y))$. Ann. Polon. Math. **18**, 115–119 (1966)
52. Wołodźko, S.: Solution générale de l'équation fonctionnelle $f[x+yf(x)] = f(x)f(y)$. Aequationes Math. **2**, 12–29 (1968)
53. Wołodźko, S.: O ciągłych rozwiązaniach równania funkcyjnego $f[x+yf(x)] = f(x)f(y)$ w zespolonej przestrzeni liniowej unormowanej. Rocznik Nauk.–Dydakt. Prace Mat. **6**, 199–205 (1970)

# Chapter 33
# One-parameter Groups of Formal Power Series of One Indeterminate

**Wojciech Jabłoński**

*Dedicated to the memory of S.M. Ulam*
*on the 100th anniversary of his birth*

**Abstract** The aim of the paper is to give a survey of results and open problems concerning one-parameter groups of formal power series in one indeterminate. A one-parameter group of formal power series is a homomorphism

$$G \ni t \mapsto \sum_{k=1}^{\infty} c_k(t)X^k \in \Gamma^{\infty}$$

of some group $G$ into the group $\Gamma^{\infty}$ of invertible formal power series. To describe arbitrary one-parameter groups, several tools from different branches of mathematics are used (differential equations, functional equations, abstract algebra). Unfortunately, the problem of to give description of one-parameter group of formal power series for arbitrary abelian group $G$ seems to be still open.
Independently of these problems in ring of formal power series homomorphisms of some groups into differential groups $L_s^1$ and $L_\infty^1$ have been examined. Used there tools are rather simple and did not exceed simple methods of substitution and changing variables adopted form the theory of functional equations. Consequently only some homomorphisms into groups $L_s^1$ for $s \leq 5$ are known and a partial result on homomorphisms into $L_s^1$ with arbitrary $s$ is proved. It appears that all the results on one-parameter groups can be transferred onto differential groups $L_s^1$ and $L_\infty^1$. This can be done using algebraic isomorphisms between $\Gamma^{\infty}$ and $L_\infty^1$ as well as between a group of invertible truncated formal power series and $L_s^1$.

**Keywords** One-parameter group • Translation equation • Formal power series

**Mathematics Subject Classification (2000):** Primary 39B72, 13F25; Secondary 39B50, 13J05, 13H05

---

W. Jabłoński (✉)
Department of Mathematics, University of Rzeszów,
ul. Rejtana 16 A, 35 – 310 Rzeszów, Poland
e-mail: wojciech@univ.rzeszow.pl

Th.M. Rassias and J. Brzdęk (eds.), *Functional Equations in Mathematical Analysis*, Springer Optimization and Its Applications 52, DOI 10.1007/978-1-4614-0055-4_33,

## 33.1 Introduction

One-parameter subgroups (iteration groups) in the group of invertible formal power series transformations (formally biholomorphic mappings) was studied in the connection with the embedding of a given formal power series $\Phi(X)$ into such the iteration group (cf. [16, 19, 21, 28, 30, 33]). Recall that in general by a one-parameter subgroup of transformations is usually understood any homomorphism of an additive group $(\mathbb{K},+)$ either of the field of reals or of the field of complex numbers into the considered group of transformations. If such the group of iterations we write as $(F_t)_{t\in\mathbb{K}}$, then we obtain the celebrated translation equation

$$F_{t_1+t_2} = F_{t_1} \circ F_{t_2} \qquad \text{for } t_1, t_2 \in \mathbb{K}.$$

One of possible way to solve the embedding problem in the ring of formal power series leads to considering regular (analytic) iterations groups as solutions of a suitable formal differential equation

$$\frac{\partial F_t(X)}{\partial t} = G(F_t(X))$$

with the integral $F_t(X) = \sum_{k=1}^{\infty} c_k(t)X^k$ satisfying the initial condition $F_0(X) = X$ and such that $F_1(X) = \Phi(X)$. It appears that the formal integral $(F_t)_{t\in\mathbb{K}}$ of that formal differential equation is a one-parameter group of formal power series (cf. also Theorem 33.6).

Also, functional equations approach is possible here to obtain one-parameter groups of formal power series, but the best idea is to join both methods: differential equations as well as functional equations (cf. [10,11,13,14,22]). Another possibility is to consider formal functional equations and formal differential equations, as it is proposed in [2,3,5,6,22].

Motivated these results, we consider one-parameter groups of formal power series indexed elements of an arbitrary abelian group $(G,+)$. We give here a structure and the detailed description of some one-parameter groups of formal power series in that general case (to give detailed description of all one-parameter groups of formal power series seems to be still open).

## 33.2 Ring of Formal Power Series

Let $\mathbb{K} \in \{\mathbb{R}, \mathbb{C}\}$, where $\mathbb{R}$ and $\mathbb{C}$ denote fields of real and complex numbers, respectively. By $\mathbb{K}[\![X]\!]$ we denote the ring of all formal power series $\sum_{j=0}^{\infty} c_j X^j$ of one indeterminate with coefficients $c_k \in \mathbb{K}$ for $k \in \mathbb{N}\cup\{0\}$ ($\mathbb{N}$ stands here for the set of all positive integers) and with the usual operations of addition and multiplication given by

$$\sum_{j=0}^{\infty} a_j X^j + \sum_{j=0}^{\infty} b_j X^j = \sum_{j=0}^{\infty} (a_j + b_j) X^j,$$

and

$$\left(\sum_{j=0}^{\infty} a_j X^j\right) \cdot \left(\sum_{j=0}^{\infty} b_j X^j\right) = \sum_{j=0}^{\infty} \left(\sum_{k=0}^{j} a_k b_{j-k}\right) X^j.$$

For a formal power series $\sum_{j=0}^{\infty} c_j X^j$ with $c_k \neq 0$ for some $k \in \mathbb{N} \cup \{0\}$ we define

$$\operatorname{ord}\left(\sum_{j=0}^{\infty} c_j X^j\right) = \min\{k \in \mathbb{N} \cup \{0\} : c_k \neq 0\},$$

with the additional assumption that $(\sum_{k=0}^{\infty} 0 X_k) = \infty$.

Let us consider the set $\Gamma^{\infty} = \{f(X) \in \mathbb{K}[\![X]\!] : \operatorname{ord} f(X) = 1\}$. The substitution $\circ$ defined by

$$\left(\sum_{j=0}^{\infty} a_j X^j\right) \circ \left(\sum_{j=0}^{\infty} b_j X^j\right) = \sum_{j=0}^{\infty} a_j \left(\sum_{k=0}^{\infty} b_k X^k\right)^j$$

is then a binary operation on $\Gamma^{\infty}$ and a pair $(\Gamma^{\infty}, \circ)$ is a group. It is known that $\Gamma^{\infty}$ consists of all invertible formal power series in $\mathbb{K}[\![X]\!]$. A very good reference for this topic is a monograph by Henrici [7, pp. 9–51]. In the analytic description of the operation of substitution the crucial point is the following multinomial theorem.

**Theorem 33.1 (multinomial theorem, cf. [4]).** *Let $R$ be a commutative ring, $k, m \in \mathbb{N}$ and let $x_1, \ldots, x_m \in R$. Then*

$$(x_1 + x_2 + \cdots + x_m)^k = \sum_{\substack{u_1,\ldots u_m \geq 0 \\ u_1+\ldots+u_m=k}} \frac{k!}{u_1!\, u_2!\, \ldots\, u_m!} x_1^{u_1} x_2^{u_2} \ldots x_m^{u_m}.$$

Using the multinomial theorem one can show that if

$$\sum_{i=1}^{\infty} a_i X^i,\ \sum_{i=1}^{\infty} b_i X^i \in \Gamma^{\infty} \quad \text{and} \quad \sum_{k=1}^{\infty} a_k \left(\sum_{l=1}^{\infty} b_l X^l\right)^k = \sum_{n=1}^{\infty} d_n X^n,$$

then

$$d_n = \sum_{k=1}^{n} a_k \sum_{\overline{u}_n \in U_{n,k}} B_{\overline{u}_n} \prod_{j=1}^{n} b_j^{u_j} \quad \text{for } n \in \mathbb{N},$$

where

$$U_{n,k} = \left\{\overline{u}_n = (u_1, \ldots, u_n) \in |0,k|^n : \sum_{j=1}^{n} u_j = k,\ \sum_{j=1}^{n} j u_j = n\right\}, \quad B_{\overline{u}_n} = \frac{k!}{\prod_{j=1}^{n} u_j!}.$$

## 33.3 The Groups $L_s^1$ and $L_\infty^1$

Let us consider the family $\mathscr{D}_0$ of all real local $C^\infty$-diffeomorphisms defined on neighborhoods of 0 and mapping 0 to 0. For a fixed positive integer $s$ we may introduce in $\mathscr{D}_0$ the equivalence relations $j^s$ and $j^\infty$ by

$$(f,g) \in j^s \text{ if and only if } (f-g)^{(k)} = 0 \text{ for } k \in \{1,\dots,s\},\ f,g \in \mathscr{D}_0,$$
$$(f,g) \in j^\infty \text{ if and only if } (f-g)^{(k)} = 0 \text{ for } k \in \mathbb{N},\ f,g \in \mathscr{D}_0.$$

Then the sets $J_s\mathbb{R}$ and $J_\infty\mathbb{R}$ of all equivalence classes $j^s f$ and $j^\infty f$, respectively, with binary operations

$$(j^s f)\cdot(j^s g) = (j^s(f\circ g)), \tag{33.1}$$
$$(j^\infty f)\cdot(j^\infty g) = (j^\infty(f\circ g)), \tag{33.2}$$

are groups. Elements $j^s f$ of $J_s\mathbb{R}$ are called jets of a function $f$.

The set $L_s^1 := J_s\mathbb{R}$ is a Lie group, for which a natural mapping

$$J_s\mathbb{R} \ni j^s f \mapsto (x_1,\dots,x_s) \in (\mathbb{R}\setminus\{0\}) \times \mathbb{R}^{s-1},$$

where $x_k := f^{(k)}(0)$ for $k \in \{1,\dots,s\}$, is a coordinate system on whole $J_s\mathbb{R}$ and it makes a one-element atlas on $J_s\mathbb{R}$. For fixed $s,r \in \mathbb{N}$, $r \le s$, let us consider a homomorphism $\pi_r^s : J_s\mathbb{R} \to J_r\mathbb{R}$ (called a restriction of a jet) given by

$$\pi_r^s(j^s f) = j^r f \quad \text{for } j^s f \in J_s\mathbb{R}.$$

It satisfies $\pi_p^q \circ \pi_q^r = \pi_p^r$ for all $p,q,r \in \mathbb{N}$, $p \le q \le r$; hence, $L_\infty^1 := J_\infty\mathbb{R}$ may also be considered as a projective limit of the Lie groups $L_s^1$, i.e.,

$$\varprojlim L_s^1 = L_\infty^1.$$

Thus, with every $j^\infty f \in L_\infty^1$ we coincide a sequence $(x_1,x_2,\dots) \in (\mathbb{R}\setminus\{0\}) \times \mathbb{R}^\infty$, where $x_k := f^{(k)}(0)$ for $k \in \mathbb{N}$. To give analytic description of groups $L_s^1$ and $L_\infty^1$, we will need the famous Faá di Bruno formulae which express the $n$-th derivative of the composition of functions.

**Theorem 33.2 (Faa di Bruno formulae, cf. [15, 29]).** *If $f$ and $g$ are functions for which all the necessary derivatives are defined, then*

$$(f\circ g)^{(n)}(x) = \sum_{k=1}^{n} f^{(k)}(g(x)) \sum_{\substack{u_1,\dots,u_n \ge 0 \\ u_1+\dots+u_n=k \\ u_1+2u_2+\dots+nu_n=n}} \frac{n!}{u_1!\ \dots\ u_n!} \prod_{j=1}^{n} \left(\frac{g^{(j)}(x)}{j!}\right)^{u_j}.$$

Hence, from (33.1) and (33.2), on account of Theorem 33.2, we may give the following, equivalent description of groups $L_s^1$ and $L_\infty^1$.

**Definition 33.1.** The groups $L_s^1$ and $L_\infty^1$ we call the sets

$$Z_s = \{\overline{x}_s = (x_1, \dots, x_s) \in \mathbb{R}^s : x_1 \neq 0\},$$
$$Z_\infty = \{\overline{x}_\infty = (x_1, x_2, \dots) \in \mathbb{R}^\infty : x_1 \neq 0\}$$

with the operations

$$\overline{x}_s \cdot \overline{y}_s = \overline{z}_s,$$
$$\overline{x}_\infty \cdot \overline{y}_\infty = \overline{z}_\infty,$$

defined by

$$z_n = \sum_{k=1}^{n} x_k \sum_{\overline{u}_n \in U_{n,k}} A_{\overline{u}_n} \prod_{j=1}^{n} y_j^{u_j} \quad \text{for } n \in \mathbb{N},$$

where

$$U_{n,k} = \left\{ \overline{u}_n = (u_1, \dots, u_n) \in |0,k|^n : \sum_{j=1}^{n} u_j = k, \ \sum_{j=1}^{n} j u_j = n \right\},$$

$$A_{\overline{u}_n} = \frac{n!}{\prod\limits_{j=1}^{n} (u_j!(j!)^{u_j})}.$$

## 33.4 The Ring of Truncated Formal Power Series

As it has been announced, we will show, that it can be possible to transform results concerning one-parameter groups of formal power series to homomorphisms into groups $L_s^1$ and $L_\infty^1$. As we will see, the group $L_\infty^1$ will be isomorphic to $\Gamma^\infty$ in the case, when $\mathbb{K} = \mathbb{R}$. To obtain results for homomorphisms into $L_s^1$ we will need a notion of the ring of truncated formal power series. To this purpose let us fix a positive integer $s \in \mathbb{N}$. The set

$$I_s = X^{s+1}\mathbb{K}[X] = \{f(X) \in \mathbb{K}[X] : \operatorname{ord} f(X) \geq s+1\}$$

is an ideal in the ring $\mathbb{K}[X]$. Define a congruence modulo $X^{s+1}$ in the following way: we will say that $f_1(X), f_2(X) \in \mathbb{K}[X]$ are congruent modulo $X^{s+1}$ (which will be written as $(f_1(X) \equiv f_2(X) \mod X^{s+1})$ provided

$$(f_1 - f_2)(X) = f_1(X) - f_2(X) \in I_s.$$

This is clearly equivalent with the condition that $X^{s+1}$ divides the difference

$$(f_1 - f_2)(X).$$

Let $\mathbb{K}[X]/I_s$ be the quotient ring of all cosets $[f(X)]_s = f(X) + I_s$. With each coset $f(X) + I_s$ ($f(X) = \sum_{k=0}^{\infty} c_k X^k \in \mathbb{K}[X]$) we may associate the $s$-truncation of a formal power series $f(X)$ given by

$$f^{[s]}(X) := \sum_{k=0}^{s} c_k X^k \in \mathbb{K}[X]_s.$$

In the set $\mathbb{K}[X]_s$ (which may be treated as the set of all formal power series $\sum_{k=0}^{\infty} c_k X^k$ with coefficients $c_k = 0$ for $k \geq s+1$), we introduce, in a natural way, an addition of truncated formal power series. However, the multiplication and the substitution of truncated formal power series must be defined in a specific way that $\mathbb{K}[X]_s$ should be closed under them. For $f(X), g(X) \in \mathbb{K}[X]_s$ define

$$(fg)(X) := (fg)^{[s]}(X),$$

and, in the case when $\operatorname{ord} g(X) \geq 1$,

$$(f \circ g)(X) := (f \circ g)^{[s]}(X).$$

The ring $(\mathbb{K}[X]_s, +, \cdot)$ is then isomorphic with $\mathbb{K}[X]/I_s$.

For truncated formal power series we also have that the set $\Gamma^s := \{f(X) \in \mathbb{K}[X]_s : \operatorname{ord} f(X) = 1\}$ is a group under substitution defined by

$$\left(\sum_{j=0}^{s} a_j X^j\right) \circ \left(\sum_{j=0}^{s} b_j X^j\right) = \sum_{j=0}^{s} a_j \left(\sum_{k=0}^{s} b_k X^k\right)^j \mod X^{s+1}.$$

Also here the analytic description of substitution is possible. If

$$\sum_{i=1}^{s} a_i X^i, \sum_{i=1}^{s} b_i X^i \in \Gamma^s$$

and

$$\sum_{k=1}^{s} a_k \left(\sum_{l=1}^{s} b_l X^l\right)^k = \sum_{n=1}^{s} d_n X^n \mod X^{s+1},$$

then

$$d_n = \sum_{k=1}^{n} a_k \sum_{\overline{u}_n \in U_{n,k}} B_{\overline{u}_n} \prod_{j=1}^{n} b_j^{u_j} \quad \text{for } n \in \{1, 2, \ldots, s\},$$

where $U_{n,k}$ and $b_{\overline{u}_n}$ are defined as for formal power series.

Similarly, as for jets, for fixed $s, r \in \mathbb{N}$, $r \leq s$, we may consider a natural homomorphism $\pi_s^r : \Gamma^s \to \Gamma^r$ given by

$$\pi_s^r(f(X)) = f^{[r]}(X) \quad \text{for } f(X) \in \Gamma^s.$$

It is easy to observe, that also here we have $\pi_p^q \circ \pi_q^r = \pi_p^r$ for all $p,q,r \in \mathbb{N}$, $p \le q \le r$. Thus, the group $\Gamma^\infty$ we may also consider as a projective limit of the groups $\Gamma^s$, that is

$$\lim_{\longleftarrow} \Gamma^s = \Gamma^\infty.$$

## 33.5 Homomorphisms into Differential Groups $L_s^1$ and $L_\infty^1$

In classical differential geometry, a one-parameter group (a one-parameter subgroup) is called any homomorphism of the additive group $(\mathbb{R},+)$ into a Lie group. Such one-parameter groups $\mathbb{R} \ni t \mapsto \Phi(t) \in L_s^1$ for $s \le 5$ have been determined in [17]. Moszner [18] gave an example of homomorphism $\mathbb{R} \ni t \mapsto \Phi(t) \in L_\infty^1$. Also homomorphisms of a group $L_r^1$ into $L_s^1$ are determined (cf. [1,8,9]). It appears (this is a subject of the paper in preparation) that almost all (all but a finite number) homomorphisms $\Phi : L_r^1 \to L_s^1$ are of the form

$$\Phi = \Phi_1 \circ \pi_1^r,$$

where $\Phi_1 : L_1^1 \to L_s^1$ is a homomorphism. Since $L_1^1$ is isomorphic to the multiplicative group $(\mathbb{R} \setminus \{0\}, \cdot)$, we need to determine homomorphisms $\Phi_1 : \mathbb{R} \setminus \{0\} \to L_s^1$. In general we will be interested in finding homomorphisms of an abelian group $(G,+)$ into $L_s^1$.

Let $s$ be a positive integer or $s = \infty$. For integers $k,l$ let $|k,l|$ denote the set of all integers $n$ such that $k \le n \le l$ as well as by $|k,\infty|$ we will mean the set of all integers $n \ge k$. Consider a function $\Phi_s : G \to Z_s$,

$$\Phi_s = (f_j)_{j\in|1,s|},$$

where $f_1 : G \to \mathbb{R} \setminus \{0\}$, $f_n : G \to \mathbb{R}$ for $n \in |2,s|$. The function $\Phi_s$ is a homomorphism if and only if

$$\Phi_s(x+y) = \Phi_s(x) \cdot \Phi_s(y) \quad \text{for } x,y \in G,$$

i.e. the functions $f_n$ solve the following system of functional equations

$$\begin{cases} f_1(x+y) = f_1(x) f_1(y), \\ f_2(x+y) = f_1(x) f_2(y) + f_2(x) f_1(y)^2, \\ f_n(x+y) = f_1(x) f_n(y) \\ \qquad + \sum\limits_{k=2}^{n-1} f_k(x) \sum\limits_{\overline{u}_n \in U_{n,k}} A_{\overline{u}_n} \prod\limits_{j=1}^{n-1} f_j(y)^{u_j} + f_n(x) f_1(y)^n, \ n \in |3,s| \end{cases} \tag{33.3}$$

for $x,y \in G$.

As it is easy to see, $f_1$ must be then en exponential function. Two cases appears in general:

*case* 1. $f_1 = 1$, there is an integer $p \geq 0$ such that $f_j = 0$ for $j \in |2, p+1|$, $f_{p+2} \neq 0$;
*case* 2. $f_1 \neq 1$.

We are able to prove the following results using classical methods of changing of variables. All that we can prove, in general, in the first case reads as follows:

**Theorem 33.3 (cf. [10, Proposition 1]).** *Let $(G,+)$ be an abelian group which admits a nonzero additive function into $(\mathbb{R},+)$ (then G must be infinite). There exists a sequence $(w_n^{p+2})_{n\in|p+3,s-p-1|}$ of polynomials with $w_n^{p+2} \in \mathbb{R}[X; d_{p+3},\ldots,d_{n-1}]$ for $n \in |p+3, s-p-1|$ such that for every homomorphism $\Phi_s = (f_j)_{j\in|1,s|}$ with $f_1 = 1$, $f_j = 0$ for $j \in |2, p+1|$ and some integer $p \geq 0$, $f_{p+2} \neq 0$, there exists a sequence of constants $(d_n)_{n\in|p+3,s-p-1|}$ such that*

*(i) $f_{p+2}$ is a nonzero additive function;*

*(ii) $f_n = d_n f_{p+2} + w_n^{p+2}(f_{p+2}; d_{p+3},\ldots,d_{n-1})$ for every $n \in |p+2, s-p-1|$.*

To show the methods used in the proof we give a sketch of it.

*Proof.* In the considered case, the system (33.3) can be reduced to

$$\begin{cases} f_{p+2}(x+y) = f_{p+2}(x) + f_{p+2}(y), \\ f_n(x+y) = f_n(x) + f_n(y) \\ \qquad + \sum\limits_{k=p+2}^{n-p-1} f_k(x) \sum\limits_{\overline{u}_n\in U_{n,k}^p} A_{\overline{u}_n} \prod\limits_{j=p+2}^{n-k+1} f_j(y)^{u_j}, \quad n \in |p+3, s|, \end{cases} \tag{33.4}$$

for $x, y \in G$, where $U_{n,k}^p := \{\overline{u}_n \in U_{n,k} : \forall_{i\in|2,p+1|}\ u_i = 0\}$.

As one can see, $f_{p+2}$ is a nonzero additive function. To find a necessary form of the function $f_n$, let us assume that for some $n$ with $n+p+1 \leq s$ we have determined functions $f_{p+2},\ldots,f_{n-1}$ and consider the equation of the system (33.4) for the function $f_r$ with $r = n+p+1$, i.e.,

$$f_r(x+y) = f_r(x) + \sum_{k=p+2}^{n} f_k(x) \sum_{\overline{u}_r\in U_{r,k}^p} A_{\overline{u}_r} \prod_{j=p+2}^{r-k+1} f_j(y)^{u_j} + f_r(y).$$

Interchanging $x$ and $y$ and using the symmetry of the left hand side of above equality we get

$$\sum_{k=p+2}^{n} f_k(x) \sum_{\overline{u}_r\in U_{r,k}^p} A_{\overline{u}_r} \prod_{j=p+2}^{r-k+1} f_j(y)^{u_j} = \sum_{k=p+2}^{n} f_k(y) \sum_{\overline{u}_r\in U_{r,k}^p} A_{\overline{u}_r} \prod_{j=p+2}^{r-k+1} f_j(x)^{u_j}.$$

Using several technical facts concerning properties of the group action in $L_s^1$ (cf. [10, proof of Proposition 1]) we obtain

$$\binom{r}{p+1} f_{p+2}(x) f_n(y) + \sum_{k=p+2}^{n-1} f_k(x) \sum_{\overline{u}_r \in \widetilde{U}_{r,k}^p} A_{\overline{u}_r} \prod_{j=p+2}^{r-k+1} f_j(y)^{u_j}$$
$$+ \binom{r}{p+2} f_n(x) f_{p+2}(y) = \binom{r}{p+1} f_{p+2}(y) f_n(x)$$
$$+ \sum_{k=p+2}^{n-1} f_k(y) \sum_{\overline{u}_r \in \widetilde{U}_{r,k}^p} A_{\overline{u}_r} \prod_{j=p+2}^{r-k+1} f_j(x)^{u_j}$$
$$+ \binom{r}{p+2} f_n(y) f_{p+2}(x)$$

with some $\widetilde{U}_{r,k}^p \subset U_{r,k}^p$. We know that the sum

$$\sum_{k=p+2}^{n-1} f_k(x) \sum_{\overline{u}_r \in \widetilde{U}_{r,k}^p} A_{\overline{u}_r} \prod_{j=p+2}^{r-k+1} f_j(y)^{u_j}$$

depend on the functions $f_{p+2}, \ldots, f_{n-1}$ only. Thus, form the above equality we derive

$$f_n(x) f_{p+2}(y) = f_{p+2}(x) f_n(y) + \frac{n!\,(p+2)!}{(n+p+1)!\,(n-p-2)} \sum_{k=p+2}^{n-1} \sum_{\overline{u}_r \in \widetilde{U}_{r,k}^p}$$
$$A_{\overline{u}_r} \left[ W_k(f_{p+2}(y); d_{p+3}, \ldots, d_k) \prod_{j=p+2}^{r-k+1} W_j(f_{p+2}(x); d_{p+3}, \ldots, d_j)^{u_j} \right.$$
$$\left. - W_k(f_{p+2}(x); d_{p+3}, \ldots, d_k) \prod_{j=p+2}^{r-k+1} W_j(f_{p+2}(y); d_{p+3}, \ldots, d_j)^{u_j} \right],$$

which for a fixed $y \in G$ such that $f_{p+2}(y) \neq 0$ allows us to write $f_n(x) = d_n f_{p+2}(x) + w_n^{p+2}(f_{p+2}(x); d_{p+3}, \ldots, d_{n-1})$ with suitably defined polynomial $w_n^{p+2}$. □

In the second case we can prove the following:

**Theorem 33.4 (cf. [13, Proposition 1]).** *Assume that the abelian group $(G,+)$ admits nonconstant exponential functions with infinite image. There exists a sequence $(v_n)_{n\in|2,s|}$ of universal polynomials $v_n \in \mathbb{R}[X; p_2, \ldots, p_{n-1}]$ with $v_2 = 0$ such that for each homomorhpism $\Phi_s = (f_j)_{j\in|1,s|}$ from $G$ to $L_s^1$ with an exponential function $f_1$ taking infinitely many values, there is a sequence of constants $(p_n)_{n\in|2,s|}$ such that*

$$f_n = (f_1^2 - f_1)\left( p_n \sum_{l=0}^{n-2} f_1^l + v_n(f_1; p_2, \ldots, p_{n-1}) \right) \quad \textit{for } n \in |2,s|.$$

The proof of that theorem is similar to the proof of Theorem 33.3. Let us present a sketch of it.

*Proof.* Assume that $f_1$ is an exponential function with infinite image. From the second equation of the system (33.3), using symmetry of the left hand side of it, we get

$$f_1(x)f_2(y)+f_1(y)^2f_2(x)=f_1(y)f_2(x)+f_1(x)^2f_2(y),$$

and thus $f_2(x)(f_1(y)^2-f_1(y))=f_2(y)(f_1(x)^2-f_1(x))$. Let $y_0\in G$ be such that $f_1(y_0)^2-f_1(y_0)\neq 0$ (this is possible by our assumption on infinite image of $f_1$). Then $f_2(x)=p_2(f_1(x)^2-f_1(x))$, where

$$p_2:=\frac{f_2(y_0)}{f_1(y_0)^2-f_1(y_0)}.$$

To determine $f_n$, let us assume that $f_2,\dots,f_{n-1}$ have required form. From the $n$-th equation of the system (33.3), using the symmetry of the left hand side of it, we obtain

$$f_1(x)f_n(y)+\sum_{k=2}^{n-1}f_k(x)\sum_{\bar{u}_n\in U_{n,k}}A_{\bar{u}_n}\prod_{j=1}^{n-k+1}f_j(y)^{u_j}+f_1(y)^nf_n(x)$$

$$=f_1(y)f_n(x)+\sum_{k=2}^{n-1}f_k(y)\sum_{\bar{u}_n\in U_{n,k}}A_{\bar{u}_n}\prod_{j=1}^{n-k+1}f_j(x)^{u_j}+f_1(x)^nf_y(x)$$

and hence

$$f_n(x)(f_1(y)^n-f_1(y))=f_n(y)(f_1(x)^n-f_1(x))$$
$$+\sum_{k=2}^{n-1}\sum_{\bar{u}_n\in U_{n,k}}A_{\bar{u}_n}\left[f_k(y)\prod_{j=1}^{n-k+1}f_j(x)^{u_j}-f_k(x)\prod_{j=1}^{n-k+1}f_j(y)^{u_j}\right].$$

Thus, for a fixed $y\in G$ with $f_1(y_0)^n-f_1(y_0)\neq 0$, using some properties of the binary operation in $L_s^1$ (cf. [13, proof of Theorem 1]) we get

$$f_n=(f_1^2-f_1)\left(p_n\sum_{l=0}^{n-2}f_1^l+v_n(f_1;p_2,\dots,p_{n-1})\right)$$

with some polynomial $v_n$. □

Above theorems contain the best results, up to our knowledge, which can be proved using the presented methods. Moreover, in the case when $f_1=1$ and $s$ is finite, we cannot determine the necessary form of all functions $f_1,\dots,f_s$, but we only have it for functions $f_1,\dots,f_{s-p-1}$. What is more, even when either $f_1=1$ and $s=\infty$ or $f_1\neq 1$, we are not able to verify, whether the functions determined in above theorems, really satisfy the system of equations (33.3). However, it appears that we can solve analogous problems in the rings of formal power series and truncated formal power series, and then we can transform the obtained results onto groups $L_s^1$ and $L_\infty^1$. This can be made by using of the following theorem.

**Theorem 33.5 (cf. [10, Theorem 2]).** *Let $s$ be either a positive integer or $s = \infty$. The groups $L_s^1$ and $(\Gamma^s, \circ)$ are isomorphic. A natural isomorphism from $L_s^1$ to $(\Gamma^s, \circ)$ is given by $\Psi : Z_s \to \Gamma^s$,*

$$\Psi(\bar{x}_s)(X) = \sum_{k=1}^{s} \frac{x_k}{k!} X^k. \tag{33.5}$$

## 33.6 One-Parameter Groups of Formal Power Series

**Definition 33.2.** Let $s \in \mathbb{N}$ or $s = \infty$, and let $(G, +)$ be a group. By a one-parameter group of formal power series we mean a homomorphism $\Theta_G^s : G \to \Gamma^s$, i.e. every function $\Theta_G^s$ satisfying

$$\Theta_G^s(t_1 + t_2)(X) = (\Theta_G^s(t_1) \circ \Theta_G^s(t_2))(X) \quad \text{for } t_1, t_2 \in G. \tag{33.6}$$

If $\Theta_G^s : G \to \Gamma^s$ is a one-parameter group of formal power series, then the family $\left(F_t^{[s]}(X)\right)_{t \in G}$, $F_t^{[s]}(X) = F^{[s]}(t, X) = \Theta_G^s(t)(X)$ will also be called a one-parameter group of formal power series. Hence, (33.6) implies

$$\begin{cases} F_{t_1+t_2}^{[s]} = F_{t_1}^{[s]} \circ F_{t_2}^{[s]} & \text{for } t_1, t_2 \in G, \\ F_0^{[s]} = \mathrm{id}. \end{cases} \tag{33.7}$$

Let $\Theta_G^s(t)(X) = F^{[s]}(t, X) = \sum_{k=1}^{s} c_k(t) X^k$ with $c_1 : G \to \mathbb{K} \setminus \{0\}$ and $c_k : G \to \mathbb{K}$ for $k \in |2, s|$. Then (33.6) and (33.7) are equivalent to the following system of functional equations

$$\begin{cases} c_1(t_1 + t_2) = c_1(t_1) c_1(t_2), \\ c_2(t_1 + t_2) = c_1(t_1) c_2(t_2) + c_2(t_1) c_1(t_2)^2, \\ c_n(t_1 + t_2) = c_1(t_1) c_n(t_2) \\ \quad + \sum_{k=2}^{n-1} c_k(t_1) \sum_{\bar{u}_n \in U_{n,k}} B_{\bar{u}_n} \prod_{j=1}^{n-k+1} c_j(t_2)^{u_j} + c_n(t_1) c_1(t_2)^n, \quad n \in |3, s| \end{cases} \tag{33.8}$$

for $t_1, t_2 \in G$.

### 33.6.1 Regular One-Parameter Groups of Formal Power Series

As we pointed out, the usual simple methods of substituting and changing of variables are not sufficient to describe the general solution (and hence all one-parameter groups of formal power series) of the system (33.8). We use here much more advanced methods adopted from the theory of formal power series. First, we determine all regular one-parameter groups $(F^{[\infty]}(t,X))_{t\in\mathbb{K}}$ of formal power series. Using them, we are able to describe one-parameter groups $(F^{[s]}(t,X))_{t\in G}$ such that either $c_1 = 1$ or $c_1(G)$ is an infinite set or both $c_1(G)$ and $(F^{[s]}(t,X))_{t\in G}$ are finite sets.

To this purpose we introduce natural in those settings operation of differentiation, which is defined purely algebraically. Let and $c_k : G \to \mathbb{K}$ for $k \in |1,s|$ and

$$F(t,X) = \sum_{k=1}^{s} c_k(t)X^k.$$

Define

$$\frac{\partial F}{\partial X}(t,X) = \sum_{k=1}^{s} kc_k(t)X^{k-1}.$$

We will also need an operator of differentiation with respect to the "time"-variable, which in the case when $G = \mathbb{K}$ and the coefficient functions $c_k$ are differentiable, is given by

$$\frac{\partial F}{\partial t}(t,X) = \sum_{k=1}^{s} c_k'(t)X^k.$$

For $G = \mathbb{K}$ the following theorem allow us to describe the all regular one-parameter groups of formal power series.

**Theorem 33.6.** *(i) If a family $(F^{[\infty]}(t,X))_{t\in\mathbb{K}}$ is a regular one-parameter group of formal power series, then there exists a formal power series $H(X) \in \mathbb{K}[\![X]\!]$ such that*

$$\begin{cases} \dfrac{\partial F^{[\infty]}}{\partial t}(t,X) = H(F(t,X)) & \text{for } t \in \mathbb{K}, \\ F^{[\infty]}(0,X) = X. \end{cases} \tag{33.9}$$

*(ii) For each $H(X) \in \mathbb{K}[\![X]\!]$, $\operatorname{ord} H \geq 1$, the family $(F^{[\infty]}(t,X))_{t\in\mathbb{K}}$ defined by* (33.9) *is a regular one-parameter group of formal power series.*

*(iii) The series $H$ is uniquely determined by $(F^{[\infty]}(t,X))_{t\in\mathbb{K}}$. It is given by the formula*

$$H(X) := \frac{\partial F^{[\infty]}}{\partial t}(0,X),$$

*in particular,* $\operatorname{ord} H \geq 1$.

Crucial for our purposes here is the part (ii) which states that the solutions of (33.9) *are the only* regular one-parameter groups of formal power series.

Theorem 33.6 may be derived from [27, 28], where one-parametr groups of formal power series in $n$ indeterminates are considered, but a simple and direct proof is also possible. In [14], it is proved, following some ideas from [30], that the general solution of (33.9) may be given by

$$F^{[\infty]}(t,X) = \mathrm{e}^{tH(X)\,\partial/\partial X}X := \sum_{k=0}^{\infty} \frac{t^k\left(H(X)\frac{\partial}{\partial X}\right)^k}{k!}X \quad \text{for } t \in \mathbb{K}.$$

Let

$$F^{[\infty]}(t,X) = \sum_{k=1}^{\infty} c_k(t)X^k$$

be a regular one-parameter group of formal power series. Then it satisfies (33.9), which means that coefficient function $c_k$ satisfy the following system of differential equations

$$\begin{cases} c_1'(t) = c_1'(0)c_1(t), \; c_1(0) = 1, \\ c_2'(t) = c_1'(0)c_2(t) + c_2'(0)c_1(t)^2, \; c_2(0) = 0, \\ c_n'(t) = c_1'(0)c_n(t) + c_n'(0)c_1(t)^n \\ \qquad + \sum\limits_{k=2}^{n-1} c_k'(0) \sum\limits_{\overline{u}_n \in U_{n,k}} B_{\overline{u}_n} \prod\limits_{j=1}^{n-k+1} c_j(t)^{u_j}, \; c_n(0) = 0, \quad n \geq 3. \end{cases} \tag{33.10}$$

Let us consider the following cases:

1. $c_1'(0) = \cdots = c_{p+1}'(0)$ and $c_{p+2}'(0) \neq 0$ for some integer $p \geq 0$;
2. $c_1'(0) \neq 0$.

In the first case, from (33.10) we get $c_1(t) = 1, c_2(t) = \cdots = c_{p+1}(t) = 0$ for all $t \in \mathbb{K}$. Hence, with $h_k := c_k'(0)$ for $k \geq p+2$, $h_{p+2} \neq 0$, the system (33.10) can be reduced to

$$\begin{cases} c_{p+2}'(t) = h_{p+2}, \; c_{p+2}(0) = 0, \\ c_n'(t) = h_n + \sum\limits_{k=p+2}^{n-p-1} h_k \sum\limits_{\overline{u}_n \in U_{n,k}^p} B_{\overline{u}_n} \prod\limits_{j=p+2}^{n-k+1} c_j(t)^{u_j}, \; c_n(0) = 0, \quad n \geq p+3. \end{cases} \tag{33.11}$$

Thus, $c_{p+2}(t) = h_{p+2}t$ for $t \in \mathbb{K}$, an we need to solve recursively simple ordinary differential equations.

In the second case, with the sequence $(\lambda_n)_{n\geq 2}$, defined by:

$$\lambda_1 := c_1'(0) \quad \text{and} \quad (k-1)\lambda_1\lambda_k := c_k'(0) \quad \text{for } k \geq 2,$$

we obtain

$$\begin{cases} c_1'(t) = \lambda_1 c_1(t)\,, \;\; c_1(0) = 1, \\ c_2'(t) = \lambda_1 c_2(t) + \lambda_1\lambda_2 c_1(t)^2\,, \;\; c_2(0) = 0, \\ c_n'(t) = \lambda_1 c_n(t) + (n-1)\lambda_1\lambda_n c_1(t)^n \\ \qquad + \sum_{k=2}^{n-1}(k-1)\lambda_1\lambda_k \sum_{\overline{u}_n\in U_{n,k}} B_{\overline{u}_n} \prod_{j=1}^{n-k+1} c_j(t)^{u_j}\,, \;\; c_n(0) = 0, \quad n \geq 3. \end{cases} \tag{33.12}$$

Thus, $c_1(t) = \mathrm{e}^{\lambda_1 t}$ for $t \in \mathbb{K}$, and in this case we have to solve recursively simple ordinary differential equation of Bernoulli's type.

The form of all regular one-parameter groups of formal power series is described in the following

**Theorem 33.7 (cf. [10, Theorem 4], [13, Theorem 6] and [14, Theorem 1]).** *Fix an integer $p \geq 0$. There are sequences of polynomials $\left(L_n^p\right)_{n\geq p+2}$ and $(P_n)_{n\geq 2}$ given by*

$$\begin{cases} L_{p+2}^p(X) = 0, \\ L_n^p(X;(h_l)_{l\in|p+2,n-p-1|}) = \sum_{k=p+2}^{n-p-1} h_k \sum_{\overline{u}_n\in U_{n,k}^p} B_{\overline{u}_n} \\ \qquad \times \int_0^X \left( \prod_{j=p+2}^{n-k+1} \left(h_j v + L_j^p(v;(h_l)_{l\in|p+2,j-p-1|})\right)^{u_j} \right) dv\,, \quad n \geq p+3, \end{cases} \tag{33.13}$$

*with $U_{n,k}^p := \left\{\overline{u}_n \in U_{n,k} : \; \forall_{j\in|2,p+1|,j\leq n-1}\, u_j = 0\right\}$ and by*

$$\begin{cases} P_2(X) = 0, \quad R_2(X;\lambda_2) = \lambda_2 X - \lambda_2\,, \\ P_n(X;(\lambda_l)_{l\in|2,n-1|}) = \sum_{k=2}^{n-1}(k-1)\lambda_k \sum_{\overline{u}_n\in U_{n,k}} B_{\overline{u}_n} \\ \qquad \times \int_1^X v^{k-2} \prod_{j=2}^{n-k+1} \left(R_j(v;(\lambda_l)_{l\in|2,j|})\right)^{u_j} \mathrm{d}v\,, \quad n \geq 3, \\ R_n(X;(\lambda_l)_{l\in|2_n|}) = \lambda_n(X^{n-1}-1) + P_n(X;(\lambda_l)_{l\in|2,n-1|}), \end{cases} \tag{33.14}$$

*such that coefficient functions of every nontrivial regular one-parameter group of formal power series $F(t,X) = \sum_{k=1}^{\infty} c_k(t)X^k$ are given by*

$$c_n(t) = h_n t + L_n^p(t;(h_k)_{k\in|p+2,n-p-1|})\,, \quad t \in \mathbb{K}\,, \; n \geq p+2, \tag{33.15}$$

*provided $c_1'(0) = 0$, $c_i'(0) = 0$ for $i \in |2, p+1|$, $c_{p+1}'(0) \neq 0$, and by*

$$\begin{cases} c_1(t) = \mathrm{e}^{\lambda_1 t}\,, \quad t \in \mathbb{K}, \\ c_n(t) = \lambda_n\left(\mathrm{e}^{n\lambda_1 t} - \mathrm{e}^{\lambda_1 t}\right) + \mathrm{e}^{\lambda_1 t} P_n\left(\mathrm{e}^{\lambda_1 t};(\lambda_k)_{k\in|2,n-1|}\right)\,, \quad t \in \mathbb{K}\,, \; n \geq 2, \end{cases} \tag{33.16}$$

*if $c_1'(0) \neq 0$, where $(h_n)_{n\geq p+2}$ and $(\lambda_n)_{n\geq 1}$ with $h_{p+2} \neq 0$, $\lambda_1 \neq 0$ are arbitrary sequences of constants.*

### 33.6.2 One-Parameter Groups with Either $c_1 = 1$ or Infinite $c_1(G)$ or Finite Both $c_1(G)$ and $(F^{[s]}(t,X))_{t\in G}$

Now using regular one-parameter groups of formal power series we will be able to construct some one-parameter groups without regularity assumptions. This is possible by the simply observation, that we obtain in considered cases either systems of inhomogeneous Cauchy equations or inhomogeneous equations of some type.

The following theorem describes one-parameter groups in considered cases.

**Theorem 33.8. (cf. [10, Theorem 4], [13, Theorem 6] and [14, Theorems 4,5, and Corollary 6]).** *Every one-parameter group*

$$\left(F^{[s]}(t,X)\right)_{t\in G}, \qquad F^{[s]}(t,X) = \sum_{k=1}^{s} c_k(t)X^k,$$

*$c_1 : G \to \mathbb{K}\setminus\{0\}$, $c_k : G \to \mathbb{K}$ for $k \in |2,s|$, of formal power series such that either $c_1 = 1$ or $c_1(G)$ is an infinite set or $c_1(G)$ and $\left(F^{[s]}(t,X)\right)_{t\in G}$ are both finite sets, is given either by*

$$\begin{cases} c_1(t) = 1, \quad t \in G, \\ c_n(t) = 0, \quad t \in G,\ n \in |2,p+1|, \\ c_{p+2}(t) = a(t), \quad t \in G, \\ c_n(t) = h_n a(t) + \overline{L}_n^p(a(t);(h_j)_{j\in|p+3,n-p-1|}), \quad t \in G,\ n \in |p+3,s-p-1|, \\ c_n(t) = a_n(t) + \overline{L}_n^p(a(t);(h_j)_{j\in|p+3,n-p-1|})), \quad t \in G,\ n \in |s-p,s|, \end{cases}$$

*where $p \in \mathbb{N}\cup\{0\}$, $a : G \to \mathbb{K}$ and $a_n : G \to \mathbb{K}$ for $|s-p,s|$ are arbitrary additive functions and $(h_n)_{n\in|p+3,s-p-1|}$is arbitrary sequence of constants, or by*

$$\begin{cases} c_1(t) = c(t), \quad t \in G, \\ c_n(t) = \lambda_n\left(c(t)^n - c(t)\right) + c(t)P_n\left(c(t);(\lambda_l)_{l\in|2,n-1|}\right), \quad t \in G,\ n \in |2,s|, \end{cases}$$

*where $(\lambda_n)_{n\in|2,s|}$ is an arbitrary sequence of constants and $c : G \to \mathbb{K}\setminus\{0\}$ is an arbitrary nonconstant exponential function.*

*Proof.* Let us consider similar cases as for regular one-parameter groups of formal power series, that is

1. $c_1 = 1, c_2 = \cdots = c_{p+1} = 0$ and $c_{p+2} \neq 0$ for some integer $p \geq 0$;
2. $c_1 \neq 1$ and either $c_1(G)$ is infinite or both $c_1(G)$ and $(F^{[s]}(t,X))_{t\in G}$ are finite.

In the first case, with either finite $s$ or $s = \infty$, we get the following system of functional equations:

$$\begin{cases} c_{p+2}(t_1+t_2)=c_{p+2}(t_1)+c_{p+2}(t_2), \\ c_n(t_1+t_2)=c_n(t_1)+c_n(t_2) \\ \quad +\sum\limits_{k=p+2}^{n-p-1} c_k(t_1) \sum\limits_{\overline{u}_n\in U^p_{n,k}} B_{\overline{u}_n} \prod\limits_{j=p+2}^{n-k+1} c_j(t_2)^{u_j}, \quad n\in|p+3,s|, \end{cases} \tag{33.17}$$

for $t_1,t_2 \in G$. Thus, $c_{p+2}$ is nonzero additive mapping. Let $a=c_{p+2}$ and define a polynomial $\overline{L}^p_{p+2}(X)=0$. Assume that for some $n\in|p+3,s-p-1|$ we have

$$c_j(t)=h_j a(t)+\overline{L}^p_j(a(t);(h_l)_{l\in|p+3,j-p-1|}),\ t\in G, j\in|p+2,n-1| \tag{33.18}$$

with

$$\overline{L}^p_n(X;(h_l)_{l\in|p+3,n-p-1|}) := L^p_n(X;(h_l)_{l\in|p+2,n-p-1|})$$

and $h_{p+2}=1$. Let us consider the equation

$$c_n(t_1+t_2)=c_n(t_1)+c_n(t_2) + \sum_{k=p+2}^{n-p-1} c_k(t_1) \sum_{\overline{u}_n\in U^p_{n,k}} B_{\overline{u}_n} \prod_{j=p+2}^{n-k+1} c_j(t_2)^{u_j} \quad \text{for } t_1,t_2\in G, \tag{33.19}$$

with functions $c_j$ given by (33.18) for $j\in|p+2,n-1|$. Observe, that (33.19) is, in fact, inhomogeneous Cauchy equation. Hence, every solution $c_n$ of (33.19) is a sum of particular solution of (33.19) and an additive function (every two solutions of (33.19) differs on an additive function). What we need now to prove (all details can be find in [14]) is that $\overline{L}^p_n(a(t);(h_l)_{l\in|p+3,n-p-1|})$ is a solution of (33.19). Then

$$c_n(t)=a_n(t)+\overline{L}^p_n(a(t);(h_l)_{l\in|p+3,n-p-1|}) \quad \text{for } t\in G,$$

with an additive function $a_n : G\to\mathbb{K}$. Finally, considering the equation of the system (33.17) with $c_{n+p+1}$ on the left hand side, we are able to prove (cf. [14] for details) that $a_n(t)=h_n a(t)$ for $t\in G$ with some $h_n\in\mathbb{K}$.

To see that with obtained functions $c_k$ the family

$$F^{[\infty]}(t,X)=X+\sum_{k=p+2}^{\infty} c_k(t)X^k$$

is really the one-parameter group of formal power series it is enough to observe that $F(t,X)=\overline{F}(a(t),X)$, where

$$\overline{F}(t,X)=X+tX^{p+2}+\cdots$$

is a regular one-parameter group of formal power series with $h_{p+2}=1$ (cf. Theorem 33.7). Thus, in the case $s=\infty$, the proof is completed.

Let us consider now the case $s \in \mathbb{N}$. In the same way as above we prove that

$$c_n(t) = h_n a(t) + \overline{L}_n^p\left(a(t);(h_l)_{l\in|p+3,n-p-1|}\right) \quad \text{for } t \in G,$$

for all $n \in |p+2, s-p-1|$. For $n \in |s-p, s|$ we obtain (see [14])

$$c_n(t) = a_n(t) + \overline{L}_n^p\left(a(t);(h_l)_{l\in|p+3,n-p-1|}\right) \quad \text{for } t \in G,$$

with arbitrary additive functions $a_n$ for $n \in |s-p, s|$. The justification that with these functions the family $F^{[s]}(t,X)$ is the one-parameter group of truncated formal power series is similar as in the case $s = \infty$, but a little bit more complicated. In this way, we obtain the description of the solutions of (33.17) in the first considered case.

For the case $c \neq 1$ we give the sketch of the proof only when $c_1(G)$ is an infinite set (the other subcase is much more complicated and uses some algebraic considerations. In that case we need to consider the system (33.8) directly, that is

$$\begin{cases} c_1(t_1+t_2) = c_1(t_1)c_1(t_2), \\ c_2(t_1+t_2) = c_1(t_1)c_2(t_2) + c_2(t_1)c_1(t_2)^2, \\ c_n(t_1+t_2) = c_1(t_1)c_n(t_2) + c_n(t_1)c_1(t_2)^n \\ \qquad + \sum\limits_{k=2}^{n-1} c_k(t_1) \sum\limits_{\bar{u}_n\in U_{n,k}} B_{\bar{u}_n} \prod\limits_{j=1}^{n-k+1} c_j(t_2)^{u_j}, \quad n \in |3,s| \end{cases} \tag{33.20}$$

for $t_1, t_2 \in G$.

From the first equation of that system we see that $c_1$ is en exponential function. Moreover, from the second equation of (33.20), using commutativity of the left hand side one can easily derive

$$c_1(t_1)c_2(t_2) + c_2(t_1)c_1(t_2)^2 = c_1(t_2)c_2(t_1) + c_2(t_2)c_1(t_1)^2,$$

whence $c_2(t_1)(c_1(t_2)^2 - c_1(t_2)) = c_2(t_2)(c_1(t_1)^2 - c_1(t_1))$. Hence we find $t_0 \in G$ with $c_1(t_0)^2 - c_1(t_0) \neq 0$ ($c_1(G)$ is the infinite set), which implies

$$c_2(t) = \lambda_2(c_1(t)^2 - c_1(t)) \quad \text{for } t \in G,$$

where $\lambda_2 := c_2(t_0)/(c_1(t_0)^2 - c_1(t_0))$.

Put $P_2(X) = 0$ and assume that for some $n \in |3,s|$ we have

$$c_j(t) = \lambda_n\left(c_1(t)^j - c_1(t)\right) + c_1(t)P_j\left(c_1(t);(\lambda_l)_{l\in|2,j-1|}\right),$$

for $t \in G$, $j \in |2, n-1|$, and consider the equation of the system (33.20) with the function $c_n$ on the left hand side, i.e.,

$$\begin{aligned} c_n(t_1+t_2) &= c_1(t_1)c_n(t_2) \\ &\quad + \sum_{k=2}^{n-1} c_k(t_1) \sum_{\bar{u}_n\in U_{n,k}} B_{\bar{u}_n} \prod_{j=1}^{n-k+1} c_j(t_2)^{u_j} + c_n(t)c_1(t')^n \end{aligned} \tag{33.21}$$

for $t_1, t_2 \in G$. Note that if functions $c_n^1, c_n^2 : G \to \mathbb{K}$ satisfy (33.21) then the difference $f = c_n^2 - c_n^1$ satisfies the equation

$$f(t_1 + t_2) = c_1(t_1) f(t_2) + c_1(t_2)^n f(t_1) \quad \text{for } t_1, t_2 \in G,$$

which general solution (cf. the solution of the second equation of system (33.20)) is given by $f(t) = \lambda_n(c_1(t)^n - c_1(t))$ with arbitrary $\lambda_n \in \mathbb{K}$. If we show (cf. [14]) that $c_1(t) P_n\left(c_1(t); (\lambda_l)_{l \in |2, n-1|}\right)$ is a particular solution of (33.21), then all solutions of (33.21) must be given by

$$c_n(t) = \lambda_n(c_1(t)^n - c_1(t)) + c_1(t) P_n\left(c_1(t); (\lambda_l)_{l \in |2, n-1|}\right) \quad \text{for } t \in G.$$

To see that these functions give us one-parameter groups of (truncated) formal power series let us compare that description with the description of regular one-parameter group of formal power series. It is easy to see that instant exponential function $\mathbb{K} \ni t \mapsto e^t$ we have arbitrary exponential function $G \ni t \mapsto c_1(t)$. □

### 33.6.3 "Translation Equation's-Type" One-Parameter Groups of Formal Power Series

If we consider (33.7), which is an equivalent definition of one-parameter group of formal power series, for the first look it is a translation equation with the initial condition. Hence, we may ask whether we can give (maeby not in general) the description of its solutions, which is similar to regular solutions of "true" translation equation. It appears that the possible description is given by the following

**Theorem 33.9 (cf. [12]).** *Every one-parameter group*

$$\left(F^{[s]}(t, X)\right)_{t \in G}, \qquad F^{[s]}(t, X) = \sum_{k=1}^{s} c_k(t) X^k,$$

*with $c_1 : G \to \mathbb{K} \setminus \{0\}$, $c_k : G \to \mathbb{K}$ for $k \in |2, s|$, of formal power series such that either*

*(i) $c_1(G)$ is an infinite group*

*or*

*(ii) $c_1(G)$ and $\left(F^{[s]}(t, X)\right)_{t \in G}$ are finite groups*

*is given by*

$$F(t, X) = S^{-1}(c_1(t) S(X)) \quad \text{for } t \in G,$$

*where $S(X) \in \Gamma_1^s$ is arbitrary.*

### 33.6.4 One-Parameter Groups of Formal Power Series: Algebraic Approach

Previous sections contain all the results which can be proved, up to our knowledge, using differential equations and functional equations only. To attack the last case we will apply some algebraic tools. Unfortunately we are able to give only a characterization of one-parameter group in that remaining case without detailed formulas. This is caused by the fact that in our characterization we use iterative roots of formal power series (problem stil open in general, cf. [20,21,23–26,31,32]) which in addition need to commute with some known one-parameter groups of formal power series. Our characterization reads as follows.

For arbitrary exponential function $c_1 : G \to \mathbb{K}\setminus\{0\}$ with $c_1(G) = E_m$, $m \geq 2$ and for $t_0 \in G$ with $c_1(t_0)$ being a primitive root of order $m$ we have

$$G/\ker c_1 = \{lt_0 + \ker c_1 : 0 \leq l \leq m-1\}, \qquad mt_0 \in \ker c_1.$$

Moreover, $G \ni t = lt_0 + t'$ with unique $0 \leq l \leq m-1$ and $t' \in \ker c_1$.

**Theorem 33.10 (cf. [14, Proposition 1]).** *Let $(G,+)$ be a commutative group and let $c_1 : G \to \mathbb{K}\setminus\{0\}$ be an exponential function such that $c_1(G) = E_m$ with an integer $m \geq 2$. Fix $t_0 \in G$ such that $c_1(t_0)$ is a primitive root of 1 of order m. Finally, let s be a positive integer or $s = \infty$.*

*If*

$$\Theta_s : G \to \Gamma^s, \qquad \Theta_s(t)(X) = \sum_{k=1}^{s} c_k(t)X^k,$$

*with $c_k : G \to \mathbb{K}$ for $k \in |2,s|$, is a one-parameter group of formal power series, then*

$$\Theta_s|_{\ker c_1} = \overline{\Theta}_s : \ker c_1 \to \Gamma_1^s, \qquad \overline{\Theta}_s(t)(X) = X + \sum_{k=2}^{s} \overline{c}_k(t)X^k$$

*is a one-parameter group of formal power series, and there exists a formal power series*

$$P(X) = \Theta_s(t_0)(X) = \sum_{k=1}^{s} d_k X^k$$

*($d_1 = c_1(t_0)$ is a primitive root of 1 of order m) such that*

$$\left(P \circ \overline{\Theta}_s(t)\right)(X) = \left(\overline{\Theta}_s(t) \circ P\right)(X) \quad \text{for } t \in \ker c_1, \tag{33.22}$$

$$P^m(X) = \overline{\Theta}_s(mt_0)(X), \tag{33.23}$$

$$\Theta_s(lt_0 + t')(X) = \left(P^l \circ \overline{\Theta}_s(t')\right)(X) \;\; 0 \leq l \leq m-1 \quad \text{for } t' \in \ker c_1. \tag{33.24}$$

*Conversely, for every one-parameter group of formal power series*

$$\overline{\Theta}_s : \ker c_1 \to \Gamma^s, \qquad \overline{\Theta}_s(t)(X) = X + \sum_{k=2}^{s} \overline{c}_k(t)X^k,$$

*and for arbitrary*

$$P(X) = \sum_{k=1}^{s} d_k X^k$$

*with* $d_1$ *being a primitive root of* 1 *of order m, such that conditions* (33.22) *and* (33.23) *are satisfied, the formula* (33.24) *properly defines a function*

$$\Theta_s : G \to \Gamma^s, \qquad \Theta_s(t)(X) = \sum_{k=1}^{s} c_k(t) X^k$$

*and* $\Theta_s$ *is a one-parameter group of formal power series with* $c_1(G) = E_m$.

For the convenience of the reader we give the proof of that theorem.

*Proof.* If

$$\Theta_s : G \to \Gamma^s, \qquad \Theta_s(t)(X) = \sum_{k=1}^{s} c_k(t) X^k$$

is a one-parameter group of formal power series, then

$$\Theta_s|_{\ker c_1} = \overline{\Theta}_s : \ker c_1 \to \Gamma^s, \qquad \overline{\Theta}_s(t)(X) = X + \sum_{k=2}^{s} \overline{c}_k(t) X^k$$

is a one-parameter group of formal power series. Put $P(X) = \Theta_s(t_0)(X)$. Since $mt_0 \in \ker c_1$, we get

$$P^m(X) = \Theta_s(t_0)^m(X) = \Theta_s(mt_0)(X) = \overline{\Theta}_s(mt_0)(X).$$

Commutativity of $G$ implies now

$$\begin{aligned} \left(P \circ \overline{\Theta}_s(t)\right)(X) &= (\Theta_s(t_0) \circ \Theta_s(t)) = \Theta_s(t_0 + t)(X) = \Theta_s(t + t_0)(X) \\ &= (\Theta_s(t) \circ \Theta_s(t_0))(X) = \left(\overline{\Theta}_s(t) \circ P\right)(X) \quad \text{for } t \in \ker c_1. \end{aligned}$$

Finally

$$\begin{aligned} \Theta_s(lt_0 + t')(X) &= (\Theta_s(lt_0) \circ \Theta_s(t'))(X) \\ &= \left(\Theta_s(t_0)^l \circ \overline{\Theta}_s(t')\right)(X) = \left(P^l \circ \overline{\Theta}_s(t')\right)(X) \end{aligned}$$

for every $0 \le l \le m-1$ and $t' \in \ker c_1$.

Now let us fix a one-parameter group of formal power series

$$\overline{\Theta}_s : \ker c_1 \to \Gamma_1^s, \qquad \overline{\Theta}_s(t)(X) = X + \sum_{k=2}^{s} \overline{c}_k(t) X^k$$

and formal power series

$$P(X)=\sum_{k=1}^{s} d_k X^k,$$

where $d_1$ is a primitive root of 1 of the order $m$, which satisfy conditions (33.22) and (33.23). The formula (33.24) properly defines a function $\Theta_s : G \to \Gamma^s$. We only need to show that $\Theta_s$ is then a one-parameter group of formal power series. Indeed, from (33.22) it follows that

$$\left(P^n \circ \overline{\Theta}_s(t)\right)(X) = \left(\overline{\Theta}_s(t) \circ P^n\right)(X) \quad \text{for } t \in \ker c_1, \tag{33.25}$$

for every integer $n \geq 1$. Then, for $l_1 t_0 + t_1, l_2 t_0 + t_2 \in G$, where $0 \leq l_1, l_2 \leq m-1$ and $t_1, t_2 \in \ker c_1$, using (33.25) we get

$$\begin{aligned}
\left(\left(P^{l_1} \circ \overline{\Theta}_s(t_1)\right) \circ \left(P^{l_2} \circ \overline{\Theta}_s(t_2)\right)\right)(X) &= \left(P^{l_1+l_2} \circ \overline{\Theta}_s(t_1+t_2)\right)(X) \\
&= \left(P^{l_1+l_2-[(l_1+l_2)/m]m+[(l_1+l_2)/m]m} \circ \overline{\Theta}_s(t_1+t_2)\right)(X) \\
&= \left(P^{l_1+l_2-\left[\frac{l_1+l_2}{m}\right]m} \circ \overline{\Theta}_s\left(\left[\frac{l_1+l_2}{m}\right] m t_0 + t_1 + t_2\right)\right)(X) \\
&= \Theta_s\left(\left(l_1+l_2-\left[\frac{l_1+l_2}{m}\right] m\right) t_0 + \left[\frac{l_1+l_2}{m}\right] m t_0 + t_1 + t_2\right)(X) \\
&= \Theta_s\left((l_1 t_0 + t_1) + (l_2 t_0 + t_2)\right)(X).
\end{aligned}$$

□

## 33.7 Final Remarks and Observations

Note first that in the remaining case when $c_1(G)$ is a finite set with infinite $(F(t,X))_{t\in G}$, it is possible to give the detailed description of some one-parameter groups of formal power series. Indeed, if $(F(t,X))_{t\in G}$ is infinite then the group $G$ must be infinite. The detailed description is possible in the subcase when $c_1(G) = E_m$ with some integer $m \geq 2$ and the group $G$ contains cyclic subgroup of order $m$. For example, the group $(\mathbb{R} \setminus \{0\}, \cdot)$ (then a sentence "exponential" should be replaced by "multiplicative") admits a multiplicative function $c_1 = \operatorname{sgn}$ with $c_1(\mathbb{R} \setminus \{0\}) = \{-1, 1\}$. Moreover, the group $(\mathbb{R} \setminus \{0\}, \cdot)$ contains a cyclic subgroup $\{-1, 1\}$. Unfortunately, even then the description is complicated and uses semicanonical forms of formal power series. Therefore, we decided to omit it here.

**Problem 33.1.** It is possible to give detailed formulas for coefficient functions

$$G \ni t \mapsto c_k(t) \in \mathbb{K}$$

in the case, when $c_1(G)$ is a finite group of complex roots of 1 and $(F(t,X))_{t\in G}$ is infinite?

Finally, the considered here cases fit all possibilities in classical cases when $(G,+)$ is either the additive group of reals or the additive group of complex numbers (then every exponential function is either constant and hence equal to 1 or it has infinite image). The mentioned above subcase of an infinite group having a cyclic group of order $m$ allows us to describe in addition all one-parameter groups of formal power series for $(G,+)$ being $(\mathbb{R}\setminus\{0\},\cdot)$. This follows from the fact that for arbitrary multiplicative function $c_1 : \mathbb{R}\setminus\{0\} \to \mathbb{R}\setminus\{0\}$ we have either $c_1 = 1$ or $c_1 = \text{sgn}$ or $c_1(\mathbb{R}\setminus\{0\})$ is infinite. Thus, we are able to determine all homomorphisms of $(\mathbb{R}\setminus\{0\},\cdot)$ into $L_s^1$ and $L_\infty^1$.

## References


1. Chudziak J., Drygaś P., Jabłoński W., Midura S.: On homomorphisms of the group $L_2^1$ into the groups $L_2^1$ and $L_3^1$. Demonstratio Math. **31**, 143–152 (1998)
2. Fripertinger, H., Reich, L.: The formal translation equation and formal cocycle equations for iteration groups of type I. Aequationes Math. **76**, 54–91 (2008)
3. Fripertinger, H., Reich, L.: The formal translation equation for iteration groups of type II. Aequationes Math. **79**, 111–156 (2010)
4. Gan X., Knox, N.: On Composition of Formal Power Series. Intern. J. Math. Math. Sci. **30(12)**, 761–770 (2002)
5. Gronau, D.: Two iterative functional equations for power series. Aequationes Math. **25**, 233–246 (1982)
6. Gronau, D.: Über die multiplikative Translationgleichung und Idempotente Potenzreihenvektoren. Aequationes Math. **28**, 312–320 (1985)
7. Henrici, P.: Applied and Computational Complex Analysis. Vol. I: Power Series - Integration - Conformal Mapping - Location of Zeros. John Wiley & Sons, New York–London–Sydney–Toronto (1974)
8. Jabłoński, W.: On existence of subsemigroups of the group $L_s^1$ and extensibility of homomorphisms into $L_s^1$. Aequationes Math. **74**, 219–225 (2007)
9. Jabłoński, W.: Homomorphisms of the group $L_s^1$ into the group $L_r^1$ for $r \leq 3$. Aequationes Math. **74**, 287–309 (2007)
10. Jabłoński, W., Reich, L.: On the solutions of the translation equation in rings of formal power series. Abh. Math. Sem. Univ. Hamburg **75**, 179–201 (2005)
11. Jabłoński, W., Reich, L.: On the form of homomorphisms into the differential group $L_s^1$ and their extensibility. Results Math. **47**, 61–68 (2005)
12. Jabłoński, W., Reich, L.: On the standard form of the solution of the translation equation in rings of formal power series. Math. Panon. **18**, 169–187 (2007)
13. Jabłoński, W., Reich, L.: On homomorphisms of an abelian group into the group of invertible formal power series. Publ. Math. Debrecen **73**, 25–47 (2008)
14. Jabłoński, W., Reich, L.: A new approach to the description of one-parameter groups of formal power series in one indeterminate (manuscript)
15. Johnson, W.P.: The Curious History of Faa di Bruno's Formula. Amer. Math. Monthly **109**, 217–234 (2002)
16. Lewis, D. C.: Formal power series transformation. Duke Math. J. **5**, 794–805 (1939)
17. Midura S., Wilczyński Z.: Sur les homomorphismes du groupe $(\mathbb{R},+)$ au groupe $L_s^1$ pour $s \leq 5$. Rocznik Nauk.–Dydak. WSP w Krakowie **13**, 241–258 (1993)
18. Moszner Z.: Sur un sous-groupe a un parametre du groupe $L_s^1$. Opuscula Math. **6**, 149–155 (1990)

19. Peschl, E., Reich, L.: Beispiel einer kontrahierrenden biholomorphen Abbildung, die in keine Liesche Gruppe biholomorpher Abbildungen einbettbar ist. Bayer. Akad. Wiss. Math.-Natur. Kl. Sitzungsber. **1971**, 81–92 (1972)
20. Praagman, C.: Roots and logarithms of automorphisms of complete local rings. Aequationes Math. **33**, 220–229 (1987)
21. Praagman, C.: Roots, iterations and logarithms of formal automorphisms. Aequationes Math. **33**, 251–259 (1987)
22. Reich, L.: Über die allgemeine Lösung der Translationgleichung in Potenzreihenringen. In: Second Zagreb–Graz Meeting of Mathematicians (Plitvice, 1980), Bericht **159**, 22 pp. Bericht 154–162, Forschungszentrum Graz, Graz (1981)
23. Reich, L.: On power series transformations in one indeterminate having iterative roots of a given order and with given multiplier. In: European Conference on Iteration Theory (Singapore 1991), pp. 210–216. World Sci. Publ., Singapore (1992)
24. Reich, L.: Über Gruppen von iterativen Wurzeln der formalen Potenzreihe $F(x) = x$. Results Math. **26**, 366–371 (1994)
25. Reich, L.: On iterative roots of formal power series $F(x) = x$. Various Publ. Ser. (Aarhus) **43**, 245–255 (1994)
26. Reich, L.: Iterative roots of formal power series: universal expression for the coefficients and analytic iteration. Grazer Math. Ber. **327**, 21–32 (1996)
27. Reich, L., Schwaiger, J.: Über analytische Iterierbatkeit formaler Potenzreihenvektoren. Österreich. Akad. Wiss. Math.-Natur. Kl. Sitzungsber. II **184**, 599–617 (1975)
28. Reich, L., Schwaiger, J.: Über einen Satz von Shl. Sternberg in der Theorie der analytischen Iterationen. Monatsh. Math. **83**, 207–221 (1977)
29. Roman, S.: The formula of Faa di Bruno. Amer. Math. Monthly **87**, 805–809 (1980)
30. Scheinberg, S.: Power series in one variable. J. Math. Anal. Appl. **31**, 321–333 (1970)
31. Schwaiger, J.: Roots of power series in one variable. Aequationes Math. **29**, 40–43 (1985)
32. Schwaiger, J.: On polynomials having different types of iterative roots. In: European Conference on Iteration Theory (Batschuns, 1989), pp. 313–323. World Sci. Publ., River Edge, NJ (1991)
33. Sternberg, S.: Infinite Lie groups and formal aspects of dynamical systems. J. Math. Mech. **10**, 451–474 (1961)

# Chapter 34
# On Some Problems Concerning a Sum Type Operator

**Hans-Heinrich Kairies**

*Dedicated to the Memory of S.M. Ulam*
*on the 100th anniversary of his birth*

**Abstract** Several open problems are presented that are connected with the sum type operator of the form:

$$F[\varphi](x) := \sum_{k=0}^{\infty} \frac{1}{2^k}\varphi(2^k x).$$

They concern the images and pre-images of the operator, its spectral properties, its maximal domain and possible extensions.

**Keywords** Sum type operators • Takagi and Weierstrass type and functions

**Mathematics Subject Classification (2000):** Primary: 47B38; Secondary: 26A27, 39B22

The sum type operator $F$, given by

$$F[\varphi](x) := \sum_{k=0}^{\infty} \frac{1}{2^k}\varphi(2^k x),$$

has been thoroughly discussed in the last years, see [1–12]. Nevertheless, there remained some open problems. We state here a few of them which are connected to six aspects of the operator $F$. Some of them are rather projects than particular problems.

H.-H. Kairies (✉)
Institut für Mathematik, Technische Universität Clausthal, Erzstraße 1, D-38678 Clausthal-Zellerfeld, Germany
e-mail: kairies@math.tu-clausthal.de

Th.M. Rassias and J. Brzdęk (eds.), *Functional Equations in Mathematical Analysis*, Springer Optimization and Its Applications 52, DOI 10.1007/978-1-4614-0055-4_34,

## 34.1 Images and Pre-images of $F$

To keep things simple, let us consider $F$ just on the domain

$$\mathscr{C} := \{\varphi : \mathbb{R} \to \mathbb{R}\ ;\ \varphi \text{ continuous and 1-periodic}\}.$$

Then $F$ is a Banach space automorphism. The functions $w$ and $t$, given by

$$w(x) = \cos 2\pi x, \qquad t(x) = \operatorname{dist}(x, \mathbb{Z}),$$

belong to the set $\mathscr{C}$ and generate the prominent Weierstrass resp. Takagi functions $F[w]$ and $F[t]$, both continuous and nowhere differentiable (cnd). Let

$$\mathscr{N} := \{f : \mathbb{R} \to \mathbb{R};\ f \text{ cnd and 1-periodic}\}.$$

**Problem 34.1.** Characterize $\mathscr{M} := F^{-1}[\mathscr{N}] = \{\varphi \in \mathscr{C}\ ;\ F[\varphi] \in \mathscr{N}\}$.

There are some subsets of $\mathscr{M}$ which are interesting by themselves.

**Problem 34.2.** Characterize $\mathscr{M}_1 := F^{-1}[\mathscr{N}] \cap \mathscr{N} = \{\varphi \in \mathscr{M}\ ;\ \varphi\ \textit{cnd}\}$.

**Problem 34.3.** Characterize $\mathscr{M}_2 := \{\varphi \in \mathscr{M}\ ;\ \varphi\ \textit{polygonal}\}$.

**Problem 34.4.** Characterize $\mathscr{M}_3 := \{\varphi \in \mathscr{M}\ ;\ \varphi\ \textit{convex on}\ [0,1]\}$.

The above mentioned $t$ belongs to $\mathscr{M}_2$ and to $\mathscr{M}_3$. Some partial answers to Problems 34.1–34.4 are contained in [3, 4, 11].

In [4] we discussed $\varphi_v \in \mathscr{C}$, which are generated by a real sequence $v = (v_n)$:

$$\varphi_v(x) := \sum_{n=0}^{\infty} \frac{v_n}{2^n} \cos(2\pi 2^n x),$$

and found the following sufficient conditions for $v$ to guarantee that $\varphi_v \in \mathscr{M}_1$.

**Theorem 34.1 ([4, Theorem 2]).** *Let $(v_n)$ be a bounded sequence of positive real numbers which does not converge to zero. Then $\varphi_v \in \mathscr{N}$ and $F[\varphi_v] \in \mathscr{N}$.*

**Problem 34.5.** Characterize $\mathscr{V} := \{(v_n)\ ;\ \varphi_v \in \mathscr{M}_1\}$.

The paper [4] contains also some partial information regarding the next problem.

**Problem 34.6.** Characterize $\mathscr{M}_4 := \{\varphi \in \mathscr{C}\ ;\ F^{(m)}[\varphi] \in \mathscr{N}\ \textit{for every}\ m \in \mathbb{N}\}$.

## 34.2 Spectral Properties of $F$

Spectral properties of $F$ have been investigated in [1, 5–8]. In [1, 7], the operator $F$ has been considered on sixteen different domains, all of them vector spaces. In [1], joint work with Baron, all the according eigenspaces have been described. One arithmetical problem remained open.

**Problem 34.7.** Let $K_\alpha := \alpha \cdot 2^{\mathbb{Z}} + \mathbb{D}$ ($\mathbb{Z}$ denotes the set of integers and $\mathbb{D}$ the dyadic rationals). Characterize $S := \{\alpha \in \mathbb{R}\ ;\ K_\alpha = K_{-\alpha}\}$.

This problem was already stated in Ann. Acad. Paed. Cracoviensis Studia Math. **5** (2006), p. 118. More information is provided in [1, p. 258].

In [7], all the according continuous and residual spectra have been calculated except in two cases.

**Problem 34.8.** *Let $\mathscr{C}$ be as in Section 34.1 and $\mathscr{C}' := \{\varphi \in \mathscr{C}\ ;\ \varphi$ even$\}$. Compute the continuous and the residual spectra of the Banach space automorphisms $F : \mathscr{C} \to \mathscr{C}$ and $F : \mathscr{C}' \to \mathscr{C}'$.*

## 34.3 The Maximal Domain of $F$

The set

$$\mathscr{D} := \left\{ \varphi : \mathbb{R} \to \mathbb{R}\ ;\ \sum_{k=0}^{\infty} \frac{1}{2^k}\, \varphi(2^k x) \text{ converges for every } x \in \mathbb{R} \right\}$$

is the maximal possible domain of our operator with respect to the arguments $\varphi : \mathbb{R} \to \mathbb{R}$. $\mathscr{D}$ is a real vector space containing the following two subsets:

$$\mathscr{D}_A := \{\varphi : \mathbb{R} \to \mathbb{R}\ ;\ \exists\, a : \mathbb{R} \to \mathbb{R},\ a \text{ additive} : \varphi(x) = a\,(\log|x|)\},$$

$$\mathscr{D}_B := \{\varphi : \mathbb{R} \to \mathbb{R}\ ; \exists\, b : J \to \mathbb{R}\ :\ \varphi = b \text{ on } J,\ \varphi(x) = \varphi(2x) \text{ otherwise}\},$$

where

$$J := (-2,-1] \cup [1,2).$$

Both subsets are far from exhausting the full set $\mathscr{D}$. This can be seen from [8], joint work with Volkmann, where the set $\mathscr{D}$ is characterized in the following way:

**Theorem 34.2 ([8, Satz 2]).** *Equivalent are*

(I) $\varphi \in \mathscr{D}$;

(II) $\forall t \in J \exists (a^t_\mu)_{\mu \in \mathbb{Z}} : \sum_{k=0}^{\infty} a^t_k \in \mathbb{R}$ *and* $\forall x \in \mathbb{R}\backslash\{0\} : \varphi(x) = \varphi(2^{m(x)}\tau(x)) = 2^{m(x)} a^{\tau(x)}_{m(x)}$.

Condition (II) can be replaced by the equivalent statement

(III) $\exists a_\mu : J \to \mathbb{R}\ (\mu \in \mathbb{Z})\ \forall \xi \in J : \sum_{k=0}^{\infty} a_k(\xi) \in \mathbb{R}$ *and* $\forall \xi \in J, m \in \mathbb{Z} : \varphi(2^m \xi) = 2^m a_m(\xi)$, which allows a straightforward comparision of $\mathscr{D}$ and $\mathscr{D}_B$. In fact, $\mathscr{D}_B$ is the eigenspace of $F:\mathscr{D}\to F[\mathscr{D}]$ with respect to the eigenvalue 2 and can be rewritten as

$$\mathscr{D}_{\mathscr{B}} = \{\varphi:\mathbb{R}\to\mathbb{R}\,;\ \forall x\in\mathbb{R}\ :\ \varphi(x)=\varphi(2x)\}.$$

This last representation suggests the following

**Problem 34.9.** Characterize $\mathscr{D}$ as solution set of a system of (iterative) functional equations.

Connected with the above mentioned [8, Satz 2] is

**Problem 34.10.** Which functions $a_\mu : J\to\mathbb{R}$ $(\mu\in\mathbb{Z})$ generate $\varphi\in\mathscr{A}$ ?

$\mathscr{A}\subset\mathscr{D}$ can be any relevant function set, but we have especially in mind sets like

$$\mathscr{A}_1=\mathscr{D}_{\kappa,\rho}\cap\mathscr{N}^*\quad\text{ or }\quad \mathscr{A}_2=F^{-1}[\mathscr{N}^*].$$

Here we used the notation from [7] for the sixteen subspaces $\mathscr{D}_{\kappa,\rho}\,(1\le\kappa,\rho\le 4)$ of $\mathscr{D}$ already mentioned in Sect. 34.2. $\mathscr{N}^*$ may be any suitable set of cnd functions, whose elements must not necessarily be 1-periodic or bounded (as the elements of $\mathscr{N}$ are). Problem 34.10 with $\mathscr{A}=\mathscr{A}_2$ is closely related to the problems from Sect. 34.1.

## 34.4 Characterizations of $F[\varphi]$

Assume that $\varphi\in\mathscr{B}'$, i.e., $\varphi$ is bounded, 1-periodic and even. Then $F[\varphi]$ satisfies the following system of seven functional equations:

$$f(x)-2f\left(\frac{x}{2}\right)=-2\varphi\left(\frac{x}{2}\right),\tag{34.1}$$

$$f(x)-2f\left(\frac{x+1}{2}\right)=-2\varphi\left(\frac{x+1}{2}\right),\tag{34.2}$$

$$f\left(\frac{x}{2}\right)-f\left(\frac{x+1}{2}\right)=\varphi\left(\frac{x}{2}\right)-\varphi\left(\frac{x+1}{2}\right),\tag{34.3}$$

$$f(x)-f\left(\frac{x}{2}\right)-f\left(\frac{x+1}{2}\right)=-\varphi\left(\frac{x}{2}\right)-\varphi\left(\frac{x+1}{2}\right),\tag{34.4}$$

$$f(x+1)-f(x)=0,\tag{34.5}$$

$$f(-x)-f(x)=0,\tag{34.6}$$

$$f(1-x)-f(x)=0.\tag{34.7}$$

The paper [2] contains a thorough investigation of this and the associated homogeneous system. Moreover, all possible characterizations of $F[\varphi]$ by subsystems of (34.1)–(34.7) in the frame of boundedness and of continuity are given:

**Theorem 34.3 ([2, Theorem 2]).** *Assume that $f : \mathbb{R} \to \mathbb{R}$ is bounded and $\varphi \in \mathscr{B}'$. Then $f = F[\varphi]$ iff either (34.1) or (34.2) or (34.3,34.4) holds. Any "superset" of either (34.1) or (34.2) or (34.3,34.4) is characteristic as well. No other "subset" of (34.1)–(34.7) is characteristic.*

**Theorem 34.4 ([2, Theorem 3]).** *Assume that $f : \mathbb{R} \to \mathbb{R}$ is continuous and $\varphi \in \mathscr{B}'$. Then $f = F[\varphi]$ iff either (34.1,34.2) or (34.1,34.3) or (34.1,34.4) or (34.1,34.5) or (34.1,34.7) or (34.2,34.3) or (34.2,34.4) or (34.2,34.5) or (34.2,34.6) or (34.2,34.7) or (34.3,34.4) holds. Any "superset" of either (34.1,34.2) or ... or (34.3,34.4) is characteristic as well. No other "subset" of (34.1)–(34.7) is characteristic.*

In the language of [7], we have $\varphi \in \mathscr{D}_{3,4}$ in the above Theorems. But there are in fact sixteen possible cases $\varphi \in \mathscr{D}_{\kappa,\rho}$ – in some of them only a "subset" of (34.5), (34.6), (34.7) is available – and many more possible regularity assumptions on $f$. This leads to

**Problem 34.11.** Give a list of all possible characterizations of $F[\varphi], \varphi \in \mathscr{D}_{\kappa,\rho}$, under suitable regularity assumptions on $f$.

To find out those properties of $\varphi$ resp. $f$, which give interesting modifications (extensions) of Theorems 34.3 and 34.4 above, is left to the reader. Some partial results are already given in [2].

## 34.5 A Two Parameter Extension of *F*

Let $\alpha \in (0,1)$, $\beta \in (0,\infty)$ and

$$\mathscr{B}_0 := \{\varphi : \mathbb{R} \to \mathbb{R}\ ;\ \varphi \text{ bounded}\}.$$

Then $F_{\alpha,\beta} : \mathscr{B}_0 \to \mathscr{B}_0$,

$$F_{\alpha,\beta}[\varphi](x) := \sum_{k=0}^{\infty} \alpha^k \varphi(\beta^k x),$$

provides a two parameter extension of $F$:

$$F = F_{\frac{1}{2},2}.$$

In the paper [5], $F_{\alpha,\beta}$ is defined on the Banach space $\mathscr{B}_0$ and on five closed subspaces $\mathscr{B}_1$ to $\mathscr{B}_5$. It is proved that $F_{\alpha,\beta} : \mathscr{B}_0 \to \mathscr{B}_0$ is a Banach space automorphism; the operator norms of $F_{\alpha,\beta}$ and $F^{-1}_{\alpha,\beta}$ are computed and the point spectra and eigenspaces of $F_{\alpha,\beta}$ and $F^{-1}_{\alpha,\beta}$ are given. Similar results for

$$F_{\alpha,\beta} : \mathscr{B}_n \to \mathscr{B}_n \qquad (1 \le n \le 5)$$

are obtained, but only for $\beta \in \mathbb{N}\backslash\{1\}$. Moreover, there is no complete description of the eigenspaces in the spirit of [1] and no information on the continuous and residual spectra. The paper [5] closes with stating several problems concerning images and pre-images of $F_{\alpha,\beta}$, as we did in Sect. 34.1 for $F$. Instead of formulating $\alpha,\beta$-modifications of our Problems 34.1–34.11, we rather present a final project.

**Problem 34.12.** Discuss in connection with $F_{\alpha,\beta}$:

- Images and pre-images;
- Eigenspaces, continuous and residual spectra;
- Fourier series;
- Maximal domains;
- Characterizations;
- Identifications.

Fourier series properties could extend results of [3, 4]. The last point "Identifications" takes care of the fact that particular properties of $F_{\alpha,\beta}[\varphi]$ may essentially depend on $\alpha$ and $\beta$.

*Example 34.1. Identify all $\alpha,\beta$ such that $F_{\alpha,\beta}[w]$ is cnd.*
Recall that

$$F_{\alpha,\beta}[w](x) = \sum_{k=0}^{\infty} \alpha^k \cos(2\pi\beta^k s).$$

A famous result of Hardy gives a concise solution:

$$F_{\alpha,\beta}[w] \quad \text{is cnd if and only if} \quad \alpha\beta \ge 1.$$

## 34.6 Other Extensions of $F$

Let

$$\mathscr{B} := \{\varphi : \mathbb{R} \to \mathbb{R}\,;\ \varphi \text{ bounded and 1-periodic}\},$$
$$\Delta(x) := 2\delta(x) = 2\text{dist}(x, \mathbb{Z})$$

and denote by $\Delta^{(n)}$ the $n$-th iterate of $\Delta$. Starting from the observation that

$$T(x) = \sum_{k=0}^{\infty} 2^{-k}\delta(2^k x) = \sum_{n=1}^{\infty} 2^{-n}\Delta^{(n)}(x),$$

Hata/Yamaguti (1984) introduced the Takagi class of functions

$$F^{\Delta}[(\alpha_n)] := \sum_{n=1}^{\infty} \alpha_n \Delta^{(n)}. \tag{34.8}$$

$F^{\Delta} : l^1 \to \mathscr{B}$ is linear and continuous. Kôno [9] proved the beautiful characterization

$$F^{\Delta}[(\alpha_n)] \quad \text{cnd iff} \quad \limsup 2^n |\alpha_n| > 0. \tag{34.9}$$

Now define for fixed $\Phi \in \mathscr{B}$ the operator $F^{\Phi}$ by

$$F^{\Phi}[(\alpha_n)] := \sum_{n=1}^{\infty} \alpha_n \Phi^{(n)}. \tag{34.10}$$

Again, $F^{\Phi} : l^1 \to \mathscr{B}$ is linear and continuous.

Formula (34.10) can be seen from another point of view: For fixed $(\alpha_n) : \mathbb{N} \to \mathbb{R}$ let

$$F^{(\alpha_n)}[\varphi] := \sum_{n=1}^{\infty} \alpha_n \varphi^{(n)}. \tag{34.11}$$

In case $(\alpha_n) \in l^1$, we have $F^{(\alpha_n)} : \mathscr{B} \to \mathscr{B}$, however $F^{(\alpha_n)}$ is no more linear.

Kôno-type characterizations for particular $\Phi \in \mathscr{B}$ might be of interest.

**Problem 34.13.** $F^{\Phi}[(\alpha_n)]$ cnd (or absolutely continous or smooth or Lipschitz or ...) iff $(\alpha_n)$ satisfies ....

**Problem 34.14.** Find all $\Phi \in \mathscr{B}$ such that:

$$F^{\Phi}[(\alpha_n)] \quad \text{cnd iff} \quad \limsup 2^n (\alpha_n) > 0.$$

There are many open problems concerning images and pre-images of $F^{\Phi}$ resp. $F^{(\alpha_n)}$ (as for $F$ in Sect. 34.1) and many open problems concerning characterizations of particular $F^{\Phi}[(\alpha_n)]$ resp. $F^{(\alpha_n)}[\varphi]$ (as for $F[\varphi]$ in Sect. 34.4).

Let us close with the remark that very recently Tabor/Tabor [12] and Makó/Páles [10] investigated thoroughly particular operators of the form (34.8) and (34.10) in connection with approximate mid-convexity and other modifications of convexity.

## References

1. Baron, K., Kairies, H.-H.: Characterizations of eigenspaces by functional equations. Aequationes Math. **68**, 248–259 (2004)
2. Kairies, H.-H.: Functional equations and characterizations of classes of particular function series. Prace Matem. **17**, 151–164 (2000)

3. Kairies, H.-H.: On a Banach space automorphism and its connections to functional equations and continuous nowhere differentiable functions. Ann. Acad. Paed. Cracoviensis Studia Mathematica **1**, 39–48 (2001)
4. Kairies, H.-H.: On an operator preserving lacunarity of Fourier series. Aequationes Math. **62**, 1–8 (2001)
5. Kairies, H.-H.: Properties of an operator acting on the space of bounded real functions and certain subspaces. In: (eds. Z. Daróczy, Z. Palés) Functional Equations - Results and Advances, pp. 175–186, Kluwer Acad. Publishers, Boston-Dordrecht-London (2001)
6. Kairies, H.-H.: Spectra of certain operators and iterative functional equations. Ann. Acad. Paed. Cracoviensis Studia Mathematica **2**, 13–22 (2002)
7. Kairies, H.-H.: On continuous and residual spectra of operators connected with iterative functional equations. Ann. Acad. Paed. Cracoviensis Studia Mathematica **5**, 52–57 (2006)
8. Kairies, H.-H., Volkmann, P.: Ein Vektorraumisomorphismus vom Summentyp. Seminar LV **13** (2002). http://www.mathematik.uni-karlsruhe.de/~semlv
9. Kôno, N.: On generalized Takagi functions. Acta Math. Hungar. **49**, 315–324 (1987)
10. Makó, J., Páles, Z.: Approximate convexity of Takagi type functions. J. Math. Anal. Appl. **369**, 545–554 (2010)
11. Mikolás, M.: Construction des familles de fonctions partout continues non dérivables. Acta Sci. Math. Szeged **17**, 49–62 (1956)
12. Tabor, Ja., Tabor, Jó.: Takagi functions and approximate midconvexity. J. Math. Anal. Appl. **356**, 729–737 (2009)

# Chapter 35
# Priors on the Space of Unimodal Probability Measures

**George Kouvaras and George Kokolakis**

*Dedicated to the Memory of S.M. Ulam*
*on the 100th anniversary of his birth*

**Abstract** Construction of unimodal random probability measures on finite dimensional Euclidean space is considered. The approach based on Bayesian nonparametric models and Convexity Theory. Specifically, the proposed model makes use of the special properties of convex sets and Choquet's theorem. As a result, we get random probability measures that admit derivatives almost everywhere in $\Re^d$.

**Keywords** Univariate and multivariate unimodality • Convexity • Choquet's theorem • Random probability measure • Dirichlet process prior

**Mathematics Subject Classification (2000):** Primary 62C10, 62G05

## 35.1 Introduction

We present a Bayesian framework with prior distributions concentrated on the space of unimodal probability measures which are defined on finite dimensional Euclidean space. One of the simplest and most widely used nonparametric Bayesian prior distributions is the Dirichlet process (DP) prior, Ferguson (1973) [7] having weak support on an entire space of probability measures. The DP is a distribution over distributions. Thus, it can be used as a prior over distributions of random variables. Models in which the DP is used as a prior over distributions of the parameters are referred to as mixtures of Dirichlet process (DPM) models, Lo (1984) [13]. Nonparametric priorsare extensively discussed in Walker et al. (1999) [16], Hjort

G. Kouvaras (✉) • G. Kokolakis
Department of Mathematics, National Technical University of Athens,
Zografou Campus, GR-15780, Greece
e-mail: gkouv@math.ntua.gr; kokolakis@math.ntua.gr

Th.M. Rassias and J. Brzdęk (eds.), *Functional Equations in Mathematical Analysis*,
Springer Optimization and Its Applications 52, DOI 10.1007/978-1-4614-0055-4_35,

(2003) [8] and Ghosh and Ramamoorthi (2003) [14]. The present paper uses a mixture of Dirichlet process on the space of probability distributions on $\Re^d$.

The distribution function $F(x) = Pr(X \leq x)$ of a real-valued random variable $X$ is unimodal with mode (or vertex) at $m$ if $F(x)$ is convex for $x < m$ and concave for $x > m$. A consequence of the above definition is the following: if $F$ is unimodal about $m$, then apart from a possible mass at $m$, $F$ is absolutely continuous. The main problem with multidimensional unimodality is that there is not a unique extension of the notion of univariate unimodality in $\Re^d$, and different definitions of multivariate unimodality actually imply different unimodality structure in $\Re$. In this paper we analyse the class of block unimodal distributions about a given mode.

The remainder of this article is organized as follows. In Sect. 2 we summarize some basic properties of the DP and DPM priors. Section 3 is devoted to the illustration of the properties of block unimodal distributions and the Choquet-type representation theorem which enables us to write a block unimodal distribution, as a generalized mixture of uniform distributions on rectangles in $\Re^d$ containing the mode. In this sense, the Choquet's theorem can be considered as one of the possible generalizations in $\Re^d$ of Khintchin's representation theorem [1] and [5]. Our results are described in Sect. 4 and some possible issues for future research are mentioned.

## 35.2 Dirichlet Process and Mixtures of Dirichlet Process Priors

Random probability distributions play an important role in the theory of statistical inference, and in particular in nonparametric inference. The DP is one of the most prominent random probability measures due to its richness, computational ease, and interpretability.

The DP was introduced by Ferguson (1973) [7] as a prior on the space of probability measures on a given measurable space $(\mathscr{X}, \mathscr{F})$.

**Definition 35.1.** Let $M > 0$ and $G_0$ be a probability measure on $(\mathscr{X}, \mathscr{F})$. A DP on $(\mathscr{X}, \mathscr{F})$ with parameters $(M, G_0)$ is a random probability measure $G$ which assigns a number $G(A)$ to every $A \in \mathscr{F}$ such that:

- $G(A)$ is a measurable $[0,1]$-valued random variable;
- each realization of $G$ is a probability measure on $(\mathscr{X}, \mathscr{F})$;
- for each measurable finite partition $\{A_1, \ldots, A_k\}$ of $\mathscr{X}$ the joint distribution of the vector $(G(A_1), \ldots, G(A_k))$ is the Dirichlet distribution with parameters $(MG_0(A_1), \ldots, MG_0(A_k))$.

Thus, the DP is defined by two parameters, the base distribution $G_0$ and the concentration parameter $M$. The base distribution can be interpreted as the prior guess for the random distribution function and the concentration parameter as expressing the strength of the belief in the prior.

Most of the basic properties of DP arise as an extension of the properties of the Dirichlet distribution. Some useful properties of the DP are the following:

- $E(G(A)) = G_0(A)$, for every $A \in \mathscr{F}$;

- $Var(G(A)) = \frac{G_0(A)(1+G_0(A))}{1+M}$, for every $A \in \mathscr{F}$.
- If $X_1,\ldots,X_n \overset{\text{iid}}{\sim} G$ and $G|M, G_0 \sim DP(MG_0)$, then

$$G|X_1,\ldots,X_n,M,G_0 \sim DP(MG_0 + \sum_{i=1}^{n} \delta_{X_i}).$$

The major drawback of the DP is that it selects discrete distributions with probability one, Blackwell (1973) [2]. Since the posterior distribution of a $DP(MG_0)$ is also a DP with parameter $(MG_0 + \sum_{i=1}^{n} \delta_{X_i})$, we can never obtain an absolutely continuous one, no matter what the sampling distribution and the sample size are.

Since the DP selects discrete measures, this is clearly unsuitable to be used as a prior for absolutely continuous and thus for unimodal distributions. Lo (1984) [13] developed a useful construction of priors for densities. Let $\mathscr{Y}$ be the parameter set, typically a finite dimensional Euclidean space. Under the DPM setting, the generic form of the random mixture density is given by

$$f(\cdot;G) = \int k(\cdot;y)\mathrm{d}G(y), \quad y \in \mathscr{Y},$$

where $k(\cdot;y)$ is a parametric kernel (with parameter vector $y$), and the random mixing distribution $G$ is assigned a DP prior. This gives rise to the following model for observations $X_1,\ldots,X_n$ distributed according to:

$$X_i|Y_i \overset{\text{ind}}{\sim} k(x_i|y_i), \quad y_i \in \mathscr{Y},$$

$$y_i|G \overset{\text{iid}}{\sim} G,$$

$$G \sim \pi,$$

for $i = 1,\ldots,n$. Note that the latent data $Y = \{Y_1,\ldots,Y_n\}$ are unobserved. A convenient choice for $\pi$ is the Dirichlet process prior, for which there exist closed form expressions for various quantities of interest. The specification of the base distribution for the DP, which corresponds to the prior on the component parameters for infinite mixture models, is often guided by mathematical and practical convenience. For DPM models, the use of conjugate priors makes the analysis much more tractable; however, conjugate priors might sometimes fail to represent the prior beliefs.

Most of the work with DPM is based on normal kernels, either univariate or multivariate. However, there are several applications where different choices of kernels are more appropriate. Relevant examples include models for unimodal densities on the real line (cf. Brunner and Lo (1989) [3]; Brunner (1992) [4]). In this paper, we study the utility of DPM of uniform distributions. This class of DPM models emerges as a natural candidate for unimodal distributions.

## 35.3 Mixture Representations of Unimodal Probability Measures

Let us suppose, without loss of generality, that the mode $m = 0$. For univariate distributions there exists a well known representation theorem due to Khinchin (cf. [6], p.158) that refers to the classical univariate unimodality.

**Theorem 35.1.** *A real valued random variable $X$ has a unimodal density at 0 if and only if it is a product of two independent random variables $U$ and $Y$, with $U$ uniformly distributed on $(0,1)$ and $Y$ having an arbitrary distribution.*

This can be expressed in the following equivalent form (cf. Shepp (1962) [15], Brunner (1992) [4] and Kokolakis and Kouvaras (2007) [9]).

**Theorem 35.2.** *The c.d.f. $F$ is convex on the negative real line and concave on the positive real line, if and only if there exists a distribution function $G$ on $\mathfrak{R}$ such that $F$ admits the representation:*

$$F(x) = G(x) + xf(x),$$

*for all $x$ points of continuity of $G$ where $f$ is the density of $F$ on $\mathfrak{R} - \{0\}$.*

In the real line, Khinchin's representation theorem allows to consider univariate unimodal distributions as generalized mixtures of uniform distributions in intervals. This is the starting point for several extensions of unimodality in $\mathfrak{R}^d$, $d > 1$. Among the various extensions proposed, the most popular one seems to be the star-unimodality. A set $S \subseteq \mathfrak{R}^d$ is said to be star-shaped about $\mathbf{0} \in S$ if for all $s \in S$ the line connecting $s$ to $\mathbf{0}$ is contained in $S$.

**Definition 35.2.** A distribution on $\mathfrak{R}^d$ is star-unimodal about $\mathbf{0}$ if it belongs to the closed convex hull of the class of all the uniform distributions on sets in $\mathfrak{R}^d$ which are star-shaped subsets about $\mathbf{0}$, where the closure is meant with respect to weak convergence.

A special kind of a star-shaped set about $\mathbf{0}$ is a rectangle containing $\mathbf{0}$. This leads to the following definition.

**Definition 35.3.** A distribution on $\mathfrak{R}^d$ is block unimodal about $\mathbf{0}$ if it belongs to the closed convex hull of the set of all uniform distributions on rectangles in $\mathfrak{R}^d$ containing $\mathbf{0}$ and having edges parallel to the axes.

Obviously, block unimodality implies star unimodality but the converse is not always valid. However, star unimodality is not preserved under marginalisation, i.e. star unimodal multivariate distributions do not necessarily have star unimodal marginals. On the other hand, the class of block unimodal distributions is closed under marginalisation.

According to Shepp (1962) [15] one has the following Theorem.

**Theorem 35.3.** *The c.d.f. $F$ is block unimodal about $\mathbf{0}$ if and only if there exists a random vector $X = (X_1,\ldots,X_n)$ with c.d.f. $F$, such that $X = (Y_1U_1,\ldots,Y_nU_n)$, where $U = (U_1,\ldots,U_n)$ is uniformly distributed on the unit cube $(0,1)^d$ and $Y$ is independent of $U$ having an arbitrary c.d.f. $G$.*

An important property of block unimodal distributions is that the class of all distributions on $\Re^d$ which are unimodal at $\mathbf{0}$ is convex and closed. The convex structure of the set of block unimodal distributions gives a characterization of such distributions. Under some mild conditions, Choquet's theorem states that each point in a convex set of distributions can be written as a generalized mixture over the extreme points, of the convex set, that is,

$$F = \int_{\mathrm{ex}(C)} G dQ(G).$$

Here, $\mathrm{ex}(C)$ is the set of extreme points of a convex set $C$, i.e. the subset of $C$ whose elements cannot be written as a convex combination of any two other points in $C$. If $C$ is a compact subset of a vector space, then Choquet's theorem states that each element of $C$ can be written as a mixture over the extreme points.

In particular, one has the following:

**Theorem 35.4.** *The c.d.f. $F$ is block unimodal about $\mathbf{0}$ if and only if there exists a probability measure $G$ on the Borel $\sigma$-field*

$$\mathscr{B}_d = \mathscr{B}(\Re^d)$$

*such that*

$$F(A) = \int U_y(A)G(\mathrm{d}y), \quad A \in \mathscr{B}_d,$$

*where $U_y$ denotes the uniform distribution over the open rectangle with opposite vertices $\mathbf{0}$ and $y$.*

This again means that the c.d.f. $F$ belongs to the closed convex hull of the set of

$$\{U_y : \quad 0 < ||y|| < \infty\}.$$

## 35.4 Bayesian Inference for Unimodal Distributions

The purpose of this section is to make inference about the unknown unimodal distribution function $F$, given a random sample $X = (X_1,\ldots,X_n)$ from $F$, that is

$$X_i \overset{\text{iid}}{\sim} F, \quad i = 1,\ldots,n.$$

In our Bayesian model specification, we could induce a prior distribution for $F$ by the relation

$$F(A) = \int U_y(A)G(\mathrm{d}y), \quad A \in \mathscr{B}_d,$$

where $G$ is a random cumulative distribution function distributed as a Dirichlet process,

$$G \sim DP(M, G_0).$$

By Theorem 4, the distribution $F$ will be a random block unimodal c.d.f. Thus, in this way, one can obtain a prior distribution on the subspace of block unimodal c.d.f.'s.

The model can alternatively be expressed as follows:

$$X_i|Y_i \overset{\text{ind}}{\sim} U(x_i|y_i),$$
$$y_i|G \overset{\text{iid}}{\sim} G,$$
$$G \sim DP(M, G_0),$$

for $i = 1, \ldots, n$.

According to the above procedure, one can always gets a c.d.f. $F$ with a single mode at zero, no matter what the distribution $G$, one starts with, is. A possible generalization to multimodal distributions is obtained applying a"partial convexification" procedure. A partial convexification procedure of a c.d.f. $G$ in univariate case relies on using $U(\alpha, 1)$ distributions instead of $U(0,1)$, with $0 < \alpha < 1$, (cf. Kokolakis & Kouvaras (2007) [9] and Kouvaras & Kokolakis (2008) [10]).

*Open problem.* It would be of interest to extend the above methods to other prior distributions than Dirichlet process prior, such as for example Pólya tree prior, Lavine(1992, 1994) [11], [12].

## References

1. Bertin, E.M-J. Cuculescu, I., Theodorescu, R.: Unimodality of Probability Measures. Kluwer Academic Publishers (1997)
2. Blackwell, D.: Discreteness of Ferguson selections. Ann. Statist. **1**, 356–358 (1973)
3. Brunner, L. J., Lo, A. Y.: Bayes methods for a symmetric unimodal density and its mode. Ann. Statist. **17**, 1550–1566 (1989)
4. Brunner, L.J.: Bayesian nonparametric methods for data from a unimodal density. Statist. Probab. Lett. **14**, 195–199 (1992)
5. Dharmadhikari, S., Kumar, J-D..: Unimodality, Convexity and Applications. Academic Press (1988)
6. Feller, W.: An Introduction to Probability Theory and its Applications, Volume 2. Wiley, New York, 2nd edition (1971)
7. Ferguson, Th.S.: A Bayesian analysis of some nonparametric problems. Ann. Statist. **1**, 209–230 (1973)

8. Hjort, N.: Topics in nonparametric Bayesian statistics. In: Hjort, N. Green, P., Richardson, S. (eds.) Highly structured stochastic systems, pp. 455-478. Oxford, Oxford University Press (2003)
9. Kokolakis, G., Kouvaras, G.: On the multimodality of random probability measures. Bayesian Anal. **2**, 213–220 (2007)
10. Kouvaras, G., Kokolakis, G.: Random multivariate multimodal distributions. In: Skiadas, Ch. (ed.) Recent Advances in Stochastic Modelling and Data Analysis, pp. 68–75. World Scientific Publishing Co. (2008)
11. Lavine, M.: Some aspects of Pólya tree distributions for statistical modelling. Ann. Statist. **20**, 1222–1235 (1992)
12. Lavine, M.: More aspects of Pólya tree distributions for statistical modelling. Ann. Statist. **22**, 1161–1176 (1994)
13. Lo, A.Y.: On a class of nonparametric estimates: I. Density estimates. Ann. Statist. **12**, 351–357 (1984)
14. Ghosh, J., Ramamoorthi, R.: Bayesian Nonparametrics. New York, Springer-Verlag, Inc (2003)
15. Shepp, L.A.: Symmetric random walk. Trans. Amer. Math. Soc. **104**, 144–153 (1962)
16. Walker, S. Damien, P., Laud, P.W., Smith, A.F.M.: Bayesian nonparametric inference for random distributions and related functions. J. Roy. Statist. Soc. Ser. B **61**, 485–527 (1999)

# Chapter 36
# Generalized Weighted Arithmetic Means

**Janusz Matkowski**

*Dedicated to the memory of S.M. Ulam*
*on the 100th anniversary of his birth*

**Abstract** Means which are the sum of single variable functions are considered. It is shown among other that if such a mean is weighted quasi-arithmetic, or subtranslative or subadditive then it must be a weighted quasi-arithmetic mean. Conditions under which the functions of the form $f(x) = ax + b$ are affine or convex with respect to such a mean are presented. Invariance of a weighted arithmetic mean with respect to the relevant mean-type mappings is considered. Some open problems are presented.

**Keywords** Mean • Generalized weighted mean • Mean-convex function mean-affine functions • Subadditivity • Subtransitivity • Functional equation • Functional inequalities

**Mathematics Subject Classification (2000):** Primary 26E30, 26A18, 39B22

## 36.1 Introduction

A function $M : I^2 \to \mathbb{R}$ is called a *mean* in an interval $I \subseteq \mathbb{R}$, if

$$\min(x,y) \le M(x,y) \le \max(x,y), \qquad x,y \in I.$$

If for all $x,y \in I, x \ne y$, these inequalities are sharp, $M$ is called *strict*; and *symmetric*, if $M(x,y) = M(y,x)$ for all $x,y \in I$. If $M$ is a mean in $I$, then $M(J^2) = J$, for every

J. Matkowski (✉)
Faculty of Mathematics Computer Science and Econometrics, University of Zielona Góra, Podgórna 50, PL-65-246 Zielona Góra, Poland

Institute of Mathematics, Silesian University, Bankowa 14, PL-42-007 Katowice, Poland
e-mail: J.Matkowski@wmie.uz.zgora.pl

Th.M. Rassias and J. Brzdęk (eds.), *Functional Equations in Mathematical Analysis*, Springer Optimization and Its Applications 52, DOI 10.1007/978-1-4614-0055-4_36,

subinterval $J \subseteq I$. Moreover, $M$ is *reflexive*, i.e.

$$M(x,x) = x, \qquad x \in I.$$

Every reflexive function $M : I^2 \to \mathbb{R}$ which is increasing with respect to each variable is a mean in $I$. If $I = (0,\infty)$ and

$$M(tx,ty) = tM(x,y), \qquad t,x,y > 0,$$

then $M$ is positively homogeneous.

For a strictly monotonic function $\gamma : I \to \mathbb{R}$ and $p \in (0,1)$, the function $M : I^2 \to I$, $M = \mathscr{A}_p^{[\gamma]}$,

$$\mathscr{A}_p^{[\gamma]}(x,y) := \gamma^{-1}\left(p\gamma(x) + (1-p)\gamma(y)\right), \qquad x,y \in I,$$

is a mean, and it is called *quasi-arithmetic weighted mean.* The function $\gamma$ is called a *generator* of the mean and $p$ its *weight*. The symmetric mean

$$\mathscr{A}^{[\gamma]} := \mathscr{A}_{1/2}^{[\gamma]}$$

is called *quasi-arithmetic* mean.

Taking here $\gamma(x) = ax + b$, $a \neq 0$, we get

$$\mathscr{A}_p^{[\gamma]}(x,y) = \mathscr{A}_p(x,y) := px + (1-p)y, \qquad x,y \in I.$$

If $p = 1/2$ we write $\mathscr{A} = \mathscr{A}_{1/2}$.

Note that the weighted arithmethic mean is sum of two functions of single variable.

In this paper, we examine the means which are of the form $M(x,y) = \varphi(x) + \psi(y)$ for some functions $\varphi, \psi : I \to \mathbb{R}$. In Sect. 36.2, we observe that $M$ is a mean iff $\psi(x) = x - \varphi(x)$ for all $x \in I$ and both $\varphi$ and $\psi$ are increasing – in particular $\varphi$ and $\psi$ must be nonexpansive, i.e. Lipschitz continuous with a Lipschitz constant 1. We denote this mean by $\mathscr{W}^{[\varphi]}$. For obvious reason, the means of this type can be treated as a generalization of the classical weighted arithmetic means. We show that if $\mathscr{W}^{[\varphi]} \leq \mathscr{W}^{[\gamma]}$ then $\mathscr{W}^{[\varphi]} = \mathscr{W}^{[\gamma]}$, and $\mathscr{W}^{[\varphi]}$ is symmetric iff $\mathscr{W}^{[\varphi]} = \mathscr{A}$.

In Sect. 36.3, we prove that a mean $\mathscr{W}^{[\varphi]}$ defined on $[0,\infty)^2$ is weighted quasi-arithmetic iff it is weighted arithmetic. The positive homogeneity of $\mathscr{W}^{[\varphi]}$ is also considered.

In Sects. 36.4 and 36.5, we consider conditions under which the functions of the form $f(x) = ax + b$ are affine or convex with respect to $\mathscr{W}^{[\varphi]}$. It turns out that if the function $f(x) = ax$ is $\mathscr{W}^{[\varphi]}$-convex, then it is $\mathscr{W}^{[\varphi]}$-affine and, under the differentiability of $\varphi$ at 0 and the condition $a \neq 1$, the mean $\mathscr{W}^{[\varphi]}$ must be a weighted arithmetic. Some related results concerning general convexity are also given.

Recall that the notion of a convex function with respect to two given means was introduced by Aumann [3] and, for the first time was considered by Aczél [1] (cf. also Matkowski and Rätz [9]).

In Sect. 36.6, we show that $\mathscr{W}^{[\varphi]}$ is subtranslative iff it is weighted arithmetic. In Sect. 36.7, we deal with subadditivity of $\mathscr{W}^{[\varphi]}$. In Sect. 36.8, we consider the invariance of a weighted arithmetic mean with respect to the mean-type mapping of the form $\left(\mathscr{W}^{[\varphi]},\mathscr{W}^{[\psi]}\right)$ and we apply the obtained result in solving a functional equation. In Sect. 36.9 we propose a generalization of the weighted quasi-arithmetic means. In Sect. 36.10 the finite dimensional counterparts of the means $\mathscr{W}^{[\varphi]}$ are discussed.

At the end we present some open problems related to Daróczy–Páles identity.

## 36.2 A Generalization of Weighted Arithmetic Mean

We begin with

**Theorem 36.1.** *Let $I$ be an interval and $\varphi,\psi : I \to \mathbb{R}$. The function $M : I^2 \to \mathbb{R}$,*

$$M(x,y) := \varphi(x) + \psi(y), \qquad x,y \in I, \tag{36.1}$$

*is a (strict) mean if, and only if, the functions $\varphi$ and $\psi$ are (strictly) increasing and*

$$\varphi(x) + \psi(x) = x, \qquad x \in I.$$

*Proof.* Suppose that $M$ given by (1) is a mean. Setting $y = x$ in (1) we get $\varphi(x) + \psi(x) = x$, whence $\psi(x) = x - \varphi(x)$ for $x \in I$. Hence, taking $x,y \in I$, $x \le y$, from the definition of the mean we have

$$x = \min(x,y) \le \varphi(x) + y - \varphi(y) \le \max(x,y) = y,$$

i.e. $\varphi(x) \le \varphi(y)$ and $x - \varphi(x) \le y - \varphi(y)$,which proves that $\varphi$ and $\psi$ are increasing. Conversely, suppose that $\varphi$ and $\psi$ are increasing and such that $\varphi(x) + \psi(x) = x$ for $x \in I$. Take arbitrary $x,y \in I$, assume that $x \le y$. Then we have $\varphi(x) \le \varphi(y)$ and $x - \varphi(x) \le y - \varphi(y)$, whence $x \le \varphi(x) + y - \varphi(y) \le y$, which means that $\min(x,y) \le M(x,y) \le \max(x,y)$. In the case when $M$ is a strict mean we argue analogously. □

**Corollary 36.1.** *Let $I$ be an interval and $\varphi : I \to \mathbb{R}$. Then*

*(i) the functions $\varphi$ and $\mathrm{id}_I - \varphi$ are nondecreasing iff the function $\mathscr{W}^{[\varphi]} : I^2 \to I$ defined by*

$$\mathscr{W}^{[\varphi]}(x,y) := \varphi(x) + y - \varphi(y), \qquad x,y \in I, \tag{36.2}$$

*is a mean;*

*(ii) the functions $\varphi$ and $\mathrm{id}_I - \varphi$ are strictly increasing iff $\mathscr{W}^{[\varphi]}$ is a strict mean.*

*Remark 36.1.* Let $I$ be an interval and $\varphi : I \to \mathbb{R}$. Then

(i) $\mathscr{W}^{[\varphi]}$ is a mean iff $\varphi$ is nondecreasing and non-expansive, i.e.

$$x \le y \Longrightarrow 0 \le \varphi(y) - \varphi(x) \le y - x;$$

(ii) $\mathscr{W}^{[\varphi]}$ is a strict mean iff $\varphi$ is strictly increasing and contractive, i.e.

$$x < y \Longrightarrow 0 < \varphi(y) - \varphi(x) < y - x.$$

Note the following easy to verify

*Property 36.1.* The following conditions are equivalent

(i) $\mathscr{W}^{[\varphi]}$ is symmetric;
(ii) there is a $c \in \mathbb{R}$ such that

$$\varphi(x) = \frac{x}{2} + c, \qquad x \in I;$$

(iii) $\mathscr{W}^{[\varphi]} = \mathscr{A}$, where $\mathscr{A}(x,y) = \frac{x+y}{2}$ for $x, y \in I$.

*Property 36.2.* Let $I$ be an interval and assume that $\varphi, \gamma : I \to \mathbb{R}$, $\mathrm{id}_I - \varphi$ and $\mathrm{id}_I - \gamma$ are nondecreasing. Then the following conditions are equivalent:

(i) $\mathscr{W}^{[\varphi]} \le \mathscr{W}^{[\gamma]}$;
(ii) $\mathscr{W}^{[\varphi]} = \mathscr{W}^{[\gamma]}$;
(iii) there is a $c \in \mathbb{R}$ such that $\gamma = \varphi + c$.

*Proof.* By definitions of the means $\mathscr{W}^{[\varphi]}$ and $\mathscr{W}^{[\gamma]}$, the inequality $\mathscr{W}^{[\varphi]} \le \mathscr{W}^{[\gamma]}$ is equivalent to the inequality $\varphi(x) - \gamma(x) \le \varphi(y) - \gamma(y)$ for all $x, y \in I$, which implies that $\varphi - \gamma$ is a constant function. □

## 36.3 When $\mathscr{W}^{[\varphi]}$ is Weighted Quasi Arithmetic or Homogeneous

The main result of this section reads as follows

**Theorem 36.2.** *Let $I \subset \mathbb{R}$ be an interval and suppose that $\varphi : I \to \mathbb{R}$ and $\mathrm{id}_I - \varphi$ are nondecreasing. Then $\mathscr{W}^{[\varphi]}$ defined by (2) is a weighted quasi-arithmetic mean if, and only if, there are $p \in (0,1)$ and $c \in \mathbb{R}$, $c \ne 0$, such that $\varphi(x) = px + c$ for $x \in I$, that is iff,*

$$\mathscr{W}^{[\varphi]}(x,y) = px + (1-p)y, \qquad x, y \in I.$$

*Proof.* If $\mathscr{W}^{[\varphi]}$ is a weighted quasi-arithmetic mean, then there is a continuous strictly monotonic function $\gamma : I \to \mathbb{R}$ and $p \in (0,1)$ such that

$$\mathscr{W}^{[\varphi]}(x,y) = \varphi(x) + \psi(x) = \gamma^{-1}\left(p\gamma(x) + (1-p)\gamma(y)\right), \qquad x, y \in I, \tag{36.3}$$

where

$$\psi(x) = x - \varphi(x), \qquad x \in I.$$

Put $J := \gamma(I)$ and take arbitrary $u, v \in J$. Setting $x := \gamma^{-1}(u)$, $y := \gamma^{-1}(b)$ in (36.3) gives

$$\gamma^{-1}(pu + (1-p)v) = \varphi \circ \gamma^{-1}(u) + \psi \circ \gamma^{-1}(v), \qquad u, v \in J.$$

Hence, setting

$$f := \gamma^{-1}, \qquad g := \varphi \circ \gamma^{-1}, \qquad h := \psi \circ \gamma^{-1},$$

we obtain the functional equation

$$f(pu + (1-p)v) = g(u) + h(v), \qquad u, v \in J.$$

Interchanging $u$ and $v$ we hence get

$$f((1-p)u + pv) = g(v) + h(u), \qquad u, v \in J.$$

Adding the respective sides of these two equations we obtain

$$f(pu + (1-p)v) + f((1-p)u + pv) = [g(u) + h(u)] + [g(v) + h(v)]$$

for all $u, v \in J$. Taking here $v = u$ gives

$$g(u) + h(u) = f(u), \qquad u, v \in J,$$

whence

$$f(pu + (1-p)v) + f((1-p)u + pv) = f(u) + f(v), \qquad u, v \in J,$$

that is $f$ is $p$-Wright affine. In view of Ng's theorem [11], there is an additive function $\alpha : \mathbb{R} \to \mathbb{R}$ and $b \in \mathbb{R}$ such that $f(u) = \alpha(u) + b$ for all $u \in J$. Since $f = \gamma^{-1}$ is continuous, we have $\alpha(u) = au$ for all $u \in \mathbb{R}$. Hence $\gamma^{-1}(u) = au + b$ for all $u \in J$ and, obviously, $a \neq 0$. From (36.2) we obtain our result. □

Let us also note the following easy to prove

**Theorem 36.3.** *Let $\varphi : I \to \mathbb{R}$ and $\mathrm{id}_{(0,\infty)} - \varphi$ be nondecreasing. The mean $\mathscr{W}^{[\varphi]}$ is homogeneous iff there is $p \in [0,1]$ such that*

$$\mathscr{W}^{[\varphi]}(x, y) = px + (1-p)y, \qquad x, y > 0.$$

## 36.4 $\mathscr{W}^{[\varphi]}$-Affine Functions

Recall the following

**Definition 36.1.** (cf. [9]) Let $I, J \subset \mathbb{R}$, $J \subset I$ be intervals, and let $M : I^2 \to I$ be a mean. A function $f : J \to I$ is said to be affine with respect $M$, briefly, $M$-affine, if

$$f(M(x,y)) = M(f(x), f(y)), \qquad x, y \in J.$$

*Remark 36.2.* It easy to see that, under the assumptions of the above definition, the following three statements are valid:

(i) if $f$ is one-to-one, onto and $M$-affine, then $f^{-1}$ is $M$-affine;
(ii) if $J = I$ and $f, g : I \to I$ are $M$-affine then so is $f \circ g$;
(iii) the function $f = \mathrm{id}_{(0,\infty)}$ and the constant function $f = c$ for $c \in I$ are $M$-affine.

**Theorem 36.4.** *Let $I = \mathbb{R}$ or $I = [0, \infty)$. Suppose that $\varphi : I \to \mathbb{R}$ and $\mathrm{id}_{(0,\infty)} - \varphi$ are strictly increasing.*

*(i) If $I = \mathbb{R}$ and $f : \mathbb{R} \to \mathbb{R}$ is $\mathscr{W}^{[\varphi]}$-affine, then there exist an additive function $\alpha : \mathbb{R} \to \mathbb{R}$ and $b \in \mathbb{R}$ such that*

$$f(x) = ax + b, \qquad x \in \mathbb{R}.$$

*If, moreover, $f$ is bounded from above (from below) or measurable in a neighborhood of a point, then $f(x) = ax + b$ for some $a, b \in \mathbb{R}$ and for all $x \in \mathbb{R}$.*

*(ii) If $I = [0, \infty)$ and $f : [0, \infty) \to [0, \infty)$ is $\mathscr{W}^{[\varphi]}$-affine, then there are $a, b \in \mathbb{R}$ such that*

$$f(x) = \alpha x + b, \qquad x \geq 0.$$

*Proof.* 1. We can assume without any loss of generality that $\varphi(0) = 0$. Suppose that $f : \mathbb{R} \to \mathbb{R}$ is $\mathscr{W}^{[\varphi]}$-affine. From Definition 36.1 we have

$$f(\varphi(x) + y - \varphi(y)) = \varphi(f(x)) + f(y) - \varphi(f(y)), \qquad x, y \in \mathbb{R}.$$

Taking here $y = 0$ we get

$$f(\varphi(x)) = \varphi(f(x)) + f(0) - \varphi(f(0)), \qquad x \in \mathbb{R}.$$

and taking $x = 0$ we get

$$f(y - \varphi(y)) = \varphi(f(0)) + f(0) - \varphi(f(y)), \qquad y \in \mathbb{R}.$$

Hence, setting $b := f(0)$, and adding the respective sides of these two equations, we obtain

$$f(\varphi(x) + y - \varphi(y)) = f(\varphi(x)) - b + f(y - \varphi(y)), \qquad x, y \in \mathbb{R}.$$

that is,

$$f\left(\varphi(x)+\psi(y)\right)-b=\left[f(\varphi(x))-b\right]+\left[f(\psi(y))-b\right], \qquad x,y\in\mathbb{R},$$

where $\psi(y)=y-\varphi(y)$ for $y\in\mathbb{R}$. As $\mathscr{W}^{[\varphi]}$ is a strict mean (cf. Corollary 36.1 (*ii*)), we have $\varphi(\mathbb{R})=(r_1,r_2)$, $\psi(\mathbb{R})=(s_1,s_2)$ where $-\infty\leq r_1<r_2\leq\infty$. It follows that the function $\alpha:\mathbb{R}\to\mathbb{R}$ defined by

$$\alpha(u):=f(u)-c, \qquad u\in\mathbb{R},$$

satisfies the functional equation

$$\alpha(u+v)=\alpha(u)+\alpha(v), \qquad u\in(r_1,r_2), v\in(s_1,s_2).$$

Since at least one of the numbers $r_1,r_2,s_1,s_2$ is not finite, it follows that

$$\alpha(u+v)=\alpha(u)+\alpha(v), \qquad u,v\in\mathbb{R},$$

which shows that $\alpha$ is additive. The remaining part of the theorem is well known (cf. Aczél [2], Kuczma [5]) and the proof of part (i) is completed.

In case (ii), the argument showing the additivity of $\alpha$ is analogous. The nonnegativity of $f$ implies that $f$ is of the desired form. □

**Theorem 36.5.** *Let $I\subset\mathbb{R}$ be a closed interval and $\varphi:I\to\mathbb{R}$ a nondecreasing function such that $\mathrm{id}_{(0,\infty)}-\varphi$ is nondecreasing. Suppose that $f:I\to I$, $f(x)=ax+b$ $(x\in I)$ for some $a,b\in\mathbb{R}$. Then*

*(i) $f$ is $\mathscr{W}^{[\varphi]}$-affine iff the function*

$$I\ni x\longrightarrow\varphi(ax+b)-a\varphi(x) \text{ is constant;}$$

*(ii) if $f$ is $\mathscr{W}^{[\varphi]}$-affine, $a\neq1$ and $\varphi$ is differentiable at the point $x_0=\frac{b}{1-a}$, then*

$$f(x)=px+c, \qquad x\in I,$$

*for some $p\in[0,1]$, $c\in\mathbb{R}$; moreover*

$$\mathscr{W}^{[\varphi]}=px+(1-p)y, \qquad x,y\in I;$$

*(iii) assuming that $f$ is $\mathscr{W}^{[\varphi]}$-affine and $a=1$, if either*

$$b>0 \text{ and } \lim_{x\to\infty}\frac{\varphi(x)}{x} \text{ exists,}$$

*or*

$$b<0 \text{ and } \lim_{x\to-\infty}\frac{\varphi(x)}{x} \text{ exists,}$$

*then the conclusion of (ii) remains valid.*

*Proof.* The $\mathscr{W}^{[\varphi]}$-affinity of $f$, that is the equality

$$a(\varphi(x)+y-\varphi(y))=\varphi(ax+b)+ay+b-\varphi(ay+b), \qquad x,y\in I,$$

simplifies to

$$\varphi(ax+b)-a\varphi(x)=\varphi(ay+b)-a\varphi(y), \qquad x,y\in I,$$

that is equivalent to the fact that, for some $C\in\mathbb{R}$,

$$\varphi(ax+b)-a\varphi(x)=C, \qquad x\in I.$$

To prove $(ii)$ note that this equation can be written in the form

$$\left[\varphi(ax+b)-\frac{C}{1-a}\right]=a\left[\varphi(x)-\frac{C}{1-a}\right], \qquad x\in I.$$

Setting $\phi(x):=\varphi(x)-\frac{C}{1-a}$ we get

$$\phi(ax+b)=a\gamma(x), \qquad x\in I.$$

Taking into account that $f$ maps $I$ into itself and considering the four possible cases $I=\mathbb{R}$, $I$ is unbounded from below, $I$ is unbounded from above, and $I$ is bounded, it is easy to see that, without any loss of generality, we can assume that $|a|<1$. It follows that $x_0:=\frac{b}{1-a}$, the only fixed point of $f$ (which of course belongs to $I$) is contractive. Moreover, setting $x=x_0$ in the above equation gives $\phi(x_0)=0$. Setting

$$\gamma(x):=\frac{\phi(x)}{x-x_0}, \qquad x\in I,\ x\neq x_0,$$

in the above functional equation gives

$$\gamma(x)=\gamma(ax+b), \qquad x\in I,\ x\neq x_0,$$

whence, by induction,

$$\gamma(x)=\gamma\left(a^nx+b\frac{1-a^n}{1-a}\right), \qquad x\in I,\ x\neq x_0,\ n\in\mathbb{N}.$$

The differentiability of $\varphi$ at $x_0$ implies the existence of $p:=\lim_{x\to x_0}\gamma(x)$. Therefore, letting $n\to\infty$ in the above formula, we obtain $\gamma(x)=p$ for $x\in I\setminus\{x_0\}$, and the proof of $(ii)$ is completed.

We omit similar reasoning in the case $(iii)$. □

*Remark 36.3.* According to Corollary 36.1, the function $\varphi$ in the above result is Lipschitzian and, consequently, it is absolutely continuous. Nevertheless, in the second part of this theorem, the assumption of differentiability of $\varphi$ at the fixed point of $f$ is essential. In fact, we have the following.

**Proposition 36.1.** *Let $a \in (0,1)$ be fixed. Suppose that $\varphi_0 : [a,1] \to \mathbb{R}$ is increasing, nonexpansive and such that $\varphi_0(a) = a\varphi_0(1)$. Then there exists a unique increasing nonexpansive function $\varphi : (0,\infty) \to \mathbb{R}$ such that $\varphi\big|_{[a,1]} = \varphi_0$, and the function $f : (0,\infty) \to (0,\infty)$, $f(x) = ax$, is $\mathscr{W}^{[\varphi]}$-affine.*

*Proof.* Since

$$\bigcup_{k\in\mathbb{Z}} \left(a^{k+1}, a^k\right] = (0,\infty), \qquad \left(a^{k+1}, a^k\right] \cap \left(a^{m+1}, a^m\right] = \emptyset,\ k,m \in \mathbb{Z},\ k \neq m,$$

(where $\mathbb{Z}$ stands for the set of integers), the function $\varphi : (0,\infty) \to \mathbb{R}$,

$$\varphi(x) := a^k \varphi_0\left(a^{-k}x\right), \qquad x \in \left(a^{k+1}, a^k\right],\ k \in \mathbb{Z},$$

is correctly defined. Note that

$$\varphi(a^k) := a^k \varphi_0(1), \qquad k \in \mathbb{Z},$$

and, for every $k \in \mathbb{Z}$, the function $\varphi$ is left-continuous at the point $a^k$. Since $\varphi_0(a) = a\varphi_0(1)$, we have

$$\varphi(a^k+) = \lim_{x\to a^k+} \varphi_0\left(a^{a-k+1}x\right) = a^{k-1}\varphi_0(a) = a^k\varphi_0(1) = \varphi\left(a^k\right),$$

so the function $\varphi$ is continuous at each point $a^k$ and, consequently, $\varphi$ is continuous. The increasing monotonicity of $\varphi_0$ implies that, for every $k \in \mathbb{Z}$, the function $\varphi$ is increasing on the interval $\left(a^{k+1}, a^k\right]$. The continuity of $\varphi$ implies that $\varphi$ is increasing on $(0,\infty)$. Since $\varphi_0$ is nonexpansive, for a fixed $k \in \mathbb{Z}$, and arbitrary $x, y \in \left(a^{k+1}, a^k\right]$, $x < y$, we have

$$\begin{aligned} 0 \le \varphi(y) - \varphi(x) &= a^k\left[\varphi_0\left(a^{-k}y\right) - \varphi_0\left(a^{-k}x\right)\right] \\ &\le a^k\left[\left(a^{-k}y\right) - \left(a^{-k}x\right)\right] = y - x, \end{aligned}$$

which proves that $\varphi$ is nonexpansive on every interval $\left(a^{k+1}, a^k\right]$. The continuity of $\varphi$ at every point $a^k$ implies hat $\varphi$ is nonexpansive on $(0,\infty)$. Thus, we have shown that the function $\varphi$ satisfies condition (i) of Remark 36.1 and, consequently, $\mathscr{W}^{[\varphi]}$ is a mean.

Now take arbitrary $x>0$. There is a unique $k\in\mathbb{Z}$ such that $x\in\left(a^{k+1},a^{k}\right]$. Then $ax\in\left(a^{k+2},a^{k+1}\right]$ and, by the definition of $\varphi$ we have

$$\varphi(ax)=a^{k+1}\varphi_0(a^{-k+1}(ax))=a(a^{k}\varphi_0(a^{-k}x))=a\varphi(x)$$

which proves that that mean $\varphi(ax)-a\varphi(x)$ is constant on $(0,\infty)$ and, in view of Theorem 36.5$(i)$, the function $f(x)=ax$ for $x>0$, is $\mathscr{W}^{[\varphi]}$-affine. □

*Remark 36.4.* Proposition 36.1 gives a construction of all $\varphi:(0,\infty)\to\mathbb{R}$ such that the function $f:(0,\infty)\to(0,\infty)$, $f(x)=f(1)x$, is $\mathscr{W}^{[\varphi]}$-affine. Moreover, it shows that the converse of Theorem 36.5 is false.

Let us note the following

**Lemma 36.1.** *Let $M:(0,\infty)^2\to(0,\infty)$ be a continuous mean. If there are $a,b>0$, $a\neq 1\neq b$, such that* $\log(b-a)$ *is irrational, and the functions $f,g:(0,\infty)\to(0,\infty)$, $f(x)=ax$, $g(x)=bx$ are $M$-affine, then $M$ is positively homogeneous.*

*Proof.* By assumption we have $M(ax,ay)=aM(x,y)$, $M(bx,by)=bM(x,y)$ for all $x,y>0$, whence, by induction,

$$M(a^{k}x,a^{k}y)=a^{k}M(x,y),\quad M(b^{m}x,b^{m}y)=b^{m}M(x,y),\quad k,m\in\mathbb{Z},\ x,y>0,$$

that is

$$M(tx,ty)=tM(x,y),\quad t\in D,\ x,y>0,$$

where $D:=\{a^{k}b^{m}:k,m\in\mathbb{Z}\}$. Since $\log(b-a)$ is irrational, by Kronecker theorem, $D$ is a dense in $(0,\infty)$. The continuity of $M$ implies that $M(tx,ty)=tM(x,y)$ for $t,x,y>0$. □

As an immediate consequence of this lemma and Theorem 36.4 we obtain

**Theorem 36.6.** *Let $I=(0,\infty)$ and suppose that $\varphi:I\to\mathbb{R}$ and $\mathrm{id}_I-\varphi$ are nondecreasing. If there are $a,b>0$, $a\neq 1\neq b$, such that $\frac{\log b}{\log a}$ is irrational, and the functions $f,g:(0,\infty)\to(0,\infty)$, $f(x)=ax$, $g(x)=bx$ are $\mathscr{W}^{[\varphi]}$-affine, then $\mathscr{W}^{[\varphi]}$ is a weighted arithmetic mean.*

**Theorem 36.7.** *Let $I=\mathbb{R}$ and suppose that $\varphi:I\to\mathbb{R}$ and $\mathrm{id}_I-\varphi$ are nondecreasing. If there are $a,b>0$, $a\neq 1\neq b$, such that $a/b$ is irrational, and the functions $f,g:\mathbb{R}\to\mathbb{R}$, $f(x)=x+a$, $g(x)=x+b$ are $\mathscr{W}^{[\varphi]}$-affine, then $\mathscr{W}^{[\varphi]}$ is a weighted arithmetic mean.*

In connection with the above two results note that the conditions under which the simultaneous system of functional equations

$$\varphi(Ax+\alpha)=a\varphi(x),\qquad \varphi(Bx+\beta)=b\varphi(x)$$

has a nontrivial continuous solution is given in [8].

## 36.5 $\mathscr{W}^{[\varphi]}$-Convex Functions

**Definition 36.2 (Aczél [1], Aumann [3], also [9]).** Let $I, J \subset \mathbb{R}$, $I \subset J$, be intervals and $M : I^2 \to I$ be a mean. A function $f : J \to I$ is said to be

[$2^0$]

$1^0$ convex with respect to $M$, briefly $M$-convex, if

$$f(M(x,y)) \leq M(f(x), f(y)), \qquad x, y \in J,$$

$2^0$ $M$-concave, if

$$f(M(x,y)) \geq M(f(x), f(y)), \qquad x, y \in J.$$

**Proposition 36.2.** *Let $I, J \subset \mathbb{R}$, $I \subset J$, be intervals and $M, N : I^2 \to I$ some means.*

*$1^0$ If a decreasing function $f : J \to I$ is $M$-convex and $M \leq N$, then $f$ is $N$-convex.*
*$2^0$ If a decreasing function $f : J \to I$ is $M$-convex and $M \leq N$, then $f$ is $N$-convex.*

*Proof.* Applying in turn: the inequality $M \leq N$ and the decreasing monotonicity of $f$, the $M$-convexity of $f$ and again the inequality $M \leq N$, we get, for all $x, y \in J$,

$$f(N(x,y)) \leq f(M(x,y)) \leq M(f(x), f(y)) \leq N(f(x), f(y)),$$

which proves $1^0$. The proof of $2^0$ is similar. □

From Property 36.1(ii) we obtain

**Corollary 36.2.** *Let $I, J \subset \mathbb{R}$, $I \subset J$, be intervals. Suppose that $\varphi : I \to \mathbb{R}$ , $\mathrm{id}_I - \varphi$ are nondecreasing and there is $p \in (0,1)$ such that either*

$$\mathscr{W}^{[\varphi]}(x,y) \leq px + (1-p)y, \qquad x, y \in I,$$

*or*

$$\mathscr{W}^{[\varphi]}(x,y) \geq px + (1-p)y, \qquad x, y \in I.$$

*If $f : J \to I$ is $\mathscr{W}^{[\varphi]}$-convex ($\mathscr{W}^{[\varphi]}$-concave), then $f$ is Jensen-convex (resp. Jensen concave).*

*Proof.* If $\mathscr{W}^{[\varphi]}(x,y) \leq px + (1-p)y$ for all $x, y \in I$, then, in view of Property 36.2(ii) with $\gamma(x) = px$ ($x \in I$), we have $\mathscr{W}^{[\varphi]}(x,y) = px + (1-p)y$ for all $x, y \in I$. It follows that

$$f(px + (1-p)y) \leq pf(x) + (1-p)f(y), \qquad x, y \in I.$$

By Kuhn's theorem [6], $f$ satisfies the inequality with $p$ replaced by $1/2$, i.e. $f$ is Jensen convex (for a simple proof of Kuhn's result see Daróczy and Páles [4]. □

**Theorem 36.8.** *Let $I = (0, \infty)$ and suppose that $\varphi : I \to \mathbb{R}$ , $\mathrm{id}_I - \varphi$ are nondecreasing.*

*(i) The function $f : (0,\infty) \to (0,\infty)$, $f(x) = f(1)x$, is $\mathscr{W}^{[\varphi]}$-convex iff it is $\mathscr{W}^{[\varphi]}$-afine.*

(ii) *Let $C_\varphi$ denote the set of all $a > 0$ such that the function $f(x) = ax$ $(x > 0)$ is $\mathscr{W}^{[\varphi]}$-convex. Then*

$$\text{either } C_\varphi = (0,\infty) \quad \text{or} \quad C_\varphi = \{d^k : k \in \mathbb{Z}\} \text{ for some } d > 0.$$

*Proof.* 1. Put $a := f(1)$. By the definition of $\mathscr{W}^{[\varphi]}$ the function $f(x) = ax$ $(x \in I)$ is $\mathscr{W}^{[\varphi]}$-convex iff

$$a(\varphi(x) + y - \varphi(y)) \le \varphi(ax) + ay - \varphi(ay), \qquad x, y \in I,$$

that is, iff

$$a\varphi(x) - \varphi(ax) \le a\varphi(y) - \varphi(ay), \qquad x, y \in I.$$

Obviously, this inequality is equivalent to the equality

$$a\varphi(x) - \varphi(ax) = a\varphi(y) - \varphi(ay), \qquad x, y \in I,$$

which holds iff

$$a(\varphi(x) + y - \varphi(y)) = \varphi(ax) + ay - \varphi(ay), \quad x, y \in I,$$

that is, iff $f$ is $\mathscr{W}^{[\varphi]}$-afine. Now the first part follows from Theorem 36.5$(i)$.

2. Suppose that $a, b \in C_\varphi$. Then, according to what has been already shown, the functions $f(x) = ax, g(x) = bx$ $(x > 0)$ are $\mathscr{W}^{[\varphi]}$-affine. In view of Remark 36.2, the function $f \circ g^{-1}$ is also $\mathscr{W}^{[\varphi]}$-affine and, consequently, $ab^{-1} \in C_\varphi$. It follows that $C_\varphi$ is a subgroup of the multiplicative group $(0,\infty)$. It is obvious that every such a subgroup is either dense or discrete. If $C_\varphi$ is discrete then, which is a well-known fact, it must be a cyclic subgroup of $(0,\infty)$, i.e. there is a number $d > 0$ such that $C_\varphi = \{d^k : k \in \mathbb{Z}\}$. □

## 36.6 Translativity, Subtranslativity and Supertranslativity of $\mathscr{W}^{[\varphi]}$

Recall that a mean $M : (0,\infty)^2 \to (0,\infty)$ is *translative* (cf. Aczél [2, p. 234]) if

$$M(x+t, y+t) = M(x,y) + t, \qquad x, y, t > 0.$$

We call a mean $M : (0,\infty)^2 \to (0,\infty)$ *subtranslative* if

$$M(x+t, y+t) \le M(x,y) + t, \qquad x, y, t > 0;$$

and *supertranslative* if the reversed inequality holds true.

**Theorem 36.9.** *Let $I = (0,\infty)$ and suppose that $\varphi : I \to \mathbb{R}$, $\mathrm{id}_I - \varphi$ are nondecreasing. The following conditions are equivalent:*

*(i)* $\mathscr{W}^{[\varphi]}$ *is subtranslative;*
*(ii)* $\mathscr{W}^{[\varphi]}$ *is supertranslative;*
*(iii)* $\mathscr{W}^{[\varphi]}$ *is translative;*
*(iv)* $\mathscr{W}^{[\varphi]}$ *is a weighted arithmetic mean.*

*Proof.* Suppose that $\mathscr{W}^{[\varphi]}$ is subtranslative. Then from the definition of $\mathscr{W}^{[\varphi]}$ we easily get

$$\varphi(x+t)+\varphi(y) \le \varphi(y+t)+\varphi(x), \qquad x,y,t>0.$$

Interchanging $x$ and $y$ we infer that

$$\varphi(x+t)+\varphi(y) = \varphi(y+t)+\varphi(x), \qquad x,y,t>0,$$

whence

$$\varphi(x+t)-\varphi(x) = \varphi(y+t)-\varphi(y), \qquad x,y,t>0.$$

It follows that

$$\varphi(x+t)-\varphi(x) = \beta(t), \qquad x,t>0,$$

for some function $\beta : (0,\infty) \to \mathbb{R}$. Solving this Pexider equation we conclude that, for some real constant $c$, the function $\varphi - c$ is additive. The conditions on $\varphi$ imply that

$$\varphi(x) = px - c \quad (x>0)$$

for some $p \in [0,1]$. Now all the remaining statements are obvious. □

## 36.7 Subadditivity and Superadditivity of $\mathscr{W}^{[\varphi]}$

The subadditivity and superadditivity of a mean plays important role in the theory of convexity with respect to the mean. Suppose, for instance, that $M : (0,\infty)^2 \to (0,\infty)$ is a superadditive mean and $I \subset (0,\infty)$ is an open interval. It is shown in [9] that if $f,g : I \to (0,\infty)$ are $M$-convex in $I$, then $f+g$ is $M$-convex.

**Theorem 36.10.** *Let $I = (0,\infty)$ and suppose that $\varphi : I \to \mathbb{R}$, $\mathrm{id}_I - \varphi$ are nondecreasing. The following conditions are equivalent*

*(i) the mean* $\mathscr{W}^{[\varphi]}$ *is subadditive (or superadditive);*
*(ii) there are $p \in [0,1]$ and $c \in \mathbb{R}$ such that*

$$\varphi(x) = px + c, \qquad x>0;$$

*(iii) there is $p \in [0,1]$ such that*

$$\mathscr{W}^{[\varphi]}(x,y) = px + (1-p)y, \qquad x,y>0.$$

*Proof.* Suppose that $\mathscr{W}^{[\varphi]}$ is subadditive, i.e.

$$\mathscr{W}^{[\varphi]}(x_1+x_2,y_1+y_2) \le \mathscr{W}^{[\varphi]}(x_1,y_1) + \mathscr{W}^{[\varphi]}(x_2,y_2), \qquad x_1,x_2,y_1,y_2>0.$$

Hence, by the definition of $\mathscr{W}^{[\varphi]}$,

$$\varphi(x_1+x_2)-\varphi(x_1)-\varphi(x_2) \le \varphi(y_1+y_2)-\varphi(y_1)-\varphi(y_2), \qquad x_1,x_2,y_1,y_2>0,$$

which, obviously implies that there is a constant $c \in \mathbb{R}$ such that

$$\varphi(x_1+x_2)-\varphi(x_1)-\varphi(x_2) = -c, \qquad x_1,x_2,y_1,y_2>0.$$

Writing this equality in the form

$$\varphi(x_1+x_2)-c = [\varphi(x_1)-c]+[\varphi(x_2)-c], \quad x_1,x_2,y_1,y_2>0,$$

we conclude that $\varphi - c$ is additive. Since the remaining statements are obvious, the proof is complete. □

## 36.8 Invariance of the Arithmetic Mean with Respect to the Mean-Type Mappings of $\mathscr{W}^{[\varphi]}$-type

Let $M,N : I^2 \to I$ be means. A mean $K : I^2 \to I$ is called *invariant with respect to the mean-type mapping* $(M,N) : I^2 \to I^2$ (briefly, $K$ is $(M,N)$*-invariant),* if

$$K(M(x,y),N(x,y)) = K(x,y), \qquad x,y \in I.$$

The invariant mean is useful when we are looking for the limits of the sequence of iterates of the mean-type mapping $(M,N) : I^2 \to I^2$ (cf. [10]).

Let us note that the proportion

$$x : \frac{x+y}{2} = \frac{2xy}{x+y} : y,$$

the base of the theory of harmony made by Pythagorean school, can be written in the form

$$\sqrt{\frac{x+y}{2}\cdot\frac{2xy}{x+y}} = \sqrt{xy}.$$

Setting

$$A(x,y) = \frac{x+y}{2}, \quad H(x,y) = \frac{2xy}{x+y}, \quad G(x,y) = \sqrt{xy}$$

for the arithmetic, harmonic and geometric mean, respectively, we hence get

$$G \circ (A,H) = G$$

which says that the geometric mean $G$ is $(A,H)$-invariant. This fact allows to determine effectively the limit of the sequence of iterates $((A,H)^n)_{n\in\mathbb{N}}$ of the mean-type mapping $(A,H) : (0,\infty)^2 \to (0,\infty)^2$.

We prove the following

**Theorem 36.11.** *Let $I \subset \mathbb{R}$ be an interval, $p \in (0,1)$, and $M : I^2 \to I$ be a mean. Suppose that $\varphi : I \to \mathbb{R}$ and $\mathrm{id}_I - \varphi$ are nondecreasing. The weighted arithmetic mean $\mathscr{A}_p$ is invariant with respect to the mean-type mapping $\left(\mathscr{W}^{[\varphi]}, M\right)$, i.e.*

$$\mathscr{A}_p \circ \left(\mathscr{W}^{[\varphi]}, M\right) = \mathscr{A}_p,$$

*if, and only if,*

$$M = \mathscr{W}^{[\psi]}$$

*where, for some $c \in \mathbb{R}$,*

$$\psi(x) = \frac{p}{1-p}(x - \varphi(x)) + c, \qquad x \in I.$$

*Proof.* If $\mathscr{A}_p$ is $\left(\mathscr{W}^{[\varphi]}, M\right)$-invariant then, by the definition of the involved means, we have

$$p\,[\varphi(x) + y - \varphi(y)] + (1-p)M(x,y) = px + (1-p)y, \qquad x,y \in I,$$

whence, after simple calculations,

$$M(x,y) = \frac{p}{1-p}(x - \varphi(x)) + \left[y - \frac{p}{1-p}(y - \varphi(y))\right], \qquad x,y \in I,$$

and, consequently,

$$M(x,y) = \mathscr{W}^{[\psi]}(x,y), \qquad x,y \in I,$$

where, by Property 36.2$(iii)$, there is $c \in \mathbb{R}$ such that

$$\psi(x) = \frac{p}{1-p}(x - \varphi(x)) + c, \qquad x \in I.$$

The converse implication is easy to verify. □

Hence, applying the main result of [10] we obtain

**Corollary 36.3.** *Let $I \subset \mathbb{R}$ be an interval, $p \in (0,1)$. Suppose that $\varphi : I \to \mathbb{R}$ and $\mathrm{id}_I - \varphi$ are nondecreasing. If*

$$\psi(x) = \frac{p}{1-p}(x - \varphi(x)) + c, \qquad x \in I,$$

*then $\psi : I \to \mathbb{R}$ and $\mathrm{id}_I - \psi$ are nondecreasing, and the sequence of iterates*

$$\left(\left(\mathscr{W}^{[\varphi]}, \mathscr{W}^{[\psi]}\right)^n\right)_{n\in\mathbb{N}}$$

*of the mean-type mapping $(\mathscr{W}^{[\varphi]}, \mathscr{W}^{[\psi]})$ converges to the mean-type mapping $(\mathscr{A}_p, \mathscr{A}_p)$.*

**Corollary 36.4.** *Let $I \subset \mathbb{R}$ be an interval. Suppose that $\varphi : I \to \mathbb{R}$ and $\mathrm{id}_I - \varphi$ are nondecreasing and $F : I^2 \to \mathbb{R}$ is continuous on the diagonal $\{(x,x) : x \in I\}$. Then the function $F$ satisfies the functional equation*

$$F\left(\varphi(x) + y - \varphi(y), x - \varphi(x) + \varphi(y)\right) = F(x,y), \qquad x,y \in I,$$

*if, and only if, there is a continuous function $f : I \to \mathbb{R}$ such that*

$$F(x,y) = f\left(\frac{x+y}{2}\right), \qquad x,y \in I.$$

*Proof.* Putting $\psi(x) = x - \varphi(x)$ for $x \in I$, we can write the above functional equation in the form

$$F(x,y) = F\left(\mathscr{W}^{[\varphi]}(x,y), \mathscr{W}^{[\psi]}(x,y)\right), \qquad x,y \in I,$$

whence, by induction,

$$F(x,y) = F\left(\left(\mathscr{W}^{[\varphi]}(x,y), \mathscr{W}^{[\psi]}(x,y)\right)^n\right), \qquad x,y \in I, n \in \mathbb{N},$$

where

$$\left(\mathscr{W}^{[\varphi]}(x,y), \mathscr{W}^{[\psi]}(x,y)\right)^n = \left(\mathscr{W}^{[\varphi]}, \mathscr{W}^{[\psi]}\right)^n(x,y)$$

and $\left(\mathscr{W}^{[\varphi]}, \mathscr{W}^{[\psi]}\right)^n$ is the $n$th iterate of the mean-type mapping $(\mathscr{W}^{[\varphi]}, \mathscr{W}^{[\psi]})$. Letting $n \to \infty$ and, applying the previous corollary for $p = 1/2$ and making use of the continuity of $F$, we obtain

$$F(x,y) = F\left(\frac{x+y}{2}, \frac{x+y}{2}\right) = f\left(\frac{x+y}{2}\right), \qquad x,y \in I,$$

where $f(x) := F(x,x)$ for $x \in I$. The converse implication is easy to verify. □

## 36.9 A Generalization of the Weighted Quasi-Arithmetic Means

We begin this section with the following easy to verify

*Remark 36.5.* Let $J \subset \mathbb{R}$ be an interval and $\gamma : J \to \mathbb{R}$ be a continuous strictly monotonic function and let $I = \gamma(J)$. Suppose that $\varphi : I \to \mathbb{R}$ and and $\mathrm{id}_I - \varphi$ are nondecreasing. Then the function $M_{\gamma,\varphi} : J^2 \to J$,

$$M_{\gamma,\varphi}(x,y) := \gamma^{-1}(\varphi(\gamma(x)) + \gamma(y) - \varphi(\gamma(y))), \qquad x,y \in J,$$

is a mean.

For $\varphi(x) = px$ with $p \in (0,1)$, the mean $M_{\gamma,\varphi}$ reduces to the weighted quasi-arithmetic mean.

We prove the following

**Theorem 36.12.** *Let $\gamma : [0,\infty) \to \mathbb{R}$ be differentiable strictly monotonic and suppose that a differentiable $\varphi : I \to \mathbb{R}$ and $\mathrm{id}_I - \varphi$ are nondecreasing in $I := \gamma([0,\infty))$. Then the mean $M_{\gamma,\varphi}$ is homogeneous if, and only if, there is $p \in [0,1]$ such that*

$$M_{\gamma,\varphi}(x,y) = px + (1-p)y, \qquad x,y \geq 0.$$

*Proof.* Suppose that $M_{\gamma,\varphi}$ is homogeneous, i.e.

$$M_{\gamma,\varphi}(tx,ty) = tM_{\gamma,\varphi}(x,y), \qquad x,y,t \geq 0.$$

By the definition of $M_{\gamma,\varphi}$ we can write this equality in the form

$$\gamma^{-1}(\varphi(\gamma(tx)) + \gamma(ty) - \varphi(\gamma(ty))) = t\gamma^{-1}(\varphi(\gamma(x)) + \gamma(y) - \varphi(\gamma(y)))$$

for $x,y,t \geq 0$. Differentiating both sides with respect to $t$ and then setting $t = 0$ and $p := \varphi'[\gamma(0]$ gives $\gamma^{-1}(\varphi(\gamma(x)) + \gamma(y) - \varphi(\gamma(y))) = px + (1-p)y$ for $x,y \geq 0$. □

## 36.10 Finite Dimensional Counterparts of $\mathscr{W}^{[\varphi]}$

In a similar way as Theorem 36.1, we can prove the following

**Theorem 36.13.** *Let $I \subset \mathbb{R}$ be an interval and $\varphi_1, \ldots, \varphi_k : I \to \mathbb{R}$. Then the function $M : I^k \to \mathbb{R}$ defined by*

$$M(x_1, \ldots, x_k) := \varphi_1(x_1) + \cdots + \varphi_k(x_k)$$

*is a mean, i.e.*

$$\min(x_1,\dots,x_k) \le M(x_1,\dots,x_k) \le \max M(x_1,\dots,x_k), \qquad x_1,\dots,x_k \in I,$$

*if, and only if, the functions $\varphi_1,\dots,\varphi_k$ are nondecreasing and*

$$\varphi_1(x) + \cdots + \varphi_k(x) = x, \qquad x \in I.$$

Thus, if the nondecreasing functions $\varphi_1,\dots,\varphi_k : I \to \mathbb{R}$ are summing up to the identity, the function

$$\mathscr{W}^{[\varphi_1,\dots,\varphi_{k-1}]}(x_1,\dots,x_k) = \sum_{i=1}^{k-1} \varphi_i(x_i) + x_k - \sum_{i=1}^{k-1} \varphi_i(x_k), \qquad x_1,\dots,x_k \in I,$$

is a mean a $k$-variable mean in $I$.

Let us note the following easy to show

**Theorem 36.14.** *Let $I \subset \mathbb{R}$ be an interval. Suppose that the functions $\varphi_i : I \to \mathbb{R}$, $i = 1,\dots,k-1$, and*

$$\mathrm{id}\big|_I - \sum_{i=1}^{k-1} \varphi_i,$$

*are nondecreasing. Then the following conditions are equivalent*

*(i) the mean $\mathscr{W}^{[\varphi_1,\dots,\varphi_{k-1}]}$ is symmetric;*
*(ii) the mean $\mathscr{W}^{[\varphi_1,\dots,\varphi_{k-1}]}$ is subadditive;*
*(iii) there are $p_i \ge 0$, $i = 1,\dots,k$, such that $\sum_{i=1}^k p_i = 1$ and*

$$\mathscr{W}^{[\varphi_1,\dots,\varphi_{k-1}]}(x_1,\dots,x_k) = \sum_{i=1}^{k} p_i x_i, \qquad x_1,\dots,x_k \in I.$$

*Remark 36.6.* Let $I \subset \mathbb{R}$ be an interval. Suppose that the functions $\varphi_i, \psi_i : I \to \mathbb{R}$, $i = 1,\dots,k-1$, and

$$\mathrm{id}\big|_I - \sum_{i=1}^{k-1} \varphi_i, \qquad \mathrm{id}\big|_I - \sum_{i=1}^{k-1} \psi_i$$

are nondecreasing. Then the following conditions are equivalent

(i) $\mathscr{W}^{[\varphi_1,\dots,\varphi_{k-1}]} \le \mathscr{W}^{[\psi_1,\dots,\psi_{k-1}]}$;
(ii) $\mathscr{W}^{[\varphi_1,\dots,\varphi_{k-1}]} = \mathscr{W}^{[\psi_1,\dots,\psi_{k-1}]}$.

*Proof.* Assume that inequality (i) holds true. From the definition of $\mathscr{W}^{[\varphi_1,\dots,\varphi_{k-1}]}$ we have

$$\mathscr{W}^{[\varphi_1,\dots,\varphi_{k-1}]}(x_1,\dots,x_k) = \sum_{i=1}^{k-1} [\varphi_i(x_i) - \varphi_i(x_k)] + x_k, \qquad x_1,\dots,x_k \in I,$$

and, similarly,

$$\mathscr{W}^{[\psi_1,\dots,\psi_{k-1}]}(x_1,\dots,x_k) = \sum_{i=1}^{k-1}[\psi_i(x_i) - \psi_i(x_k)] + x_k, \qquad x_1,\dots,x_k \in I.$$

Let us fix arbitrarily $j \in \{1,\dots,k-1\}$ and take $x, y \in I$. Hence, putting $x_i = x_k = y$ for all $i \in \{1,\dots,k-1\}$, $i \neq j$, in inequality (i) we get

$$\varphi_j(x) - \varphi_j(y) \leq \psi_j(x) - \psi_j(y), \qquad x, y \in I,$$

whence

$$\varphi_j(x) - \psi_j(x) \leq \varphi_j(y) - \psi_j(y), \qquad x, y \in I.$$

Since $x, y \in I$ are arbitrary, it follows that

$$\varphi_j(x) - \psi_j(x) = \varphi_j(y) - \psi_j(y), \qquad x, y \in I,$$

for any $j \in \{1,\dots,k-1\}$ and $j \in \{1,\dots,k-1\}$. □

## 36.11 Final Remark and Some Questions

For a fixed $p \in (0,1)$ denote the weighted arithmetic mean

$$\mathscr{A}_p(x,y) = px + (1-p)y, \qquad x, y \in \mathbb{R}.$$

Daróczy and Páles [4] observed the following identity

$$\mathscr{A}_p\left(\mathscr{A}_p\left(\frac{x+y}{2}, x\right), \mathscr{A}_p\left(y, \frac{x+y}{2}\right)\right) = \frac{x+y}{2}, \qquad x, y \in \mathbb{R},$$

which appears very useful in the theory of convex functions.

Note that if $\gamma$ is continuous strictly increasing in an interval $I$ and $p \in (0,1)$, then

$$\mathscr{A}_p^{[\gamma]}\left(\mathscr{A}_p^{[\gamma]}\left(\mathscr{A}^{[\gamma]}, x\right), \mathscr{A}_p^{[\gamma]}\left(y, \mathscr{A}^{[\gamma]}\right)\right) = \mathscr{A}_p^{[\gamma]}(x,y), \qquad x, y \in I,$$

which reduces to the previous identity for $\gamma = id|_I$.

In this connection, the following problems arise.

Let $I \subset \mathbb{R}$ be an interval and let $\varphi : I \to \mathbb{R}$ and $\mathrm{id}_I - \varphi$ be strictly increasing.

**Problem 36.1.** Suppose that the following equation

$$\mathscr{W}^{[\varphi]}\left(\mathscr{W}^{[\varphi]}\left(\frac{x+y}{2},x\right),\mathscr{W}^{[\varphi]}\left(y,\frac{x+y}{2}\right)\right)=\frac{x+y}{2},\qquad x,y\in I,$$

is satisfied. Does it imply that $\mathscr{W}^{[\varphi]}$ coincides with $\mathscr{A}_p$ for some $p\in(0,1)$?

**Problem 36.2.** Do there exist $\mathscr{W}^{[\varphi]}$, being not weighted quasi-arithmetic, and a quasi-arithmetic mean $\mathscr{A}^{[\gamma]}$ such that

$$\mathscr{W}^{[\varphi]}\left(\mathscr{W}^{[\varphi]}\left(\mathscr{A}^{[\gamma]}(x,y),x\right),\mathscr{W}^{[\varphi]}\left(y,\mathscr{A}^{[\gamma]}(x,y)\right)\right)=\mathscr{A}^{[\gamma]}(x,y),\qquad x,y\in I\,?$$

## References

1. Aczél, J.: A generalization of the notion of convex functions. D.K.N.V.S Forh **19**, 87–90 (1946)
2. Aczél, J.: Functional equations and their applications. Academic Press, New York–London (1966)
3. Aumann, G.: Aufbau von Mittelwerten mehrerer Argumente I. Math. Ann. **109**, 235–253 (1934)
4. Daróczy, Z., Páles, Zs.: Convexity with given infinite weight sequences. Stochastica **11**, 5–12 (1987)
5. Kuczma, M.: An Introduction to the Theory of Functional Equations and Inequalities. Cauchy's Equations and Jensen Inequality. Uniwersytet Śląski–PWN, Warszawa–Kraków–Katowice (1985)
6. Kuhn, N.: A note on *t*–convex functions In: General Inequalities 4, Internat. Ser. Numer. Math. **71**, 269–276 (1984)
7. Matkowski, J.: On *a*–Wright convexityand the converse of Minkowski's inequality. Aequationes Math. **43**, 106–112 (1992)
8. Matkowski, J.: On a system of simultaneous iterative functional equations. Ann. Math. Siles. **9**, 123–135 (1995)
9. Matkowski, J., Rätz, J.: Convex functions with respect to an arbitrary mean. Internat. Ser. Numer. Math. **123**, 249–259 (1997)
10. Matkowski, J.: Iterations of mean–type mappings and invariant means. Ann. Math. Siles. **13**, 211–226 (1999)
11. Ng, C.T.: On midconvex functions with midconcave bounds. Proc. Amer. Math. Soc. **102**, 538–540 (1988).

# Chapter 37
# On Means Which are Quasi-Arithmetic and of the Beckenbach–Gini Type

**Janusz Matkowski**

*Dedicated to the memory of S.M. Ulam*
*on the 100th anniversary of his birth*

**Abstract** The class of quasi-arithmetic means and the class of Beckenbach–Gini means are essentially different. The problem of characterization of the means which belong to both classes leads to a composite functional equation for two unknown functions. We solve this functional equation assuming that a generator of quasi-arithmetic mean is once continuously differentiable.

**Keywords** Quasi-arithmetic mean • Beckenbach–Gini mean • Functional equation

**Mathematics Subject Classification (1991):** Primary 30B12, 26E60

## 37.1 Introduction

Let $I \subset \mathbb{R}$ be an interval. By a *mean* we mean any function $M : I^2 \to I$ such that

$$\min(x,y) \leq M(x,y) \leq \max(x,y), \qquad x,y \in I.$$

If for all $x,y \in I, x \neq y$, these inequalities are sharp, $M$ is called a *strict mean*. If $M : I^2 \to I$ is a mean then, of course, $M$ is *reflexive*, i.e.,

$$M(x,x) = x, \qquad x \in I.$$

J. Matkowski (✉)
Faculty of Mathematics Computer Science and Econometrics, University of Zielona Góra, Podgórna 50, PL-65-246 Zielona Góra, Poland

Institute of Mathematics, Silesian University, Bankowa 14, PL-42-007 Katowice, Poland
e-mail: J.Matkowski@wmie.uz.zgora.pl

Th.M. Rassias and J. Brzdęk (eds.), *Functional Equations in Mathematical Analysis*, Springer Optimization and Its Applications 52, DOI 10.1007/978-1-4614-0055-4_37,

Every reflexive function $M : I^2 \to I$ which is increasing with respect to each variable is a mean. In this paper we call such functions the *increasing means*.

If $\varphi : I \to \mathbb{R}$ is continuous and strictly monotonic and $f : I \to (0,\infty)$ an arbitrary function then $M : I^2 \to I$ given by

$$M(x,y) := \varphi^{-1}\left(\frac{\varphi(x)f(x)+\varphi(y)f(y)}{f(x)+f(y)}\right), \qquad x,y \in I,$$

is a mean and it is called *weighted quasi-arithmetic.*

Losonczi [5] considered the problem of equality of two quasi-arithmetic weighted means. Assuming among others that some of the involved functions are sixth times differentiable, reduces the problem to a sixth order differential equation. Next, employing the software package Maple V, he got 32 families of solutions. (For the historical background of the problem cf. [5].)

In the present paper, we consider a special case of the problem considered by Losonczi. Namely, we examine when the classical *quasi-arithmetic means* and *Beckenbach–Gini means* (cf. Beckenbach [2], Gini [4]; also: Bullen, Mitrinović, Vasić [3, Chap. III, p. 189]) coincide. However, we assume that functions involved are only continuously differentiable, and we apply elementary methods.

Recall that a mean $M = M^{[\varphi]} : I^2 \to I$ is quasi-arithmetic if there is a continuous and strictly monotonic function $\varphi : I \to \mathbb{R}$, called a *generator* of $M$, such that

$$M^{[\varphi]}(x,y) = \varphi^{-1}\left(\frac{\varphi(x)+\varphi(y)}{2}\right), \qquad x,y \in I.$$

Of course, every quasi-arithmetic mean is increasing and continuous. A mean $M = M_f$ is called a Beckenbach–Gini mean if there is a function $f : I \to (0,\infty)$, a generator of the mean, such that

$$M_f(x,y) = \frac{xf(x)+yf(y)}{f(x)+f(y)}, \qquad x,y \in I,$$

(note that one can assume that $f$ is of a constant sign). In general, a Beckenbach–Gini mean need be increasing and continuous. Clearly, both quasi-arithmetic and Beckenbach–Gini means are strict.

In Sect. 2 of the present paper, we examine when $M^{[\varphi]} = M_f$, i.e. when the Beckenbach–Gini and quasi-arithmetic means coincide. More exactly, we consider the functional equation

$$\varphi^{-1}\left(\frac{\varphi(x)+\varphi(y)}{2}\right) = \frac{xf(x)+yf(y)}{f(x)+f(y)}, \qquad x,y \in I,$$

when $\varphi : I \to \mathbb{R}$ and $f : I \to (0,\infty)$ are the unknown functions. Under the assumption of the continuous differentiability of the unknown functions we show (Theorem 37.1) that the functions $\varphi$ and $f$ satisfy this equation if, and only if,

$$\varphi'(x) = cf(x)^2, \qquad x \in I,$$

and

$$f(x) = \frac{1}{\sqrt{|Ax^2 + Bx + C|}}, \qquad x \in I,$$

for some $A, B, C, c \in \mathbb{R}$, $C \neq 0 \neq c$. Applying this result we determine all families of pairs of functions $(\varphi, f)$ satisfying the considered functional equation (Corollary 37.1) as well as the relevant means. In some cases, the generator $\varphi$ can be the log composed with a homographic function or arctan composed with an affine function.

The case $I = \mathbb{R}$ is considered. In this context, the generalized geometric and harmonic means appear. Note that these means can be treated as the Beckenbach–Gini ones but they are not quasi-arithmetic. The problem of extension of quasi-arithmetic means is considered. We end this section with some remarks on the three parameter family of Beckenbach–Gini means $\{M_{A,B,C} : A, B, C \in \mathbb{R}, A^2 + B^2 + C^2 > 0\}$, with $M_{A,B,C} := M_f$, where $f$ is given by

$$f(x) = \frac{1}{\sqrt{|Ax^2 + Bx + C|}},$$

indicating some interesting relations among these means and the generalized harmonic and geometric means.

In Sect. 3, we consider a higher dimensional version of equation $M^{[\varphi]} = M_f$. More exactly, we examine the functional equation

$$\varphi^{-1}\left(\frac{\sum_{j=1}^{k} \varphi(x_k)}{k}\right) = \frac{\sum_{j=1}^{k} x_k f(x_k)}{\sum_{j=1}^{k} f(x_k)}, \qquad x_1, \ldots, x_k \in I,$$

where $k \geq 3$ is a fixed positive integer. We show that a strictly monotonic continuously differentiable function $\varphi : I \to \mathbb{R}$ and a continuously differentiable function $f : I \to (0,\infty)$ satisfy this equation if, and only if, either

$$\varphi(x) = ax + b, \qquad f(x) = c, \qquad x \in I,$$

or

$$\varphi(x) = \frac{a}{x+d} + b, \qquad f(x) = \frac{c}{x+d}, \qquad x \in I,$$

for some $a, b, c, d \in \mathbb{R}$, $a \neq 0 \neq c$ (and the joint forms of both means $M^{[\varphi]}$ and $M_f$ are given).

Let us mention that if one of the means $M^{[\varphi]}$ or $M_f$ is positively homogeneous then the equation $M^{[\varphi]} = M_f$ is easy to solve (cf. [6, Theorem 6 and Corollary 3]).

## 37.2 The Functional Equation $M^{[\varphi]} = M_f$

Some properties of Beckenbach–Gini and quasi-arithmetic means can be found in [3, Chaps. III and IV]. We begin with recalling the following easy to verify

*Remark 37.1.* Let $I \subset \mathbb{R}$ be an interval and $f, g : I \to (0,\infty)$.Then $M_f = M_g$ if, and only if, $g = cf$ for some $c > 0$.

*Remark 37.2.* Let $I \subset \mathbb{R}$ be an interval and $\varphi, \psi : I \to \mathbb{R}$ continuous and strictly monotonic. Then $M^{[\varphi]} = M^{[\psi]}$ if, and only if, $\psi = af + b$ for some $a, b \in \mathbb{R}$, $a \neq 0$.

Let us also note that every mean has the following *mean property:* if $M : I^2 \to I$ is a mean and $J \subset I$ is an interval, then $M$ restricted to the set $J^2$ is a mean on $J^2$ and, consequently, $M(J^2) = J$.

The main result reads as follows

**Theorem 37.1.** *Let $I \subset \mathbb{R}$ be an open interval. A strictly monotonic continuously differentiable function $\varphi : I \to \mathbb{R}$ and a function $f : I \to (0,\infty)$ satisfy the functional equation $M^{[\varphi]} = M_f$ in $I$, i.e.,*

$$\varphi^{-1}\left(\frac{\varphi(x)+\varphi(y)}{2}\right) = \frac{xf(x)+yf(y)}{f(x)+f(y)}, \qquad x, y \in I, \tag{37.1}$$

*if, and only if,*

$$\varphi'(x) = cf(x)^2, \qquad x \in I,$$

*and*

$$f(x) = \frac{1}{\sqrt{|Ax^2+Bx+C|}}, \qquad x \in I,$$

*for some $A, B, C, c \in \mathbb{R}$, $C \neq 0 \neq c$.*

*Proof.* Suppose that a strictly monotonic and continuously differentiable function $\varphi : I \to \mathbb{R}$ and a function $f : I \to (0,\infty)$ satisfy equation 37.1.

Assume first that

$$\varphi'(x) \neq 0, \qquad x \in I.$$

Then (37.1) easily implies that $f$ is also continuously differentiable in $I$. Write (37.1) in the form

$$\varphi(x)+\varphi(y) = 2\varphi\left(\frac{xf(x)+yf(y)}{f(x)+f(y)}\right), \qquad x, y \in I. \tag{37.2}$$

Differentiating both sides of (37.2) with respect to $x$ and $y$, respectively, gives

$$\varphi'(x) = 2\varphi'\left(\frac{xf(x)+yf(y)}{f(x)+f(y)}\right)\frac{f(x)^2+f(x)f(y)+xf'(x)f(y)-yf'(x)f(y)}{(f(x)+f(y))^2} \tag{37.3}$$

and

$$\varphi'(y) = 2\varphi'\left(\frac{xf(x)+yf(y)}{f(x)+f(y)}\right)\frac{f(y)^2+f(y)f(x)+yf'(y)f(x)-xf'(y)f(x)}{(f(x)+f(y))^2} \tag{37.4}$$

for all $x,y \in I$. Subtracting these equations by sides and then dividing by $x-y$, for all $x,y \in I$, $x \neq y$, we obtain

$$\frac{\varphi'(x)-\varphi'(y)}{x-y} = 2\varphi'\left(\frac{xf(x)+yf(y)}{f(x)+f(y)}\right)\frac{\dfrac{f(x)^2-f(y)^2}{x-y}+f'(x)f(y)+f(x)f(y)'}{(f(x)+f(y))^2}.$$

Letting here $y \to x$ we see that the limit of the right hand side exists and, consequently,

$$\varphi''(x) = 2\varphi'(x)\frac{f'(x)}{f(x)}, \qquad x \in I.$$

It follows that the function $\varphi$ is twice continuously differentiable in $I$. Writing this relation in the form

$$\frac{\varphi''(x)}{\varphi'(x)} = 2\frac{f'(x)}{f(x)}, \qquad x \in I,$$

we infer that there exists a number $c \in \mathbb{R}$ such that

$$\varphi'(x) = cf(x)^2, \qquad x \in I. \tag{37.5}$$

Note that $c \neq 0$, as, by assumptions, $f$ has positive values and $\varphi$ is strictly monotonic. It follows that

$$\varphi'(x) \neq 0, \qquad x \in I,$$

and, by (37.1), the function $f$ is also twice continuously differentiable in $I$.

Now from (37.3) and (37.4) we have

$$\frac{\varphi'(x)}{\varphi'(y)} = \frac{f(x)^2+f(x)f(y)+xf'(x)f(y)-yf'(x)f(y)}{f(y)^2+f(y)f(x)+yf'(y)f(x)-xf'(y)f(x)}, \qquad x,y \in I.$$

Hence, making use of (37.5), we obtain

$$\frac{f(x)^2}{f(y)^2} = \frac{f(x)^2+f(x)f(y)+xf'(x)f(y)-yf'(x)f(y)}{f(y)^2+f(y)f(x)+yf'(y)f(x)-xf'(y)f(x)}, \qquad x,y \in I,$$

which is equivalent to the relation

$$\begin{aligned} f(x)f(y)\left[f(x)^2 - f(y)^2\right] \\ = (x-y)\left[f'(y)f(x)^3 - f'(x)f(y)^3\right], \qquad x,y \in I. \end{aligned} \tag{37.6}$$

Without any loss of generality, we can assume that

$$0 \in I .$$

By Remark 1 we can also assume that

$$f(0) = 1 .$$

Now, setting $y = 0$ in (37.5) gives

$$f'(x) = \frac{f(x)^3 - f(x)}{x} + af(x)^3 , \qquad x \in I, x \neq 0 . \tag{37.7}$$

Differentiating both sides of (37.6) with respect to $y$ we obtain, for all $x, y \in I$,

$$\begin{aligned} & f(x)f'(y)\left[f(x)^2 - f(y)^2\right] - 2f(x)f(y)^2f'(y) \\ & \quad = -\left[f'(y)f(x)^3 + f'(x)f(y)^3\right] + (x-y)\left[f''(y)f(x)^3 + 3f'(x)f(y)^2f'(y)\right] . \end{aligned}$$

Putting here $y := 0$ gives

$$(3ax - 1)f'(x) = (2a - bx)f(x)^3 - 3af(x) , \qquad x \in I ,$$

where

$$a := f'(0), \qquad b := f''(0) .$$

Making use of (37.7), we hence get, for all $x \in I$, $x \neq 0$,

$$(3ax - 1)\left[\frac{f(x)^3 - f(x)}{x} + af(x)^3\right] = (2a - bx)f(x)^3 - 3af(x) ,$$

which, as $f$ is positive and continuous, implies that

$$(3ax - 1)\left[f(x)^2 - 1 + axf(x)^2\right] = (2a - bx)xf(x)^2 - 3ax , \qquad x \in I ,$$

and, consequently,

$$f(x)^2 = \frac{1}{(3a^2 - b)x^2 - 2ax + 1} , \qquad x \in I .$$

Since, by the assumption, $f$ is positive, we get

$$f(x) = \frac{1}{\sqrt{|(3a^2 - b)x^2 - 2ax + 1|}} , \qquad x \in I .$$

Thus, we have shown that $f$ must be of the form

$$f(x) = \frac{1}{\sqrt{|Ax^2+Bx+C|}}, \qquad x \in I, \tag{37.8}$$

for some $A,B,C \in \mathbb{R}$, such that $C \neq 0$.

To finish this part of the proof put

$$Z := \{x \in I : \varphi'(x) = 0\}.$$

The continuity of $\varphi'$ implies that $Z$ is closed. By the strict monotonicity of $\varphi$, the interior of $Z$ is empty. Moreover, there exists at most countable set $S$ such that

$$I \setminus Z = \bigcup_{s \in S} I_s,$$

where $\{I_s : s \in S\}$ is a family of open (in $I$) and disjoint intervals. If $Z$ were not empty then there would exist an $s \in S$ such that

$$I_s = (x_0, x_1), x_0 < x_1, \text{ with } x_0 \in Z \text{ or } x_1 \in Z.$$

Assume, for instance, that $x_0 \in Z$. Since $\varphi' \neq 0$ in $I_s$, according to what we have already shown,

$$\varphi'(x) = cf(x)^2, \qquad x \in (x_0, x_1).$$

Equation (37.1) and the assumptions of $\varphi$ imply that $f$ is continuous in $I$. In particular, $f$ is continuous at $x_0$. Letting $x \to x_0$ we hence get

$$0 = \varphi'(x_0) = cf(x_0)^2,$$

and, consequently, $f(x_0) = 0$. This contradiction shows that the set $Z$ is empty, and completes the proof of the "only if" part of the theorem.

Making use formula (37.5), and considering several obvious cases (cf. Corollary 37.1 below), it is easy to verify that the converse implication holds true. □

*Remark 37.3.* Note that a function $f : I \to (0,\infty)$ satisfies (37.6) if, and only if, $f$ is of the form (37.8).

Applying Theorem 37.1 we obtain

**Corollary 37.1.** *Let $I \subset \mathbb{R}$ be an interval. A strictly monotonic continuously differentiable function $\varphi : I \to \mathbb{R}$ and a continuously differentiable function $f : I \to (0,\infty)$ satisfy the functional equation*

$$M^{[\varphi]}(x,y) = M_f(x,y), \qquad x,y \in I,$$

*if, and only if, one of the following cases occurs:*

1. *there are* $p,q,r,a,b\in\mathbb{R}$, $p<q$, $r>0$, $a\neq 0$, *such that* $I\subset(p,q)$,

$$f(x)=\frac{r}{\sqrt{(x-p)(q-x)}},\qquad \varphi(x)=a\log\frac{x-p}{q-x}+b,\qquad x\in I;$$

*moreover*

$$M^{[\varphi]}(x,y)=\frac{x\sqrt{(y-p)(q-y)}+y\sqrt{(x-p)(q-x)}}{\sqrt{(y-p)(q-y)}+\sqrt{(x-p)(q-x)}}=M_f(x,y),\qquad x,y\in I;$$

2. *there are* $p,q,r,a,b\in\mathbb{R}$, $p<q$, $r>0$, $a\neq 0$, *such that either* $I\subset(-\infty,p)$ *or* $I\subset(q,\infty)$, *and*

$$f(x)=\frac{r}{\sqrt{(x-p)(x-q)}},\qquad \varphi(x)=a\log\frac{x-p}{q-x}+b,\qquad x\in I;$$

*moreover*

$$M^{[\varphi]}(x,y)=\frac{x\sqrt{(y-p)(y-q)}+y\sqrt{(x-p)(x-q)}}{\sqrt{(y-p)(y-q)}+\sqrt{(x-p)(x-q)}}=M_f(x,y),\qquad x,y\in I;$$

3. *there are* $p,r,a,b\in\mathbb{R}$, $r>0$, $a\neq 0$, *such that either* $I\subset(-\infty,p)$ *or* $I\subset(p,\infty)$, *and*

$$f(x)=\frac{r}{\sqrt{(x-p)^2}}=\frac{r}{|x-p|},\qquad \varphi(x)=\frac{a}{x-p}+b,\qquad x\in I;$$

*moreover*

$$M^{[\varphi]}(x,y)=p+\frac{2(x-p)(y-p)}{(x-p)+(y-p)}=M_f(x,y),\qquad x,y\in I;$$

4. *there are* $p,q,r,a,b\in\mathbb{R}$, $p^2-4q<0$, $r>0$, $a\neq 0$, *such that*

$$f(x)=\frac{r}{\sqrt{x^2+px+q)}},\qquad \varphi(x)=a\arctan\frac{2x+q}{\sqrt{4q-p^2}}+b,\qquad x\in I;$$

*moreover*

$$M^{[\varphi]}(x,y)=\frac{x\sqrt{y^2+py+q}+y\sqrt{x^2+px+q}}{\sqrt{y^2+py+q}+\sqrt{x^2+px+q}}=M_f(x,y),\qquad x,y\in I;$$

5. *there are* $p,r,a,b\in\mathbb{R}$, $r>0$, $a\neq 0$, *such that either* $I\subset(-\infty,p)$,

$$f(x)=\frac{r}{\sqrt{p-x}},\qquad \varphi(x)=a\log(p-x)+b,\qquad x\in I,$$

*and moreover*

$$M^{[\varphi]}(x,y) = -\sqrt{(x-p)(y-p)}+p = \frac{x\sqrt{p-y}+y\sqrt{p-x}}{\sqrt{p-y}+\sqrt{p-x}} = M_f(x,y)\,, \quad x,y \in I\,;$$

*or* $I \subset (p,\infty)$,

$$f(x) = \frac{r}{\sqrt{x-p}}\,, \qquad \varphi(x) = a\log(x-p)+b\,, \qquad x \in I\,,$$

*and moreover*

$$M^{[\varphi]}(x,y) = \sqrt{(x-p)(y-p)}+p = \frac{x\sqrt{y-p}+y\sqrt{x-p}}{\sqrt{y-p}+\sqrt{x-p}} = M_f(x,y),\ \ x,y \in I\,;$$

6. *there are* $r$, $a,b \in \mathbb{R}$, $r > 0$, $a \neq 0$, *such* $f(x) = r$ *and* $\varphi(x) = ax+b$ *for* $x \in I$; *moreover,*

$$M^{[\varphi]}(x,y) = \frac{x+y}{2} = M_f(x,y)\,, \qquad x,y \in I\,.$$

*Remark 37.4.* Note the following facts.

1. *In the cases 4 and 6 of Corollary 37.1, taking* $I = \mathbb{R}$ *gives the suitable Beckenbach–Gini and quasi-arithmetic means which are globally defined on* $\mathbb{R}^2$.
2. *Let*

$$g(x) := \frac{1}{\sqrt{|(x-p)(x-q)|}}\,, \qquad x \in \mathbb{R}\backslash\{p,q\}\,,$$

*where* $p \neq q$, *and define* $F_{p,q} : (\mathbb{R}^2\backslash A) \to \mathbb{R}$ *where* $A := \{p,q\} \times \mathbb{R} \cup \mathbb{R} \times \{p,q\}$ *by*

$$F_{p,q}(x,y) := \frac{x\sqrt{|(y-p)(y-q)|}+y\sqrt{|(x-p)(x-q)|}}{\sqrt{|(y-p)(y-q)|}+\sqrt{|(x-p)(x-q)|}}\,.$$

*Since, for every* $y \in \mathbb{R}\backslash\{p,q\}$,

$$\lim_{x\to p} F_{p,q}(x,y) = \lim_{x\to p} F(x,y) = p\,, \quad \lim_{x\to q} F_{p,q}(x,y) = \lim_{x\to q} F(x,y) = q\,, \tag{37.9}$$

*the function* $F_{p,q}$ *is not extendable to a continuous mean on* $\mathbb{R}^2$. *Note that the Beckenbach–Gini means mentioned in the cases 1 and 2 of Corollary 37.1 coincide with* $F_{p,q}$ *on some proper subsets of* $\mathbb{R}^2$.
3. *The mean considered in case 3 (where* $p = q$*), which is a translated harmonic one, is extendable to a generalized harmonic mean* $H_p : \mathbb{R}^2 \to \mathbb{R}$ *defined by* $H_p(p,p) := p$ and

$$H_p(x,y) := p + \frac{(x-p)\,|y-p| + (y-p)\,|x-p|}{|y-p|+|x-p|}\,, \quad (x,y) \neq (p,p)\,.$$

*This mean is continuous on* $\mathbb{R}^2$ *and* $H_p(x,y) = p$ *for all* $x,y$ *with* $xy \le 0$. *Since*

$$H_p(x,y) := \frac{x|y-p|+y|x-p|}{|y-p|+|x-p|}, \quad (x,y) \ne (p,p),$$

$H_p$ *can be treated as a Beckenbach–Gini mean of the generator* $f(x) = 1/|x-p|$.

4. *Similarly, the mean considered in case 5 (where* $p = q$*), a translated geometric one, is extendable to a generalized geometric mean* $G_p : \mathbb{R}^2 \to \mathbb{R}$ *defined by* $H_p(p,p) := p$ and

$$H_p(x,y) := p + \frac{(x-p)\sqrt{|y-p|}+(y-p)\sqrt{|x-p|}}{\sqrt{|y-p|}+\sqrt{|x-p|}}, \quad (x,y) \ne (p,p).$$

*This mean is continuous on* $\mathbb{R}^2$ *and* $G_p(x,y) = 0$ *for all* $x,y$ *with* $xy \le 0$. *Since*

$$H_p(x,y) := \frac{x\sqrt{|y-p|}+y\sqrt{|x-p|}}{\sqrt{|y-p|}+\sqrt{|x-p|}}, \qquad (x,y) \ne (p,p),$$

so $H_p$ *can be treated as a Beckenbach–Gini mean of the generator* $f(x) = 1/\sqrt{|x-p|}$.

*Remark 37.5.* For every $p \in \mathbb{R}$, the generalized harmonic and geometric means, $H_p$ and $G_p$, are unique increasing means which coincide with the translated harmonic and geometric means, respectively.

*Proof.* Since for all $x,y$ such that $(x-p)+(y-p) > 0$,

$$H_p(x,y) = p + \frac{2(x-p)(y-p)}{(x-p)+(x-p)}, \qquad G_p(x,y) := p + \sqrt{(x-p)(y-p)},$$

and for all $x,y$ such that $(x-p)+(y-p) < 0$,

$$H_p(x,y) = p + \frac{2(p-x)(p-y)}{(p-x)+(p-y)}, \qquad G_p(x,y) := p + \sqrt{(p-x)(p-y)},$$

the means $H_p$ and $G_p$ on the sets $(p,\infty)^2$ and $(-\infty,p)^2$ coincide, respectively, with the translated harmonic and geometric means.

Now suppose that $M : \mathbb{R}^2 \to \mathbb{R}$ is an increasing mean which coincides with the harmonic translated means on $(p,\infty)^2$ and $(-\infty,p)^2$. Then for all $x,y$ such that $x \le p \le y$ we have

$$p = H_p(x,p) = M(x,p) \le M(x,y) \le M(p,y) = H_p(p,y) = p,$$

which proves that $M(x,y) = p$ for all $x \le p \le y$. □

As we have observed, the generalized harmonic and geometric means are of the Beckenbach–Gini type. However, we have the following.

*Remark 37.6.* The generalized harmonic and geometric means are not quasi-arithmetic ones.

This is a consequence of the subsequent

**Proposition 37.1.** *Let $I, J \subset \mathbb{R}$ be intervals such that $I = (a,b)$, $J = (b,c)$ for some $a,b,c$, $-\infty \le a < b < c \le \infty$. Suppose that $M^{[\psi]} : I^2 \to I$ and $M^{[\gamma]} : J^2 \to J$ are quasi-arithmetic means of the suitable generators $\psi : I \to \mathbb{R}$ and $\gamma : J \to \mathbb{R}$. The following two conditions are equivalent:*

1. *there exists a quasi-arithmetic mean $M^{[\varphi]} : (a,c)^2 \to (a,c)$ such that*

$$M^{[\varphi]}(x,y) = M^{[\psi]}(x,y), \quad x,y \in I; \qquad M^{[\varphi]}(x,y) = M^{[\gamma]}(x,y), \quad x,y \in J;$$

2. *the limits*

$$\psi(b-) := \lim_{x \to b-} \psi(x), \qquad \gamma(b+) := \lim_{x \to b+} \gamma(x)$$

*exist and are finite.*

*Proof.* It is an immediate consequence of the definition of a quasi-arithmetic mean and Remark 37.2. □

Take arbitrarily fixed $A, B, C \in \mathbb{R}$ such that $A^2 + B^2 + C^2 > 0$, and put $M_{A,B,C} := M_f$, where $f$ is given by $f(x) = 1/\sqrt{|Ax^2 + Bx + C|}$. The next remark shows an interesting relations of the family of Beckenbach–Gini means $M_{A,B,C}$ and the generalized harmonic and geometric means for $p = 0$.

*Remark 37.7.* The three-parameter family of means

$$\{M_{A,B,C} : A, B, C \in \mathbb{R},\ A^2 + B^2 + C^2 > 0\}$$

has the following properties: for all $x, y \in \mathbb{R}$,

$$\lim_{A \to \infty} M_{A,B,C}(x,y) = \lim_{A \to -\infty} M_{A,B,C}(x,y) = \frac{x|y| + y|x|}{|y| + |x|} = H_0(x,y),$$

$$\lim_{B \to \infty} M_{A,B,C}(x,y) = \lim_{B \to -\infty} M_{A,B,C}(x,y) = \frac{x\sqrt{|y|} + y\sqrt{|x|}}{\sqrt{|y|} + \sqrt{|x|}} = G_0(x,y),$$

$$\lim_{C \to \infty} M_{A,B,C}(x,y) = \lim_{C \to -\infty} M_{A,B,C}(x,y) = \frac{x+y}{2}.$$

## 37.3 A Higher Dimensional Case

In this section, we prove the following.

**Theorem 37.2.** *Let $I \subset \mathbb{R}$ be an interval and $k \in \mathbb{N}$, $k \geq 3$, fixed. A strictly monotonic continuously differentiable function $\varphi : I \to \mathbb{R}$ and a continuously differentiable function $f : I \to (0,\infty)$ satisfy the equation $M^{[\varphi]} = M_f$ in $I$, i.e.*

$$\varphi^{-1}\left(\frac{\sum_{j=1}^{k}\varphi(x_j)}{j}\right) = \frac{\sum_{j=1}^{k} x_j f(x_j)}{\sum_{j=1}^{k} f(x_j)}, \qquad x_1,\dots,x_k \in I, \tag{37.10}$$

*if, and only if, either, for some $a,b,c \in \mathbb{R}$, $a \neq 0 \neq c$,*

$$\varphi(x) = ax+b, \qquad f(x) = c, \qquad x \in I,$$

*and then both means $M^{[\varphi]}$ and $M_f$ are equal to the arithmetic one; or for some $a,b,c,d \in \mathbb{R}$, $a \neq 0 \neq c$,*

$$\varphi(x) = \frac{a}{x+d} + b, \qquad f(x) = \frac{c}{x+d}, \qquad x \in I,$$

*and then both means $M^{[\varphi]}$ and $M_f$ are equal to the d-translated harmonic one, i.e., for all $x_1,\dots,x_k \in I$,*

$$M^{[\varphi]}(x_1,\dots,x_k) = \frac{k}{1/(x+d)+\dots+1/(x+d)} - d = M_f(x_1,\dots,x_k)\,.$$

*Proof.* For the simplicity of notations assume that $k = 3$. Suppose $f : I \to (0,\infty)$ and $\varphi : I \to \mathbb{R}$ are continuously differentiable and satisfy (37.10). Then we have

$$\varphi\left(\frac{xf(x)+yf(y)+zf(z)}{f(x)+f(y)+f(z)}\right) = \frac{\varphi(x)+\varphi(y)+\varphi(z)}{3}, \qquad x,y,z \in I\,.$$

Differentiating this equation, first with respect to $x$ and next with respect to $y$, gives

$$\begin{aligned}\varphi'(x) &= 3\varphi'\left(M_f(x,y,z)\right)\\ &\quad\times \frac{[f(x)+xf'(x)]\,[f(x)+f(y)+f(z)] - f'(x)\,[xf(x)+yf(y)+zf(z)]}{[f(x)f(y)+f(z)]^2}\,,\end{aligned}$$

$$\begin{aligned}\varphi'(y) &= 3\varphi'\left(M_f(x,y,z)\right)\\ &\quad\times \frac{[f(y)+yf'(y)]\,[f(x)+f(y)+f(z)] - f'(y)\,[xf(x)+yf(y)+zf(z)]}{[f(x)f(y)+f(z)]^2}\end{aligned}$$

for all $x,y,z \in I$.

Since $\varphi$ is strictly monotonic, the set $U := \{x \in I : \varphi'(x) \neq 0\}$ is non-empty and open. Let $J \subset U$ be a nontrivial maximal interval. Since $M_f$ is a mean, we have $M_f(J^3) \subset J$. Dividing these two equations by sides we get

$$\frac{\varphi'(x)}{\varphi'(y)} = \frac{[f(x)+xf'(x)]\,[f(x)+f(y)+f(z)] - f'(x)\,[xf(x)+yf(y)+zf(z)]}{[f(y)+yf'(y)]\,[f(x)+f(y)+f(z)] - f'(y)\,[xf(x)+yf(y)+zf(z)]} \tag{37.11}$$

for all $x,y,z \in J$, which proves that the right hand side does not depend on $z$. Hence, writing the derivative of the right hand side with respect to $z$, we get

$$\begin{aligned} 0 = &\left\{\left[f(x)+xf'(x)\right] f'(z) - f'(x)\left[f(z)+zf'(z)\right]\right\} \\ &\times \left\{\left[f(y)+yf'(y)\right][f(x)+f(y)+f(z)] - f'(y)\,[xf(x)+yf(y)+zf(z)]\right\} \\ &- \left\{\left[f(y)+yf'(y)\right] f'(z) - f'(y)\,[f(z)+zf(z)]\right\} \\ &\times \left\{\left[f(x)+xf'(x)\right][f(x)+f(y)+f(z)] - f'(x)\,[xf(x)+yf(y)+zf(z)]\right\} \end{aligned}$$

for all $x,y,z \in J$. Setting here $y := z$ gives

$$\begin{aligned} 0 = &\left\{\left[f(x)+xf'(x)\right] f'(z) - f'(x)\left[f(z)+zf'(z)\right]\right\} \\ &\times \left\{\left[f(z)+zf'(z)\right][f(x)+2f(z)] - f'(z)\,[xf(x)+2zf(z)]\right\} = 0 \end{aligned}$$

for all $x,z \in J$. For $z = x$ the factor in the second brace reduces to $3f(x)^2$. Since, by assumption, $f$ is positive, it follows that for every $x \in J$ there is a non-empty open interval $J_x$ such that $x \in J_x$ and for all $z \in J_x$, the second factor is different than 0. Now the above equation implies that, for every $x \in J$,

$$\left[f(x)+xf'(x)\right] f'(z) - f'(x)\left[f(z)+zf'(z)\right] = 0\,, \qquad z \in J_x\,. \tag{37.12}$$

Treating here $x$ as fixed we get

$$\frac{f'(z)}{f(z)} = \frac{\beta}{z+d}\,, \qquad z \in J_x\,,$$

where $d$ and $\beta$ are some constant (depending on fixed $x$) and, consequently,

$$f(z) = c(z+d)^{\beta}\,, \qquad z \in I_x\,, \tag{37.13}$$

for some constant $c > 0$. Assume first that $\beta \neq 0$. Setting the function (37.13) into relation (37.12), and performing simple calculations leads to $\beta(z-x) = x - z$ for

$z \in J_x$, whence $\beta = -1$. Now the differentiability of $f$ easily implies that $d$ and $c$ must be constant independent on $x$. Thus, we have shown that, for some $d \in \mathbb{R}$ and $c > 0$,

$$f(x) = \frac{c}{x+d}, \qquad x \in J. \tag{37.14}$$

If $d = 0$, then

$$M_f(x,y,z) = \frac{xf(x)+yf(y)+zf(z)}{f(x)+f(y)+f(z)} = \frac{3}{1/x+1/y+1/z}, \quad x,y,z \in J,$$

is the harmonic mean. Applying Remark 37.2 we infer that, for some $a, b \in \mathbb{R}$, $a \neq 0$, $\varphi(x) = a/x + b$ for $x \in J$. Now we are going to determine the form of the function $\varphi$ in the case $d \neq 0$. Setting the function (37.14) into (37.11) and fixing arbitrarily $y$ ,$z \in J$ gives

$$\varphi'(x) = \frac{Ax^2+Bx+C}{Dx^2+Ex+F}, \qquad x \in J, \tag{37.15}$$

for some real constant $A,B,C,D,E,F$. Replacing the values of $\varphi'$, $f$ and $f'$ in (37.11) by the suitable expressions given by the formulas (37.14) and (37.15), we obtain an equality of two polynomials of the variables $x$ and $y$ (the variable $z$ disappears). Comparing the coefficients at the relevant monomials leads to the following system of equations for the unknown numbers $A,B,C,D,E,F$ :

$$\begin{gathered}
AF = 0, \qquad AE = 0, \qquad AD = 0, \\
D(2Ad+B) = 0, \qquad E(2Ad+B) = 0, \qquad F(2Ad+B) = 0, \\
AFd^2 + 2BFd + C(F - Dd^2) = 0, \qquad BFd + C(2F - Ed) = 0, \\
AEd^2 + Bd(2E - Dd) + C(E - 2Dd) = 0.
\end{gathered}$$

If $A \neq 0$ then, according to the first three equations of this system, we would have $D = E = F = 0$ which, in view of formula (37.15), is impossible. Thus, $A = 0$ and this algebraic system of equations simplifies to the following one:

$$\begin{gathered}
DB = 0, \qquad EB = 0, \qquad FB = 0, \\
2BFd + C(F - Dd^2) = 0, \qquad BFd + C(2F - Ed) = 0, \\
Bd(2E - Dd) + C(E - 2Dd) = 0.
\end{gathered}$$

Note that $B = 0$ (otherwise we would again have $D = E = F = 0$ which, by (37.15), is impossible). Now the algebraic system of equations implies that $C(F - Dd^2) = 0$, $C(2F - Ed) = 0$, and $C(E - 2Dd) = 0$. According to our assumption, $\varphi' \neq 0$ in $J$. Since $A = B = 0$, in view of (37.15), we have $C \neq 0$. Therefore $F = Dd^2$ and $E = 2Dd$, whence

$$\varphi'(x) = \frac{C}{Dx^2 + 2D\mathrm{d}x + Dd^2} = \frac{C}{D(x+d)^2}, \qquad x \in J.$$

Hence, making use of the continuous differentiability of $\varphi$ and the maximality of $J$ we infer that $J = I$, and

$$\varphi(x) = \frac{a}{x+d} + b, \qquad x \in I,$$

where $a := -C/D$ and $b \in \mathbb{R}$. Note that this formula includes also the case $d = 0$.

Assume now that there is an $x \in J$ such that $\beta = 0$ in (37.13). Then the function $f$ is constant in in the interval $J_x$. Now it is a consequence of the previous reasoning that $f$ is constant in $I$, i.e. $f(x) = c$ for all $x \in I$. In this case we have

$$M_f(x,y,z) = \frac{xf(x)+yf(y)+zf(z)}{f(x)+f(y)+f(z)} = \frac{x+y+z}{3}, \qquad x,y,z \in I.$$

It follows from (37.9) that, for some $a, b \in \mathbb{R}$, $a \neq 0$, $\varphi(x) = ax + b$ for $x \in I$. The verification of the converse implication is a matter of easy calculations. □

*Remark 37.8.* Let us mention that the problem of equality of $k$-variable weighted quasi-arithmetic means of the form

$$\varphi^{-1}\left(\frac{\sum_{j=1}^{k} g(x_j)\varphi(x_j)}{g(x_j)}\right)$$

with $k \geq 3$, under the assumptions that the involved functions are twice continuously differentiable, was considered by Bajraktarević [1].

## References

1. Bajraktarević, M.: Sur une équation fonctionelle aux valeurs moyennes. Glasnik Mat.-Fiz. Astr. **13**, 243–248 (1958)
2. Beckenbach, E.F.: A class of mean value functions. Amer. Math. Monthly **57**, 1–6 (1950)
3. Bullen, P.S., Mitrinivić, D.S., Vasić, P.M.: Means and Their Inequalities. D. Reidel Publishing Company, Dordrecht-Boston-Lancaster-Tokyo (1988)
4. Gini, C. : Di una formula comprensiva delle medie. Metron **13**(2), 3–22 (1938)
5. Losonczi, L.: Equality of two variable weighted means: reduction to differential equations. Aequationes Math. **58**, 223–241 (1999)
6. Matkowski, J.: On invariant generalized Beckenbach–Gini means. In: Daróczy, Z., Páles, Zs. (eds.) Functional Equations – Results and Advances, pp. 219–230. Kluwer Academic Press, Boston-Dordrecht-London (2002)

# Chapter 38
# Scalar Riemann–Hilbert Problem for Multiply Connected Domains

**Vladimir V. Mityushev**

*Dedicated to the memory of S.M. Ulam*
*on the 100th anniversary of his birth*

**Abstract** We solve the scalar Riemann–Hilbert problem for circular multiply connected domains. The method is based on the reduction of the boundary value problem to a system of functional equations (without integral terms). In the previous works, the Riemann–Hilbert problem and its partial cases such as the Dirichlet problem were solved under geometrical restrictions to the domains. In the present work, the solution is constructed for any circular multiply connected domain in the form of modified Poincaré series.

**Keywords** Boundary value problem • Multiply connected domain • Schwartz operator • Poincaré series • Dirichlet problem • Harmonic measure • Green function

**Mathematics Subject Classification (2000):** Primary 30E25

## 38.1 Introduction

Various boundary value problems are reduced to singular integral equations [13,32,38]. Only some of them can be solved in closed form. In the present paper we follow the lines of the book [30] and describe application of the functional equations method to the scalar linear Riemann–Hilbert problem for multiply connected domains.

This problem can be considered as a generalization of the classical Dirichlet and Neumann problems for harmonic functions. It includes as a partial case the mixed

---

V.V. Mityushev (✉)
Department of Computer Science and Computer Methods, Pedagogical University, Podchorążych 2, 30-084 Kraków, Poland
e-mail: vmityu@yahoo.com

Th.M. Rassias and J. Brzdęk (eds.), *Functional Equations in Mathematical Analysis*, Springer Optimization and Its Applications 52, DOI 10.1007/978-1-4614-0055-4_38,

boundary value problem. One knows the famous Poisson formula which solves the Dirichlet problem for a disk. The exact solution of the Dirichlet problem for a circular annulus is also known due to Villat–Dini (see [8, p. 119] and [3, p.169]). Formulae from Theorems 38.4 and 38.5 presented below can be considered as a generalization of the Poisson and Villat–Dini formulae to arbitrary circular multiply connected domains.

In the present paper, we propose complete solution to the scalar Riemann–Hilbert problem for any multiply connected domain using functional equations. By *functional equations* we mean iterative functional equations [21, 30] with shift into domain. Hence, we do not use traditional integral equations and infinite systems of linear algebraic equations. We give exact formulae for the solution when the unknown function is given explicitly in terms of the coefficients of the problem and geometric parameters of the domain.

Despite the solution is given by exact formulae, its structure is not elementary. More precisely, it is represented in the form of integrals involving the Abelian functions [5] (Poincaré series or their counterparts [30]). The reason, why the solution in general is not presented by integrals involving elementary kernels, has the topological nature. In order to explain this, we shortly recall the scheme of the solution to the Riemann–Hilbert problem

$$\phi(t)+G(t)\overline{\phi(t)}=g(t)\,,\quad t\in\partial\mathbb{C}^+, \tag{38.1}$$

for the upper half-plane $\mathbb{C}^+$ following [13, 32]. Define the function $\phi^-(z):=\overline{\phi(\overline{z})}$ analytic in the lower half-plane. Then the Riemann–Hilbert problem (38.1) becomes the $\mathbb{C}$-linear problem (Riemann problem)

$$\phi^+(t)+G(t)\phi^-(t)=g(t)\,,\quad t\in\partial\mathbb{C}^+. \tag{38.2}$$

The latter problem is solved in terms of the Cauchy type integrals (see details in [13, 32]).

Let us look at this scheme from another point of view [41]. Introduce a copy of the upper half-plane $\mathbb{C}^+$ with the local complex coordinate $\overline{z}$ and glue it with $\mathbb{C}^+$ along the real axis. Define the function $\phi^-(\overline{z}):=\overline{\phi(z)}$ analytic on the copy of $\mathbb{C}^+$. Then we again arrive at the $\mathbb{C}$-linear problem (38.2) but on the double of $\mathbb{C}^+$ which is conformally equivalent to the Riemann sphere $\hat{\mathbb{C}}$. The fundamental functionals of $\hat{\mathbb{C}}$ are expressed by means of meromorphic functions which produces the Cauchy type integrals. The same scheme holds for any $n$-connected domain $D$. In result, we arrive at the problem (38.2) on the Schottky double of $D$, the Riemann surface of genus $(n-1)$, where the life is more complicated than on the plane considered as the Riemann sphere of zero genus. It does not described by meromorphic functions. Therefore, if one tries to solve the problem (38.2) on the double of $D$, he has to use meromorphic analogies of the Cauchy kernel on Riemann surfaces, i.e., the Abelian functions. In the case $n=2$, the double of $D$ becomes a torus in which meromorphic functions are replaced by the classical elliptic functions [3]. Crowdy [9–11] used the Schottky–Klein prime function associated with the Schottky double of $D$ to solve many different problems for multiply connected domains.

The paper is organized as follows. First, we state the problem, describe the known results, discuss the $\mathbb{R}$-linear and the Schwarz problems. In Sect. 38.2, we discuss functional equations and prove the convergence of the method of successive approximations for these equations. In Sect. 38.3, the harmonic measures of the circular multiply connected domains and the Schwarz operator are constructed by the method which can be outlined as follows. At the beginning the Schwarz problem is written as an $\mathbb{R}$-linear problem. Then we reduce it to functional equations. Application of the method of successive approximations yields the solution in the form of the Poincaré series of weight 2. As a sequence we obtain the almost uniform convergence of the Poincaré series for any multiply connected domain.

Further, by application of the standard factorization method we reduce the problem (38.3) to the Riemann–Hilbert problem with constant coefficients in each component of $\partial D$. The latter problem is also solved by application of successive approximations to functional equations.

### 38.1.1 Statement of the Riemann–Hilbert Problem

Let $D$ be a multiply connected domain on the complex plane whose boundary $\partial D$ consists of $n$ simple closed Jordan curves. The positive orientation on $\partial D$ leaves $D$ to the left.

The scalar linear Riemann–Hilbert problem for $D$ is stated as follows. Given Hölder continuous functions $\lambda(t) \neq 0$ , $f(t)$ on $\partial D$. To find a function $\phi(z)$ analytic in $D$, continuous in the closure of $D$ with the boundary condition

$$\operatorname{Re} \overline{\lambda(t)}\phi(t) = f(t) , \quad t \in \partial D . \tag{38.3}$$

This condition can be also written in the form (38.1).

The problem (38.3) had been completely solved for simply connected domains ($n = 1$). Its solution and general theory of boundary value problems is presented in the classic books by Gakhov [13], Muskhelishvili [32] and Vekua [38]. In 1975 Bancuri [6] had solved the Riemann–Hilbert problem for circular annulus ($n = 2$).

First results concerning the Riemann–Hilbert problem for general multiply connected domains were obtained by Kveselava [20] in 1945. He reduced the problem to an integral equation. From 1952, Vekua and later Bojarski began to extensively study this problem. Their results are presented in the book [38]. This Georgian attack to the problem supported by young Polish mathematician were successful. Due to Kveselava, Vekua and Bojarski, we have a theory of solvability of the problem (38.3) based on integral equations and estimations of its defect numbers, $l$, the number of linearly independent solutions and $p$, the number of linearly independent conditions of solvability on $f(t)$. Here, $\varkappa = \operatorname{wind}_{\partial D}\lambda$ is the

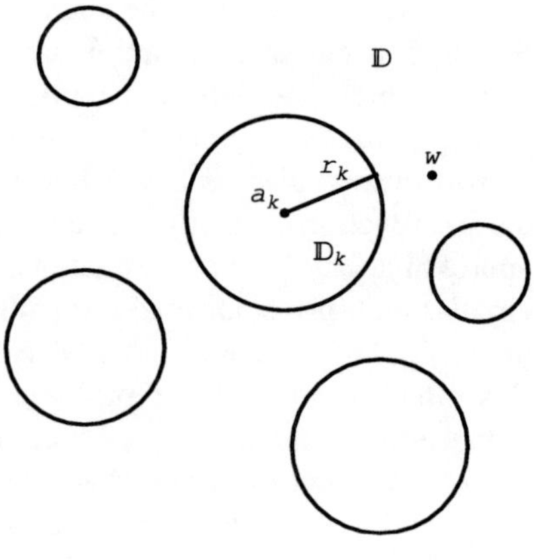


**Fig. 38.1** Multiply connected circular domain $\mathbb{D}$ and disks $\mathbb{D}_k$. A point $w$ is fixed in $\mathbb{D}\setminus\{\infty\}$

index of the problem. In particular, Bojarski obtained the exact estimation $l_\varkappa \leq \varkappa + 1$. In the special case $0 < \varkappa < n-2$, Bojarski shown that solvability of the problem depends on a system of linear algebraic equations with $2\varkappa$ unknowns. It was also demonstrated that the rank of this system differs from $2\varkappa$ on the set of zeros of an analytic function of few variables. Hence, almost always

$$l = \max(0, 2\varkappa - n + 2).$$

In 1971, Zverovich [41] developed the theory by reduction the problem (38.3) to the $\mathbb{C}$-linear problem on the Riemann surface (38.2) and shown that the solution of the problem is expressed in terms of the fundamental functionals of the double of $D$. The complete solution to the problem (38.3) had been given in [26, 29, 30] by the method of functional equations. It is presented in Sect. 38.4 with some corrections and modifications.

Any multiply connected domain $D$ can be conformally mapped onto a circular multiply connected domain ([17, p. 235]). Hence, it is sufficient to solve the problem (38.3) for a circular domain and after to write the solvability conditions and solution using the conformal mapping.

Let us consider mutually disjointed disks (Fig. 38.1)

$$\mathbb{D}_k := \{z \in \mathbb{C} : \ |z - a_k| < r_k\} \quad (k = 1, 2, ..., n)$$

in the complex plane $\mathbb{C}$. Let $\mathbb{D}$ be the complement of the closed disks $|z - a_k| \leq r_k$ to the extended complex plane $\widehat{\mathbb{C}} = \mathbb{C} \cup \{\infty\}$, i.e.,

$$\mathbb{D} := \widehat{\mathbb{C}} \setminus \bigcup_{k=1}^{n} (\mathbb{D}_k \cup \partial \mathbb{D}_k) .$$

It is assumed that $\mathbb{T}_k \cap \mathbb{T}_m = \emptyset$ for $k \neq m$. The circles $\mathbb{T}_k := \{t \in \mathbb{C} : \ |t - a_k| = r_k\}$ leaves $\mathbb{D}$ to the left.

### 38.1.2 $\mathbb{R}$-Linear Problem

Let $D$ be a multiply connected domain described above. Let $D_k$ $(k = 1, 2, \ldots, n)$ be simply connected domains complementing $D$ to the extended complex plane. The $\mathbb{R}$-*linear conjugation problem* or simply $\mathbb{R}$-*linear problem* is stated as follows. Given Hölder continuous functions $a(t) \neq 0$, $b(t)$ and $c(t)$ on $\partial D$. To find a function $\phi(z)$ analytic in $\bigcup_{k=1}^{n} D_k \cup D$, continuous in $D_k \cup \partial D_k$ and in $D \cup \partial D$ with the conjugation condition

$$\phi^+(t) = a(t)\phi^-(t) + b(t)\overline{\phi^-(t)} + c(t)\,, \quad t \in \partial D. \tag{38.4}$$

Here, $\phi^+(t)$ is the limit value of $\phi(z)$ when $z \in D$ tends to $t \in \partial D$, $\phi^-(t)$ is the limit value of $\phi(z)$ when $z \in D_k$ tends to $t \in \partial D$. In the case $|a(t)| \equiv |b(t)|$ the $\mathbb{R}$-linear problem is reduced to the Riemann–Hilbert problem (38.3) [24].

In the case of the smooth boundary $\partial D$, the homogeneous $\mathbb{R}$-linear problem with constant coefficients

$$\phi^+(t) = a\phi^-(t) + b\overline{\phi^-(t)}\,, \quad t \in \partial D. \tag{38.5}$$

is equivalent to the *transmission problem* from the theory of harmonic functions

$$u^+(t) = u^-(t), \quad \lambda^+ \frac{\partial u^+}{\partial n}(t) = \lambda^- \frac{\partial u^-}{\partial n}(t)\,, \quad t \in \partial D\,. \tag{38.6}$$

Here, the real function $u(z)$ is harmonic in $D$ and continuously differentiable in $D_k \cup \partial D_k$ and in $D \cup \partial D$, $\frac{\partial}{\partial n}$ is the normal derivative to $\partial D$. The conjugation conditions express the perfect contact between materials with different conductivities $\lambda^+$ and $\lambda^-$. The functions $\phi(z)$ and $u(z)$ are related by the equalities

$$u(z) = \mathrm{Re}\,\phi(z)\,, \quad z \in D\,, \tag{38.7}$$

$$u(z) = \frac{\lambda^- + \lambda^+}{2\lambda^+}\mathrm{Re}\,\phi(z)\,, \quad z \in D_k \ (k = 1, 2, \ldots, n)\,.$$

The coefficients are related by formulae

$$a = 1\,, \quad b = \frac{\lambda^- - \lambda^+}{\lambda^- + \lambda^+}\,. \tag{38.8}$$

For details see [30, Sect. 2.12]. Let us note that for positive $\lambda^+$ and $\lambda^-$ we arrive at the elliptic case $|b| < |a|$ in accordance with Mikhajlov's terminology [24].

The non-homogeneous problem (38.4) with real coefficients $a(t)$ and $b(t)$ can be written as a transmission problem (38.6). If $a(t)$ and $b(t)$ are complex the transmission problem takes more complicated form [24].

In 1932, having used the theory of potentials Muskhelishvili [31] (see also [33, p. 522]) reduced the problem (38.6) to a Fredholm integral equation and proved that it has a unique solution in the case $\lambda^{\pm} > 0$, the most interesting in applications. In 1933 I. N. Vekua and Ruhadze [36, 37] constructed a solution of (38.6) in closed form for annulus and ellipse (see also papers by Ruhadze quoted in [33]). Hence, the paper [31] published in 1932 is the first result on solvability of the $\mathbb{R}$-linear problem, [36] and [37] published in 1933 are the first papers devoted to exact solution to the the $\mathbb{R}$-linear problem for annulus and ellipse. A little bit later Golusin [16] considered the $\mathbb{R}$-linear problem in the form (38.6) by use of the functional equations for analytic functions (see below Sect. 38.1.4). Therefore, Golusin's paper [16] published in 1935 is the first paper which concerns constructive solution to the the $\mathbb{R}$-linear problem for special circular multiply connected domains. In the further works these first results were not associated to the $\mathbb{R}$-linear problem even by their authors.

In 1946 Markushevich [22] had stated the $\mathbb{R}$-linear problem in the form (38.4) and studied it in the case $a(t) = 0$, $b(t) = 1$, $c(t) = 0$ when (38.4) is not a Nöther problem. Latter Muskhelishvili [32] (p. 455 in Russian edition) did not determined whether (38.4) was his problem (38.6) discussed in 1932 in terms of harmonic functions. In 1952 Vekua [39] established that the vector–matrix problem (38.4) is Nötherian if $\det a(t) \neq 0$.

In 1960 Bojarski [7] shown that in the case $|b(t)| < |a(t)|$ with $a(t)$, $b(t)$ belonging to the Hölder class $H^{1-\varepsilon}$ with sufficiently small $\varepsilon$, the $\mathbb{R}$-linear problem (38.4) qualitatively is similar to the $\mathbb{C}$-linear problem

$$\phi^{+}(t) = a(t)\phi^{-}(t) + c(t)\,, \quad t \in \partial D\,. \tag{38.9}$$

More precisely, Bojarski proved the following theorem for simply connected domains. His proof is also valid for multiply connected domains. Let $\mathrm{wind}_{\partial D} a(t)$ denote the winding number (index) of $a(t)$ along $\partial D$

**Theorem 38.1.** *[Bojarski [7]] Let the coefficients of the problem (38.4) satisfies the inequality*

$$|b(t)| < |a(t)|\,. \tag{38.10}$$

*If $\varkappa = \mathrm{wind}_{\partial D}\, a(t) \geq 0$, the problem (38.4) is solvable and the homogeneous problem (38.4) ($f(t) = 0$) has $2\varkappa$ $\mathbb{R}$-linearly independent solutions vanishing at infinity. If $\varkappa < 0$, the problem (38.4) has a unique solution if and only if $|2\varkappa|$ $\mathbb{R}$-linearly independent conditions on $f(t)$ are fulfilled.*

Later Mikhajlov [24] (first published in [23]) developed this result to continuous coefficients $a(t)$ and $b(t)$; $f(t) \in \mathscr{L}^{p}(\partial D)$. The case $|b(t)| < |a(t)|$ was called the elliptic case. It corresponds to the partial case of the real constant coefficients $a$ and $b$ considered by Muskhelishvili [31].

Mikhajlov [24] reduced the problem (38.4) to an integral equation and justified the absolute convergence of the method of successive approximation for the later equation in the space $\mathscr{L}^p(\partial D)$ under the restrictions $\mathrm{wind}_{\partial D} a(t) = 0$ and

$$(1+S_p)|b(t)| < 2|a(t)|\,, \tag{38.11}$$

where $S_p$ is the norm of the singular integral in $\mathscr{L}^p(\partial D)$.

### *38.1.3 Schwarz Problem*

As we noted above, the Riemann–Hilbert problem (38.3) is a partial case of the $\mathbb{R}$-linear problem. Latter we will need this fact in the case $a = 1, b = -1$.

**Theorem 38.2.** *The problem*

$$\mathrm{Re}\,\phi(t) = c(t)\,, \quad t \in \partial D \tag{38.12}$$

*is equivalent to the problem*

$$\phi^+(t) = \phi^-(t) - \overline{\phi^-(t)} + c(t)\,, \quad t \in \partial D\,, \tag{38.13}$$

*i.e., the problem (38.12) is solvable if and only if (38.13) is solvable. If (38.12) has a solution $\phi(z)$, it is a solution of (38.13) in D and solution of (38.13) in $D_k$ can be found from the following simple problem for the simply connected domain $D_k$ with respect to function* $2\mathrm{Im}\,\phi^-(z)$ *harmonic in $D_k$*

$$2\mathrm{Im}\,\phi^-(t) = \mathrm{Im}\,\phi^+(t) - c(t)\,, \quad t \in \partial D\,. \tag{38.14}$$

*The problem (38.14) has a unique solution up to an arbitrary additive real constant.*

The proof of the theorem is evident. We call the problem (38.12) by the *Schwarz problem* for the domain $D$. Along similar lines, (38.14) is called the Schwarz problem for the domain $D_k$. The operator solving the Schwarz problem is called the *Schwarz operator* (in appropriate functional space). The function $v(z) = 2\mathrm{Im}\,\phi(z)$ is harmonic in $D_k$. Therefore, the Schwarz problem (38.14) is equivalent to the Dirichlet problem

$$v(t) = \mathrm{Im}\,\phi^+(t) - c(t)\,, \quad t \in \partial D\,.$$

For multiply connected domains $D$, the Schwarz problem (38.12) is not equivalent to a Dirichlet problem for harmonic functions, since any function harmonic in $D$ is represented as the real part of a single-valued analytic function plus logarithmic terms (see for instance (38.28)).

The problem

$$\mathrm{Re}\,\phi(t) = c(t) + c_k\,, \quad t \in \partial D_k,\ k = 1,2,\ldots,n\,, \tag{38.15}$$

with undetermined constants $c_k$ is called the *modified Schwarz problem*. The problem (38.15) always has a unique solution up to an arbitrary additive complex constant [25].

### 38.1.4 Functional Equations

In 1934, Golusin [14–16] reduced the Dirichlet problem for multiply connected circular domains to a system of functional equations and applied the method of successive approximations to obtain its solution under some geometrical restrictions. Such a restriction can be roughly presented in the following form: each hole $\mathbb{D}_k$ lies sufficiently far away from all other holes $\mathbb{D}_m$ $(m \neq k)$. Golusin's approach were developed in [4, 12, 40]. Aleksandrov and Sorokin [4] extended Golusin's method to arbitrary multiply connected circular domains. However, the analytic form of the Schwarz operator was lost. More precisely, the Schwarz problem was reduced via functional equations to an infinite system of linear algebraic equations. Application of the method of truncation to this infinite system was justified.[1]

We also reduce the problem to functional equations which are similar to Golusin's ones. The main advantage of our modified functional equations is based on the possibility to solve them without any geometrical restriction by successive approximations. It is worth noting that this solution produce the Poincaré series discussed below.

The same story repeats with the alternating Schwarz method, which we call for non-overlapping domains by the *generalized Schwarz method* [15, 25]. It is also known as a *decomposition method* [35]. Miklin [25] developed the alternating Schwarz method to the Dirichlet problem for multiply connected domains and proved its convergence under some geometrical restrictions coinciding with Golusin's restrictions for circular domains. Having modified this method we obtained a method convergent for any multiply connected domain (for details see [27, 30]).

### 38.1.5 Poincaré Series

Let us consider mutually disjointed disks $\mathbb{D}_k$ in the complex plane $\mathbb{C}$. Let

$$z_{(k)}^{*} = \frac{r_k^2}{\overline{z - a_k}} + a_k$$

[1]Despite the method of truncation can be effective in numeric computations, one can hardly accept that this method yields a closed form solution. Any way it depends on the definition of the term "closed form solution". By my private definition, reduction of a boundary value problem to an integral equation does not yields a closed form solution. A regular infinite system [18] can be considered as an equation with compact operator, i.e., it is no more than a discreet form of an Fredholm integral equation.

be the inversion with respect to the circle $\mathbb{T}_k$. It is known that if $\Phi(z)$ is analytic in the disk $|z-a_k| < r_k$ and continuous in its closure, $\overline{\Phi(z^*_{(k)})}$ is analytic in $|z-a_k| > r_k$ and continuous in $|z-a_k| \geq r_k$.

Introduce the composition of successive inversions with respect to the circles $\mathbb{T}_{k_1}, \mathbb{T}_{k_2}, \ldots \mathbb{T}_{k_p}$:

$$z^*_{(k_p k_{p-1} \ldots k_1)} := \left( z^*_{(k_{p-1} \ldots k_1)} \right)^*_{(k_p)} . \tag{38.16}$$

In the sequence $k_1, k_2, \ldots, k_p$, no two neighbouring numbers are equal. The number $p$ is called the *level* of the mapping. When $p$ is even, these are Möbius transformations. If $p$ is odd, we have anti-Möbius transformations, i.e., Möbius transformations in $\overline{z}$. Thus, these mappings can be written in the form

$$\begin{aligned} \gamma_j(z) &= (e_j z + b_j)/(c_j z + d_j) , \quad p \in 2\mathbb{Z} , \\ \gamma_j(\overline{z}) &= (e_j \overline{z} + b_j)/(c_j \overline{z} + d_j) , \quad p \in 2\mathbb{Z}+1 , \end{aligned} \tag{38.17}$$

where $e_j d_j - b_j c_j = 1$. Here $\gamma_0(z) := z$ (identical mapping with the level $p = 0$), $\gamma_1(\overline{z}) := z^*_{(1)}$ ,…, $\gamma_n(\overline{z}) := z^*_{(n)}$ ($n$ simple inversions, $p = 1$), $\gamma_{n+1}(z) := z^*_{(12)}$, $\gamma_{n+2}(z) := z^*_{(13)}$,…, $\gamma_{n^2}(z) := z^*_{(n,n-1)}$ ($n^2 - n$ pairs of inversions, $p = 2$), $\gamma_{n^2+1}(\overline{z}) := z^*_{(121)}$,… and so on. The set of the subscripts $j$ of $\gamma_j$ is ordered in such a way that the level $p$ is increasing. The functions (38.17) generate a Schottky group $\mathscr{K}$. Thus, each element of $\mathscr{K}$ is presented in the form of the composition of inversions (38.16) or in the form of linearly ordered functions (38.17). Let $\mathscr{K}_m$ be such a subset of $\mathscr{K} \backslash \{\gamma_0\}$ that the last inversion of each element of $\mathscr{K}_m$ is different from $z^*_{(m)}$, i.e.,

$$\mathscr{K}_m = \left\{ z^*_{(k_p k_{p-1} \ldots k_1)} : k_p \neq m \right\} .$$

Let $H(z)$ be a rational function. The following series is called the Poincaré series

$$\theta_{2q}(z) := \sum_{j=0}^{\infty} H(\gamma_j(z))(c_j z + d_j)^{-2q} , \quad q \in \mathbb{Z}/2 , \tag{38.18}$$

associated with the subgroup $\mathscr{K}$.

**Definition 38.1.** A point $z$ is called a limit point of the group $\mathscr{K}$ if $z$ is a point of accumulation of the sequence $\gamma_j(z)$ for some $z \in \widehat{\mathbb{C}}$. A point which is not a limit point is called an ordinary point.

In other words, if $z$ runs over the extended complex plane, then the accumulation points of the sequence $\gamma_j(z)$ generate the limit set $\Lambda(\mathscr{K})$. It is assumed that in the formula (38.18) $z \in B := \widehat{\mathbb{C}} \backslash (B_1 \cup \Lambda(\mathscr{K}))$, $B_1$ is the set of poles of all $H(\gamma_j(z))$ and $\gamma_j(z)$. Ordinary points are characterized by the following property.

**Lemma 38.1.** *A point $z$ be a regular point of $\mathscr{K}$ if there exist numbers $k_1, k_2, \ldots, k_m$ such that $z_{(k_m k_{m-1} \ldots k_1)}$ belongs to $\mathbb{D} \cup \partial \mathbb{D}$.*

The points $z_1$ and $z_2$ are called congruent if there exists such $\gamma_j \in \mathscr{K}$ that $\gamma_j(z_1) = z_2$. All limit points of the Schottky group $\mathscr{K}$ lie within the disks $\mathbb{D}_k$. In the neighbourhood of a limit point $\varsigma$ there is an infinite number of distinct points congruent to any point of $\widehat{\mathbb{C}}$, at most, exception of $\varsigma$ itself and of one other point. The limit set $\Lambda(\mathscr{K})$ is transformed into itself by any $\gamma_j \in \mathscr{K}$; $\Lambda(\mathscr{K})$ is closed and dense itself.

Poincaré [34] created the series (38.18) and investigated its absolute convergence using algebraic–geometric ideas. When $q = 1$ the series (38.18) can be either absolutely convergent or absolutely divergent. It depends on the properties of $\mathscr{K}$. The main method of the proof of absolute convergence is concluded in the estimation of the $\theta_2$-series (38.18) by the number series $\sum_{j=1}^{\infty} |c_j|^{-2}$, where $c_j$ is a coefficient of $\gamma_j(z)$. The $c_j$ are the radii of the so-called isometric circles $|c_j z + d_j| = 1$. If the series $\sum_{j=1}^{\infty} |c_j|^{-2}$ converges, then (38.18) for $q = 1$ converges absolutely and uniformly in each compact subset of $B$. Absolute convergence immediately yields the automorphic property of the Poincaré $\theta_2$-series, since it is possible to change the order of summation.

Necessary and sufficient conditions for absolute and uniform convergence of the series have been found in [1,2] in terms of the Hausdorff dimension of $\Lambda(\mathscr{K})$. This result is based on the study of the series $\sum_{j=1}^{\infty} |c_j|^{-2}$. Let us note that absolute and uniform convergence was not studied separately in the previous works [1, 2, 34].

In the present paper, we deal only with uniform convergence. This approach allows us to extend knowledge about the Poincaré series. The uniform convergence does not directly imply the automorphy relation (invariance under some discsrete group of transformations), since it is forbidden to change the order of summation. But this difficulty can be easily overcome by using of functional equations. As a result, the Poincaré series satisfies the required automorphy relation and written in each fundamental domain with a prescribed summation depending on this domain [28].

## 38.2 Linear Functional Equations

### *38.2.1 Homogeneous Equation*

Let $G$ be a domain on the extended complex plane whose boundary $\partial G$ consists of simple closed Jordan curves. Introduce the Banach space $\mathscr{C}(\partial G)$ of functions continuous on the curves of $\partial G$ with the norm $\|f\| = \max_{1 \le k \le n} \max_{\partial G} |f(t)|$. Let us consider a closed subspace $\mathscr{C}_{\mathscr{A}}(G)$ of $\mathscr{C}(\partial G)$ consisting of the functions analytically continued into all disks $G$. Further, we usually take $\cup_{k=0}^{n} \mathbb{D}_k$ and sometimes $\mathbb{D}$ as the domain $G$ (not necessary connected). For brevity, the notation $\mathscr{C}_{\mathscr{A}}$ for $\mathscr{C}_{\mathscr{A}}\left(\cup_{k=0}^{n} \mathbb{D}_k\right)$ is used.

Hereafter, a point $w \in \mathbb{D} \backslash \{\infty\}$ is fixed.

**Lemma 38.2.** *Let given numbers $\nu_k$ have the form $\nu_k := \exp(-i\mu_k)$ with $\mu_k \in \mathbb{R}$. Consider the system of functional equations with respect to the functions $\phi_k(z)$ analytic in $\mathbb{D}_k$*

$$\phi_k(z) = -\nu_k \sum_{m \neq k} \overline{\nu_m} \left[ \overline{\phi_m\left(z^*_{(m)}\right)} - \overline{\phi_m\left(w^*_{(m)}\right)} \right], \quad |z-a_k| \leq r_k, \qquad (38.19)$$

$$k = 1, 2, ..., n.$$

*This system has only the trivial solution.*

*Proof.* Let $\phi_m(z)$ $(m = 1, 2, ..., n)$ be a solution of (38.19). Then the right hand part of (38.19) implies that the function $\phi_k(z)$ is analytic in $|z-a_k| \leq r_k$ $(k = 1, 2, ..., n)$. Introduce the function

$$\psi(z) := -\sum_{m=1}^{n} \overline{\nu_m} \left[ \overline{\phi_m\left(z^*_{(m)}\right)} - \overline{\phi_m\left(w^*_{(m)}\right)} \right],$$

analytic in the closure of $\mathbb{D}$. Then the functions $\psi$, $\phi_k$ satisfy the $\mathbb{R}$-linear boundary conditions

$$\nu_k \psi(t) = \phi_k(t) - \overline{\phi_k(t)} + \overline{\phi_k\left(w^*_{(k)}\right)}, \quad |t-a_k| = r_k, \ k = 1, \ldots, n.$$

One can write the latter relations in the following form

$$\text{Re } \nu_k \psi(t) = c_k, \quad |t-a_k| = r_k, \ k = 1, \ldots, n, \qquad (38.20)$$

$$2\text{Im } \phi_k(t) = \text{Im } \nu_k \psi(t) + d_k, \quad |t-a_k| = r_k, \ k = 1, \ldots, n. \qquad (38.21)$$

Here, $\phi_k\left(w^*_{(k)}\right) = c_k + i d_k$. One may consider equalities (38.20) as a boundary value problem with respect to the function $\psi(z)$ analytic in $\mathbb{D}$ and continuous in its closure, i.e., $\psi \in \mathscr{C}_{\mathscr{A}}(\mathbb{D})$. The real constants $c_k$ have to be determined. We prove that the problem (38.20) has only constant solutions: $\psi(z) \equiv c$, $c_k = \text{Re } \nu_k c$. Denote by

$$\psi(\mathbb{D}) := \left\{ \varsigma \in \widehat{\mathbb{C}} : z \in \mathbb{D}, \ \varsigma = \psi(z) \right\}$$

the image of $\mathbb{D}$ under mapping $\psi$. It follows from Boundary Correspondence Principle for conformal mapping that the boundary of $\psi(\mathbb{D})$ consists of the segments $\text{Re } \nu_k \varsigma = c_k$ $(k = 1, 2, ..., n)$. But in this case the point $\varsigma = \infty \in \psi(\mathbb{D})$ corresponds to a point of $\mathbb{D}$. It contradicts to boundedness of the function $\psi(z)$ in the closure of $\mathbb{D}$. Hence, $\psi(z) = \textit{constant}$ and equalities (38.21) imply that $\phi_k(t) = \textit{constant}$ [13]. Using (38.19) we have $\phi_k(z) \equiv 0$. □

### 38.2.2 Non-homogeneous Equation

**Lemma 38.3.** *Let $h \in \mathscr{C}_{\mathscr{A}}$, $|\nu_k| = 1$. Then the system of functional equations*

$$\phi_k(z) = -\nu_k \sum_{m \neq k} \overline{\nu_m} \left[ \overline{\phi_m\left(z^*_{(m)}\right)} - \overline{\phi_m\left(w^*_{(m)}\right)} \right] + h_k(z) \,, \tag{38.22}$$

$$|z - a_k| \leq r_k, \; k = 1, 2, \ldots, n \,,$$

*has a unique solution $\Phi \in \mathscr{C}_{\mathscr{A}}$. Here*

$$\Phi(z) := \phi_k(z) \,, \quad |z - a_k| \leq r_k, \; k = 1, 2, \ldots, n \,.$$

*This solution can be found by the method of successive approximations. The approximations are converging in $\mathscr{C}_{\mathscr{A}}$.*

*Proof.* Rewrite the system (38.22) on $\mathbb{T}_k$ in the form of a system of integral equations

$$\phi_k(t) = -\nu_k \sum_{m \neq k} \overline{\nu_m} \overline{\frac{1}{2\pi i} \int_{\mathbb{T}_m^-} \phi_m(\tau) \left( \frac{1}{\tau - t^*_{(m)}} - \frac{1}{\tau - w^*_{(m)}} \right) d\tau} + h_k(t) \,,$$

$$|t - a_k| = r_k, \; k = 1, 2, \ldots, n \,. \tag{38.23}$$

The orientation on $\mathbb{T}_m^-$ leaves $\mathbb{D}_m$ to the left. The system (38.23) can be written as an equation in the space $\mathscr{C}\left(\cup_{k=1}^n \mathbb{T}_k\right)$:

$$\Phi = \mathbf{A}\Phi + h \,. \tag{38.24}$$

The integral operators from (38.23) are compact in $\mathscr{C}(\mathbb{T}_k)$; multiplication by $\overline{\nu_m}$ and complex conjugation are bounded operators in $\mathscr{C}$. Then $\mathbf{A}$ is a compact operator in $\mathscr{C}$. Since $\Phi$ is a solution of (38.24) in $\mathscr{C}$, $\Phi \in \mathscr{C}_{\mathscr{A}}$ (see Pumping principle from [30], Sect. 2.3). This follows from the properties of the Cauchy integral and the condition $h \in \mathscr{C}_{\mathscr{A}}$. Therefore, (38.24) in $\mathscr{C}$ and (38.22) in $\mathscr{C}_{\mathscr{A}}$ are equivalent when $h \in \mathscr{C}_{\mathscr{A}}$. It follows from Lemma 38.2 that the homogeneous equation $\Phi = \mathbf{A}\Phi$ has only trivial solution. Then the Fredholm theorems imply that (38.24) or the system (38.22) has a unique solution.

Let us show the convergence of the method of successive approximations. By virtue of Successive Approximation Theorem (see [19] and [30], Sect. 2.3) it is sufficient to prove the inequality $\rho(\mathbf{A}) < 1$, where $\rho(\mathbf{A})$ is the spectral radius of the operator $\mathbf{A}$. The inequality $\rho(\mathbf{A}) < 1$ is satisfied if for all complex numbers $\lambda$ such that $|\lambda| \leq 1$ equation

$$\Phi = \lambda \mathbf{A} \Phi$$

has only trivial solution. This equation can be rewritten in the form

$$\phi_k(z) = -\lambda \nu_k \sum_{m \neq k} \overline{\nu_m} \left[ \overline{\phi_m\left(z^*_{(m)}\right)} - \overline{\phi_m\left(w^*_{(m)}\right)} \right], \quad |z - a_k| \leq r_k \,. \tag{38.25}$$

Consider the case $|\lambda| < 1$. Introduce the function analytic in the closure of $\mathbb{D}$

$$\psi(z) = -\lambda \sum_{m=0}^{n} \overline{\nu_m} \left( \overline{\phi_m\left(z^*_{(m)}\right)} - \overline{\phi_m\left(w^*_{(m)}\right)} \right).$$

Then $\psi(z)$ and $\phi_k(z)$ satisfy the $\mathbb{R}$-linear problem

$$\nu_k \psi(t) = \phi_k(t) - \lambda \overline{\phi_k(t)} + \gamma_k\,, \quad |t - a_k| = r_k,\ k = 1,2,...,n\,,$$

where

$$\gamma_k := \lambda \overline{\phi_k\left(w^*_{(k)}\right)}\,.$$

It can be written in the form

$$\nu_k \psi_0(t) = \phi_k(t) - \lambda \overline{\phi_k(t)} + \gamma_k - \nu_k \psi(\infty)\,, \quad |t - a_k| = r_k,\ k = 1,2,...,n\,, \tag{38.26}$$

where $\psi_0(z) = \psi(z) - \psi(\infty)$. Theorem 38.1 implies that the problem (38.26) has the unique solution

$$\psi_0(z) = 0\,, \quad \phi_k(z) = \frac{\gamma_k - \nu_k \psi(\infty) + \lambda \left( \overline{\gamma_k - \nu_k \psi(\infty)} \right)}{|\lambda|^2 - 1}\,, \qquad k = 1,2,...,n\,.$$

Hence, $\phi_k(z) = constant$. Then (38.25) yields $\phi_k(z) \equiv 0$.

Consider the case $|\lambda| = 1$. Then by substituting $\omega_k(z) = \phi_k(z)/\sqrt{\lambda}$ the system (38.25) is reduced to the same system with $\lambda = 1$. It follows from Lemma 38.2 that $\omega_k(z) = \phi_k(z) = 0$. Hence, $\rho(\mathbf{A}) < 1$.

This inequality proves the lemma. □

## 38.3 Schwarz Operator

### *38.3.1 Harmonic Measures*

In the present section, the number $s$ is chosen from $1, 2, \ldots, n$ and fixed. The *harmonic measure* $\alpha_s(z)$ of the circle $\mathbb{T}_s$ with respect to $\partial\mathbb{D}$ is a function harmonic in $\mathbb{D}$, continuous in its closure, satisfying the boundary conditions

$$\alpha_s(t) = \delta_{sk}\,, \quad |t - a_k| = r_k,\ k = 1,2,...,n\,, \tag{38.27}$$

where $\delta_{sk}$ is the Kronecker symbol. The functions $\alpha_s$ is infinitely $\mathbb{R}$-differentiable in the closure of $\mathbb{D}$ (see [30], Sect. 2.7.2). Using Logarithmic Conjugation Theorem [30] we look for $\alpha_s(z)$ in the form

$$\alpha_s(z) = \operatorname{Re} \phi(z) + \sum_{m=1}^{n} A_m \ln|z - a_m| + A\,, \tag{38.28}$$

where $A_m$ and $A$ are real constants,

$$\sum_{m=1}^{n} A_m = 0\,. \tag{38.29}$$

The latter condition follows from the limit in (38.28) as $z$ tends to infinity. Using the boundary condition (38.27) and the representation (38.28) we arrive at the following boundary value problem

$$\operatorname{Re} \phi(t) + \sum_{m=1}^{n} A_m \ln|t - a_k| + A = \delta_{sk}\,, \quad |t - a_k| = r_k\,. \tag{38.30}$$

This problem is equivalent to the $\mathbb{R}$ -linear problem (see Introduction)

$$\phi(t) = \phi_k(t) - \overline{\phi_k(t)} + f_k(t)\,, \quad |t - a_k| = r_k,\ k = 1, 2, \ldots, n\,, \tag{38.31}$$

where the unknown functions $\phi \in \mathscr{C}_{\mathscr{A}}(\mathbb{D})$, $\phi_k \in \mathscr{C}_{\mathscr{A}}(\mathbb{D}_k)$,

$$\phi(w) = 0\,, \tag{38.32}$$

$$f_k(z) := \delta_{sk} - A - A_k \ln r_k - \sum_{m \neq k} A_m \ln(z - a_m)\,, \quad z \in \mathbb{D}_k\,. \tag{38.33}$$

The branch of $\ln(z - a_m)$ is fixed in such a way that the cut connecting the points $z = a_m$ and $z = \infty$ does not intersect the circles $\mathbb{T}_k$ for $k \neq m$ and does not pass through the point $z = w$. The function $f_k(z)$ satisfies the boundary condition

$$\operatorname{Re} f_k(t) := \delta_{sk} - A - \sum_{m=1}^{n} A_m \ln|t - a_m|\,, \quad |t - a_k| = r_k$$

and belongs to $\mathscr{C}_{\mathscr{A}}(\mathbb{D}_k)$.

*Remark 38.1.* More precisely, the functions $\phi$, $\phi_k$ and $f_k$ are infinitely $\mathbb{C}$-differentiable in the closures of the domains considered.

Let us introduce the function

$$\Phi(z) := \begin{cases} \phi_k(z) + \sum\limits_{m \neq k} \left[ \overline{\phi_m\left(z^*_{(m)}\right)} - \overline{\phi_m\left(w^*_{(m)}\right)} \right] - \overline{\phi_k\left(w^*_{(k)}\right)} + f_k(z); \\ \hfill |z - a_k| \leq r_k, \\ \phi(z) + \sum\limits_{m=1}^{n} \left[ \overline{\phi_m\left(z^*_{(m)}\right)} - \overline{\phi_m\left(w^*_{(m)}\right)} \right], \quad z \in \mathbb{D}. \end{cases}$$

Calculate the jump across the circle $\mathbb{T}_k$

$$\Delta_k := \Phi^+(t) - \Phi^-(t), \quad t \in \mathbb{T}_k,$$

where

$$\Phi^+(t) := \lim_{z \to t\ z \in \mathbb{D}} \Phi(z), \quad \Phi^-(t) := \lim_{z \to t\ z \in \mathbb{D}_k} \Phi(z).$$

Using (38.31), (38.33) we get $\Delta_k = 0$. It follows from Analytic Continuation Principle that $\Phi(z)$ is analytic in the extended complex plane. Then Liouville's theorem implies that $\Phi(z)$ is a constant. Using (38.32) we calculate $\Phi(w) = 0$ and hence $\Phi(z) \equiv 0$. The definition of $\Phi(z) \equiv 0$ in $|z - a_k| \leq r_k$ yields the following system of functional equations

$$\phi_k(z) = -\sum_{m \neq k} \left[ \overline{\phi_m\left(z^*_{(m)}\right)} - \overline{\phi_m\left(w^*_{(m)}\right)} \right] - \delta_{sk} + A$$
$$+ A_k \ln r_k + \sum_{m \neq k} A_m \ln(z - a_m) + \overline{\phi_k\left(w^*_{(k)}\right)}, \quad |z - a_k| \leq r_k, \tag{38.34}$$

with respect to the functions $\phi_k(z) \in \mathscr{C}_{\mathscr{A}}(\mathbb{D}_k)$. The branches of logarithms are chosen in the same way as in (38.33).

The system of functional equations (38.34) is the main point to construct the harmonic measure $\alpha_s$ via the analytic function $\phi(z)$ by formula (38.28). If $\phi_k(z)$ are known, the required function $\phi(z)$ has the form

$$\phi(z) = -\sum_{m=1}^{n} \left[ \overline{\phi_m\left(z^*_{(m)}\right)} - \overline{\phi_m\left(w^*_{(m)}\right)} \right], \quad z \in \mathbb{D} \cup \partial\mathbb{D}. \tag{38.35}$$

It is convenient to represent $\phi_k(z)$ in the following form

$$\phi_k(z) = \varphi_k^{(0)}(z) + \sum_{m=1}^{n} A_m \varphi_k^{(m)}(z), \tag{38.36}$$

where $\varphi_k^{(0)}(z)$ satisfies

$$\varphi_k^{(0)}(z) = -\sum_{m\neq k}\left[\overline{\varphi_m^{(0)}\left(z_{(m)}^*\right)} - \overline{\varphi_m^{(0)}\left(w_{(m)}^*\right)}\right] - \delta_{sk} + A + A_k \ln r_k$$

$$+\sum_{m\neq k} A_m \ln(w - a_m) + \overline{\phi_k\left(w_{(k)}^*\right)}, \quad |z-a_k| \leq r_k,\ k = 1,\ldots,n, \tag{38.37}$$

$\varphi_k^{(m)}(z)$ satisfies

$$\varphi_k^{(m)}(z) = -\sum_{k_1\neq k}\left[\overline{\varphi_{k_1}^{(m)}\left(z_{(k_1)}^*\right)} - \overline{\varphi_{k_1}^{(m)}\left(w_{(k_1)}^*\right)}\right] + \delta'_{km}\ln\frac{z-a_m}{w-a_m},$$

$$|z-a_k| \leq r_k, \quad k = 1,\ldots,n,\ m = 1,\ldots,n, \tag{38.38}$$

In (38.38), $n$ systems of functional equations are written, $m$ is the number of the system,

$$\delta'_{km} = 1 - \delta_{km},$$

where $\delta_{km}$ is the Kronecker symbol. It is assumed that the constants $A$, $A_k$ and

$$\overline{\phi_k\left(w_{(k)}^*\right)}$$

are fixed in (38.37). After, these constant will be found. According to Lemma 38.3 functional equations (38.37)–(38.38) can be solved by the method of successive approximations. The method of successive approximations applied to (38.38) yields

$$\varphi_k^{(m)}(z) = \delta'_{km}\ln\frac{z-a_m}{w-a_m} - \sum_{k_1\neq k}\delta'_{k_1 m}\ln\frac{\overline{z_{(k_1)}^* - a_m}}{\overline{w_{(k_1)}^* - a_m}}$$

$$+\sum_{k_1\neq k}\sum_{k_2\neq k_1}\delta'_{k_2 m}\ln\frac{z_{(k_2k_1)}^* - a_m}{w_{(k_2k_1)}^* - a_m} - \cdots, \tag{38.39}$$

where the sums $\sum_{k_j\neq k_{j-1}}$ contains the terms with $k_j = 1,2,\ldots,n$; $k_j \neq k_{j-1}$. By virtue of Lemma 38.3 with $v_m = 1$ the series (38.39) converges uniformly in $|z-a_k| \leq r_k$. It follows from (38.37) that $\phi_k^{(0)}(z)$ $(k = 1,2,...,n)$ are constants, since the zero-th approximation is a constant and the operator from the right-hand side of (38.37) produces constants.

One can see from (38.35) that the constants $\varphi_k^{(0)}(z)$ does not impact on $\phi(z)$, hence using (38.36) we have

$$\phi(z) = -\sum_{m=1}^{n} A_m \sum_{k=1}^{n}\left[\overline{\varphi_k^{(m)}\left(z_{(k)}^*\right)} - \overline{\varphi_k^{(m)}\left(w_{(k)}^*\right)}\right]. \tag{38.40}$$

Substitution of (38.39) into (38.40) yields

$$\phi(z) = -\sum_{m=1}^{n} A_m \sum_{k=1}^{n} \delta'_{km} \ln \overline{\frac{z^*_{(k)} - a_m}{w^*_{(k)} - a_m}}$$

$$+\sum_{m=1}^{n} A_m \sum_{k=1}^{n} \sum_{k_1 \neq k} \delta^{(m)}_{k_1} \ln \frac{z^*_{(k_1 k)} - a_m}{w^*_{(k_1 k)} - a_m}$$

$$-\sum_{m=1}^{n} A_m \sum_{k=1}^{n} \sum_{k_1 \neq k} \sum_{k_2 \neq k_1} \delta^{(m)}_{k_2} \ln \overline{\frac{z^*_{(k_2 k_1 k)} - a_m}{w^*_{(k_2 k_1 k)} - a_m}} + \ldots, \quad z \in \mathbb{D} \cup \partial\mathbb{D}. \quad (38.41)$$

Using the properties of $\delta'_{km}$ one can rewrite (38.41) in the form

$$\phi(z) = -\sum_{m=1}^{n} A_m \sum_{k \neq m} \ln \overline{\frac{z^*_{(k)} - a_m}{w^*_{(k)} - a_m}} + \sum_{m=1}^{n} A_m \sum_{k=1}^{n} \sum_{k_1 \neq k,m} \ln \frac{z^*_{(k_1 k)} - a_m}{w^*_{(k_1 k)} - a_m}$$

$$-\sum_{m=1}^{n} A_m \sum_{k=1}^{n} \sum_{k_1 \neq k} \sum_{k_2 \neq k_1,m} \ln \overline{\frac{z^*_{(k_2 k_1 k)} - a_m}{w^*_{(k_2 k_1 k)} - a_m}} + \cdots, \quad z \in \mathbb{D} \cup \partial\mathbb{D} (38.42)$$

In order to write (38.42) in more convenient form we use the following

**Lemma 38.4.** *There holds the equality*

$$\sum_{k=1}^{n} \sum_{k_1 \neq k} \cdots \sum_{k_s \neq k_{s-1},m} A_m R_{mk_s k_{s-1} \ldots k_1 k}$$

$$= \sum_{k_1 \neq m} \sum_{k_2 \neq k_1} \cdots \sum_{k_s \neq k_{s-1}} \sum_{k \neq k_s} A_m R_{mk_1 k_2 \ldots k_s k} \,. \quad (38.43)$$

*Proof.* It is sufficiently to demonstrate that the both parts of equality (38.43) contain the same terms. First, replace $k_1 k_2 \ldots k_s$ by $k_s k_{s-1} \ldots k_1$ in the right hand part of (38.43) which becomes

$$\sum_{k_s \neq m} \sum_{k_{s-1} \neq k_s} \cdots \sum_{k_1 \neq k_2} \sum_{k \neq k_1} A_m R_{mk_s k_{s-1} \ldots k_1 k} \,. \quad (38.44)$$

The left hand part of (38.43) can be written as the sum

$$\sum_{k=1}^{n} \sum_{k_1=1}^{n} \cdots \sum_{k_s=1}^{n} \delta'_{k_1 k} \delta'_{k_2 k_1} \cdots \delta'_{k_s k_{s-1}} \delta'_{k_s m} A_m R_{mk_s k_{s-1} \ldots k_1 k} \,, \quad (38.45)$$

where $\delta'_{lk} = 1 - \delta_{lk}$, $\delta_{lk}$ is the Kronecker symbol. One can see that the sum (38.44) written in the similar form (38.45) contains the same product of the complimentary Kronecker symbols.

This proves the lemma. □

Applying Lemma 38.4 to (38.42) we obtain

$$\phi(z) = -\sum_{m=1}^{n} A_m \sum_{k \neq m} \ln \overline{\frac{\overline{z_{(k)}^* - a_m}}{w_{(k)}^* - a_m}}$$

$$+ \sum_{m=1}^{n} A_m \sum_{k_1 \neq m} \sum_{k \neq k_1} \ln \frac{z_{(k_1 k)}^* - a_m}{w_{(k_1 k)}^* - a_m}$$

$$- \sum_{m=1}^{n} A_m \sum_{k_1 \neq m} \sum_{k_2 \neq k_1} \sum_{k \neq k_2} \ln \overline{\frac{\overline{z_{(k_1 k_2 k)}^* - a_m}}{w_{(k_1 k_2 k)}^* - a_m}} + \cdots, \quad z \in \mathbb{D} \cup \partial \mathbb{D}. \quad (38.46)$$

It can be also written in the form

$$\phi(z) = \sum_{m=1}^{n} A_m \psi_m(z),$$

where

$$\psi_m(z) = \ln \prod_{k \neq m} \overline{\frac{\overline{w_{(k)}^* - a_m}}{z_{(k)}^* - a_m}} + \ln \prod_{k_1 \neq m} \prod_{k \neq k_1} \frac{z_{(k_1 k)}^* - a_m}{w_{(k_1 k)}^* - a_m}$$

$$+ \ln \prod_{k_1 \neq m} \prod_{k_2 \neq k_1} \prod_{k \neq k_2} \overline{\frac{\overline{w_{(k_1 k_2 k)}^* - a_m}}{z_{(k_1 k_2 k)}^* - a_m}} + \cdots, \quad z \in \mathbb{D} \cup \partial \mathbb{D}. \quad (38.47)$$

Let us rewrite (38.47) in terms of the group $\mathscr{K}$

$$\psi_m(z) = \ln \left[ \prod_{j \in \mathscr{K}_m}^{\infty} \psi_m^{(j)}(z) \right], \quad (38.48)$$

where

$$\psi_m^{(j)}(z) = \begin{cases} \dfrac{\gamma_j(z) - a_m}{\gamma_j(w) - a_m}, & \text{if level of } \gamma_j \text{ is even;} \\ \overline{\dfrac{\overline{\gamma_j(\overline{w}) - a_m}}{\gamma_j(\overline{z}) - a_m}}, & \text{if level of } \gamma_j \text{ is odd.} \end{cases}$$

The numeration on $j$ in (38.48) is fixed with increasing level.

In order to determine the constants $A$ and $A_m$ substitute $z = w_{(k)}^*$ in the real parts of (38.34)

$$0 = -\sum_{m \neq k} \operatorname{Re} \left[ \overline{\varphi_m\left( \left( w_{(k)}^* \right)_{(m)}^* \right) - \overline{\varphi_m\left( w_{(m)}^* \right)}} \right] - \delta_{sk} \quad (38.49)$$

$$+ A + A_k \ln r_k + \sum_{m \neq k} A_m \ln \left| w_{(k)}^* - a_m \right|, \quad k = 1, 2, \ldots, n.$$

The function $\varphi_m$ has the form (38.39) and linearly depends on the unknown constants $A_m$. The equalities (38.29), (38.49) generate a system of $n+1$ linear algebraic equations with respect to $n+1$ unknowns $A, A_1, ..., A_n$. This system has a unique solution, since in the opposite case it contradicts to uniqueness of the solution of Dirichlet problem.

**Theorem 38.3.** *The harmonic measures have the form*

$$\alpha_s(z) = \sum_{m=1}^{n} A_m \left[\text{Re } \psi_m(z) + \ln|z - a_m|\right] + A \,, \tag{38.50}$$

*where $\psi_m(z)$ is given in (38.48). The infinite product (38.48) converges uniformly on each compact subset of $\mathbb{D} \setminus \{\infty\}$. The real constants $A$ and $A_m$ are uniquely determined by the system (38.29), (38.49).*

*Proof.* Exact formulae for the harmonic measures were deduced in the formal way. In order to justify them it is necessary to prove the change of the summation in (38.42) to obtain (38.46).

Using the designations of Sect. 38.2 we write the series (38.39) in the form

$$\Phi = \sum_{k=1}^{\infty} \mathbf{A}^k h \,, \tag{38.51}$$

where

$$h(z) = \delta'_{km} \ln \frac{z - a_m}{w - a_m} \,, \quad \mathbf{A}h(z) = -\sum_{k_1 \neq k} \delta'_{k_1 m} \ln \overline{\frac{z^*_{(k_1)} - a_m}{w^*_{(k_1)} - a_m}} \,,$$

$$\mathbf{A}^2 h(z) = \sum_{k_1 \neq k} \sum_{k_2 \neq k_1} \delta'_{k_2 m} \ln \frac{z^*_{(k_2 k_1)} - a_m}{w^*_{(k_2 k_1)} - a_m} - \cdots \,. \tag{38.52}$$

It is possible to change the order of summation in each term $\mathbf{A}^k h$ of the successive approximations which contains inversions only of the level $k$. Therefore, it is possible to change the order of summation in the series (38.41) in the terms having the same level. The series (38.46) is obtained from (38.42) by application of this rule.

This proves the theorem. □

*Remark 38.2.* The constants $A$ and $A_m$ depend on the choice of $w$.

*Remark 38.3.* The logarithmic terms in (38.50) can be included into infinite product (38.47). Then (38.50) becomes

$$\alpha_s(z) = \sum_{m=1}^{n} A_m \ln \prod_{j \in \mathscr{K}_m \cup \{0\}} |\psi_m^{(j)}(z)| + A_0 \,,$$

where $A_0 := A - \sum_{m=1}^{n} A_m \ln|w - a_m|$.

### 38.3.2 Schwarz Operator

Following the previous section we construct the complex Green function $M(z,\zeta)$ and the Schwarz operator for the circular multiply connected domain $\mathbb{D}$. One can find general properties of the Schwarz operator in [25, 30].

Let $z$ and $\zeta$ belong to the closure of $\mathbb{D}$. The real *Green function*

$$G(z,\zeta) = g(z,\zeta) - \ln|z-\zeta|$$

is introduced via the function $g(z,\zeta)$ harmonic in $\mathbb{D}$ satisfying the Dirichlet problem

$$g(t,\zeta) - \ln|t-\zeta| = 0\,, \quad |t-a_k| = r_k,\ k = 1,2,\ldots,n \tag{38.53}$$

with respect to the first variable. If $G(z,\zeta)$ is known, the solution of the Dirichlet problem

$$u(t) = f(t), \quad |t-a_k| = r_k,\ k = 1,2,\ldots,n \tag{38.54}$$

has the form

$$u(z) = \frac{1}{2\pi}\sum_{k=1}^{n}\int_{\mathbb{T}_k} f(\zeta)\frac{\partial G}{\partial \nu}(z,\zeta)\mathrm{d}\sigma\,, \tag{38.55}$$

where $\nu$ is the outward (in sense of orientation) normal vector at the point $\zeta \in \partial\mathbb{D}$.

The *complex Green function* $M(z,\zeta)$ is defined by the formula

$$M(z,\zeta) = G(z,\zeta) + iH(z,\zeta)\,, \tag{38.56}$$

where the function $H(z,\zeta)$ is harmonically conjugated to $G(z,\zeta)$ on the variable $z$. It has the form

$$H(z,\zeta) = \int_{w}^{z} -\frac{\partial G}{\partial y}\mathrm{d}x + \frac{\partial G}{\partial x}\mathrm{d}y$$

with $z = x + \mathrm{i}y$.

Introduce the *Schwarz kernel* (see [30, Sect. 2.7.2])

$$T(z,\zeta) = \frac{\partial M}{\partial \nu}(z,\zeta), \quad \zeta \in \mathbb{D}\,. \tag{38.57}$$

In accordance with (38.54)–(38.57) the function

$$F(z) = \frac{1}{2\pi}\sum_{k=1}^{n}\int_{\mathbb{T}_k} f(\zeta)T(z,\zeta)\mathrm{d}\sigma \tag{38.58}$$

satisfies the boundary value problem

$$\operatorname{Re} F(t) = f(t), \quad |t - a_k| = r_k,\ k = 1,2,\dots,n\,. \tag{38.59}$$

Here, $u(z)$ from (38.55) is the real part of the analytic function $F(z)$ having in general multi-valued imaginary part in the multiply connected domain $\mathbb{D}$. If we are looking for single-valued $F(z)$ by (38.59), we arrive at the Schwarz problem in accordance with the terminology introduced in the Introduction.

We use the representation for the Green function (see [30, Sect. 2.7.2])

$$M(z,\zeta) = M_0(z,\zeta) + \sum_{k=1}^{n} \alpha_k(\zeta)\ln(z - a_k) - \ln(\zeta - z) + A(\zeta)\,, \tag{38.60}$$

where $\alpha_k$ is a harmonic measure of $\mathbb{D}$, $A(\zeta)$ is a real function in $\zeta$. The point $w$ and the branches of $\ln(z - a_k)$ are fixed as in the previous section. Using (38.57), (38.60) we obtain

$$T(z,\zeta) = \frac{\partial M_0}{\partial \nu}(z,\zeta) + \sum_{m=1}^{n} \frac{\partial \alpha_m}{\partial \nu}(\zeta)\ln(z - a_m) - \frac{1}{\zeta - z}\frac{\partial \zeta}{\partial \nu} + \frac{\partial A}{\partial \nu}(\zeta)\,. \tag{38.61}$$

The function $M_0(z,\zeta)$ is infinitely $\mathbb{C}$-differentiable in the closure of $\mathbb{D}$ in $z$ and satisfies the boundary value problem which follows from (38.53) and (38.56)

$$\operatorname{Re}\left[M_0(t,\zeta) + \sum_{k=1}^{n} \alpha_k(\zeta)\ln(t - a_k) - \ln(\zeta - t) + A(\zeta)\right] = 0\,,$$

$$|t - a_k| = r_k,\ k = 1,2,\dots,n;\ M_0(w,\zeta) = 0\,. \tag{38.62}$$

The problem (38.62) has a unique solution. It is reduced to the following system of functional equations

$$\phi_k(z) = -\sum_{m\neq k}\left[\overline{\phi_m\left(z^*_{(m)}\right) - \phi_m\left(w^*_{(m)}\right)}\right] - \ln(\zeta - z) + A + \alpha_k(\zeta)\ln r_k$$

$$+\sum_{m\neq k} \alpha_m(\zeta)\ln(z - a_m) + \overline{\phi_k\left(w^*_{(k)}\right)}\,, \quad |z - a_k| \le r_k,\ k = 1,\dots,n\,. \tag{38.63}$$

where $\phi_k(z)$ belongs to $\mathscr{C}_{\mathscr{A}}(\mathbb{D}_k)$ and $\mathbb{C}$-infinitely differentiable in the closure of $\mathbb{D}_k$. Here, $\zeta$ is considered as a parameter fixed in the closure of $\mathbb{D}$. The required function $M_0(z,\zeta)$ is related to the auxiliary functions $\phi_k(z)$ by equality

$$M_0(z,\zeta) = -\sum_{k=1}^{n}\left[\overline{\phi_k\left(z^*_{(k)}\right) - \phi_k\left(w^*_{(k)}\right)}\right], \quad z \in \mathbb{D}\cup\partial\mathbb{D}\setminus\{\zeta\}\,. \tag{38.64}$$

In order to solve (38.63) we consider two auxiliary systems of functional equations

$$\Psi_k(z) = -\sum_{m\neq k}\left[\overline{\Psi_m(z_m^*)} - \overline{\Psi_m(w_m^*)}\right] + A$$

$$+\alpha_k(\zeta)\ln r_k + \sum_{m\neq k}\alpha_m(\zeta)\ln(z-a_m) + \overline{\phi_m\left(w_{(k)}^*\right)},$$

$$\Omega_k(z) = -\sum_{m\neq k}\left[\overline{\Omega_m(z_m^*)} - \overline{\Omega_m(w_m^*)}\right] - \ln(\zeta - z), \quad |z-a_k|\leq r_k,\ k=1,\dots,n.$$

The first system coincides with the system (38.23) ($\nu_k = 1$), and thus can be solved by the method of successive approximations (cf. Lemma 38.3). Let us consider the second system. If $|\zeta - z| \leq r_s$ for some $s$, the right hand part of the second system $-\ln(\zeta - z)$ does not belong to $\mathscr{C}_{\mathscr{A}}(\mathbb{D}_s)$. But by introducing a new unknown function

$$\Omega_s^{(0)}(z) := \Omega_s(z) - \ln(\zeta - z)$$

we get an equation in the space $\mathscr{C}_{\mathscr{A}}(\mathbb{D}_s)$. Therefore, the method of successive approximations can be applied to the second system too, and

$$\overline{\Psi_k\left(z_{(k)}^*\right)} - \overline{\Psi_k\left(w_{(k)}^*\right)} = \sum_{m\neq k}\alpha_m(\zeta)\ln\frac{\overline{z_{(k)}^* - a_m}}{\overline{w_{(k)}^* - a_m}}$$

$$-\sum_{k_1\neq k}\sum_{m\neq k_1}\alpha_m(\zeta)\ln\frac{z_{(k_1k)}^* - a_m}{w_{(k_1k)}^* - a_m}$$

$$+\sum_{k_1\neq k}\sum_{k_2\neq k_1}\sum_{m\neq k_2}\alpha_m(\zeta)\ln\frac{\overline{z_{(k_2k_1k)}^* - a_m}}{\overline{w_{(k_2k_1k)}^* - a_m}} - \cdots, \tag{38.65}$$

$$\overline{\Omega_k\left(z_{(k)}^*\right)} - \overline{\Omega_k\left(w_{(k)}^*\right)} = \ln\frac{\overline{\zeta - w_{(k)}^*}}{\overline{\zeta - z_{(k)}^*}} + \sum_{k_1\neq k}\ln\frac{\zeta - z_{(k_1k)}^*}{\zeta - w_{(k_1k)}^*}$$

$$+\sum_{k_1\neq k}\sum_{k_2\neq k_1}\ln\frac{\overline{\zeta - z_{(k_2k_1k)}^*}}{\overline{\zeta - w_{(k_2k_1k)}^*}} + \cdots, \quad z\in\mathbb{D}\cup\partial\mathbb{D}. \tag{38.66}$$

The series (38.65), (38.66) converge uniformly in every compact subset of $\mathbb{D}\cup\partial\mathbb{D}\setminus\{\zeta\}$. We have $\phi_k(z) = \Psi_k(z) + \Omega_k(z)$; hence, the values

$$\overline{\phi_k\left(z_{(k)}^*\right)} - \overline{\phi_k\left(w_{(k)}^*\right)} = \overline{\Psi_k\left(z_{(k)}^*\right)} - \overline{\Psi_k\left(w_{(k)}^*\right)} + \overline{\Omega_k\left(z_{(k)}^*\right)} - \overline{\Omega_k\left(w_{(k)}^*\right)} \tag{38.67}$$

are completely determined. It follows from (38.64) that

$$M_0(z,\zeta) = \sum_{m=1}^{n} \alpha_m(\zeta)\psi_m(z) - \omega(z,\zeta)\,,$$

where the functions $\psi_m(z)$ have the form (38.47) or (38.48 ), $\alpha_m(\zeta)$ are given in Theorem 38.3,

$$\omega(z,\zeta) = \ln\left(\prod_{k=1}^{n} \overline{\frac{\zeta - z^*_{(k)}}{\zeta - w^*_{(k)}}}\right)\left(\prod_{k=1}^{n}\prod_{k_1\neq k} \frac{\zeta - w^*_{(k_1 k)}}{\zeta - z^*_{(k_1 k)}}\right)$$

$$\times\left(\prod_{k=1}^{n}\prod_{k_1\neq k}\prod_{k_2\neq k_1} \overline{\frac{\zeta - z^*_{(k_2 k_1 k)}}{\zeta - w^*_{(k_2 k_1 k)}}}\right)\cdots. \tag{38.68}$$

This infinite product can be represented in the form

$$\omega(z,\zeta) = \ln\prod_{j=1}^{\infty} \omega_j(z,\zeta)\,, \tag{38.69}$$

where

$$\omega_j(z,\zeta) = \begin{cases} \dfrac{\zeta-\gamma_j(z)}{\zeta-\gamma_j(w)}\,, & \text{if level of } \gamma_j \text{ is even;} \\[2ex] \overline{\dfrac{\zeta-\gamma_j(\overline{w})}{\zeta-\gamma_j(\overline{z})}}\,, & \text{if level of } \gamma_j \text{ is odd.} \end{cases}$$

In order to find $A(\zeta)$ we substitute $w^*_{(k)}$ in the real part of (38.63) and obtain

$$0 = -\sum_{m\neq k} \operatorname{Re}\left[\overline{\phi_m\left(\left(w^*_{(k)}\right)^*_{(m)}\right)} - \overline{\phi_m\left(w^*_{(m)}\right)}\right] - \ln\left|\zeta - w^*_{(k)}\right|$$

$$+A(\zeta) + \alpha_k(\zeta)\ln r_k + \sum_{m\neq k} \alpha_m(\zeta)\ln\left|w^*_{(k)} - a_m\right|\,, \quad k = 1,\ldots,n\,. \tag{38.70}$$

The harmonic measures satisfy the equality

$$\sum_{m=1}^{n} \alpha_m(\zeta) = 1\,. \tag{38.71}$$

One can consider (38.70), (38.71) as a system of $n+1$ real linear algebraic equations with respect to $n+1$ real unknowns $\alpha_1(\zeta)$, $\alpha_2(\zeta)$,..., $\alpha_n(\zeta)$, $A(\zeta)$. The systems (38.70), (38.71) and (38.29), (38.49) have the same homogeneous part. Therefore, the system (38.70), (38.71) has a unique solution. We may at the beginning look for the complex Green function $M(z,\zeta)$ with undetermined periods $\alpha_k(\zeta)/2\pi$, find $\alpha_k(\zeta)$ from (38.70), (38.71) and after assert that $\alpha_k(\zeta)$ is a harmonic measure. In order to determine $A(\zeta)$, we fix for instance $k=n$ in (38.70) and obtain

$$A(\zeta) = \sum_{m=1}^{n-1} \mathrm{Re} \left[ \overline{\phi_m\left(\left(w^*_{(n)}\right)^*_{(m)}\right) - \phi_m\left(w^*_{(m)}\right)} \right]$$

$$+ \ln\left|\zeta - w^*_{(n)}\right| - \alpha_k(\zeta)\ln r_k - \sum_{m=1}^{n-1} \alpha_m(\zeta)\ln\left|w^*_{(k)} - a_m\right| , \quad (38.72)$$

where $\left[\overline{\phi_m\left(z^*_{(m)}\right) - \phi_m\left(w^*_{(m)}\right)}\right]$ has the form (38.65)–(38.67).

The function (38.58) is single-valued in $\mathbb{D}$ if and only if

$$\sum_{k=1}^{n} \int_{\mathbb{T}_k} f(\zeta)\frac{\partial \alpha_m}{\partial \nu}(\zeta)\mathrm{d}\sigma = 0 , \quad m = 1,2,...,n . \quad (38.73)$$

Note that one of the relation (38.73) follows from the other ones. For instance, let (38.73) be valid for $m = 1,2,...,n-1$. Then ( 38.73) for $m = n$ is fulfilled, since

$$\sum_{k=1}^{n} \int_{\mathbb{T}_k} f(\zeta)\frac{\partial \alpha_n}{\partial \nu}(\zeta)\mathrm{d}\sigma = -\sum_{m=1}^{n-1}\sum_{k=1}^{n} \int_{\mathbb{T}_k} f(\zeta)\frac{\partial \alpha_m}{\partial \nu}(\zeta)\mathrm{d}\sigma = 0 .$$

Here the identity (38.71) is used. With the help of (38.61) the single- and the multi-valued components of the Schwarz operator can be separated

$$T(z,\zeta) = T_s(z,\zeta) + T_m(z,\zeta) ,$$

$$T_s(z,\zeta) = \sum_{m=1}^{n} \frac{\partial \alpha_m}{\partial \nu}(\zeta)\left[\psi_m(z) + \ln(z-a_m)\right] ,$$

$$T_m(z,\zeta) = \frac{\partial \omega}{\partial \nu}(z,\zeta) - \frac{1}{\zeta - z}\frac{\partial \zeta}{\partial \nu} + \frac{\partial A}{\partial \nu}(\zeta) .$$

We now proceed to calculate the normal derivatives in the latter formulae. One can see that

$$\frac{\partial f}{\partial \nu}\mathrm{d}\sigma = -\frac{1}{i}\left[\frac{\partial f}{\partial \zeta} + \left(\frac{r_k}{\zeta - a_k}\right)^2 \frac{\partial f}{\partial \overline{\zeta}}\right]\mathrm{d}\tau , \quad |\zeta - a_k| = r_k , \quad (38.74)$$

for any $f \in \mathscr{C}^1(\partial\mathbb{D})$. Recall that we deal with the outward normal to $\mathbb{D}$. In order to apply (38.74) to $\omega(z,\zeta)$ we find from (38.68)

$$\frac{\partial\omega}{\partial\zeta}(z,\zeta) = \sum_{k=1}^{n}\sum_{k_1\neq k}\left(\frac{1}{\zeta - w^*_{(k_1k)}} - \frac{1}{\zeta - z^*_{(k_1k)}}\right)$$
$$+\sum_{k=1}^{n}\sum_{k_1\neq k}\sum_{k_2\neq k_1}\sum_{k_3\neq k_2}\left(\frac{1}{\zeta - w^*_{(k_3k_2k_1k)}} - \frac{1}{\zeta - z^*_{(k_3k_2k_1k)}}\right)$$
$$+\cdots = \sum_{j=1}^{\infty}{}''\left(\frac{1}{\zeta-\gamma_j(w)} - \frac{1}{\zeta-\gamma_j(z)}\right), \tag{38.75}$$

where the terms in the latter sum are ordered due to increasing even level. We also have

$$\frac{\partial\omega}{\partial\overline{\zeta}}(z,\tau) = \sum_{j=1}^{\infty}{}'\left(\frac{1}{\overline{\zeta-\gamma_j(\overline{z})}} - \frac{1}{\overline{\zeta-\gamma_j(\overline{w})}}\right), \tag{38.76}$$

where elements $\gamma_j$ have the odd level. Substituting (38.75), (38.76) into (38.61), (38.58) we arrive at the following

**Theorem 38.4.** *The Schwarz operator of* $\mathbb{D}$ *has the form*

$$\phi(z) = \frac{1}{2\pi i}\sum_{k=1}^{n}\int_{\mathbb{T}_k} f(\zeta)\left\{\sum_{j=2}^{\infty}{}''\left[\frac{1}{\zeta-\gamma_j(w)} - \frac{1}{\zeta-\gamma_j(z)}\right]\right.$$
$$\left.+\left(\frac{r_k}{\zeta-a_k}\right)^2\sum_{j=1}^{\infty}{}'\left[\frac{1}{\overline{\zeta-\gamma_j(\overline{z})}} - \frac{1}{\overline{\zeta-\gamma_j(\overline{w})}}\right] - \frac{1}{\zeta-z}\right\}\mathrm{d}\zeta$$
$$+\frac{1}{2\pi i}\sum_{k=1}^{n}\int_{\mathbb{T}_k} f(\zeta)\frac{\partial A}{\partial\nu}(\zeta)\mathrm{d}\sigma + \sum_{m=1}^{n} A_m\left[\ln(z-a_m) + \psi_m(z)\right] + i\varsigma\,, \tag{38.77}$$

*where*

$$A_m := \frac{1}{2\pi i}\sum_{k=1}^{n}\int_{\mathbb{T}_k} f(\zeta)\frac{\partial\alpha_m}{\partial\nu}(\zeta)\mathrm{d}\sigma\,, \quad m = 1,2,\ldots,n\,,$$

*$A(\zeta)$ has the form (38.72), The functions $\alpha_m(\zeta)$ and $\psi_m(z)$ are derived in Theorem 38.3, $\varsigma$ is an arbitrary real constant, $\sum'$ contains $\gamma_j$ of odd level, $\sum''$ – of even level. The series converges uniformly in each compact subset of $\mathbb{D}\cup\partial\mathbb{D}\setminus\{\infty\}$.*

The single-valued part of the Schwarz operator can be determined by solution to the modified Dirichlet problem (see [30, Sect. 2.7.2]):

$$\mathrm{Re}\,\phi(t) = f(t) + c_k\,, \quad t\in\mathbb{T}_k,\ k = 1,2,\ldots,n\,, \tag{38.78}$$

where a given function $f \in \mathscr{C}(\partial\mathbb{D})$, $c_k$ are undetermined real constants. If one of the constants $c_k$ is fixed arbitrary, the remaining ones are determined uniquely and $\phi(z)$ is determined up to an arbitrary additive purely imaginary constant (see [30, Sect. 2.7.2]). Thus, we have

**Theorem 38.5.** *The single-valued part of the Schwarz operator of $\mathbb{D}$ corresponding to the modified Dirichlet problem (38.78) has the form*

$$\phi(z) = \frac{1}{2\pi i}\sum_{k=1}^{n}\int_{\mathbb{T}_k}(f(\zeta)+c_k)\left\{\sum_{j=2}^{\infty}{}''\left[\frac{1}{\zeta-\gamma_j(w)}-\frac{1}{\zeta-\gamma_j(z)}\right]\right.$$
$$\left.+\left(\frac{r_k}{\zeta-a_k}\right)^2\sum_{j=1}^{\infty}{}'\left[\frac{1}{\overline{\zeta-\gamma_j(\overline{z})}}-\frac{1}{\overline{\zeta-\gamma_j(\overline{w})}}\right]-\frac{1}{\zeta-z}\right\}d\zeta$$
$$+\frac{1}{2\pi i}\sum_{k=1}^{n}\int_{\mathbb{T}_k}f(\zeta)\frac{\partial A}{\partial \nu}(\zeta)d\sigma+i\varsigma\,. \tag{38.79}$$

*One of the real constants $c_k$ can be fixed arbitrary, the remaining ones are determined uniquely from the linear algebraic system*

$$\sum_{k=1}^{n}\int_{\mathbb{T}_k}(f(\zeta)+c_k)\frac{\partial\alpha_m}{\partial\nu}(\zeta)d\sigma=0\,,\quad m=1,2,\dots,n-1\,. \tag{38.80}$$

For brevity we write (38.79) in the following form $\phi = \mathbf{T}f + i\varsigma$. For the definiteness it is assumed that the operator $\mathbf{T}$ contains the constants $c_k$ as parameters, i.e., the conditions (38.80) on $c_k$ are satisfied and $\mathbf{T}f$ is single-valued.

## 38.4 Solution to the Riemann–Hilbert Problem

Applying the method of factorization we first reduce the problem (38.3) for circular multiply connected domain $\mathbb{D}$ with arbitrary $\lambda(t)$ to the case when $\lambda(t)$ is constant in each circle $\mathbb{T}_k$. This possibility were also used by Bojarski (see Addition to [38]). However, now we can do it explicitly. More precisely, the factorization function is written in terms of the Schwarz operator constructed in the previous section.

Let $\varkappa_k$ be the index of $\lambda(t)$ along the curve $\mathbb{T}_k$

$$\varkappa_k = \operatorname{wind}_{\mathbb{T}_k}\lambda(t) := \frac{1}{2\pi i}\int_{\mathbb{T}_k} d\ln\lambda(t)\,.$$

The value $\varkappa := \sum_{k=1}^{n}\varkappa_k$ is called the *index of the problem*. Consider the case $\varkappa > 0$. Introduce the new unknown function from $\mathscr{C}_{\mathscr{A}}(\mathbb{D})$

$$\psi(z) := \frac{\phi(z)}{R(z)} - \sum_{s=1}^{\varkappa}\delta_s z^s\,,$$

where

$$R(z) := \prod_{m=1}^{n} (z - a_m)^{-\varkappa_m} ,$$

$\sum_{s=1}^{\varkappa} \delta_s z^s$ is the principal part of the function $\phi(z)/R(z)$ at infinity. If $\varkappa \leq 0$, the sum $\sum_{s=1}^{\varkappa}$ is identically equal to zero. The boundary value problem (38.3) becomes

$$\operatorname{Re} R(t)\overline{\lambda(t)}\psi(t) = f(t) - \operatorname{Re}\left(\overline{\lambda(t)}\sum_{s=1}^{\varkappa} \delta_s t^s R(t)\right), \quad t \in \partial\mathbb{D} , \tag{38.81}$$

with respect to $\psi(z) \in \mathscr{C}_{\mathscr{A}}(\mathbb{D})$. The index of this problem is equal to zero but the right-hand side contains $\varkappa$ undetermined constants $\delta_s$.

Following [13] we apply the factorization method to the problem (38.81). Introduce the function

$$P(t) := \overline{R(t)}\lambda(t)\left(R(t)\overline{\lambda(t)}\right)^{-1}, \quad t \in \partial\mathbb{D} .$$

First, note that $\ln P(t)$ is a continuous function on every $\mathbb{T}_k$, since $\operatorname{wind}_{\mathbb{T}_k} P(t) = 0$. Consider the modified Schwarz problem

$$X_1(t) - \overline{X_1(t)} = \ln P(t) - 2\mathrm{i}\mu_k , \quad |t - a_k| = r_k,\ k = 1, 2, ..., n , \tag{38.82}$$

with respect to the function $X_1(z) \in \mathscr{C}_{\mathscr{A}}(\mathbb{D})$ with undetermined real constants $\mu_k$. The problem (38.82) is solved in Theorem 38.5

$$X_1(z) = \mathrm{i}\left(\mathbf{T}\frac{\ln P}{2i}\right)(z) ,$$

where $\mathbf{T}$ denote the single–valued part of Schwarz's operator for $\mathbb{D}$, the parameters $c_k$ from Theorem 38.5 here have the form $c_k = -\mu_k$.

The function $X(z) := \exp X_1(z)$ is analytic in $\mathbb{D}$, Hölder continuous in its closure and does not vanish in the closure of $\mathbb{D}$. Therefore, (38.82) yields

$$\overline{\nu(t)}X(t)\left[\nu(t)\overline{X(t)}\right]^{-1} = P(t) , \quad t \in \partial\mathbb{D} ,$$

where $\nu(t) = \nu_k := \exp(-\mathrm{i}\mu_k)$ on $|t - a_k| = r_k$. Then the boundary condition (38.81) can be rewritten in the form

$$\operatorname{Re} \nu(t)\omega(t) = h(t) + \sum_{s=1}^{2\varkappa} p_s\beta_s(t) , \quad t \in \partial\mathbb{D} , \tag{38.83}$$

where the unknown function

$$\omega(z) := \psi(z)\,[X(z)]^{-1}$$

belongs to $\mathscr{C}_{\mathscr{A}}(\mathbb{D})$ and

$$h(t) := f(t)\nu(t)\left[R(t)\overline{\lambda(t)}X(t)\right]^{-1}$$

is a known real function. The real constants $p_s$ and functions $\beta_s(t)$ are defined by the relations

$$-\mathrm{Re}\left(\nu(t)\,[X(t)]^{-1}\sum_{s=1}^{\varkappa}\delta_s t^s\right) = \sum_{s=1}^{2\varkappa} p_s\beta_s(t)\,, \quad t\in\partial\mathbb{D}\,,$$

where

$$\delta_s = p_s - \mathrm{i}p_{s+\varkappa},\ \nu(t)\,[X(t)]^{-1}t^s = -\left(\beta_s(t)+\mathrm{i}\beta_{s+\varkappa}(t)\right)\,,$$
$$s=1,2,\ldots,\varkappa,\ k=1,2,\ldots,n\,. \tag{38.84}$$

Thus, the problem (38.81), hence (38.3), is reduced to the problem (38.83) with piece-constant coefficient $\nu(t)$. Following [13] we consider two cases for (38.83): a) all numbers $\nu_k$ are equal ($\nu_k=\nu$); b) $\nu_k\neq\nu_m$ for some $k\neq m$. In the case a), the problem (38.83) is solved in terms of Schwarz's operator $\mathbf{T}$ constructed in Theorem 38.5, namely

$$\omega(z) = \nu^{-1}\left[(\mathbf{T}h)(z)+\sum_{s=1}^{2\varkappa}p_s(\mathbf{T}\beta_s)(z)+\mathrm{i}\varsigma\right]\,, \tag{38.85}$$

where $\varsigma$ is an arbitrary real constant. Conditions (38.73) yield

$$\sum_{s=1}^{2\varkappa}p_s\sum_{m=1}^{n}\int_{\mathbb{T}_m}\beta_s(\zeta)\frac{\partial\alpha_k}{\partial n_\zeta}(\zeta)\mathrm{d}s = -\sum_{m=1}^{n}\int_{\mathbb{T}_m}h(\zeta)\frac{\partial\alpha_k}{\partial n_\zeta}(\zeta)\mathrm{d}s\,,$$
$$k=1,2,...,n-1\,. \tag{38.86}$$

The relation (38.86) with $k=n$ follows from others, since the condition

$$\sum_{k=1}^{n}\alpha_k(\zeta)\equiv 1$$

implies

$$\sum_{k=1}^{n}\frac{\partial\alpha_k}{\partial n_\zeta}(\zeta)\equiv 0\,,\quad |\zeta-a_m| = r_m,\ m=1,2,...,n\,.$$

Therefore, (38.86) is a system of $n-1$ real linear algebraic equations with respect to $2\varkappa$ unknowns $p_1,p_2,...,p_{2\varkappa}$. This is the system predicted by Bojarski [38].

It corresponds to solvability of the Riemann–Hilbert problem in the case a). If the system (38.86) is solvable then the solution of (38.3) has the form

$$\phi(z) = R(z)\left[X(z)\omega(z) + \sum_{s=1}^{\varkappa} \delta_s z^s\right], \quad z \in \mathbb{D} \cup \partial\mathbb{D}, \tag{38.87}$$

where $\delta_s$ are defined by (38.84).

Consider now the case (b). The problem (38.83) can be reduced to a system of functional equations in the following way. First, write the problem (38.83) as the $\mathbb{R}$-linear problem

$$\nu_k \omega(t) = \phi_k(t) - \overline{\phi_k(t)} + f_k(t), \quad |t - a_k| = r_k,\ k = 1, 2, \ldots, n.$$

Here, the unknown functions $\phi_k(z)$ belong to $\mathscr{C}_{\mathscr{A}}(\mathbb{D}_k)$,

$$f_k(z) := -h_k(z) - \sum_{s=1}^{2\varkappa} p_s \beta_{sk}(z),$$

and the function $h_k(z)$ is a solution of the Schwarz problem for the disk $\mathbb{D}_k$

$$\operatorname{Re} h_k(t) = h(t),\ |t - a_k| = r_k,$$

with respect to $h_k(z) \in \mathscr{C}_{\mathscr{A}}(\mathbb{D}_k)$. According to formula (2.7.2) from [30]

$$h_k(z) = -\frac{1}{\pi \mathrm{i}} \int_{\mathbb{T}_k} \frac{h(\tau)}{\tau - z} \mathrm{d}\tau + \frac{1}{2\pi \mathrm{i}} \int_{\mathbb{T}_k} \frac{h(\tau)}{\tau - a_k} \mathrm{d}\tau.$$

The functions $\beta_{sk}(z)$ are defined through $\beta_s(t)$ by the same formulae.

Introduce the function

$$\Phi(z) := \begin{cases} \overline{\nu_k}\phi_k(z) + \sum_{m \neq k} \overline{\nu_m}\left[\overline{\phi_m\left(z^*_{(m)}\right) - \phi_m\left(w^*_{(m)}\right)}\right] \\ \qquad -\overline{\nu_k}\left(\overline{\phi_k(w^*_{(k)}) - f_k(z)}\right), & |z - a_k| \leq r_k; \\ \omega(z) + \sum_{m=1}^{n} \overline{\nu_m}\left[\overline{\phi_m\left(z^*_{(m)}\right)} - \overline{\phi_m\left(w^*_{(m)}\right)}\right], & z \in \mathbb{D}. \end{cases}$$

Taking into account Analytic Continuation Principle and Liouville's Theorem we obtain that $\Phi(z)$ is a constant:

$$\Phi(z) = Q + \sum_{m=1}^{n} \overline{\nu_m \phi_m\left(w^*_{(m)}\right)},$$

where $Q = \omega(w)$. The definition of $\Phi(z)$ in $|z-a_k| \leq r_k$ yields the system of functional equations

$$\phi_k(z) = -\nu_k \sum_{m\neq k} \overline{\nu_m}\left[\overline{\phi_m\left(z^*_{(m)}\right)} - \overline{\phi_m\left(w^*_{(m)}\right)}\right]$$

$$+\overline{\phi_k\left(w^*_{(k)}\right)} - f_k(z) + \nu_k Q, \quad |z-a_k| \leq r_k,\ k=1,2,...,n\,, \tag{38.88}$$

with respect to the functions

$$\phi_k(z) \in \mathscr{C}_{\mathscr{A}}(\mathbb{D}_k)\,.$$

The function $\omega(z)$ is related to the auxiliary functions by the formula

$$\omega(z) = Q - \sum_{m=1}^{n} \overline{\nu_m}\left[\overline{\phi_m\left(z^*_{(m)}\right)} - \overline{\phi_m\left(w^*_{(m)}\right)}\right]. \tag{38.89}$$

The system (38.88) is solved in Lemma 38.3

$$\phi_k(z) = \left(\mathbf{A}_k h^+\right)(z) + \sum_{s=1}^{2\varkappa} p_s(\mathbf{A}_k\beta_s^+)(z)$$

$$+\nu_k Q + \overline{\phi_k\left(w^*_{(k)}\right)}, \quad |z-a_k| \leq r_k,\ k=1,2,\ldots,n\,. \tag{38.90}$$

The operators $\mathbf{A}_k$ are defined by the following formula

$$(\mathbf{A}_k F)(z) = F(z)$$

$$+\nu_k \sum_{m=0}^{\infty} \sum_{k_1\neq k} \sum_{k_2\neq k_1} \cdots \sum_{k_m\neq k_{m-1}} (-1)^m \mathbf{C}^m \nu_{k_m}\left[F\left(z^*_{(k_m k_{m-1}\ldots k_1)}\right)\right.$$

$$\left.-F\left(w^*_{(k_m k_{m-1}\ldots k_1)}\right)\right], \quad |z-a_k| \leq r_k,\ k=1,2,\ldots,n\,. \tag{38.91}$$

Here, $F(z)$ is a function from $\mathscr{C}_{\mathscr{A}}(\cup_{k=0}^{n}\mathbb{D}_k)$, $\mathbf{C}$ is the operator of complex conjugation.

Calculate the values

$$R_{km} := \overline{\phi_m\left(w^*_{(mk)}\right)} - \overline{\phi_m\left(w^*_{(m)}\right)} = \overline{(\mathbf{A}_k f_k)\left(w^*_{(mk)}\right)} - \overline{(\mathbf{A}_k f_k)\left(w^*_{(m)}\right)},$$

$$k=1,2,\ldots,n;\ m=1,2,\ldots,n;\ m\neq k\,.$$

Substitution of $z = w^*_{(k)}$ into (38.88) and taking the real part yield

$$\mathrm{Re}\left(-\sum_{s=1}^{2\varkappa} p_s\beta^+_{sk}(w^*_{(k)}) + \nu_k Q\right) = \xi_k, \quad k=1,2,\ldots,n\,, \tag{38.92}$$

where

$$\xi_k = \mathrm{Re}\left( \nu_k \sum_{m \neq k} \overline{\nu_m} R_{km} + h_k\left(w^*_{(k)}\right) \right). \tag{38.93}$$

Let for definiteness $\nu_n \neq \nu_{n-1}$. Then

$$\Delta = \frac{\nu_n}{\nu_{n-1}} - \frac{\nu_{n-1}}{\nu_n} \neq 0$$

and

$$Q = \frac{2}{\Delta}\left[ \frac{\xi_n}{\nu_{n-1}} - \frac{\xi_{n-1}}{\nu_n} + \sum_{s=1}^{2\varkappa} p_s \left( \frac{\beta^+_{sn}\left(w^*_{(n)}\right)}{\nu_{n-1}} - \frac{\beta^+_{s,n-1}\left(w^*_{(n-1)}\right)}{\nu_n} \right) \right]. \tag{38.94}$$

Therefore, (38.92) becomes

$$\sum_{s=1}^{2\varkappa} p_s \,\mathrm{Re}\left[ -\beta^+_{sk}(w^*_{(k)}) + \frac{2}{\Delta}\left( \frac{\nu_k}{\nu_{n-1}} \beta^+_{sn}(w^*_{(n)}) - \frac{\nu_k}{\nu_n} \beta^+_{s,n-1}(w^*_{(n-1)}) \right) \right]$$
$$= \xi_k - \frac{2}{\Delta}\left( \frac{\xi_n}{\nu_{n-1}} - \frac{\xi_{n-1}}{\nu_n} \right), \quad k = 1, 2, \ldots, n-2. \tag{38.95}$$

The relation (38.95) is a system of $n-2$ real linear algebraic equations with respect to $2\varkappa$ real unknowns $p_1, p_2, ..., p_{2\varkappa}$. This is Bojarski's system [38] which corresponds to solvability of the Riemann–Hilbert problem. Equality (38.95) gives the necessary and sufficient solvability condition of the Riemann–Hilbert problem (38.3) in the case (b). If (38.95) is solvable, $\phi(z)$ has the form (38.87), (38.89), (38.90). It is worth noting that $\overline{\phi_k(w^*_{(k)})}$ remains undetermined. But it follows from (38.89) that we do not need to determine the constants $\overline{\phi_k(w^*_{(k)})}$ $(k = 1, 2, \ldots, n)$ to find the function $\omega(z)$.

**Theorem 38.6.** $(\varkappa \geq 0)$

*(a) If all numbers $\nu_k$ are equal, the Riemann–Hilbert boundary value problem (38.3) is solvable if and only if the system (38.86) is solvable.*

*If (38.86) is fulfilled, the general solution to the problem (38.3) has the form (38.87), (38.85).*

*(b) If at least two numbers $\nu_k$ are non-equal, the Riemann–Hilbert boundary value problem (38.3) is solvable if and only if the system (38.95) is solvable.*

*If (38.95) is satisfied, the general solution to the problem (38.3) has the form (38.87), (38.89), (38.90).*

The case of the negative index $\varkappa$ is studied in the same way. In this case, we introduce the auxiliary unknown function $\omega(z) := \phi(z)\,[R(z)X(z)]^{-1}$ from the space $\mathscr{C}_{\mathscr{A}}(\mathbb{D})$ with the following additional condition

$$\omega(z) \text{ has zero at } z = \infty \text{ of the order } |\varkappa|. \tag{38.96}$$

Thus, the function $\omega(z)$ has to satisfy the following $\mathbb{R}$-linear problem

$$\nu_k \omega(t) = \phi_k(t) - \overline{\phi_k(t)} + f_k(t), \quad |t-a_k| = r_k, \; k=1,2,\ldots,n,$$

with the additional condition (38.96). Application of the same arguments yields

**Theorem 38.7.** $(\varkappa < 0)$

*(a) If all numbers $\nu_k$ are equal, the Riemann–Hilbert boundary value problem (38.3) with respect to the function $\phi(z)$ having the order $|\varkappa|$ at infinity has a solution if and only if $(n-1)$ linearly independent conditions are fulfilled*

$$\sum_{m=1}^{n} \int_{\mathbb{T}_m} h(\zeta) \frac{\partial \alpha_k}{\partial n_\zeta}(\zeta) ds = 0, \quad k = 1,2,\ldots,n-1. \tag{38.97}$$

*If these conditions are valid, the general solution of the problem (38.3) has the form*

$$\phi(z) := \omega(z)R(z)X(z) = \nu^{-1}[(\mathbf{T}h)(z) + i\varsigma]R(z)X(z)$$

*and contains one arbitrary real constant $\varsigma$.*

*(b) If at least two numbers $\nu_k$ are non-equal, the Riemann–Hilbert boundary value problem (38.3) with respect to the function $\phi(z)$ having the order $|\varkappa|$ at infinity has a solution if and only if $(n-2)$ linearly independent conditions are fulfilled*

$$\xi_k - \frac{2}{\Delta}\left(\frac{\xi_n}{\nu_{n-1}} - \frac{\xi_{n-1}}{\nu_n}\right) = 0, \quad k = 1,2,\ldots,n-2, \tag{38.98}$$

*where $\xi_k = \mathrm{Re}\, h_k(w^*_{(k)})$ (see (38.93)). If these conditions are valid, the unique solution of the problem (38.3) has the form $\phi(z) := \omega(z)R(z)X(z)$.*

*In all cases, the solution $\phi(z)$ is regular at infinity if and only if the function $R(z)X(z)$ satisfies the condition (38.96).*

Let $l$ be the number of $\mathbb{R}$-linear independent solutions of the homogeneous Riemann–Hilbert problem

$$\mathrm{Re}\, \overline{\lambda(t)}\phi(t) = 0, \quad t \in \partial D. \tag{38.99}$$

Let $p$ be the number of $\mathbb{R}$-linear independent solvability conditions of the inhomogeneous Riemann–Hilbert problem (38.3)

$$\mathrm{Re}\, \overline{\lambda(t)}\phi(t) = f(t), \quad t \in \partial D. \tag{38.100}$$

Let the rank of the system (38.86) is equal to $r$. Then the homogeneous system (38.86) has $2\varkappa - r$ $\mathbb{R}$-linearly independent solutions. Along similar lines, the homogeneous system (38.95) has $2\varkappa - d$ $\mathbb{R}$-linearly independent solutions, where $d$ is the rank of the system. One can see that the ranks depend only on the coefficient

$\lambda(t)$. Moreover, $r \leq n-1$ and $d \leq n-2$. According to Theorems 38.6 and 38.7, we have in all cases

$$l - p = 2\varkappa - n + 2\,, \tag{38.101}$$

where in the case of negative index, $|\varkappa|$ complex conditions to provide that $\phi$ has the zero order at infinity are taken into account. Formula (38.101) coincided to formula (37.50′) from Gakhov's book [13].

## References


1. Akaza, T.: Singular sets of some Klainian groups. Nagoya Math. J. **26**, 127–143 (1966)
2. Akaza, T., Inoue, K.: Limit sets of geometrically finite free Klainian groups. Tohoku Math. J. **36**, 1–16 (1984)
3. Akhiezer, N.I.: Elements of the Theory of Elliptic Functions. Nauka, Moscow, 1970 (in Russian); English transl.: AMS, Providence, Rhode Island (1990)
4. Aleksandrov, I.A., Sorokin, A.S.: The problem of Schwarz for multiply connected domains. Sib. Math. Zh. **13**, n. 5, 971–1001 (1972) (in Russian)
5. Baker, H.F.: Abelian Functions. Cambridge University Press, Cambridge (1996)
6. Bancuri, R.D.: On the Riemann–Hilbert problem for doubly connected domains. Soobsch. AN GruzSSR **80**, n. 4, 549–552 (1975) (in Russian)
7. Bojarski, B.: On generalized Hilbert boundary value problem. Soobsch. AN GruzSSR **25**, n. 4, 385–390 (1960) (in Russian)
8. Koppenfels, V., Schtalman, Tz.: Practice of conformal mappings, Moscow, IL (1963) (in Russian)
9. Crowdy, D.: Explicit solution of a class of Riemann–Hilbert problems. Annales Universitatis Paedagogicae Cracoviensis, Studia Mathematica VIII (2009) http,://mat.ap.krakow.pl/aapc/index.php/studmath/article/view/77/58 (online version)
10. Crowdy, D.: The Schwarz problem in multiply connected domains and the SchottkyKlein prime function. Complex Variables and Elliptic Equations **53**, No. 3, 221–236 (2008)
11. Crowdy, D.: Geometric function theory: a modern view of a classical subject. Nonlinearity **21**, T205–T219 (2008) http,://www.iop.org/EJ/abstract/0951-7715/21/10/T04/
12. Dunduchenko, L.E.: On the Schwarz formula for an $n$-connected domain. Dopovedi AN URSR **5**, 1386–1389 (1966) (in Ukrainian)
13. Gakhov, F.D.: Boundary Value Problems. Nauka, Moscow (1977) (3rd edition; in Russian); Engl. transl. of 1st ed.: Pergamon Press, Oxford (1966)
14. Golusin, G.M.: Solution of basic plane problems of mathetical physics for the case of Laplace equation and multiply connected domains bounded by circles (method of functional equations). Math. Zb. **41:2**, 246–276 (1934)
15. Golusin, G.M.: Solution of spatial Dirichlet problem for Laplace equation and for domains enbounded by finite number of spheres. Math. Zb. **41:2**, 277–283 (1934)
16. Golusin, G.M.: Solution of plane heat conduction problem for multiply connected domains enclosed by circles in the case of isolated layer. Math. Zb. **42:2**, 191–198 (1935)
17. Golusin, G.M.: Geometric Theory of Functions of Complex Variable. Nauka, Moscow (1966) (2nd ed.; in Russian); Engl. transl. by AMS, Providence, RI (1969)
18. Kantorovich, L.V., Krylov, V.I.: Approximate methods of higher analysis, Groningen, Noordhoff (1958)
19. Krasnosel'skii, M.A., Vainikko, G.M., Zabreiko, P.P., Rutickii, Ja.B., Stecenko, V.Ja.: Approximate Methods for Solution of Operator Equations. Nauka, Moscow (1969) (in Russian); Engl. transl.: Wolters-Noordhoff Publ., Groningen (1972)

20. Kveselava, D.A.: Riemann–Hilbert problem for multiply connected domain. Soobsch. AN GruzSSR **6**, n. 8, 581–590 (1945) (in Russian)
21. Kuczma, M., Choczewski, B., Ger, R.: Iterative functional equations. Encyclopedia Math. Appl. 32, Cambridge University Press (1990)
22. Markushevich, A.I.: On a boundary value problem of analytic function theory. Uch. Zapiski MGU **1**, vyp. 100, 20–30 (1946) (in Russian)
23. Mikhailov, L.G.: On a boundary value problem. DAN SSSR **139**, 294–297 (1961) (in Russian)
24. Mikhailov, L.G.: New Class of Singular Integral Equations and its Applications to Differential Equations with Singular Coefficients. AN TadzhSSR, Dushanbe (1963) (in Russian); English transl.: Akademie Verlag, Berlin (1970)
25. Mikhlin, S.G.: Integral Equations. Pergamon Press, New York (1964)
26. Mityushev, V.V.: Solution of the Hilbert boundary value problem for a multiply connected domain. Slupskie Prace Mat.-Przyr. **9a**, 37–69 (1994)
27. Mityushev, V.V.: Generalized method of Schwarz and addition theorems in mechanics of materials containing cavities. Arch. Mech. **47**, 1169–1181 (1995)
28. Mityushev, V.V.: Convergence of the Poincaré series for classical Schottky groups. Proc. Amer. Math. Soc. **126**, 2399–2406 (1998)
29. Mityushev, V.V.: Hilbert boundary value problem for multiply connected domains. Complex Variables **35**, 283–295 (1998)
30. Mityushev, V.V., Rogosin, S.V.: Constructive methods to linear and non-linear boundary value problems for analytic function. Theory and Applications. Chapman & Hall / CRC, Monographs and Surveys in Pure and Applied Mathematics, Boca Raton etc. (2000)
31. Muskhelishvili, N.I.: To the problem of torsion and bending of beams constituted from different materials. Izv. AN SSSR **7**, 907–945 (1932) (in Russian)
32. Muskhelishvili, N.I.: Singular Integral Equations. Nauka, Moscow (1968) (3rd edition; in Russian); English translation of the 1st ed.: Noordhoff, Groningen (1946)
33. Muskhelishvili, N.I.: Some Basic Problems of Mathematical Elasticity Theory. Nauka, Moscow (1966) (5th edition: in Russian); English translation of the 1st ed.: Noordhoff, Groningen (1953)
34. Poincaré, H.: Oeuvres. Gauthier-Villart, Paris, v. **2** (1916); v. **4** (1950); v. **9** (1954)
35. Smith, B., Björstad, P., Gropp, W.: Domain Decomposition. Parallel Multilevel Methods for Elliptic Partial Differential Equations. Cambridge University Press (1996)
36. Vekua, I.N., Rukhadze, A.K.: The problem of the torsion of circular cylinder reinforced by transversal circular beam. Izv. AN SSSR **3**, 373–386 (1933)
37. Vekua, I.N., Rukhadze, A.K.: Torsion and transversal bending of the beam compounded by two materials restricted by confocal ellipces. Prikladnaya Matematika i Mechanika (Leningrad) **1**, n. 2, 167–178 (1933)
38. Vekua, I.N.: Generalized Analytic Functions. Nauka, Moscow (1988) (in Russian)
39. Vekua, N.P.: Systems of Singular Integral Equations. Noordhoff, Groningen (1967)
40. Zmorovich, V.A.: On a generalization of the Schwarz integral formula on n-connected domains. Dopovedi URSR **5**, 489–492 (1958) (in Ukranian)
41. Zverovich, E.I.: Boundary value problems of analytic functions in Hölder classes on Riemann surfaces. Uspekhi Mat. Nauk **26**(1), 113–179 (1971) (in Russian)

# Chapter 39
# Hodge Theory for Riemannian Solenoids

**Vicente Muñoz and Ricardo Pérez Marco**

*Dedicated to the memory of S.M. Ulam*
*on the 100th anniversary of his birth*

**Abstract** A measured solenoid is a compact laminated space endowed with a transversal measure. The De Rham $L^2$-cohomology of the solenoid is defined by using differential forms which are smooth in the leafwise directions and $L^2$ in the transversal direction. We develop the theory of harmonic forms for Riemannian measured solenoids, and prove that this computes the De Rham $L^2$-cohomology of the solenoid. This implies in particular a Poincaré duality result.

**Keywords** Solenoids • Harmonic forms • Cohomology • Hodge theory

**Mathematics Subject Classification (2000):** Primary 58A14; Secondary 57R30, 58A12

## 39.1 Introduction

In this article, we continue the geometric study of solenoids initiated in [6]. Solenoids are compact topological spaces which are locally modeled by the product of a $k$-dimensional ball by some transversal space and admit a transversal measure invariant by holonomy.

---

V. Muñoz (✉)
Facultad de Matemáticas, Universidad Complutense de Madrid,
Plaza Ciencias 3, 28040 Madrid, Spain
e-mail: vicente.munoz@mat.ucm.es

R.P. Marco
CNRS, LAGA UMR 7539, Université Paris XIII, 99, Avenue J.-B. Clément,
93430-Villetaneuse, France
e-mail: ricardo@math.univ-paris13.fr

Th.M. Rassias and J. Brzdęk (eds.), *Functional Equations in Mathematical Analysis*, Springer Optimization and Its Applications 52, DOI 10.1007/978-1-4614-0055-4_39,

First, we review how the De Rham cohomology theory, fundamental classes, and singular cohomology do extend to solenoids and preserve most of the functorial properties. The theory of bundles, connections and Chern classes also goes through.

For a solenoid $S$ endowed with a transversal measure $\mu$, we have a well defined De Rham $L^2$-cohomology theory, which is suited to implement classical techniques from Harmonic and Functional Analysis. The De Rham $L^2$-cohomology is defined by using forms which are smooth in the leaf-wise directions, and are $L^2$-integrable with respect to $\mu$ in the transversal direction. We introduce the *reduced* De Rham $L^2$-cohomology $\bar{H}^*_{DR}(S_\mu)$ as the quotient of the De Rham $L^2$-cohomology $H^*_{\mathrm{DR}}(S_\mu)$ with the closure of $\{0\}$ (making it a Hausdorff topological space). See [5] for these notions.

Most of the formal theory of pseudo-differential operators can be carried out for spaces of sections of bundles which are $L^2$-transversally. In order to extend classical Hodge theory as developed in [12], we define Sobolev spaces of sections using the $L^2$ transversal structure. Sobolev regularity lemma holds in this general setting but Rellich compact embedding does not in general. With these tools at hand we can develop harmonic theory for solenoids.

We obtain a Hodge theorem giving an isomorphism of the reduced De Rham cohomology with the space of harmonic forms. We prove the main results:

**Theorem 39.1.** *Let $S_\mu$ be a compact oriented solenoid endowed with a transversal measure $\mu$. Let $\mathscr{K}^p(S_\mu)$ be the space of p-forms which are harmonic in the leaf-wise directions and $L^2$-transversally (with respect to $\mu$). Then there is an isomorphism:*

$$\bar{H}^p_{\mathrm{DR}}(S_\mu) \cong \mathscr{K}^p(S_\mu).$$

**Corollary 39.1.** *The $*$-Hodge operator gives an isomorphism (Poincaré duality)*

$$* : \bar{H}^p_{\mathrm{DR}}(S_\mu) \to \bar{H}^{k-p}_{\mathrm{DR}}(S_\mu).$$

Contrary to the classical situation for compact manifolds where Rellich theorem holds, here the spaces of harmonic forms are not necessarily finite dimensional. Actually, the situation for solenoids is more similar to that of Hodge theory for $L^2$-forms for complete non-compact manifolds (see e.g. [1]). However, when the transversal measure $\mu$ is ergodic (see Definition 39.3), we may expect finite-dimensionality (see Question 39.2). We discuss at the end the simple example of Kronecker foliations on the torus.

## 39.2 Solenoids

We review the basic notions on solenoids introduced in [6].

**Definition 39.1.** Let $k \geq 0$, $r \geq 1$, $s \geq 0$ with $r \geq s$. A $k$-solenoid of class $C^{r,s}$ is a compact Hausdorff space endowed with an atlas of flow-boxes $\mathscr{A} = \{(U_i, \varphi_i)\}$,

$$\varphi_i : U_i \to D^k \times K(U_i),$$

where $D^k$ is the $k$-dimensional open ball, and $K(U_i) \subset \mathbb{R}^l$ is an open set, the transversal of the flow-box. The changes of charts $\varphi_{ij} = \varphi_i \circ \varphi_j^{-1}$ are of the form $\varphi_{ij}(x,y) = (X(x,y), Y(y))$, where $X(x,y)$ is of class $C^{r,s}$ and $Y(y)$ is of class $C^s$.

We shall always use *good* flow-boxes. By this, we mean a flow-box

$$U = D^k \times K(U)$$

whose closure is contained in another flow-box $V = D^k \times K(V)$, where $\overline{K(U)} \subset K(V)$ is compact.

Let $S$ be a $k$-solenoid, and $U \cong D^k \times K(U)$ be a flow-box for $S$. The sets $L_y = D^k \times \{y\}$ are called the (local) leaves of the flow-box. A leaf $l \subset S$ of the solenoid is a connected $k$-dimensional manifold whose intersection with any flow-box is a collection of local leaves. The solenoid is oriented if the leaves are oriented (in a transversally continuous way).

A transversal for $S$ is a subset $T$ which is a finite union of transversals of flow-boxes. Given two local transversals $T_1$ and $T_2$ and a path contained in a leaf from a point of $T_1$ to a point of $T_2$, there is a well-defined germ of holonomy map at this point, $h$, from $T_1$ to $T_2$.

**Definition 39.2.** Let $S$ be a $k$-solenoid. A transversal measure $\mu = (\mu_T)$ for $S$ associates to any local transversal $T$ a locally finite measure $\mu_T$ supported on $T$, which are invariant by the holonomy, i.e. if $h$ is a germ of holonomy map, then $h_*\mu_{T_1} = \mu_{T_2}$.

We denote by $S_\mu$ a $k$-solenoid $S$ endowed with a transversal measure $\mu = (\mu_T)$. We refer to $S_\mu$ as a measured solenoid. Observe that for any transversal measure $\mu = (\mu_T)$ the scalar multiple $c\,\mu = (c\,\mu_T)$, where $c > 0$, is also a transversal measure. Notice that there is no natural scalar normalization of transversal measures.

**Definition 39.3.** A measured solenoid $S_\mu$ is *ergodic* if for any transveral $T$, and any subset $A \subset T$ invariant by the holonomy, either $\mu_T(A) = 0$ or $\mu_T(T - A) = 0$.

A solenoid $S$ is *uniquely ergodic* if it has a unique (up to scalars) transversal measure $\mu$ and its support is the whole of $S$.

A Riemannian solenoid is a solenoid endowed with a Riemannian metric in the tangent spaces of the leaves, and with smoothness of class $C^{r,s}$. Note that a Riemannian metric defines a volume form in each leaf. A *daval* measure $\nu$ on $S$ ([6]) is a finite Borel measure on the solenoid which in any flow-box $U = D^k \times K(U)$, it decomposes as volume along leaves, $\nu = \mathrm{vol}_{D^k} \times \mu_{K(U)}$. Such a measure defines a tranversal measure and moreover there is a one-to-one correspondence between daval measures and transversal measures. In particular, if a Riemannian solenoid is uniquely ergodic, then there is a *unique* daval measure with total mass 1.

Now let $M$ be a smooth manifold of dimension $n$ (and class $C^r$). An immersion of a $k$-solenoid $S$ into $M$, with $k < n$, is a smooth map $f : S \to M$ (of class $C^{r,s}$) such that the differential restricted to the tangent spaces of leaves has rank $k$ at every point of $S$. If $M$ is a Riemannian manifold, this endows $S$ with the pull-back Riemannian structure.

Denote by $\mathscr{C}_k(M)$ the space of compactly supported currents of dimension $k$ on $M$. We have the following definition.

**Definition 39.4.** Let $S_\mu$ be an oriented measured $k$-solenoid. An immersion $f : S \to M$ defines a generalized Ruelle–Sullivan current $(S_\mu, f) \in \mathscr{C}_k(M)$ as follows. Let $\{U_i\}$ be a finite covering of $S$ by flow-boxes, and a partition of unity $\{\rho_i\}$ associated to it. For $\omega \in \Omega^k(M)$, we define

$$\langle (S_\mu, f), \omega \rangle = \sum_i \int_{K(U_i)} \left( \int_{L_y} \rho_i f^* \omega \right) d\mu_{K(U_i)}(y),$$

where $L_y$ denotes the horizontal disk of the flow-box.

In [6], it is proved that $(S_\mu, f)$ is a closed current. Therefore, it defines a real homology class

$$[S_\mu, f] \in H_k(M, \mathbb{R}). \tag{39.1}$$

Ruelle and Sullivan defined in [11] this notion for the restricted class of solenoids embedded in $M$. In [8], it is proved that if $a \in H_k(M, \mathbb{R})$ is any real homology class, then there is an immersed oriented $k$-solenoid $f : S_\mu \to M$ such that $a = [S_\mu, f]$. Moreover, $S$ can be chosen to be uniquely ergodic and with holonomy generated by a single map. In [9], we prove that the set of currents $(S_\mu, f)$ for immersed oriented uniquely ergodic $k$-solenoids with $a = [S_\mu, f]$ is actually dense in the space of closed currents $\alpha \in \mathscr{C}_k(M)$ representing $a$.

## 39.3 Cohomology of Solenoids

In general, and for the remainder of the article, we shall consider solenoids of class $C^{\infty,0}$.

### *39.3.1 De Rham Cohomology*

Let $S$ be a solenoid (here, we allow $S$ to be a *non-compact* solenoid). The space of $p$-forms $\Omega^p(S)$ consist of $p$-forms on leaves with function coefficients that are smooth on leaves and partial derivatives of all orders continuous transversally. Using the differential $d$ in the leaf-wise directions, we obtain the De Rham differential complex $(\Omega^*(S), d)$. The De Rham cohomology groups of the solenoid are defined as the quotients

$$H^p_{\mathrm{DR}}(S) := \frac{\ker(d : \Omega^p(S) \to \Omega^{p+1}(S))}{\mathrm{im}(d : \Omega^{p-1}(S) \to \Omega^p(S))}. \tag{39.2}$$

We can also consider the spaces $\Omega_m^p(S)$ of differential forms with function coefficients that are smooth on leaves and measurable transversally (then the partial derivatives are automatically measurable transversally). Then define in the same way the De Rham measurable cohomology groups $H^p_{\mathrm{DR}m}(S)$ using the complex $(\Omega_m^*(S),d)$. Note the natural map $H^p_{\mathrm{DR}}(S) \to H^p_{\mathrm{DR}m}(S)$.

**Proposition 39.1.** *Let $\underline{\mathbb{R}}_c$ and $\underline{\mathbb{R}}_m$ be respectively the sheaf of functions which are locally constant on leaves and transversally continuous, resp. measurable. Then we have isomorphisms*

$$H^p_{\mathrm{DR}}(S) \cong H^p(S,\underline{\mathbb{R}}_c),$$

*and*

$$H^p_{\mathrm{DR}m}(S) \cong H^p(S,\underline{\mathbb{R}}_m).$$

*Proof.* The proofs are similar. We prove the first isomorphism. It follows from the existence of the sheaf resolution

$$\underline{\mathbb{R}}_c \to \Omega^0 \xrightarrow{d} \Omega^1 \xrightarrow{d} \dots .$$

The exactness of this complex of sheaves is a Poincaré lemma: for a small open set $U = D^k \times K(U)$, the complex $0 \to \underline{\mathbb{R}}_c(U) \to \Omega^0(U) \to \Omega^1(U) \to \dots$ is exact. If $\mathrm{d}_x\alpha(x,y) = 0$ then $\alpha(x,y) = \mathrm{d}_x\beta(x,y)$, for a collection of forms $\beta(x,y)$, $y \in K(U)$. We can choose $\beta(x,y)$ to depend continuously (or measurably for the proof of the second isomorphism) on $y$, as can be seen by the usual construction. Note that, clearly, if $f(x,y)$ is a function with $\mathrm{d}f = 0$ then $f(x,y)$ is locally constant on $x$. □

*Remark 39.1.* The spaces $\Omega^p(S)$ are topological vector spaces. Therefore, the De Rham cohomology (39.2) inherits a natural topology. In general, these spaces are infinite dimensional (even for compact solenoids). In some references, it is customary to take the closure of the spaces $\operatorname{im} d$ in definition (39.2), obtaining the *reduced De Rham cohomology groups*

$$\bar{H}^p_{\mathrm{DR}}(S) = \frac{\ker d|_{\Omega^p}}{\overline{\operatorname{im} d|_{\Omega^{p-1}}}} .$$

This is equivalent to quotienting $H^p_{\mathrm{DR}}(S)$ by $\overline{\{0\}}$, obtaining thus Hausdorff vector spaces.

We shall list some basic properties of the De Rham cohomology:

1. Functoriality. Let $S_1$, $S_2$ be two solenoids. A smooth map $f : S_1 \to S_2$ is a map sending leaves to leaves and transversally continuous. $f$ defines a map on De Rham cohomologies, $f^* : H^p_{\mathrm{DR}}(S_2) \to H^p_{\mathrm{DR}}(S_1)$, by $f^*[\omega] = [f^*\omega]$. This applies in particular to an immersion of a solenoid into a smooth manifold $f : S \to M$, or to the inclusion of a leaf $i : l \to S$.

2. Mayer–Vietoris sequence. Let $U,V$ be two open subsets of a solenoid $S$. There is a short exact sequence of complexes:

$$\Omega^\bullet(U\cup V)\to\Omega^\bullet(U)\oplus\Omega^\bullet(V)\to\Omega^\bullet(U\cap V).$$

The only non-trivial point is the surjectivity of the last map, but if follows from the existence of a partition of unity $\{\rho_U,\rho_V\}$ subordinated to $\{U,V\}$: any $\omega\in\Omega^\bullet(U\cap V)$ is the image of $(\rho_V\omega,-\rho_U\omega)$. Taking the associated long exact sequence, we get the Mayer–Vietoris exact sequence

$$\cdots\to H^p_{\mathrm{DR}}(U\cup V)\to H^p_{\mathrm{DR}}(U)\oplus H^p_{\mathrm{DR}}(V)$$
$$\to H^p_{\mathrm{DR}}(U\cap V)\to H^{p+1}_{\mathrm{DR}}(U\cup V)\to\cdots.$$

3. Homotopy. A homotopy between two maps $f_0,f_1:S_1\to S_2$ is a map $F:S_1\times[0,1]\to S_2$ (where $S_1\times[0,1]$ is given the solenoid structure with leaves $l\times[0,1]$, for $l\subset S_1$ a leaf of $S_1$) such that $F(x,0)=f_0(x)$ and $F(x,1)=f_1(x)$. We say that the maps $f_0,f_1$ are homotopic, written $f_0\sim f_1$. In this case $f_0^*=f_1^*:H^p_{\mathrm{DR}}(S_2)\to H^p_{\mathrm{DR}}(S_1)$.

   We prove this as follows: factor $f_0=F\circ i_0$, and $f_1=F\circ i_1$, where $i_t:S_1\to S_1\times[0,1]$ is given as $i_t(x)=(x,t)$. Then $f_0^*=f_1^*$ follows from $i_0^*=i_1^*$. Let us check this. Consider $\pi:S_1\times[0,1]\to S_1$. Then $\pi\circ i_t=Id$. Let us see that $i_t^*:H^p_{\mathrm{DR}}(S\times I)\to H^p_{\mathrm{DR}}(S)$ is an isomorphism inverse to $\pi^*$. It is enough to see that $h^*=(i_t\circ\pi)^*:H^p_{\mathrm{DR}}(S\times I)\to H^p_{\mathrm{DR}}(S\times I)$ is the identity, $h(x,s)=(x,t)$, $t\in I$ fixed. For a closed $p$-form $\omega=\omega_1(x,s)+\omega_2(x,s)\wedge ds$, note that

$$0=d\omega=d_x\omega_1+(-1)^p\frac{\partial\omega_1}{\partial s}\wedge ds+d_x\omega_2\wedge ds=0$$

implies that $d_x\omega_1=0$ and $\frac{\partial\omega_1}{\partial s}=(-1)^{p+1}d_x\omega_2$. Now $\pi^*i_t^*\omega=\omega_1(x,t)$. Hence

$$\begin{aligned}(Id-\pi^*i_t^*)\omega&=\omega_1(x,s)-\omega_1(x,t)+\omega_2(x,s)\wedge ds\\&=\int_t^s\frac{\partial\omega_1}{\partial s}(x,u)du+\omega_2(x,s)\wedge ds\\&=(-1)^{p+1}\left(\int_t^s d_x\omega_2(x,u)du+(-1)^{p-1}\omega_2(x,s)\wedge ds\right)\\&=(-1)^{p+1}d\left(\int_t^s\omega_2(x,u)du\right),\end{aligned}$$

as required.

4. We say that two solenoids $S_1$, $S_2$ are of the same homotopy type if there are maps $f:S_1\to S_2$, $g:S_2\to S_1$, such that $f\circ g\sim Id_{S_2}$, $g\circ f\sim Id_{S_1}$. Then the De Rham cohomology groups of $S_1$ and $S_2$ are isomorphic.

### 39.3.2 Fundamental Classes

Let $S$ be an *oriented* compact $k$-solenoid. The De Rham cohomology groups do not depend on any measure of $S$. If $\mu = (\mu_T)$ is a transversal measure, then the integral $\int_{S_\mu}$ descends to cohomology giving a map [6]

$$\int_{S_\mu} : H^k_{\mathrm{DR}}(S) \to \mathbb{R}. \tag{39.3}$$

We define the solenoidal homology as

$$H_p(S, \underline{\mathbb{R}}_c) := H^p(S, \underline{\mathbb{R}}_c)^* = H^p_{\mathrm{DR}}(S)^* .$$

Then the map (39.3) defines a homology class $[S_\mu] \in H^k(S, \underline{\mathbb{R}}_c)^* = H_k(S, \underline{\mathbb{R}}_c)$. We shall call this element the *fundamental class* of $S_\mu$.

Any map $f : S_1 \to S_2$ defines a map $f^* : H^p_{\mathrm{DR}}(S_2) \to H^p_{\mathrm{DR}}(S_1)$ and hence, by dualizing, a map $f_* : H_p(S_1, \underline{\mathbb{R}}_c) \to H_p(S_2, \underline{\mathbb{R}}_c)$. Applying this to an immersion $f : S_\mu \to M$ of an oriented, measured, compact solenoid into a smooth manifold, then we have the equality

$$f_*[S_\mu] = [S_\mu, f],$$

with the generalized Ruelle–Sullivan class defined in (39.1).

Note that if $S$ has a dense leaf (in particular when $S$ it is minimal, i.e. all leaves are dense), then $H_0(S, \underline{\mathbb{R}}_c) = \mathbb{R}$. On the other hand, the dimension of the top degree homology counts the number of mutually singular tranverse measures on $S$.

**Theorem 39.2.** *Let $S$ be a compact, oriented $k$-solenoid. Then $H_k(S, \underline{\mathbb{R}}_c)$ is isomorphic to the real vector space generated by all transversal measures.*

*Proof.* There is a well-defined linear map

$$\Psi : \mu \mapsto [S_\mu] \in H_k(S, \underline{\mathbb{R}}_c)$$

which sends each signed transversal measure (with finite total mass in each transversal) to the associated fundamental class: Decompose $\mu = \mu_+ - \mu_-$ and integrate with respect to each measure. We prove first that $\Psi$ is a bijection.

If $\Psi(\mu) = 0$ then

$$\int_{S_\mu} \omega = 0$$

for all $k$-forms $\omega \in \Omega^k(S)$. Let $U = D^k \times K(U)$ be a flow-box and let $\omega$ be a compactly supported $k$-form with integral 1 on $D^k$. For any function $\phi$ with $\operatorname{supp} \phi \subset K(U)$, we have

$$0 = \int_{S_\mu} \phi\, \omega = \int_{K(U)} \phi \left( \int_{L_y} \omega \right) \mathrm{d}\mu_{K(U)}(y) = \int_{K(U)} \phi\, \mathrm{d}\mu_{K(U)}(y).$$

Therefore, $\mu_{K(U)} = 0$. So $\mu = 0$.

We prove that $\Psi$ is onto. Let $C \in H_k(S, \mathbb{R}_c)$ be given. By definition $C : H^k_{\mathrm{DR}}(S) \to \mathbb{R}$, so for any $k$-form $\omega$, we have $C(\omega) \in \mathbb{R}$ and $C(\mathrm{d}\eta) = 0$, for $\eta \in \Omega^{k-1}(S)$. Let $U = D^k \times K(U)$ be a flow-box, and fix a compactly supported $k$-form $\omega$ with integral 1 on $D^k$. Let $\phi$ be a continuous function with $\operatorname{supp}\phi \subset K(U)$ and let $\tilde{\phi}$ be the corresponding function on the flow box $U = D^k \times K(U)$ constant on leaves. Now the map

$$\phi \mapsto C(\tilde{\phi}\,\omega),$$

is a continuous linear functional: if $\phi_n \to \phi$ in $C^0$ then $\tilde{\phi}_n\,\omega \to \tilde{\phi}\,\omega$ in $\Omega^k(S)$, so $C(\tilde{\phi}_n\,\omega) \to C(\tilde{\phi}\,\omega)$. Therefore, by Riesz representation theorem it is represented by a signed measure $\mu_{K(U)}$ with finite total mass,

$$C(\tilde{\phi}\,\omega) = \int_{K(U)} \phi\, \mathrm{d}\mu_{K(U)}.$$

The measure $\mu_{K(U)}$ does not depend on the choice of $\omega$. Taking another $\omega'$ we have $\omega' - \omega = \mathrm{d}\eta$, with $\eta$ compactly supported in $D^k$. Hence, $\tilde{\phi}\,\omega' - \tilde{\phi}\,\omega = d(\tilde{\phi}\,\eta)$ and $C(\tilde{\phi}\,\omega') = C(\tilde{\phi}\,\omega)$.

Also, the constructed measure does not depend on the coordinates of the flow-box. Taking a change of chart $\Phi : D^k \times K(U) \to D^k \times K(U)$, we have that $C(\Phi^*(\tilde{\phi}\,\omega)) = C(\tilde{\phi}\,\omega)$ since $\Phi^*(\tilde{\phi}\,\omega) - \tilde{\phi}\,\omega = \mathrm{d}\eta$ (both are compactly supported forms with leaf-wise integral 1). Finally, the $(\mu_T)$ are invariant by the holonomy. We only need to check the invariance by local holonomy, and this follows from the previous remark. □

*Remark 39.2.* There is no Poincaré duality for $H^*_{\mathrm{DR}}(S)$ in general. Moreover, these spaces may be infinite dimensional (even for uniquely ergodic solenoids): if $S$ is an $n$-torus foliated by irrational lines, then $H^1_{\mathrm{DR}}(S)$ can be infinite-dimensional as the example in Sect. 39.8 shows.

### *39.3.3 Singular Cohomology*

We consider the space $\mathrm{Map}(I^n, S)$ of continuous maps $T : I^n \to S$ mapping into a leaf, and endow it with the uniform convergence topology. The degenerate maps (see [4]) form a closed subspace; therefore, the quotient, $\mathrm{Map}'(I^n, S)$, has a natural quotient topology. The space of singular chains $C_n(S)$ is the free abelian group generated by $\mathrm{Map}'(I^n, S)$. There is a natural boundary map $\partial : C_n(S) \to C_{n-1}(S)$.

Let $G$ be any topological abelian group. Define the cochains

$$C^n(S, G) = \mathrm{Hom}_{\mathrm{cont}}(C_n(S), G)$$

as the continous homomorphisms. That is, $\varphi : C_n(S) \to G$ such that if $T_k : I^n \to S$ are maps which converge to $T_o : I^n \to S$ in the uniform topology, then $\varphi(T_k) \to \varphi(T_o)$. Define the differential $\delta : C^n(S) \to C^{n+1}(S)$ by $\delta\varphi(T) = \varphi(\partial T)$. The solenoid singular cohomology of $S$ with coefficients in $G$ is defined as:

$$H^n(S,G) := \frac{\ker(\delta : C^n(S,G) \to C^{n+1}(S,G))}{\mathrm{im}(\delta : C^{n-1}(S,G) \to C^n(S,G))}\,.$$

We have some basic properties:

1. Functoriality. Let $f : S_1 \to S_2$ be a solenoid map. Then there is a map $f_* : C_n(S_1) \to C_n(S_2)$, $f_*(T) = f \circ T$, and a map $f^* : C^n(S_2,G) \to C^n(S_1,G)$, $f^*(\varphi) = \varphi \circ f$. Clearly, $f^*\delta = \delta f^*$, so the map descends to cohomology: $f^* : H^n(S_2,G) \to H^n(S_1,G)$.
2. Homotopy. Suppose that $f,g : S_1 \to S_2$ are two homotopic solenoid maps. The usual construction yields a chain homotopy $H$ between $f^*$ and $g^*$ (one only have to check that this map sends continuous cochains into continuous cochains). Therefore, $f^* = g^* : H^n(S_2,G) \to H^n(S_1,G)$.
3. If $S_1$, $S_2$ are of the same homotopy type, then $H^n(S_1,G) \cong H^n(S_2,G)$.
4. If $U = D^k \times K(U)$ is a flow-box, then $U$ is of the same homotopy type than $\{*\} \times K(U)$. Therefore, $H^n(U) = 0$ for $n > 0$, and $H^0(U) = \mathrm{Map}_{\mathrm{cont}}(K(U),G)$. In particular, this implies that
$$\underline{\mathbb{R}}_c \to C^0(-,\mathbb{R}) \xrightarrow{\delta} C^1(-,\mathbb{R}) \xrightarrow{\delta} \cdots$$
is a resolution. Therefore, there is an isomorphism $H^n(S,\mathbb{R}) \cong H^n(S,\underline{\mathbb{R}}_c)$.
5. Mayer–Vietoris. For two open sets $U,V$ with $S = U \cup V$, define $C_n(S;U,V)$ as the subcomplex generated by those singular chains completely contained in either $U$ or $V$. Define accordingly $C^n(S;U,V)$. It is not difficult to see that the restriction $C^n(S,G) \to C^n(S,G;U,V)$ is chain homotopy equivalence (by a process of subdivision of simplices, as in [4]). Therefore, the exact sequence $0 \to C^n(S,G;U,V) \to C^n(U,G) \oplus C^n(V,G) \to C^n(U \cap V,G) \to 0$ gives rise to a long exact sequence:
$$\cdots \to H^p(U \cup V,G) \to H^p(U,G) \oplus H^p(V,G) \\ \to H^p(U \cap V,G) \to H^{p+1}(U \cup V,G) \to \cdots\,.$$

## 39.4 De Rham $L^2$-Cohomology

Now consider a $k$-solenoid $S$ with a transversal measure $\mu$. There is a notion of cohomology which takes into account the transversal measure structure. For this, we work with forms which are $L^2$-transversal relative to $\mu$.

**Definition 39.5.** A function $f$ is $L^2(\mu)$*-transversally smooth* if in any (good) flow-box $U = D^k \times K(U)$ all partial derivatives on the first variable exist and are in $L^2(\mu_{K(U)})$, i.e. if we write $f$ as $f(x,y)$ then for all $r \geq 0$,

$$\int_{K(U)} \|f(\cdot,y)\|^2_{C^r} \, \mathrm{d}\mu_{K(U)}(y) < \infty.$$

We consider the space of forms

$$\Omega^p_{L^2(\mu)}(S)$$

which are $L^2(\mu)$-transversally smooth, i.e. locally these are forms $\alpha = \sum f_I(x,y)\mathrm{d}x_I$, where $f_I$ are $L^2(\mu)$-transversally smooth functions. There exists a well-defined differential along leaves $d : \Omega^p_{L^2(\mu)}(S) \to \Omega^{p+1}_{L^2(\mu)}(S)$ which defines the complex $(\Omega^*_{L^2(\mu)}(S), d)$. We define the *De Rham $L^2$-cohomology* vector space as the quotients

$$H^p_{\mathrm{DR}}(S_\mu) := \frac{\ker\left(d : \Omega^p_{L^2(\mu)}(S) \to \Omega^{p+1}_{L^2(\mu)}(S)\right)}{\mathrm{im}\left(d : \Omega^{p-1}_{L^2(\mu)}(S) \to \Omega^p_{L^2(\mu)}(S)\right)}.$$

We also introduce the reduced De Rham $L^2$-cohomology:

$$\bar{H}^p_{\mathrm{DR}}(S_\mu) := \frac{\ker d}{\overline{\overline{\mathrm{im}\, d}}}.$$

Note that there are natural maps

$$H^p_{\mathrm{DR}}(S) \to H^p_{\mathrm{DR}}(S_\mu) \to H^p_{\mathrm{DR}m}(S),$$

since $C^{\infty,0}$-functions are $L^2(\mu)$-transversally smooth. The integration map $\int_{S_\mu}$ is well-defined for forms in $\Omega^k_{L^2(\mu)}$, since a $L^2(\mu)$-transversally smooth $k$-form is automatically $L^1(\mu)$-transversal (all measures are finite measures on compact transversals). So we have

$$\int_{S_\mu} : H^k_{\mathrm{DR}}(S_\mu) \to \mathbb{R}.$$

Let $\underline{\mathbb{R}}_\mu$ be the sheaf of measurable functions which are locally constant on leaves and $L^2(\mu)$-transversally. A standard Poincaré lemma shows that there is a resolution of sheaves

$$\underline{\mathbb{R}}_\mu \to \Omega^0_{L^2(\mu)} \to \Omega^1_{L^2(\mu)} \to \cdots \to \Omega^k_{L^2(\mu)}.$$

So we get a natural isomorphism

$$H^p_{\mathrm{DR}}(S_\mu) \cong H^p(S, \underline{\mathbb{R}}_\mu).$$

**Definition 39.6.** The ergodic dimension of $\mu$ is $1 \leq d(\mu) \leq +\infty$ defined to be the maximal number of mutually singular non-zero transversal measures $\mu_i$ such that

$$\mu \geq \sum_{i=1}^{d} c_i \mu_i,$$

with $c_i > 0$.

By classical ergodic theory, if the ergodic dimension $d(\mu)$ is finite then $\mu$ is a linear combination of exactly $d(\mu)$ ergodic transversal measures.

**Lemma 39.1.** *The ergodic dimension of $\mu$ is equal to the dimension of $H^0_{\mathrm{DR}}(S_\mu)$,*

$$d(\mu) = \dim_{\mathbb{R}} H^0_{\mathrm{DR}}(S_\mu).$$

*In particular, $\mu$ is ergodic if and only if $H^0_{\mathrm{DR}}(S_\mu) \cong \mathbb{R}$.*

*Proof.* Let $d \geq 1$ be finite and $d \leq d(\mu)$. By definition, $\mu \geq \sum_{i=1}^{d} c_i \mu_i$, where $\mu_i$ are ergodic measures and $c_i > 0$. Consider disjoint measurable subsets $S_i \subset S$ which are leaf-saturated, such that $S_i$ is of total measure for $\mu_i$ and of zero measure for any $\mu_j$ with $j \neq i$ in any transversal. Then the characteristic functions $\chi_{S_i}$ are independent in $\Omega^0_{L^2(\mu)}(S)$, since $\mu(S_i) > 0$ for all $i$: if $f = \sum \lambda_i \chi_{S_i} = 0$ then $0 = \|f\|^2 = \int_{S_\mu} |f|^2 = \sum \lambda_i^2 \mu(S_i) \implies \lambda_i = 0$ for all $i$. (Here, we fix an auxiliary Riemannian metric and consider the daval measures corresponding to the transversal measures.)

Thus, $\dim_{\mathbb{R}} H^0_{\mathrm{DR}}(S_\mu) = \dim_{\mathbb{R}} H^0(S, \underline{\mathbb{R}}_\mu) \geq d$. This proves the result when $d(\mu) = +\infty$.

Now assume that $d(\mu)$ is finite. Write $\mu = \sum_{i=1}^{d} c_i \mu_i$, where $\mu_i$ are ergodic measures, $d = d(\mu)$. Let $S_i$ be as before. Let us prove that the $(\chi_{S_i})$ do generate $H^0(S, \underline{\mathbb{R}}_\mu)$.

Let $f$ be any measurable function that is locally constant on leaves and $L^2(\mu)$-transversally. Note that $f$ is automatically $L^2(\mu_j)$-transversally. For a transversal $T$, the real function $x \mapsto \mu_{j,T}(T \cap f^{-1}((-\infty, x]))$ is a non-decreasing Heaviside function taking the values 0 and 1 (by ergodicity of $\mu_j$). So this function has a jump at some $x_j \in \mathbb{R}$ that is independent of the transversal $T$ (as holonomy shows). Then $F_{x_j} = f^{-1}(x_j)$ has total $\mu_j$-measure. So $F_{x_j} = S_j$ up to a set of $\mu_j$-measure zero, thus of $\mu$-measure zero. That means that $f$ is constant along each $S_j$ up to a set of $\mu$-measure zero. So $f = \sum x_j \chi_{S_j}$ in $L^2(\mu)$ (note that $S - (\cup S_j)$ is of zero $\mu$-measure). □

We review basic properties of the De Rham $L^2$-cohomology:

1. There is not cup product, and therefore the $H^*_{\mathrm{DR}}(S_\mu)$ are just vector spaces (not rings).
2. Functoriality. If $f : S_1 \to S_2$ is a solenoidal map, then we require that $\mu_2 = f_* \mu_1$. This means that for any local transversal $T_1$ of $S_1$, $f(T_1)$ is a local transversal of $S_2$ and the transported measure $f_* \mu_1$ is a constant multiple of $\mu_2$ on the transversal.

Note that this is automatic when the solenoids are uniquely ergodic. Then for any form $\omega$ which is $L^2(\mu_2)$-transversally smooth we have that $f^*\omega$ is $L^2(\mu_1)$-transversally smooth.
3. Mayer–Vietoris. It holds exactly as in Sect. 39.3.1.
4. Poincaré duality. We shall see that it holds for the reduced $L^2$-cohomology for oriented ergodic solenoids (see Corollary 39.3).

## 39.5 Bundles Over Solenoids

Let $S$ be a $k$-solenoid. A vector bundle of rank $n$ over $S$ consists of a $(k+n)$-solenoid $E$ and a projection map $\pi : E \to S$ satisfying the following condition: there is an open covering $U_\alpha$ for $S$, and solenoid isomorphisms $\psi_\alpha : E_\alpha = \pi^{-1}(U_\alpha) \stackrel{\cong}{\to} U_\alpha \times \mathbb{R}^n = D^k \times K(U_\alpha) \times \mathbb{R}^n$, such that $\pi = pr_1 \circ \psi_\alpha$, where $pr_1 : U_\alpha \times \mathbb{R}^n \to U_\alpha$ denotes the projection, and the transition functions

$$\psi_\alpha \circ \psi_\beta^{-1} : (U_\beta \cap U_\alpha) \times \mathbb{R}^n \to (U_\beta \cap U_\alpha) \times \mathbb{R}^n$$

are of the form $(x,y,v) \mapsto (x,y,g_{\alpha\beta}(x,y)(v))$, where $g_{\alpha\beta}$ is a $C^{\infty,0}$-smooth function from $U_\alpha \cap U_\beta$ to $\mathrm{GL}(n,\mathbb{R})$.

Some points are easy to check:

1. The usual constructions of vector bundles remain valid here: direct sums, tensor products, symmetric and anti-symmetric products. Also, there are notions of sub-bundle and of quotient bundle.
2. A section of a bundle $\pi : E \to S$ is a map $s : S \to E$ such that $\pi \circ s = Id$. We denote the space of sections as $\Gamma(E)$. By definition these are maps of class $C^{\infty,0}$.
3. If $S_\mu$ is a measured solenoid, and $E \to S$ is a vector bundle, then we have the notion of sections which are $L^2(\mu)$-transversally smooth. Locally, in a chart $E_\alpha = D^k \times K(U) \times \mathbb{R}^n \to U_\alpha = D^k \times K(U)$, the section is written $s(x,y) = (x,y,v(x,y))$. We require that $v$ is $C^\infty$ on $x$ and $L^2(\mu)$ on $y$. This does not depend on the chosen trivialization.
4. If $f : S_1 \to S_2$ is a solenoid map, and $\pi : E \to S_2$ is a vector bundle, then the pull-back $f^*E = \{(p,v) \in S_1 \times E \mid f(p) = \pi(v)\}$ is naturally a vector bundle over $S_1$.
5. The tangent bundle $TS$ of $S$ is an example of vector bundle. We have bundles of $(p,q)$-tensors $TS^{\otimes p} \otimes (TS^*)^{\otimes q}$ on any solenoid $S$. In particular, we have bundles of $p$-forms (anti-symmetric contravariant tensors) $\bigwedge^p T^*S$. Its sections are the $p$-forms $\Omega^p(S)$.
6. A metric on a bundle $E$ is a section of $\mathrm{Sym}^2(E^*)$ which is positive definite at every point. A metric on $S$ is a metric on the tangent bundle. An orientation of a bundle $E$ is a continuous choice of orientation for each of the fibers of $E$. An orientation of $S$ is an orientation of its tangent bundle.

We define $\Omega^p(E) = \Gamma(\bigwedge^p T^*S \otimes E)$. A connection on a vector bundle $E \to S$ is a map

$$\nabla : \Gamma(E) \to \Omega^1(E),$$

such that $\nabla(f \cdot s) = f\nabla s + df \wedge s$. Consider a local trivialization in a flow-box $U_\alpha$ with coordinates $(x,y)$. Then $\nabla|_{U_\alpha} = d + a_\alpha$, where $a_\alpha \in \Omega^1(U_\alpha, \mathrm{End}\, E)$. Under a change of trivialization $g_{\alpha\beta}$, for two trivializing open subsets $U_\alpha, U_\beta$, we have the usual formula $a_\beta = g_{\alpha\beta}^{-1} a_\alpha g_{\alpha\beta} + g_{\alpha\beta}^{-1} dg_{\alpha\beta}$.

A partition of unity argument proves that there are always connections on a vector bundle $E \to S$. The space of connections is an affine space over $\Omega^1(\mathrm{End}\, E)$.

Given a connection $\nabla$ on $E$, there is a unique map $d_\nabla : \Omega^p(E) \to \Omega^{p+1}(E)$, $p \geq 0$, such that $d_\nabla s = \nabla s$ for $s \in \Gamma(E)$, and $d_\nabla(\alpha \wedge \beta) = \mathrm{d}\alpha \wedge \beta + (-1)^p \alpha \wedge d_\nabla \beta$, for $\alpha \in \Omega^p(S)$, $\beta \in \Omega^q(E)$. It is easy to see that $\hat{F}_\nabla : \Gamma(E) \to \Omega^2(E)$, given by $\hat{F}_\nabla(s) = d_\nabla d_\nabla s$, has a tensorial character (i.e., it is linear on functions). Therefore, there is a $F_\nabla \in \Omega^2(\mathrm{End}\, E)$, called *curvature* of $\nabla$, such that $\hat{F}_\nabla(s) = F_\nabla \cdot s$. Locally, on a trivialization $U_\alpha$, we have the formula $F_\nabla = da_\alpha + a_\alpha \wedge a_\alpha$.

Given connections on vector bundles, there are induced connections on associated bundles (dual bundle, tensor product, direct sum, symmetric product, pull-back under a solenoid map, etc.). This follows in a straightforward way from the standard theory. In particular, if $l \hookrightarrow S$ is a leaf of a solenoid $S$, then we can perform the pull-back of the bundle and connection to the leaf, which consists on restricting them to $l$. This gives a bundle and connection of a complete $k$-dimensional manifold. Also, if $f : S \to M$ is an immersion of a solenoid in a smooth $n$-manifold, and $E \to M$ is a bundle with connection, then the pull-back construction produces a bundle with connection on $S$.

Consider a vector bundle $E \to S$ endowed with a metric. We say that a connection $\nabla$ is compatible with the metric if it satisfies

$$d\langle s,t\rangle = \langle \nabla s, t\rangle + \langle s, \nabla t\rangle.$$

In the particular case of the tangent bundle $TS$ of a Riemannian solenoid $S$, we have the Levi–Civita connection $\nabla^{\mathrm{LC}}$, which is the unique connection compatible with the metric and with torsion $T_\nabla(X,Y) = \nabla_X Y - \nabla_Y X = 0$. This is the Levi–Civita connection on each leaf, and the transversal continuity follows easily.

## 39.6 Chern Classes

We can also define a complex vector bundle over a solenoid, by using $\mathbb{C}^n$ as fiber, and taking the transition functions with values in $\mathrm{GL}(n,\mathbb{C})$. An Hermitian metric on a complex vector bundle is a positive definite Hermitian form in each fiber with smoothness of type $C^{\infty,0}$ on any local trivialization.

Let $E \to S$ be a complex vector bundle over a solenoid of rank $n$. Put a Hermitian structure on $E$, and consider any Hermitian connection $\nabla$ on $E$. Then the curvature $F_\nabla$ is a 2-form with values in $\operatorname{End} E$, i.e. $F_\nabla \in \Omega^2(\operatorname{End} E)$. The Bianchi identity says

$$d_\nabla F_\nabla = 0.$$

This holds leaf-wise, so it holds on the solenoid.

Consider the elementary functions: $\operatorname{Tr}_i : M_{r\times r} \to \mathbb{C}$, given by $\operatorname{Tr}_i(A) = \operatorname{Tr}(\bigwedge^i A)$. Then the Chern classes are

$$c_i(E) = \left[\operatorname{Tr}_i\left(\frac{\sqrt{-1}}{2\pi} F_\nabla\right)\right] \in H^{2i}_{\mathrm{DR}}(S).$$

These classes are well defined (since the forms inside are closed, which again follows by working on leaves) and do not depend on the connection (different connections give forms differing by exact forms), see [12, Chap. III].

We have some facts:

1. If $M$ is a manifold, we recover the usual Chern classes.
2. If $f : S_1 \to S_2$ is a solenoid map, then $f^* c_i(E) = c_i(f^* E)$. In particular,
   - If $f : S \to M$ is an immersion of a solenoid in a manifold and $E|_S = f^* E$, then $c_i(E|_S) = f^* c_i(E)$.
   - If $j : l \to S$ is the inclusion of a leaf, then $c_i(E|_l) = j^* c_i(E)$.

*Question 39.1.* Are the Chern classes defined as elements in $H^{2i}(S, \underline{\mathbb{Z}})$?

This question has an affirmative answer for the case of line bundles (complex vector bundles of rank 1). A line bundle $L \to S$ is given by its transition functions $g_{\alpha\beta} : U_\alpha \cap U_\beta \to S^1$ which are of class $C^{\infty,0}$. Therefore, the line bundles on $S$ are parametrized by $H^1(S, C^{\infty,0}(S^1))$, where $C^{\infty,0}(S^1)$ is the sheaf which assigns $U \mapsto C^{\infty,0}(U, S^1)$.

Note that there is an exact sequence of sheaves:

$$0 \to \underline{\mathbb{Z}} \to C^{\infty,0}(\mathbb{R}) \to C^{\infty,0}(S^1) \to 0.$$

Here, the sheaf $\underline{\mathbb{Z}}$ is the locally constant sheaf. As $C^{\infty,0}(\mathbb{R})$ is a fine sheaf (it has partitions of unity), it is acyclic. So the map

$$\delta : H^1(S, C^{\infty,0}(S^1)) \to H^2(S, \underline{\mathbb{Z}})$$

is an isomorphism. There is a natural map $\alpha : H^2(S, \underline{\mathbb{Z}}) \to H^2(S, \underline{\mathbb{R}}_c)$. It is easy to see that $\alpha(\delta([L])) = c_1(L)$, by using the transition functions $g_{\alpha\beta}$ to construct a suitable connection on $L$ with which we compute the curvature (as in the case of manifolds, see [12]).

## 39.7 Hodge Theory

### *39.7.1 Sobolev Norms*

Let $S_\mu$ be a compact Riemannian $k$-solenoid which is oriented and endowed with a transversal measure. We denote the associated (finite) daval measure also by $\mu$. Now consider a vector bundle $E \to S$ and endow it with a metric. The space of sections of class $C^{\infty,0}$ is denoted $\Gamma(S,E)$. The space of $L^2(\mu)$-transversally smooth sections (sections of class $C^\infty$ along leaves and $L^2$ in the transversal directions) is denoted by $\Gamma_{L^2(\mu)}(S,E)$.

Now let us introduce suitable completions of these spaces of sections. Fix a connection $\nabla$ for $E$ and the Levi–Civita connection for $TS$. There is an $L^2$-norm on sections of $E$, given by

$$(s,t)_E = \int_S \langle s,t\rangle \, d\mu.$$

We can complete the spaces of sections to obtain spaces of $L^2$-sections $L^2_\mu(S,E)$. Do not confuse with $\Gamma_{L^2(\mu)}(S,E)$, which refers to sections smooth along leaves. We consider also Sobolev norms $W^{l,2}$ as follows. Take $s$ a section of $E$. Then we set

$$\|s\|^2_{W^{l,2}_\mu} = \int_S \left( \sum_{i=0}^{l} |\nabla^i s|^2 \right) d\mu.$$

Completing with respect to this norm gives a Hilbert space consisting of sections with regularity $W^{l,2}$ on leaves and $L^2(\mu)$-transversally, denoted $W^{l,2}_\mu(S,E)$. These spaces do not depend on the choice of metrics and connections.

For future use, we also introduce the norms $C^r_\mu$, which give spaces of sections with $C^r$-regularity on leaves and $L^2(\mu)$-transversally. Take $s$ a section of $E$. Assume it has support in a flow box $U = D^k \times K(U)$, and assume that $E$ has been trivialized by an orthonormal frame. Then

$$\|s\|^2_{C^r_\mu} = \int_{K(U)} \|s(\cdot,y)\|^2_{C^r} \, d\mu_{K(U)}(y).$$

These norms are patched (via partitions of unity, in a non-canonical way) to get a norm on the spaces of sections on the whole solenoid. The topology defined by this norm is independent of the partition of unity. The spaces of sections are denoted $C^r_\mu(S,E)$. Note that $\bigcap_{r\geq 0} C^r_\mu(S,E) = \Gamma_{L^2(\mu)}(S,E)$.

We can define the norm $W^{l,2}_\mu$ by using Fourier transforms. For this we have to restrict to a flow-box $U = D^k \times K(U)$. We Fourier-transform the section $s(x,y)$ in the leaf-wise directions, to get $\hat{s}(\xi,y)$, and then take the integral

$$\int_{K(U)} \left( \int (1+|\xi|^2)^l |\hat{s}(\xi,y)|^2 d\xi \right) d\mu_{K(U)}(y).$$

**Proposition 39.2 (Sobolev).** *$W_\mu^{s,2}(S,E) \subset C_\mu^p(S,E)$, for $s > [k/2]+p+1$.*

This is similar to Proposition 1.1 in Chap. IV of [12]. The proof carries over to the solenoid situation verbatim. As a consequence,

$$\bigcap_{r\geq 0} W_\mu^{r,2}(S,E) = \Gamma_{L^2(\mu)}(S,E).$$

### 39.7.2 Pseudodifferential Operators

Let $E,F$ be two vector bundles over $S$ of ranks $n,m$ respectively. A differential operator $L$ of order $l$ is an operator

$$L : \Gamma(S,E) \to \Gamma(S,F)$$

which locally on a flow-box $U = D^k \times K(U)$ is of the form

$$L(s) = \sum_{|\alpha|\leq l} A_\alpha(x,y) D^\alpha s,$$

where $\alpha = (\alpha_1,\dots,\alpha_k)$ is a multi-index, with $|\alpha| = \sum \alpha_i$,

$$D^\alpha = \frac{\partial^{|\alpha|}}{\partial^{\alpha_1}x_1\dots\partial^{\alpha_k}x_k},$$

and $A_\alpha$ are $(n\times m)$-matrices of functions (with regularity $C^{\infty,0}$). Note that a differential operator gives rise to differential operators on each leaf. Moreover, $L$ extends to

$$L : W_\mu^{p,2}(S,E) \to W_\mu^{p-l,2}(S,F).$$

The usual properties, like the existence of adjoints, extend to this setting.

The symbol of a differential operator on a solenoid is defined in the same fashion as for the case of manifolds, and coincides with the symbol of the differential operator on the leaves. We recall that the symbol $\sigma_l(L) \in \mathrm{Hom}(\pi^*E,\pi^*F)$, $\pi : TS \to S$, has the form

$$\sigma_l(L)(x,y,v) = \sum_{|\alpha|=l} A_\alpha(x,y) v_1^{\alpha_1}\dots v_k^{\alpha_k}.$$

The properties of the symbol map, such as the rule of the symbol of the composition of differential operators, or the symbol of the adjoint, hold here. This is just the fact that they can be done leaf-wise, and the continuous transversality is easy to check.

Differential operators can be generalized to pseudodifferential operators as in the case of manifolds. A pseudodifferential operator of order $l$ on a flow-box $U = D^k \times K(U)$ is an operator

$$L(p) : \Gamma_c(U,E) \to \Gamma(U,F)$$

which sends a (compactly supported) section $s(x,y)$ to

$$L(p)s(x,y) = \int p(x,\xi,y)\hat{s}(\xi,y)e^{i\langle x,\xi\rangle}d\xi,$$

where $\hat{s}(\xi,y)$ is the (leaf-wise) Fourier transform, and $p(x,\xi,y)$ is a function defined in $D^k \times \mathbb{R}^k \times K(U)$, smooth on $x$ and $\xi$, continuous on $y$, and satisfying:

- $|D_x^\beta D_\xi^\alpha p(x,\xi,y)| \leq C_{\alpha\beta l}(1+|\xi|)^{l-|\alpha|}$, for constants $C_{\alpha\beta l}$
- the limit $\sigma_l(p)(x,\xi,y) = \lim_{\lambda\to\infty} p(x,\lambda\xi,y)\lambda^{-l}$ exists
- $p(x,\xi,y) - \sigma_l(p)(x,\xi,y)$ should be of order $\leq l-1$ for $|\xi| \geq 1$

A pseudodifferential operator of order $l$ on $S$ is an operator

$$L : \Gamma(S,E) \to \Gamma(S,F)$$

which is locally of the form $L(p_U)$ for some $p_U$ as above. The symbol of $L$ is $\sigma_l(L) = \sigma_l(p_U)$ for a local representative $L|_U = L(p_U)$. This symbol is well-defined and independent of choices, which follows from the case of manifolds, since the symbol can be defined leaf-wise (see [12]). The usual properties of the symbol map (composition, adjoint) hold here.

A pseudodifferential operator of order $l$ is an operator of order $l$, i.e., it extends as a continuous map to

$$L : W_\mu^{p,2}(S,E) \to W_\mu^{p-l,2}(S,E).$$

This is done as in Theorem 3.4 of [12, Chap. IV], by noting that

$$||L(p)s(\cdot,y)||_{W_\mu^{p-l,2}} \leq C||s(\cdot,y)||_{W_\mu^{p,2}},$$

where $C$ is a constant depending on $C_{\alpha\beta l}$.

The key of the theory is the fact that we can construct a pseudodifferential operator given a symbol $\sigma_l(L)$.

**Proposition 39.3.** *Let S be a compact solenoid. Then there is an exact sequence*

$$0 \to \mathrm{OP}_{l-1}(E,F) \to \mathrm{PDiff}_l(E,F) \to \mathrm{Symb}_l(E,F) \to 0,$$

*where* $\mathrm{OP}_{l-1}(E,F)$ *is the space of operators of order* $l-1$, $\mathrm{PDiff}_l(E,F)$ *the space of pseudodifferential operators of order l, and* $\mathrm{Symb}_l(E,F)$ *the space of symbols of order l.*

### *39.7.3 Elliptic Operator Theory for Solenoids*

We say that a pseudodifferential operator $L : E \to F$ of order $l$ is elliptic if the symbol $\sigma_l(L)$ satisfies that $\sigma_l(L)(x,v) : E_x \to F_x$ is an isomorphism for each $x \in S$, $v \in T_xS$, $v \neq 0$.

**Theorem 39.3.** *Let L be an elliptic pseudodifferential operator of order l. Then there exists a pseudo-inverse, a pseudodifferential operator $\tilde{L}$ of order $-l$ such that $L \circ \tilde{L} = Id + K_1$ and $\tilde{L} \circ L = Id + K_2$, where $K_1, K_2$ are operators of order $-1$.*

This is done as in Theorem 4.4 of [12, Ch. IV]. The basic idea is to construct a pseudo-inverse by using Proposition 39.3. Note that $K_1, K_2$ are not usually compact operators (this is due to the failure of the Rellich lemma in our situation), so we will not have finite-dimensionality of the kernel and cokernel of elliptic operators.

**Corollary 39.2.** *Let L be an elliptic pseudodifferential operator of order l, and let*

$$\mathscr{K}_{L_s} = \ker(L : W_\mu^{s,2}(S,E) \to W_\mu^{s-l,2}(S,F)).$$

*Then $\mathscr{K}_{L_s} \subset \Gamma_{L^2(\mu)}(S,E)$, and it is independent of s.*

*Proof.* Let $\sigma \in W_\mu^{2,s}(S,E)$ such that $L\sigma = 0$. Then $\sigma = (\tilde{L} \circ L - K_2)(\sigma) = -K_2\sigma \in W_\mu^{2,s+1}(S,E)$. Working inductively,

$$\sigma \in \bigcap_{r \geq 0} W_\mu^{2,r}(S,E) = \Gamma_{L^2(\mu)}(S,E).$$

The second assertion is clear. □

An operator $L : \Gamma(E) \to \Gamma(E)$ is called self-adjoint if $L^* = L$. If $L$ is an elliptic self-adjoint operator, then there is a pseudo-inverse $G$ which is self-adjoint (just take the pseudo-inverse $\tilde{L}$ provided by Theorem 39.3 and let $G = (\tilde{L} + \tilde{L}^*)/2$). Then we have that $L \circ G = G \circ L$, because

$$\langle (L \circ G - G \circ L)s, s \rangle = \langle Gs, Ls \rangle - \langle Ls, Gs \rangle = 0.$$

In particular, $K_1 = K_2$ in Theorem 39.3.

For self-adjoint operators, we have the following result

**Theorem 39.4.** *Let L be an elliptic self-adjoint operator of order l. Then*

$$W_\mu^{s,2}(S,E) = \ker L \oplus \overline{\operatorname{im} L}\,.$$

*and an analogous result for $\Gamma_{L^2(\mu)}(S,E)$.*

*Proof.* Clearly, $\ker L$ is a closed subspace. Let $s_1 \in \ker L$ and $s_2 \in \mathrm{im}\, L$, say $s_2 = L(t)$. From this we have $\langle s_1, s_2 \rangle = \langle s_1, L(t) \rangle = \langle L(s_1), t \rangle = 0$. The decomposition $\ker L \oplus \overline{\mathrm{im}\, L}$ is therefore orthogonal.

It remains to prove that $(\ker L)^{\perp} \subset \overline{\mathrm{im}\, L}$. Equivalently, $(\mathrm{im}\, L)^{\perp} \subset \ker L$. Let $s \in (\mathrm{im}\, L)^{\perp}$. Then $\langle L(s), s_2 \rangle = \langle s, L(s_2) \rangle = 0$, for all $s_2$. Hence $L(s) = 0$. □

A complex of differential operators is a sequence

$$\Gamma(E_0) \xrightarrow{L_0} \Gamma(E_1) \xrightarrow{L_1} \cdots \xrightarrow{L_{m-1}} \Gamma(E_m),$$

where $E_i$ are vector bundles, and $L_i$ are differential operators such that $L_i \circ L_{i-1} = 0$. The complex is called elliptic if the sequence of symbols

$$\pi^* E_0 \xrightarrow{\sigma(L_0)} \pi^* E_1 \xrightarrow{\sigma(L_1)} \cdots \xrightarrow{\sigma(L_{m-1})} \pi^* E_m,$$

is exact for each $v \neq 0$. We define the cohomology of the complex as

$$H^q(S,E) = \frac{\ker(L_q : \Gamma(E_q) \to \Gamma(E_{q+1}))}{\mathrm{im}\,(L_{q-1} : \Gamma(E_{q-1}) \to \Gamma(E_q))},$$

and the $L^2$-cohomology by

$$H^q(S_\mu,E) = \frac{\ker(L_q : \Gamma_{L^2(\mu)}(E_q) \to \Gamma_{L^2(\mu)}(E_{q+1}))}{\mathrm{im}(L_{q-1} : \Gamma_{L^2(\mu)}(E_{q-1}) \to \Gamma_{L^2(\mu)}(E_q))}.$$

The *reduced* $L^2$-cohomology is

$$\bar{H}^q(S_\mu,E) = \frac{\ker L_q}{\overline{\mathrm{im}\, L_{q-1}}}.$$

This is the group $H^q(S_\mu,E)$ quotiented by the closure of $\{0\}$, making it a Hausdorff space.

We construct the Laplacian operators of the elliptic complex as follows:

$$\Delta_j = L_j^* L_j + L_{j-1} L_{j-1}^* : \Gamma_{L^2(\mu)}(E_j) \to \Gamma_{L^2(\mu)}(E_j).$$

These are self-adjoint elliptic operators. There is an associated operator $G$ given by Theorem 39.4. Denote

$$\mathscr{H}^j(E) = \ker \Delta_j.$$

And note that $\Delta_j s = 0$ if and only if $L_j s = 0$ and $L_j^* s = 0$. We remove the subindex $j$ from now on.

**Theorem 39.5.** *We have the following:*

1. $\overline{\operatorname{im}\Delta} = \overline{\operatorname{im}L} \oplus \overline{\operatorname{im}L^*}$, *and it is an orthogonal decomposition.*
2. $\Gamma_{L^2(\mu)}(S,E_j) = \mathscr{H}^j(E) \oplus \overline{\operatorname{im}L} \oplus \overline{\operatorname{im}L^*}$.
3. *There is a canonical isomorphism* $\mathscr{H}^j(E) \simeq \bar{H}^j(S_\mu,E)$.

*Proof.* 1. The inclusion $\subset$ is clear. Note that the decomposition is orthogonal: $\langle L(s_1), L^*(s_2)\rangle = \langle LL(s_1), s_2\rangle = 0$, for all $s_1, s_2$.

For the reverse inclusion, let us check that $\operatorname{im}L \subset \overline{\operatorname{im}\Delta}$ (the other inclusion is similar). This is equivalent to proving that $\ker\Delta \subset (\operatorname{im}L)^\perp$. But $\ker\Delta = \ker L \cap \ker L^* \subset \ker L^* \subset (\operatorname{im}L)^\perp$, so the result follows.

2. It follows from 1) and Theorem 39.4.
3. Consider the map $\mathscr{H}^j(E) \to \bar{H}^j(S_\mu,E)$. This is well defined because if $\Delta s = 0$, $s \in \Gamma_{L^2(\mu)}(S,E_j)$ then $0 = \langle s, \Delta s\rangle = \langle s, (LL^* + L^*L)s\rangle = \langle L^*s, L^*s\rangle + \langle Ls, Ls\rangle \Longrightarrow Ls = L^*s = 0$.

   It is injective. If $s = Lt$ and $\Delta s = 0$ then $0 = L^*s = L^*Lt$, so $\|Lt\|^2 = \langle L^*Lt, t\rangle = 0$, so $s = Lt = 0$.

   It is surjective. Take a class $[s] \in \bar{H}^j(S_\mu,E)$. Decompose $s = s_1 + s_2 + s_3$, where $s_1 \in \mathscr{H}^j(E)$, $s_2 \in \overline{\operatorname{im}L}$ and $s_3 \in \overline{\operatorname{im}L^*}$. Now $0 = Ls = Ls_3$, so $s_3 \perp \operatorname{im}L^*$. Therefore $s_3 = 0$. So $s = s_1 + s_2$ and $[s] = [s_1]$ in $\bar{H}^j(S_\mu,E)$, where $s_1 \in \mathscr{H}^j(E)$. □

### *39.7.4 Harmonic Theory*

The Riemannian metric and the orientation give rise to a natural volume form along leaves $\mathrm{vol} \in \Omega^k(S)$. The usual Hodge-$*$ operator (see [12]) can be defined for forms on $S$, actually, it is the $*$ operator on leaves. This operator $* : \Omega^p(S) \to \Omega^{k-p}(S)$ is defined by

$$\alpha \wedge *\beta = (\alpha, \beta)\,\mathrm{vol},$$

for $\alpha, \beta \in \Omega^p(S)$, where $(\cdot,\cdot)$ is the point-wise metric induced on forms. Note that $*$ extends to $* : \Omega^p_{L^2_\mu}(S) \to \Omega^{k-p}_{L^2_\mu}(S)$, since it is leaf-wise isometric. Note that $\mathrm{vol} = *1$.

It is easy to check that $d^* = \pm * d *$. The Laplacian is defined as $\Delta = dd^* + d^*d$. Note that if $\Delta s = 0$ then $(s, \Delta s) = (s, dd^*s) + (s, d^*ds) = (d^*s, d^*s) + (ds, ds) = \|d^*s\|^2 + \|ds\|^2$. So $d^*s = 0$ and $ds = 0$. We define the space of harmonic forms:

$$\mathscr{K}^j(S_\mu) = \mathscr{H}_\Delta(\wedge^j T^*S).$$

Then the theory of elliptic operators says the following

**Theorem 39.6.** *We have:*

- *The space of harmonic sections* $\mathscr{K}^j(S_\mu) \subset \Omega^j_{L^2(\mu)}(S)$.
- *There is a natural isomorphism* $\bar{H}^j_{\mathrm{DR}}(S_\mu) \cong \mathscr{K}^j(S_\mu)$.

**Corollary 39.3.** *Poincaré duality:*

$$* : \mathscr{K}^p(S_\mu) \to \mathscr{K}^{k-p}(S_\mu)$$

*is an isomorphism.*

*If $S_\mu$ is ergodic, then $H^0_{\mathrm{DR}}(S_\mu) \cong H^k_{\mathrm{DR}}(S_\mu) \cong \mathbb{R}$ (with the isomorphism given by integration $\int_{S_\mu}$. Therefore*

$$\int_{S_\mu} : \bar{H}^p_{\mathrm{DR}}(S_\mu) \otimes \bar{H}^{k-p}_{\mathrm{DR}}(S_\mu) \to \mathbb{R}$$

*is a perfect pairing.*

In general, the spaces $\mathscr{K}^p(S_\mu)$ are not in general finite dimensional. For instance, take a solenoid which is a fibration, i.e., $S$ is a compact $(n+k)$-manifold such that there is a submersion $\pi : S \to B$ onto an $n$-dimensional manifold, and the transversal measure is induced by a measure $\mu$ on $B$. Then we have a fiber bundle $\mathbb{H}^p \to B$ such that $\mathbb{H}^p_y = H^p(F_y)$, $F_y = \pi^{-1}(y)$, $y \in B$. Then $\mathscr{K}^p(S_\mu) \cong L^2_\mu(\mathbb{H}^p)$.

Nonetheless, we propose the following:

*Question 39.2.* If $S_\mu$ is an ergodic solenoid with controlled growth, are the spaces $\mathscr{K}^p(S_\mu)$ of finite dimension?

A controlled solenoid has transversal measures and is defined in [7]. For such a solenoid, any leaf has an exhaustion $K_n$ by compact sets such that for any flow-box in a finite covering, the number of local leaves which intersect partially $K_n$ is negligible. In this case, the normalized measure supported on $K_n$ converges to a daval measure (giving rise to a transversal measure, as usual). If the solenoid $S_\mu$ is ergodic, then for almost all $\mu$-leaves, this limit coincides with $\mu$. So in this situation, we can understand the behaviour of harmonic forms by restricting to a leaf, and then study the spaces of harmonic forms on this leaf (which is a complete manifold with a quasi-periodic behaviour).

The example in Sect. 39.8 shows that the question has a negative answer if we ask instead for $H^p_{\mathrm{DR}}(S_\mu)$.

## 39.8 Example: Kronecker Solenoids

Let us develop the example of Kronecker solenoids, i.e. the flat torus with a linear foliation. Let $\mathbb{T}^n = \mathbb{R}^n/\mathbb{Z}^n$ with a flat Euclidean metric on $\mathbb{T}^n$. Consider a foliation given by a $k$-dimensional linear subspace $W \subset \mathbb{R}^n$. That is, fix an orthonormal basis $w_1, \dots, w_n$, so that $W = \langle w_1, \dots, w_k \rangle$. Consider the solenoid $S$ whose leaves are the images of $k$-planes of $\mathbb{R}^n$ parallel to $W$ under $\pi : \mathbb{R}^n \to \mathbb{T}^n$.

The solenoid is minimal with all leaves dense in $\mathbb{T}^n$ if there is no hyperplane $H = \sum m_i x_i = 0$ with $m_i \in \mathbb{Z}$ so that $W \subset H$. Equivalently, for any integer vector $m \in \mathbb{Z}^n$, we have that $\langle m, w_j \rangle \neq 0$ for some $j = 1, \dots, k$. Equivalently

$$\sum \langle m, w_j \rangle^2 \neq 0. \tag{39.4}$$

Notice that all leaves have the same topological type and are simply connected and diffeomorphic to $\mathbb{R}^k$ if and only if $W \cap \mathbb{Z}^n = \{0\}$. We shall suppose that the solenoid is minimal, but we allow $W \cap \mathbb{Z}^n \neq \{0\}$.

In this situation, we have affine transversals $T \subset W^\perp = \langle w_{k+1}, \dots, w_n \rangle$ and the holonomy are irrational translations, therefore by Haar theorem the unique transversal measure is the Lebesgue measure on the transversal. Thus, $S$ is uniquely ergodic, and the unique daval measure is the Haar measure on $\mathbb{T}^n$, image of the Lebesgue measure on $\mathbb{R}^n$.

We can characterize the space of functions $C^\infty$ on leaves and $L^2$-transversally by their Fourier expansion. All such functions are in particular $L^2$ on the torus, so they have a Fourier expansion ( $x = (x_1, \dots, x_n) \in \mathbb{T}^n$, $m = (m_1, \dots, m_n) \in \mathbb{Z}^n$)

$$f(x) = \sum_{m \in \mathbb{Z}^n} a_m \, \mathrm{e}^{2\pi \mathrm{i} \langle m, x \rangle},$$

where $\sum |a_m|^2 < \infty$.

To be smooth on leaves is equivalent to be $W^{l,2}$ on leaves for all $l$. Let us take a derivative along some $w_j$, $j = 1, \dots, k$.

$$D_{w_j} f = \sum_{m \in \mathbb{Z}^n} \langle m, w_j \rangle \, a_m \, \mathrm{e}^{2\pi \mathrm{i} \langle m, x \rangle}$$

So the condition $C^\infty$ on leaves and $L^2$-transversally is

$$\sum_{m \in \mathbb{Z}^n} \left| \prod_{j=1}^{k} \langle m, w_j \rangle^{r_j} a_m \right|^2 < \infty$$

for any $r_1, \dots, r_k \geq 0$.

Note that the Laplacian on functions is the usual Laplacian on the leaves direction, so that

$$\Delta f = -\sum_{j=1}^{k} D_{w_j}^2 f = -\sum_{m \in \mathbb{Z}^n} \left( \sum_{j=1}^{k} \langle m, w_j \rangle^2 \right) a_m \, \mathrm{e}^{2\pi \mathrm{i} \langle m, x \rangle}.$$

So $\Delta f = 0 \iff f = \text{constant}$, using (39.4).

Now consider $p$-forms, $0 \leq p \leq k$. A $p$-form $\omega = \sum_{|J|=p} f_J \, d\varpi_J$, where $\varpi_1, \dots, \varpi_k$ are coordinates on the leaves, dual to the basis $w_1, \dots, w_k$, and $J = (j_1, \dots, j_p)$,

$1 \leq j_1 < \ldots < j_p \leq k$, $\mathrm{d}\varpi_J = \mathrm{d}\varpi_{j_1} \wedge \ldots \wedge \mathrm{d}\varpi_{j_p}$. For the flat metric, $\Delta\omega = \sum_{|J|=p} \Delta f_J \, \mathrm{d}\varpi_J$. So $\Delta\omega = 0 \iff f_J = \text{constant}$.

In conclusion,

$$\mathscr{H}^p(S_\mu) = \bigwedge^p W^* ,$$

which is of dimension $\binom{k}{p}$.

Note that the $*$-Hodge operator satisfies $*\varpi_J = \pm\varpi_{J^c}$, where $J^c$ is the complement $\{1,\ldots,k\} - J$. So $* : \mathscr{H}^p(S_\mu) \to \mathscr{H}^{k-p}(S_\mu)$ is the natural isomorphism $\bigwedge^p W^* \cong \bigwedge^{k-p} W^*$.

## *The Case of the Kronecker* 1*-Solenoid*

Let us see the particular case of the Kronecker 1-solenoid with $w_1 = \alpha = (\alpha_1, \ldots, \alpha_n)$ and

$$\dim_{\mathbb{Q}} \langle \alpha_1, \ldots, \alpha_n \rangle = n.$$

As above, $\mathscr{H}^0(S_\mu) = \mathbb{R}$ and $\mathscr{H}^1(S_\mu) = \mathbb{R}$.

Let us compute $H^0_{\mathrm{DR}}(S_\mu)$ and $H^1_{\mathrm{DR}}(S_\mu)$. For a function $f = \sum a_m \mathrm{e}^{2\pi \mathrm{i}\langle m,x\rangle}$,

$$df = \sum \langle m, \alpha \rangle \, a_m \, \mathrm{e}^{2\pi \mathrm{i}\langle m,x\rangle} .$$

So $df = 0 \iff f = \text{constant}$. Then $H^0_{\mathrm{DR}}(S_\mu) = \mathbb{R}$.

Now a 1-form $\omega = g \, \mathrm{d}\varpi_1$, $g = \sum b_m \, \mathrm{e}^{2\pi \mathrm{i}\langle m,x\rangle}$, is exact if and only if

$$g = d\mathrm{f} = \sum \langle m, \alpha \rangle \, a_m \, \mathrm{e}^{2\pi \mathrm{i}\langle m,x\rangle} ,$$

i.e.

$$a_m = \frac{b_m}{\langle m, \alpha \rangle} , \qquad m \in \mathbb{Z}^n ,$$

should be in $L^2$. So

$$H^1_{\mathrm{DR}}(S_\mu) = \frac{\{(b_m) | \sum |\langle m, \alpha \rangle^r b_m|^2 < \infty, \forall r \geq 0\}}{\{(b_m) | \sum |\langle m, \alpha \rangle^r b_m|^2 < \infty, \forall r \geq -1\}} . \tag{39.5}$$

This space is always infinite dimensional. Using Minkowski's diophantine approximation theorem, there are infinitely many integer vectors $m \in \mathbb{Z}^n$ such that

$$|\langle m, \alpha \rangle| < \frac{1}{\|m\|} .$$

Then we can choose an infinite number of *disjoint* sequences $(m_k) \subset \mathbb{Z}^n$ such that $\langle m_k, \alpha \rangle \to 0$ and $\sum \frac{1}{\|m\|^2} < +\infty$, when $k \to +\infty$. For each such sequence take a

Fourier series with support on it (that is $b_m = 0$ if $m \neq m_k$) such that $\sum_k b_{m_k}^2 < +\infty$ but

$$\sum_k \left( \frac{b_{m_k}}{\langle m_k, \alpha \rangle} \right)^2 = +\infty.$$

We can take for example $b_{m_k} = \langle m_k, \alpha \rangle$. Then clearly, $(b_{m_k})$ lies in (39.5). In this way we get infinitely many independent elements in (39.5).

As an aside, note that the De Rham cohomology $H^1_{\mathrm{DR}}(S)$ is also of infinite dimension (as it is proved in [3] by sheaf theoretic methods). However, if we use forms with regularity $C^{\infty,\infty}$, using the criterion characterising $C^\infty$ as those with polynomially decaying Fourier coefficients, we can compute that

$$H^1(S, \underline{\mathbb{R}}_\infty) \cong \frac{\{(b_m) \mid \|b_m\| = O((1 + \|m\|)^{-r}), \forall r > 0\}}{\{(b_m) \mid \|\langle m, \alpha \rangle^{-1} b_m\| = O((1 + \|m\|)^{-r}), \forall r > 0\}},$$

(here $\underline{\mathbb{R}}_\infty$ is the sheaf of functions locally constant on leaves and $C^\infty$ transversally). A vector $\alpha$ is diophantine if there exist $\tau > 0$ and $\gamma > 0$ such that for all $m \in \mathbb{Z}^n$ we have

$$|\langle m, \alpha \rangle| \geq \frac{\gamma}{\|m\|^{n+\tau}}.$$

(The set of diophantine vectors is of full Lebesgue measure.) Then it follows that when $\alpha$ is diophantine, $H^1(S, \underline{\mathbb{R}}_\infty)$ is one dimensional, and otherwise, when $\alpha$ is Liouville, it is infinite dimensional [2, 10]. To impose such type of transversal regularity in this problem does not seem to be natural.

**Acknowledgement** Partially supported through grant MEC (Spain) MTM2007-63582.

## References


1. Dodziuk, J.: Sobolev spaces of differential forms and De Rham-Hodge isomorphism. J. Diff. Geom. **16**, 63–73 (1981)
2. Heitsch, J.L.: A cohomology of foliated manifolds. Comment. Math. Helvetici **50**, 197–218 (1975)
3. Macias, E.: Continuous cohomology of linear foliations on $T^2$. Rediconti di Matematica Serie VII (Roma) **11**, 523–528 (1991)
4. Massey, W.S.: Singular Homology Theory. Graduate Texts in Mathematics 70, Springer-Verlag (1980)
5. Moore, C., Schochet, C.: Global analysis on foliated spaces. Mathematical Sciences Research Institute Publications 9, Springer-Verlag (1988)
6. Muñoz, V., Pérez-Marco, R.: Ergodic solenoids and generalized currents. Revista Matemática Complutense, **24**(2), 493–525 (2011)
7. Muñoz, V., Pérez-Marco, R.: Schwartzman cycles and ergodic solenoids. To appear in Essays in Mathematics and its Applications. Dedicated to Stephen Smale (eds. P. Pardalos and Th.M. Rassias), Springer.
8. Muñoz, V., Pérez-Marco, R.: Ergodic solenoidal homology: Realization theorem. Comm. Math. Phys. **302**(3), 737–753 (2011)

9. Muñoz, V., Pérez-Marco, R.: Ergodic solenoidal homology II: Density of ergodic solenoids. Aust. J. Math. Anal. Appl. **6**(1), Article 11, 1–8 (2009)
10. Reinhart, B.L.: Harmonic integrals on almost product manifolds. Trans. Amer. Math. Soc. **88**, 243–276 (1958)
11. Ruelle, D., Sullivan, D.: Currents, flows and diffeomorphisms. Topology **14**, 319–327 (1975)
12. Wells, R.O.: Differential analysis on complex manifolds. GTM **65**, Second Edition, Springer-Verlag (1979)

# Chapter 40
# On Solutions of a Generalization of the Gołąb–Schinzel Functional Equation

**Anna Mureńko**

*Dedicated to the memory of S.M. Ulam*
*on the 100th anniversary of his birth*

**Abstract** Under some additional assumptions we characterize solutions of the functional equation

$$f(x+M(f(x))y) = f(x) \circ f(y),$$

where $f, M : \mathbb{R} \to \mathbb{R}$, $\circ : \mathbb{R}^2 \to \mathbb{R}$ are unknown functions and $f$ is continuous at a point.

**Keywords** Gołąb–Schinzel functional equation • Addition formulas

**Mathematics Subject Classification (2000):** Primary 39B22; Secondary 20N02

## 40.1 Introduction

Let $\mathbb{Z}$ and $\mathbb{R}$ denote the sets of integers and reals. Under some additional assumptions we give a description of solutions of the functional equation

$$f(x+M(f(x))y) = f(x) \circ f(y). \tag{40.1}$$

Equation (40.1) is a common generalization of the Gołąb–Schinzel equation

$$f(x+f(x)y) = f(x)f(y)$$

and of equations of the form

$$f(x+y) = f(x) \circ f(y)$$

A. Mureńko (✉)
Department of Mathematics, University of Rzeszów, Rejtana 16 A,
35–959 Rzeszów, Poland
e-mail: aniam@univ.rzeszow.pl

Th.M. Rassias and J. Brzdęk (eds.), *Functional Equations in Mathematical Analysis*, Springer Optimization and Its Applications 52, DOI 10.1007/978-1-4614-0055-4_40,

(see e.g. [1, pp. 49–81]) known as addition formulas.

The Gołąb–Schinzel functional equation and its generalizations have been considered by many authors in several classes of functions (see e.g. [2–4, 7–17,20,21]). Among others Brzdęk (see [5,6]) has obtained some results concerning the following functional equations

$$f(x+f(x)^n y) = f(x)f(y) \tag{40.2}$$

and

$$f(x+f(x)^n y) = tf(x)f(y). \tag{40.3}$$

In his survey paper [11], he raised the problem which results obtained for (40.2) or (40.3) can be carried over to the case of (40.1). Brzdęk has solved (40.2) in the class of functions $f : \mathbb{R} \to \mathbb{R}$ that are continuous at a point. It seems to be interesting that every solution of (40.2) continuous at a point $a$ with $f(a) \neq 0$ is continuous. It is easy to prove the analogous statement for (40.3). We show that the case of (40.1) is more complicated.

We consider (40.1) as an equation of three unknown functions $f, M$ and $\circ$ under the following assumptions:

$(A)$ $f, M : \mathbb{R} \to \mathbb{R}$ and $\circ : \mathbb{R}^2 \to \mathbb{R}$;
$(B)$ $M^{-1}(\{0\}) = \{0\}$;
$(C)$ $\circ : \mathbb{R}^2 \to \mathbb{R}$ is commutative and associative.

We use the notation $T := f^{-1}(\{f(0)\})$ and $W = f(\mathbb{R}) \setminus \{0\}$.

## 40.2 Auxiliary Lemmas

First, let us recall some already known results.

**Lemma 40.1 (see [19, Lemmas 1 and 2]).** *Let conditions $(A)$–$(B)$ be valid, $f, M, \circ$ satisfy (40.1) and the operation $\circ$ be commutative. Then*

*$(i)$ if $f(0) = 0$, then $f \equiv 0$;*
*$(ii)$ $T \setminus \{0\}$ is the set of periods of $f$;*
*$(iii)$ $T$ is an additive subgroup of $\mathbb{R}$;*
*$(iv)$ $y - x \in T$ for every $x, y \in \mathbb{R}$ with $f(x) = f(y) \neq 0$.*

**Lemma 40.2 (see [19, Lemma 3]).** *Let conditions $(A)$–$(C)$ be valid, $f, M, \circ$ satisfy (40.1) and $f \not\equiv 0$. Then*

$$M(a)T = T \quad \textit{for } a \in W.$$

**Lemma 40.3 (see [19, Lemma 4]).** *Assume that conditions $(A)$–$(C)$ hold and $f, M, \circ$ satisfy (40.1). Then*

$$f(x) \circ f(y) = 0 \Leftrightarrow f(x)f(y) = 0 \quad \textit{for } x, y \in \mathbb{R}. \tag{40.4}$$

**Definition 40.1.** We say that a function $f : \mathbb{R} \to \mathbb{R}$ is trivial if and only if

$$f(\mathbb{R} \setminus \{0\}) = \{0\}.$$

**Lemma 40.4 ((see [19, Lemma 4 and Theorem 5]).** *Let conditions (A)–(C) hold, $f$ be not trivial, $M(f(\mathbb{R})) \setminus \{0,1\} \neq \emptyset$ and $T = \{0\}$. Then $f, M, \circ$ satisfy (40.1) if and only if there exist a multiplicative subgroup $D \neq \{1\}$ of $\mathbb{R} \setminus \{0\}$, $c \in \mathbb{R} \setminus \{0\}$ and an injective function $h : D \cup \{0\} \to \mathbb{R}$ such that*

$$h(0) = 0\,; \tag{40.5}$$

$$M(y) = h^{-1}(y) \quad \text{for } y \in h(D \cup \{0\})\,; \tag{40.6}$$

$$f(x) = \begin{cases} h(cx+1)\,, & \text{if } cx+1 \in D\,, \\ 0\,, & \text{otherwise}\,, \end{cases} \quad \text{for } x \in \mathbb{R}\,; \tag{40.7}$$

$$a \circ b = h(h^{-1}(a)h^{-1}(b)) \quad \text{for } a, b \in h(D \cup \{0\}). \tag{40.8}$$

**Lemma 40.5 (see [18, Lemma 9]).** *Let conditions (A)–(C) hold, $f, M, \circ$ satisfy (40.1), $M(W) \subset \{-1,1\}$ and $M(W) \neq \{1\}$. Then*

*(i)* $f^{-1}(M^{-1}(\{1\})) \subset \frac{1}{2}T$;
*(ii)* $x - y \in \frac{1}{2}T$ *for* $x, y \in f^{-1}(M^{-1}(\{-1\}))$.

We also need the following auxiliary lemmas.

**Lemma 40.6.** *Assume that conditions (A)–(C) are fulfilled, $f, M, \circ$ satisfy (40.1), $M(f(\mathbb{R})) \setminus \{0,1\} \neq \emptyset$, $T \setminus \{0\} \neq \emptyset$ and the function $f$ is continuous at a point $a \in \mathbb{R}$ such that $f(U) \neq \{0\}$ for each neighbourhood $U$ of $a$. Then the set of periods of $f$ is dense in $\mathbb{R}$.*

*Proof.* First, suppose that there exists $w \in W$ with $|M(w)| \neq 1$. Then, according to Lemmas 40.1 $(ii)$, $(iii)$ and 40.2 the set of periods of $f$ is dense in $\mathbb{R}$.

Now assume that $M(W) \subset \{-1,1\}$ and $M(f(y)) = -1$ for some $y \in \mathbb{R}$. By the assumption we conclude that there exists a sequence $(a_n)$ with

$$a = \lim_{n \to \infty} a_n$$

and $f(a_n) \neq 0$ for $n \in \mathbb{N}$. Clearly, $M(f(a_n)) \in \{-1,1\}$ for $n \in \mathbb{N}$.

Two cases may occur:

1. There exists a subsequence $(b_n)$ of $(a_n)$ such that $M(f(b_n)) = 1$ for $n \in \mathbb{N}$.
2. There exists a subsequence $(b_n)$ of $(a_n)$ such that $M(f(b_n)) = -1$ for $n \in \mathbb{N}$.'

In the first case, by (40.1) and Lemma 40.3, we get

$$f(b_n + y) = f(b_n) \circ f(y) = f(y) \circ f(b_n) = f(y - b_n) \neq 0 \quad \text{for } n \in \mathbb{N}.$$

Thus, in view of Lemma 40.1$(iii),(iv)$, $2b_n \in T$ for $n \in \mathbb{N}$ and consequently $2b_n - 2b_m \in T$ for $m,n \in \mathbb{N}$. Moreover

$$a = \lim_{n\to\infty} b_n,$$

so $T$ is a dense subset of $\mathbb{R}$. Consequently, according to Lemma 40.1$(ii)$, we get our assertion.

In the latter case, using (40.1) and Lemma 40.3, we have

$$f(b_n - y) = f(b_n) \circ f(y) = f(y) \circ f(b_n) = f(y - b_n) \neq 0 \quad \text{for } n \in \mathbb{N}.$$

Similarly, as before, we get $2y - 2b_n \in T$ for $n \in \mathbb{N}$, hence that $2b_n - 2b_m \in T$ for $m,n \in \mathbb{N}$, and finally that the set of periods of $f$ is dense. □

**Lemma 40.7.** *Assume that condition (A) is valid, $f, M, \circ$, satisfy (40.1), $f$ is continuous at a point $a \in \mathbb{R}$ and the set of periods of $f$ is dense in $\mathbb{R}$. Then there exists $c \in \mathbb{R}$ such that $f \equiv c$ and $c \circ c = c$.*

*Proof.* Since a function $f$ is continuous at a point $a$ and the set of periods of $f$ is dense in $\mathbb{R}$, there exists a constant $c \in \mathbb{R}$ such that $f \equiv c$. This together with (40.1) yield $c \circ c = c$. □

**Lemma 40.8.** *Assume that (A)–(C) are valid, $M(f(\mathbb{R})) \setminus \{0,1\} \neq \emptyset$, $T \setminus \{0\} \neq \emptyset$ and $f(U) = \{0\}$ for some neighbourhood $U$ of some point $a \in \mathbb{R}$. Then functions $f, M, \circ$ satisfy functional (40.1) if and only if one of the following four assertions holds.*

$(i)$ *There exist $d_0, d_1 \in \mathbb{R} \setminus \{0\}$ with $d_0 \neq d_1$ and $k > 0$ such that*

$$f(x) = \begin{cases} d_0, & \text{if } x \in k\mathbb{Z}; \\ d_1, & \text{if } x \in k(\frac{1}{2} + \mathbb{Z}); \\ 0, & \text{otherwise}, \end{cases} \qquad \text{for } x \in \mathbb{R}; \tag{40.9}$$

$$M(d_0) = -1 \quad \text{and} \quad M(d_1) \in \{-1,1\}; \tag{40.10}$$

$$d_0 \circ d_0 = d_0; \tag{40.11}$$

$$d_1 \circ d_0 = d_1; \tag{40.12}$$

$$d_1 \circ d_1 = d_0; \tag{40.13}$$

$$u \circ 0 = 0 \quad \text{for } u \in \{d_0, d_1, 0\}. \tag{40.14}$$

$(ii)$ *There exist $d_0, d_1 \in \mathbb{R} \setminus \{0\}$ with $d_0 \neq d_1$, $k > 0$ and $x_0 \in (0,k)$ such that conditions (40.11)–(40.14) are fulfilled and*

$$f(x)=\begin{cases} d_0\,, & \text{if } x\in k\mathbb{Z};\\ d_1\,, & \text{if } x\in x_0+k\mathbb{Z};\\ 0\,, & \text{otherwise}, \end{cases}\qquad \text{for } x\in\mathbb{R}\,; \tag{40.15}$$

$$M(d_0)=1 \quad \text{and} \quad M(d_1)=-1. \tag{40.16}$$

(*iii*) *There exist* $d_0,d_1,d_2,d_3\in\mathbb{R}\setminus\{0\}$ *with* $d_i\neq d_j$ *for* $i,j\in\{0,1,2,3\}$, $i\neq j$, $k>0$ *and* $x_0\in(0,k)\setminus\{\frac{1}{2}k\}$ *such that*

$$f(x)=\begin{cases} d_0\,, & \text{if } x\in k\mathbb{Z};\\ d_1\,, & \text{if } x\in k\left(\frac{1}{2}+\mathbb{Z}\right);\\ d_2\,, & \text{if } x\in x_0+k\mathbb{Z};\\ d_3\,, & \text{if } x\in x_0+k\left(\frac{1}{2}+\mathbb{Z}\right);\\ 0\,, & \text{otherwise}, \end{cases}\qquad \text{for } x\in\mathbb{R}\,; \tag{40.17}$$

$$M(d_0)=M(d_1)=1 \quad \text{and} \quad M(d_2)=M(d_3)=-1\,; \tag{40.18}$$

$$d_2\circ d_1=d_3\,; \tag{40.19}$$

$$d_3\circ d_1=d_2\,; \tag{40.20}$$

$$d_3\circ d_2=d_1\,; \tag{40.21}$$

$$d_1\circ d_1=d_2\circ d_2=d_3\circ d_3=d_0\,; \tag{40.22}$$

$$d_t\circ d_0=d_t \quad \text{for } t\in\{0,1,2,3\}\,; \tag{40.23}$$

$$u\circ 0=0 \quad \text{for } u\in\{d_0,d_1,d_2,d_3,0\}. \tag{40.24}$$

(*iv*) *There exist* $d_0\in\mathbb{R}\setminus\{0\}$ *and* $k>0$ *such that*

$$f(x)=\begin{cases} d_0\,, & \text{if } x\in k\mathbb{Z};\\ 0\,, & \text{otherwise}, \end{cases}\qquad \text{for } x\in\mathbb{R}\,; \tag{40.25}$$

$$M(d_0)=-1\,; \tag{40.26}$$

$$d_0\circ d_0=d_0\,; \tag{40.27}$$

$$u\circ 0=0 \quad \text{for } u\in\{d_0,0\}. \tag{40.28}$$

*Proof.* Assume that $f,M,\circ$ satisfy (40.1). According to Lemma 40.3, condition (40.4) is valid. By the assumption there exists a neighbourhood $U$ of a point $a$ such that $f(U)=\{0\}$, moreover it is easy to see that $f\not\equiv 0$. Therefore, the set of periods of $f$ is not dense. Thus, in view of Lemma 40.1(*ii*), (*iii*) and the assumption that $T\setminus\{0\}\neq\emptyset$, we conclude that there exists $k>0$ with

$$T=k\mathbb{Z}. \tag{40.29}$$

This and Lemma 40.2 imply

$$M(W) \subset \{-1,1\}. \tag{40.30}$$

By assumption we have $M(W) \neq 1$. Consequently, from (40.29) and Lemma 40.5, we have

$$x \in \frac{1}{2}k\mathbb{Z} \quad \text{for } x \in f^{-1}(M^{-1}(\{1\})) \tag{40.31}$$

and

$$x - y \in \frac{1}{2}k\mathbb{Z} \quad \text{for } x, y \in f^{-1}(M^{-1}(\{-1\})). \tag{40.32}$$

Hence, making use of Lemma 40.1 $(ii)$, (40.29) and (40.30), we deduce that

$$\operatorname{card} W \leq 4. \tag{40.33}$$

On account of Lemma 40.1 $(i)$ we have $f(0) \neq 0$, so (40.30) gives

$$M(f(0)) \in \{-1,1\}.$$

We consider five cases.

*Case* 1: $\operatorname{card} W = 1$. Then $W = \{f(0)\}$. Thus, by (40.29) and Lemma 40.1 $(ii)(iv)$, we get (40.25). The assumption

$$M(f(\mathbb{R})) \setminus \{0,1\} \neq \emptyset$$

and $(B)$ yield (40.26). Conditions (40.27), (40.28) follow from (40.25), (40.26) and (40.1).

*Case* 2: $\operatorname{card} W = 2$ and $M(f(0)) = 1$. Then there exits $z_0 \in \mathbb{R}$ such that $f(z_0) \notin \{f(0),0\}$. Condition (40.30) implies $M(f(z_0)) \in \{-1,1\}$, thus by $(B)$ and the assumption that $M(f(\mathbb{R})) \setminus \{0,1\} \neq \emptyset$ we obtain that $M(f(z_0)) = -1$. Next, using Lemma 40.1 $(ii)(iv)$ and (40.29) we get

$$k\mathbb{Z} = f^{-1}(\{f(0)\})$$

and

$$z_0 + k\mathbb{Z} = f^{-1}(\{f(z_0)\}).$$

Clearly, $z_0 \notin k\mathbb{Z}$. Now, it is easily seen that there exists $x_0 \in (0,k)$ such that $f(x_0) = f(z_0)$, whence

$$x_0 + k\mathbb{Z} = z_0 + k\mathbb{Z}.$$

Consequently, we get (40.15) and (40.16). These conditions together with (40.1) imply (40.11)–(40.14).

*Case* 3: $\operatorname{card} W = 2$ and $M(f(0)) = -1$. Similarly, as in Case 2, we get that there exists $z_0 \in \mathbb{R}$ such that $f(z_0) \notin \{f(0),0\}$ and $M(f(z_0)) \in \{-1,1\}$. In view of (40.31) and (40.32), we conclude that

$$x_0 \in \frac{1}{2}k\mathbb{Z}.$$

Consequently, (40.29) and Lemma 40.1$(ii)(iv)$ yield (40.9). It is clear that (40.10) is valid. Conditions (40.11)–(40.14) are a consequence of (40.9), (40.10) and (40.1).

*Case* 4: $\operatorname{card} W = 3$. Then $W = \{d_0, d_1, d_2\}$ where $d_i \neq d_j$ for $i, j \in \{0,1,2\}$, $i \neq j$. If $M(f(0)) = -1$, then by (40.30)–(40.32) we would have

$$f^{-1}(W) \subset \frac{1}{2}k\mathbb{Z},$$

so using (40.29) and Lemma 40.1$(ii)$, we would obtain $\operatorname{card} W \leq 2$. This means that $M(f(0)) = 1$. Moreover, by (40.30), $(B)$ and the assumption $M(f(\mathbb{R})) \setminus \{0,1\} \neq \emptyset$, we deduce that there exists $z_0 \in \mathbb{R}$ with $M(f(z_0)) = -1$. Without loss of generality we can assume that $d_0 = f(0)$ and $d_1 = f(z_0)$. So, making use of (40.1), we have

$$d_1 \circ d_1 = f(z_0) \circ f(z_0) = f(z_0 - z_0) = f(0) = d_0. \tag{40.34}$$

Observe that

$$d_0 \circ f(x) = f(0 + M(f(0))x) = f(0 + x) = f(x) \quad \text{for } x \in \mathbb{R}$$

and, for $x \in \mathbb{R} \setminus \{0\}$,

$$f(x) \circ f\left(\frac{-x}{M(f(x))}\right) = f\left(x + M(f(x))\frac{-x}{M(f(x))}\right) = f(0) = d_0.$$

This together with $(C)$, (40.1) and (40.4) show that $(W, \circ|_{W^2})$ is a commutative group. Thus, since $d_1 \neq d_0$, we get $d_1 \circ d_2 \neq d_0 \circ d_2 = d_2$. We deduce analogously that $d_1 \circ d_2 \neq d_1$. Finally, we must have $d_1 \circ d_2 = d_0$. Hence, in view of (40.34), $d_1 = d_2$. This brings a contradiction with $d_1 \neq d_2$, thus the case $\operatorname{card} W = 3$ may not occur.

*Case* 5: $\operatorname{card} W = 4$. Then, on account of (40.30), $(B)$ and the assumption

$$M(f(\mathbb{R})) \setminus \{0,1\} \neq \emptyset,$$

there exists $z_0 \in \mathbb{R}$ with $M(f(z_0)) = -1$. Hence, in view of (40.30)–(40.32), we get

$$f^{-1}(W) \subset \frac{1}{2}k\mathbb{Z} \cup \left(z_0 + \frac{1}{2}k\mathbb{Z}\right).$$

Therefore, using Lemma 40.1$(ii)$, we obtain (40.17). It is easy to see that (40.18) holds. Further, conditions (40.19)–(40.24) follow from (40.17), (40.18) and (40.1).

□

We give an example of functions $f, M, \circ$ satisfying condition $(i)$ of Lemma 40.8.

*Example 40.1.* Let

$$f(x) = \begin{cases} 7, & \text{if } x \in 3\mathbb{Z}; \\ -4, & \text{if } x \in 3\left(\frac{1}{2}+\mathbb{Z}\right); \\ 0, & \text{otherwise}, \end{cases} \qquad \text{for } x \in \mathbb{R};$$

$$M(y) = \begin{cases} -1, & \text{if } y \in \{-4,7\}; \\ 0, & \text{otherwise}, \end{cases} \qquad \text{for } y \in \mathbb{R};$$

$$7 \circ 7 = -4 \circ -4 = 7;$$
$$-4 \circ 7 = 7 \circ -4 = -4;$$

and

$$s \circ t = 0 \quad \text{if } s \in \mathbb{R} \setminus \{-4,7\} \text{ or } t \in \mathbb{R} \setminus \{-4,7\}.$$

Then $f, M, \circ$ satisfy condition $(i)$ of Lemma 40.8 with $d_0 = 7$, $d_1 = -4$ and $k = 3$. Since the assumptions of Lemma 40.8 are fulfilled, so $f, M, \circ$ satisfy (40.1).

## 40.3 The Main Results

Now we are in a position to prove the main results of this paper.

**Theorem 40.1.** *Assume that conditions (A)–(C) are valid, the function $f$ is not trivial and continuous at a point $a \in \mathbb{R}$. Then $f, M, \circ$ satisfy (40.1) if and only if one of the following five assertions holds.*

1° *There exists $c \in \mathbb{R}$ such that $f \equiv c$ and $c \circ c = c$.*

2° *There exist a multiplicative subgroup $D \neq \{1\}$ of $\mathbb{R} \setminus \{0\}$, $c \in \mathbb{R} \setminus \{0\}$ and an injective function $h : D \cup \{0\} \to \mathbb{R}$ such that conditions (40.5)–(40.8) are valid.*

3° *There exist an additive subgroup $G \neq \{0\}$ of $\mathbb{R}$ and an injection $h : G \to \mathbb{R} \setminus \{0\}$ such that*

$$M(h(G)) = \{1\};$$

$$f(x) = \begin{cases} h(x), & \text{if } x \in G; \\ 0, & \text{otherwise}; \end{cases} \qquad \text{for } x \in \mathbb{R};$$

*and*

$$s \circ t = h(h^{-1}(s) + h^{-1}(t)) \quad \text{for } s,t \in h(G).$$

*Moreover, if $G \neq \mathbb{R}$, then*

$$u \circ 0 = 0 \quad \text{for } u \in h(G) \cup \{0\}.$$

4° *There exist an additive subgroup $G \neq \{0\}$ of $\mathbb{R}$, $k \in G \cap (0,\infty)$ and an injective function $h : (G \cap [0,k)) \to \mathbb{R} \setminus \{0\}$ such that*

$$M(h(G\cap[0,k)))=\{1\};$$

$$f(x+km)=\begin{cases} h(x), & \text{if } x\in(G\cap[0,k));\\ 0, & \text{otherwise},\end{cases} \qquad \text{for } x\in[0,k),\ m\in\mathbb{Z};$$

*and*

$$s\circ t=f(h^{-1}(s)+h^{-1}(t)) \quad \text{for } s,t\in h(G\cap[0,k)).$$

*Moreover, if* $G\neq\mathbb{R}$, *then*

$$u\circ 0=0 \quad \text{for } u\in h(G\cap[0,k))\cup\{0\}.$$

5° *One of conditions* $(i)$–$(iv)$ *of Lemma 40.8 is fulfilled.*

*Proof.* Assume that $f,M,\circ$ satisfy (40.1) and the function $f$ is not trivial. According to Lemma 40.3, condition (40.4) is fulfilled.

If

$$M(f(\mathbb{R}))\setminus\{0,1\}\neq\emptyset$$

and $T=\{0\}$, then in virtue of Lemma 40.4, we get 2°. Next, in the case

$$M(f(\mathbb{R}))\setminus\{0,1\}\neq\emptyset$$

and

$$T\setminus\{0\}\neq\emptyset,$$

using Lemmas 40.6–40.8, we obtain 1° or 5°.

Now assume that $M(f(\mathbb{R}))\subset\{0,1\}$ and $f\not\equiv 0$. First, we show that

$$G:=f^{-1}(M^{-1}(\{1\}))$$

is an additive subgroup of $\mathbb{R}$. Observe that

$$G=f^{-1}(W)$$

and take $x,y\in G$. Using (40.1), (40.4) and $(B)$ we get

$$f(x+y)=f(x+M(f(x))y)=f(x)\circ f(y)\neq 0,$$

hence, $x+y\in G$. As $f\not\equiv 0$, it follows from Lemma 40.1$(i)$ that $f(0)\neq 0$. Furthermore, notice that

$$f(x)\circ f(-x)=f(x+M(f(x))(-x))=f(x-x)=f(0)\neq 0,$$

whence $-x\in G$. If the function $f$ is injective, we put

$$h:=f|_G.$$

Making use of $(B)$ and (40.1) we obtain condition 3°. Now consider the case where the function $f$ is not injective. Then, on account of Lemma 40.1$(iii),(iv)$, $T$ is a nonzero additive subgroup of $\mathbb{R}$. Therefore, $T$ is dense in $\mathbb{R}$ or $T = k\mathbb{Z}$ with some $k > 0$. In the first case, it results from Lemma 40.7 that 1° holds. So, assume that $T = k\mathbb{Z}$. According to Lemma 40.1 $(ii),(iv)$ the function

$$h := f|_{[0,k)}$$

is injective and

$$f(x+km) = h(x)$$

for $x \in [0,k)$, $m \in \mathbb{Z}$. Now, it is easily seen that condition 4° is valid.

Conversely, if condition 2° is valid, then in view of Lemma 40.4, $f, M, \circ$ satisfy (40.1). In the other cases it is easy to check that $f, M, \circ$ fulfil (40.1). □

*Example 40.2.* Let

$$f(x) = \begin{cases} \exp x, & \text{if } x \in \mathbb{Z}; \\ 0, & \text{otherwise}, \end{cases} \qquad \text{for } x \in \mathbb{R};$$

$$M(y) = \begin{cases} 1, & \text{if } y \in \exp(\mathbb{Z}); \\ 0, & \text{otherwise}, \end{cases} \qquad \text{for } y \in \mathbb{R};$$

and

$$s \circ t = \begin{cases} \exp(\ln s + \ln t), & \text{if } s,t \in \exp(\mathbb{Z}); \\ 0, & \text{otherwise}, \end{cases} \qquad \text{for } s,t \in \mathbb{R}.$$

Then $f, M, \circ$ satisfy condition 3° of Theorem 40.1 with $G = \mathbb{Z}$ and $h : \mathbb{Z} \to \mathbb{R} \setminus \{0\}$ defined by $h(x) = \exp x$. Consequently, on account of Theorem 40.1, $f, M, \circ$ satisfy (40.1).

In the case where $f(a) \neq 0$, we can give clearer description of solutions of (40.1).

**Theorem 40.2.** *Assume that conditions $(A)$–$(C)$ are valid and the function $f$ is continuous at a point $a \in \mathbb{R}$ with $f(a) \neq 0$. Then $f, M, \circ$ satisfy (40.1) if and only if one of the following five assertions holds.*

$(i)$ *There exists $c \in \mathbb{R} \setminus \{0\}$ such that $f \equiv c$ and $c \circ c = c$.*

$(ii)$ *There exist $c \in \mathbb{R} \setminus \{0\}$ and an injective, continuous at the point $ca+1$ function $h : [0,\infty) \to \mathbb{R}$ such that*

$$h(0) = 0;$$

$$M(y) = h^{-1}(y) \quad \textit{for } y \in h([0,\infty));$$

$$f(x) = h(\max\{0, cx+1\}) \quad \textit{for } x \in \mathbb{R};$$

$$s \circ t = h(h^{-1}(s)h^{-1}(t)) \quad \textit{for } s,t \in h([0,\infty)).$$

(*iii*) *There exist* $c \in \mathbb{R} \setminus \{0\}$ *and an injective, continuous at the point* $ca+1$ *function* $h : \mathbb{R} \to \mathbb{R}$ *such that*

$$h(0) = 0\,;$$

$$M(y) = h^{-1}(y) \quad \text{for } y \in h(\mathbb{R})\,;$$

$$f(x) = h(cx+1) \quad \text{for } x \in \mathbb{R}\,;$$

$$s \circ t = h(h^{-1}(s)h^{-1}(t)) \quad \text{for } s,t \in h(\mathbb{R}).$$

(*iv*) $M(f(\mathbb{R})) = \{1\}$, *the function* $f$ *is injective and*

$$s \circ t = f(f^{-1}(s) + f^{-1}(t)) \quad \text{for } s,t \in f(\mathbb{R}).$$

(*v*) $M(f(\mathbb{R})) = \{1\}$, *there exist* $k > 0$ *and an injective function* $h : [0,k) \to \mathbb{R} \setminus \{0\}$ *such that*

$$f(x+km) = h(x) \quad \text{for } x \in [0,k),\ m \in \mathbb{Z}\,;$$

$$s \circ t = f(h^{-1}(s) + h^{-1}(t)) \quad \text{for } s,t \in h([0,k)).$$

*Proof.* Assume that $f, M, \circ$ satisfy (40.1). Then one of conditions 1°– 5° of Theorem 40.1 is valid. If condition 1° is fulfilled, then since $f(a) \neq 0$, we obtain (*i*).

Now assume that 2° holds. Then the continuity of $f$ at a point $a$ with $f(a) \neq 0$ implies that $f(I) \subset \mathbb{R} \setminus \{0\}$ for some interval $I \subset \mathbb{R}$. Thus, according to (40.7), the multiplicative subgroup $D$ contains an interval. Consequently $D = \mathbb{R} \setminus \{0\}$ or $D = \mathbb{R}_+$. Moreover, since $f$ is continuous at a point $a$, $h$ is continuous at the point $ca+1$. Finally, we get (*ii*) or (*iii*).

Next, consider the case where condition 3° or 4° holds. Then from the continuity of $f$ at a point $a$ with $f(a) \neq 0$ it follows that $G = \mathbb{R}$, whence we get (*iii*) or (*iv*).

Clearly, if condition 5° is valid, then the function $f$ is not continuous at any point $x$ with $f(x) \neq 0$.

The converse results from Theorem 40.1. □

## References


1. Aczél, J.: Lectures on Functional Equations and Their Applications. Academic Press, New York–London (1966)
2. Aczél, J., Schwaiger, J.: Continuous solutions of the Gołąb-Schinzel equation on the nonnegative reals and on related domains. Sitzungsber. Österreich. Akad. Wiss. Math.-Natur. Kl. Abt. II **208**, 171–177 (1999)
3. Baron, K.: On the continuous solutions of the Gołąb-Schinzel equation. Aequationes Math. **38**, 155–162 (1989)
4. Brillouët-Belluot, N.: On some functional equations of Gołąb-Schinzel type. Aequationes Math. **42**, 239–270 (1991)
5. Brzdęk, J.: O rozwiązaniach równania funkcyjnego $f(x+f(x)^n y) = f(x)f(y)$. Ph.D. Thesis, Silesian University, Katowice (1990)
6. Brzdęk, J.: Some remarks on solutions of the functional equation $f(x+f(x)^n y) = tf(x)f(y)$. Publ. Math. Debrecen **43**, 147–160 (1993)

7. Brzdęk, J.: On solutions of the Gołąb-Schinzel functional equation. Publ. Math. Debrecen **44**, 235–241 (1994)
8. Brzdęk, J.: The Christensen measurable solutions of a generalization of the Gołąb-Schinzel functional equation. Ann. Polon. Math. **64**, 195–205 (1996)
9. Brzdęk, J.: On the continuous solutions of a generalization of the Gołąb-Schinzel equation. Publ. Math. Debrecen **63**, 421–429 (2003)
10. Brzdęk, J.: A generalization of addition formulae. Acta Math. Hungar. **101**, 281–291 (2003)
11. Brzdęk, J.: The Gołąb-Schinzel equation and its generalization. Aequationes Math. **70**, 14–24 (2005)
12. Chudziak, J.: Continuous solutions of a generalization of the Gołąb-Schinzel equation. Aequationes Math. **61**, 63–78 (2001)
13. Chudziak, J.: Continuous solutions of a generalization of the Gołąb-Schinzel equation II. Aequationes Math. **71**, 115–123 (2006)
14. Gołąb, S., Schinzel, A.: Sur l'équation fonctionnelle $f(x+yf(x)) = f(x)f(y)$. Publ. Math. Debrecen **6**, 113–125 (1959)
15. Jabłońska, E.: On solutions of a generalization of the Gołąb-Schinzel equation. Aequationes Math. **71**, 269–279 (2006)
16. Jabłońska, E.: Functions having the Darboux property and satisfying some functional equation. Colloq. Math. **114**, 113–118 (2009)
17. Kahlig, P., Matkowski, J.: A modified Gołąb-Schinzel equation on a restricted domain (with applications to meteorology and fluid mechanics). Österreich. Akad. Wiss. Math.-Natur. Kl. Sitzungsber. II **211**, 117–136 (2002)
18. Mureńko, A.: On solutions of a common generalization of the Gołąb-Schinzel equation and of the addition formulae. J. Math. Anal. Appl. **341**, 1236–1240 (2008)
19. Mureńko, A.: On the general solution of a generalization of the Gołąb-Schinzel equation. Aequationes Math. **77**, 107–118 (2009)
20. Reich, L.: Über die stetigen Lösungen der Gołąb-Schinzel-Gleichung auf $\mathbb{R}_{\geq 0}$. Sitzungsber. Österreich. Akad. Wiss. Math.-Natur. Kl. Abt. II **208**, 165–170 (1999)
21. Sablik, M.: A conditional Gołąb-Schinzel equation. Anz. Österreich. Akad. Wiss. Math.-Natur. Kl. Abt. II **137**, 11–15 (2000)

# Chapter 41
# On a Functional Equation Containing an Indexed Family of Unknown Mappings

**Prem Nath and Dhiraj Kumar Singh**

*Dedicated to the memory of S.M. Ulam*
*on the 100th anniversary of his birth*

**Abstract** In this paper, without imposing any regularity conditions on the mappings under considerations, all possible general solutions of a functional equation have been obtained.

**Keywords** Additive mapping • Multiplicative mapping

**Mathematics Subject Classification (2000):** Primary: 39B22, 39B52

## 41.1 Introduction

For $n = 1,2,3,\ldots$, let

$$\Gamma_n = \left\{ (p_1,\ldots,p_n) : p_i \geq 0, i = 1,\ldots,n; \sum_{i=1}^{n} p_i = 1 \right\}$$

denote the set of all $n$-component discrete complete probability distributions with non-negative elements. Let $I = \{x \in \mathbb{R} : 0 \leq x \leq 1\} = [0,1]$, $\mathbb{R}$ denoting the set of all real numbers.

Losonczi [3] considered thefunctional equation

$$\sum_{i=1}^{n} \sum_{j=1}^{m} F_{ij}(p_i q_j) = \sum_{i=1}^{n} G_i(p_i) + \sum_{j=1}^{m} H_j(q_j) + \lambda \sum_{i=1}^{n} G_i(p_i) \sum_{j=1}^{m} H_j(q_j) \quad (41.1)$$

---

P. Nath (✉) • D.K. Singh
Department of Mathematics, University of Delhi, Delhi – 110007, India
e-mail: pnathmaths@gmail.com; dksingh@maths.du.ac.in; dhiraj426@rediffmail.com

Th.M. Rassias and J. Brzdęk (eds.), *Functional Equations in Mathematical Analysis*, Springer Optimization and Its Applications 52, DOI 10.1007/978-1-4614-0055-4_41,

with $(p_1,\dots,p_n)\in\Gamma_n$, $(q_1,\dots,q_m)\in\Gamma_m$, $\lambda\neq 0$, $\lambda\in\mathbb{R}$, $F_{ij}:I\to\mathbb{R}$, $G_i:I\to\mathbb{R}$, $H_j:I\to\mathbb{R}$ as unknown mappings. He found the measurable (in the sense of Lebesgue) solutions of (41.1) for all $(p_1,\dots,p_n)\in\Gamma_n$, $(q_1,\dots,q_m)\in\Gamma_m$ by taking $n\geq 3$, $m\geq 3$ as fixed integers; $i=1,\dots,n$; $j=1,\dots,m$; in [3, Theorem 6, p. 69]. For the last more than two decades, the general solutions of (41.1) for all $(p_1,\dots,p_n)\in\Gamma_n$, $(q_1,\dots,q_m)\in\Gamma_m$, $n\geq 3$, $m\geq 3$ being fixed integers, without imposing any regularity condition on any of the unknown mappings, *are still not known so far*.

The object of this paper is to present the general solutions of the functional equation

$$\sum_{i=1}^{n}\sum_{j=1}^{m}F(p_iq_j)=\sum_{i=1}^{n}G(p_i)+\sum_{j=1}^{m}H_j(q_j)+\lambda\sum_{i=1}^{n}G(p_i)\sum_{j=1}^{m}H_j(q_j) \tag{41.2}$$

with $F:I\to\mathbb{R}$, $G:I\to\mathbb{R}$, $H_j:I\to\mathbb{R}$, $j=1,\dots,m$; $\lambda\neq 0$, $(p_1,\dots,p_n)\in\Gamma_n$, $(q_1,\dots,q_m)\in\Gamma_m$, $n\geq 3$, $m\geq 3$ being fixed integers.

The functional equation (41.1) contains $nm+n+m$ unknown mappings. In particular, if $n=m=3$, then (41.1) contains fifteen unknown mappings, a number significantly large. The functional equation (41.2) contains five unknown mappings when $m=3$. Even this number is also significantly large. But the authors have succeeded in finding its general solutions which is, indeed, first step towards finding the general solutions of (41.1). After finding the general solutions of (41.2), it seems that its one solution is useful. This has been pointed out in the discussion undertaken at the end of the paper.

It is worth mentioning that if $H_j=F=G$, $j=1,\dots,m$ and $\lambda=2^{1-\alpha}-1\neq 0$, then the resulting form of (41.2), that is,

$$\sum_{i=1}^{n}\sum_{j=1}^{m}F(p_iq_j)=\sum_{i=1}^{n}F(p_i)+\sum_{j=1}^{m}F(q_j)+(2^{1-\alpha}-1)\sum_{i=1}^{n}F(p_i)\sum_{j=1}^{m}F(q_j)$$

is useful in characterizing the nonadditive entropy of degree $\alpha$ (see [2]),

$$H_n^{\alpha}(p_1,\dots,p_n)=(2^{1-\alpha}-1)^{-1}\left(\sum_{i=1}^{n}p_i^{\alpha}-1\right),\quad \alpha\neq 1$$

in information theory.

Let us define the mappings $f:I\to\mathbb{R}$, $g:I\to\mathbb{R}$, $h_j:I\to\mathbb{R}$, $j=1,\dots,m$ as

$$f(x)=x+\lambda F(x)\,;\quad g(x)=x+\lambda G(x)\,;\quad h_j(x)=x+\lambda H_j(x) \tag{41.3}$$

for all $x\in I$. Then (41.2) reduces to the Pexider type functional equation

$$\sum_{i=1}^{n}\sum_{j=1}^{m}f(p_iq_j)=\sum_{i=1}^{n}g(p_i)\sum_{j=1}^{m}h_j(q_j)\,. \tag{41.4}$$

The process of finding the general solutions of (41.4), for fixed integers $n \geq 3, m \geq 3$, needs determining the general solutions of the functional equation

$$\sum_{i=1}^{n} \sum_{j=1}^{m} \psi(p_i q_j) = \sum_{i=1}^{n} \phi(p_i) \sum_{j=1}^{m} \psi(q_j) + m(n-1)\,\psi(0) \tag{41.5}$$

where $\psi : I \to \mathbb{R}$, $\phi : I \to \mathbb{R}$ and $n \geq 3$, $m \geq 3$ are fixed integers. This task has been accomplished in Sect. 41.3. The corresponding general solutions of (41.4) and (41.2) have been investigated in Sects. 41.4 and 41.5, respectively. Section 41.2 contains some known definitions and results needed for the subsequent development of this paper.

## 41.2 Some Preliminary Definitions and Results

In this section, we mention some known definitions and results.

A mapping $a : I \to \mathbb{R}$ is said to be additive on $I$ or on the unit triangle

$$\Delta = \{(x,y) : 0 \leq x \leq 1, 0 \leq y \leq 1, 0 \leq x+y \leq 1\}$$

if it satisfies the equation $a(x+y) = a(x) + a(y)$ for all $(x,y) \in \Delta$. A mapping $A : \mathbb{R} \to \mathbb{R}$ is said to be additive on $\mathbb{R}$ if it satisfies the equation $A(x+y) = A(x) + A(y)$ for all $x \in \mathbb{R}$, $y \in \mathbb{R}$. It is known [1] that if a mapping $a : I \to \mathbb{R}$ is additive on $I$, then it has a unique additive extension $A : \mathbb{R} \to \mathbb{R}$ in the sense that $A$ is additive on $\mathbb{R}$ and $A(x) = a(x)$ for all $x \in I$.

**Lemma 41.1 ([4]).** *Let $f : I \to \mathbb{R}$ be a mapping which satisfies the equation*

$$\sum_{i=1}^{n} f(p_i) = c$$

*for all $(p_1, \ldots, p_n) \in \Gamma_n$; $c$ a given constant and $n \geq 3$ a fixed integer. Then there exists an additive mapping $b : \mathbb{R} \to \mathbb{R}$ such that*

$$f(p) = b(p) - \frac{1}{n} b(1) + \frac{c}{n}$$

*for all $p \in I$.*

**Lemma 41.2 ([3]).** *Let $n \geq 3$ be a fixed integer, $c \in \mathbb{R}$ a given constant and $f_i : I \to \mathbb{R}$, $i = 1, \ldots, n$ be mappings which satisfy the functional equation*

$$\sum_{i=1}^{n} f_i(p_i) = c$$

*for all* $(p_1,\ldots,p_n) \in \Gamma_n$. *Then there exists an additive mapping* $A : \mathbb{R} \to \mathbb{R}$ *and constants* $b_i$ $(i = 1,\ldots,n)$ *such that* $f_i(p) = A(p) + b_i$ *with*

$$A(1) + \sum_{i=1}^{n} b_i = c\,.$$

**Lemma 41.3 ([5]).** *If a mapping* $f : I \to \mathbb{R}$ *satisfies the functional equation*

$$\sum_{i=1}^{n}\sum_{j=1}^{m} f(p_i q_j) = 0 \tag{41.6}$$

*for all* $(p_1,\ldots,p_n) \in \Gamma_n$, $(q_1,\ldots,q_m) \in \Gamma_m$, $n \geq 3$, $m \geq 3$ *fixed integers, then there exists an additive mapping* $e_2 : \mathbb{R} \to \mathbb{R}$ *such that*

$$f(p) = e_2(p) - \frac{1}{nm} e_2(1) \tag{41.7}$$

*for all* $p \in I$.

**Definition 41.1.** A mapping $M : I \to \mathbb{R}$ is said to be multiplicative on $I$ if $M(0) = 0, M(1) = 1$ and

$$M(pq) = M(p)M(q)$$

for all $p \in ]0,1[$, $q \in ]0,1[$ where $]0,1[ = \{x \in \mathbb{R} : 0 < x < 1\}$.

**Lemma 41.4 ([6]).** *Let* $n \geq 3$, $m \geq 3$ *be fixed integers and* $\varphi : I \to \mathbb{R}$ *be a mapping which satisfies the functional equation*

$$\sum_{i=1}^{n}\sum_{j=1}^{m} \varphi(p_i q_j) = \sum_{i=1}^{n} \varphi(p_i) \sum_{j=1}^{m} \varphi(q_j)$$
$$+(n-m)\varphi(0)\sum_{i=1}^{n} \varphi(p_i) + n(m-1)\varphi(0) \tag{41.8}$$

*for all* $(p_1,\ldots,p_n) \in \Gamma_n$, $(q_1,\ldots,q_m) \in \Gamma_m$. *Then* $\varphi$ *is of the form*

$$\varphi(p) = \overline{b}_1(p) + \varphi(0) \tag{41.9}$$

*where* $\overline{b}_1 : \mathbb{R} \to \mathbb{R}$ *is an additive mapping with*

$$\overline{b}_1(1) = \begin{cases} -n\varphi(0), & \text{if } \varphi(1) + (n-1)\varphi(0) \neq 1; \\ 1 - n\varphi(0), & \text{if } \varphi(1) + (n-1)\varphi(0) = 1, \end{cases} \tag{41.10}$$

*or*

$$\varphi(p) = M(p) - \overline{b}(p) + \varphi(0)\,, \tag{41.11}$$

*where $\overline{b} : \mathbb{R} \to \mathbb{R}$ is an additive mapping with*

$$\overline{b}(1) = n\varphi(0) \tag{41.12}$$

*and $M : I \to \mathbb{R}$ is a nonconstant nonadditive mapping which is multiplicative in the sense of Definition* 41.1.

## 41.3 The Functional Equation (41.5)

We prove:

**Theorem 41.1.** *Let $n \geq 3$, $m \geq 3$ be fixed integers and $\psi : I \to \mathbb{R}$, $\varphi : I \to \mathbb{R}$ be mappings which satisfy the functional equation* (41.5) *for all $(p_1, \ldots, p_n) \in \Gamma_n$, $(q_1, \ldots, q_m) \in \Gamma_m$. Then, any general solution $(\psi, \varphi)$ of* (41.5) *is of the form (for all $p \in I$)*

$$\begin{cases} \text{(i)} \ \ \psi(p) = b(p) + \psi(0), \\ \text{(ii)} \ \ \varphi(p) = b_1(p) - \dfrac{1}{n} b_1(1) + \dfrac{1}{n}, \end{cases} \tag{41.13}$$

*or*

$$\begin{cases} \text{(i)} \ \ \psi(p) = b(p) + \psi(0), \ \ b(1) = -m\psi(0), \\ \text{(ii)} \ \ \varphi \textit{ arbitrary}, \end{cases} \tag{41.14}$$

*or*

$$\begin{cases} \text{(i)} \ \ \psi(p) = [\psi(1) + (m-1)\,\psi(0)][M(p) - \overline{b}(p)] + B_1(p) + \psi(0), \\ \qquad\qquad\qquad\qquad\qquad\qquad\qquad\qquad \psi(1) + (m-1)\,\psi(0) \neq 0, \\ \text{(ii)} \ \ \varphi(p) = M(p) - \overline{b}(p) + \varphi(0), \end{cases} \tag{41.15}$$

*where $b : \mathbb{R} \to \mathbb{R}$, $b_1 : \mathbb{R} \to \mathbb{R}$, $\overline{b} : \mathbb{R} \to \mathbb{R}$ and $B_1 : \mathbb{R} \to \mathbb{R}$ are additive mappings such that $\overline{b}(1) = n\varphi(0)$, $B_1(1) = [\psi(1) + (m-1)\,\psi(0)]n\,\varphi(0) - m\psi(0)$ and the mapping $M : I \to \mathbb{R}$ is nonconstant, nonadditive, and multiplicative in the sense of Definition* 41.1.

*Proof.* We divide the discussion into two cases:

*Case* 1: $\sum_{i=1}^{n} \varphi(p_i) - 1$ vanishes identically on $\Gamma_n$. In this case, the equation

$$\sum_{i=1}^{n} \varphi(p_i) = 1 \tag{41.16}$$

holds for all $(p_1,\ldots,p_n) \in \Gamma_n$, $n \geq 3$ a fixed integer. By Lemma 41.1, there exists an additive mapping $b_1 : \mathbb{R} \to \mathbb{R}$ such that $\varphi(p)$ is of the form (ii) in (41.13). Making use of (41.16) in (41.5), we obtain the equation

$$\sum_{i=1}^{n} \sum_{j=1}^{m} \psi(p_i q_j) = \sum_{j=1}^{m} \psi(q_j) + m(n-1)\,\psi(0) \qquad (41.17)$$

valid for all $(p_1,\ldots,p_n) \in \Gamma_n$, $(q_1,\ldots,q_m) \in \Gamma_m$, $n \geq 3$, $m \geq 3$ being fixed integers. The substitutions $q_1 = 1$, $q_2 = \cdots = q_m = 0$ in (41.17) yield the functional equation

$$\sum_{i=1}^{n} \psi(p_i) = \psi(1) + (n-1)\,\psi(0)$$

valid for all $(p_1,\ldots,p_n) \in \Gamma_n$, $n \geq 3$ a fixed integer. Making use of Lemma 41.1, there exists an additive mapping $b : \mathbb{R} \to \mathbb{R}$ such that $\psi(p)$ is of the form (i) in (41.13). Thus, we have obtained the solution (41.13) of (41.5).

*Case* 2: $\sum_{i=1}^{n} \varphi(p_i) - 1$ does not vanish identically on $\Gamma_n$. In this case, there exists a probability distribution $(p_1^*,\ldots,p_n^*) \in \Gamma_n$ such that

$$\sum_{i=1}^{n} \varphi(p_i^*) - 1 \neq 0\,.$$

Let us write (41.5) in the form

$$\sum_{j=1}^{m} \left\{ \sum_{i=1}^{n} \psi(p_i q_j) - \psi(q_j) \sum_{i=1}^{n} \varphi(p_i) \right\} = m(n-1)\,\psi(0)\,.$$

By Lemma 41.1, there exists a mapping $A_1 : \Gamma_n \times \mathbb{R} \to \mathbb{R}$, additive in the second variable, such that

$$\sum_{i=1}^{n} \psi(p_i q) - \psi(q) \sum_{i=1}^{n} \varphi(p_i) = A_1(p_1,\ldots,p_n;q) - \psi(0)\left[\sum_{i=1}^{n} \varphi(p_i) - n\right] \qquad (41.18)$$

valid for all $(p_1,\ldots,p_n) \in \Gamma_n$ and $q \in I$ with

$$A_1(p_1,\ldots,p_n;1) = m\,\psi(0)\left[\sum_{i=1}^{n} \varphi(p_i) - 1\right]. \qquad (41.19)$$

Let $x \in I$ and $(r_1,\ldots,r_n) \in \Gamma_n$. Putting successively $q = xr_t$, $t = 1,\ldots,n$ in (41.18), adding the resulting $n$ equations so obtained and then substituting the value of $\sum_{t=1}^{n} \psi(xr_t)$ calculated from (41.18), we get the equation

$$\sum_{i=1}^{n}\sum_{t=1}^{n}\psi(xp_ir_t) - [\psi(x)-\psi(0)]\sum_{i=1}^{n}\phi(p_i)\sum_{t=1}^{n}\varphi(r_t)-n^2\psi(0)$$
$$= A_1(r_1,\ldots,r_n;x)\sum_{i=1}^{n}\varphi(p_i)+A_1(p_1,\ldots,p_n;x)\,. \quad (41.20)$$

The left side of (41.20) does not undergo any change if we interchange $r_t$ and $p_i$, $t = 1,\ldots,n$; $i = 1,\ldots,n$. So, the right hand side of (41.20) must also remain unchanged on interchanging $r_t$ and $p_i$, $t = 1,\ldots,n$; $i = 1,\ldots,n$. This gives rise to the symmetry equation

$$A_1(p_1,\ldots,p_n;x)\left[\sum_{t=1}^{n}\varphi(r_t)-1\right] = A_1(r_1,\ldots,r_n;x)\left[\sum_{i=1}^{n}\varphi(p_i)-1\right]. \quad (41.21)$$

Setting $p_i = p_i^*$, $i = 1,\ldots,n$ in (41.21) and using the fact that

$$\sum_{i=1}^{n}\varphi(p_i^*)-1\neq 0$$

we obtain the equation

$$A_1(r_1,\ldots,r_n;x) = B(x)\left[\sum_{t=1}^{n}\varphi(r_t)-1\right], \quad (41.22)$$

where $B:\mathbb{R}\to\mathbb{R}$ is defined as

$$B(x) = A_1(p_1^*,\ldots,p_n^*;x)\left[\sum_{i=1}^{n}\varphi(p_i^*)-1\right]^{-1}$$

for all $x\in\mathbb{R}$. It can be easily verified that $B:\mathbb{R}\to\mathbb{R}$ is an additive mapping with $B(1) = m\psi(0)$. By taking $r_1 = p_1,\ldots,r_n = p_n$ in (41.22), using (41.18) and the additivity of $B:\mathbb{R}\to\mathbb{R}$, it follows that

$$\sum_{i=1}^{n}\psi_1(p_iq)-\psi_1(q)\sum_{i=1}^{n}\varphi(p_i) = 0 \quad (41.23)$$

where $\psi_1 : I\to\mathbb{R}$ is defined as

$$\psi_1(x) = \psi(x)+B(x)-\psi(0) \quad (41.24)$$

for all $x\in I$.
*Case* 2.1: $\psi_1(q)\equiv 0$ on $I$. In this case, it is obvious, from (41.23), that $\varphi$ is as in (41.14)(ii). Also, since $\psi_1(q) = 0$ for all $q\in I$, (41.24) gives $\psi(p) = -B(p)+\psi(0)$ for all $p\in I$. Define $b:\mathbb{R}\to\mathbb{R}$ as $b(x) = -B(x)$ for all $x\in I$. Then (41.14)(i) follows as $b(1) = -B(1) = -m\psi(0)$.

*Case* 2.2: $\psi_1(q)$ does not vanish identically on $I$. In this case, there exists an element $q_0 \in I$ such that $\psi_1(q_0) \neq 0$. But, from (41.24), it is easy to see that $\psi_1(0) = 0$. Hence, $0 < q_0 \leq 1$. Also, putting $x = 1$ in (41.24) and using $B(1) = m\,\psi(0)$, we obtain $\psi_1(1) = \psi(1) + (m-1)\psi(0)$.

Let us consider the case $\psi_1(1) = 0$. Then (41.23) gives the equation

$$\sum_{i=1}^{n} \psi_1(p_i) = 0$$

valid for all $(p_1, \ldots, p_n) \in \Gamma_n$, $n \geq 3$ a fixed integer. Now, from (41.24), it follows that

$$\sum_{i=1}^{n} \psi(p_i) = (n-m)\psi(0)$$

as $B(1) = m\,\psi(0)$. By Lemma 41.1, there exists an additive mapping $b : \mathbb{R} \to \mathbb{R}$ such that $\psi(p) = b(p) + \psi(0)$ for all $p \in I$ with $b(1) = -m\psi(0)$. Consequently

$$\sum_{j=1}^{m} \psi(q_j) = 0\,.$$

Now, from (41.5), it follows that $\varphi$ is, indeed, an arbitrary real-valued mapping. Thus, in this case, we again get the solution (41.14).

Now, let us consider the case when $\psi_1(1) \neq 0$. This means that

$$[\psi(1) + (m-1)\psi(0)] \neq 0\,.$$

Let us put $q = 1$ in (41.23) and use (41.24). We obtain the equation

$$\sum_{i=1}^{n} \{\psi(p_i) - [\psi(1) + (m-1)\psi(0)]\varphi(p_i)\} = (n-m)\psi(0)\,.$$

By Lemma 41.1, there exists an additive mapping $B_1 : \mathbb{R} \to \mathbb{R}$ such that

$$\psi(p) = [\psi(1) + (m-1)\psi(0)][\varphi(p) - \varphi(0)] + B_1(p) + \psi(0) \tag{41.25}$$

with

$$B_1(1) = [\psi(1) + (m-1)\psi(0)]n\varphi(0) - m\psi(0)\,. \tag{41.26}$$

Let us choose $p_1 = 1, p_2 = \cdots = p_n = 0$, $q = q_0$ in (41.23), use $\psi_1(0) = 0$ and $\psi_1(q_0) \neq 0$. We obtain $\varphi(1) + (n-1)\varphi(0) = 1$.

From (41.5), (41.25), (41.26) and using $\psi(1) + (m-1)\psi(0) \neq 0$, we obtain the functional equation (41.8). Keeping in view the fact that $\varphi(1) + (n-1)\varphi(0) = 1$, $\varphi$ is of the form (41.9) for all $p \in I$ in which $\bar{b}_1 : \mathbb{R} \to \mathbb{R}$ is an additive mapping with $\bar{b}_1(1) = 1 - n\varphi(0)$ or $\varphi$ is of the form (41.11) in which $\bar{b} : \mathbb{R} \to \mathbb{R}$ is an additive mapping with $\bar{b}(1) = n\varphi(0)$; $M : I \to \mathbb{R}$ being a nonconstant nonadditive mapping which is also multiplicative in the sense of Definition 41.1. In the former case, making use of the additivity of $\bar{b}_1 : \mathbb{R} \to \mathbb{R}$ and $\bar{b}_1(1) = 1 - n\varphi(0)$, it follows that

$$\sum_{i=1}^{n} \varphi(p_i) - 1 = 0$$

for all $(p_1, \ldots, p_n) \in \Gamma_n$ contradicting

$$\sum_{i=1}^{n} \varphi(p_i^*) - 1 \neq 0\,.$$

Hence, we ignore this case. In the latter case, using the fact that $\varphi(p) - \varphi(0) = M(p) - \bar{b}(p)$ for all $p \in I$, (41.25) gives

$$\psi(p) = [\psi(1) + (m-1)\,\psi(0)][M(p) - \bar{b}(p)] + B_1(p) + \psi(0)$$

for all $p \in I$ with $B_1(1)$ given by (41.26) and $\bar{b}(1) = n\varphi(0)$. Thus, we have obtained the solution (41.15) of (41.5). □

## 41.4 The Functional Equation (41.4)

We prove:

**Theorem 41.2.** *Let $f : I \to \mathbb{R}$, $g : I \to \mathbb{R}$ and $h_j : I \to \mathbb{R}$, $j = 1, \ldots, m$ be mappings which satisfy* (41.4) *for all $(p_1, \ldots, p_n) \in \Gamma_n$, $(q_1, \ldots, q_m) \in \Gamma_m$; $n \geq 3$, $m \geq 3$ being fixed integers. Then, the general solution of* (41.4) *(for all $p \in I$) is of the form*

$$\begin{cases} f(p) = e_2(p) - \dfrac{1}{nm}\, e_2(1)\,, \\ g(p) = e_1(p) - \dfrac{1}{n}\, e_1(1)\,, \\ h_j \ \text{an arbitrary}, \quad j = 1, \ldots, m, \end{cases} \tag{41.27}$$

*or*

$$\begin{cases} f(p) = e_2(p) - \dfrac{1}{nm}\, e_2(1)\,, \\ g \ \text{arbitrary}\,, \\ h_j(p) = e_3(p) + h_j(0)\,, \quad j = 1, \ldots, m, \end{cases} \tag{41.28}$$

*or*

$$\begin{cases} f(p) = B(p) + f(0)\,, \\ g(p) = [g(1) + (n-1)g(0)]\left[b_1(p) - \frac{1}{n}b_1(1) + \frac{1}{n}\right], \\ h_j(p) = [g(1) + (n-1)g(0)]^{-1}B(p) + B^*(p) + h_j(0)\,, \end{cases} \tag{41.29}$$

*or*

$$\begin{cases} f(p) = [f(1) + (nm-1)\,f(0)][M(p) - \overline{b}(p)] + \overline{B}(p) + f(0)\,, \\ \qquad\qquad f(1) + (nm-1)\,f(0) \neq 0\,, \\ g(p) = [g(1) + (n-1)\,g(0)][M(p) - \overline{b}(p)] + g(0)\,, \\ h_j(p) = [g(1) + (n-1)\,g(0)]^{-1}\{[f(1) + (nm-1)f(0)] \\ \qquad\qquad \times [M(p) - \overline{b}(p)] + \overline{B}(p)\} + B^*(p) + h_j(0)\,, \end{cases} \tag{41.30}$$

*where* $e_i : \mathbb{R} \to \mathbb{R}$ $(i = 1,2,3)$*,* $b_1 : \mathbb{R} \to \mathbb{R}$*,* $\overline{b} : \mathbb{R} \to \mathbb{R}$*,* $B : \mathbb{R} \to \mathbb{R}$*,* $\overline{B} : \mathbb{R} \to \mathbb{R}$*,* $B^* : \mathbb{R} \to \mathbb{R}$ *are additive mappings with*

$(\alpha_1)$ $e_3(1) = -\sum_{j=1}^{m} h_j(0)\,,$

$(\alpha_2)$ $\overline{b}(1) = n\,[g(1) + (n-1)\,g(0)]^{-1}\,g(0)\,,$

$(\alpha_3)$ $\overline{B}(1) = n[f(1) + (nm-1)f(0)][g(1) + (n-1)g(0)]^{-1}g(0) - nmf(0)\,,$

$(\alpha_4)$ $B^*(1) = nm[g(1) + (n-1)\,g(0)]^{-1}\,f(0) - \sum_{j=1}^{m} h_j(0)\,,$

*and* $M : I \to \mathbb{R}$ *is a nonconstant nonadditive mapping which is multiplicative in the sense of Definition 41.1.*

Before giving the proof of this theorem, we need to prove the following:

**Lemma 41.5.** *The mappings* $g : I \to \mathbb{R}$*,* $h_j : I \to \mathbb{R}$*,* $j = 1,\ldots,m$*, appearing in* (41.4)*, satisfy the equation*

$$[g(1) + (n-1)g(0)]\sum_{i=1}^{n}\sum_{j=1}^{m} h_j(p_i q_j) - (n-1)[g(1) + (n-1)\,g(0)] \times \sum_{j=1}^{m} h_j(0) = \sum_{i=1}^{n} g(p_i) \sum_{j=1}^{m} h_j(q_j) \tag{41.31}$$

*for all* $(p_1,\ldots,p_n) \in \Gamma_n$*,* $(q_1,\ldots,q_m) \in \Gamma_m$*,* $n \geq 3$*,* $m \geq 3$ *being fixed integers.*

*Proof.* Let us write (41.4) in the form

$$\sum_{i=1}^{n}\left\{\sum_{j=1}^{m} f(p_i q_j) - g(p_i)\sum_{j=1}^{m} h_j(q_j)\right\} = 0\,.$$

By Lemma 41.1, there exists a mapping $B_2 : \mathbb{R} \times \Gamma_m \to \mathbb{R}$, additive in the first variable, such that

$$\sum_{j=1}^{m} f(pq_j) - g(p)\sum_{j=1}^{m} h_j(q_j) = B_2(p;q_1,\ldots,q_m) \\ -g(0)\sum_{j=1}^{m} h_j(q_j) + mf(0) \qquad (41.32)$$

for all $p \in I$ and $(q_1,\ldots,q_m) \in \Gamma_m$ with

$$B_2(1;q_1,\ldots,q_m) = ng(0)\sum_{j=1}^{m} h_j(q_j) - nmf(0)\,. \qquad (41.33)$$

Putting $p = 1$ in (41.32), making use of (41.33) and the Lemma 41.2, it follows that there exists an additive mapping $A : \mathbb{R} \to \mathbb{R}$ such that

$$f(p) = [g(1) + (n-1)g(0)][h_j(p) - h_j(0)] + A(p) + f(0) \qquad (41.34)$$

with

$$A(1) = -nmf(0) + [g(1) + (n-1)g(0)]\sum_{j=1}^{m} h_j(0)\,. \qquad (41.35)$$

Equation (41.31) now follows from (41.4), (41.34) and (41.35). □

*Proof of Theorem* 41.2. We divide the discussion into three cases:

*Case* 1: $\sum_{i=1}^{n} g(p_i)$ vanishes identically on $\Gamma_n$. In this case,

$$\sum_{i=1}^{n} g(p_i) = 0$$

for all $(p_1,\ldots,p_n) \in \Gamma_n$. By Lemma 41.1, there exists an additive mapping $e_1 : \mathbb{R} \to \mathbb{R}$ such that

$$g(p) = e_1(p) - \frac{1}{n}e_1(1)$$

for all $p \in I$. Since $\sum_{i=1}^{n} g(p_i) = 0$, it follows from (41.4) that $h_j$ $(j = 1,\ldots,m)$ are, indeed, arbitrary real-valued mappings and also (41.4) reduces to (41.6) for all $(p_1,\ldots,p_n) \in \Gamma_n$, $(q_1,\ldots,q_m) \in \Gamma_m$. Consequently, by Lemma 41.3, $f$ is of the form (41.7) for all $p \in I$. Thus, we have obtained the solution (41.27) of (41.4).

*Case* 2: $\sum_{j=1}^{m} h_j(q_j)$ vanishes identically on $\Gamma_m$. In this case,

$$\sum_{j=1}^{m} h_j(q_j) = 0$$

for all $(q_1,\ldots,q_m) \in \Gamma_m$. By Lemma 41.2, there exists an additive mapping $e_3 : \mathbb{R} \to \mathbb{R}$ such that $h_j(p) = e_3(p) + h_j(0)$ for all $j = 1,\ldots,m$ with $e_3(1)$ given by $(\alpha_1)$. Now, proceeding as in Case 1, the solution (41.28) can be obtained.

*Case* 3: Neither $\sum_{i=1}^{n} g(p_i)$ vanishes identically on $\Gamma_n$ nor $\sum_{j=1}^{m} h_j(q_j)$ vanishes identically on $\Gamma_m$. Then there exist a probability distribution $(p_1^*,\ldots,p_n^*) \in \Gamma_n$ and a probability distribution $(q_1^*,\ldots,q_m^*) \in \Gamma_m$ such that

$$\sum_{i=1}^{n} g(p_i^*) \neq 0 \quad \text{and} \quad \sum_{j=1}^{m} h_j(q_j^*) \neq 0\,.$$

Then

$$\sum_{i=1}^{n} g(p_i^*) \sum_{j=1}^{m} h_j(q_j^*) \neq 0\,.$$

Now let us pay attention to (41.31). We claim that $[g(1) + (n-1)\,g(0)] \neq 0$. On the contrary, suppose that $g(1) + (n-1)\,g(0) = 0$. Then (41.31) reduces to

$$\sum_{i=1}^{n} g(p_i) \sum_{j=1}^{m} h_j(q_j) = 0$$

valid for all probability distributions $(p_1,\ldots,p_n) \in \Gamma_n$ and $(q_1,\ldots,q_m) \in \Gamma_m$. In particular

$$\sum_{i=1}^{n} g(p_i^*) \sum_{j=1}^{m} h_j(q_j^*) = 0\,,$$

a contradiction. Hence, $[g(1) + (n-1)\,g(0)] \neq 0$. Now, (41.34) gives

$$h_j(p) = [g(1) + (n-1)\,g(0)]^{-1}\Big[f(p) - f(0)\Big] + B^*(p) + h_j(0) \qquad (41.36)$$

where $B^* : \mathbb{R} \to \mathbb{R}$ is an additive mapping such that

$$B^*(p) = -\,[g(1) + (n-1)\,g(0)]^{-1}A(p)$$

and $B^*(1)$ is given by $(\alpha_4)$. Now, from (41.4), (41.36) and $(\alpha_4)$, the equation

$$\sum_{i=1}^{n}\sum_{j=1}^{m} f(p_i q_j) = [g(1)+(n-1)\,g(0)]^{-1}\sum_{i=1}^{n} g(p_i)\left[\sum_{j=1}^{m} f(q_j)+m(n-1)\,f(0)\right] \tag{41.37}$$

follows. This equation is valid for all $(p_1,\ldots,p_n)\in\Gamma_n$ and $(q_1,\ldots,q_m)\in\Gamma_m$, $n\geq 3$, $m\geq 3$ being fixed integers. Define the mappings $\varphi: I\to\mathbb{R}$ and $\psi: I\to\mathbb{R}$ as

$$\varphi(x)=[g(1)+(n-1)\,g(0)]^{-1}g(x) \tag{41.38}$$

and

$$\psi(x)=f(x)+m(n-1)\,f(0)x \tag{41.39}$$

for all $x\in I$. Then (41.37) reduces to the functional equation (41.5) with

$$\varphi(1)+(n-1)\,\varphi(0)=1\,.$$

So, we need to consider only those solutions of (41.5) which satisfy the requirement $\varphi(1)+(n-1)\,\varphi(0)=1$.

Referring to (41.14), we can choose $\varphi$ to be any arbitrary mapping which satisfies the condition

$$\varphi(1)+(n-1)\,\varphi(0)=1\,.$$

Then, from (41.38), (41.39) and (41.14), we obtain the solution (of (41.4))

$$\begin{cases} f(p)=b(p)-m(n-1)f(0)\,p+f(0)\,,\\ g \text{ arbitrary}\,,\\ h_j(p)=[g(1)+(n-1)\,g(0)]^{-1}[b(p)-m(n-1)f(0)\,p]\\ \qquad\qquad +B^*(p)+h_j(0) \end{cases} \tag{41.40}$$

with $b(1)=-mf(0)$. But this solution is contained in (41.28) if we set

$$e_2(p)=b(p)-m(n-1)\,f(0)p$$

and

$$e_3(p)=[g(1)+(n-1)\,g(0)]^{-1}[b(p)-m(n-1)\,f(0)p]+B^*(p).$$

The solutions (41.13) and (41.15), of (41.5), also satisfy the condition

$$\varphi(1)+(n-1)\,\varphi(0)=1\,.$$

Making use of (41.38), (41.39), (41.36), (41.13) and (41.15), the solutions (41.29) and (41.30) can be obtained in which $\overline{b}(1)$, $\overline{B}(1)$ are given, respectively, by $(\alpha_2)$ and $(\alpha_3)$. □

## 41.5 The Functional Equation (41.2)

We prove:

**Theorem 41.3.** *Let* $F:I\to\mathbb{R}$, $G:I\to\mathbb{R}$ *and* $H_j:I\to\mathbb{R}$, $j=1,\ldots,m$ *be mappings which satisfy the equation* (41.2) *for all* $(p_1,\ldots,p_n)\in\Gamma_n$, $(q_1,\ldots,q_m)\in\Gamma_m$; $n\geq 3$, $m\geq 3$ *being fixed integers and* $\lambda\neq 0$. *Then the general solution of* (41.2) *(for all* $p\in I$*) is of the form*

$$\begin{cases} F(p)=\dfrac{1}{\lambda}\{e_2(p)-p-\dfrac{1}{nm}e_2(1)\}, \\ G(p)=\dfrac{1}{\lambda}\{e_1(p)-p-\dfrac{1}{n}e_1(1)\}, \\ H_j \quad arbitrary, \qquad j=1,\ldots,m, \end{cases} \tag{41.41}$$

*or*

$$\begin{cases} F(p)=\dfrac{1}{\lambda}\{e_2(p)-p-\dfrac{1}{nm}e_2(1)\}, \\ G \quad arbitrary, \\ H_j(p)=\dfrac{1}{\lambda}\{e_3(p)-p\}+H_j(0), \end{cases} \tag{41.42}$$

*or*

$$\begin{cases} F(p)=\dfrac{1}{\lambda}\{B(p)-p\}+F(0), \\ G(p)=\dfrac{1}{\lambda}\{[\lambda(G(1)+(n-1)\,G(0))+1][b_1(p)-\dfrac{1}{n}b_1(1)+\dfrac{1}{n}]-p\}, \\ H_j(p)=\dfrac{1}{\lambda}\{[\lambda(G(1)+(n-1)G(0))+1]^{-1}B(p)+B^*(p)-p\}+H_j(0), \end{cases} \tag{41.43}$$

*or*

$$\begin{cases} F(p)=\frac{1}{\lambda}\{[\lambda(F(1)+(nm-1)F(0))+1][M(p)-\overline{b}(p)]+\overline{B}(p)-p\}+F(0)\,, \\ \qquad\qquad\qquad\qquad [\lambda(F(1)+(nm-1)F(0))+1\neq 0]\,, \\ G(p)=\frac{1}{\lambda}\{[\lambda(G(1)+(n-1)G(0))+1][M(p)-\overline{b}(p)]-p\}+G(0)\,, \\ H_j(p)=\frac{1}{\lambda}\{[\lambda(G(1)+(n-1)G(0))+1]^{-1}\{[\lambda(F(1)+(nm-1)F(0))+1] \\ \qquad\qquad \times[M(p)-\overline{b}(p)]+\overline{B}(p)\}+B^*(p)-p\}+H_j(0)\,, \end{cases} \tag{41.44}$$

*where* $e_i:\mathbb{R}\to\mathbb{R}$ $(i=1,2,3)$, $b_1:\mathbb{R}\to\mathbb{R}$, $\overline{b}:\mathbb{R}\to\mathbb{R}$, $B:\mathbb{R}\to\mathbb{R}$, $\overline{B}:\mathbb{R}\to\mathbb{R}$, $B^*:\mathbb{R}\to\mathbb{R}$ *are additive mappings with*

$(\beta_1)$ $e_3(1)=-\lambda\sum_{j=1}^{m}H_j(0)\,,$

$(\beta_2)$ $\overline{b}(1)=\lambda n[\lambda(G(1)+(n-1)G(0))+1]^{-1}G(0)\,,$

$(\beta_3)$ $\overline{B}(1)=\lambda n[\lambda(F(1)+(nm-1)F(0))+1]$
$\times[\lambda(G(1)+(n-1)G(0))+1]^{-1}G(0)-\lambda nmF(0)\,,$

$(\beta_4)$ $B^*(1)=\lambda nm[\lambda(G(1)+(n-1)G(0))+1]^{-1}F(0)-\lambda\sum_{j=1}^{m}H_j(0)\,,$

*and* $M:I\to\mathbb{R}$ *is a nonconstant nonadditive mapping which is multiplicative in the sense of Definition* 41.1.

*Proof.* From (41.3) and the solutions of the functional equation (41.4) i.e., (41.27)–(41.30) with $(\alpha_1),(\alpha_2),(\alpha_3),(\alpha_4)$, we obtain, respectively, the solutions (41.41)–(41.44) with $(\beta_1),(\beta_2),(\beta_3),(\beta_4)$; of the functional equation (41.2). The details are omitted. □

*Remark 41.1.* Out of the four solutions (41.41)–(41.44), the solution (41.44) is of importance from information-theoretic point of view. From this solution, it can be easily computed that

$$\sum_{i=1}^{n}F(p_i)=\frac{1}{\lambda}\left\{d_1\sum_{i=1}^{n}M(p_i)-1\right\}-d_2\,,$$

$$\sum_{i=1}^{n}G(p_i)=\frac{1}{\lambda}\left\{d_3\sum_{i=1}^{n}M(p_i)-1\right\},$$

and

$$\sum_{j=1}^{m} H_j(q_j) = \frac{1}{\lambda}\left\{d_4 \sum_{j=1}^{m} M(q_j) - 1\right\},$$

where

$$d_1 = [\lambda(F(1) + (nm-1)F(0)) + 1],$$
$$d_2 = n(m-1)F(0),$$
$$d_3 = [\lambda(G(1) + (n-1)G(0)) + 1],$$
$$d_4 = [\lambda(G(1) + (n-1)G(0)) + 1]^{-1}[\lambda(F(1) + (nm-1)F(0)) + 1].$$

If $d_1 = 1$, $d_2 = 0$, $d_3 = 1$ and $d_4 = 1$, then

$$\sum_{i=1}^{n} F(p_i) = L_n^{\lambda}(p_1, \ldots, p_n),$$

$$\sum_{i=1}^{n} G(p_i) = L_n^{\lambda}(p_1, \ldots, p_n)$$

and

$$\sum_{j=1}^{m} H_j(q_j) = L_m^{\lambda}(q_1, \ldots, q_m),$$

where (see Nath and Singh [5])

$$L_t^{\lambda}(x_1, \ldots, x_t) = \frac{1}{\lambda}\left[\sum_{i=1}^{t} M(x_i) - 1\right]. \tag{41.45}$$

The non-additive measure of entropy

$$H_t^{\alpha}(x_1, \ldots, x_t) = (2^{1-\alpha} - 1)^{-1}\left(\sum_{i=1}^{t} x_i^{\alpha} - 1\right),$$

with $\alpha \neq 1$, is a particular case of (41.45) when $\lambda = 2^{1-\alpha} - 1$, $\alpha > 0$, $\alpha \neq 1$ and $M : I \to \mathbb{R}$ is of the form $M(p) = p^{\alpha}$, $p \in I$, $\alpha \neq 1$, $\alpha > 0$, $0^{\alpha} := 0$, $1^{\alpha} := 1$.

## References

1. Daróczy, Z., Losonczi, L.: Über die Erweiterung der auf einer Punktmenge additiven Funktionen. Publ. Math. Debrecen **14**, 239–245 (1967)
2. Havrda, J., Charvát, F.: Quantification method of classification process, concept of structural $\alpha$-entropy. Kybernetika (Prague) **3**, 30–35 (1967)

3. Losonczi, L: Functional equations of sum form. Publ. Math. Debrecen **32**, 57–71 (1985)
4. Losonczi, L., Maksa, Gy.: On some functional equations of the information theory. Acta Math. Acad. Sci. Hungar. **39**, 73–82 (1982)
5. Nath, P., Singh, D.K.: On a multiplicative type sum form functional equation and its role in information theory. Appl. Math. **51** (5), 495–516 (2006)
6. Nath, P., Singh, D.K.: A sum form functional equation and its relevance in information theory. Aust. J. Math. Anal. Appl. **5**(1), Article 9, 1–18 (2008)

# Chapter 42
# Two-Step Iterative Method for Nonconvex Bifunction Variational Inequalities

**Muhammad Aslam Noor, Khlaida Inayat Noor, and Eisa Al-Said**

*Dedicated to the memory of S.M. Ulam*
*on the 100th anniversary of his birth*

**Abstract** In this paper, we suggest and analyze a two-step iterative method for solving nonconvex bifunction variational inequalities. We also discuss the convergence of the iterative method under partially relaxed strongly monotonicity, which is a weaker condition than cocoerciveness.

**Keywords** Bifunction variational inequalities • Nonconvex sets • Monotone operators • Iterative method • Convergence

**Mathematics Subject Classification (2000):** Primary 49J40; Secondary 90C33

## 42.1 Introduction

Variational inequalities theory, which was introduced by Stampacchia [23], provides us with a simple, general and unified framework to study a wide class of problems arising in pure and applied sciences. For the applications, physical formulation, numerical methods and other aspects of variational inequalities, see [1–23] and the references therein. It is well known that the variational inequalities represent

M.A. Noor (✉)
Mathematics Department, COMSATS Institute of Information Technology, Islamabad, Pakistan

Mathematics Department, College of Science, King Saud University, Riyadh, Saudi Arabia
e-mail: mnoor.c@ksu.edu.sa

K.I. Noor
Mathematics Department, COMSATS Institute of Information Technology, Islamabad, Pakistan
e-mail: khalidanoor@hotmail.com

E. Al-Said
Mathematics Department, College of Science, King Saud University, Riyadh, Saudi Arabia
e-mail: eisasaid@ksu.edu.sa

Th.M. Rassias and J. Brzdęk (eds.), *Functional Equations in Mathematical Analysis*, Springer Optimization and Its Applications 52, DOI 10.1007/978-1-4614-0055-4_42,

the optimality condition for the differentiable convex functions on the convex sets in a normed space. For the directionally differentiable convex functions, we have the bifunction variational inequalities. For the applications, numerical methods and formulation of the bifunction variational inequalities, see [3–7, 11, 14] and references therein. We would like to point out that all the work assumed that the underlying choice set is convex set. In many practical situations, the choice set may not be convex one.

Recently, Noor [17] has introduced and considered a new class of variational inequalities, called nonconvex variational inequalities on the uniformly prox-regular sets. It is well-known that the uniformly prox-regular sets are nonconvex and include the convex sets as a special case, see [2, 22]. We would like to point out that the projection-type methods and their invariant forms can not be used for solving the bifunction variational inequalities. To overcome this drawback, one usually uses the auxiliary principle technique, which is due to Glowinski, Lions and Tremolieres [10]. The main idea in this technique is to consider a suitable auxiliary problem related to the original problem. This way, one defines a mapping connecting the solutions of these problems. In this case, one has shown that the map connecting the solutions is a contraction mapping and consequently, it has fixed point satisfying the original problem. This technique has been used to suggest and analyze several methods for solving variational inequalities and variational inequalities. It has been shown that a substantial number of numerical methods can be obtained as special cases from this technique, see [13–18, 20]. In this paper, we again use the auxiliary principle technique to suggest and analyze a two-step predictor–corrector method for solving nonconvex bifunction variational inequalities. We also consider the convergence of the new iterative method under partially relaxed strongly monotonicity, which is a weaker condition than cocoerciveness. In this sense, our result represents an improvement and refinement of the known results. Our results can be considered as a novel and important application of the auxiliary principle technique. The interested readers are advised to discover novel and other new applications of the nonconvex bifunction variational inequalities in Engineering, Physical, Optimization, Regional, Operations Research and Mathematical Sciences. It is interesting to consider the implementation of the proposed iterative method for solving the bifunction variational inequalities. This is an open problem for further research.

## 42.2 Preliminaries

Let $H$ be a real Hilbert space whose inner product and norm are denoted by $\langle \cdot, \cdot \rangle$ and $\|\cdot\|$, respectively. Let $K$ be a nonempty, closed and convex set in $H$. The basic concepts and definitions used in this paper are exactly the same as in Noor [17].

Poliquin et al. [22] and Clarke et al. [2] have introduced and studied a new class of nonconvex sets, which are called uniformly prox-regular sets.

**Definition 42.1.** For a given $r \in ]0,\infty]$, a subset $K_r$ is said to be normalized uniformly $r$-prox-regular, if and only if every nonzero proximal normal to $K_r$ can be realized by an $r$-ball, that is,$\forall u \in K_r$, and $0 \neq \xi \in N^P_{K_r}(u)$, one has

$$\langle (\xi)/\|\xi\|, v-u\rangle \leq (1/2r)\|v-u\|^2, \qquad \forall v \in K_r.$$

It is clear that the class of normalized uniformly prox-regular sets is sufficiently large to include the class of convex sets, $p$-convex sets, $C^{1,1}$submanifolds (possibly with boundary) of $H$, the images under a $C^{1,1}$ diffeomorphism of convex sets and many other nonconvex sets; see [1,2,22]. Obviously, for $r = \infty$, the uniformly prox-regularity of $K_r$ is equivalent to the convexity of $K$. This class of uniformly prox-regular sets have played an important part in many nonconvex applications such as optimization, dynamic systems and differential inclusions. It is known that if $K_r$ is a uniformly prox-regular set, then the proximal normal cone $N^P_{K_r}(u)$ is closed as a set-valued mapping.

We now recall the well known proposition which summarizes some important properties of the uniformly prox-regular sets $K_r$.

**Lemma 42.1.** *Let $K$ be a nonempty closed subset of $H$, $r \in ]0,\infty]$ and*

$$K_r = \{u \in H : d_K(u) < r\}.$$

*If $K_r$ is uniformly prox-regular, then*

*(i) $P_{K_r}(u) \neq \emptyset$ for $u \in K_r$;*
*(ii) for each $r' \in ]0,r[$, $P_{K_r}$ is Lipschitz continuous with constant $r/(r-r')$ on $K_{r'}$.*

For given bifunction $B(\cdot,\cdot)$, we consider the problem of finding $u \in K_r$, such that

$$B(u, v-u) \geq 0, \qquad \forall v \in K_r, \tag{42.1}$$

which is called the *nonconvex bifunction variational inequality.*

If $k_r \equiv K$, the convex set, then problem (42.1) is equivalent to finding $u \in K$ such that

$$B(u, v-u) \geq 0, \qquad \forall v \in K, \tag{42.2}$$

which is called the bifunction variational inequality. Noor [14] has shown that bifunction variational inequalities represent the optimality condition of a directionally differentiable convex functions on a convex set in a normed space. For the applications, numerical methods and formulation of the bifunction variational inequalities, see [3–7,11,14] and the references therein.

If $B(u, v-u) = \langle Tu, v-u\rangle$, where $T$ is a nonlinear operator, then problem (42.1) reduces to finding $u \in K_r$ such that

$$\langle Tu, v-u\rangle \geq 0, \qquad \forall v \in K_r, \tag{42.3}$$

which is called the nonconvex variational inequality, introduced and studied by Noor [16–18] in recent years.

It is well-known [2, 18, 22] that the union of two disjoint intervals $[a,b]$ and $[c,d]$ is a prox-regular set with $r = (c-b)/2$. We also consider the following simple examples to give an idea of the importance of the nonconvex sets. These examples are due to Noor [18].

*Example 42.1.* Let $u = (x,y)$ and $v = (t,z)$ belong to the real Euclidean plane and consider $Tu = (2x, 2(y-1))$. Let $K = t^2 + (z-2)^2 \geq 4,\ -2 \leq t \leq 2$, and $z \geq -2$ be a subset of the Euclidean plane. Then one can easily show that the set $K$ is a prox-regular set $K_r$. It is clear that nonconvex variational inequality (42.1) has no solution.

*Example 42.2.* Let $u = (x,y) \in \mathbb{R}^2,\ v = (t,z) \in \mathbb{R}^2$ and let $Tu = (-x, 1-y)$. Let the set $K$ be the union of 2 disjoint squares, say $A$ and $B$ having respectively, the vertices in the points $(0,1),(2,1),(2,3),(0,3)$ and in the points $(4,1),(5,2),(4,3),(3,2)$. The fact that $K$ can be written in the form:

$$\{(t,z) \in \mathbb{R}^2 : \max\{|t-1|,|z-2|\} \leq 1\} \cup \{|t-4| + |z-2| \leq 1\}\}$$

shows that it is a prox-regular set in $\mathbb{R}^2$ and the nonconvex variational inequality (42.1) has a solution on the square $B$. We note that the operator $T$ is the gradient of a strictly concave function. This shows that the square $A$ is redundant.

We note that, if $K_r \equiv K$, the convex set in $H$, then problem (42.3) is equivalent to finding $u \in K$, such that

$$\langle Tu, v-u \rangle \geq 0, \qquad \forall v \in K. \tag{42.4}$$

Inequality of type (42.4) is called the *variational inequality*, which was introduced and studied by Stampacchia [23] in 1964. It turned out that a number of unrelated obstacle, free, moving, unilateral and equilibrium problems arising in various branches of pure and applied sciences can be studied via variational inequalities, see [1–23] and the references therein.

**Definition 42.2.** A bifunction $B(\cdot,\cdot) : H \times H \to H$ is said to be *partially relaxed strongly monotone,* provided

$$B(u, v-u) + B(v, z-v) \leq -\alpha \|u-z\|^2, \qquad \forall u,v,z \in H.$$

It is well known that the cocoerciveness implies partially relaxed strongly monotonicity, but the converse is not true.

## 42.3 Main Results

In this section, we suggest and analyze some iterative methods for solving the nonconvex bifunction variational inequality (42.1) using the auxiliary principle technique of Glowinski, Lions and Tremolieres [10] as developed by Noor [13–18].

For a given $u \in K_r$ satisfying (42.1), find $w \in K_r$ such that

$$\rho B(u, v-u) + \langle w-u, v-w \rangle \geq 0, \qquad \forall v \in K_r, \tag{42.5}$$

where $\rho > 0$ is a constant. Problem (42.5) is known as the auxiliary nonconvex bifunction variational inequality. We note that if $w = u$, then clearly $w$ is a solution of the (42.1). This observation enables us to suggest and analyze the following iterative method for solving (42.1).

*Example 42.3.* (**Algorithm 1**) For a given $u_0 \in H$, compute the approximate solution $u_{n+1}$ by the iterative scheme

$$\rho B(w_n, v-w_n) + \langle u_{n+1} - w_n, v - u_{n+1} \rangle \geq 0, \qquad \forall v \in K_r, \tag{42.6}$$

$$\rho B(u_n, v-u_n) + \langle w_n - u_n, v - w_n \rangle \geq 0, \qquad \forall v \in K_r, \tag{42.7}$$

where $\rho > 0$ and $\beta > 0$ are constants. This algorithm is called the two-step predictor–corrector method for solving the nonconvex bifunction variational inequality (42.1).

If $B(u, v-u) = \langle Tu, v-u \rangle$, then Algorithm 1 collapses to the following.

*Example 42.4.* (**Algorithm 2**) For a given $u_0 \in H$, compute the approximate solution by the iterative scheme

$$\rho \langle Tw_n, v - u_{n+1} \rangle + \langle u_{n+1} - w_n, v - u_{n+1} \rangle \geq 0, \qquad \forall v \in K_r,$$

$$\rho \langle Tu_n, v - w_n \rangle + \langle w_n - u_n, v - w_n \rangle \geq 0, \qquad \forall v \in K_r,$$

for solving nonconvex variational inequalities (42.3). For the convergence analysis of this algorithm see Noor [18].

We now consider the convergence analysis of Algorithm 1 and this is the main motivation of our next result.

**Theorem 42.1.** *Let $u \in K_r$ be a solution of (42.1) and let $u_{n+1}$ be the approximate solution obtained from Algorithm 1. If the bifunction $B(\cdot,\cdot)$ is partially relaxed strongly monotone with constant $\alpha > 0$, then*

$$\|u_{n+1} - u\|^2 \leq \|w_n - u\|^2 - (1 - 2\alpha\rho)\|u_{n+1} - w_n\|^2, \tag{42.8}$$

$$\|w_n - u\|^2 \leq \|u_n - u\|^2 - (1 - 2\alpha\rho)\|w_n - u_n\|^2. \tag{42.9}$$

*Proof.* Let $u \in K_r$ be a solution of (42.1). Then

$$B(u, v-u) \geq 0, \quad \forall v \in K_r. \tag{42.10}$$

Take $v = w_n$ in (42.10); we have

$$B(u, u_{n+1} - u) \geq 0. \tag{42.11}$$

Taking $v = u$ in (42.6) and using (42.11), we have

$$\begin{aligned}\langle w_n - u_n, u - w_n\rangle &\geq \rho\{\langle B(u_n, u - w_n) + B(u, w_n - u)\}\\ &\geq -\alpha\rho\|u_n - w_n\|^2,\end{aligned} \tag{42.12}$$

since $B(\cdot,\cdot)$ is partially relaxed strongly monotone with constant $\alpha > 0$.
From (42.13), we have

$$\|u - w_n\|^2 \leq \|u - u_n\|^2 - (1 - 2\alpha\rho)\|w_n - u_n\|^2,$$

the required result (42.9).
Taking $v = u$ in (42.6), we have

$$\rho B(w_n, u - u_{n+1}) + \langle u_{n+1} - w_n, u - u_{n+1}\rangle \geq 0. \tag{42.13}$$

From (42.11), (42.12) and using the partially relaxed strongly monotonicity $TB(\cdot,\cdot)$ with constant $\alpha > 0$, we have

$$\|u_{n+1} - u\|^2 \leq \|w_n - u\|^2 - (1 - 2\alpha\rho)\|u_{n+1} - w_n\|^2,$$

the required result (42.8). □

**Theorem 42.2.** *Let $u \in K_r$ be a solution of (42.1) and let $u_{n+1}$ be the approximate solution obtained from Algorithm 1. If $H$ is a finite dimensional space and $0 < \rho < \frac{1}{2\alpha}$, then*

$$\lim_{n\to\infty} u_n = u.$$

*Proof.* Let $\bar{u} \in K_r$ be a solution of (42.1). Then, the sequence $\{\|u_n - \bar{u}\|\}$ is nonincreasing and bounded, and

$$\sum_{n=0}^{\infty}(1 - 2\alpha\rho)\|u_{n+1} - w_n\|^2 \leq \|w_0 - u\|^2,$$

$$\sum_{n=0}^{\infty}(1 - 2\alpha\rho)\|w_n - u_n\|^2 \leq \|u_0 - u\|^2,$$

which implies

$$\lim_{n\to\infty}\|u_{n+1} - w_n\| = 0, \qquad \lim_{n\to\infty}\|w_n - u_n\| = 0.$$

Thus

$$\lim_{n\to\infty}\|u_{n+1} - u_n\| = \lim_{n\to\infty}\|u_{n+1} - w_n\| + \lim_{n\to\infty}\|w_n - u_n\| = 0. \tag{42.14}$$

Let $\hat{u}$ be a cluster point of $\{u_n\}$; there is a subsequence $\{u_{n_i}\}$ that converges to $\hat{u}$. Replacing $u_{n+1}$ by $u_{n_i}$ in (42.6), $w_n$ by $u_{n_i}$ in (42.7), and taking the limits, and using (42.14), we have

$$B(\hat{u}, v-\hat{u}) \geq 0, \quad \forall v \in K_r.$$

This shows that $\hat{u} \in K_r$ solves (42.1) and $\|u_{n+1}-\hat{u}\|^2 \leq \|u_n-\hat{u}\|^2$, which implies that the sequence $\{u_n\}$ has a unique cluster point and $\hat{u} = \lim_{n\to\infty} u_n$ is the solution of (42.1). □

**Acknowledgement** I would like to express my sincere gratitude to Prof. Dr. Themistocles M. Rassias for the kind invitation and Dr. S.M. Junaid Zaidi, Rector, CIIT, Islamabad, Pakistan, for providing excellent research facilities.

This research is supported by the Visiting Professor Program of King Saud University, Riyadh, Saudi Arabia under grant NO. KSU.VPP. 108.

## References

1. Bounkhel, M., Tadj, L., Hamdi, A.: Iterative schemes to solve nonconvex variational problems. J. Inequal. Pure Appl. Math. **4**, 1–14 (2003)
2. Clarke, F.H., Ledyaev, Y.S., Wolenski, P.R.: Nonsmooth Analysis and Control Theory. Springer–Verlag, Berlin (1998)
3. Crespi, G.P., Ginchev, J., Rocca, M.: Minty variational inequalities, increase along rays property and optimization. J. Optim. Theory Appl. **123**, 479–496 (2004)
4. Crespi, G.P., Ginchev, J., Rocca, M.: Existence of solutions and star-shapedness in Minty variational inequalities. J. Global Optim. **32**, 485–494 (2005)
5. Crespi, G.P., Ginchev, J., Rocca, M.: Increasing along rays property for vector functions. J. Nonconvex Anal. **7**, 39–50 (2006)
6. Crespi, G.P., Ginchev, J., Rocca, M.: Some remarks on the Minty vector variational principle. J. Math. Anal. Appl. **345**, 165–175 (2008)
7. Fang, Y.P., Hu, R.: Parametric well-posedness for variational inequalities defined by bifunction. Comput. Math. Appl. **53**, 1306–1316 (2007)
8. Giannessi, F., Maugeri, A., Pardalos, P.M.: Equilibrium Problems: Nonsmooth Optimization and Variational Inequality Models. Kluwer Academics Publishers, Dordrecht (2001)
9. Gilbert, R.P., Panagiotopoulos, P.D., Pardalos, P. M.: From Convexity to Nonconvexity. Kluwer Academic Publishers, Dordrecht (2001)
10. Glowinski, R., Lions, J.L., Tremolieres, R.: Numerical Analysis of Variational Inequalities. North-Holland, Amsterdam (1981)
11. Lalitha, C.S., Mehra, M.: Vector variational inequalities with cone-pseudomonotone bifunction. Optimization **54**, 327–338 (2005)
12. Noor, M.A.: General variational inequalities. Appl. Math. Lett. **1**, 119–121 (1988)
13. Noor, M.A.: New approximation schemes for general variational inequalities. J. Math. Anal. Appl. **251**, 217–229 (2000)
14. Noor, M.A.: Some new classes of nonconvex functions. Nonlinear Funct. Anal. Appl **11**, 165–171 (2006)
15. Noor, M.A.: Some developments in general variational inequalities, Appl. Math. Comput. **152**, 199–277 (2004)

16. Noor, M.A.: Implicit iterative methods for nonconvex variational inequalities. J. Optim. Theory Appl. **143**, 619–624 (2009)
17. Noor, M.A.: Projection methods for nonconvex variational inequalities, Optim. Lett. **3**, 411–418 (2009)
18. Noor, M.A.: On an implicit method for nonconvex variational inequalities. J. Optim. Theory Appl. **147**, 411–417 (2010)
19. Noor, M.A., Noor, K.I.: Iterative schemes for trifunction variational inequalities, to appear
20. Noor, M.A., Noor, K.I., Rassias, Th.M.: Some aspects of variational inequalities. J. Comput. Appl. Math. **47**, 285–312 (1993)
21. Pardalos, P.M., Rassias, Th.M., Khan, A.A.: Nonlinear Analysis and Variational Problems. Springer (2010)
22. Poliquin, R.A., Rockafellar, R.T., Thibault, L.: Local differentiability of distance functions. Trans. Amer. Math. Soc. **352**, 5231–5249 (2000)
23. Stampacchia, G.: Formes bilineaires coercitives sur les ensembles convexes. C. R. Math. Acad. Sci. Paris **258**, 4413–4416 (1964)

# Chapter 43
# On a Sincov Type Functional Equation

**Prasanna K. Sahoo**

*Dedicated to the memory of S.M. Ulam*
*on the 100th anniversary of his birth*

**Abstract** The present work aims to find the general solution $f_1, f_2, f_3 : G^2 \to H$ and $f : G \to H$ of the Sincov type functional equation $f_1(x,y) + f_2(y,z) + f_3(z,x) = f(x+y+z)$ for all $x, y, z \in G$ without any regularity assumption. Here, $G$ and $H$ are additive abelian groups, and the division by 2 is uniquely defined in $H$.

**Keywords** Additive Abelian group • Additive function • Biadditive function • Quadratic function • Sincov functional equation

**Mathematics Subject Classification (2000):** Primary 39B52

## 43.1 Introduction

In 1903, the Ukrainian mathematician Sincov (see [13, 14]) studied the functional equation

$$f(x,y) + f(y,z) = f(x,z) \tag{43.1}$$

for all $x, y, z \in \mathbb{R}$ (the set of real numbers). Others like Moritz Cantor [3] and Gottlob Frege [7] (see also [8, 9]) treated this functional equation before Sincov. The most general solution $f : \mathbb{R}^2 \to \mathbb{R}$ of (43.1) is given by $f(x,y) = \phi(y) - \phi(x)$, where $\phi : \mathbb{R} \to \mathbb{R}$ is an arbitrary function. The functional equation (43.1) is known as the Sincov functional equation. The multiplicative version of (43.1) is the following: $f(x,y)\,f(y,z) = f(x,z)$ for all $x, y, z \in \mathbb{R}$; the most general solution $f : \mathbb{R}^2 \to \mathbb{R}$ of it are $f(x,y) = 0$ or $f(x,y) = \phi(y)/\phi(x)$, where $\phi : \mathbb{R} \to \mathbb{R}$ is an arbitrary nonzero function. The Sincov functional equation has many applications. The Sincov functional equation arises in the axiomatic characterization of the

P. K. Sahoo (✉)
Department of Mathematics, University of Louisville, Louisville, KY 40292, USA
e-mail: sahoo@louisville.edu

Th.M. Rassias and J. Brzdęk (eds.), *Functional Equations in Mathematical Analysis*, Springer Optimization and Its Applications 52, DOI 10.1007/978-1-4614-0055-4_43, 

maximum entropy principle (see Shore and Johnson [12]). This functional equation plays an important role in the characterization of information measures on open domains (see Ebanks, Sahoo and Sander [5]). It also arises in the study of subgroup consistent poverty indices in economics (see Foster and Shorrocks [6]). Further, the Sincov functional equation has found applications in actuarial mathematics (see Shiu [11]). In their book, Dodson and Poston [4] define an affine space as follows: *An affine space is a triple $(S,V,f)$ where $S$ is a set, $V$ a vector space and $f : S^2 \to V$ such that $f(x,y) + f(y,z) = f(x,z)$ for $x,y,z \in S$, and for all $x \in S$, the map $f_x(y) = f(x,y)$ is a bijection.* In this definition, we again encounter the Sincov functional equation. The Sincov functional equation becomes the additive Cauchy functional equation $g(x+y) = g(x) + g(y)$ if $f(x,y) = g(y-x)$, that is $f(x,y)$ is a function of $(y-x)$. Since every solution $f(x,y)$ of the Sincov functional equation is skew symmetric, therefore the Sincov functional equation (43.1) can be rewritten as

$$f(x,y) + f(y,z) + f(z,x) = 0 \tag{43.2}$$

for all $x,y,z \in \mathbb{R}$.

Let $G$ and $H$ be additive Abelian groups, and $f_1, f_2, f_3 : G^2 \to H$ and $f : G \to H$. The functional equation

$$f_1(x,y) + f_2(y,z) + f_3(z,x) = f(x+y+z) \tag{43.3}$$

holding for all $x,y,z \in G$ is a generalization of the Sincov functional equation (43.1). The functional equation (43.3) arises in certain type of aggregation problems. If $f = 0$, then the functional equation (43.3) is the pexiderized Sincov functional equation. The functional equation (43.3) is also a generalization of a functional equation studied by Baker [2].

In this paper, we determine the most general solution of the functional equation (43.3) without any regularity assumptions on the unknown functions. The method used for determining the general solution of (43.3) is elementary and interesting.

## 43.2 Preliminaries

A group $G$ is said to be 2-*divisible* if for every element $y \in G$, there exists an element $x \in G$ such that $2x = y$. If this element $x$ is unique, then $G$ is said to be *uniquely 2-divisible*. In a uniquely divisible group, this unique element $x$ is denoted by $\frac{y}{2}$.

Let $G$ and $H$ be Abelian groups. A function $a : G \to H$ is said to be additive if and only if $a(x+y) = a(x) + a(y)$ for all $x,y \in G$. If $G = H = \mathbb{R}$, then a continuous additive function $a : G \to H$ is of the form $a(x) = kx$, where $k$ is a real constant (see [1] and also [10]). A function $b : G^2 \to H$ is biaditive if it is additive in each variable [10]. If $G = H = \mathbb{R}$, then a continuous biadditive function $b : G^2 \to H$ is of the form $b(x,y) = mxy$, where $m$ is a real constant. A function $q : G \to H$ is said to be quadratic if and only if $q(x+y) + q(x-y) = 2q(x) + 2q(y)$ for all $x,y \in G$. A continuous quadratic function $q : \mathbb{R} \to \mathbb{R}$ has the form $q(x) = nx^2$ where $n$ is a real constant. It is well known that a quadratic function $q(x)$ can be represented as the diagonal of a symmetric biadditive map $b : G^2 \to H$, that is $q(x) = b(x,x)$. In the sequel, we will write an abelian group $G$ in additive notation so that 0 will denote the identity element of $G$.

The following lemma will be useful in establishing the main theorem in this paper.

**Lemma 43.1.** *Let $G$ and $H$ be groups written in additive notation. The functions $f,g : G^2 \to H$ satisfy the functional equation*

$$f(x,y) + g(y,z) - f(x,z) = 0 \tag{43.4}$$

*holding for all $x,y,z \in G$ if and only if*

$$f(x,y) = \phi(x) - \psi(y) \qquad \text{and} \qquad g(x,y) = \psi(x) - \psi(y) \tag{43.5}$$

*where $\phi, \psi : G \to H$ are arbitrary functions.*

*Proof.* It is easy to verify that (43.5) satisfies (43.4). Hence, we prove that (43.5) is only the most general solution of (43.4). From (43.4) we have

$$f(x,y) = f(x,z) - g(y,z) \tag{43.6}$$

for all $x,y,z \in G$. Letting $z = 0$, we obtain

$$f(x,y) = f(x,0) - g(y,0) \tag{43.7}$$

for all $x,y \in G$. From relation (43.7), we obtain

$$f(x,y) = \phi(x) - \psi(y) \tag{43.8}$$

where

$$\phi(x) = f(x,0) \qquad \text{and} \qquad \psi(y) = g(y,0). \tag{43.9}$$

Similarly, letting $x = 0$ in (43.4) we have

$$g(y,z) = f(0,z) - f(0,y) \tag{43.10}$$

for all $y,z \in G$. Hence

$$g(y,z) = \delta(z) - \delta(y) \tag{43.11}$$

where $\delta(x) = f(0,x)$. Next substituting (43.8) and (43.11) into (43.4), we obtain

$$\delta(z) + \psi(z) = \psi(y) + \delta(y)$$

for all $y, z \in G$. Therefore

$$\delta(z) = -\psi(z) + c \tag{43.12}$$

where $c$ is a constant. Using (43.12) in (43.11) we obtain the asserted solution (43.5). The proof of the lemma is now complete. □

The proof of the following lemma is similar to the proof of Lemma 43.1.

**Lemma 43.2.** *Let $G$ and $H$ be groups written in additive notation. The functions $f, g : G^2 \to H$ satisfy the functional equation*

$$f(x,y) + g(y,z) - g(x,z) = 0 \tag{43.13}$$

*holding for all $x, y, z \in G$ if and only if*

$$f(x,y) = \phi(x) - \phi(y) \qquad \text{and} \qquad g(x,y) = \phi(x) - \psi(y) \tag{43.14}$$

*where $\phi, \psi : G \to H$ are arbitrary functions.*

The proof of the following Lemma is also similar to the proof of Lemma 43.1.

**Lemma 43.3.** *Let $G$ and $H$ be groups written in additive notation. The functions $f_1, f_2, f_3 : G^2 \to H$ satisfy the functional equation*

$$f_1(x,y) + f_2(y,z) + f_3(z,x) = 0 \tag{43.15}$$

*for all $x, y, z \in G$ if and only if*

$$f_1(x,y) = \alpha(x) - \beta(y), \; f_2(x,y) = \beta(x) - \gamma(y), \; \text{and} \; f_3(x,y) = \gamma(x) - \alpha(y), \tag{43.16}$$

*where $\alpha, \beta, \gamma : G \to H$ are arbitrary functions.*

## 43.3 Solution of (43.3)

**Theorem 43.1.** *Suppose $G$ and $H$ are abelian groups. Moreover, suppose the division by 2 is uniquely defined in $H$. The functions $f_1, f_2, f_3 : G^2 \to H$ and $f : G \to H$ satisfy the functional equation (43.3) for all $x, y, z \in G$ if and only if*

$$f(x) = 2a(x) + 2q(x) + c_1 + c_2 + c_3, \tag{43.17}$$

$$f_1(x,y) = \alpha(x) - \beta(y) + a(x) + a(y) + 2q(x+y) - q(x) - q(y) + c_1, \tag{43.18}$$

$$f_2(x,y) = \beta(x) - \gamma(y) + a(x) + a(y) + 2q(x+y) - q(x) - q(y) + c_2, \tag{43.19}$$

$$f_3(x,y) = \gamma(x) - \alpha(y) + a(x) + a(y) + 2q(x+y) - q(x) - q(y) + c_3, \tag{43.20}$$

*where $\alpha, \beta, \gamma : G \to H$ are arbitrary functions, $q : G \to H$ is a quadratic function, $a : G \to H$ is an additive function, and $c_1, c_2, c_3$ are arbitrary elements in $H$.*

*Proof.* It is easy to check that the functions enumerated in (43.17)–(43.20) is the solution of (43.3). Next, we show that the only solution of (43.3) is (43.17)–(43.20).

Interchanging $x$ with $y$ in (43.3) we have

$$f_1(y,x)+f_2(x,z)+f_3(z,y)=f(x+y+z) \tag{43.21}$$

for all $x,y,z \in G$. Subtracting (43.21) from (43.3) we obtain

$$f_1(x,y)-f_1(y,x)+f_2(y,z)-f_3(z,y)+f_3(z,x)-f_2(x,z)=0 \tag{43.22}$$

for all $x,y,z \in G$. Equation (43.22) can be rewritten as

$$F_1(x,y)+F_2(y,z)-F_2(x,z)=0 \tag{43.23}$$

for all $x,y,z \in G$, where

$$F_1(x,y)=\frac{1}{2}\left[f_1(x,y)-f_1(y,x)\right], \tag{43.24}$$

$$F_2(x,y)=\frac{1}{2}\left[f_2(x,y)-f_3(y,x)\right]. \tag{43.25}$$

From Lemma (43.2), we obtain

$$F_1(x,y)=\phi(x)-\phi(y) \tag{43.26}$$

and

$$F_2(x,z)=\phi(x)-\psi(z) \tag{43.27}$$

where $\phi,\psi : G\to H$ are arbitrary functions.

Next adding (43.21) to (43.3) we obtain

$$f_1(x,y)+f_1(y,x)+f_2(y,z)+f_3(z,y)+f_2(x,z)+f_3(z,x)=2f(x+y+z) \tag{43.28}$$

for all $x,y,z \in G$. The functional equation (43.28) reduces to

$$L_1(x,y)+L_2(y,z)+L_2(x,z)=f(x+y+z) \tag{43.29}$$

for all $x,y,z \in G$, where

$$L_1(x,y)=\frac{1}{2}\left[f_1(x,y)+f_1(y,x)\right], \tag{43.30}$$

$$L_2(x,y)=\frac{1}{2}\left[f_2(x,y)+f_3(y,x)\right]. \tag{43.31}$$

From (43.30) we see that $L_1$ is a symmetric function, that is

$$L_1(x,y)=L_1(y,x) \tag{43.32}$$

for all $x,y \in G$.

Interchanging $y$ with $z$ in (43.29) we get

$$L_1(x,z)+L_2(z,y)+L_2(x,y)=f(x+y+z) \tag{43.33}$$

for all $x,y,z \in G$. Subtracting (43.33) from (43.29) we have

$$L_1(x,y)-L_2(x,y)+L_2(y,z)-L_2(z,y)+L_2(x,z)-L_1(x,z)=0 \tag{43.34}$$

for all $x,y,z \in G$. From Lemma 43.1 we obtain

$$L_1(x,y)-L_2(x,y)=m(x)-\ell(y), \tag{43.35}$$

$$L_2(x,y)-L_2(y,x)=\ell(x)-\ell(y), \tag{43.36}$$

where $m,\ell : G \to H$ are arbitrary functions.

Interchanging $x$ with $y$ in (43.35) we see that

$$L_1(y,x)-L_2(y,x)=m(y)-\ell(x). \tag{43.37}$$

Next using the symmetry of $L_1$ (that is using (43.32)) in (43.37), we obtain

$$L_1(x,y)-L_2(y,x)=m(y)-\ell(x). \tag{43.38}$$

Subtracting (43.38) from (43.35) we have

$$L_2(y,x)-L_2(x,y)=m(x)+\ell(x)-m(y)-\ell(y) \tag{43.39}$$

for all $x,y \in G$. Comparing (43.39) with (43.36) and simplifying we obtain

$$m(x)+2\ell(x)=m(y)+2\ell(y) \tag{43.40}$$

for all $x,y \in G$. Hence, from (43.40) we have

$$m(x)=-2\ell(x)+d \tag{43.41}$$

where $d$ is a constant. Thus, (43.41) in (43.35) yields

$$L_1(x,y)-L_2(x,y)=-2\ell(x)-\ell(y)+d. \tag{43.42}$$

Letting $z=0$ in (43.29) we obtain

$$\begin{aligned} L_1(x,y) &= f(x+y)-L_2(y,0)-L_2(x,0) \\ &= f(x+y)-r(x)-r(y), \end{aligned} \tag{43.43}$$

where

$$r(x)=L_2(x,0). \tag{43.44}$$

Letting $y = 0$ in (43.36) we see that

$$L_2(0,x) = L_2(x,0) - \ell(x) + \ell(0). \tag{43.45}$$

Letting $x = 0$ in (43.29) we have

$$L_2(y,z) = f(y+z) - L_1(0,y) - L_2(0,z). \tag{43.46}$$

Using (43.32) and (43.45) in (43.46) we obtain

$$L_2(y,z) = f(y+z) - L_1(y,0) - L_2(z,0) + \ell(z) - \ell(0) \tag{43.47}$$

for all $y, z \in G$. Using (43.42) in (43.47) we have

$$L_2(y,z) = f(y+z) - L_2(y,0) + 2\ell(y) + \ell(z) - d - L_2(z,0). \tag{43.48}$$

Using (43.44) in (43.48) we have

$$L_2(y,z) = f(y+z) - r(y) - r(z) + 2\ell(y) + \ell(z) - d \tag{43.49}$$

for all $y, z \in G$. Again from (43.47), (43.45) and (43.43) we see that

$$\begin{aligned} L_2(y,z) &= f(y+z) - L_1(0,y) - L_2(0,z) \\ &= f(y+z) - f(y) + r(y) - r(z) + \ell(z) + r(0) - \ell(0) \end{aligned} \tag{43.50}$$

for all $y, z \in G$. Comparing (43.49) and (43.50) we have

$$f(y) = 2r(y) - 2\ell(y) - \ell(0) + r(0) + d. \tag{43.51}$$

Using (43.51) in (43.49) we obtain

$$L_2(y,z) = 2r(y+z) - r(y) - r(z) - 2\ell(y+z) + 2\ell(y) + \ell(z) - \ell(0) + r(0). \tag{43.52}$$

Using (43.43), (43.51) and (43.52) in (43.29) we obtain

$$\begin{aligned} r(x+y+z) - \ell(x+y+z) &= r(x+y) + r(y+z) + r(z+x) - r(x) - r(y) - r(z) + r(0) \\ &\quad - \ell(x+y) - \ell(y+z) - \ell(z+x) + \ell(x) + \ell(y) + \ell(z) - \ell(0) \end{aligned} \tag{43.53}$$

for all $x, y, z \in G$. Define $g : G \to H$ by

$$g(x) = r(x) - r(0) - \ell(x) + \ell(0) \tag{43.54}$$

for all $x \in G$. Then $g(0) = 0$. Using (43.54) in (43.53) we have

$$g(x+y+z) = g(x+y) + g(y+z) + g(z+x) - g(x) - g(y) - g(z) \tag{43.55}$$

for all $x, y, z \in G$.

Substituting $z = -y$ in (43.55) and using the fact that $g(0) = 0$, we get

$$g(x+y) + g(x-y) = 2g(x) + g(y) + g(-y) \tag{43.56}$$

for all $x, y \in G$. Replacing $y$ with $-y$ and $x$ with $-x$ in (43.56) we obtain

$$g(-x-y) + g(-x+y) = 2g(-x) + g(-y) + g(y) \tag{43.57}$$

for all $x, y \in G$. Adding (43.56) and (43.57) we have

$$\begin{aligned} g(x+y) + g(-x-y) + g(x-y) + g(-x+y) \\ = 2g(x) + 2g(-x) + 2g(y) + 2g(-y) \end{aligned} \tag{43.58}$$

for all $x, y \in G$. The functional equation (43.58) reduces to

$$q(x+y) + q(x-y) = 2q(x) + 2q(y) \tag{43.59}$$

for all $x, y \in G$ where

$$q(x) = \frac{1}{2}\left[g(x) + g(-x)\right]. \tag{43.60}$$

In view of (43.59), the function $q : G \to H$ is a quadratic function.

Next, substituting $z = -x - y$ in (43.55) and using the fact that $g(0) = 0$, we obtain

$$g(x+y) - g(-x-y) = g(x) - g(-x) + g(y) - g(-y) \tag{43.61}$$

for all $x, y \in G$. Define a function $a : G \to H$ by

$$a(x) = \frac{1}{2}\left[g(x) - g(-x)\right]. \tag{43.62}$$

Then (43.62) in (43.61) yields

$$a(x+y) = a(x) + a(y) \tag{43.63}$$

for all $x, y \in G$. Therefore, $a : G \to H$ is an additive mapping on $G$.

From (43.62), (43.60) and (43.54), we obtain

$$r(x) = a(x) + q(x) + \ell(x) + r(0) - \ell(0). \tag{43.64}$$

Letting (43.64) into (43.51) we get

$$f(x) = 2a(x) + 2q(x) + 3r(0) - 3\ell(0) + d. \tag{43.65}$$

Next using (43.64) and (43.65) in (43.43) we get

$$L_1(x,y) = a(x) + a(y) + 2q(x+y) - q(x) - q(y)$$
$$- \ell(x) - \ell(y) + r(0) - \ell(0) + d. \tag{43.66}$$

Similarly, using (43.64) in (43.52) we obtain

$$L_2(x,y) = a(x) + a(y) + 2q(x+y) - q(x) - q(y) + \ell(x) + r(0) - \ell(0). \tag{43.67}$$

Next, from (43.24), (43.25), (43.30) and (43.31) we see that

$$f_1(x,y) = L_1(x,y) + F_1(x,y), \tag{43.68}$$
$$f_2(x,y) = L_2(x,y) + F_2(x,y), \tag{43.69}$$
$$f_3(x,y) = L_2(x,y) - F_2(x,y). \tag{43.70}$$

Using (43.66) and (43.26) in (43.68) we obtain

$$f_1(x,y) = \alpha(x) - \beta(y) + a(x) + a(y) + 2q(x+y) - q(x) - q(y) + c_1, \tag{43.71}$$

where

$$\alpha(x) = \phi(x) - \ell(x) - \alpha_0, \qquad \beta(y) = \phi(y) + \ell(y) - \beta_0 \tag{43.72}$$

and

$$c_1 = d + r(0) - \ell(0) + \alpha_0 - \beta_0$$

for some constants $\alpha_0, \beta_0$ in $H$. Using (43.67) and (43.27) in (43.69) we have

$$f_2(x,y) = \beta(x) - \gamma(y) + a(x) + a(y) + 2q(x+y) - q(x) - q(y) + c_2, \tag{43.73}$$

where $\gamma(y) = \psi(y) - \gamma_0$ and

$$c_2 = r(0) - \ell(0) + \beta_0 - \gamma_0.$$

Similarly, letting (43.67) and (43.27) in (43.70) we obtain

$$f_3(x,y) = \gamma(x) - \alpha(y) + a(x) + a(y) + 2q(x+y) - q(x) - q(y) + c_2, \tag{43.74}$$

where

$$c_3 = r(0) - \ell(0) + \gamma_0 - \alpha_0.$$

Now inserting (43.65), (43.71), (43.73) and (43.74) into the functional equation (43.3), we see that

$$3r(0) - 3\ell(0) + d = c_1 + c_2 + c_3.$$

Replacing the constant $3r(0)-3\ell(0)+d$ by $c_1+c_2+c_3$ in (43.65) we have the asserted solution (43.17)–(43.20). Now the proof of the theorem is complete.

The following remark is in order.

*Remark 43.1.* The general solution (43.17)–(43.20) of the functional equation (43.3) can also be represented as follows:

$$\begin{aligned} f(x) &= 2a(x)+2b(x,x)+c_1+c_2+c_3,\\ f_1(x,y) &= \alpha(x)-\beta(y)+a(x)+a(y)+b(x,x)+4b(x,y)+b(y,y)+c_1,\\ f_2(x,y) &= \beta(x)-\gamma(y)+a(x)+a(y)+b(x,x)+4b(x,y)+b(y,y)+c_2,\\ f_3(x,y) &= \gamma(x)-\alpha(y)+a(x)+a(y)+b(x,x)+4b(x,y)+b(y,y)+c_3,\end{aligned}$$

where $\alpha,\beta,\gamma,c_1,c_2,c_3$ are as in Theorem 43.1 and $b: G^2\to H$ is a biadditive function.

The following corollaries easily follow from Theorem 43.1.

**Corollary 43.1.** *The continuous functions $f_1,f_2,f_3:\mathbb{R}^2\to\mathbb{R}$ and $f:\mathbb{R}\to\mathbb{R}$ satisfy the functional equation* (43.3) *for all $x,y,z\in\mathbb{R}$ if and only if*

$$\begin{aligned} f(x) &= 2ax+2bx^2+c_1+c_2+c_3,\\ f_1(x,y) &= \alpha(x)-\beta(y)+ax+ay+bx^2+4bxy+by^2+c_1,\\ f_2(x,y) &= \beta(x)-\gamma(y)+ax+ay+bx^2+4bxy+by^2+c_2,\\ f_3(x,y) &= \gamma(x)-\alpha(y)+ax+ay+bx^2+4bxy+by^2+c_3,\end{aligned}$$

*where $\alpha,\beta,\gamma:\mathbb{R}\to\mathbb{R}$ are arbitrary continuous functions, and $a,b,c_1,c_2,c_3$ are arbitrary real constants.*

**Corollary 43.2.** *Let $G$ and $H$ be abelian groups, and the division by 2 is uniquely defined in $H$. The functions $f: G^2\to H$ and $g: G\to H$ satisfy the functional equation*

$$f(x_1,x_2)+f(x_2,x_3)+\cdots+f(x_n,x_1)=g(x_1+x_2+\cdots+x_n) \tag{43.75}$$

*holding for all $x_1,x_2,\ldots,x_n\in G$ and $n\geq 3$ if and only if*

$$g(x)=2a(x)+2b(x,x)+nc,$$

$$f(x,y)=\begin{cases}\alpha(x)-\alpha(y)+a(x)+a(y)+b(x,x)\\ \qquad\qquad +4b(x,y)+b(y,y)+c, & \text{if } n=3;\\ \alpha(x)-\alpha(y)+a(x)+a(y)+c, & \text{if } n>3,\end{cases}$$

*where $\alpha: G\to H$ is an arbitrary function, $b: G^2\to H$ is a biadditive function, $a: G\to H$ is an additive function, and $c$ is an arbitrary element in $H$.*

*Proof.* Letting $x_4 = x_5 = \cdots = x_n = 0$ in (43.75) we get

$$\begin{aligned} f(x_1,x_2) + f(x_2,x_3) + f(x_3,0) + (n-4)f(0,0) \\ + f(0,x_1) = g\big(x_1+x_2+x_3\big) \end{aligned} \tag{43.76}$$

for all $x_1, x_2, x_3 \in G$. Define

$$h(x_3,x_1) = f(x_3,0) + f(0,x_1) + (n-4)f(0,0). \tag{43.77}$$

Then (43.76) by (43.77) reduces to

$$f(x_1,x_2) + f(x_2,x_3) + h(x_3,x_1) = g\big(x_1+x_2+x_3\big) \tag{43.78}$$

for all $x_1, x_2, x_3 \in G$. Now applying the Theorem 43.1 we have

$$g(x) = 2a(x) + 2b(x,x) + 3c \tag{43.79}$$

and

$$f(x,y) = \alpha(x) - \alpha(y) + a(x) + a(y) + b(x,x) + 4b(x,y) + b(y,y) + c, \tag{43.80}$$

where $c_1 = c_2 = c_3 = c$. If $n = 3$, then it is easy to check that (43.79) and (43.80) are solution of (43.75) and we have the asserted solution for this case. If $n > 3$, then again substituting (43.79) and (43.80) into (43.75) we obtain $b(x,y) = 0$ for all $x, y \in G$. Hence, we have the asserted solution in this case.

**Corollary 43.3.** *Let $G$ and $H$ be Abelian groups. Further, suppose $G$ is uniquely $n$-divisible and $H$ is uniquely 2-divisible. Let $\bar{x} = (x_1 + x_2 + \cdots + x_n)/n$. The function $f : G^2 \to H$ satisfies the functional equation*

$$f(x_1,x_2) + f(x_2,x_3) + \cdots + f(x_n,x_1) = n f(\bar{x},\bar{x}) \tag{43.81}$$

*holding for all $x_1, x_2, \ldots, x_n \in G$ and $n \geq 3$ if and only if*

$$f(x,y) = \begin{cases} \alpha(x) - \alpha(y) + a(x) + a(y) & \\ \qquad\qquad + b(x,x) + 4b(x,y) + b(y,y) + c, & \text{if } n = 3; \\ \alpha(x) - \alpha(y) + a(x) + a(y) + c, & \text{if } n > 3, \end{cases}$$

*where $\alpha : G \to H$ is an arbitrary function, $b : G^2 \to H$ is a biadditive function, $a : G \to H$ is an additive function, and $c$ is an arbitrary element in $H$.*

**Acknowledgement** The work was partially supported by an IRI grant from the Office of the VP for Research, and an intramural grant from the College of Arts and Sciences of the University of Louisville.

## References

1. Aczél, J.: Lectures on Functional Equations and Their Applications. Dover Publications, New York (2006)
2. Baker, J.: A functional equation, Solution to problem E 2607. Amer. Math. Monthly **10**, 824–825 (1977)
3. Cantor, M.: Funktionalgleichungen mit drei von einander unabhängigen veränderlichen. Z. Math. Phys. **41**, 161–163 (1896)
4. Dobson, C.T.J., Poston, T.: Tensor Geometry. Pitman, London (1977)
5. Ebanks, B., Sahoo, P., Sander, W.: Characterizations of Information Measures. World Scientific Publishing Co., NJ (1998)
6. Foster, J.E., Shorrocks, A.F.: Subgroup consistent poverty indices. Econometrica **59**, 697–709 (1991)
7. Frege, G.: Rechungsmethoden, die sich auf eine Erweiterung des Grössebegriffes gründen. Verlag Friedrich Frommann, Jena (1874)
8. Gronau, D.: A remark on Sincov's functional equation. Not. S. Afr. Math. Soc. **31**, 1–8 (2000)
9. Gronau, D.: Gottlob Frege, a pioneer in iteration theory. In: Iteration Theory (ECIT 94) (Opava), pp. 105–119. Grazer Math. Ber. **334** (1997)
10. Sahoo, P.K., Riedel, T.R.: Mean Value Theorems and Functional Equations. World Scientific Publishing Co., NJ (1998)
11. Shiu, E.S.W.: Some functional equations in actuarial mathematics. Actuarial Research Clearing House **2**, 191–208 (1988)
12. Shore, J.E., Johnson, R.W.: Axiomatic derivation of the principle of maximum entropy and the principle of minimum cross-entropy. IEEE Trans. Inform. Theory **26**, 26–37 (1980)
13. Sincov, D.M.: Notes sur la calcul fonctionnel (in Russian). Bull. Soc. Phys.–Math. Kazam. **13**, 48–72 (1903)
14. Sincov, D.M.: Über eine funktionalgleichung. Arch. Math. Phys. **6**, 216–217 (1903)

# Chapter 44
# Invariance in Some Families of Means

**Gheorghe Toader, Iulia Costin, and Silvia Toader**

*Dedicated to the memory of S.M. Ulam*
*on the 100th anniversary of his birth*

**Abstract** A mean $P$ is $(M,N)$-invariant if $P(M,N) = P$. In the same time, the mean $N$ is called complementary to $M$ with respect to $P$. For the determination of complementaries, three methods have been used: the direct calculation, the methods of functional equations, and the series expansion of means. In the current paper, we consider the method of series expansion of means to study the invariance in the family of extended logarithmic means.

**Keywords** Extended logarithmic mean • Complementary mean • Invariance in a class of means

**Mathematics Subject Classification (2000):** Primary 26E60

## 44.1 Introduction

A *mean* is a function $M : \mathbb{R}_+^2 \to \mathbb{R}_+$, with the property

$$\min(a,b) \leq M(a,b) \leq \max(a,b), \qquad \forall a,b > 0.$$

Each mean is *reflexive*, that is

$$M(a,a) = a, \qquad \forall a > 0.$$

This isalso used as the definition of $M(a,a)$.

---

G. Toader (✉) • S. Toader
Department of Mathematics, Technical University of Cluj-Napoca, Romania
e-mail: Gheorghe.Toader@math.utcluj.ro; Silvia.Toader@math.utcluj.ro

I. Costin
Department of Computer Sciences, Technical University of Cluj-Napoca, Romania
e-mail: Iulia.Costin@cs.utcluj.ro

Th.M. Rassias and J. Brzdęk (eds.), *Functional Equations in Mathematical Analysis*, Springer Optimization and Its Applications 52, DOI 10.1007/978-1-4614-0055-4_44,

A mean is *symmetric* if

$$M(b,a) = M(a,b), \qquad \forall a,b > 0;$$

it is *strict* if

$$[M(a,b) - a]\,[M(a,b) - b] \neq 0, \qquad \text{for } a \neq b.$$

A reflexive function $M : \mathbb{R}_+^2 \to \mathbb{R}_+$ is called also a *pre-mean* (or a *generalized mean*).

We shall refer here to the following families of means:

– the weighted power means $\mathscr{P}_{p;\lambda}$, defined by

$$\mathscr{P}_{p;\lambda}(a,b) = [\lambda a^p + (1-\lambda) b^p]^{1/p}, \lambda \in (0,1), \qquad p \neq 0;$$

– the weighted arithmetic means $\mathscr{A}_\lambda = \mathscr{P}_{1;\lambda}$;
– the weighted Lehmer means $\mathscr{C}_{p;\lambda}$, defined by

$$\mathscr{C}_{p;\lambda}(a,b) = \frac{\lambda a^p + (1-\lambda) b^p}{\lambda a^{p-1} + (1-\lambda) b^{p-1}}, \qquad \lambda \in (0,1);$$

– the extended logarithmic means $\mathscr{L}_p$ defined by

$$\mathscr{L}_p(a,b) = \left( \frac{1}{p} \frac{a^p - b^p}{\ln a - \ln b} \right)^{\frac{1}{p}}, \qquad p \neq 0, a \neq b.$$

The symmetric means $\mathscr{P}_{p;1/2}, \mathscr{A}_{1/2}$ and $\mathscr{C}_{p;1/2}$ are writen simply as $\mathscr{P}_p$, $\mathscr{A}$ and $\mathscr{C}_p$, respectively. The special case $\mathscr{L}_1 = \mathscr{L}$ is the logarithmic mean, while the limiting case

$$\lim_{p \to 0} \mathscr{L}_p(a,b) = \mathscr{L}_0(a,b) = \sqrt{ab} = \mathscr{G}(a,b),$$

is the geometric mean.

Given three functions $M, N, P : \mathbb{R}_+^2 \to \mathbb{R}_+$, we can compose them, obtaining a new function $P(M,N) : \mathbb{R}_+^2 \to \mathbb{R}_+$, defined by

$$P(M,N)(a,b) = P(M(a,b), N(a,b)), \qquad \forall a,b > 0.$$

If $M, N, P$ are means (pre-means) then $P(M,N)$ is also a mean (pre-mean).

**Definition 44.1.** The function $P$ is called $(M,N)$-*invariant* if it verifies

$$P(M,N) = P.$$

In the same case, the function $N$ is called complementary to $M$ with respect to $P$ (or $P$-*complementary* to $M$).

These properties are related to the following problem. Given two functions $M, N : \mathbb{R}_+^2 \to \mathbb{R}_+$ and $a_0, b_0 \in \mathbb{R}_+$, we can define a (Gaussian) *double sequence* by:

$$a_{n+1} = M(a_n, b_n), b_{n+1} = N(a_n, b_n), \qquad \forall n > 0. \tag{44.1}$$

If $M,N$ are means which have some properties (for instance,one of them is continuous and strict (see [13]), the sequences $(a_n)_{n\geq 0}$ and $(b_n)_{n\geq 0}$ are convergent to a common limit $P(a_0,b_0)$. Moreover, $P$ also defines a mean. K. F. Gauss was the first author who related the problem of determination of the common limit of the double sequences, to the invariance of the mean $P$ with respect to $(M,N)$, in the special case in which $M$ is the arithmetic mean while $N$ is the geometric mean. A general *invariance principle* was proved in [1]. It was generalized for pre-means in [15]:

**Theorem 44.1.** *Let $P$ be a continuous pre-mean and $M$ and $N$ be two functions such that $P$ is $(M,N)$-invariant. If the sequences $(a_n)_{n\geq 0}$ and $(b_n)_{n\geq 0}$ defined by (44.1) are convergent to a common limit $l$, then $l = P(a_0,b_0)$.*

As it is known from the above mentioned classical example of the arithmetic-geometric mean of Gauss, the determination of a $(M,N)$-invariant mean $P$ is a very difficult problem (see [1]). That is why we study the (equivalent) problem of finding a mean $N$ which is complementary to $M$ with respect to $P$. For the determination of complementaries, three methods have been used: the direct calculation (see [14]), the methods of functional equations (see [9]), and the series expansion of means (see [8]). In the current paper, we use the method of series expansion of means.

Given a fixed mean $M$, we construct a family of means by taking for every bijection $f : \mathbb{R}_+ \to \mathbb{R}_+$, the mean $M(f)$ defined by

$$M(f)(a,b) = f^{-1}(M(f(a), f(b))), \qquad \forall a,b > 0.$$

If we take $M = \mathscr{A}_\lambda$, we get the family of weighted quasi-arithmetic means. A simpler case is obtained by taking the sub-family given by $f(x) = x^p$, thus

$$M_p(a,b) = [M(a^p,b^p)]^{1/p}, \qquad \forall a,b > 0, p \neq 0.$$

The family of weighted power means is of this type. In what follows we study the problem of invariance for the class $\mathscr{L}_p$ which is also of this type with $M = \mathscr{L}$.

## 44.2 Invariance in a Family of Means

Given a family $\mathscr{X}$ of means, we can consider three problems of invariance:

- A first problem is that of the study of *invariance* of a given mean $P$ *with respect to the family* $\mathscr{X}$. This means the determination of all the pairs of means $(M,N)$ from $\mathscr{X}$ such that $P$ is $(M,N)$-invariant.
- A second problem is named *invariance in the family* $\mathscr{X}$. It consists in determining all the triples of means $(P,M,N)$ from $\mathscr{X}$ such that $P$ is $(M,N)$-invariant.

– A third type of problem was called as *reproducing identities* and assumes the determination of quadruples of means $(P,M,N,Q)$ from $\mathscr{X}$ such that

$$P(M,N) = Q.$$

This problem has the trivial solution $P(M,M) = M$, the solutions of the invariance problem $P(M,N) = P$, but it also can have other solutions.

Many problems of the first type were formulated as functional equations. The first one was related to the invariance of arithmetic mean $\mathscr{A}$ with respect to the family of quasi-arithmetic means $\mathscr{A}(f)$. It was solved in [11, 12] for analytic functions $f$ and in [9] for the second order continuously differentiable functions $f$. It was called the Matkowski–Sutô problem (see [5]). Many other problems of this type were also solved.

The first problem of the second type was solved in [8] for the family of symmetric Lehmer means $\mathscr{C}_p$, finding three solutions. The problem was studied in [4] for the general case of weighted Lehmer means $\mathscr{C}_{p;\lambda}$. Fifteen solutions were found, including pre-means (for $\lambda \notin [0,1]$). For the family of weighted power means $\mathscr{P}_{p;\lambda}$, the problem of invariance was solved in [3]. There are ten solutions, some of them trivial or involve pre-means. To solve these last two problems, were used the coefficients of the series development of the means up to the degree five. The invariance in the class of weighted quasi-arithmetic means was solved in [7], while that of generalized Beckenbach–Gini means in [10], using functional equations methods.

The first reproducing identities problem was studied in [2] for the families of Lehmer means and for that of power means.

## 44.3 Series Expansion of Means

For the study of some problems related to means, in [8] the power series expansion of the normalized function $M(1,1-t)$ is used. When it is impossible to determine all the coefficients, a recurrence relation for the coefficients is very useful. Such a formula was given by Euler (see [6]) in the following:

**Theorem 44.2.** *If $p$ is a real number and the function $f$ has the Taylor series*

$$f(x) = \sum_{n=0}^{\infty} a_n\, x^n, \qquad [f(x)]^p = \sum_{n=0}^{\infty} b_n\, x^n,$$

*then we have the recurrence relation*

$$\sum_{k=0}^{n} [k(p+1)-n]\, a_k\, b_{n-k} = 0, \qquad n \geq 0.$$

We need the series expansion of the extended logarithmic mean.

**Theorem 44.3.** *The first coefficients of the series expansion of the extended logarithmic mean $\mathscr{L}_p$ (with $p \neq 0$) are*

$$\begin{aligned}\mathscr{L}_p(1,1-x) =& 1 - \frac{1}{2}x + \frac{p-3}{24}x^2 + \frac{p-3}{48}x^3 \\ &- \frac{(p-5)\left(2p^2+5p-45\right)}{5760}x^4 \\ &- \frac{6p^3-15p^2-90p+315}{11520}x^5 + \cdots.\end{aligned}$$

*Proof.* Denote

$$\mathscr{L}_p(1,1-x) = \sum_{n=0}^{\infty} h_n x^n = \left(\frac{1-(1-x)^p}{-p\ln(1-x)}\right)^{\frac{1}{p}} = \left(\sum_{n=0}^{\infty} l_n x^n\right)^{\frac{1}{p}}.$$

This gives

$$1 - \sum_{n=0}^{\infty}(-1)^n \binom{p}{n} x^n = p\left(\sum_{n=0}^{\infty} l_n x^n\right)\left(\sum_{n=1}^{\infty} \frac{x^n}{n}\right),$$

or

$$\frac{(-1)^n}{n+1}\binom{p-1}{n} = \sum_{k=0}^{n} \frac{l_{n-k}}{k+1},$$

which allows the step by step determination of the coefficients $l_n$,

$$l_0 = 1, \quad l_1 = -\frac{p}{2}, \quad l_2 = \frac{p}{12}(2p-3), \quad l_3 = -\frac{p}{24}(p-2)^2,$$

$$\begin{aligned}l_4 &= \frac{p}{720}\left(6p^3-45p^2+110p-90\right), \\ l_5 &= -\frac{p}{1440}(p-4)^2\left(2p^2-8p+9\right), \ldots.\end{aligned}$$

Using Euler's formula for $l_n$ and $h_n$,

$$h_n = \frac{1}{n}\sum_{k=1}^{n}\left[k\left(1+\frac{1}{p}\right)-n\right] l_k h_{n-k},$$

we get the desired result. □

*Remark 44.1.* For $p = 0$ we get the first coefficients of the series expansion of the geometric mean $\mathscr{G} = \mathscr{L}_0$.

## 44.4 Complementaries with Respect to $\mathscr{L}_p$

Denote the $\mathscr{L}_p$-complementary of the mean $M$ by ${}^{\mathscr{L}(p)}M$. We establish the following theorem.

**Theorem 44.4.** *If the mean M has the series expansion*

$$M(1,1-x) = 1 + \sum_{n=1}^{\infty} \alpha_n x^n,$$

*then the first coefficients of the series expansion of* ${}^{\mathscr{L}(p)}M$ *are*

$$\begin{aligned}
{}^{\mathscr{L}(p)}M(1,1-x) =& 1 - (\alpha_1+1)x + \frac{1}{3}\left[\alpha_1(\alpha_1+1)(3-p) - 3\alpha_2\right]x^2 \\
&+ \frac{1}{18}\Big[-2(p-3)^2\alpha_1^3 + 3\left(5p-6-p^2\right)\alpha_1^2 + p(3-p)\alpha_1 \\
&+ 12(3-p)\alpha_1\alpha_2 - 18\alpha_3 + 6(3-p)\alpha_2\Big]x^3 \\
&+ \frac{1}{540}\Big[\left(540-525p+180p^2-19p^3\right)\alpha_1^4 \\
&+ \left(540-690p+300p^2-38p^3\right)\alpha_1^3 \\
&+ p\left(-135+120p-21p^2\right)\alpha_1^2 + 2p\left(15-p^2\right)\alpha_1 \\
&+ 180(3-p)\alpha_2^2 + 30p(3-p)\alpha_2 - 180(3-p)^2\alpha_1^2\alpha_2 \\
&- 180(3-p)(2-p)\alpha_1\alpha_2 + 360(3-p)\alpha_1\alpha_3 \\
&+ 180(3-p)\alpha_3 - 540\alpha_4\Big]x^4 + \cdots.
\end{aligned}$$

*Proof.* Given the coefficients $\alpha_n$ we deduce the coefficients $a_n$ of $M^p$ by Euler's formula:

$$a_n = \frac{1}{n}\sum_{k=1}^{n}\left[k(p+1)-n\right]\alpha_k a_{n-k}, \qquad n \geq 1.$$

Denoting ${}^{\mathscr{L}(p)}M = N$ we have the condition $\mathscr{L}_p(M,N) = \mathscr{L}_p$, thus

$$\frac{M^p - N^p}{\mathscr{L}_p^p} = \ln M^p - \ln N^p.$$

For

$$N^p(1,1-x) = 1 + \sum_{n=1}^{\infty} b_n x^n,$$

we have

$$\frac{\sum_{n=0}^{\infty} a_n x^n - \sum_{n=0}^{\infty} b_n x^n}{\left(\sum_{n=0}^{\infty} h_n x^n\right)^p} = \ln\left(\sum_{n=0}^{\infty} a_n x^n\right) - \ln\left(\sum_{n=0}^{\infty} b_n x^n\right).$$

Substituting

$$\frac{\sum_{n=0}^{\infty} a_n x^n - \sum_{n=0}^{\infty} b_n x^n}{\left(\sum_{n=0}^{\infty} h_n x^n\right)^p} = \sum_{n=0}^{\infty} c_n x^n,$$

and using the notations from the previous proof, we obtain

$$\sum_{n=1}^{\infty} (a_n - b_n) x^n = \left(\sum_{n=0}^{\infty} l_n x^n\right)\left(\sum_{n=0}^{\infty} c_n x^n\right),$$

thus

$$c_0 = 0, c_1 = a_1 - b_1, c_n = a_n - b_n - \sum_{k=1}^{n-1} l_k c_{n-k}, \qquad n > 1.$$

Differentiating the equality

$$\sum_{n=1}^{\infty} c_n x^n = \ln\left(\sum_{n=0}^{\infty} a_n x^n\right) - \ln\left(\sum_{n=0}^{\infty} b_n x^n\right),$$

we deduce

$$\sum_{n=1}^{\infty} n c_n x^{n-1} = \frac{\sum_{n=1}^{\infty} n a_n x^{n-1}}{\sum_{n=0}^{\infty} a_n x^n} - \frac{\sum_{n=1}^{\infty} n b_n x^{n-1}}{\sum_{n=0}^{\infty} b_n x^n},$$

thus

$$\left(\sum_{n=0}^{\infty} a_n x^n\right)\left(\sum_{n=0}^{\infty} b_n x^n\right)\left[\sum_{n=0}^{\infty} (n+1) c_{n+1} x^n\right]$$
$$= \left(\sum_{n=0}^{\infty} b_n x^n\right)\left[\sum_{n=0}^{\infty} (n+1) a_{n+1} x^n\right] - \left(\sum_{n=0}^{\infty} a_n x^n\right)\left[\sum_{n=0}^{\infty} (n+1) b_{n+1} x^n\right],$$

or

$$\sum_{n=0}^{\infty} \left[d_0 (n+1) c_{n+1} + d_1 n c_n + \cdots + d_n c_1\right] x^n = \sum_{n=0}^{\infty} [b_0 (n+1) a_{n+1}$$
$$+ b_1 n a_n + \cdots + b_n a_1 - a_0 (n+1) b_{n+1} - a_1 n b_n - \cdots - a_n b_1] x^n,$$

where $d_n = a_0 b_n + a_1 b_{n-1} + \cdots + a_n b_0$. The equality of the coefficients of the same power of $x$ allows the step by step determination of $b_n$. Given the coefficients $b_n$ we deduce the coefficients $\beta_n$ of $N$ again by Euler's formula:

$$\beta_n = \frac{1}{n} \sum_{k=1}^{n} \left[ k\left(\frac{1}{p}+1\right) - n \right] b_k \beta_{n-k}, \qquad n \geq 1.$$

□

*Remark 44.2.* For $p = 0$ we obtain the first coefficients of the series expansion of

$$^{\mathscr{L}(p)}M = ^{\mathscr{G}} M.$$

**Theorem 44.5.** *The first terms of the series expansion of the $\mathscr{L}_p$-complementary of $\mathscr{L}_q$ are*

$$\begin{aligned} ^{\mathscr{L}(p)}\mathscr{L}_q(1,1-x) =& 1 - \frac{x}{2} + (2p-q-3)\,\frac{x^2}{48}(2+x) \\ & - \left[22p^3 - 20(q+3)p^2 + 10\left(q^2+6q-42\right)p \right. \\ & \left. - 3\left(2q^3 + 5q^2 - 70q - 225\right)\right] \frac{x^4}{17280} + \cdots. \end{aligned}$$

**Corollary 44.1.** *We have $^{\mathscr{L}(p)}\mathscr{L}_q = \mathscr{L}_r$ if and only if $p = 0, r = -q$.*

*Proof.* Equating the coefficients of $x^2$ we get the condition: $2p = q + r$. Then, equating the coefficients of $x^k, k = 0, 1, \ldots, 5$, we get the solution. □

*Remark 44.3.* The problem of invariance in the class of extended logarithmic means has the only solution $\mathscr{L}_0(\mathscr{L}_q, \mathscr{L}_{-q}) = \mathscr{L}_0$.

**Acknowledgement** These researches are supported by the Project PN2-Partenership Nr. 11018 MoDef.

## References

1. Borwein, J.M., Borwein, P.B.: Pi and the AGM - a Study in Analytic Number Theory and Computational Complexity. John Wiley & Sons, New York (1987)
2. Brenner, J.L., Mays, M.E.: Some reproducing identities for families of mean values. Aequationes Math. **33**, 106–113 (1987)
3. Costin, I.: Invariance in the class of weighted power means. In: Negru, V., Jebelean, T., Petcu, D., Zaharie, D. (eds.) Ninth International Symposium on Symbolic and Numeric Algorithms for Scientific Computing (SYNASC 2007), Timisoara, Romania, 2007, pp. 131–133. IEEE Computer Society, Los Alamos (2007)
4. Costin, I., Toader, G.: Invariance in the class of weighted Lehmer means. J. Ineq. Pure Appl. Math. **9**, Issue 2, Article 54, 7 pp. (2008)

5. Daróczy, Z., Páles, Zs.: Gauss-composition of means and the solution of the Matkowski-Sutô problem. Publ. Math. Debrecen **61**, 157–218 (2002)
6. Gould, H.W.: Coefficient identities for powers of Taylor and Dirichlet series. Amer. Math. Monthly **81**, 3–14 (1974)
7. Jarczyk, J.: Invariance in the class of weighted quasi-arithmetic means with continuous generators. Publ. Math. Debrecen **71**, 279–294 (2007)
8. Lehmer, D.H.: On the compounding of certain means. J. Math. Anal. Appl. **36**, 183–200 (1971)
9. Matkowski, J.: Invariant and complementary quasi-arithmetic means. Aequationes Math. **57**, 87–107 (1999)
10. Matkowski, J.: On invariant generalized Beckenbach-Gini means. In: Daróczy Z., Páles, Zs. (eds.) Functional Equations - Results and Advances, Advances in Mathematics, vol. 3, pp. 219–230. Kluwer Acad. Publ., Dordrecht (2002)
11. Sutô, O.: Studies on some functional equations I. Tôhoku Math. J. **6**, 1–15 (1914)
12. Sutô, O.: Studies on some functional equations II. Tôhoku Math. J. **6**, 82–101 (1914)
13. Toader, G.: Some remarks on means. Anal. Numér. Théor. Approx. **20**, 97–109 (1991)
14. Toader, G., Toader, S.: Greek Means and the Arithmetic-Geometric Mean. RGMIA Monographs, Victoria University, (2005) http://staff.vu.edu.au/RGMIA/monographs/toader.htm
15. Toader, G., Toader, S.: Means and generalized means. J. Ineq. Pure Appl. Math. **8**, Issue 2, Article 45, 6 pp. (2007)

# Chapter 45
# On a Hilbert-Type Integral Inequality

**Bicheng Yang**

*Dedicated to the memory of S.M. Ulam*
*on the 100th anniversary of his birth*

**Abstract** By introducing some parameters and using the way of weight function, a new Hilbert-type integral inequality with a combination kernel is given, which is a best extension of Hilbert's integral inequality. As applications, the equivalent form is considered.

**Keywords** Hilbert's integral inequality • Weight function • Parameter

**Mathematics Subject Classification (2000):** Primary 26D15

## 45.1 Introduction

If

$$0 < \int_0^\infty f^2(x)\mathrm{d}x < \infty \qquad \text{and} \qquad 0 < \int_0^\infty g^2(x)\mathrm{d}x < \infty,$$

then we have [1]:

$$\int_0^\infty \int_0^\infty \frac{f(x)g(y)}{x+y}\mathrm{d}x\mathrm{d}y < \pi \left( \int_0^\infty f^2(x)\mathrm{d}x \int_0^\infty g^2(x)\mathrm{d}x \right)^{1/2}, \tag{45.1}$$

$$\int_0^\infty \int_0^\infty \frac{f(x)g(y)}{\max\{x,y\}}\mathrm{d}x\mathrm{d}y < 4 \left( \int_0^\infty f^2(x)\mathrm{d}x \int_0^\infty g^2(x)\mathrm{d}x \right)^{1/2}, \tag{45.2}$$

where the constant factors $\pi$ and 4 are all the best possible. We call (45.1) Hilbert's integral inequality. Both (45.1) and (45.2) are important in analysis and its applications [1, 2]. In recent years, by using the way of weight function, a number

B. Yang (✉)
Department of Mathematics, Guangdong Education Institute, Guangzhou, 510303, China
e-mail: bcyang@pub.guangzhou.gd.cn

Th.M. Rassias and J. Brzdęk (eds.), *Functional Equations in Mathematical Analysis*, Springer Optimization and Its Applications 52, DOI 10.1007/978-1-4614-0055-4_45,

of extensions of (45.1) and (45.2) were given by Yang et al. [3–5]. In 2006, Li et al. [6] gave the following inequality with a kernel relating (45.1) and (45.2):

$$\int_0^\infty \int_0^\infty \frac{f(x)g(y)}{x+y+\max\{x,y\}}\mathrm{d}x\mathrm{d}y < c\left(\int_0^\infty f^2(x)\mathrm{d}x \int_0^\infty g^2(x)\mathrm{d}x\right)^{1/2}, \tag{45.3}$$

where the constant factor

$$c = 2\sqrt{2}\arctan\left(\frac{1}{\sqrt{2}}\right)$$

is the best possible. In 2007, Xie [7] gave a best extension of (45.3), and Guo [8] gave a similar form of (45.3) as

$$\int_0^\infty \int_0^\infty \frac{f(x)g(y)\mathrm{d}x\mathrm{d}y}{x^\lambda + y^\lambda + \min\{x^\lambda, y^\lambda\}} < c_\lambda \left(\int_0^\infty x^{1-\lambda} f^2(x)\mathrm{d}x \int_0^\infty x^{1-\lambda} g^2(x)\mathrm{d}x\right)^{1/2},$$

with the best possible constant factor

$$c_\lambda = \frac{2\sqrt{2}}{\lambda}\arctan\sqrt{2} \qquad (\lambda > 0).$$

In this paper, by using the way of weight function, we give a new Hilbert-type integral inequality with the combination kernel of two classes of mean value as

$$\frac{1}{(x^\lambda + By^\lambda) + 2A(xy)^{\lambda/2}} \qquad (\lambda, B > 0, A > -\sqrt{B})$$

and a best constant factor, which is an extension of (45.1). As applications, the equivalent form is obtained.

## 45.2 Some Lemmas

First, we prove some auxiliary lemmas.

**Lemma 45.1.** *If $\lambda > 0$, $B > 0$ and $A > -\sqrt{B}$, then*

$$k_\lambda(A,B) := \int_0^\infty \frac{1}{u^\lambda + B + 2Au^{\lambda/2}} u^{\lambda/2-1}\mathrm{d}u$$

$$= \begin{cases} \dfrac{2}{\lambda\sqrt{B-A^2}}\left(\dfrac{\pi}{2} - \arctan\dfrac{A}{\sqrt{B-A^2}}\right), & -\sqrt{B} < A < \sqrt{B}; \\ \dfrac{2}{\lambda\sqrt{B}}, & A = \sqrt{B}; \\ \dfrac{2}{\lambda\sqrt{A^2-B}}\ln\dfrac{A+\sqrt{A^2-B}}{\sqrt{B}}, & A > \sqrt{B}. \end{cases}$$

*Proof.* Setting $v = u^{\lambda/2}$, we find

$$k_\lambda(A,B) = \frac{2}{\lambda}\int_0^\infty \frac{1}{v^2+2Av+B}\mathrm{d}v$$

$$= \begin{cases} \dfrac{2}{\lambda\sqrt{B-A^2}}\left[\arctan\dfrac{v+A}{\sqrt{B-A^2}}\right]_0^\infty, & -\sqrt{B}<A<\sqrt{B}; \\ \dfrac{-2}{\lambda(v+\sqrt{B})}\Big|_0^\infty, & A=\sqrt{B}; \\ \dfrac{1}{\lambda\sqrt{A^2-B}}\left[\ln\dfrac{v+A-\sqrt{A^2-B}}{v+A+\sqrt{A^2-B}}\right]_0^\infty, & A>\sqrt{B}. \end{cases}$$

Hence, we obtain the statement. □

**Lemma 45.2.** *Let $p > 1$, $q > 1$, $\lambda, B > 0$, $A > -\sqrt{B}$, and $0 < \varepsilon < \frac{\lambda}{2}\min\{p,q\}$. Then*

$$\int_0^1 \frac{u^{\lambda/2+\varepsilon/q-1}\mathrm{d}u}{u^\lambda+B+2Au^{\lambda/2}} = \int_0^1 \frac{u^{\lambda/2-1}\mathrm{d}u}{u^\lambda+B+2Au^{\lambda/2}} + o_1(1)(\varepsilon\to 0^+), \tag{45.4}$$

$$\int_1^\infty \frac{u^{\lambda/2-\varepsilon/p-1}\mathrm{d}u}{u^\lambda+B+2Au^{\lambda/2}} = \int_1^\infty \frac{u^{\lambda/2-1}\mathrm{d}u}{u^\lambda+B+2Au^{\lambda/2}} + o_2(1)(\varepsilon\to 0^+), \tag{45.5}$$

$$\int_0^1 \frac{u^{\lambda/2-\varepsilon/p-1}\mathrm{d}u}{u^\lambda+B+2Au^{\lambda/2}} = \int_0^1 \frac{u^{\lambda/2-1}\mathrm{d}u}{u^\lambda+B+2Au^{\lambda/2}} + o_3(1)(\varepsilon\to 0^+). \tag{45.6}$$

*Proof.* Since $A+\sqrt{B}>0$, there exists $\delta\in(0,1)$, such that $A+\sqrt{B(1-\delta)}\geq 0$. Now, it is enough to note that, in view of the assumption, for $\varepsilon\to 0^+$, we have

$$0 < \int_0^1 \frac{u^{\lambda/2-1}\mathrm{d}u}{u^\lambda+B+2Au^{\lambda/2}} - \int_0^1 \frac{u^{\lambda/2+\varepsilon/q-1}\mathrm{d}u}{u^\lambda+B+2Au^{\lambda/2}}$$

$$\leq \int_0^1 \frac{u^{\lambda/2-1}(1-u^{\varepsilon/q})\mathrm{d}u}{2\sqrt{B}u^{\lambda/2}+2Au^{\lambda/2}} \leq \frac{1}{2(\sqrt{B}+A)}\left(\frac{2}{\lambda}-\frac{1}{\lambda/2+\varepsilon/q}\right)\to 0,$$

$$0 < \int_1^\infty \frac{u^{\lambda/2-1}\mathrm{d}u}{u^\lambda+B+2Au^{\lambda/2}} - \int_1^\infty \frac{u^{\lambda/2-\varepsilon/p-1}\mathrm{d}u}{u^\lambda+B+2Au^{\lambda/2}}$$

$$= \int_1^\infty \frac{u^{\lambda/2-1}(1-u^{-\varepsilon/p})\mathrm{d}u}{\delta u^\lambda+[(1-\delta)u^\lambda+B]+2Au^{\lambda/2}} \leq \int_1^\infty \frac{u^{\lambda/2-1}(1-u^{-\varepsilon/p})\mathrm{d}u}{\delta u^\lambda+2[\sqrt{B(1-\delta)}+A]u^{\lambda/2}}$$

$$\leq \int_1^\infty \frac{u^{\lambda/2-1}(1-u^{-\varepsilon/p})\mathrm{d}u}{\delta u^\lambda} = \frac{1}{\delta}\left(\frac{2}{\lambda}-\frac{1}{\lambda/2+\varepsilon/p}\right)\to 0,$$

$$0<\int_0^1 \frac{u^{\lambda/2-\varepsilon/p-1}\mathrm{d}u}{u^\lambda+B+2Au^{\lambda/2}}-\int_0^1 \frac{u^{\lambda/2-1}\mathrm{d}u}{u^\lambda+B+2Au^{\lambda/2}}$$
$$\leq \int_0^1 \frac{u^{\lambda/2-1}(u^{-\varepsilon/p}-1)\mathrm{d}u}{2\sqrt{B}u^{\lambda/2}+2Au^{\lambda/2}}\leq \frac{1}{2(\sqrt{B}+A)}\left(\frac{1}{\lambda/2-\varepsilon/p}-\frac{2}{\lambda}\right)\to 0.$$ □

## 45.3 Main Results and Applications

**Theorem 45.1.** *Let $p>1$, $1/p+1/q=1$, $\lambda, B>0$, $A>-\sqrt{B}$, $\phi_r(x)=x^{r(1-\lambda/2)-1}$ $(r=p,q)$, $f,g\geq 0$,*

$$0<\|f\|_{p,\phi_p}=\int_0^\infty x^{p(1-\lambda/2)-1}f^p(x)\mathrm{d}x<\infty$$

*and*

$$0<\|g\|_{q,\phi_q}=\int_0^\infty x^{q(1-\lambda/2)-1}g^q(x)\mathrm{d}x<\infty.$$

*Then we have the following two equivalent inequalities:*

$$I_\lambda:=\int_0^\infty y^{(p\lambda)/2-1}\left[\int_0^\infty \frac{f(x)\mathrm{d}x}{x^\lambda+By^\lambda+2A(xy)^{\lambda/2}}\right]^p \mathrm{d}y<k_\lambda^p(A,B)\|f\|_{p,\phi_p}^p, \tag{45.7}$$

$$J_\lambda:=\int_0^\infty\int_0^\infty \frac{f(x)g(y)}{x^\lambda+By^\lambda+2A(xy)^{\lambda/2}}\mathrm{d}x\mathrm{d}y<k_\lambda(A,B)\|f\|_{p,\phi_p}\|g\|_{q,\phi_q}. \tag{45.8}$$

*Proof.* Setting $u=x/y$, we find that

$$\varpi_\lambda(y):=\int_0^\infty \frac{y^{\lambda/2}x^{\lambda/2-1}}{x^\lambda+By^\lambda+2A(xy)^{\lambda/2}}\mathrm{d}x=k_\lambda(A,B), \tag{45.9}$$

$$\omega_\lambda(x):=\int_0^\infty \frac{x^{\lambda/2}y^{\lambda/2-1}}{x^\lambda+By^\lambda+2A(xy)^{\lambda/2}}\mathrm{d}y=k_\lambda(A,B). \tag{45.10}$$

By Hölder's inequality with weight and (45.9), for $y\in(0,\infty)$, we obtain

$$\left[\int_0^\infty \frac{f(x)\mathrm{d}x}{x^\lambda+By^\lambda+2A(xy)^{\lambda/2}}\right]^p$$
$$=\left\{\int_0^\infty \frac{1}{x^\lambda+By^\lambda+2A(xy)^{\lambda/2}}\left[\frac{x^{(1-\lambda/2)/q}}{y^{(1-\lambda/2)/p}}f(x)\right]\left[\frac{y^{(1-\lambda/2)/p}}{x^{(1-\lambda/2)/q}}\right]\mathrm{d}x\right\}^p$$

$$\leq \int_0^\infty \frac{x^{(1-\lambda/2)(p-1)}y^{\lambda/2-1}f^p(x)}{x^\lambda+By^\lambda+2A(xy)^{\lambda/2}}\mathrm{d}x\left[\int_0^\infty \frac{y^{(1-\lambda/2)(q-1)}x^{\lambda/2-1}}{x^\lambda+By^\lambda+2A(xy)^{\lambda/2}}\mathrm{d}x\right]^{p-1}$$

$$= k_\lambda{}^{p-1}(A,B)y^{1-(p\lambda)/2}\int_0^\infty \frac{x^{(1-\lambda/2)(p-1)}y^{\lambda/2-1}}{x^\lambda+By^\lambda+2A(xy)^{\lambda/2}}f^p(x)\mathrm{d}x. \tag{45.11}$$

By (45.11) and (45.10), in view of Fubini's Theorem, it follows that

$$I_\lambda \leq k_\lambda^{p-1}(A,B)\int_0^\infty\int_0^\infty \frac{x^{(1-\lambda/2)(p-1)}y^{\lambda/2-1}f^p(x)}{x^\lambda+By^\lambda+2A(xy)^{\lambda/2}}\mathrm{d}x\mathrm{d}y$$

$$= k_\lambda^{p-1}(A,B)\int_0^\infty \omega_\lambda(x)\phi_p(x)f^p(x)\mathrm{d}x = k_\lambda^p(A,B)\|f\|_{p,\phi_p}^p. \tag{45.12}$$

If (45.11) takes the form of equality, then [8] for any $y \in (0,\infty)$, there exist non-zero constants $A$ and $B$, with $Ax^{(1-\lambda/2)(p-1)}y^{\lambda/2-1}f^p(x) = By^{(1-\lambda/2)(q-1)}x^{\lambda/2-1}$ a.e. in $(0,\infty)$. It follows

$$Ax^{p(1-\lambda/2)}f^p(x) = By^{q(1-\lambda/2)}, \quad \textit{a.e. in } (0,\infty).$$

Hence

$$x^{p(1-\lambda/2)-1}f^p(x) = \frac{B}{Ax}y^{q(1-\lambda/2)}$$

a.e. in $(0,\infty)$, which contradicts

$$0 < \int_0^\infty x^{p(1-\lambda/2)-1}f^p(x)\mathrm{d}x < \infty.$$

Then (45.11) keeps the form of strict inequality; so does (45.12). Inequality (45.10) is valid. By Hölder's inequality, we find

$$J_\lambda = \int_0^\infty \left[\int_0^\infty \frac{y^{-1/p+\lambda/2}f(x)\mathrm{d}x}{x^\lambda+By^\lambda+2A(xy)^{\lambda/2}}\right]\left[y^{1/p-\lambda/2}g(y)\right]\mathrm{d}y \leq I_\lambda^{1/p}\|g\|_{q,\phi_q}. \tag{45.13}$$

In view of (45.7), we have (45.8). On the other hand, suppose that (45.8) is valid. We find $I_\lambda > 0$, since $\|f\|_{p,\phi_p} > 0$. If $I_\lambda = \infty$, then (45.7) is not valid, since $\|f\|_{p,\phi_p} < \infty$. Hence, $0 < I_\lambda < \infty$. Setting

$$g(y) := y^{(p\lambda)/2-1}\left[\int_0^\infty \frac{f(x)\mathrm{d}x}{x^\lambda+By^\lambda+2A(xy)^{\lambda/2}}\right]^{p-1}, \qquad y \in (0,\infty),$$

by (45.8), we find

$$0 < \int_0^\infty y^{q(1-\lambda/2)-1} g^q(y) \mathrm{d}y = I_\lambda = J_\lambda$$

$$< k_\lambda(A) \|f\|_{p,\phi_p} \left\{ \int_0^\infty y^{q(1-\lambda/2)-1} g^q(y) \mathrm{d}y \right\}^{1/q} < \infty, \tag{45.14}$$

$$I_\lambda = \int_0^\infty y^{q(1-\lambda/2)-1} g^q(y) \mathrm{d}y < k_\lambda^p(A) \|f\|_{p,\phi_p}^p. \tag{45.15}$$

Thus, we have (45.7), which is equivalent to (45.8). □

**Theorem 45.2.** *All the constant factors in inequalities (45.7) and (45.8) of Theorem 45.1 are the best possible.*

*Proof.* For $0 < 2\varepsilon < \lambda \min\{p,q\}$, we define $f_\varepsilon, g_\varepsilon$ by: $f_\varepsilon(x) = g_\varepsilon(x) = 0$ for each $x \in (0,1)$ and

$$f_\varepsilon(x) = x^{\lambda/2-\varepsilon/p-1}, \qquad g_\varepsilon(x) = x^{\lambda/2-\varepsilon/q-1}$$

for $x \in [1,\infty)$. If there exists a constant $0 < k \le k_\lambda(A,B)$ such that (45.8) is still valid when we replace $k_\lambda(A,B)$ by $k$, then we have in particular

$$k = \varepsilon k \|f_\varepsilon\|_{p,\phi_p} \|g_\varepsilon\|_{q,\phi_q} > \varepsilon \int_0^\infty \int_0^\infty \frac{f_\varepsilon(x) g_\varepsilon(y) \mathrm{d}x\mathrm{d}y}{x^\lambda + By^\lambda + 2A(xy)^{\lambda/2}}$$

$$= \varepsilon \int_1^\infty x^{\lambda/2-\varepsilon/p-1} \left[ \int_1^\infty \frac{y^{\lambda/2-\varepsilon/q-1}}{x^\lambda + By^\lambda + 2A(xy)^{\lambda/2}} \mathrm{d}y \right] \mathrm{d}x.$$

Setting $u = x/y$ in the above integral, by Fubini's Theorem, we obtain

$$k > \varepsilon \int_1^\infty x^{-\varepsilon-1} \left[ \int_0^x \frac{u^{\lambda/2+\varepsilon/q-1}}{u^\lambda + B + 2Au^{\lambda/2}} \mathrm{d}u \right] \mathrm{d}x$$

$$= \int_0^1 \frac{u^{\lambda/2+\varepsilon/q-1} \mathrm{d}u}{u^\lambda + B + 2Au^{\lambda/2}} + \varepsilon \int_1^\infty x^{-\varepsilon-1} \left[ \int_1^x \frac{u^{\lambda/2+\varepsilon/q-1} \mathrm{d}u}{u^\lambda + B + 2Au^{\lambda/2}} \right] \mathrm{d}x$$

$$= \int_0^1 \frac{u^{\lambda/2+\varepsilon/q-1} \mathrm{d}u}{u^\lambda + B + 2Au^{\lambda/2}} + \varepsilon \int_1^\infty \frac{(\int_u^\infty x^{-\varepsilon-1} \mathrm{d}x) u^{\lambda/2+\varepsilon/q-1}}{u^\lambda + B + 2Au^{\lambda/2}} \mathrm{d}u$$

$$= \int_0^1 \frac{u^{\lambda/2+\varepsilon/q-1} \mathrm{d}u}{u^\lambda + B + 2Au^{\lambda/2}} + \int_1^\infty \frac{u^{\lambda/2-\varepsilon/p-1}}{u^\lambda + B + 2Au^{\lambda/2}} \mathrm{d}u.$$

For $\varepsilon \to 0^+$, in view of (45.4) and (45.5), we find $k \ge k_\lambda(A,B)$. Hence, $k = k_\lambda(A,B)$ is the best constant factor of (45.8). If the constant factor in (45.7) is not optimal, then by (45.13), we get a contradiction that the constant factor in (45.8) is not optimal. □

*Remark 45.1.* For $p = q = 2, \lambda = B = 1$ and $A = 0$ in (45.8), we have (45.1). For $B = 1, A = 1/2$ in (45.8), we obtain a new inequality:

$$\int_0^\infty \int_0^\infty \frac{f(x)g(y)}{x^\lambda + y^\lambda + (xy)^{\lambda/2}} \mathrm{d}x\mathrm{d}y < \frac{4\sqrt{3}}{9\lambda}\pi \|f\|_{p,\phi_p} \|g\|_{q,\phi_q}. \tag{45.16}$$

## References


1. Hardy, G.H., Littlewood, J.E., Polya, G.: Inequalities. Cambridge Univ. Press, Cambridge (1952)
2. Mintrinovic, D.S., Pecaric, J.E., Kink, A.M.: Inequalities Involving Functions and Their Integrals and Derivertives. Kluwer Academic Publishers, Boston (1991)
3. Yang, B.: On an extension of Hilbert's integral inequality with some parameters. Austr. J. Math. Anal. Appl. **1**, Article 11, 1–8 (2004)
4. Yang, B.: On a generalization of the Hilbert's type inequality and its applications. Chinese J. Engrg. Math. **21**, 821–824 (2004)
5. Yang, B., Brnetic, I., Krnic, M., Pecaric, J.: Generalization of Hilbert and Hardy-Hilbert integral inequalities. Math. Inequal. Appl. **8**, 259–272 (2005)
6. Li, Y., He, B.: A new Hilbert–type integral inequality and the equivalent form. Internat. J. Math. Math. Sci. **2006**, Article ID 457378, 1–6 (2006)
7. Xie, Ch.: Best generalization of a new Hilbert–type inequality. J. Jinan Univ. Nat. Sci. **28**, 24–27 (2007)
8. Ge, X.: New extension of a Hilbert–type integral inequality. J. Jinan Univ. Nat. Sci. **28**, 447–450 (2007)

# Chapter 46
# An Extension of Hardy–Hilbert's Inequality

**Bicheng Yang**

*Dedicated to the memory of S.M. Ulam*
*on the 100th anniversary of his birth*

**Abstract** By using the way of weight coefficient and the improved Euler–Maclaurin's summation formula, a new extension of Hardy–Hilbert's inequality with multi-parameter and a best constant factor is obtained, and the equivalent form is considered.

**Keywords** Hardy–Hilbert's inequality • Multi-parameter • Weight coefficient • Equivalent form

**Mathematics Subject Classification (2000):** Primary 26D15

## 46.1 Introduction

In 1908, Weyl [1] published the well known Hilbert's inequality:

$$\sum_{n=1}^{\infty}\sum_{m=1}^{\infty}\frac{a_m b_n}{m+n} < \pi\left(\sum_{n=1}^{\infty}a_n^2\sum_{n=1}^{\infty}b_n^2\right)^{1/2}, \tag{46.1}$$

for real sequences $\{a_n\}_{n=1}^{\infty}$ and $\{b_n\}_{n=1}^{\infty}$ with

$$0<\sum_{n=1}^{\infty}a_n^2<\infty,$$

$$0<\sum_{n=1}^{\infty}b_n^2<\infty,$$

B. Yang (✉)
Department of Mathematics, Guangdong Education Institute, Guangzhou, 510303, China
e-mail: bcyang@pub.guangzhou.gd.cn

Th.M. Rassias and J. Brzdęk (eds.), *Functional Equations in Mathematical Analysis*, Springer Optimization and Its Applications 52, DOI 10.1007/978-1-4614-0055-4_46,

where the constant factor $\pi$ is the best possible. In 1925, Hardy [2] gave the following extension of (46.1) with a pair of conjugate exponents $(p,q)$ $(1/p+1/q=1)$:

$$\sum_{n=1}^{\infty}\sum_{m=1}^{\infty}\frac{a_m b_n}{m+n}<\frac{\pi}{\sin(\pi/p)}\left(\sum_{n=1}^{\infty}a_n^p\right)^{1/p}\left(\sum_{n=1}^{\infty}b_n^q\right)^{1/q}, \qquad (46.2)$$

for $p>1$, $a_n,b_n\geq 0$, with

$$0<\sum_{n=1}^{\infty}a_n^p<\infty$$

and

$$0<\sum_{n=1}^{\infty}b_n^q<\infty,$$

where the constant factor $\pi/\sin(\pi/p)$ is the best possible. We give (46.2) the name of Hardy–Hilbert's inequality.

In 1934, Hardy et al. [3] gave some applications of (46.1) and (46.2); namely, introducing a pair of non-conjugate exponents $(p,q)$ in (46.1), Hardy et al. proved that:

$$\sum_{n=1}^{\infty}\sum_{m=1}^{\infty}\frac{a_m b_n}{(m+n)^{\lambda}}\leq K(p,q)\left(\sum_{n=1}^{\infty}a_n^p\right)^{1/p}\left(\sum_{n=1}^{\infty}b_n^q\right)^{1/q}, \qquad (46.3)$$

for $p,q>1$, $1/p+1/q\geq 1$, $0<\lambda=2-(1/p+1/q)\leq 1$, where the constant factor $K(p,q)$ is optimal only for $\lambda=1$. In 1951, Bonsall [4] considered (46.3) in the case of general kernel. But the constants are not the best possible unless $(p,q)$ is a pair of conjugate exponents. In 1991, Mitrinovic et al. [5] summarized the above results.

In 1997–1998, by using the way of weight coefficient, Yang and Gao [6, 7] gave a strengthened version of (46.2) as:

$$\sum_{n=1}^{\infty}\sum_{m=1}^{\infty}\frac{a_m b_n}{m+n}<\left\{\sum_{n=1}^{\infty}\left[\frac{\pi}{\sin(\pi/p)}-\frac{1-\gamma}{n^{1/p}}\right]a_n^p\right\}^{1/p}$$
$$\times\left\{\sum_{n=1}^{\infty}\left[\frac{\pi}{\sin(\pi/p)}-\frac{1-\gamma}{n^{1/q}}\right]b_n^q\right\}^{1/q}, \qquad (46.4)$$

where $1-\gamma=0.42278433^{+}$ ($\gamma$ is the Euler constant).

In 2001, Yang [8] gave an extension of (46.1) by using the way of weight coefficient and the improved Euler–Maclaurin's summation formula and introducing an independent parameter $0<\lambda\leq 4$ as follows

$$\sum_{n=1}^{\infty}\sum_{m=1}^{\infty}\frac{a_m b_n}{(m+n)^{\lambda}}<B\left(\frac{\lambda}{2},\frac{\lambda}{2}\right)\left(\sum_{n=1}^{\infty}n^{1-\lambda}a_n^2\sum_{n=1}^{\infty}n^{1-\lambda}b_n^2\right)^{1/2}, \qquad (46.5)$$

where the constant factor $B(\lambda/2,\lambda/2)$ is the best possible (B(u,v) is the Beta function). In 2004, Yang [9] published the dual form of (46.2) as follows:

$$\sum_{n=1}^{\infty}\sum_{m=1}^{\infty}\frac{a_m b_n}{m+n}<\frac{\pi}{\sin(\pi/p)}\left(\sum_{n=1}^{\infty}n^{p-2}a_n^p\right)^{1/p}\left(\sum_{n=1}^{\infty}n^{q-2}b_n^q\right)^{1/q}, \tag{46.6}$$

where $\pi/\sin(\pi/p)$ is the best possible constant. For $p=q=2$, both (46.6) and (46.2) reduce to (46.1). It means that there are two different best extensions of (46.1). In 2005, Yang [10] gave an extension of (46.2) and (46.6) with two pairs of conjugate exponents $(p,q)$, $(r,s)$ $(p,r>1)$ and two parameters $\alpha,\lambda>0$ $(\alpha\lambda\le\min\{r,s\})$ as

$$\sum_{n=1}^{\infty}\sum_{m=1}^{\infty}\frac{a_m b_n}{(m^{\alpha}+n^{\alpha})^{\lambda}}<k_{\alpha\lambda}(r)\left\{\sum_{n=1}^{\infty}n^{p(1-(\alpha\lambda)/r)-1}a_n^p\right\}^{1/p}$$
$$\times\left\{\sum_{n=1}^{\infty}n^{q(1-(\alpha\lambda)/s)-1}b_n^q\right\}^{1/q}, \tag{46.7}$$

where the constant factor $k_{\alpha\lambda}(r)=B(\lambda/r,\lambda/s)/\alpha$ is the best possible. In particular, for $\alpha=1$ and $0<\lambda\le\min\{r,s\}$, inequality (46.7) reduces to

$$\sum_{n=1}^{\infty}\sum_{m=1}^{\infty}\frac{a_m b_n}{(m+n)^{\lambda}}<B\left(\frac{\lambda}{r},\frac{\lambda}{s}\right)\left\{\sum_{n=1}^{\infty}n^{p(1-\lambda/r)-1}a_n^p\right\}^{1/p}$$
$$\times\left\{\sum_{n=1}^{\infty}n^{q(1-\lambda/s)-1}b_n^q\right\}^{1/q}. \tag{46.8}$$

*Remark 46.1.* For $\lambda=1$, $r=q$, $s=p$, (46.8) reduces to (46.2); for $\lambda=1$, $r=p$, $s=q$, (46.8) reduces to (46.6); for $p=q=r=s=2$, (46.8) reduces to (46.5) only for $0<\lambda\le 2$. But for $2<\lambda\le 4$, inequality (46.8) is not an extension of (46.5).

In this paper, by using the way of weight coefficient and the improved Euler–Maclaurin's summation formula as in [8, 10], we obtain inequality (46.8) with the extended parameter $0<\lambda\le 2\min\{r,s\}$, which is a best extension of (46.5) in the case $0<\lambda\le 4$. The equivalent form is considered.

## 46.2 Some Lemmas

**Lemma 46.1 (the improved Euler-Maclaurin's summation formula, cf. [11]).**

*1. If $f(x)\in C^1[1,\infty)$, then we have*

$$\sum_{k=1}^{\infty}f(k)=\int_1^{\infty}f(x)\mathrm{d}x+\frac{1}{2}f(1)+\int_1^{\infty}P_1(x)f'(x)\mathrm{d}x, \tag{46.9}$$

*where* $P_1(x) = x - [x] - 1/2$ *is the Bernoulli function of the 1st order.*

2. *If* $g(x) \in C^2[1,\infty)$, *then*

$$\int_1^\infty P_1(x)g(x)\mathrm{d}x = -\frac{1}{12}g(1) + \frac{1}{6}\int_1^\infty P_3(x)g''(x)\mathrm{d}x, \qquad (46.10)$$

*where* $P_3(x)$ *is the Bernoulli function of the 3rd order.*

3. *If* $G \in C^3[1,\infty)$, $(-1)^i G^{(i)}(x) > 0$, *and* $G^{(i)}(\infty) = 0$ $(i = 0,1,2,3)$, *then*

$$-\frac{1}{12}G(1) < \int_1^\infty P_1(x)G(x)\mathrm{d}x < 0$$
$$< \int_1^\infty P_3(x)G(x)\mathrm{d}x < \frac{1}{120}G(1). \qquad (46.11)$$

**Lemma 46.2.** *If* $r > 1$,

$$\frac{1}{r} + \frac{1}{s} = 1,$$

*and*

$$0 < \lambda \leq 2\min\{r,s\},$$

*then setting the weigh coefficient* $\varpi_\lambda(s,m)$ *as*

$$\varpi_\lambda(s,m) := \sum_{n=1}^{\infty} \frac{m^{\lambda/r}}{(m+n)^\lambda n^{1-\lambda/s}} \qquad (m \in \mathbf{N}) \qquad (46.12)$$

*(*$\mathbf{N}$ *is the set of positive integer numbers), we have*

$$B\left(\frac{\lambda}{r},\frac{\lambda}{s}\right)[1-\theta_\lambda(r,m)] = \varpi_\lambda(s,m) < B\left(\frac{\lambda}{r},\frac{\lambda}{s}\right), \qquad (46.13)$$

*where*

$$\theta_\lambda(r,m) := 1 - \frac{1}{B(\lambda/r,\lambda/s)}\varpi_\lambda(s,m) = O\left(\frac{1}{m^{\lambda/s}}\right) \qquad (m \to \infty).$$

*Proof.* For fixed $s > 1$, $m \in \mathbf{N}$ and $\lambda \in (0, 2\min\{r,s\})$, set

$$f(x) := \frac{1}{(m+x)^\lambda x^{1-\lambda/s}}, \qquad (x \in (0,\infty)).$$

Then by (46.9), it follows that

$$\begin{aligned}\varpi_\lambda(s,m) &= m^{\lambda/r}\sum_{n=1}^{\infty} f(n)\\ &= m^{\lambda/r}\left[\int_1^\infty f(x)\mathrm{d}x + \frac{1}{2}f(1) + \int_1^\infty P_1(x)f'(x)\mathrm{d}x\right]\\ &= m^{\lambda/r}\int_0^\infty f(x)\mathrm{d}x - m^{\lambda/r}\rho_\lambda(s,m),\end{aligned}\tag{46.14}$$

and

$$\rho_\lambda(s,m) := \int_0^1 f(x)\mathrm{d}x - \frac{1}{2}f(1) - \int_1^\infty P_1(x)f'(x)\mathrm{d}x.\tag{46.15}$$

We find $f(1) = (m+1)^{-\lambda}$ and, integrating by parts,

$$\begin{aligned}\int_0^1 f(x)\mathrm{d}x &= \frac{s}{\lambda}\int_0^1 \frac{\mathrm{d}x^{\lambda/s}}{(m+x)^\lambda}\\ &= \frac{s}{\lambda(m+1)^\lambda} + s\int_0^1 \frac{x^{\lambda/s}\mathrm{d}x}{(m+x)^{\lambda+1}}\\ &= \frac{s}{\lambda(m+1)^\lambda} + \frac{s^2}{s+\lambda}\int_0^1 \frac{\mathrm{d}x^{\lambda/s+1}}{(m+x)^{\lambda+1}}\\ &= \frac{s}{\lambda(m+1)^\lambda} + \frac{s^2}{(s+\lambda)(m+1)^{\lambda+1}}\\ &\quad + \frac{s^2(\lambda+1)}{(s+\lambda)}\int_0^1 \frac{x^{\lambda/s+1}\mathrm{d}x}{(m+x)^{\lambda+2}},\end{aligned}$$

whence

$$\begin{aligned}\int_0^1 f(x)\mathrm{d}x &> \frac{s}{\lambda(m+1)^\lambda} + \frac{s^2}{(s+\lambda)(m+1)^{\lambda+1}}\\ &\quad + \frac{s^3(\lambda+1)}{(s+\lambda)(2s+\lambda)(m+1)^{\lambda+2}}\\ &\geq \frac{s}{\lambda(m+1)^\lambda} + \frac{s}{3(m+1)^{\lambda+1}} + \frac{s(\lambda+1)}{12(m+1)^{\lambda+2}},\end{aligned}$$

$$f'(x) = \frac{-\lambda}{(m+x)^{\lambda+1}x^{1-\lambda/s}} - \frac{(1-\lambda/s)}{(m+x)^{\lambda}x^{2-\lambda/s}}$$
$$= \frac{-(1+\lambda/r)}{(m+x)^{\lambda}x^{2-\lambda/s}} + \frac{m\lambda}{(m+x)^{\lambda+1}x^{2-\lambda/s}}.$$

Since

$$0 < \lambda \leq 2\min\{r,s\} \leq 4,$$

by (46.10) and (46.11), we find

$$-\int_1^{\infty} P_1(x)f'(x)\mathrm{d}x = \int_1^{\infty} P_1(x)\frac{(1+\lambda/r)\mathrm{d}x}{(m+x)^{\lambda}x^{2-\lambda/s}}$$
$$-\int_1^{\infty} P_1(x)\frac{m\lambda\mathrm{d}x}{(m+x)^{\lambda+1}x^{2-\lambda/s}}$$
$$> \frac{-(1+\lambda/r)}{12(m+1)^{\lambda}} + \frac{m\lambda}{12(m+1)^{\lambda+1}}$$
$$-\frac{m\lambda}{6}\int_1^{\infty} P_3(x)\left[\frac{1}{(m+x)^{\lambda+1}x^{2-\lambda/s}}\right]''\mathrm{d}x$$
$$> \frac{-(1+\lambda/r)}{12(m+1)^{\lambda}} + \frac{(m+1)\lambda-\lambda}{12(m+1)^{\lambda+1}}$$
$$-\frac{4(m+1)}{720}\left[\frac{(\lambda+1)(\lambda+2)}{(m+1)^{\lambda+3}}\right.$$
$$\left.+\frac{2(\lambda+1)(2s-\lambda)}{s(m+1)^{\lambda+2}} + \frac{(2s-\lambda)(3s-\lambda)}{s^2(m+1)^{\lambda+1}}\right]$$
$$\geq \left(\frac{-1}{12}+\frac{\lambda}{12s}\right)\frac{1}{(m+1)^{\lambda}} - \frac{\lambda}{12(m+1)^{\lambda+1}}$$
$$-\frac{1}{180}\left[\frac{6(\lambda+1)}{(m+1)^{\lambda+2}} + \frac{20s}{(m+1)^{\lambda+1}} + \frac{6}{(m+1)^{\lambda}}\right]. \quad (46.16)$$

Hence, by (46.15), it follows

$$\rho_{\lambda}(s,m) > \left(\frac{s}{\lambda} - \frac{37}{60} + \frac{\lambda}{12s}\right)\frac{1}{(m+1)^{\lambda}}$$
$$+\left(\frac{2s}{9}-\frac{\lambda}{12}\right)\frac{1}{(m+1)^{\lambda+1}} + \left(\frac{s}{12}-\frac{1}{30}\right)\frac{(\lambda+1)}{(m+1)^{\lambda+2}}. \quad (46.17)$$

Since for $\lambda \leq 2s$ and $s > 1$,

$$\left(\frac{2s}{9} - \frac{\lambda}{12}\right) \geq \left(\frac{2s}{9} - \frac{2s}{12}\right) = \frac{s}{18} > 0, \qquad \frac{s}{12} - \frac{1}{30} > \frac{1}{12} - \frac{1}{30} > 0,$$

setting $h(\lambda) := 5\lambda^2 - 37s\lambda + 60s^2$, by (46.17), we find

$$\rho_\lambda(s,m) > \frac{h(\lambda)}{60\lambda s}\frac{1}{(m+1)^\lambda}.$$

Next, since $h'(\lambda) := 10\lambda - 37s \leq 20s - 37s < 0$, we get $h(\lambda) \geq h(2s) = 6s^2 > 0$ and $\rho_\lambda(s,m) > 0$. Setting $u = x/m$, we obtain

$$m^{\lambda/r}\int_0^\infty f(x)\mathrm{d}x = \int_0^\infty \frac{1}{(u+1)^\lambda}u^{\lambda/s-1}\mathrm{d}u = B\left(\frac{\lambda}{r}, \frac{\lambda}{s}\right). \tag{46.18}$$

Consequently, by (46.14), we have

$$\varpi_\lambda(s,m) < B\left(\frac{\lambda}{r}, \frac{\lambda}{s}\right).$$

Since

$$\int_0^1 f(x)\mathrm{d}x = \int_0^1 \frac{\mathrm{d}x}{(m+x)^\lambda x^{1-\lambda/s}} \leq \int_0^1 \frac{\mathrm{d}x}{m^\lambda x^{1-\lambda/s}} = \frac{s}{\lambda m^\lambda},$$

and, by (46.16) and (46.11),

$$-\int_1^\infty P_1(x)f'(x)\mathrm{d}x = \left(1+\frac{\lambda}{r}\right)\int_1^\infty P_1(x)\frac{\mathrm{d}x}{(m+x)^\lambda x^{2-\lambda/s}}$$
$$-m\lambda\int_1^\infty P_1(x)\frac{\mathrm{d}x}{(m+x)^{\lambda+1}x^{2-\lambda/s}} < \frac{m\lambda}{12(m+1)^{\lambda+1}},$$

so, by (46.15), it follows

$$0 < m^{\lambda/r}\rho_\lambda(s,m) < m^{\lambda/r}\left[\frac{s}{\lambda m^\lambda} - \frac{1}{2(m+1)^\lambda} + \frac{m\lambda}{12(m+1)^{\lambda+1}}\right]$$
$$< m^{\lambda/r}\left[\frac{s}{\lambda m^\lambda} + \frac{m\lambda}{12m^{\lambda+1}}\right] = \left(\frac{s}{\lambda} + \frac{\lambda}{12}\right)\frac{1}{m^{\lambda/s}}.$$

Hence, by (46.14) and (46.18), we have

$$\theta_\lambda(r,m) = 1 - \frac{\varpi_\lambda(s,m)}{B(\lambda/r,\lambda/s)} = \frac{1}{B(\lambda/r,\lambda/s)}m^{\lambda/r}\rho_\lambda(s,m) = O\left(\frac{1}{m^{\lambda/s}}\right).$$

□

**Lemma 46.3.** *If $p, r > 1$, $1/p+1/q = 1$, $1/r+1/s = 1$, $0 < \lambda \le 2\min\{r,s\}$, and $0 < \varepsilon < (q\lambda)/s$, then*

$$\sum_{n=1}^{\infty} \frac{m^{\lambda/r+\varepsilon/q} n^{\lambda/s-\varepsilon/q-1}}{(m+n)^{\lambda}} = B\left(\frac{\lambda}{r}+\frac{\varepsilon}{q}, \frac{\lambda}{s}-\frac{\varepsilon}{q}\right)\left[1-O\left(\frac{1}{m^{\lambda/s-\varepsilon/q}}\right)\right]$$

$$(m \to \infty). \qquad (46.19)$$

*Proof.* Let $(r', s')$ be such that

$$\frac{1}{r'} = \frac{1}{r} + \frac{\varepsilon}{q\lambda} > 0, \qquad \frac{1}{s'} = \frac{1}{s} - \frac{\varepsilon}{q\lambda} > 0.$$

Then $1/r' + 1/s' = 1$ and $r' > 1$. Hence by (46.12) and (46.13),

$$\sum_{n=1}^{\infty} \frac{m^{\lambda/r+\varepsilon/q}\, n^{\lambda/s-\varepsilon/q-1}}{(m+n)^{\lambda}} = \varpi_{\lambda}(r', m) = B\left(\frac{\lambda}{r'}, \frac{\lambda}{s'}\right)[1-\theta_{\lambda}(r', m)]$$

$$= B\left(\frac{\lambda}{r}+\frac{\varepsilon}{q}, \frac{\lambda}{s}-\frac{\varepsilon}{q}\right)[1-\theta_{\lambda}(r', m)],$$

where

$$\theta_{\lambda}(r', m) = O\left(\frac{1}{m^{\lambda/s'}}\right) = O\left(\frac{1}{m^{\lambda/s-\varepsilon/q}}\right) \qquad (m \to \infty).$$

Consequently, (46.19) is valid. □

## 46.3 Main Result and the Equivalent Form

**Theorem 46.1.** *Let $p, r > 1$, $1/p+1/q = 1$, $1/r+1/s = 1$, $0 < \lambda \le 2\min\{r,s\}$, $a_n, b_n \ge 0$, and*

$$0 < \sum_{n=1}^{\infty} n^{p(1-\lambda/r)-1} a_n^p < \infty, \qquad 0 < \sum_{n=1}^{\infty} n^{q(1-\lambda/s)-1} b_n^q < \infty.$$

*Then inequality (46.8) is valid, which is equivalent to*

$$J := \sum_{n=1}^{\infty} n^{(p\lambda)/s-1}\left[\sum_{m=1}^{\infty} \frac{a_m}{(m+n)^{\lambda}}\right]^p < \left[B\left(\frac{\lambda}{r}, \frac{\lambda}{s}\right)\right]^p \sum_{n=1}^{\infty} n^{p(1-\lambda/r)-1}\, a_n^p, \quad (46.20)$$

*where the constant factors $B(\lambda/r, \lambda/s)$ and $[B(\lambda/r, \lambda/s)]^p$ are the best possible.*

*Proof.* By the symmetric property, it is obvious that $\varpi_\lambda(r,n) < B(\lambda/r,\lambda/s)$. By Hölder's inequality [12], we find

$$\begin{aligned}
\left[\sum_{m=1}^{\infty}\frac{a_m}{(m+n)^\lambda}\right]^p &= \left\{\sum_{m=1}^{\infty}\frac{a_m}{(m+n)^\lambda}\frac{m^{(1-\lambda/r)/q}}{n^{(1-\lambda/s)/p}}\frac{n^{(1-\lambda/s)/p}}{m^{(1-\lambda/r)/q}}\right\}^p \\
&\le \left[\sum_{m=1}^{\infty}\frac{1}{(m+n)^\lambda}\frac{m^{[(1-\lambda/r)p]/q}}{n^{1-\lambda/s}}a_m^p\right] \\
&\quad\times\left[\sum_{m=1}^{\infty}\frac{1}{(m+n)^\lambda}\frac{n^{[(1-\lambda/s)q]/p}}{m^{1-\lambda/r}}\right]^{p-1} \\
&= \varpi_\lambda^{p-1}(r,n)n^{1-(p\lambda)/s}\sum_{m=1}^{\infty}\frac{1}{(m+n)^\lambda}\frac{m^{[(1-\lambda/r)p]/q}}{n^{1-\lambda/s}}a_m^p \\
&\le \left[B\left(\frac{\lambda}{r},\frac{\lambda}{s}\right)\right]^{p-1} n^{1-(p\lambda)/s}\sum_{m=1}^{\infty}\frac{1}{(m+n)^\lambda}\frac{m^{[(1-\lambda/r)p]/q}}{n^{1-\lambda/s}}a_m^p.
\end{aligned}$$

Hence, by (46.13), we obtain

$$\begin{aligned}
J &= \sum_{n=1}^{\infty} n^{(p\lambda)/s-1}\left[\sum_{m=1}^{\infty}\frac{a_m}{(m+n)^\lambda}\right]^p \\
&\le \left[B\left(\frac{\lambda}{r},\frac{\lambda}{s}\right)\right]^{p-1}\sum_{n=1}^{\infty}\sum_{m=1}^{\infty}\frac{1}{(m+n)^\lambda}\frac{m^{[(1-\lambda/r)p]/q}}{n^{1-\lambda/s}}a_m^p \\
&= \left[B\left(\frac{\lambda}{r},\frac{\lambda}{s}\right)\right]^{p-1}\sum_{m=1}^{\infty}\varpi_\lambda(s,m)m^{p(1-\lambda/r)-1}a_m^p \\
&< \left[B\left(\frac{\lambda}{r},\frac{\lambda}{s}\right)\right]^{p}\sum_{m=1}^{\infty} m^{p(1-\lambda/r)-1}a_m^p. \qquad (46.21)
\end{aligned}$$

Therefore, (46.20) is valid. By Hölder's inequality, we find

$$\begin{aligned}
\sum_{n=1}^{\infty}\sum_{m=1}^{\infty}\frac{a_m b_n}{(m+n)^\lambda} &= \sum_{n=1}^{\infty}\left[n^{\lambda/s-1/p}\sum_{m=1}^{\infty}\frac{a_m}{(m+n)^\lambda}\right]n^{-\lambda/s+1/p}b_n \\
&\le J^{1/p}\left\{\sum_{n=1}^{\infty} n^{q(1-\lambda/s)-1}b_n^q\right\}^{1/q}. \qquad (46.22)
\end{aligned}$$

So, by (46.20), we have (46.8).

On the other-hand, suppose (46.8) is valid. Setting

$$b_n := n^{(p\lambda)/s-1}\left[\sum_{m=1}^{\infty}\frac{a_m}{(m+n)^{\lambda}}\right]^{p-1}, \qquad (n\in\mathbf{N}),$$

we get

$$\sum_{n=1}^{\infty} n^{q(1-\lambda/s)-1}\, b_n^q = J.$$

By (46.21), we deduce that $J<\infty$. If $J=0$, then (46.20) is naturally valid; if $0< J<\infty$, then by (46.8), we find

$$\begin{aligned}\sum_{n=1}^{\infty} n^{q(1-\lambda/s)-1}\, b_n^q &= J = \sum_{n=1}^{\infty}\sum_{m=1}^{\infty}\frac{a_m b_n}{(m+n)^{\lambda}}\\ &< B\left(\frac{\lambda}{r},\frac{\lambda}{s}\right)\left\{\sum_{n=1}^{\infty} n^{p(1-\lambda/r)-1}\, a_n^p\right\}^{1/p}\left\{\sum_{n=1}^{\infty} n^{q(1-\lambda/s)-1}\, b_n^q\right\}^{1/q},\end{aligned}$$

and

$$\left\{\sum_{n=1}^{\infty} n^{q(1-\lambda/s)-1}\, b_n^q\right\}^{1/p} = J^{1/p} < B\left(\frac{\lambda}{r},\frac{\lambda}{s}\right)\left\{\sum_{n=1}^{\infty} n^{p(1-\lambda/r)-1}\, a_n^p\right\}^{1/p},$$

and then (46.20) is valid, which is equivalent to (46.8).

For $0<\varepsilon<(q\lambda)/s$, we set

$$\widetilde{a} = \left\{\widetilde{a}_n\right\}_{n=1}^{\infty}, \qquad \widetilde{b} = \left\{\widetilde{b}_n\right\}_{n=1}^{\infty}$$

where $\widetilde{a}_n = n^{\lambda/r-\varepsilon/p-1}$ and $\widetilde{b}_n = n^{\lambda/s-\varepsilon/q-1}$ for $n\in\mathbf{N}$. If there exists a constant

$$0<k\le B\left(\frac{\lambda}{r},\frac{\lambda}{s}\right)$$

such that (46.8) is still valid when we replace $B(\lambda/r,\lambda/s)$ by $k$, then we find

$$\begin{aligned}\widetilde{I} &:= \sum_{n=1}^{\infty}\sum_{m=1}^{\infty}\frac{\widetilde{a}_m\widetilde{b}_n}{(m+n)^{\lambda}}\\ &< k\left\{\sum_{n=1}^{\infty} n^{p(1-\lambda/r)-1}\, \widetilde{a}_n^p\right\}^{1/p}\left\{\sum_{n=1}^{\infty} n^{q(1-\lambda/s)-1}\, \widetilde{b}_n^q\right\}^{1/q}\\ &= k\sum_{n=1}^{\infty}\frac{1}{n^{1+\varepsilon}},\end{aligned}$$

and

$$\widetilde{I} = \sum_{n=1}^{\infty} \left[ \sum_{m=1}^{\infty} \frac{1}{(m+n)^{\lambda}} m^{\lambda/r-\varepsilon/p-1} \right] n^{\lambda/s-\varepsilon/q-1}$$
$$= \sum_{m=1}^{\infty} \frac{1}{m^{1+\varepsilon}} \sum_{n=1}^{\infty} \frac{m^{\lambda/r+\varepsilon/q}}{(m+n)^{\lambda}} n^{\lambda/s-\varepsilon/q-1}.$$

Write

$$\widehat{B} := B(\lambda/r+\varepsilon/q, \lambda/s-\varepsilon/q).$$

Then, by (46.19) and the above inequalities, it follows that

$$k \sum_{n=1}^{\infty} \frac{1}{n^{1+\varepsilon}} > \widetilde{I} = \widehat{B} \sum_{m=1}^{\infty} \frac{1}{m^{1+\varepsilon}} [1 - \theta_{\lambda}(r', m)]$$
$$= \widehat{B} \left[ \sum_{m=1}^{\infty} \frac{1}{m^{1+\varepsilon}} - \sum_{m=1}^{\infty} \frac{1}{m^{1+\varepsilon}} \theta_{\lambda}(r', m) \right]$$
$$= \widehat{B} \sum_{n=1}^{\infty} \frac{1}{n^{1+\varepsilon}} \left[ 1 - \left( \sum_{m=1}^{\infty} \frac{1}{m^{1+\varepsilon}} \right)^{-1} \sum_{m=1}^{\infty} \frac{1}{m^{1+\varepsilon}} O\left( \frac{1}{m^{\lambda/s-\varepsilon/q}} \right) \right],$$

whence

$$k > \widehat{B} \left[ 1 - \left( \sum_{m=1}^{\infty} \frac{1}{m^{1+\varepsilon}} \right)^{-1} \sum_{m=1}^{\infty} \frac{1}{m^{1+\varepsilon}} O\left( \frac{1}{m^{\lambda/s-\varepsilon/q}} \right) \right].$$

Therefore, $k \geq B(\lambda/r, \lambda/s)$ $(\varepsilon \to 0^+)$, and $k = B(\lambda/r, \lambda/s)$ is the best value of (46.8). Note that $[B(\lambda/r, \lambda/s)]^p$ is the best value of (46.20), because otherwise we get a contradiction by (46.22) that the constant factor in (46.8) is not the best possible. □

**Acknowledgement** This work is supported by the Emphases Natural Science Foundation of Guangdong Institution, Higher Learning, College and University (No. 05Z026), and Guangdong Natural Science Foundation (No. 7004344).

## References

1. Weyl, H.: Singulare Integral Gleichungen mit Besonderer Berucksichtigung des Fourierschen Integral Theorems. Inaugeral-Dissertation, Gottingen (1908)
2. Hardy, G.H.: Note on a theorem of Hilbert concerning series of positive term. Proc. Lond. Math. Society **23**, 45–46 (1925)

3. Hardy, G.H., Littlewood, J.E., Polya, G.: Inequalities. Cambridge University Press, Cambridge (1934)
4. Bonsall, F.F.: Inequalities with non-conjugate parameter. Q. J. Math. **2**, 135–150 (1951)
5. Mitrinovic, D.S., Pecaric, J.E., Fink, A.M.: Inequalities Involving Functions and Their Integrals and Derivatives. Kluwer Acaremic Publishers, Boston (1991)
6. Yang, B., Gao, M.: On a best value of Hardy-Hilbert's inequality. Adv. Math. **26**, 159–164 (1997)
7. Gao, M., Yang, B.: On the extended Hilbert's inequality. Proc. Amer. Math. Soc. **126**, 751–759 (1998)
8. Yang, B.: On a generalization of Hilbert double series theorem. Nanjing Univ. J. Math. Biquarterly **18**, 145–152 (2001)
9. Yang, B.: On new extensions of Hilbert's inequality. Acta Math. Hungar. **104**, 291–299 (2004)
10. Yang, B.: On best extensions of Hardy–Hilbert's inequality with two parameters. JIPAM. J. Inequal. Pure Appl. Math. **6**, Art. 81, 1–15 (2005)
11. Yang, B.: On a strengthened version of the more accurate Hardy-Hilbert's inequality. Acta Math. Sin. (Chin. Ser.) **42**, 1103–1110 (1999)
12. Kuang, J.: Applied Inequalities. Shangdong Science Technic Press, Jinan (2004)

# Chapter 47
# A Relation to Hilbert's Integral Inequality and a Basic Hilbert-Type Inequality

**Bicheng Yang and Themistocles M. Rassias**

*Dedicated to the memory of S.M. Ulam*
*on the 100th anniversary of his birth*

**Abstract** By using the way of weight function, a new integral inequality with certain parameters and a best constant factor is proved which provides a relation of Hilbert's integral inequality and a basic Hilbert-type integral inequality. Both the equivalent form as well as the reverse form are considered.

**Keywords** Basic Hilbert-type integral inequality • Parameter • Weight function

**Mathematics Subject Classification (2000):** Primary 26D15

## 47.1 Introduction

If

$$0 < \int_0^\infty f^2(x)\mathrm{d}x < \infty, \qquad 0 < \int_0^\infty g^2(x)\mathrm{d}x < \infty$$

and $f, g \geq 0$, then we have [2, 8]:

$$\int_0^\infty \int_0^\infty \frac{f(x)g(y)}{x+y}\mathrm{d}x\mathrm{d}y < \pi \left\{ \int_0^\infty f^2(x)\mathrm{d}x \int_0^\infty g^2(x)\mathrm{d}x \right\}^{1/2}, \tag{47.1}$$

$$\int_0^\infty \int_0^\infty \frac{|\ln(x/y)| f(x)g(y)}{x+y}\mathrm{d}x\mathrm{d}y < c_0 \left\{ \int_0^\infty f^2(x)\mathrm{d}x \int_0^\infty g^2(x)\mathrm{d}x \right\}^{1/2}, \tag{47.2}$$

B. Yang (✉)
Department of Mathematics, Guangdong Education Institute, Guangzhou 510303, China
e-mail: bcyang@pub.guangzhou.gd.cn

Th.M. Rassias
Department of Mathematics, National Technical University of Athens, Zografou Campus, 15780 Athens, Greece
e-mail: trassias@math.ntua.gr

Th.M. Rassias and J. Brzdęk (eds.), *Functional Equations in Mathematical Analysis*, Springer Optimization and Its Applications 52, DOI 10.1007/978-1-4614-0055-4_47,

where the constant factors $\pi$ and $c_0(=\sum_{k=1}^{\infty}(8(-1)^{k-1})/(2k-1)^2 = 7.3277^+)$ are the best possible. Inequality (47.1) is thewell known Hilbert's integral inequality and (47.2) is the basic Hilbert-type inequality. In 2005, Hardy–Riesz gave a best extension of (47.1) by introducing $(p,q)$-parameter $(p>1,\ 1/p+1/q=1)$ as [1]

$$\int_0^\infty\int_0^\infty \frac{f(x)g(y)}{x+y}\mathrm{d}x\mathrm{d}y < \frac{\pi}{\sin(\pi/p)}\left\{\int_0^\infty f^p(x)\mathrm{d}x\right\}^{1/p}\left\{\int_0^\infty g^q(x)\mathrm{d}x\right\}^{1/q}. \quad (47.3)$$

Inequality (47.3) is named Hardy-Hilbert's integral inequality, which is important in analysis and its applications [5]. In 1998, Yang gave a best extension of (47.1) by introducing an independent parameter $\lambda > 0$ as [6, 7]

$$\int_0^\infty\int_0^\infty \frac{f(x)g(y)}{(x+y)^\lambda}\mathrm{d}x\mathrm{d}y$$

$$< B\left(\frac{\lambda}{2},\frac{\lambda}{2}\right)\left\{\int_0^\infty x^{1-\lambda}f^2(x)\mathrm{d}x\int_0^\infty x^{1-\lambda}g^2(x)\mathrm{d}x\right\}^{1/2}, \quad (47.4)$$

where the Beta function $B(u,v)$ is defined by [9]:

$$B(u,v) := \int_0^\infty \frac{1}{(1+t)^{u+v}}t^{u-1}\mathrm{d}t \qquad (u,v>0). \quad (47.5)$$

In recent years, by introducing two pairs of conjugate exponents and an independent parameter, Yang et al. [10, 11] proved the following two different best extensions of (47.1) and (47.3):

$$\int_0^\infty\int_0^\infty \frac{f(x)g(y)}{x^\lambda+y^\lambda}\mathrm{d}x\mathrm{d}y < \frac{\pi}{\lambda\sin(\pi/r)}\|f\|_{p,\phi}\|g\|_{q,\psi}, \quad (47.6)$$

$$\int_0^\infty\int_0^\infty \frac{f(x)g(y)}{(x+y)^\lambda}\mathrm{d}x\mathrm{d}y < B\left(\frac{\lambda}{r},\frac{\lambda}{s}\right)\|f\|_{p,\phi}\|g\|_{q,\psi} \quad (47.7)$$

for $p,r>1,\ 1/p+1/q=1,\ 1/r+1/s=1,\ \lambda>0,$

$$\phi(x) = x^{p(1-\lambda/r)-1},$$

$$\psi(x) = x^{q(1-\lambda/s)-1},$$

$f,g \geq 0$, and

$$0 < \|f\|_{p,\phi} := \left\{\int_0^\infty x^{p(1-\lambda/r)-1}f^p(x)\mathrm{d}x\right\}^{1/p} < \infty,$$

$$0 < \|g\|_{q,\psi} := \left\{\int_0^\infty x^{q(1-\lambda/s)-1}g^q(x)\mathrm{d}x\right\}^{1/q} < \infty.$$

Yang [12] also considered the reverse of (47.6) and (47.7).

In this paper, by using the way of weight function and some basic real analysis, a new integral inequality with the homogeneous kernel of $-\lambda$ degree

$$k_\lambda(x,y) = \frac{|\ln(x/y)|^\beta}{(x+y)^{\lambda-\alpha}(\min\{x,y\})^\alpha}$$

with $\lambda > 0$, $\alpha < \lambda \min\{1/r, 1/s\}$, $\beta > -1$, is given, which is a relation between Hilbert's integral inequality and the basic Hilbert-type inequality (47.2). The equivalent form and the reverse forms are considered. All the new inequalities possess the best constant factors.

## 47.2 Three Lemmas

We introduce the following Gamma function [9]:

$$\Gamma(s) = \int_0^\infty \mathrm{e}^{-t} t^{s-1} \mathrm{d}t \qquad (s > 0). \tag{47.8}$$

**Lemma 47.1.** *For $a, b > 0$, it follows*

$$\int_0^1 x^{a-1}(-\ln x)^{b-1}\mathrm{d}x = \frac{1}{a^b}\Gamma(b) = \int_1^\infty y^{-a-1}(\ln y)^{b-1}\mathrm{d}y. \tag{47.9}$$

*Proof.* Setting $x = \mathrm{e}^{-t/a}$ in the first integral of (47.9), by calculation and (47.8), we obtain the first equality of (47.9). Setting $y = 1/x$ in the first integral of (47.9), we obtain the second equality of (47.9). □

**Lemma 47.2.** *If $r > 1$, $1/r + 1/s = 1$, $\lambda > 0$, $\alpha < \lambda \min\{1/r, 1/s\}$ and $\beta > -1$. Define the weight functions $\varpi_\lambda(s,x)$ by*

$$\varpi_\lambda(s,x) := x^{\lambda/r} \int_0^\infty \frac{|\ln(x/y)|^\beta y^{\lambda/s-1}}{(x+y)^{\lambda-\alpha}(\min\{x,y\})^\alpha}\mathrm{d}y \qquad (x \in (0,\infty)). \tag{47.10}$$

Then

$$\varpi_\lambda(s,x) = k_\lambda(r) := \int_0^\infty \frac{|\ln u|^\beta u^{\lambda/r-1}}{(u+1)^{\lambda-\alpha}(\min\{u,1\})^\alpha}\mathrm{d}u, \tag{47.11}$$

where $k_\lambda(r)$ is a positive number and

$$k_\lambda(r) = \Gamma(\beta+1)\sum_{k=0}^\infty \binom{\alpha-\lambda}{k} \times \left[\frac{1}{(k+\lambda/r-\alpha)^{\beta+1}} + \frac{1}{(k+\lambda/s-\alpha)^{\beta+1}}\right]. \tag{47.12}$$

*Proof.* Setting $u = x/y$ in (47.10), by simplification, we get (47.11). Next, by (47.11) and (47.9),

$$\begin{aligned} 0 < k_\lambda(r) &= \int_0^1 \frac{(-\ln u)^\beta u^{\lambda/r-\alpha-1}}{(u+1)^{\lambda-\alpha}} du + \int_1^\infty \frac{(\ln u)^\beta u^{\lambda/r-1}}{(u+1)^{\lambda-\alpha}} du \\ &\leq \int_0^1 (-\ln u)^{(\beta+1)-1} u^{\lambda/r-\alpha-1} du \\ &\quad + \int_1^\infty (\ln u)^{(\beta+1)-1} u^{-(\lambda/s-\alpha)-1} du \\ &= \left[\left(\frac{r}{\lambda - r\alpha}\right)^{\beta+1} + \left(\frac{s}{\lambda - s\alpha}\right)^{\beta+1}\right] \Gamma(\beta+1) < \infty. \end{aligned}$$

Hence, $k_\lambda(r)$ is a positive number. Using a property of power series, we find

$$\begin{aligned} k_\lambda(r) &= \int_0^1 \frac{(-\ln u)^\beta u^{\lambda/r-\alpha-1}}{(u+1)^{\lambda-\alpha}} du \\ &\quad + \int_1^\infty \frac{(\ln u)^\beta u^{-\lambda/s+\alpha-1}}{(1+u^{-1})^{\lambda-\alpha}} du \\ &= \int_0^1 \sum_{k=0}^\infty \binom{\alpha-\lambda}{k} (-\ln u)^\beta u^{\lambda/r-\alpha+k-1} du \\ &\quad + \int_1^\infty \sum_{k=0}^\infty \binom{\alpha-\lambda}{k} (\ln u)^\beta u^{-\lambda/s+\alpha-k-1} du \\ &= \sum_{k=0}^\infty \binom{\alpha-\lambda}{k} \left[\int_0^1 (-\ln u)^\beta u^{\lambda/r-\alpha+k-1} du \right. \\ &\quad \left. + \int_1^\infty (\ln u)^\beta u^{-(\lambda/s-\alpha+k)-1} du \right]. \end{aligned}$$

Then in view of (47.9), we have (47.13). □

**Lemma 47.3.** *Let* $p > 0$ $(p \neq 1)$, $r > 1$, $1/p + 1/q = 1$, $1/r + 1/s = 1$, $\lambda > 0$, $\alpha < \lambda \min\{1/r, 1/s\}$, $\beta > -1$ *and* $n \in \mathbb{N}$, $n > r/(|q|\lambda)$. *Then for* $n \to \infty$, *one has*

$$\begin{aligned} I_n :&= \frac{1}{n} \int_1^\infty \int_1^\infty \frac{|\ln(x/y)|^\beta x^{\lambda/r-1/(np)-1} y^{\lambda/s-1/(nq)-1}}{(x+y)^{\lambda-\alpha}(\min\{x,y\})^\alpha} \mathrm{d}x\mathrm{d}y \\ &= k_\lambda(r) + o(1). \end{aligned} \tag{47.13}$$

*Proof.* Setting $u = y/x$, by Fubini's theorem [3], we obtain

$$\begin{aligned}
I_n &= \frac{1}{n}\int_1^\infty \left[\int_1^\infty \frac{|\ln(x/y)|^\beta x^{\lambda/r-1/(np)-1} y^{\lambda/s-1/(nq)-1}}{(x+y)^{\lambda-\alpha}(\min\{x,y\})^\alpha}\mathrm{d}x\right]\mathrm{d}y \\
&= \frac{1}{n}\int_1^\infty y^{-1/n-1}\left[\int_0^y \frac{|\ln u|^\beta u^{\lambda/s+1/(np)-1}}{(1+u)^{\lambda-\alpha}(\min\{1,u\})^\alpha}\mathrm{d}u\right]\mathrm{d}y \\
&= \frac{1}{n}\int_1^\infty y^{-1/n-1}\left[\int_0^1 \frac{(-\ln u)^\beta u^{\lambda/s+1/(np)-1}}{(1+u)^{\lambda-\alpha}u^\alpha}\mathrm{d}u \right.\\
&\quad \left. + \int_1^y \frac{(\ln u)^\beta u^{\lambda/s+1/(np)-1}}{(1+u)^{\lambda-\alpha}}\mathrm{d}u\right]\mathrm{d}y \\
&= \int_0^1 \frac{(-\ln u)^\beta u^{\lambda/s+1/(np)-1}}{(1+u)^{\lambda-\alpha}u^\alpha}\mathrm{d}u \\
&\quad + \frac{1}{n}\int_1^\infty y^{-1/n-1}\left[\int_1^y \frac{(\ln u)^\beta u^{\lambda/s+1/(np)-1}}{(1+u)^{\lambda-\alpha}}\mathrm{d}u\right]\mathrm{d}y \\
&= \int_0^1 \frac{(-\ln u)^\beta u^{\lambda/s+1/(np)-1}}{(1+u)^{\lambda-\alpha}u^\alpha}\mathrm{d}u \\
&\quad + \frac{1}{n}\int_1^\infty \left(\int_u^\infty y^{-1/n-1}\mathrm{d}y\right)\frac{(\ln u)^\beta u^{\lambda/s+1/(np)-1}}{(1+u)^{\lambda-\alpha}}\mathrm{d}u \\
&= \int_0^1 \frac{(-\ln u)^\beta u^{\lambda/s-\alpha+1/(np)-1}}{(1+u)^{\lambda-\alpha}}\mathrm{d}u \\
&\quad + \int_1^\infty \frac{(\ln u)^\beta u^{\lambda/s-1/(nq)-1}}{(1+u)^{\lambda-\alpha}}\mathrm{d}u.
\end{aligned} \tag{47.14}$$

(i) If $p > 0$ $(p \neq 1)$ and $q > 0$, by Levi's theorem [3], we find

$$\int_0^1 \frac{(-\ln u)^\beta u^{\lambda/s-\alpha+1/(np)-1}}{(1+u)^{\lambda-\alpha}}\mathrm{d}u = \int_0^1 \frac{(-\ln u)^\beta u^{\lambda/s-\alpha-1}}{(1+u)^{\lambda-\alpha}}\mathrm{d}u + o_1(1),$$

$$\int_1^\infty \frac{(\ln u)^\beta u^{\lambda/s-1/(nq)-1}}{(1+u)^{\lambda-\alpha}}\mathrm{d}u = \int_1^\infty \frac{(\ln u)^\beta u^{\lambda/s-1}}{(1+u)^{\lambda-\alpha}}\mathrm{d}u + o_2(1) \quad (n \to \infty).$$

(ii) If $q < 0$, take $n_0 \in \mathbb{N}$ with

$$n_0 > \frac{r}{|q|\lambda}, \qquad \frac{1}{s'} = \frac{1}{s} - \frac{1}{n_0 q\lambda}, \qquad \frac{1}{r'} = \frac{1}{r} + \frac{1}{n_0 q\lambda}.$$

Then for $n \geq n_0$ we have

$$\int_1^\infty \frac{(\ln u)^\beta u^{\lambda/s-1/(nq)-1}}{(1+u)^{\lambda-\alpha}} du \leq \int_1^\infty \frac{(\ln u)^\beta u^{\lambda/s-1/(n_0 q)-1}}{(1+u)^{\lambda-\alpha}} du \leq k_\lambda(r'),$$

and then by Lebesgue's control convergence theorem, we find

$$\int_1^\infty \frac{(\ln u)^\beta u^{\lambda/s-1/(nq)-1}}{(1+u)^{\lambda-\alpha}} du = \int_1^\infty \frac{(\ln u)^\beta u^{\lambda/s-1}}{(1+u)^{\lambda-\alpha}} du + o_3(1) \qquad (n \to \infty).$$

Hence, by the above results and (47.14), we have (47.14). □

## 47.3 Main Results

**Theorem 47.1.** *Assume that* $p > 0,\ p \neq 1,\ r > 1,\ 1/p + 1/q = 1,\ 1/r + 1/s = 1,$ $\lambda > 0,\ \alpha < \lambda \min\{1/r, 1/s\},\ \beta > -1,\ \phi(x) = x^{p(1-\lambda/r)-1},\ \psi(x) = x^{q(1-\lambda/s)-1}$ *for* $x \in (0,\infty),\ f, g \geq 0,$ *and*

$$0 < \|f\|_{p,\phi} = \left\{\int_0^\infty x^{p(1-\lambda/r)-1} f^p(x) \mathrm{d}x\right\}^{1/p} < \infty, \qquad 0 < \|g\|_{q,\psi} < \infty.$$

*Then,*

*(i) for* $p > 1$, *one has the following inequality:*

$$I := \int_0^\infty \int_0^\infty \frac{|\ln(x/y)|^\beta f(x) g(y)}{(x+y)^{\lambda-\alpha}(\min\{x,y\})^\alpha} \mathrm{d}x\mathrm{d}y < k_\lambda(r) \|f\|_{p,\phi} \|g\|_{q,\psi}, \qquad (47.15)$$

*where the constant factor* $k_\lambda(r)$ *expressed by (47.13) is the best possible;*

*(ii) for* $0 < p < 1$, *the reverse of (47.15) holds with the best constant factor* $k_\lambda(r)$.

*Proof.* (i) By Hólder's inequality with weight [4], in view of (47.10), we find

$$I = \int_0^\infty \int_0^\infty \frac{|\ln(x/y)|^\beta}{(x+y)^{\lambda-\alpha}(\min\{x,y\})^\alpha} \frac{x^{(1-\lambda/r)/q}}{y^{(1-\lambda/s)/p}} f(x) \frac{y^{(1-\lambda/s)/p}}{x^{(1-\lambda/r)/q}} g(y)\, \mathrm{d}x\mathrm{d}y$$

$$\leq \left\{\int_0^\infty \int_0^\infty \frac{|\ln(x/y)|^\beta}{(x+y)^{\lambda-\alpha}(\min\{x,y\})^\alpha} \frac{x^{(1-\lambda/r)(p-1)}}{y^{1-\lambda/s}} f^p(x) \mathrm{d}x\mathrm{d}y\right\}^{1/p}$$

$$\times \left\{ \int_0^\infty \int_0^\infty \frac{|\ln(x/y)|^\beta}{(x+y)^{\lambda-\alpha}(\min\{x,y\})^\alpha} \frac{y^{(1-\lambda/s)(q-1)}}{x^{1-\lambda/r}} g^q(y)\mathrm{d}x\mathrm{d}y \right\}^{1/q}$$

$$= \left\{ \int_0^\infty \varpi_\lambda(s,x)\phi(x) f^p(x)\mathrm{d}x \right\}^{1/p} \left\{ \int_0^\infty \varpi_\lambda(r,y)\psi(y) g^q(y)\mathrm{d}y \right\}^{1/q}. \quad (47.16)$$

We conform that (47.16) preserves the form of strict inequality. Otherwise, there exist constants $A$ and $B$, such that they are not all zero and [4]

$$A\frac{x^{(1-\lambda/r)(p-1)}}{y^{1-\lambda/s}} f^p(x) = B\frac{y^{(1-\lambda/s)(q-1)}}{x^{1-\lambda/r}} g^q(y) \qquad \text{a.e. in } (0,\infty)\times(0,\infty).$$

It follows that $Ax^{p(1-\lambda/r)} f^p(x) = By^{q(1-\lambda/s)} g^q(y)$ a.e. in $(0,\infty)\times(0,\infty)$. If $A \neq 0$, then there is $y > 0$, such that

$$x^{p(1-\lambda/r)-1} f^p(x) = [By^{q(1-\lambda/s)} g^q(y)] \frac{1}{Ax} \qquad \text{a.e. in } x \in (0,\infty).$$

This contradicts the fact that $0 < \|f\|_{p,\phi} < \infty$. So, by (47.11) and (47.13), (47.15) is valid.

For $n \in \mathbb{N}$, $n > r/(|q|\lambda)$, we set

$$f_n(x) := \begin{cases} 0, & 0 < x \le 1; \\ x^{\lambda/r - 1/(np) - 1}, & x > 1, \end{cases} \qquad g_n(x) := \begin{cases} 0, & 0 < x \le 1; \\ x^{\lambda/r - 1/(nq) - 1}, & x > 1, \end{cases}$$

if there exists a constant factor $0 < k \le k_\lambda(r)$, such that (47.15) is valid when we replace $k_\lambda(r)$ by $k$. Then by (47.14), we have

$$k_\lambda(r) + o(1) = I_n = \frac{1}{n} \int_0^\infty \int_0^\infty \frac{|\ln(x/y)|^\beta f_n(x) g_n(y)}{(x+y)^{\lambda-\alpha}(\min\{x,y\})^\alpha} \mathrm{d}x\mathrm{d}y$$

$$< \frac{1}{n} k \|f_n\|_{p,\phi} \|g_n\|_{q,\psi} = k,$$

and $k_\lambda(r) \le k$ $(n \to \infty)$. Hence, $k = k_\lambda(r)$ is the best constant factor of (47.15).

(ii) For $0 < p < 1$, by the reverse Hölder's inequality with weight [4], in view of (47.10), we find the reverse of (47.16), which still preserves the strict form. Then by (47.11) and (47.13), we obtain the reverse of inequality (47.15). By (47.14) and applying a similar technique to the one just used above one can prove that the constant factor in the reverse inequality of (47.15) is still the best possible. □

**Theorem 47.2.** *Assume that* $p > 0$, $p \neq 1$, $r > 1$, $1/p + 1/q = 1$, $1/r + 1/s = 1$, $\lambda > 0$, $\alpha < \lambda \min\{1/r, 1/s\}$, $\beta > -1$, $\phi(x) = x^{p(1-\lambda/r)-1}$, $\psi(x) = x^{q(1-\lambda/s)-1}$ *for every* $x \in (0,\infty)$, $f \ge 0$, *and* $0 < \|f\|_{p,\phi} < \infty$. *Then,*

(i) *for* $p > 1$, *the following inequality is equivalent to (47.15):*

$$J := \int_0^\infty y^{(p\lambda)/s-1}\left[\int_0^\infty \frac{|\ln(x/y)|^\beta f(x)\mathrm{d}x}{(x+y)^{\lambda-\alpha}(\min\{x,y\})^\alpha}\right]^p \mathrm{d}y$$
$$< k_\lambda^p(r)\|f\|_{p,\phi}^p, \tag{47.17}$$

*where the constant factor* $k_\lambda^p(r)$ *is the best possible;*

(ii) *for* $0 < p < 1$, *one has the reverse inequality of (47.18), which is equivalent to the reverse of (47.15) with the best constant factor* $k_\lambda^p(r)$.

*Proof.* (i) For $p > 1, x > 0$, setting a bounded measurable function as

$$[f(x)]_n := \min\{f(x),n\} = \begin{cases} f(x), & f(x) < n; \\ n, & f(x) \geq n, \end{cases}$$

we get

$$\int_{1/n}^n \phi(x)[f(x)]_n^p \mathrm{d}x > 0 \qquad (n \geq n_0),$$

for some $n_0 \in \mathbb{N}$, because $\|f\|_{p,\phi} > 0$. Setting $\widetilde{g}_n(y)$ as

$$\widetilde{g}_n(y) := y^{(p\lambda)/s-1}\left[\int_{1/n}^n \frac{|\ln(x/y)|^\beta}{(x+y)^{\lambda-\alpha}(\min\{x,y\})^\alpha}[f(x)]_n \mathrm{d}x\right]^{p-1} \tag{47.18}$$

for $y \in (1/n,n)$, $n \geq n_0$, by (47.15), we find

$$0 < \int_{1/n}^n \psi(y)\widetilde{g}_n^q(y)\mathrm{d}y$$
$$= \int_{1/n}^n y^{(p\lambda)/s-1}\left[\int_{1/n}^n \frac{|\ln(x/y)|^\beta [f(x)]_n \mathrm{d}x}{(x+y)^{\lambda-\alpha}(\min\{x,y\})^\alpha}\right]^p \mathrm{d}y$$
$$= \int_{1/n}^n\int_{1/n}^n \frac{|\ln(x/y)|^\beta [f(x)]_n \widetilde{g}_n(y)}{(x+y)^{\lambda-\alpha}(\min\{x,y\})^\alpha}\mathrm{d}x\mathrm{d}y$$
$$< k_\lambda(r)\left\{\int_{1/n}^n \phi(x)[f(x)]_n^p \mathrm{d}x\right\}^{1/p}\left\{\int_{1/n}^n \psi(y)\widetilde{g}_n^q(y)\mathrm{d}y\right\}^{1/q} < \infty; \tag{47.19}$$

$$0 < \int_{1/n}^n \psi(y)\widetilde{g}_n^q(y)\mathrm{d}y < k_\lambda^p(r)\int_0^\infty \phi(x)f^p(x)\mathrm{d}x < \infty. \tag{47.20}$$

It follows $0 < \|g\|_{q,\psi} < \infty$. For $n \to \infty$, by (47.15), both (47.20) and (47.20) still keep the forms of strict inequality. Hence, inequality (47.18) follows.

On the other hand, suppose (47.18) is valid. By Hölder's inequality, we obtain

$$I = \int_0^\infty y^{-1/p+\lambda/s} \int_0^\infty \frac{|\ln(x/y)|^\beta f(x)\mathrm{d}x}{(x+y)^{\lambda-\alpha}(\min\{x,y\})^\alpha} y^{1/p-\lambda/s} g(y)\mathrm{d}y$$
$$\leq J^{1/p}\|g\|_{q,\psi}. \tag{47.21}$$

In view of (47.18), inequality (47.15) follows, which is equivalent to (47.18). We conform that the constant factor in (47.18) is the best possible. Otherwise, we may get a contradiction by (47.21) that the constant factor in (47.15) is not the best possible.

(ii) For $0 < p < 1$, since $\|f\|_{p,\phi} > 0$, we conform that $J > 0$. If $J = \infty$, then the reverse of (47.18) is naturally valid. Suppose $0 < J < \infty$. Setting

$$g(y) := y^{(p\lambda)/s-1} \left[ \int_0^\infty \frac{|\ln(x/y)|^\beta}{(x+y)^{\lambda-\alpha}(\max\{x,y\})^\alpha} f(x)\mathrm{d}x \right]^{p-1},$$

by the reverse of (47.15), we obtain

$$\infty > \|g\|_{q,\psi}^q = J = I > k_\lambda(r)\|f\|_{p,\phi}\|g\|_{q,\psi} > 0$$

and

$$J^{1/p} = \|g\|_{q,\psi}^{q-1} > k_\lambda(r)\|f\|_{p,\phi}.$$

Hence, we have the reverse of (47.18).

On the other hand, suppose the reverse of (47.18) is valid. By the reverse Hölder's inequality, we can get the reverse of (47.21). Hence, in view of the reverse of (47.18), we obtain the reverse of (47.15), which is equivalent to the reverse of (47.18). We conform that the constant factor in the reverse of (47.18) is the best possible. Otherwise, we may get a contradiction by the reverse of (47.21) that the constant factor in the reverse of (47.15) is not the best possible. □

*Remark 47.1.* Setting $\alpha = \beta = 0$ in (47.15), we have (47.7). For $p = r = 2$ in (47.15) setting $\alpha = \beta = 0$, $\lambda = 1$, we have (47.1); setting $\alpha = 0$, $\beta = \lambda = 1$, we have (47.2). Hence, inequality (47.15) is a relation between (47.1) and (47.2).

## References


1. Hardy, G.H.: Note on a theorem of Hilbert concerning series of positive terms. Proc. London Math. Soc. **23**, Records of Proc. xlv-xlvi (1925)
2. Hardy, G.H., Littlewood, J.E., Polya, G.: Inequalities. Cambridge Univ. Press, Cambridge (1952)
3. Kuang, J.: Introduction to Real Analysis. Hunan Education Press, Changsha (1996)

4. Kuang, J.: Applied Inequalities. Shangdong Science Press, Jinan (2004)
5. Mintrinovic, D.S., Pecaric, J.E., Kink, A.M.: Inequalities Involving Functions and Their Integrals and Derivertives. Kluwer Academic Publishers, Boston (1991)
6. Yang, B.: On Hilbert's integral inequality. J. Math. Anal. Appl. **220**, 778–785 (1998)
7. Yang, B.: A note on Hilbert's integral inequality. Chin. Quart. J. Math. **13**, 4, 83–86 (1998)
8. Yang, B.: On a Base Hilbert-type integral inequality and extensions. College Mathematics **24**, 2, 87–92 (2008)
9. Wang, Z., Guo, D.: Introduction to Special Functions. Science Press, Beijing (1979)
10. Yang, B.: On an extension of Hilbert's integral inequality with some parameters. Aust. J. Math. Anal. Appl. **1**, Article 11, 1–8 (2004)
11. Yang, B., Brnetic, I., Krnic, M., Pecaric, J.: Generalization of Hilbert and Hardy–Hilbert integral inequalities. Math. Inequal. Appl. **8**, 259–272 (2005)
12. Yang, B.: On a reverse Hardy-Hilbert's inequality Kyungpook Math. J. **47**, 411–423 (2007)

Springer Optimization and Its Applications Vol. 52

ISBN 978-1-4614-0055-4 © Springer Science + Business Media, LLC 2012

Themistocles M. Rassias and Janusz Brzdęk (Eds.)

# Functional Equations in Mathematical Analysis

## Erratum

The correct order of the authors in Chapter 25, Stability of the Quadratic—Cubic Functional Equation in Quasi–Banach Spaces, should be Zhihua Wang and Wanxiong Zhang.

Th.M. Rassias and J. Brzdęk (eds.), *Functional Equations in Mathematical Analysis*, Springer Optimization and Its Applications 52, DOI 10.1007/978-1-4614-0055-4_25,


---

**DOI 10.1007/978-1-4614-0055-4_48**